WILLIAM F. MAAG LIBRARY
YOUNGSTOWN STATE UNIVERSITY

ANNUAL REVIEW OF PHYSICAL CHEMISTRY

EDITORIAL COMMITTEE (1989)

C. AUSTEN ANGELL
GERALD T. BABCOCK
JOSEPH E. DEMUTH
JAMES L. KINSEY
TERRY A. MILLER
C. BRADLEY MOORE
ROBERT J. SILBEY
HERBERT L. STRAUSS

Responsible for the organization of Volume 40
(Editorial Committee, 1987)

HANS C. ANDERSEN
C. AUSTEN ANGELL
GERALD T. BABCOCK
JOSEPH E. DEMUTH
JAMES L. KINSEY
DONALD H. LEVY
C. BRADLEY MOORE
HERBERT L. STRAUSS
NEIL R. KESTNER (Guest)
WAYNE L. MATTICE (Guest)

Production Editor — SUZANNE COPENHAGEN
Indexing Coordinator — MARY A. GLASS
Subject Indexer — STEVEN SORENSEN

ANNUAL REVIEW OF PHYSICAL CHEMISTRY

VOLUME 40, 1989

HERBERT L. STRAUSS, *Editor*
University of California, Berkeley

GERALD T. BABCOCK, *Associate Editor*
Michigan State University

C. BRADLEY MOORE, *Associate Editor*
University of California, Berkeley

ANNUAL REVIEWS INC 4139 EL CAMINO WAY P.O. BOX 10139 PALO ALTO, CALIFORNIA 94303-0897

ANNUAL REVIEWS INC.
Palo Alto, California, USA

COPYRIGHT © 1989 BY ANNUAL REVIEWS INC., PALO ALTO, CALIFORNIA, USA. ALL RIGHTS RESERVED. The appearance of the code at the bottom of the first page of an article in this serial indicates the copyright owner's consent that copies of the article may be made for personal or internal use, or for the personal or internal use of specific clients. This consent is given on the condition, however, that the copier pay the stated per-copy fee of $2.00 per article through the Copyright Clearance Center, Inc. (21 Congress Street, Salem, MA 01970) for copying beyond that permitted by Section 107 or 108 of the US Copyright Law. The per-copy fee of $2.00 per article also applies to the copying, under the stated conditions, of articles published in any *Annual Review* serial before January 1, 1978. Individual readers, and nonprofit libraries acting for them, are permitted to make a single copy of an article without charge for use in research or teaching. This consent does not extend to other kinds of copying, such as copying for general distribution, for advertising or promotional purposes, for creating new collective works, or for resale. For such uses, written permission is required. Write to Permissions Dept., Annual Reviews Inc., 4139 El Camino Way, P.O. Box 10139, Palo Alto, CA 94303-0897 USA.

International Standard Serial Number: 0066-426X
International Standard Book Number: 0-8243-1040-3
Library of Congress Catalog Card Number: A-51-1658

Annual Review and publication titles are registered trademarks of Annual Reviews Inc.

∞ The paper used in this publication meets the minimum requirements of American National Standard for Information Sciences—Permanence of Paper for Printed Library Materials, ANSI Z39.48-1984.

Annual Reviews Inc. and the Editors of its publications assume no responsibility for the statements expressed by the contributors to this *Review*.

TYPESET BY AUP TYPESETTERS (GLASGOW) LTD., SCOTLAND
PRINTED AND BOUND IN THE UNITED STATES OF AMERICA

PREFACE

This volume presents the customary wide range of topics in physical chemistry. A continuing feature of our series has been the inclusion of contributions by authors from many different countries. This wide participation reflects the internationalization of science, caused partly by the increasing availability of rapid means of exchanging information.

The use of standard units and symbols becomes more important as the authorship and readership of the *Annual Review of Physical Chemistry* becomes more global. A comprehensive compilation of units and symbols, together with much ancillary information, has just been published. This volume, *Quantities, Units, and Symbols in Physical Chemistry*, edited by I. Mills et al (1988. Oxford: Blackwell) and prepared under the auspices of the International Union of Pure and Applied Chemistry (IUPAC), provides an indispensable reference for notation and units.

We thank the authors and the members of the editorial staff, each of whom has made a unique contribution to the continued excellence of this volume.

THE EDITORIAL COMMITTEE

ANNUAL REVIEWS INC. is a nonprofit scientific publisher established to promote the advancement of the sciences. Beginning in 1932 with the *Annual Review of Biochemistry*, the Company has pursued as its principal function the publication of high quality, reasonably priced *Annual Review* volumes. The volumes are organized by Editors and Editorial Committees who invite qualified authors to contribute critical articles reviewing significant developments within each major discipline. The Editor-in-Chief invites those interested in serving as future Editorial Committee members to communicate directly with him. Annual Reviews Inc. is administered by a Board of Directors, whose members serve without compensation.

1989 Board of Directors, Annual Reviews Inc.

Dr. J. Murray Luck, Founder and Director Emeritus of Annual Reviews Inc.
  Professor Emeritus of Chemistry, Stanford University
Dr. Joshua Lederberg, President of Annual Reviews Inc.
  President, The Rockefeller University
Dr. James E. Howell, Vice President of Annual Reviews Inc.
  Professor of Economics, Stanford University
Dr. Winslow R. Briggs, *Director, Carnegie Institution of Washington, Stanford*
Dr. Sidney D. Drell, *Deputy Director, Stanford Linear Accelerator Center*
Dr. Sandra M. Faber, *Professor of Astronomy, University of California, Santa Cruz*
Dr. Eugene Garfield, *President, Institute for Scientific Information*
Mr. William Kaufmann, *President, William Kaufmann, Inc.*
Dr. D. E. Koshland, Jr., *Professor of Biochemistry, University of California, Berkeley*
Dr. Gardner Lindzey, *Director, Center for Advanced Study in the Behavioral Sciences, Stanford*
Dr. William F. Miller, *President, SRI International*
Dr. Charles Yanofsky, *Professor of Biological Sciences, Stanford University*
Dr. Richard N. Zare, *Professor of Physical Chemistry, Stanford University*
Dr. Harriet A. Zuckerman, *Professor of Sociology, Columbia University*

Management of Annual Reviews Inc.

John S. McNeil, Publisher and Secretary-Treasurer
William Kaufmann, Editor-in-Chief
Mickey G. Hamilton, Promotion Manager
Donald S. Svedeman, Business Manager
Ann B. McGuire, Production Manager

ANNUAL REVIEWS OF
Anthropology
Astronomy and Astrophysics
Biochemistry
Biophysics and Biophysical Chemistry
Cell Biology
Computer Science
Earth and Planetary Sciences
Ecology and Systematics
Energy
Entomology
Fluid Mechanics
Genetics
Immunology
Materials Science
Medicine
Microbiology
Neuroscience
Nuclear and Particle Science
Nutrition
Pharmacology and Toxicology
Physical Chemistry
Physiology
Phytopathology
Plant Physiology and
  Plant Molecular Biology
Psychology
Public Health
Sociology

SPECIAL PUBLICATIONS

Excitement and Fascination
  of Science, Vols. 1, 2, and 3

Intelligence and Affectivity,
  by Jean Piaget

A detachable order form/envelope is bound into the back of this volume.

CONTENTS

DIELECTRICS IN PHYSICAL CHEMISTRY, *Robert H. Cole*	1
ORIENTATIONAL GLASSES, *A. Loidl*	29
THE METAL-INSULATOR TRANSITION IN EXPANDED FLUID METALS, *F. Hensel and H. Uchtmann*	61
ELECTROLYTES DISSOLVED IN POLYMERS, *J. M. G. Cowie and S. H. Cree*	85
DYNAMICS OF SOLVATION AND CHARGE TRANSFER REACTIONS IN DIPOLAR LIQUIDS, *Biman Bagchi*	115
PICOSECOND VIBRATIONAL ENERGY TRANSFER STUDIES OF SURFACE ADSORBATES, *E. J. Heilweil, M. P. Casassa, R. R. Cavanagh, and J. C. Stephenson*	143
FUNDAMENTAL MECHANISMS OF DESORPTION AND FRAGMENTATION INDUCED BY ELECTRONIC TRANSITIONS AT SURFACES, *Phaedon Avouris and Robert E. Walkup*	173
COMPUTER SIMULATIONS OF GLOBULAR PROTEIN FOLDING AND TERTIARY STRUCTURE, *Jeffrey Skolnick and Andrzej Kolinski*	207
PHYSICAL CHEMISTRY AT ULTRAHIGH PRESSURES AND TEMPERATURES, *Raymond Jeanloz*	237
THEORY OF ADSORBATE INTERACTIONS, *Peter J. Feibelman*	261
TRANSITION METAL OXIDES, *C. N. R. Rao*	291
OPTICAL SECOND HARMONIC GENERATION AT INTERFACES, *Y. R. Shen*	327
RUBBER-LIKE ELASTICITY, *B. Erman and J. E. Mark*	351
VECTOR CORRELATIONS IN PHOTODISSOCIATION DYNAMICS, *G. E. Hall and P. L. Houston*	375
SPECTROSCOPY OF THE DIATOMIC $3d$ TRANSITION METAL OXIDES, *A. J. Merer*	407
TRANSPORT OF ELECTRONS IN NONPOLAR FLUIDS, *Richard A. Holroyd and Werner F. Schmidt*	439
THEORETICAL METHODS FOR ROVIBRATIONAL STATES OF FLOPPY MOLECULES, *Zlatko Bačić and John C. Light*	469

(*Continued*)

CONTENTS (*continued*)

HOLE-BURNING SPECTROSCOPY, *Silvia Völker*	499
ATOMIC-RESOLUTION SURFACE SPECTROSCOPY WITH THE SCANNING TUNNELING MICROSCOPE, *R. J. Hamers*	531
ORIENTED MOLECULE BEAMS VIA THE ELECTROSTATIC HEXAPOLE: PREPARATION, CHARACTERIZATION, AND REACTIVE SCATTERING, *David H. Parker and Richard B. Bernstein*	561
MEASUREMENT OF FORCES BETWEEN SURFACES IN POLYMER FLUIDS, *Sanjay S. Patel and Matthew Tirrell*	597
VACUUM UV PHOTOPHYSICS AND PHOTOIONIZATION SPECTROSCOPY, *Tomas Baer*	637
PROTON TRANSLOCATION IN PROTEINS, *Robert A. Copeland and Sunney I. Chan*	671
INDEXES	
Author Index	699
Subject Index	725
Cumulative Index of Contributing Authors, Volumes 36–40	738
Cumulative Index of Chapter Titles, Volumes 36–40	740

SOME RELATED ARTICLES IN OTHER *ANNUAL REVIEWS*

From the *Annual Review of Astronomy and Astrophysics*, Volume 27 (1989):

> A New Component of the Interstellar Matter: Small Grains and Large Aromatic Molecules, J. L. Puget and A. Léger
>
> Kinematics, Chemistry, and Structure of the Galaxy, G. Gilmore, R. F. G. Wyse, and K. Kuijken
>
> Chemical Analyses of Cool Stars, B. Gustafsson

From the *Annual Review of Biochemistry*, Volume 58 (1989):

> Two-Dimensional NMR and Protein Structure, A. Bax

From the *Annual Review of Biophysics and Biophysical Chemistry*, Volume 18 (1989):

> Free Energy Via Molecular Simulation: Applications to Chemical and Biomolecular Systems, D. L. Beveridge and F. M. DiCapua
>
> Time-Resolved Macromolecular Crystallography, K. Moffat
>
> The Study of Lipid Phase Transition Kinetics by Time-Resolved X-Ray Diffraction, M. Caffrey
>
> Expanding Roles of Computers and Robotics in Biological Macromolecular Research, A Wada, S.-i. Kidokoro, and S. Endo
>
> Thermodynamic Problems of Protein Structure, P. L. Privalov
>
> The Electrostatic Properties of Membranes, S. McLaughlin

From the *Annual Review of Earth and Planetary Sciences*, Volume 17 (1989):

> Raman Spectroscopy in Mineralogy and Geochemistry, P. F. McMillan

From the *Annual Review of Materials Science*, Volume 19 (1989):

> Use of Laser Techniques to Study the Dynamics of Molecular Surface Interaction, J. Häger and H. Walther
> Fractal Phenomena in Disordered Systems, R. Orbach
> Polymer Interdiffusion, H. H. Kausch and M. Tirrell
> Crystal Chemistry and Properties of Mixed Valence Copper Oxides, B. Raveau and C. Michel
> Synthesis, Stabilization, and Electronic Structure of Quantum Semiconductor Nanoclusters, M. L. Steigerwald and L. E. Brus

RELATED ARTICLES (*continued*)

From the *Annual Review of Nuclear and Particle Science*, Volume 39 (1989):

Spin and Isospin Excitations at Intermediate Energies, W. P. Alford
Inclusive Electron-Nucleus Scattering and Scaling, T. W. Donnelly, J. S. McCarthy, I. Sick, and D. Day
Solar Neutrino Experiments, R. Davis, Jr., L. Wolfenstein, and A. Mann
Tests of Perturbative QCD, G. Altarelli

DIELECTRICS IN PHYSICAL CHEMISTRY

Robert H. Cole

Department of Chemistry, Brown University, Providence, Rhode Island 02912

INTRODUCTION

When I was invited to contribute this prefatory chapter, I felt both honored and concerned about my qualifications, the latter when I remembered some questionable credentials and a piece of advice from P. Debye.

The credentials in question or lacking go back to the 1930s. My undergraduate training included only a single course in chemistry. This was not entirely my fault: I had thought of taking physical chemistry at Oberlin, if only for cultural reasons, but had neither time nor inclination to meet the inviolable prerequisite of qualitative analysis while concentrating in physics and mathematics. I managed only a little better, and not for the record, when I was a graduate student in physics at Harvard, as my only instruction in chemistry then was auditing G. B. Kistiakowsky's course in thermodynamics.

As accounts that follow suggest, my developing research interests and activities increasingly exposed me to physical chemistry and physical chemists. So I was not entirely taken aback when one of them I had come to know well in World War II years, Paul Cross, was able with support from Donald Hornig and J. S. (Spike) Coles to offer me a position in the chemistry department at Brown. Not long after accepting and beginning to learn more by teaching and doing, C. A. Kraus told me I should join the American Chemical Society. When I innocently asked why, I was merely given a form to fill out and expose my pitiful lack of qualifications. The application was summarily rejected. When I reported this to the King, as Kraus was described by his students, he called Alden Emery, then long-time ACS executive secretary, and told him to admit me forthwith, which he did.

The directions of research that I hoped would give me further respectability were then as now largely in the field of dielectric behavior, perhaps more generally and better called "electrophysical chemistry." This is not common usage, but there is a distinguished precedent: the entry for Lars Onsager in *American Men and Women of Science* (12th edition, 1972) listed a primary interest as "electrophysics."

I hope the second description is apt enough to make these recollections and thoughts of sufficiently general interest, but I have occasional qualms when I remember some fatherly advice from Debye in 1963. I had helped to organize an ACS symposium in his honor and had contributed a somewhat tedious talk on electric dipole interactions to the occasion, but at a party afterward he took me aside and suggested that I think about something more worthwhile, like NMR. This shook me a bit, and I continued to remember it whenever a current line of interest loses some of its initial zest, but something new and exciting has always come along to which I could relate without major disruption. May this account and future developments justify my belief that there is now as ever much of interest and much to be learned.

THE INFERNAL FIELD PROBLEM

The traditional problem in dielectrics has long been to account for experimentally observed relations between the macroscopic polarization or electric moment density P and macroscopic electric field E. A stumbling block of equally long standing has been the problem of how adequately to describe the underlying fields at microscopic and molecular levels in tractable ways. As a substitute for many body problems with long-range orientation-dependent forces, local or mean field approximations derived from macroscopic quantities have of necessity been introduced, debated, and often found wanting, to the point that they have been called "infernal fields" (by a research student at Oxford in 1962).

A mainstay for many years and still useful in its proper place is the Lorentz local field E_{LOR}. In 1909, Lorentz (1) considered isotropically and linearly polarizable atoms on a cubic lattice or in an isotropic continuum, and showed that for a spherical sample in a field E_0 external to it, the sum at any one of the induced dipole fields of all the others vanishes and the field at any site is $E_{LOR} = E_0 = E + (4\pi/3)P$, where E is the macroscopic average in the sample and the relation $P = (\varepsilon-1)E/4\pi$ if the problem is linear gives $E_{LOR} = (\varepsilon+2/3)E$ where E is the dielectric constant (or relative permittivity). This led to the appearance of the ratio $(\varepsilon-1)/(\varepsilon+2)$ in many subsequent developments to relate polarization and field to charge displacements at the molecular level, with disastrous results for polar molecules in liquids and solids.

Debye's classic work (2, 3) in relating dielectric constant to polarizability α and permanent dipole moment μ of polar molecules opened up the whole field of dielectric studies for chemical purposes. His formula in simplest form is

$$(\varepsilon-1)/(\varepsilon+2) = (4\pi/3)(N/V)[\alpha+\mu^2/(3kT)] \qquad 1.$$

in which N/V is the number density of molecules and the Lorentz local field, or Clausius-Mossotti function, has been incorporated. The disaster for polar molecules just mentioned, which Van Vleck (4) called the $4\pi/3$ catastrophe, is evident from this formula predicting an infinite value of ε for the right-hand side equal to unity. This is easily realized for only moderate values of μ and T, in complete contradiction to experiment as shown schematically in Figure 1, where $\varepsilon-1$ is plotted against $(N/V)(\mu^2/3kT)$ for a variety of polar liquids at temperatures in the normal liquid range. The figure also shows a common pattern of behavior, with the only major deviations being for hydrogen bonding, or otherwise associated, liquids.

Much work since the 1920s has been devoted to modifying or embellishing the basic Lorentz field picture, with little or no success for even moderately polar media, and a variety of empirical functions have been proposed to describe the pattern of behavior suggested by Figure 1. One of the latter, proposed by Jefferies Wyman in 1936 (5), was apparently an incentive for the breakthrough in Onsager's famous 1936 paper (6), which was soon to be followed by Kirkwood's equally famous 1939 paper (7).

The gist of Onsager's contribution was in the observation that the electric field at a polar molecule was in part a reaction field induced by its own dipole moment and as such could have no role in reorienting it, as tacitly supposed in using the Lorentz field. The result of evaluating both this field and the field in a cavity volume occupied by the molecule for an electrostatic continuum model was Onsager's equation. The prediction of the equation for an average value of the contribution from polarizability is shown as the *solid curve* in Figure 1. Thus the $4\pi/3$ catastrophe from the bootstrap effect in the Clausius-Mossotti function was averted and a remarkably good account given of average behavior in the absence of specific short-range orientation effects not accounted for by the simple model. The Kirkwood treatment based on separation of specific short-range effects and correlations from long-range dipolar (Coulomb) forces, with only the latter treated macroscopically, was to appear shortly. With this, a basis became available for characterization and modeling of hydrogen bonding or other associative effects, molecular shapes, and the like to

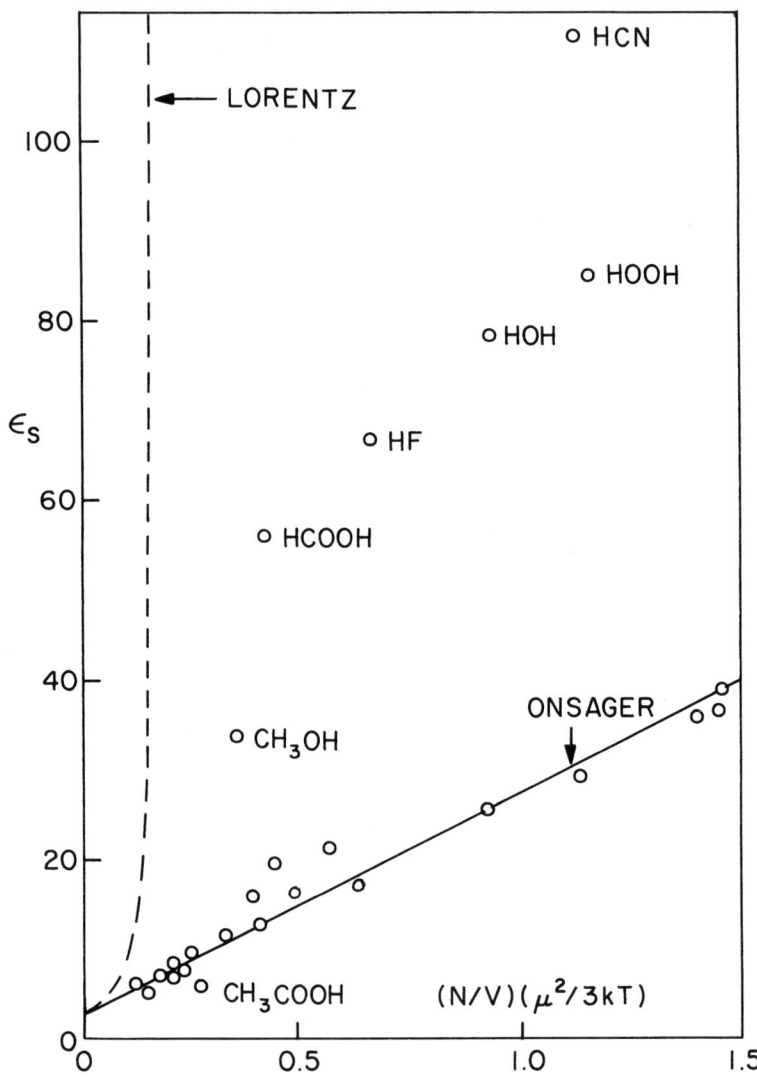

Figure 1 Permittivities of polar liquids at or near 25°C plotted as a function of relative dipole energy $(N/V)(\mu^2/3kT)$. *Points* are experimental, *curves* are from Lorentz and Onsager expressions for nominal permittivity $\varepsilon_\infty = 2.25$ of induced polarization.

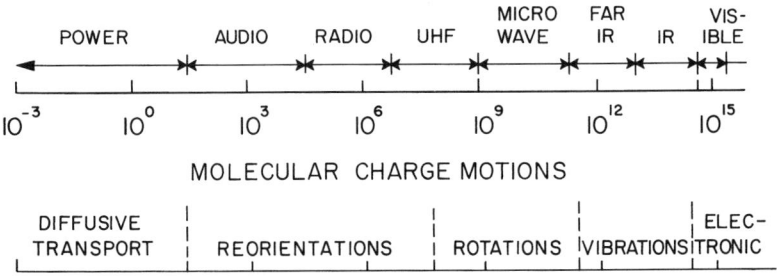
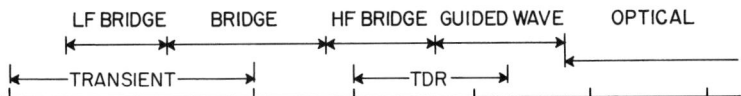

Figure 2 The dielectric regions of the electromagnetic spectrum on a logarithmic frequency scale, with representative motions of charges and experimental methods indicated.

account for gross differences and lesser deviations from the Onsager result in Figure 1. These predictions and previsions were yet to be tested, questioned, and refined, but new approaches to old problems were at hand.

These were exciting times for me. I had served a rewarding apprenticeship for several summers with my older brother, Kenneth Cole, and learned about electrical measurements and the state of physics in understanding impedance behavior of biological systems. The dielectric aspects particularly intrigued me as I worked with my brother to survey experimental evidence and theories of relaxation. I was starting graduate research with J. H. Van Vleck at Harvard as my mentor when Onsager's paper generated considerable excitement because of its obvious relevance to problems of dipole coupling and magnetic susceptibilities. It also changed my thesis problem from magnetic susceptibilities of single crystals of Tutton and (I had hoped) rare earth salts to the very different one of measuring dielectric constants of liquids and interpreting both these static results and relaxation data in the light of the new theoretical advances.

A particularly memorable occasion was a New York Academy of Sciences conference in April of 1939 (8), which featured papers by Kirkwood (9) and Van Vleck (4) among others. Van Vleck's paper summarized the state of local field treatments and discussed results of his calculations of dipole-dipole interactions on rigid lattices. The latter had considerable bearing on several theoretical and experimental problems of interest, discussed more below. Kirkwood's contribution was an outline of his statistical mechanical treatment and a first crude model for dipole correlations and the Kirkwood g-factor of liquid water that led to a static dielectric constant of 67 rather than the experimental 78 at 25°C. This was a major achievement at the time, and an incentive both for devising models of hydrogen bonding in other liquids, such as aliphatic alcohols, HCN, and several acids, and for better calculations to come as more realistic theories of liquids were developed and computer simulations became feasible. An interesting exchange in discussion, not recorded in the published account of the meeting, was between Kirkwood, insisting on spherical samples, and Onsager, equally determined to discuss a sample of arbitrary shape provided only that it was covered with tin foil. Although it was not clear to me at the time, this may have been the first mention of the Onsager "tin foil theorem," never published by him but derived much later by Felderhof (10).

All these developments gave me much to think about and also were the beginning of my conversion from a physicist to a physical chemist, at least in name and as much by accident as by design. As World War II approached in 1941, Kirkwood was commissioned by E. Bright Wilson to recruit someone who might know enough electronics to help develop instrumentation and methods for study of shock waves and other underwater explosion phenomena. Apparently on the strength of having talked with me about mutual interests and having seen me face to face with an oscillograph when visiting Harvard, Jack was responsible for shifting my activities to the National Defence Research Council (NDRC) project on explosives under G. B. Kistiakowsky as part of the war effort. It was not until 1947 that I got back to academic research, thanks to Paul Cross.

The more recent developments from theory of liquids and of dipolar crystals would require at least a chapter to do them justice, a task that I attempted in part not long ago (11). Here it is worth mentioning results of some relevant calculations that seem not to be widely known.

The first are from papers of M. Lax and co-workers (12, 13), in which an analytic theory of coupling of dipoles on a rigid lattice in an applied electric field is developed. The partition function is expressed in terms of dipole wave sums and can be evaluated to give the polarization and permittivity if the spherical model approximation is made of replacing the N constraints $\mu_i \cdot \mu_i = \mu^2$ of orientations of N dipoles μ_i by the single constraint $\Sigma \mu_i \cdot \mu_i = N\mu^2$, a procedure that has worked reasonably well for other models of cooperative interactions.

The results are of interest in several respects. First, when the discrete lattice sums for rigid dipoles are replaced by continuum integrations in the long wavelength limit, Onsager's equation is obtained from the seemingly unreasonable choice of lower boundary radius a to satisfy $a^3 = 3V/4\pi N$, i.e. the cavity volumes fill the sample volume. For face and body centered cubic lattices, however, somewhat larger permittivities are predicted (corresponding to a Kirkwood g-factor greater than one) with a phase transition predicted for a temperature $T_c \simeq 0.1(4\pi N/3V)\mu^2$, i.e. a factor 10 smaller than for the Lorentz field. A further feature is that induced moments can be included if they are represented by harmonic oscillators. The principal effect is to multiply the rigid dipole moments μ by a factor $(n^2+2)/3$, where n is the refractive index of induced polarization. A similar conclusion seems to be indicated by Wertheim's treatments for liquids using graph theory methods (14), so it may be a useful result for correcting other calculations with rigid dipoles for comparison with liquids of real polarizable dipoles. These and other features all seem to deserve further study (despite Debye's advice to me mentioned in the introduction).

Computer simulations are proving to be valuable in assessing the magnitudes and range of specific orientational correlations of simple molecules as they affect dielectric properties and serve to test the Kirkwood formulation. A stumbling block discussed in several places (14, 15) has been proper use of cyclic boundary conditions when long-range dipole interactions are involved. The calculations by Claude Brot and co-workers (16) are illuminating because they avoid the problem by using a spherical potential well or "box" to confine the simulation particles and by evaluating the dipole correlations for both an inner Kirkwood sphere of specific interactions and the entire volume.

For interactions of point dipoles and parameters corresponding roughly to CH_3F at 206 K, the results indicate that Kirkwood's formulation is approximately correct for inner sphere diameters as small as 3σ, where σ is the Lennard-Jones potential parameter used for the calculations. At the same time, however, the calculated Kirkwood g-factors are of order 2.5 whereas the experimental value is nearly unity; but this discrepancy is greatly reduced by using either off-center dipoles or quadrupole moments, indicating the importance of using more realistic descriptions of molecular charge distributions than merely point dipoles.

Obviously many points deserve further study before equilibrium properties of dipolar materials are understood really well. At the same time, the progress to date is encouraging, and, in any case, even quite simple applications of the various formulations often give semiquantitative results of respectable accuracy.

Finally in this section, it would not do to omit any mention of the very fundamental formal treatment pioneered and being implemented by Robert Fulton. In a sense, much of the controversy about local fields over the years has been confused by failures to realize that the electric field

E of the macroscopic Maxwell equations is, like the polarization P, a macroscopic average quantity. Fulton's approach (17, 18) is to recognize this at the outset and to obtain expressions for both in terms of quite arbitrary charge and current density distributions as external sources. In this way, dielectric response functions are expressed in terms of functional derivatives with respect to source fields and macroscopic E without introducing molecular cavities and the like. The general results have been shown to reduce to the Kirkwood-Fröhlich equation and Onsager-like static and relaxation functions by appropriate specialization, as well as providing a basis for evaluation of nonlinear effects. These developments have received less attention than it seems to me they deserve, no doubt in part because of formal quantum electrodynamic methods formidable to the uninitiated. The extent to which these and other approaches in the same spirit can be put together with adequate molecular dynamics to give further results remains to be seen, but the signs are encouraging.

THE RANGE AND CAPABILITIES OF DIELECTRIC SPECTROSCOPY

The term *dielectric spectroscopy* has a variety of connotations depending on one's interests and point of view, as suggested by such other terms as *impedance* or *immitance spectroscopy*. Here we take it to mean the study of relaxation processes, primarily in condensed phases, of interest to physical chemists. Defining a boundary or transition in the electromagnetic spectrum between such spectroscopy, using electronic methods primarily, and resonance spectroscopy by optical methods is arbitrary but is reasonably taken to be the region between 100 and 1000 GHz.

The distinction between the two is fairly clear-cut in terms of experimental methods: Microwave styles of measurement are currently limited to wavelengths greater than about 1 mm (frequency <300 GHz), and this is the same as the 10 cm^{-1} lower limit commonly quoted for far IR spectrometers. The change in character from IR resonance lines or bands to regions of relaxation dispersion and absorption, with line widths of a decade or more in frequency, is less well defined either conceptually or experimentally, as there is a comparative dearth of measurements in the difficult transition region.

As suggested by the schematic representation in Figure 2, the dielectric region of the spectrum extends over some 14 decades in frequency, with only a little further space on a logarithmic scale needed to accommodate the whole of resonance spectroscopy. (This obviously distorted viewpoint can of course be reversed by using a linear frequency scale to reduce the whole dielectric region to little more than a dumping ground for products of resonance excitations.) Some of the changes in character of molecular charge motions responding in different regions are suggested in Figure 2, together with an equally sketchy indication of kinds of methods available for their study.

This available range is not as enormous as it might seem to be without considering the natural line widths and shapes for relaxation processes and the changes of their characteristic relaxation true scales with temperature. The prototype of relaxation functions is of course a simple exponential approach of polarization to a new equilibrium state with time, $\exp(-t/\tau)$, after an instantaneous change in applied field. This is shown in Figure 3, together with the corresponding steady state a.c. response to applied sinusoidal field of frequency $f = \omega/2\pi$, as represented by the real (dispersive) ε' and imaginary (absorptive) ε'' parts of the complex permittivity $\varepsilon^* = \varepsilon' - i\varepsilon''$. This function by Laplace transformation of $\exp(-t/\tau)$ is

$$\varepsilon^* = \varepsilon_\infty + \frac{(\varepsilon_s - \varepsilon_\infty)}{1 + i\omega\tau} \qquad 2.$$

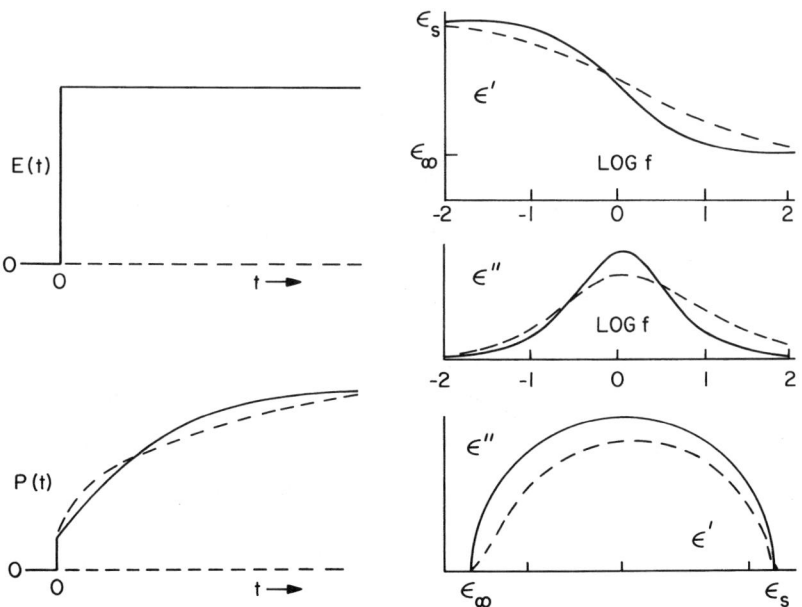

Figure 3 *Left side*: Response of polarization $P(t)$ to an applied step field $E(t)$. *Right side*: Dispersion ε' and loss ε'' as a function of log frequency and in the complex plane. *Solid curves* are for exponential (Debye) relaxation, *dashed curves* are deviations found experimentally.

where ε_∞ is the somewhat ambiguous high frequency limit of the relaxation and ε_s the low-frequency limit.

It is to be noted that the dispersion and absorption curves require at least a decade in frequency, and preferably two or more, to be reasonably well defined. Coupled with the facts that one is usually dealing with thermally activated processes with an Arrhenius or stronger temperature dependence of relaxation time scale and that observed relaxation processes are commonly more spread out in frequency or time, as indicated by the dashed curves in Figure 3, the need for broadband coverage is only too apparent.

A good deal of my research effort over the years has of necessity been devoted to developing or taking advantage of newer and better methods to meet the need. By all odds the easiest part of the frequency range is from ca. 20 Hz to 200 kHz or more, particularly by use of transformer bridges. The first of four such, built in 1948, got me into dielectrics again after World War II, and the last in 1963 is still serving faithfully and well, even though it lacks such newer amenities as automatic balancing and programmed frequency sweeping at a series of temperatures to let the equipment do all the work by itself overnight.

With such equipment, one could readily do quite definitive studies of relaxation in polar liquids of large molecules or at low temperatures, molecular crystals, and amorphous solids, as well as very precise measurements of static permitivities of gases. At the same time, however, the studies usually underscored the need to extend the range of measurement to both lower and higher frequencies; for example, on the one hand to follow further the approach of liquids to glassy behavior and on the other to study properly the dielectric aspects of dynamics of small molecules and local motions in larger ones at ordinary temperatures. The former has been taken nearly to the limits of patience or thermal noise by ultralow frequency steady state (19) and transient response methods (20, 21).

For a long time, dielectric measurements at megahertz and gigahertz frequencies were not to be undertaken lightly, but in the past 25 years or so measurements of modest accuracy at a few frequencies with any one instrument have been replaced at an increasing rate to ever higher frequencies by broad band techniques, made available largely by response to the needs of communication and information technology.

Impedance and network analyzers with a gamut of programming and data processing capabilities are becoming commonplace to frequencies upwards of 10 GHz, as are time-domain picosecond response methods using fast pulse generators and sampling oscilloscope detection. My own efforts to approach this range for looking at simpler dynamics of reasonably small molecules began as modest forays that resulted in some progress and more frustration in trying to extract molecular response functions from the obscurity of transcendental functions relating them to observables with equipment based on 50 ohms as the point of reference. This all changed for the better in 1970 when I heard Fellner-Feldegg talk about time domain reflectometry (TDR), made possible by then new developments of fast pulse and pulse sampling techniques. His somewhat deceptively simple methods (22) of getting dielectric information from a few picoseconds to nanoseconds or more all at once helped set me off on several new courses (23), about which more follows.

In the last year or two at the time of this writing, a number of new instruments have been announced or rumored that promise to open up the high-frequency range considerably. These include network analyzers to 40 GHz, time domain pulse generators and sampling systems to as high as 100 GHz, and time domain opto-electronic equipment to 350 GHz. I lack the temerity at this stage to consider very seriously the considerable problems in using these and other developments in prospect for dielectric measurements, but there are good reasons to think that such efforts will be rewarded by important results. A noteworthy example has already been provided by the work of Alain Gerschel and his Polish colleagues (24) for such liquids as halogenated methanes and chlorobenzene in the range 50–600 GHz from millimeter wave interferometer measurements that join up with far IR data above 450 GHz. These show deviations from extrapolated lower frequency patterns of behavior that are far from understood, but with various aspects attributed to correlated rotations and librations (25), ordering by short-range dipolar interactions, and dipolarons or other collective modes.

More such experiments are needed with better sources and detectors than were available for the original work. Although it is unlikely that any single approach alone will clarify the underlying dynamics, I can also believe that dielectric evidence will prove to be an important part of what is needed.

DIELECTRIC VIRIAL COEFFICIENTS OF GASES

As part of this somewhat random walk through past and present, a digression here on equilibrium dielectric properties of less than ideal gases is in keeping, as it shows how a new direction of research can develop from a combination of circumstances rather than by design and with some unexpected results.

I had begun to wonder early on about the adequacy of point dipoles, permanent or induced, for describing real molecules and their dielectric interaction effects in local field problems, Kirkwood correlation g-factors, and the relation of dipole coupling to phase transitions and relaxation in orientationally disordered molecular crystals, notably HCl, HBr, and HI. These last intrigued me because it was hard to think of any real molecules that more closely resembled the unreal spherical objects with point dipoles assumed in theory and because available evidence indicated that the crystal structures were all simple face-centered. A literature search yielded only wildly discordant information about equilibrium, let alone relaxation, dielectric behavior, and some calculations by Krieger & James (26) to suggest that interactions of molecular quadrupole moments could help to explain the multiplicity of solid phases. This was enough to start a series of investigations of the solid forms, discussed more in the next section.

My interest in gases was aroused when papers by Zwanzig (27) and Buckingham & Pople (28) appeared that pointed out collision-induced effects in gases from the field of one molecular quadrupole that induced dipole moments in neighbors during collisions. Their predictions were impossible to test and use for my purposes, however, because magnitudes of molecular quadrupole moments were virtually unknown, and the few available measurements of gas imperfections, as expressed by dielectric second virial coefficients of pair interactions, were highly uncertain about the magnitude and even the sign, let alone temperature dependence.

By this time, my research students and I had developed transformer bridge measurements to a high degree of precision for other purposes, so with realization that much more could be done it seemed entirely feasible to make the necessary electrical measurements of dielectric constants with a precision of the order of parts per million. This proved to be possible without undue difficulty, the more demanding problems proving to be in design and use of sufficiently stable capacitance cells and in sufficiently accurate determinations of gas densities.

The density problem comes about because theory leads to a virial expansion of the Clausius-Mossotti function in powers of number density N/V.

$$\frac{\varepsilon-1}{\varepsilon_2} \cdot \frac{V}{N} = A_\varepsilon + B_\varepsilon \frac{N}{V} + C_\varepsilon \left(\frac{N}{V}\right)^2 + \cdots \qquad 3.$$

where A_ε contains the single molecule contributions from polarizability and permanent dipole moments as in Eq. 1 and the second virial coefficient B_ε, the effects of pair interactions. Determination of densities from measured pressures was undesirable, because the conversion of pressure to density leads to a modified virial expansion with B_ε replaced by $B_\varepsilon - A_\varepsilon B_p$ where B_p is the second pressure virial coefficient for expansion of pressure P in powers of density. The correction term $A_\varepsilon B_p$ is typically 10–20 times larger than B_ε, and literature data for B_p of gases of interest often lacked the necessary accuracy. While recoiling from the prospect of major efforts in *PVT* measurements, I remembered the Burnett expansion method of generating a series of decreasing densities in constant ratio that could be determined by a modest number of measurements at low pressures. The basic scheme (29) was later elaborated into a sequence of expansions from one cell to two others (30), which provided valuable checks on consistency and clues to sources of error, and Roland Orcutt among others and I were finally in business.

Some early results were unexpected. Although the principal interest at the time was in quadrupolar molecules, we decided to measure some atomic gases, primarily as a test of the apparatus, as we supposed that the expected small positive B_ε values could be calculated with sufficient accuracy from Kirkwood's early theory (31) of induced dipole fluctuation effects. The experimental positive values of B_ε for argon were somewhat less than calculated, while more difficult measurements of the much smaller effect in helium (for which ε deviates from unity by 60 ppm at one atmosphere) gave negative values of B_ε. This caused us considerable concern, as it suggested the presence of systematic errors in the measurements. We could find no likely reason, and happily I heard a report of a similar anomaly in pressure-induced spectral line shifts in helium. With these reassurances, we reported these and other results (29). The effect for helium presented an interesting challenge to theorists, as one is dealing with the calculation of the polarizability of a pair of helium atoms as a function of their separation. As it has finally turned out, even such a seemingly simple problem requires very accurate wave functions, and a variety of calculations were made, not without controversy about their methods, before the matter was settled.

Meanwhile, my own work with a series of students went on to study the effects for molecules of increasing complexity. We took CO_2 as a prime example of a quadrupolar molecule, and from measurements by Tapan Bose to 150 atm of both the pure gas and mixtures with argon (32) in conjunction with theory developed by Buckingham & Pople (28) we were able to derive a rather large quadrupole moment $Q = 4.3 \times 10^{-26}$ esu cm^2. Some earlier estimates had given considerably larger values which we found to be the result of neglecting quadrupole-quadrupole interactions in the pure gas, as verified by the experiments in the mixtures with argon from which the interaction of $CO_2 - Ar$ pairs lacking the $Q - Q$ interaction could be extracted. The results were gratifying, as they demonstrated the importance of both quadrupole moment fields inducing dipoles in neighbors and their interactions with each other at fairly short range. A further satisfaction was the agreement of our value with ones from other effects, notably the one from the elegant but difficult Buckingham-Disch experiment (33) in which the anisotropy of orientations of CO_2 molecules by quadrupole interaction with a strong external field gradient was measured by the small but cumulative induced birefringence for an optical path length of ca. 1 meter.

After these and other studies of quadrupolar and octapolar molecules, our attention turned to simple polar molecules for which quadrupolar effects could also be expected, albeit added to larger effects of dipolar interactions, and such molecules as CF_3H, CHF_3, $CClF_3$, $CClH_3$, and $CClF_2H$ were chosen.

The values of B_ε obtained by Sutter (34) varied widely in magnitude and in sign as a combined result of effects of dipole and quadrupole interactions together with shape-dependent short-range repulsive forces, but obtaining consistent fits without including quadrupole effects proved impossible. Uncertainties as to best values of other parameters made derived values of quadrupole moments very approximate, but the magnitudes were in reasonable agreement with other estimates. The principal value of the work was thus to establish both the importance of quadrupole effects and the high sensitivity of dielectric effects to orientational forces, with pressure virial coefficients much less sensitive, and the temperature dependence of viscosity scarcely sensitive at all, as previously pointed out by Spurling & Mason (35).

More selective evidence as to the magnitude of quadrupole-induced electric moments during collisions is provided by the absorption spectra, in which one expects rotational lines with selection rules $\Delta J = 0, \pm 2$ rather than $\Delta J = \pm 1$ for dipole transitions. Ordinarily these are in the far infrared and too broadened because of short durations of collisions to be readily resolved, but HCl and HBr are exceptions: Because of their small moments of inertia, the quadrupole lines are widely spaced and in the near infrared. From this, Shmuel Weiss and I (36) were able to find and identify the predicted lines quite precisely and from estimated intensities obtain a quadrupole moment $Q = 5.5 \times 10^{-26}$ esu cm^2 for HBr. This was convincing evidence for the reality of quadrupole effects of sufficient magnitude to play a significant role in the structure of orientational phases and phase transitions in hydrogen halides, thus helping to resolve questions about interpretations of dielectric effects in these solid phases that had led to the studies of dielectric virial coefficients of gases in the first place.

RELAXATION OF DIPOLAR LIQUIDS AND CRYSTALS

When several students from the World War II generation and I started studies of dielectric relaxation in 1947, the initial instrumentation for audio and radio frequencies together with questions raised by studies of earlier literature suggested looking at solid phases with orientational freedom of small dipolar molecules. Two prime examples were hydrogen bromide, which we hoped was representative of simple molecules in well-defined structures, and ice, important for understanding "water substance" in any of its forms. In both cases, the experimental data from previous work were meager, with no agreements between results of different authors. We were soon to learn the hard way about the difficulties of measurements on any solids, and especially ones of high permittivity at low temperature.

The mundane problems of preparing and maintaining samples free of cracks and voids and the lack of adhesion of electrodes appeared in extreme form for HBr, with Norman Brown the chief sufferer. Solid samples grown from the liquid in an elegant coaxial cylinder cell invariably developed cracks and voids in cooling, with catastrophic results as peak permittivities near the order-disorder transition at 89 K were approached, but reproducible values were finally obtained by resorting to applying compression to parallel plate electrodes after a sample initially free of any visible defects was frozen in by judicious temperature cycling. The result was a spectacular and reproducible peak in static permittivity at $\varepsilon_s \cong 200$ (rather than 10–35 from earlier work), followed by not one but two distinct relaxations at lower temperatures of the partially ordered crystal (37). The slower and better-defined one was a beautiful depressed circular arc in the complex plane plots of $\varepsilon^* = \varepsilon' - i\varepsilon''$ expressed as $\varepsilon^* - \varepsilon_\infty = (\varepsilon_s - \varepsilon_\infty)[1 + (i\omega\tau)^{1-\alpha}]^{-1}$, so here there was a seemingly clear-cut example of the behavior my brother and I had uncovered years before (44). At the time, we did not realize that the low-temperature phase was ferroelectric rather than antiferroelectric as I supposed, with the slow relaxation due to the presence of domains. This was discovered only some 20 years later when Japanese workers reported hysteresis loops (38), which Stan Cichanowski and I confirmed and studied further (39). Other students and I in the meantime had been able to define the dielectric properties of hydrogen halides much more extensively (40, 41). The results sufficed to map some common patterns of behavior but equally brought out marked differences among HCl, HBr, and HI; these differences both dashed my hopes for any simple understanding and emphasized the need for much more refined treatments of pair and many body interactions in these supposedly simple systems.

By way of contrast, the early dielectric studies of ice that I and Bob Auty (42) performed gave very simple, clean results once the problems of void formation, electrode contacts, and electrode polarization had been diagnosed and dealt with reasonably well. The relaxation of the polycrystalline samples was accurately fitted by the simple Debye Eq. 2, with a relaxation time of 40 μs at 0°C increasing to 2 ms at -90°C and an activation energy at 13 kcal/mole. Numerous studies since have been extended to single crystals and the various polymorphic forms, but it has been gratifying to see how well the original values have been confirmed (the more so after Lars Onsager in a lecture on Bjerrum fault sites as the agents for dipole reorientations referred to me as an experimentalist whose results could be trusted).

Below $-40°C$, the relaxation in ice was too slow for measurement by available bridge methods, so with Auty and Donald Davidson, I set out to develop transient methods (43) for this and other relaxations at longer times. I wanted particularly to look for good examples of the $t^{-\alpha}(\alpha < 1)$ time dependence so often found in the literature and deviations from it at long times (44), and as a test of the new instrumentation picked glycerol for measurements by both bridge and transient measurements over a range of temperatures.

Glycerol was chosen partly because analysis of earlier measurements had seemed to show that the relaxation was described by the circular arc (Cole-Cole) relaxation function (regardless of whether glycerol could be considered representative of anything but itself), but more from the convenience that it was as slow as molasses in January and could be supercooled indefinitely with ease. This was the genesis of the skewed arc (Cole-Davidson, C-D) relaxation function (45).

When Don Davidson showed me his first results with a skewed rather than the expected symmetric circular arc in the complex plane representation, I asked him to try again with more attention to avoiding possible sources of error. After he forced me to take him seriously, I remembered the impedance characteristic of an infinite leaky cable (with capacitances charged through resistors) described by a relaxation function of the form $(1+i\omega\tau)^{-1/2}$. Changing the exponent from $-\frac{1}{2}$ to a variable parameter β with $0 < \beta < 1$ gave the skewed arc function $\varepsilon^* - \varepsilon_\infty = (\varepsilon_s - \varepsilon_\infty)[1+i\omega\tau]^{-\beta}$. For β of order 0.6–0.8 this gave beautiful fits except at the very highest frequencies with $\omega\tau \gg 1$ and set me thinking of such things as coupled diffusive mechanisms and diversity of hydrogen bonding. Going on to propylene glycol with two OH groups gave similar results except for β values nearer to value unity, whereas 1-propanol with a single OH gave the $\beta = 1$ characteristic of Debye relaxation except for evidence of small but distinct secondary relaxations at much higher frequencies. All three liquids could be supercooled quite readily to very low temperatures and long times, and all showed characteristic slowing down of relaxation time on approaching the glass transition. This could be described by the familiar non-Arrhenius law $\ln \tau = A \exp B/(T-T_0)$, with T_0 some 20–30 degrees below the glass transition temperature T_g, which I attributed to Tammann and is now commonly known as the Vogel-Tammann-Fulcher (VTF) equation.

To test whether such striking results were to be found in other than associated liquids, I naturally looked first in C. P. Smyth's book (46) with its summaries of his many pioneering exploratory studies. This soon suggested alkyl halides for further investigation, as they were normal polar liquids by usual criteria, including that of reasonable conformity with Onsager's equation for their static permittivities, and several likely candidates could be supercooled quite readily. Donald Denney had recently finished a PhD thesis on alcohols and agreed to return for a year for work on such problems and to help other students while I was on sabbatical in 1955–1956.

Any anxiety about progress in the lab during my absence was quickly dispelled by reports of very similar behavior in i-butyl chloride, i-butyl, and i-amyl bromide (47). The last was studied even more extensively in the next few years, thanks to the development of a microwave bridge method by Sivert Glarum (48) and other somewhat crude but usable methods to cover the range from 1 to 900 MHz, with the result that the relaxation spectrum of i-amyl bromide was reasonably well defined over 13 decades of frequency or time. At the same time that we were engaged in these and related studies, there were increasingly frequent reports of other examples of similar relaxation line shapes and non-Arrhenius temperature dependence in liquids and amorphous solids, including dielectric results for a variety of polymers and other systems, viscosity, and viscoelastic effects. Much of the dielectric work has been ably reviewed by Graham Williams (49), including the extensive work of his own research group, and a notable contribution of his was the finding of an empirical relaxation function variously known as the Williams-Watts, Kohlrausch-Williams-Watts (KWW), or "stretched exponential." (The addition of Kohlrausch is to recognize that he had used the same function many years before to describe time dependence of viscoelastic behavior.)

The term "stretched exponential" refers to the (macroscopic) relaxation function of the form $\gamma(t) = A \exp[-t/\bar{\tau}]^{\bar{\beta}}$ with $0 < \bar{\beta} < 1$, from which the complex permittivity ε^* can be obtained by the relation $\varepsilon^* - \varepsilon_\alpha = (\varepsilon_s - \varepsilon_\infty)[1 - i\omega L\gamma(t)]$, where L denotes the Laplace transform. Although no solution in terms of known functions exists, extensive numerical evaluations have been made and used to show that the calculated frequency dependence, although very similar to that of the skewed arc (Cole-Davidson) function above, is less asymmetric in the ε'' absorption line shape for values of β and $\bar{\beta}$ required to produce good average agreement. Numerous examples have been, or could be, cited for preferring one, the other, or neither, but while mentioning a few of these (50) I expressed my opinion that the similarities are more important than the differences. In any event, the KWW function has gained widespread acceptance and use, to the extent that it is often referred to as a universal function for relaxation processes, but so have others.

Whether or to what extent any function can be called universal is at least partly a matter of semantics. Dictionaries provide a multitude of definitions, ranging from the uncompromising "applicable to all cases" (Concise Oxford) to "any general or widely held principle, concept, or notion" (American Heritage). By the first definition, I know of no single relaxation function that is universal for liquids, let alone solids as well, whereas the second is more accommodating.

Theoretical interpretation of the widespread occurrence of characteristic patterns of behavior has until quite recently lagged far behind the accumulation of experimental evidence and empirical descriptions. A common practice has long been to beg the real questions by invoking distributions of relaxation times in a sum of individually exponential processes as an explanation. Ever since I first looked at the necessary form of such distributions to fit the data, I have been arguing that such a description may be both unnecessary and inappropriate, and I had accumulated a few examples to support the thesis drawn from electrolyte diffusion and extended electrical network theory. A more satisfying example became available when Sivert Glarum as part of his PhD thesis produced entirely on his own his "defect diffusion" model (51) of a linear chain of molecules relaxing exponentially unless and until the arrival of a defect, represented by random walk between sites, resulted in complete loss of orientational correlation. The resulting relaxation function $\gamma(t)$ is a sum of the two effects of an exponential and a cooperative diffusive process characterized by a complementary error function of time. Different ratios of the two characteristic times gave Debye relaxation in one limit, with the circular arc with $\alpha = 0.5$ in the other, and the skewed arc with $\beta = 0.5$ for equal times, with a continuous variation of breadth and asymmetry between the extremes.

Glarum's model was admittedly a highly simplified approach, and difficulties were encountered in attempting to elaborate on or modify his assumptions, but it also was instructive in showing how characteristic observed behavior could emerge from treatment of relaxation as a cooperative process. It has also evidently been seminal for recent developments: by Shlesinger & Montroll (52a,b) on non-Markovian distribution of pausing times between sequential reorientational steps, and by Anderson, Palmer, et al (53) on hierarchically constrained dynamics. Both lead for some choices of the distribution or constraints to relaxation functions essentially of the Williams-Watts form and also to predictions or inferences about their temperature dependence. Space here permits only two general comments about these very interesting developments.

The first comment is that although avowedly molecular in concept, there is no evident place in these theories for individuality of molecules, and the treatments are stochastic in nature. This comment is not meant to be unduly critical: treatment of many body dynamics at the molecular level remains formidable and, as Montroll once remarked to me, one adopts stochastic methods when the molecular going gets too rough. The second comment is that although these and other theories make connections between the onset of broadened relaxation line shapes and non-Arrhenius temperature dependence of characteristic times, there are several examples from experiment that one can perfectly well have one without the other. One already mentioned is of alcohols, which have both a principal relaxation of Debye form and relaxation time described by the VTF equation over considerable ranges of temperature. A second is of so-called conductivity relaxation in ionic glasses, where the converse is true.

A number of more specifically molecular models have also appeared, of which two for polymer chain dynamics are of interest in the present context because of their formulation and predictions. The first is the Shore & Zwanzig calculation (54) for transverse dipoles attached to polymer chain segments that are both coupled by torsional forces between segments and subject to rotational diffusion forces of the surroundings. Depending on the strength of the coupling forces relative to thermal energy kT, number of chain segments N, and number of these carrying interacting dipoles, the resulting relaxation spectra show deviations from simple Debye relaxation that are variously like the C-D, KWW, or Havriliak-Negami (55) function. (The last, which is an empirical combination of skewed and symmetrical circular arc functions, often fits polymer relaxation data better than the other functions with one less adjustable parameter.) The original paper should be consulted for further details; the above is no more than a sketch of a previously attempted summary (11) of the variety of conclusions that can be drawn about the relative importance of the assumed forces in chain dynamics.

A second kind of development by Skinner (56) invokes solitons as descriptions of localized conformational misalignments, or spins. These can perhaps be regarded as an example of the "defects" assumed by Glarum without discussion of such possible origins as localized fluctuations generated by the heat bath. Results for different constraints on the soliton dynamics also gave in this model relaxation functions of both Williams-Watts and Cole-Davidson form.

All these developments seem to show real promise for better understanding, but there are some awkward points from earlier work that have been brought back to mind by some recent preliminary results. Glarum

(51) found, for example, that experimental data for i-amyl bromide at room temperature and a limited range of microwave frequencies sufficed to show that the skewed arc representation of data at lower temperatures and frequencies was not a good fit. Rather, the absorption curve suggested the presence of two reasonably distinct peaks. These could plausibly be attributed to intramolecular conformational effects brought out by increased thermal energy, and some unpublished exploratory measurements by Elpidio Tombari using time domain (TDR) methods with high resolution to 10 GHz show what may be similar effects. These are shown in Figure 4 as complex plane loci for 1-bromopentane at $-96°$C and 1-iodoctane at $-40°$C, together with data for 3-bromopentane at $-96°$C, known from previous TDR measurements (50) to be accurately fitted by the skewed arc function, with $\beta = 0.88$ giving the *solid curve* in the figure. The deviations from similar behavior for the other two substituted alkanes are evident, but no attempt has yet been made to establish better representations.

A further interesting finding is that when the neat liquids are diluted with hexane the relaxation spectra shift to considerably higher frequencies but without any significant changes in shape. The choice of hexane as a nonpolar solvent limits the temperature range of dilution studies, but these limits are being explored by John Berberian using 3-methylpentane, known from his work to be a readily supercooled solvent. Just to tantalize me and any interested reader, he has told me of rather startling preliminary results at low temperatures that are as yet too fragmentary to be presented, much less rationalized, at this stage. Only one conclusion is clear: The old truism continues to hold that further investigations can, and often do, raise as many questions as they seek to answer and call for further work.

ELECTROLYTES AND CONDUCTING GLASSES

Although the Kraus tradition was strong in chemistry when I arrived at Brown in 1947, and I helped out a little with research in progress then, my first serious involvement with electrolytes was as usual in an unexpected and unplanned way. Ronald Gillespie wrote in 1953 to ask whether he could work in the laboratory to determine dielectric properties of neat sulfuric acid, which he needed to know better for his classic studies of ionic equilibria in this solvent. The extraordinarily high specific conductance, about that of 0.1 m KCl in water, gave us great difficulties with electrode polarization effects even at the highest frequency (<1 MHz) then available, but we did manage to obtain static permittivities of order 100, together with slight indications of dispersion at higher frequencies (57). When Stuart Lovell and I had developed methods for measurements to 250 MHz soon

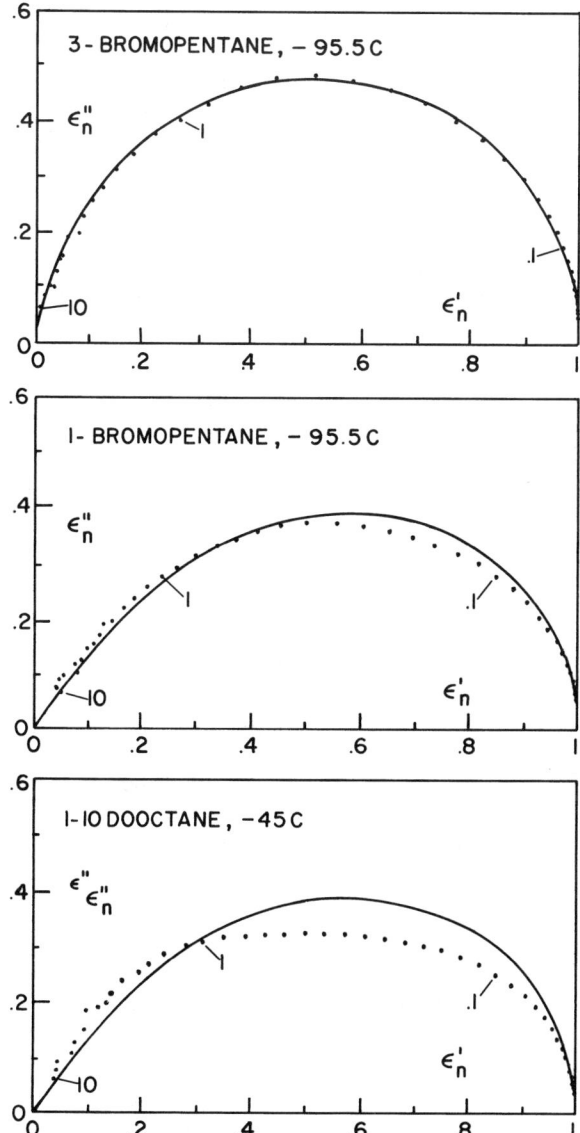

Figure 4 Complex plane plots of permittivity for three substituted alkanes. Indicated frequencies of experimental points are in GHz, *solid curves* are skewed-arc (CD) functions with $\beta = 0.88$ (3-bromopentane) and 0.60 (1-bromopentane, 1-iodo octane). The numbers along the curves are values of the frequency in GHz.

thereafter (58), sulfuric acid revisited gave some very strange results: Although the static permittivity was indeed very large as previously found, dielectric relaxation of Debye form as best we could tell was centered extraordinarily low frequency, and even slight deviations from the stoichiometric $H_2O \cdot SO_3$ composition produced large decreases of permittivity. I suspected instrumental errors, possibly as a result of higher conductivities from shifts of ionic equilibria, and refrained from publishing the results pending some grounds for more confidence in them. This finally came in 1974 on hearing a talk by Lars Onsager at Yale when Raymond Fuoss retired. My ears perked up when sulfuric acid was mentioned in the context of rate-limiting steps in proton transfer and kinetic depolarization by delayed dielectric response to changing ionic fields. I realized that the latter could explain our findings in terms of the very long solvent relaxation time, ca. 400 ps as compared to 8 ps of water and 50 ps of methanol, for example, and our results agreed with the Hubbard-Onsager prediction to the accuracy of the rather suspect measurements.

Soon thereafter, I was able to catalyze an interaction of Onsager and Hubbard with W. M. Van Beek and Michel Mandel at Leiden (59), as I had heard there about the latter's puzzlement of finding decreases of permittivity of salts in H_2O and MeOH that correlated better with conductance regardless of salt type than with theories of ion solvation. At the time, I had been developing time domain (TDR) methods for higher frequency dielectric measurements, and with David Hall returned to sulfuric acid as a prime example of dominant kinetic effects, because of the long relaxation time. The results (60), by a new method in an early stage, were gratifying as they confirmed and extended the previous results and also gave near quantitative agreement with the theory if "stick" boundary conditions for ions and solvent molecules were used in the hydrodynamic continuum model.

A series of studies of Paul Winsor of ions in other solvents (61) showed that for more representative electrolytes both the kinetic (depolarization) and static (solvation) effects had to be considered in comparison with theories of the two kinds of effects on dielectric polarization, and also on ion mobilities in conduction (62). A sticking point in comparing these and other experiments with theory is uncertainty about the roles of various complicating factors. Experimentally, for example, the classic Debye-Falkenhagen effect of ion atmosphere relaxation has yet to be observed directly and unambiguously as far as I know, and the problem of properly defining a "static" dielectric constant in a conducting medium is not trivial. Some of the theoretical difficulties have been stated succinctly by Hubbard, Colonomos & Wolynes in their paper on molecular theory of ion dynamics (63): "All equilibrium or static effects have been treated independently one from another; a complete theory should include all these effects in a self consistent form." I would also add *a fortiori*—and so should a complete dynamical theory. Ramifications as they appeared in 1980 have been ably reviewed by Wolynes (64).

Problems with the interplay of ionic charge transport and displacements are also evident in conducting glasses and the various interpretations of dielectric and conductive aspects of their behavior that have been proposed. These are suggested schematically in Figure 5 by plots of measured conductance and permittivity as a function of frequency. At low frequencies, the apparent conductance typically has a plateau taken to be a d.c. or steady state value, but with a decline at still lower frequencies accompanied by enormous increases in permittivity. These are ordinarily attributed to space charge effects at electrodes or other boundary layers rather than to intrinsic properties of the solid, but unambiguous separation of the two is even more difficult for solids than for conducting liquids.

My own attitude toward the low frequency problem over the years had been to avoid it as much as possible, but there was no escaping it entirely

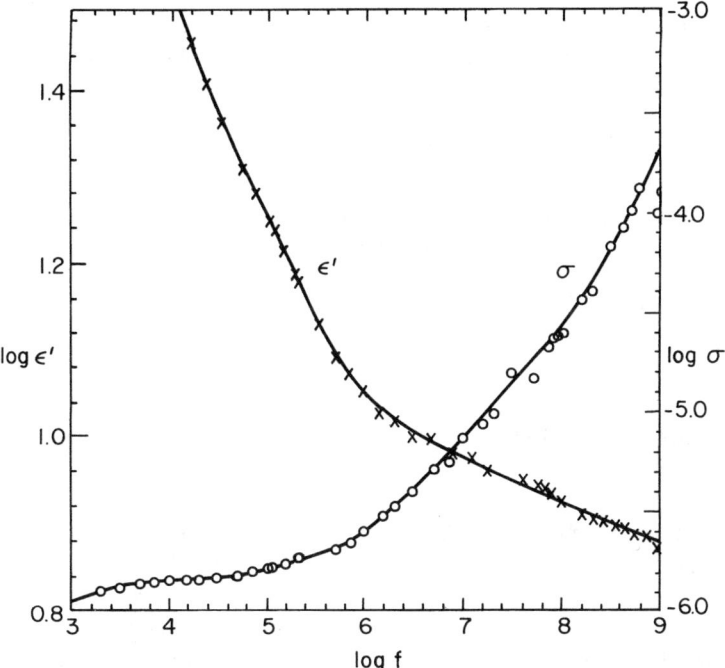

Figure 5 Logarithmic plots of permittivity ε' and conductivity σ' (mho/cm) for sodium trisilicate glass of 168°C.

when Bill Risen asked me and I agreed to look into what might be learned about fast ionic conducting glasses over the dielectric spectrum, and in particular at higher frequencies in the gap between bridge measurements below 1 MHz and far IR absorption well above 10 GHz. That there are thermally activated relaxation processes in addition to simple ion mobilities and glass network charge displacements complicated by space charge effects below 1 MHz was clear enough from the considerable and growing literature of low frequency results as typified by the plots in Figure 5. The region above 1 MHz was of particular interest to us because there seemed to be little or no indication of either conductance or permittivity approaching high frequency plateau values expected from applying either strong or weak liquid electrolyte theory to the solid state problem.

In progress to date on the problem, it has been possible by using high frequency bridge and TDR methods to show that for sodium trisilicate and lithium fluoroborate glasses conductance continues to rise and permittivity to decrease roughly as fractional powers of frequency to at least 3 GHz, and that the former may well join smoothly and continuously with the low frequency wing of far IR absorption, as fearlessly interpolated through four decades of frequency by Wong & Angell (65) in their log-log plots of absorption over 13 decades in magnitude and 12 decades in frequency. The excess of absorption seems, as Moynihan has remarked (66), "to be endemic to the solid state." Another feature that may also be common is the virtual independence of overall shape on temperature over a considerable range, despite shifts over several decades of the characteristic time scale.

All this and other evidence raises more questions than it provides answers. Are the patterns of behavior in these and other systems to be understood in terms of distinct low and high frequency processes or are they aspects of a continuum of behavior? Are they best interpreted as conductivity or polarization relaxations or (again) as aspects of a mingling of the two?

NONLINEAR DYNAMICS—SECOND AND HIGHER ORDERS

The appearance of Kubo-Green linear response theory and correlation function methods in the 1950s was another landmark for me, as it put into better perspective both the time evolution of dynamical processes generally and the relation of dielectric observables to other mechanical and thermal transport processes in particular. After Sivert Glarum introduced me to the subject as part of his thesis, I was able to use the new formalism to derive a number of earlier results obtained in other ways and to add a few new ones (67, 68). As a result, I was receptive to a proposal by Graham Williams that the possibility be explored of applying response theory and correlation function methods to analysis of the dynamics of Kerr effect birefringence more generally than for the rotational diffusion model of the classic work by Benoit (69, 70).

In Kerr effect experiments, anisotropically polarizable molecules are partially oriented by an external bias field E_b. The degree of orientation is detected by the induced birefringence of initially linearly polarized light passing through the sample at right angles to E_b. The dynamic effect as a measure of molecular response most simply consists in observing the changes in birefringence as a function of time after applying or removing a step-like pulse with sufficiently short transition time. For nonpolar molecules, only the first-order response proportional to E_b^2 is needed, and for axially symmetric molecules the first-order response is a measure of the correlation function for $P_2 (\cos \theta)$, where $\theta = \theta(t)$ is the angle between the axis and field directions and P_2 is the second Legendre polynomial. For molecules with permanent dipole moments, the analysis becomes more interesting, as dipole orientation energy proportional to E_b and $P_1 (\cos \theta)$ is also involved. This involvement changes the magnitude without affecting the P_2 character of response after E_b is removed, but it complicates the problem for response after E_b is applied because the second-order effect of dipole torques ($\sim E_b$) is required to obtain the overall response of order E_b^2.

There were intriguing differences between observed field-on and field-off responses when compared to each other and to P_1 responses from conventional dielectric measurements. Graham Williams had made me aware of these in his beautiful work on large molecules and supercooled liquids, which he has ably reviewed elsewhere (71), and I also knew of the extensive studies of biopolymers, notably by O'Konski and co-workers (72). I had supposed at the outset that a simple extension of response theory perturbation expansion in powers of E_b to second order in E_b^2 would readily yield simple expressions in terms of P_1 and P_2 correlations to account for the observations and thus increase their usefulness. The problem proved to be nontrivial, however, and it was only after considerable floundering and frustration that I was able to disentangle the nested convolutions of self and joint correlations in the field-on response to achieve the hoped for results (73).

If nothing else, this modest excursion into nonlinear dynamics gave me simple examples of several truisms. Two are that probes of two different aspects can be much more revealing than either one alone, and that higher order effects are much more than a simple extension of first order ones. A simple demonstration is that the relation of P_1 and P_2 responses is sensitive to the type of orientational dynamics, as the time constant τ_1 of P_1 response is three times τ_2 of P_2 response for small-step diffusional motion, whereas the two are the same for strong collisional loss of correlations at the other extreme, and that effects of both are observable by comparison of Kerr responses for field-on and off.

The construction of theories of nonlinear dielectric effects for molecules of arbitrary symmetry, with joint correlations, and to higher orders seems likely to encounter considerable difficulties, but at least two recent developments may help in overcoming them. The first is the approach of Fulton mentioned in the first section, the second are formal solutions, by Morita (74) and Morita & Watanabe (75), of the time evolution to arbitrary order (and increasing complexity) in powers of field strength, together with generalization of Kramers-Kronig and other relations among transient and steady state solutions.

Experimentally, new capabilities from laser techniques and high power fast electronic devices should make it possible to study the dynamics of smaller molecules and of larger ones to much higher frequencies.

CONCLUSION

The writing of any report or personal review such as this one invariably recalls unresolved questions put aside for one reason or another, brings some recognition of new ones in the light of recent developments, and adds to the store of things to think about. A "short list" of some of these, more or less in order of appearance as this is written, includes the following:

1. Small molecule dynamics and high frequency processes.
2. The roles of dipolar and shorter range forces in relaxation.
3. Unified theories of polarization and conduction.
4. Interfacial Polarization; Debye-Falkenhagen effect.
5. Nonlinear dynamics.

I am sanguine enough to believe that these and other dielectric problems, as aspects of much larger problems, will be as challenging and rewarding in the future as in the past. As Wolynes put it (64): "How a charge moves from place to place is a question that is likely to remain relevant for some time to come."

Acknowledgments

I owe thanks to many friends and colleagues for help over the years. Of them, I should mention those who have been especially helpful as I was thinking about and writing this chapter: John Berberian, Satoru Mashimo,

Bo Gestblom, Giorgio Chryssikos, Elpidio Tombari, and Andrew Burns. Support from the National Science Foundation through Grant CHE-8518365 is gratefully acknowledged.

Literature Cited

1. Lorentz, H. A. 1914. *Vortrage über die Kinetischen Theorie der Materie und der Eleckrizitat*, pp. 167–92. Leipzig: Teubner
2. Debye, P. 1912. *Physik Z.* 13: 97–100
3. Debye, P. 1929. *Polar Molecules.* New York: Chemical Catalog. 172 pp.
4. Van Vleck, J. H. 1940. *Ann. NY Acad. Sci.* 40: 293–313
5. Wyman, J. 1936. *J. Am. Chem. Soc.* 58: 1482–85
6. Onsager, L. 1936. *J. Am. Chem. Soc.* 58: 1486–93
7. Kirkwood, J. G. 1939. *J. Chem. Phys.* 7: 911–19
8. Conference papers. 1940. *Ann. NY Acad. Sci.* 40: 289–482
9. Kirkwood, J. G. 1940. See Ref. 8, pp. 315–20
10. Felderhof, B. U. 1979. *Physica A* 95: 572–80
11. Cole, R. H. 1984. In *Molecular Liquids, Dynamics and Interactions*, ed. A. J. Barnes, W. J. Orville-Thomas, J. Yarwood, pp. 59–110. Dordrecht: Reidel. 590 pp.
12. Lax, M. 1952. *J. Chem. Phys.* 20: 1351–59
13. Toupin, R. A., Lax, M. 1957. *J. Chem. Phys.* 27: 458–64
14. Wertheim, M. S. 1979. *Annu. Rev. Phys. Chem.* 30: 471–501
15. Alder, B. J., Pollock, E. L. 1981. *Annu. Rev. Phys. Chem.* 32: 311–30
16. Brot, C., Bossis, G. 1980. *Mol. Phys.* 40: 1053–72
17. Fulton, R. L. 1975. *Mol. Phys.* 29: 405–13
18. Fulton, R. L. 1983. *J. Chem. Phys.* 78: 6856–84
19. Berberian, J. G., Cole, R. H. 1969. *Rev. Sci. Instrum.* 40: 811–17
20. Davidson, D. W., Auty, R. P., Cole, R. H. 1951. *Rev. Sci. Instrum.* 22: 678–82
21. Mopsik, F. I. 1984. *Rev. Sci. Instrum.* 54: 79–87
22. Fellner-Feldegg, H. 1972. *J. Phys. Chem.* 76: 2116–22
23. Cole, R. H., Berberian, J. G., Mashimo, S., Chryssikos, G. D., Tombari, E., Burns, A. 1989. *J. Appl. Phys.* July
24. Gerschel, A., Grochulski, T., Kisiel, Z., Pszczolkowski, L., Leibler, K. 1985. *Mol. Phys.* 54: 97–117
25. Gerschel, A. 1984. See Ref. 11, pp. 163–99
26. Krieger, T. J., James, H. M. 1954. *J. Chem. Phys.* 22: 796–814
27. Zwanzig, R. W. 1956. *J. Chem. Phys.* 25: 211–16
28. Buckingham, A. D., Pople, J. A. 1955. *Trans. Faraday Soc.* 51: 1029–35; 1957. *J. Chem. Phys.* 27: 820–21
29. Johnston, D. R., Oudemanns, G. J., Cole, R. H. 1960. *J. Chem. Phys.* 33: 1310–17
30. Orcutt, R. H., Cole, R. H. 1967. *J. Chem. Phys.* 46: 697–702
31. Kirkwood, J. G. 1936. *J. Chem. Phys.* 4: 592–601
32. Bose, T. K., Cole, R. H. 1970. *J. Chem. Phys.* 52: 140–47
33. Buckingham, A. D., Disch, R. L. 1963. *Proc. R. Soc. London Ser. A* 273: 275–89
34. Sutter, H., Cole, R. H. 1970. *J. Chem. Phys.* 52: 132–39
35. Spurling, T. H., Mason, E. A. 1967. *J. Chem. Phys.* 46: 322–26
36. Weiss, S., Cole, R. H. 1967. *J. Chem. Phys.* 46: 644–49
37. Brown, N. L., Cole, R. H. 1953. *J. Chem. Phys.* 21: 1920–26
38. Hoshino, S., Shimaoka, K., Nimura, N. 1967. *Phys. Rev. Lett.* 19: 1286–88
39. Cichanowski, S. W., Cole, R. H. 1973. *J. Chem. Phys.* 59: 2420–26
40. Cole, R. H., Havriliak, S. Jr. 1957. *Discuss. Faraday Soc.* 23: 31–38
41. Groenewegen, P. P. M., Cole, R. H. 1967. *J. Chem. Phys.* 46: 1069–74
42. Auty, R. P., Cole, R. H. 1952. *J. Chem. Phys.* 20: 1309–14
43. Davidson, D. W., Auty, R. P., Cole, R. H. 1951. *Rev. Sci. Instrum.* 22: 678–82
44. Cole, K. S., Cole, R. H. 1941. *J. Chem. Phys.* 9: 341–51; 1942. *J. Chem. Phys.* 10: 98–105
45. Davidson, D. W., Cole, R. H. 1951. *J. Chem. Phys.* 19: 1484–90
46. Smyth, C. P. 1955. *Dielectric Behavior and Structure*, pp. 99–131. New York: McGraw-Hill. 441 pp.
47. Denney, D. J. 1957. *J. Chem. Phys.* 27: 259–64
48. Glarum, S. H. 1958. *Rev. Sci. Instrum.* 29: 1016–19

49. Williams, G. 1984. See Ref. 11, pp. 239–74
50. Berberian, J. G., Cole, R. H. 1986. *J. Chem. Phys.* 84: 6921–27
51. Glarum, S. H. 1960. *J. Chem. Phys.* 33: 639–43
52a. Shlesinger, M. F., Montroll, E. W. 1984. *Proc. Natl. Acad. Sci. USA* 81: 1280–83
52b. Bendler, J. T., Shlesinger, M. F. 1987. *J. Mol. Liq.* 36: 37–46
53. Palmer, R. G., Stein, D. L., Abrahams, E., Anderson, P. W. 1984. *Phys. Rev. Lett.* 53: 958–65
54. Shore, J. E., Zwanzig, R. W. 1975. *J. Chem. Phys.* 63: 5445–58
55. Havriliak, S. Jr., Negami, S. 1966. *J. Polym. Sci. Polym. Symp.* C14: 97–117
56. Skinner, J. L. 1983. *J. Chem. Phys.* 79: 1955–64
57. Gillespie, R. J., Cole, R. H. 1956. *Trans. Faraday Soc.* 52: 1325–31
58. Lovell, S. E., Cole, R. H. 1959. *Rev. Sci. Instrum.* 30: 361–62
59. Hubbard, J. B., Onsager, L., Van Beek, W. M., Mandel, M. 1977. *Proc. Natl. Acad. Sci. USA* 17: 401–4
60. Hall, D. G., Cole, R. H. 1981. *J. Chem. Phys.* 85: 1065–69
61. Winsor, P. IV, Cole, R. H. 1982. *J. Phys. Chem.* 86: 2486–94
62. Evans, D. F., Tominaga, T., Hubbard, J. B., Wolynes, P. G. 1979. *J. Phys. Chem.* 83: 2669–77
63. Hubbard, J. B., Colonomos, P., Wolynes, P. G. 1979. *J. Chem. Phys.* 71: 2652–61
64. Wolynes, P. G. 1980. *Annu. Rev. Phys. Chem.* 31: 345–76
65. Wong, J., Angell, C. A. 1976. *Glass, Structure by Spectroscopy*, Chapter 11. New York: Dekker. 864 pp.
66. Moynihan, C. T., Boesch, L. P., Laberge, N. L. 1973. *Phys. Chem. Glasses* 14: 122–25
67. Cole, R. H. 1965. *J. Chem. Phys.* 42: 637–43
68. Cole, R. H. 1981. In *Physics of Dielectric Solids*, ed. C. H. L. Goodman, Conf. Ser. No. 58, pp. 1–21. Bristol/London: Inst. Physics. 151 pp.
69. Benoit, H. 1951. *Ann. Phys. Paris* 6: 561–609
70. Benoit, H. 1952. *J. Chim. Phys.* 49: 517–21
71. Williams, G. 1984. See Ref. 11, pp. 237–74
72. O'Konski, C. T., ed. 1976. *Molecular Electro-Optics*, Vol. 1. New York: Dekker
73. Cole, R. H. 1982. *J. Phys. Chem.* 86: 4700–3
74. Morita, A. 1986. *Phys. Rev. A* 34: 1499–1504
75. Morita, A., Watanabe, H. 1987. *Phys. Rev. A* 35: 2690–96

ORIENTATIONAL GLASSES

A. Loidl

Institut für Physik, Universität Mainz, D-6500 Mainz, Federal Republic of Germany

INTRODUCTION

Orientational glasses have received considerable attention during the last decade. Orientational glasses share many properties with glasses; however, unlike canonical glasses their crystalline structure is well understood. The common signature of this new class of disordered materials is relaxation phenomena of the orientational degrees of freedom, analogous to those observed in spin glasses, canonical glasses, and polymers. In addition, orientational glasses exhibit the low-temperature thermodynamic, elastic, and dielectric properties characteristic of amorphous systems. Thus, it seems that orientational glasses are model systems of the amorphous state of condensed matter. For example, the linear term in the specific heat in the cyanide glasses has been calculated from a microscopic model. These theoretical studies were chosen as among the most outstanding achievements in solid state physics for 1985 (1).

In orientational glasses no long-range orientational order can be achieved and the orientational variables exhibit a frozen-in disordered state at low temperatures even though strong interaction forces are active. It is essential that this glassy low temperature state is established by a collective freezing process of the orientational degrees of freedom and does not occur simply because of the slowing down of thermally activated processes in the crystal field of the neighboring host atoms, i.e. a relaxation in an anisotropic crystal lattice. Well-known examples of this latter case are hexagonal ice and carbon monoxide. In these crystals the orientational disorder is frozen well above the hypothetical ordering temperature because of the high activation energy required for relaxational processes compared to the ordering energy. These systems are not considered in the present review.

The orientational variables usually are the rotational degrees of freedom of an aspherical molecule that can reorient between symmetry-equivalent directions in the crystal but can also be represented by pseudospins, e.g. off-center atoms that jump between equivalent lattice sites. In a concentrated system, the orientational degrees of freedom will exhibit long-range orientational order (assuming that finite interactions exist between the "spins" and that the hindering barriers for reorientations due to the lattice anisotropy are not too high to allow an ordering process in experimentally accessible times). Site disorder, however, e.g. a dilution of the anisotropic molecules through spherical atoms, immediately introduces competing interactions and can yield a glassy ground state. The analogy to spin glasses is straightforward. In the spin glass problem, site disorder of the spins and the Ruderman/Kittel/Kasuya/Yosida (RKKY) interaction, which changes sign with distance, yield frustrated interactions and thus a nonordered low-temperatue state. In orientational glasses, site disorder of the orientational variables and anisotropic dipolar or quadrupolar interaction forces are responsible for the orientational glass state. The interactions can be dominated by direct multipolar forces (electrostatic dipolar, quadrupolar, or octopolar forces) or can be indirect, e.g. mediated via lattice strains like in the alkali halide alkali cyanide mixtures.

The following classes of mixed molecular crystals exhibit a glassy state at low temperatures:

1. Mixed molecular systems in which the anisotropic molecules are substituted by isotropic atoms or ions: $KCl:KOH$ (2), *ortho–para* hydrogen mixtures (3), $KCN:KBr$ (4–6), $N_2:Ar$ (7–9), and $CH_4:Kr$ (10) are the most prominent examples. Off-center systems like $KTaO_3:Li$ (11), in which the thermal hopping of the off-center impurities (Li) can be described in terms of a pseudospin formalism, are also related to this group. A schematic x-T phase diagram for this type of mixed crystals is shown in Figure 1a. At high T the mixtures and the pure molecular compound exhibit a plastic phase in which the high symmetry phase is stabilized by a fast reorientational motion of the molecules (pseudospins). With decreasing temperatures, the crystals with a high molecular concentration x undergo transitions into orientationally ordered phases. For decreasing concentrations the ordering temperatures decrease because the interaction forces scale with x. However, below a critical concentration x_c, long-range orientational order is suppressed and the crystals freeze into a disordered glassy state. Some of the compounds mentioned above are characterized by a strong coupling of the orientational degrees of freedom to translational phonon modes. Hence the freezing-in of the orientational disorder strongly disturbs the long-range translational order of the center of mass lattice, and the low-temperature state can be viewed as intermediate between the crystalline and the amorphous states.

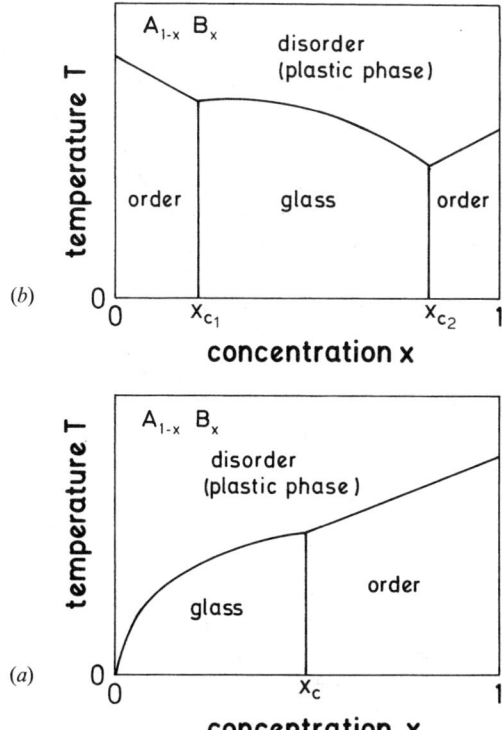

Figure 1 (a) *Lower frame*: Schematic x-T phase diagram of $A_{1-x}B_x$ mixed crystals. B denotes a molecular compound (e.g. KCN, N_2, o-H_2 etc), A a compound with no orientational degrees of freedom (e.g. KBr, Ar, p-H_2 etc). (b) *Upper frame*: Schematic x-T phase diagram of $A_{1-x}B_x$ mixed crystals. The pure compounds are elastically or electrically ordered. In the case of electric order, the high-T phase is paraelectric.

2. Mixed dipolar crystals in which the two components exhibit ferroelectric and antiferroelectric interactions: in RbH_2PO_4 : $NH_4H_2PO_4$ (RADP) (12, 13) a frustrated ground state is believed to occur via the two competing electric interactions. Glassy behavior has also been reported for the closely related system, RbH_2AsO_4 : $NH_4H_2AsO_4$ (RADA) (14). Mixtures of ferroelectric Betainphosphite [$(CH_3)_3NCH_2COO \cdot H_3PO_3$] and antiferroelectric Betainphosphate [$(CH_3)_3NCH_2COO \cdot H_3PO_4$] seem also to belong to this class of compounds (J. Albers, unpublished results). A schematic phase diagram is shown in Figure 1b. The pure compounds, A and B, exhibit ferroelectric or antiferroelectric order, respectively. However, at intermediate concentrations, long-range electric order is suppressed and a low-temperature glassy state occurs.

So far I have discussed the orientational glasses in terms of random bonds, analogous to the spin glass problem. However, another group of molecular crystals exists in which the long-range orientational order is suppressed by random fields. Well-known examples of this group are NaCN:KCN mixtures (15, 16). Both pure compounds exhibit a plastic high-temperature phase and undergo a cubic to orthorhombic phase transition. The orthorhombic structure is orientationally ordered. However, already the substitution of approximately 15% sodium by potassium (or vice versa) is enough to suppress the orientational order of the CN molecules. The phase diagram looks similar to that shown in Figure 1b. In this type of molecular crystal the sublattice of the aspherical molecules remains undiluted. Hence it is plausible that frustrated interactions are unimportant and the low-temperature state is solely determined by the strength of the random fields: in the case of NaCN:KCN, random strains are introduced by the volume difference of the impurity atoms that cause (quenched) elastic relaxations of the neighboring ions. After the importance of random fields in this system was recognized, their influence was investigated in a number of systems that previously had been treated as pure random bond problems only (17).

Frozen-in orientational disorder also occurs in undiluted molecular crystals. These substances, which are commonly termed *glassy crystals*, exhibit a stable ordered low temperature phase that can be suppressed by rapid cooling. Cyclohexanol is a paramount example (18). Another system in which the plastic phase can be supercooled is cyanoadamantane (19, 20). Both systems exhibit a thermodynamic anomaly at the glass transition (analogous to the findings in canonical glasses but contrary to the smooth specific heat at the glass transition in orientational glasses and in spin glasses). The cooling rate and the steric hindrance of the molecules with respect to reorientations are important parameters to reach the glassy state in undiluted molecular crystals. Glassy crystals are discussed in reviews by Suga (18) and by Suga & Seki (21).

The plan of this article is as follows: After a historical review I focus on the alkali halide–alkali cyanide mixed crystals. This restriction can be justified threefold: 1. some general considerations and theoretical concepts are rather unique for all orientational glasses and can be demonstrated well within one class of compounds; 2. the cyanide glasses play an outstanding role because of the strong coupling of rotational and translational degrees of freedom. A freezing-in of orientational correlations yields a breakdown of true long-range translational order. Thus alkali halide–alkali cyanide mixtures are intermediate between spin glasses and canonical glasses. 3. Reviews on *ortho*- and *para*-hydrogen mixtures (22), RADP (23), and glassy crystals (18, 21) have been published previously.

HISTORICAL SURVEY

Orientational glasses reconcile the relaxation dynamics of spin glasses and the low temperature thermodynamic properties of amorphous systems. Therefore any historical review of orientational glasses that neglects the pertinent theories and the relevant experimental results in the fields of spin glasses and amorphous systems is incomplete.

Low-Temperature Specific Heat in Glasses

In 1971 Zeller & Pohl (24) found that the low-temperature specific heat in amorphous systems can be described by

$$C = C_1(t)T + C_{exc}T^3 C_D T^3$$

where C_1 is a time-dependent term in the specific heat that is linear in temperature, C_D characterizes the normal Debye-specific heat of the crystalline state, and C_{exc} is an excess contribution to the specific heat that follows a T^3 dependence in a limited temperature range $5K \leq T \leq 20K$). This low-temperature specific heat seems to be a universal feature of the glassy state. The linear term has been successfully interpreted in terms of a tunneling model by Anderson, Halperin & Varma (25) and independently by Phillips (26). With the assumption of a constant density of tunneling states they were able to calculate the experimentally observed time and temperature dependence. On the other hand, no convincing theory exists to explain the excess T^3 term.

Freezing Process in Spin Glasses

In dilute magnetic systems like Cu:Mn, a smeared-out magnetic "phase transition" was detected by Owen et al (27) and Jacobs & Schmitt (28) and interpreted as an ordering of antiferromagnetic arrays distributed randomly in a nonmagnetic matrix. In very low fields, Cannella & Mydosh (29) found a sharp frequency-dependent magnetic cusp in the magnetic susceptibility in the dilute magnetic alloy Cu:0.9%Mn. These cusps in the susceptibilities were found to be universal features of spin glasses and were explained by two characteristic features: namely, randomness (site disordered spins) and competing interactions. In the classical spin glasses, the spins interact indirectly via a polarization (scattering) of the conduction electrons. This RKKY interaction [Ruderman & Kittel (30), Kasuya (31), and Yosida (32)] oscillates with distance. As the distances between the spins are random, the spins cannot satisfy all bonding conditions (frustrated interactions). Hence the spins undergo a collective freezing into random orientations. A detailed review on spin glasses has been given by Binder & Young (33).

Orientational Glasses

The first measurements on orientational glasses were performed by Känzig et al (2) in 1964. These authors investigated the temperature-dependent dielectric behavior of KCl:OH at low temperatures for different dipole concentrations and found concentration-dependent cusps in the real part of the dielectric constant ε'. The temperature at which the peak occurred was almost independent of frequency and increased with the OH^- concentration. The authors stressed the importance of the dipole-dipole interaction forces and explained the decrease in ε' on the low-T side of the cusps as due to dipolar interactions, not to relaxation. However, they interpreted the data in terms of a smeared-out ferroelectric phase transition. In 1965 Brout (34) gave an interpretation of the ordered state in terms of random antiferroelectric arrays reminiscent of the dilute magnetic system Mn in Cu treated by Klein & Brout (35). This paper explained the experimental results in Cu:Mn as a "random antiferromagnet" in which the low-temperature state is made up of small clusters of impurities interacting via the RKKY potential. According to this picture the cusps in the susceptibilities are due to two competing processes: namely a randomization of the clusters that results in a decrease of the susceptibility and a break-up of the internal structure of the clusters that increases the effective number of spins and thus increases the susceptibility (35), as the temperature is raised. Fisher & Klein (36) put these ideas of a phase transition of dipoles dissolved in alkali halides on a quantitative theoretical basis. In 1979 they extended their work to strain defects in alkali halides (37), where an indirect strain-mediated interaction between the defects (e.g. CN^- in KCl) is responsible for the orientational freezing. These calculations suggested a strong analogy between the low-temperature thermodynamic properties of interacting strain defects and those of glasses.

Sullivan et al (1978) (3) interpreted low-temperature NMR studies in *ortho*- and *para*-hydrogen mixtures in terms of a quadrupolar glass phase. p-H_2 has a rotational wave function that is spherical in the ground state (quantum number $J = 0$), whereas o-H_2 has a rotational quantum number $J = 1$ in the ground state and thus can be regarded as a "magnetic" species. The electrostatic quadrupolar moment of the o-H_2 gives rise to an anisotropic orientation-dependent interaction. The phase diagram of $(p\text{-}H_2)_{1-x}(o\text{-}H_2)_x$ is very similar to that shown schematically in Figure 1a. An orientationally disordered (plastic) phase is found at high temperatures with a hexagonal close-packed structure. Below 3K and for high *ortho* concentrations, the crystals undergo phase transitions into a state in which the quadrupoles are orientationally ordered. Below a critical concentration of $x_c \sim 0.55$, however, the orientational disorder is frozen-in. For this concentration range Sullivan et al (3) have reported a "spin glass" transition, a conclusion that they drew from a sudden change in the NMR line shapes. Analogous to the spin glass systems, the results were interpreted as the collective freezing process of the *ortho* molecules into random configurations.

A key experiment demonstrating the orientational glass behavior in $(KBr)_{1-x}(KCN)_x$ was performed by Satija & Wang (38) in 1978. These authors studied the temperature-dependent transverse sound waves by Brillouin scattering in a wide concentration range x and found that below a critical concentration x_c the elastic constants c_{44} pass through a minimum and increase again toward lower temperatures. The observed temperature-dependence in the elastic behavior of KBr:KCN mixed crystals corresponds to cusps in the quadrupolar susceptibility. From the fact that crystals with $x \leq x_c$ remain clear (single domain) the authors concluded that these cusps indicate no phase transition but can be interpreted as pure relaxational phenomena. They missed the importance of the possibly collective character of the freezing-in and hence the close analogy to spin glass behavior. Subsequently, Rowe et al (4) published neutron scattering results for $(KBr)_{0.5}(KCN)_{0.5}$, in which they studied the temperature-dependence of long-wavelength phonons with a sound velocity proportional to c_{44}. Analogous to the findings in the Brillouin scattering experiments, c_{44} passed through a minimum. In addition, a strong central peak intensity appeared at low T. These observations have been interpreted as possibly being due to a transition into an elastic dipole glass, as proposed by Fischer & Klein (37) using a microscopic theory of tunneling centers interacting via lattice strains and by Klenin (39) on the basis of computer simulation experiments. Later on these neutron scattering results were interpreted by Michel & Rowe (40) in terms of an orientational glass, for which the central line reflects the temperature dependence of the order parameter of the glass phase. The order parameter was defined analogous to the proposal of Edwards & Anderson (41) for magnetic impurities. In the case of KBr:KCN the CN molecules interact via elastic deformations of the host lattice. A random distribution of the impurity ions leads to a random angular distribution of the quadrupole-quadrupole interaction forces. Thus, site disorder and anisotropic interactions produce the frustration that is believed to be essential for the glassy state.

The field of orientational glasses has greatly expanded since the early 1980s. New systems have been investigated and new experimental techniques have become available. Investigations of systems like N_2:Ar (7-9), CH_4:Kr (10), $KTiO_4$:Li (11), and RbH_2PO_4:$NH_4H_2PO_4$ (12) have greatly contributed to our understanding of the orientational glass state. RADP especially has attracted considerable attention during the past few years.

Competing short-range interactions are introduced by mixing ferroelectric RbH_2PO_4 and antiferroelectric $NH_4H_2PO_4$, which form isomorphous tetragonal crystals. Thus RADP mixed crystals exhibit a glass-like phase in a broad concentration range. This result was first shown by Courtens in 1982 (12) with birefringence and dielectric techniques. Later on it was shown that the freezing process of the frustrated proton ordering follows a Vogel-Fulcher law with a Vogel-Fulcher (glass) temperature of 10K (13).

THE CYANIDE GLASSES

The cyanide glasses play an important role in the field of orientational glasses: The specific heat anomalies (42, 43), the low-temperature dielectric (44) and elastic (45) properties, were found to be essentially analogous to those in canonical glasses, like vitreous silica or PMMA (polymethyl-methacrylate). The cusps in the susceptibility indicated a freezing process with a similar dynamics as that observed in spin glasses (46, 47). In addition, the cyanide glasses are characterized by a strong coupling of the rotational and translational degrees of freedom. Thus, a collective freezing-in of the orientations breaks the long-range translational order of the crystal. Near the critical concentrations the cyanide glasses are characterized by highly anomalous line shapes of the elastically scattered neutron intensities (48).

My discussion of the cyanide glasses focuses on two systems: KBr:KCN and NaCN:KCN. In the former compounds, the aspherical CN molecules are substituted by spherical halogen ions. Hence, the quadrupoles are site disordered. In NaCN:KCN, the anisotropic species remains undiluted and the alkali metal is substituted. These mixtures exhibit no quenched site disorder of the orientational variable, and it seems plausible to assume that the glass state occurs due to random fields.

Phase and Glass Transitions

ELASTIC AND ELECTRIC PHASE TRANSITIONS The pure cyanides KCN and NaCN exhibit a cubic structure (NaCl-type) at high temperatures. This plastic phase is stabilized by a fast reorientational motion of the CN molecules between symmetry equivalent orientations. The maximum probability for the CN orientations is along $\langle 111 \rangle$ (49, 50). At 168K for KCN and at 288K for NaCN [for references see (51)] there is a first-order phase transition to an orthorhombic structure with the CN^- ions aligned along the former cubic $\langle 110 \rangle$ axis. The phase transition is accompanied by an extreme softening of the elastic constant c_{44} (shear strains along the cube axis) (52). The softening of c_{44} is driven by a strong rotation translation coupling that is mediated by elastic strains (53). The phase transition in the cyanides can be considered analogous to a cooperative Jahn-Teller effect (54). At 83K for KCN (172K for NaCN), an order–disorder phase transition yields an antiferroelectric order of the static dipoles connected to the CN^- ions (head to tail order) (51).

PHASE DIAGRAMS   Solid solutions between the cyanides NaCN, KCN, and RbCN and the alkali halides NaCl, KCl, and KBr can be formed over the entire concentration range (51). To take $(KBr)_{1-x}(KCN)_x$ mixed crystals as an example, the substitution of the aspherical CN^- ions by the spherical Br^- ions lowers the transition temperatures of both phase transitions, the elastic and the electric transition, respectively. A number of studies of the low-temperature phases have been made by X-ray and neutron diffraction, and transition temperatures have been determined by specific heat measurements [a complete list of references for the KBr:KCN system can be found in (55)]. The elastic phase transition is lowered, reaching about 90K for $x = 0.60$. For $x \leq 0.6$ no structural phase transitions can be detected and the crystals stay "pseudo-cubic" down to the lowest temperatures. In this concentration regime a new low-temperature state, a quadrupolar glass state, is observed that is indicated by frequency-dependent cusps in the quadrupolar (elastic) susceptibility. The low-temperature, elastically ordered phases in KBr:KCN pass through the sequence orthorhombic–monoclinic–rhombohedral–glass/pseudocubic as the concentration x is lowered. Five percent of Br suppresses the antiferroelectric phase transition. For concentrations $x \leq 0.95$, no tendency for dipolar order can be detected. A phase diagram (55) is shown in Figure 2. Phase diagrams for KCl:KCN, KI:KCN (56), and NaCl:NaCN (57) have been reported, which exhibit a similar polymorphism. Attempts have been made to describe the low-temperature phases in KBr:KCN theoretically by the delicate balance between the strain-rotation and the rotation-translation couplings (58) and in terms of a Landau theory (59). The various phase diagrams of the systems investigated exhibit quite different critical concentrations. This fact immediately suggests that static random strains that couple to the reorientational motion play an essential role (17).

A phase diagram for $(NaCN)_{1-x}(KCN)_x$ as determined by neutron diffraction (60) is shown in Figure 3. A preliminary phase diagram was constructed from dielectric and optical measurements, which showed that over a wide concentration range there appears no structural phase transition (15). Figure 3 shows that only close to the pure cyanides can structural phase transitions be observed, yielding two critical concentrations $x_{c1} = 0.15$ and $x_{c2} = 0.9$ with a broad glassy regime at intermediate concentrations. Also the antiferroelectric phase transition is suppressed rapidly. Again, as in KBr:KCN a striking polymorphism is found, which

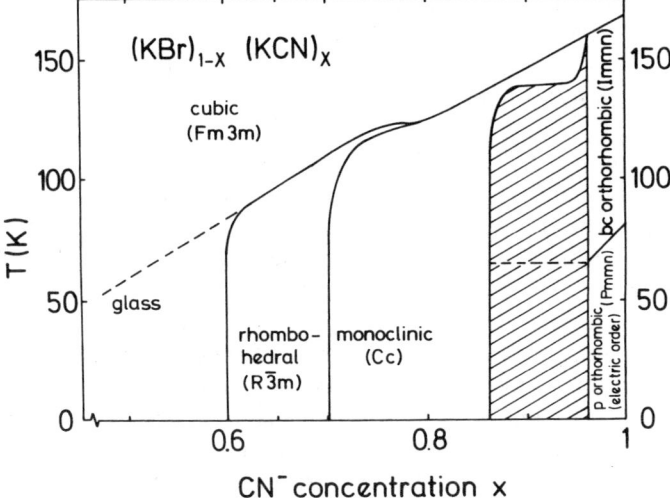

Figure 2 x-T phase diagram of $(KBr)_{1-x}(KCN)_x$. The *solid lines* give estimates of the phase boundaries between different crystallographic structures. The *hatched area* indicates a coexistance region. Figure taken from Ref. (55), reproduced with permission from *Zeitschrift für Physik* (Springer-Verlag).

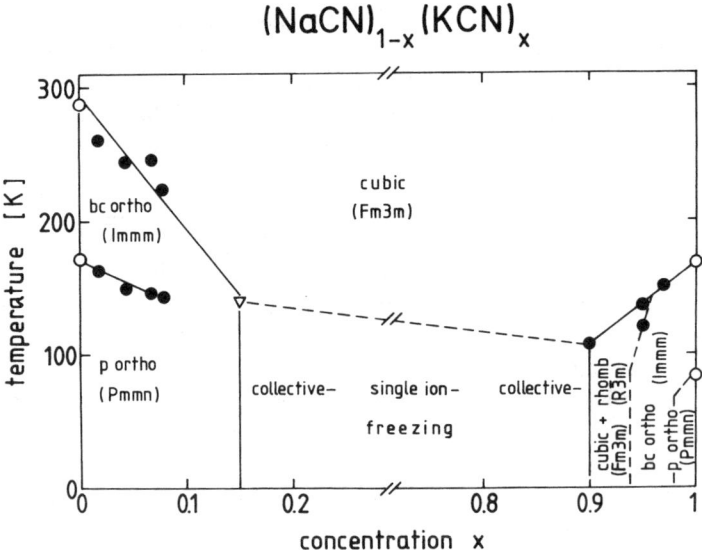

Figure 3 x-T phase diagram of $(NaCN)_{1-x}(KCN)_x$ as determined from neutron scattering experiments (60). The *solid lines* give estimates of the phase boundaries between different crystallographic structures. The *dashed lines* are rough estimates only.

is reminiscent of the structural polymorphism in a number of canonical glasses. It is still an open question, however, whether or not the tendency to form different structural phases is directly related to the tendency to form a glassy state. The ability of many covalent liquids to form a glassy state appears to be a direct consequence of the indifference of the crystalline network structure with respect to distortions (61).

THE STRUCTURE OF THE GLASS STATE In single crystal diffraction experiments the glass state is indicated by a rapid increase of diffuse scattered intensities around the reciprocal lattice points. These diffuse scattered intensities can be explained by frozen-in local shear distortions (transverse phonons with zero energy). Transverse scans through the (060) reflection in $(KBr)_{0.43}(KCN)_{0.57}$ are shown in Figure 4 to illustrate this behavior. These results have been obtained with single crystal neutron diffraction (50, 55). A strong increase of the diffuse scattered intensities and an equally strong decrease of the pure Bragg scattering contributions were found in a narrow temperature range (50). The profiles can be described by a resolution limited Bragg scattering (*empty Gaussian area*) and by an exponentially decaying diffuse contribution (*shaded area*). The exponential decay was a rather unexpected feature of the diffuse line shapes (48, 50). From a bilinear coupling of rotations and translations one would expect a $1/q^2$ dependence of the line shapes, where q is the phonon wave number (40). It was suggested that the exponential line shapes originate from higher order terms in the coupling and demonstrate the importance of anharmonic contributions (50). In $(KBr)_{0.43}(KCN)_{0.57}$, no long-range orientational order is established and coupled rotational and translational correlations are frozen-in. On the average this crystal stays cubic down to the lowest temperatures, with local shear distortions appearing for $T \leq 100K$. The observed line shapes can be interpreted as a distribution of static T_{2g} shear strains centered around the undistorted lattice. It was concluded that the glass state in the cyanide glasses is built up of clusters with a distribution of shear deformations centered around zero strain (50, 55).

Specific Heat Measurements

PHASE AND GLASS TRANSITION ANOMALIES Specific heat measurements focusing on the phase and glass transition temperature anomalies in KBr:KCN mixed crystals have been performed by Suga et al (62) and subsequently Lüty (51), Mertz & Loidl (63, 64), Moriya et al (65), and Matsuo et al (66). Representative results of the adiabatic specific heat for the mixed molecular crystals $(KBr)_{1-x}(KCN)_x$ for concentrations $x = 0.53, 0.57, 0.65, 0.73, 0.84$, and 0.93 are shown in Figure 5 (64). The specific heat anomalies for $x \geq 0.6$ indicate the phase transitions into

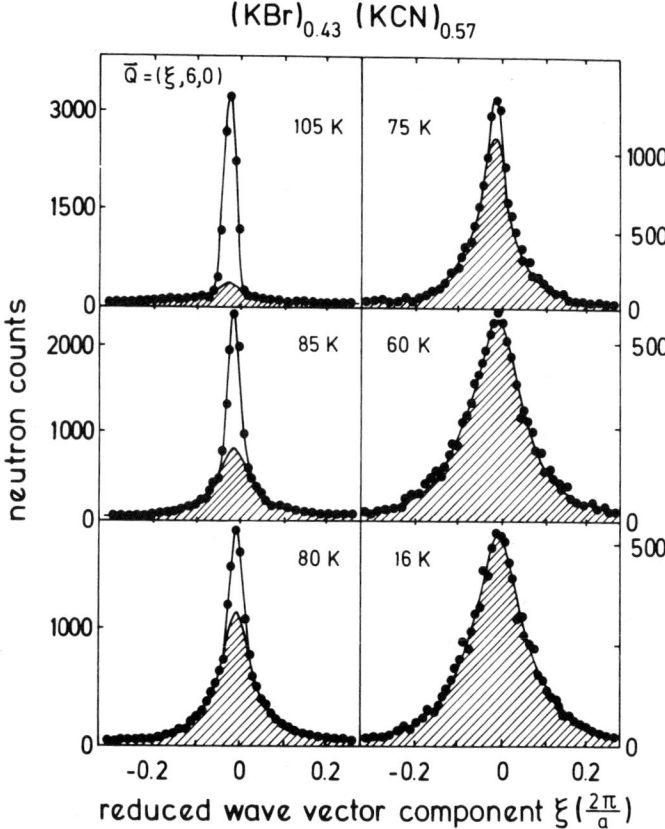

Figure 4 Transverse scans at zero energy transfer through the (060) reflection in $(KBr)_{0.43}(KCN)_{0.57}$ at various temperatures. The *shaded areas* correspond to scattering from frozen-in strain fields, the *empty areas* to Bragg scattering. Figure taken from Ref. (50) and reproduced with permission from the American Physical Society.

elastically or electrically ordered phases. For $x \leq 0.6$, no specific heat anomalies are detectable. All c_p versus T curves look smooth down to the lowest temperatures. The entropy changes have been determined from the specific heat anomalies of Figure 5 (64). The configurational entropy changes decrease strongly with decreasing concentration and extrapolate to zero at $x \approx 0.6$, the critical concentration. The absence of a specific heat anomaly at the glass transition is a characteristic feature of orientational glasses and is similar to the findings in spin glasses, where nothing happens at the freezing temperature (33). In contrast, in canonical glasses the glass transition temperature is characterized by a jump in the specific heat. The

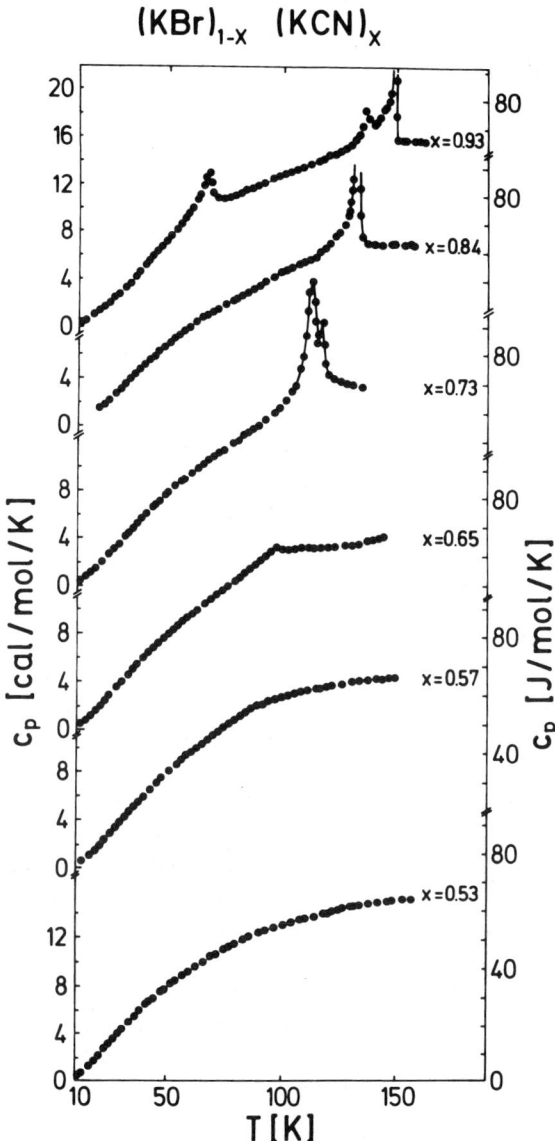

Figure 5 Adiabatic specific heat c_p for mixed molecular crystals $(KBr)_{1-x}(KCN)_x$ for concentrations $0.5 \leq x \leq 1$ and for temperatures $10K \leq T \leq 200K$ (64).

specific heat results in the glassy regime of KBr:KCN were interpreted [Mertz & Loidl (64)] by assuming that the progressive freezing of local order takes place over a large temperature range, so that the specific heat looks smooth at all temperatures. But even for concentrations above the critical concentration, the establishment of local order sets in far above the structural phase transition. The additional entropy associated with the development of long-range order is almost negligible. The situation in the orientational glasses seems to be similar to the findings in randomly diluted Ising spin systems with next nearest-neighbors interactions. In these systems, already far above the percolation limit, the specific heat appears entirely smooth. The additional entropy due to the development of long-range order is undetectably small except close to the fully concentrated systems (68a,b).

LOW-TEMPERATURE SPECIFIC HEAT As pointed out above, amorphous solids are characterized by universal low-temperature properties. Below 1K the specific heat has been found to be quasilinearly temperature dependent, while the thermal conductivity exhibits roughly a T^2 dependence. The tunneling model (25, 26) explained why glasses have a specific heat proportional to temperature in contrast to insulating crystals, in which the Debye specific heat is proportional to T^3 in this temperature regime. As a consequence of the tunneling model the effective specific heat should depend logarithmically on time. These predictions have been verified experimentally (69, 70). The tunneling center theory also correctly predicted that glasses should have a saturable ultrasonic attenuation, which also has been demonstrated experimentally (71).

Between 1 and 20K amorphous systems have another equally universal feature, namely an excess specific heat over the normal Debye behavior of the crystalline counterparts and a plateau in the thermal conductivity. It has been demonstrated experimentally that the excess T^3 term is independent of time (69), and it has been suggested that this contribution to c_p is an intrinsic property of the phonon system itself. No convincing theoretical explanation has been given so far.

The pioneering work of DeYoreo et al (42) has demonstrated that for $0.25 \leq x \leq 0.7$ the specific heat in $(KBr)_{1-x}(KCN)_x$ mixed crystals varies linearly with temperature and logarithmically with the measuring time, a characteristic that is typical for all amorphous solids. A linear term in the specific heat has also been detected in the orientational glass NaCN:KCN by Mertz et al (72). Figure 6 shows the results for $(NaCN)_{0.41}(KCN)_{0.59}$. The observed time dependence is indicated in the *insert*. In addition, it has also been demonstrated that KBr:KCN mixed crystals exhibit an excess specific heat in a wide concentration range. Some representative results

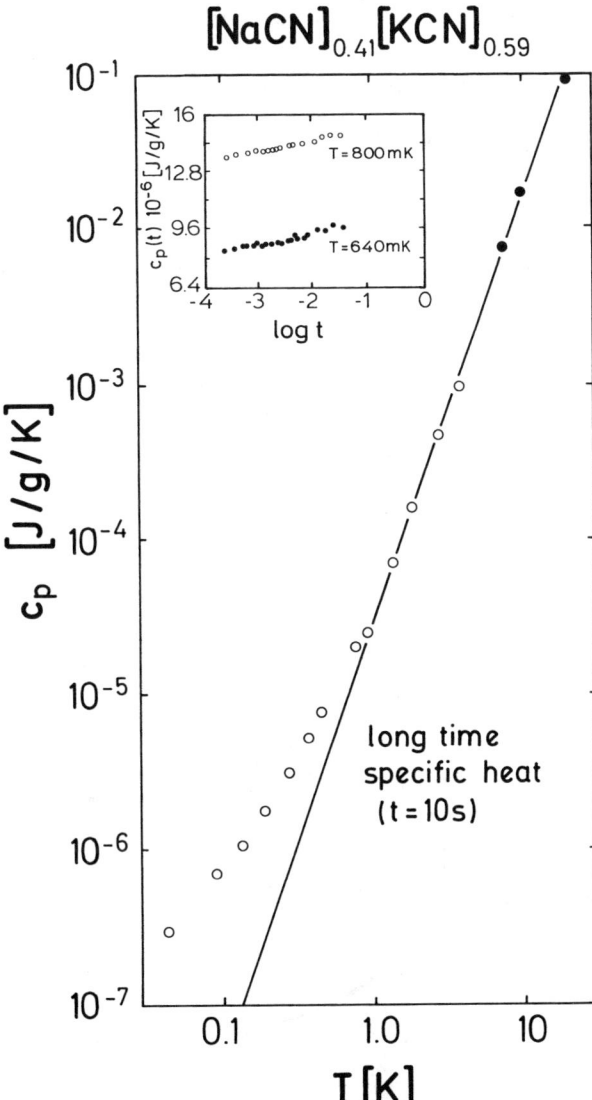

Figure 6 Long-time specific heat (10s) of $(NaCN)_{0.41}(KCN)_{0.59}$ versus temperature (72). The theoretical Debye specific heat is shown for comparison.

(73) are shown in Figure 7. It is important to note that the excess specific heat that appears as a bump in C/T^3 looks very similar for different concentrations and can even be detected in the elastically ordered systems (e.g. for $x = 0.73$). The excess specific heat almost vanishes, however, for concentrations close to the pure compounds (73).

Sethna and co-workers (74, 75) have proposed a microscopic model that assumes that the dipole flipping is the dominant process responsible for the low-temperature linear term in the specific heat. The cyanide molecules exhibit two possible orientations separated by a 180° flip. This situation can be modeled with a double well potential with a given barrier height and a given asymmetry. At low temperatures, tunneling will be the only

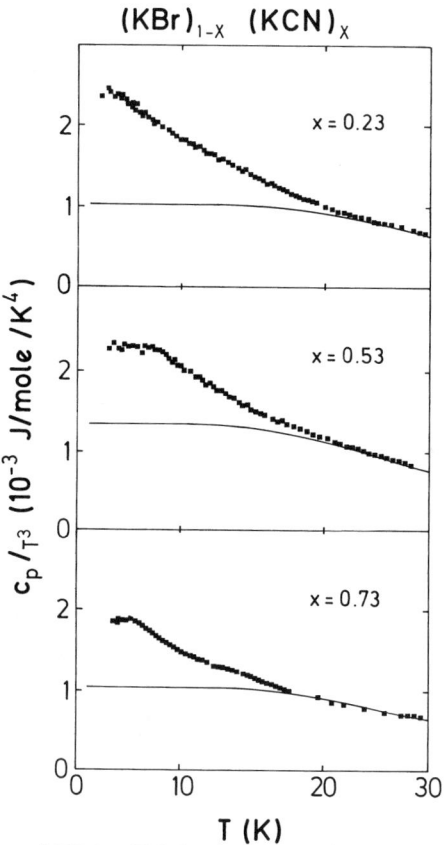

Figure 7 Specific heat of $(KBr)_{1-x}(KCN)_x$ plotted as c/T^3 versus T for $x = 0.23$, 0.53, and 0.73 (73). Note that $(KBr)_{0.27}(KCN)_{0.73}$ exhibits an elastically ordered ground state. The *solid lines* indicate the theoretical Debye specific heat.

possible reorientational channel. Due to the locally frozen-in shear strains, there exists a distribution of barrier heights that can be described by a Gaussian distribution. The barrier heights and the distribution of barrier heights can be determined experimentally with dielectric techniques (15, 51, 76). The microscopic model proposed by Sethna et al (74, 75) assumed that the low-barrier tail of the Gaussian distribution contains the active tunneling centers. The low-temperature specific heat in KBr:KCN was calculated with dielectric data and compared with the experimental specific heat (74, 77a,b). The good overall agreement for different concentrations gave strong support to the microscopic theory.

Later it was proposed that the excess specific heat and the plateau in the thermal conductivity in KBr:KCN can be explained by librational modes of the CN^- ions coupled to the lattice strains (78). From simple arguments one would expect that the energy of the librational modes scales like the square root of the energy barrier. Dielectric measurements show that the barrier height for dipolar relaxation increases linearly with the CN^- concentration. Experimentally, however, KBr:KCN exhibits bumps in C/T^3 that appear, at a first glance, to be independent of concentration (Figure 7). It is interesting to note that NaCN:KCN mixed crystals exhibit no excess specific heat in the glassy regime (72). Clearly, further experimental and theoretical work is needed to identify the microscopic origin of the excess specific heat in the cyanide glasses.

The Relaxation Dynamics at the Glass Transition

The CN molecules interact via a strain-mediated coupling. This is an effective quadrupolar interaction, whose microscopic origin is the bilinear coupling between rotational and translational degrees of freedom, and which has been extensively studied by Michel & Naudts (53) and by Sahu & Mahanti (79). In addition, the CN^- ions carry an electrostatic dipole moment of approximately 0.3 D. Both the dipole and the quadrupole moments are rigidly connected with the shape of the CN dumbbell. For reorientational jumps, the CN molecules must cross the hindering barriers produced by the Born-Mayer potential generated by the neighboring K^+ ions. Most likely, quadrupolar jumps are reorientation processes from a given [111] direction to a neighboring direction, e.g. [1$\bar{1}$1]. Dipolar reorientations are 180° flips. Certainly the dipoles still can fluctuate in a quadrupolar ordered phase (the antiferroelectric phase transition in pure KCN appears in the elastically ordered state at 83K).

QUADRUPOLAR FREEZING   The glass transition in orientational glasses is indicated by cusps in the appropriate multipolar susceptibility. The primary (glass forming) relaxation processes in the cyanide glasses are quadrupolar reorientations. To study the quadrupolar susceptibility, the temperature (T) and frequency (f) dependence of the elastic constants have to be investigated. The elastic constants c_{ij} are connected with the quadrupolar susceptibility χ^Q via

$$c_{ij}(T, f) = c_{ij}^0[1 - g^2\chi^Q(T, f)]$$

where c_{ij}^0 is the background elastic constant of the reference system, e.g. KBr, and g^2 is the rotation-translation coupling constant. The quadrupolar susceptibility is given by

$$\chi^Q(T, f) = \chi_0^Q[1 + i2\pi f \tau^Q(T)].$$

Here χ_0^Q is the static quadrupolar susceptibility and τ^Q is the quadrupolar relaxation time. As the temperature is lowered, the molecular relaxation slows down, yielding dispersion in the real part and a loss peak in the imaginary part of the orientational susceptibility. In the case of single ion freezing processes, the relaxation time can be described in terms of an Arrhenius law

$$\tau = (2\pi f_0)^{-1} \exp(E/k_B T)$$

where f_0 is the attempt frequency and E is the height of the hindering barrier against quadrupolar relaxations. The parameterization of a collective freezing process in terms of an Arrhenius law often yields unrealistically high values of attempt frequencies and hindering barriers. New concepts have been developed in the field of amorphous systems and spin glasses to describe (or to parameterize) the collective freezing processes at the glass transition: The relaxation has been described in the framework of a phenomenological Vogel (80)–Fulcher (81) law in terms of hierarchically constrained relaxations (82), by dynamic scaling laws (83), etc [for a discussion see (33)].

The first observation of cusps in the quadrupolar susceptibility, i.e. minima in the temperature dependence of the elastic constant c_{44}, have been reported by Satija & Wang (38) using Brillouin spectroscopy in KBr:KCN mixed crystals. Later the temperature dependence of c_{44} was studied with neutron scattering (4, 84), ultrasonic (85–89), and torsion pendulum techniques (90). The measuring frequencies of these techniques are rather different: Neutron scattering probes the elastic properties in the THz regime, Brillouin scattering in the GHz regime. Ultrasonic and torsion pendulum techniques operate at MHz and kHz frequencies, respectively. Thus, the quadrupolar freezing dynamics can be studied in a frequency range of almost ten decades.

Figure 8 shows the temperature dependence of the elastic constants and the damping in $(KBr)_{0.965}(KCN)_{0.035}$ as derived by Loidl et al (86) from the sound wave propagation along the [100] and [110] directions as measured at 10 MHz. All the elastic constants except c_{44} exhibit a minimum at about 16K. The attenuation peaks at somewhat lower temperatures. The signal of the c_{44} mode was lost at 22K because of a large damping, but from the analysis of all the data it was deduced that c_{44} exhibits the same quantitative temperature dependence (86). By focusing on the elastic constants with the representations T_{2g}, E_g, and A_{1g}, it was concluded that the T_{2g} mode c_{44} exhibits the strongest temperature dependence, whereas the bulk modulus (A_{1g}) is practically temperature independent (86). In addition, clear dispersion effects were detected with overtone frequencies up to 90 MHz (86).

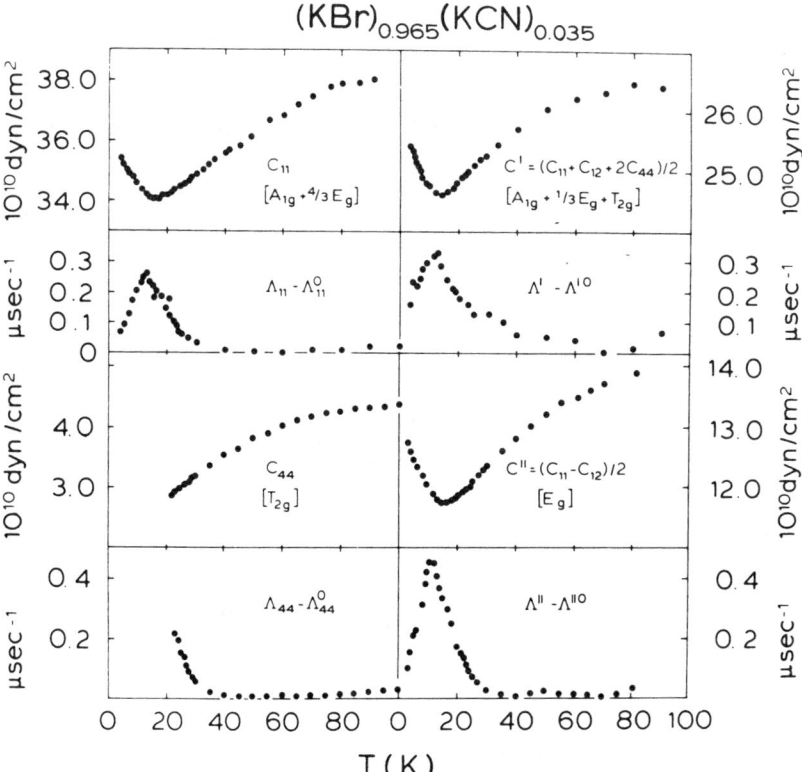

Figure 8 The temperature dependence of the elastic constants and of the sound wave attenuation for $(KBr)_{0.965}(KCN)_{0.035}$. Figure taken from Ref. (85) and reproduced with permission from *Zeitschrift für Physik* (Springer-Verlag).

The frequency dependence of the quadrupolar freezing over ten decades of frequencies as measured in $(KBr)_{0.2}(KCN)_{0.8}$ is shown in Figure 9 (91, 92). Here the temperature dependence of the elastic constant c_{44}, as measured with neutron (1 THz), Brillouin (1 GHz), ultrasonic (10 MHz), and torsion pendulum techniques (1 kHz), is compared. These results clearly demonstrate the frequency dependence of the freezing process at the glass transition. A detailed discussion of the freezing dynamics is given below.

The temperature dependence of the elastic constant c_{44} in the NaCN:KCN glasses as determined by inelastic neutron scattering techniques is shown in Figure 10 (16, 60). Again we find the well-known behavior that c_{44} softens, passes through a minimum, and recovers again toward 0K. However, the strength of the softening is very different for the three different concentrations x. Although for $x = 0.19$ and $x = 0.85$ well-defined minima in $c_{44}(T)$ were observed, the elastic shear constant is nearly temperature independent for $x = 0.59$ (60). The random field couples to the reorientational variable and weakens the rotation-translation coupling. Its strength scales with $x(1-x)$ and exhibits a maximum for intermediate concentrations. The observed temperature behavior of c_{44} can be taken as strong evidence that the NaCN:KCN glasses are dominated by random fields (60).

DIPOLAR FREEZING   Measurements of the dielectric constant probe the dipolar susceptibility $\chi^D(T, f)$:

$$\varepsilon(f) - \varepsilon_\infty = 4\pi\chi_0^D/[1 + (i2\pi f\tau^D)^{1-\alpha}].$$

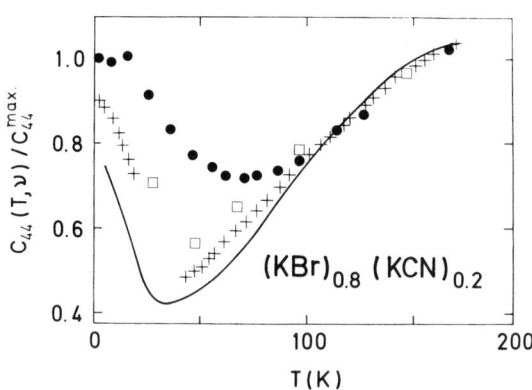

Figure 9 Normalized elastic constant $c_{44}(T, f)$ versus temperature as determined in $(KBr)_{0.8}(KCN)_{0.2}$ for different measuring frequencies: (●) inelastic neutron scattering data (THz) (84); (□) Brillouin data ($x = 0.19$) (GHz) (38); (+) ultrasonic results (10 MHz) (87); (———) torsion pendulum measurements (kHz) (90). Figure taken from Ref. (92) and reproduced with permission from the American Physical Society.

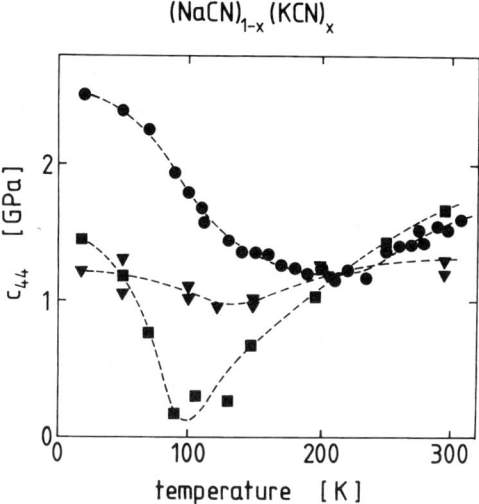

Figure 10 Temperature dependence of the elastic constants c_{44} for three different concentrations in the glassy regime of $(NaCN)_{1-x}(KCN)_x$: (●) $x = 0.19$; (▼) $x = 0.59$; (■) $x = 0.85$ (60). The lines are drawn to guide the eye.

Here χ_0 is the static dielectric susceptibility and τ^D is the most probable dipolar relaxation time. Again, the temperature dependence of τ can be described with an Arrhenius law. The introduction of the parameter α describes a symmetric broadening of the Debye loss peak and an equivalent smearing out of the dispersion step in the real part of the dielectric constant. This concept of a symmetrically broadened loss peak was introduced by Cole & Cole (93). The Cole-Cole approach introduces a distribution of the attempt frequencies. For the cyanide glasses it has been demonstrated that a Gaussian distribution of activation energies describes the experimental results equally well (76).

Detailed dielectric measurements in the alkali halide–alkali cyanide mixed crystals have been performed by Lüty and co-workers (15, 94, 95). A review of these results is given by Lüty in (51). Further dielectric measurements that focus predominantly on the glassy behavior of the cyanides can be found in Refs. (16, 46, 47, 76, 77a,b, 92, 96). The cyanide glasses are characterized by an extreme broad distribution of relaxation times. This was demonstrated by Birge et al (76), who showed that the dielectric loss in $(KBr)_{0.5}(KCN)_{0.5}$ extends over eight decades in frequency. In contrast the dielectric loss peak in KCN forms a good Debye peak with a width of 1.14 decades (94).

As a representative example of the extremely broad distributions of relaxation times that are observed in the cyanide glasses, Figure 11 shows the frequency dependence of the imaginary part of the dielectric constant in $(NaCN)_{0.15}(KCN)_{0.85}$ (*upper frame*) and in $(NaCN)_{0.41}(KCN)_{0.59}$ (*lower frame*) (16). The real and the imaginary part of ε have been fitted with a Cole-Cole distribution of relaxation times. The *solid lines* in Figure 13 are the results of these fits. For both crystals the width of the dielectric loss is anomalously broad [8–10 decades at 50K (16)] and increases with decreasing temperatures. These findings are analogous to the observed behavior of β-relaxation processes in a variety of glasses and polymers.

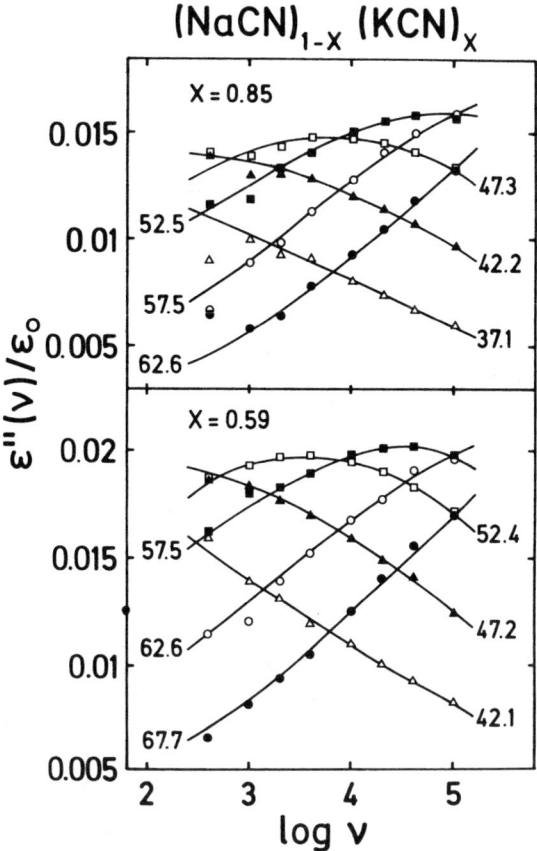

Figure 11 Normalized imaginary part of the dielectric constant $\varepsilon''(f)$ versus the logarithm of the measuring frequency in $(NaCN)_{0.15}(KCN)_{0.85}$ (*upper frame*) and $(NaCN)_{0.41}(KCN)_{0.59}$ (*lower frame*) at different temperatures. The lines are calculated by using a Cole-Cole distribution of relaxation times. Figure taken from Ref. (16) and reproduced with permission from the American Physical Society.

The concentration dependence of the mean hindering barriers E_0 and the concentration dependence of their distribution width σ is shown for $(KBr)_{1-x}(KCN)_x$ in the Figures 12 and 13. Here σ denotes the full width at half maximum (FWHM) of Gaussian-distributed activation energies. For low concentrations E_0 scales almost linearly with x. This indicates that the dipoles experience a potential determined by the quadrupolar interaction forces, or, to be more specific, the dipoles probe the local environment, which exhibits locally strong deviations from cubic symmetry due to the quadrupolar freezing transition. In the framework of a mean field theory (97), the local deformations scale linearly with the rotation-translation couplings, which in turn depend linearly on x. At higher concentrations E_0 reflects the structures of the low-temperature phases. The width parameter σ, on the other hand, is anomalously broad, even in the ordered phases. This immediately implies that domain walls occupy a

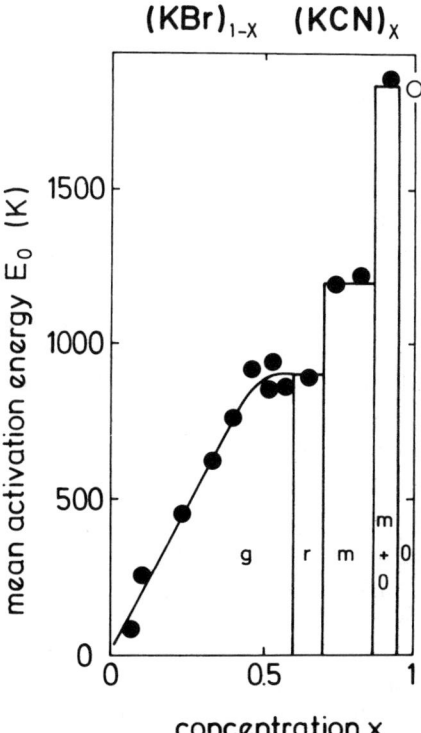

Figure 12 Concentration dependence of the mean activation energy E_0 for dipolar reorientations in $(KBr)_{1-x}(KCN)_x$: (●) (73); (○) (94). (See Figure 2 for definition of the symbols defining the phases.)

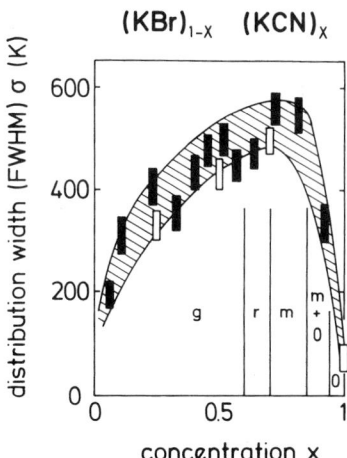

Figure 13 Concentration dependence of the distribution width of activation energies σ [full width at half maximum (FWHM)]: (■) (73); (□) (78).

large fraction of the crystal volume. Hence, reorientations within the domain wall are responsible for the observed low-temperature specific heat anomalies in the ordered compounds.

The loss peaks in the cyanide glasses shift to lower frequencies for lower temperatures. Indeed, the peak frequencies follow an Arrhenius law over the experimental range (92). Some representative results of the slowing down of the dipolar reorientations are shown in Figure 14 (92). Here the temperature dependence of the real part of the dielectric constant in $(KBr)_{1-x}(KCN)_x$ is plotted for different measuring frequencies. Clearly, Figures 9 and 14 demonstrate that the dispersion regimes for strain and polarization appear at different temperatures, thus indicating a complete decoupling of dipolar and quadrupolar relaxations.

SEQUENTIAL FREEZING OF DIPOLES AND QUADRUPOLES For a quantitative comparison of dipolar and quadrupolar freezing-in in KBr:KCN, $\ln(f)$ was plotted versus $1/T_f^D$ and $1/T_f^Q$ (92) for $(KBr)_{1-x}(KCN)_x$ for concentrations $x = 0.2$ and $x = 0.5$. Here, analogous to the definition of the freezing temperatures in the spin glass problem, the cusps in real part of the dipolar and in the quadrupolar susceptibilities were defined as freezing temperatures T_f^D and T_f^Q, respectively (92). Figure 15 (92) reveals that for both concentrations, the freezing temperatures coincide only at gigahertz frequencies and high temperatures. At even higher temperatures the dynamics of dipoles and quadrupoles should be the same. As soon as the quadrupoles start to order locally (glass transition temperature), however,

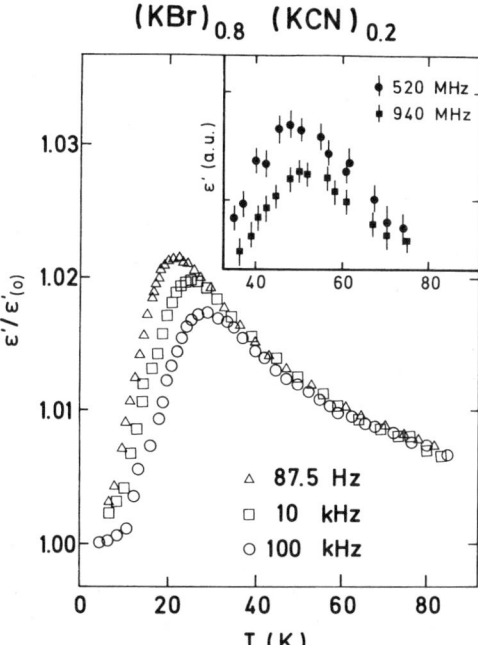

Figure 14 Real part of the dielectric constant in $(KBr)_{1-x}(KCN)_x$ versus temperature for different measuring frequencies (96): (△) 87.5 Hz; (□) 10 kHz; (○) 100 kHz. The high frequency data are shown in the *insert* (92): (●) 520 MHz; (■) 940 MHz. Figure taken from Ref. (92) and reproduced with permission from the American Physical Society.

significant differences in the relaxational behavior become apparent. The appearance of a primary (α-) and a secondary (β-) relaxation process is a well-known phenomenon in polymers and glasses (98). In these systems the β-relaxation exhibits an Arrhenius-type of behavior with a broad distribution of relaxation times, whereas the α-relaxation often is described with a Vogel-Fulcher law. Surprisingly, similar trends were detected in KBr:KCN glasses (92). In the cyanide glasses the quadrupolar relaxation describes the collective freezing process of orientations coupled to lattice strains. Hence, this relaxation is the relevant primary relaxation and describes the structural glass transition. The dipolar relaxation is a thermally activated process of single CN^- ions over the hindering barriers created by the quadrupolar interaction forces (97).

The Nature of the Glass Transition in the Cyanide Glasses

In this review orientational glasses have been treated thus far as theoretically analogous to spin glasses, and has focused on the random bond

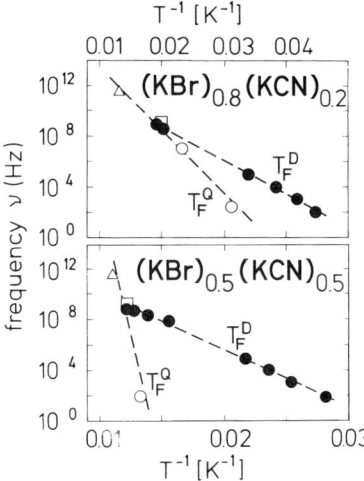

Figure 15 Arrhenius plots [log(f) versus $1/T$] of the dipolar (92, 96) (*full symbols*) and the quadrupolar freezing temperatures (*empty symbols*) in $(KBr)_{1-x}(KCN)_x$ for $x = 0.2$ and 0.5: ($\triangle$) inelastic neutron scattering results for $x = 0.2$ (84) and $x = 0.5$ (4); ($\square$) Brillouin scattering results (38); ($\bigcirc$) ultrasonic results (87) and torsion pendulum results (90). Figure taken from Ref. (92) and reproduced with permission from the American Physical Society.

disorder and has disregarded the random field completely. This approach was used in the early work by Fischer & Klein (37) and later on in the framework of infinite-range mean field models (99–102). Monte Carlo simulations have been performed by Carmesin & Binder (103) on systems with nearest neighbor interactions. Recently K. H. Michel (17) described the relaxation phenomena in the cyanide glasses within the framework of a random field model. Later he extended this theory to include collective freezing, which results in a nonergodic instability (104a,b): Near a second-order ferroelastic phase transition critical fluctuations are frozen-in due to weak random fields.

The best way to characterize the low-temperature state in disordered systems is to study the temperature dependence of the order parameter. The order parameter in orientational glasses is given by $\Psi \sim \overline{\langle Y_\alpha(n) \rangle^2}$. $Y(n)$ is the orientational coordinate at lattice site n; α labels the modes of appropriate symmetry (in the cyanide glasses, T_{2g} symmetry). The brackets $\langle \rangle$ denote the thermal average and the bar the configurational average. As a result of the rotation-translation coupling in the cyanide glasses, the order parameter is also proportional to $\overline{\langle s_i(n) \rangle^2}$ where $s_i(n)$ is a translational coordinate at site n. It is this property that leads to the existence of a static central peak in addition to the dynamic scattering law (40). According to Michels theory (104a,b), two contributions to the order parameter have to be taken into account: $\Psi = \Psi_1 + \Psi_2$. The first term results solely from the response to the static random strain field h and is a smooth function of temperature, namely $\Psi_1 \sim h^2/T^2$. Ψ_2 accounts for the nonergodic transition at the freezing temperature T_f and exists only for temperatures $T \leq T_f$.

So far only a few experiments have been reported that elucidate the nature of the glass transition in the cyanide glasses (105, 106a,b). Elschner et al (105) performed inelastic neutron scattering and NMR experiments in NaCl:NaCN. They found the order parameter to be a smooth function of temperature but with a much stronger temperature dependence than theoretically predicted. Ultrasonic results in KCl:KCN, KBr:KCN, and RbBr:RbCN (106) were analyzed within the framework of the pure random field model (17). The agreement between theory and experiment was found to be excellent.

As stated above, NaCN:KCN seems to be a model system to test the predictions of the random strain theory as proposed by Michel (17, 104a,b). Very recently inelastic neutron scattering experiments in NaCN:KCN glasses were reported that studied the temperature dependence of the central peak, which is a direct measure of the order parameter of the glass state (60). Some representative scans are presented in Figure 16 (60). Upon cooling from room temperature, the phonon side bands soften, pass through a minimum, and increase again toward 0K. In addition, a resolution-limited central peak appears at zero energy transfers. The squared frequency as determined by the center of the phonon side bands is proportional to the elastic constant c_{44}. The intensity of the central peak can be viewed as a direct measure of the glass state (17, 40). A summary of the results as obtained in the glassy crystals with concentrations $x = 0.19, 0.59$, and 0.85 is given in the Figures 10 and 17. Figure 10 shows the temperature dependence of the elastic constants c_{44}. Figure 17 documents the T dependence of the central peak intensities. Obviously, the temperature dependence of both quantities is drastically different and displays the strength of the random fields.

The logarithm of the central peak intensity was plotted versus the logarithm of temperature for a more quantitative analysis (60). The results for crystals with concentrations $x = 0.19, 0.44, 0.59, 0.85, 0.89$ are documented in Figure 18 (60) and can be summarized as follows: At high temperatures the central peak intensities $I_{cp}(T)$ exhibit a slope of two, indicative of a pure random strain system, for all crystals investigated. Saturation effects at low temperatures indicate that the order parameter is fully developed. For $x = 0.89$ the anomaly in the temperature dependence of the order parameter indicates the onset of a structural phase transition. The abrupt change of slope for $x = 0.85$ can be interpreted as nonergodic instability, with an additional order parameter appearing for temperatures below the glass transition. For crystals with concentrations $x = 0.41$ and 0.59 the temperature dependence of the order parameter is

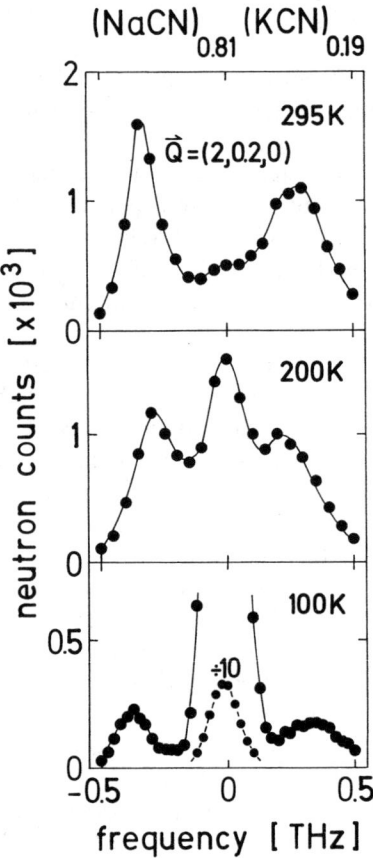

Figure 16 Neutron scattering line shapes in T_{2g} symmetry in $(NaCN)_{0.81}(KCN)_{0.19}$ at different temperatures (60). The lines are drawn to guide the eye.

always much weaker, as theoretically predicted. Obviously the theory fails to describe the case of strong random fields correctly.

CONCLUSIONS

Orientational glasses exhibit a collective freezing of the orientational degrees of freedom into random orientations. The low-temperature glass state can be viewed as an arrangement of clusters with different shear deformations distributed around the undisturbed lattice. The relaxation dynamics in the cyanide glasses is characterized by a primary (structural) relaxation followed by secondary relaxational processes.

Orientational glasses seem to be appealing model systems to study the

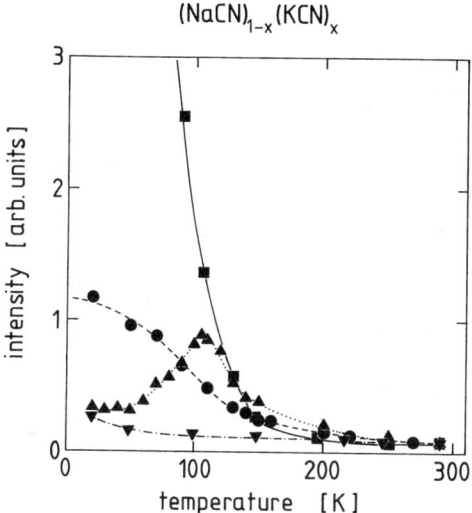

Figure 17 Temperature dependence of the central peak intensities in $(NaCN)_{1-x}(KCN)_x$ for different concentrations: (●) $x = 0.19$; (▼) $x = 0.59$; (■) $x = 0.85$. Results for $(NaCN)_{0.11}(KCN)_{0.89}$ (▲), which undergoes a structural phase transition, are included.

properties of disordered systems. Attempts have been made to describe the low-temperature glassy anomalies within a microscopic model, starting from the well-known crystalline structure.

The NaCN : KCN system is dominated by random strains. Pure random strain systems, in which the coupling of the orientational variable to the random field dominates, exhibit an order parameter that is a continuous function of temperature. Crystals, characterized by a delicate balance between the strain-rotation and rotation-translation couplings, exhibit an elastic shear constant that approaches almost zero. In these systems an additional order parameter appears for temperatures below the minimum of $c_{44}(T)$, and the glass transition can be interpreted in terms of a nonergodic instability. Concerning the KM : KCN mixtures, where M = Cl, Br, and I, an open question remains as to what extent these crystals are dominated by random fields or by random bonds. The different critical concentrations in these systems, however, which seem to scale with the volume difference of the halogen ions as compared to the CN^- ion, demonstrate that random fields, at least, play an important role.

ACKNOWLEDGMENTS

I would like to thank K. Knorr and K. H. Michel for stimulating

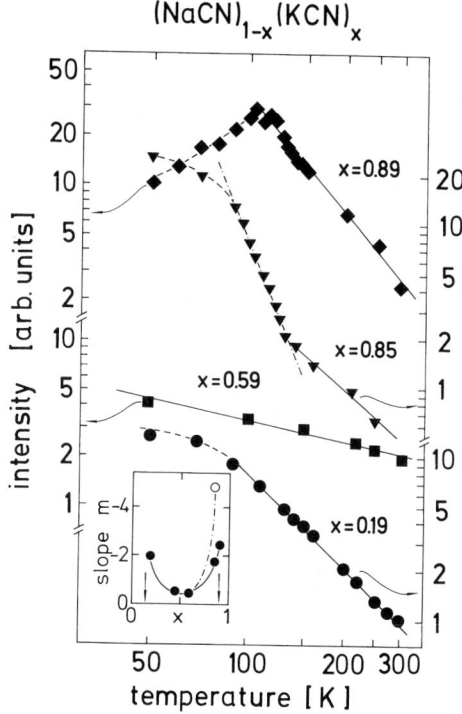

Figure 18 Log-log plot of the central peak intensities in $(NaCN)_{1-x}(KCN)_x$ for concentrations $x = 0.89$, 0.85, 0.59, and 0.19 versus temperature. The *insert* shows the slope m at high temperatures (●) and for $x = 0.85$ also below the glass transition (○). The lines are drawn to guide the eye.

discussions. I also acknowledge the invaluable help of R. Böhmer, B. Mertz, and T. Schräder.

This work was supported by the "Sonderforschungsbereich 262" (Mainz) the "Materialwissenschaftliches Forschungszentrum" (Mainz), and the German Federal Minister for Research and Technology (Bundesminister für Forschung und Technologie) under the Contract number 03-LO1MAI-0.

Literature Cited

1. Sethna, J. P. 1986. *Phys. Today* 39: S20–21
2. Känzig, W., Hart, H. R., Roberts, S. 1964. *Phys. Rev. Lett.* 13: 543–45
3. Sullivan, N. S., Devoret, M., Cowan, B. P., Urbina, C. 1978. *Phys. Rev. B* 17: 5016–24
4. Rowe, J. M., Rush, J. J., Hinks, D. G., Susman, S. 1979. *Phys. Rev. Lett.* 43: 1158–61
5. Loidl, A. 1985. *J. Chim. Phys.* 82: 305–9
6. Knorr, K. 1987. *Phys. Scr. T* 19B: 531–36

7. Press, W., Janik, B., Grimm, H. 1982. *Z. Phys. B* 49: 9–16
8. Esteve, D., Sullivan, N. S., Devoret, M. 1982. *J. Phys. Lett.* 43: 793–99
9. Klee, H., Carmesin, H. O., Knorr, K. 1988. *Phys. Rev. Lett.* 61: 1855–58
10. Grondey, S., Prager, M., Press, W., Heidemann, A. 1986. *J. Chem. Phys.* 85: 2204–10
11. Höchli, U. T. 1982. *Phys. Rev. Lett.* 48: 1494–97
12. Courtens, E. 1982. *J. Phys. Lett.* 43: 199–204
13. Courtens, E. 1984. *Phys. Rev. Lett.* 52: 69–72
14. Trybula, Z., Stankowski, J., Szczepanska, L., Blinc, R., Weiss, A., Dalal, N. S. 1988. *Physica B* 153: 143–46
15. Lüty, F., Ortiz-Lopez, J. 1986. *Phys. Rev. Lett.* 50: 1289–92
16. Loidl, A., Schräder, T., Böhmer, R., Knorr, K., Kjems, J. K., Born, R. 1986. *Phys. Rev. B* 34: 1238–49
17. Michel, K. H. 1986. *Phys. Rev. Lett.* 57: 2188–90; Michel, K. H. 1987. *Phys. Rev. B* 35: 1405–13; Michel, K. H. 1987. *Phys. Rev. B* 35: 1414–18
18. Suga, H. 1986. *Ann. NY Acad. Sci.* 484: 248–63
19. Pathamanathan, K., Johari, G. P. 1985. *J. Phys. C* 18: 6535–45
20. Descamps, M., Caucheteux, C. 1987. *J. Phys. C* 20: 5073–95
21. Suga, H., Seki, S. 1974. *J. Non-Cryst. Solids* 16: 171–94
22. Harris, A. B., Meyer, H. 1985. *Can. J. Phys.* 63: 3–23
23. Courtens, E. 1983. *Helv. Phys. Acta* 56: 705–20
24. Zeller, R. C., Pohl, R. O. 1971. *Phys. Rev. B* 4: 2029–41
25. Anderson, P. W., Halperin, B. I., Varma, C. M. 1972. *Philos. Mag.* 25: 1–9
26. Phillips, W. A. 1972. *J. Low Temp. Phys.* 7: 351–60
27. Owen, J., Browne, M., Knight, W. D., Kittel, C. 1956. *Phys. Rev.* 102: 1501–7
28. Jacobs, I. S., Schmitt, R. W. 1959. *Phys. Rev.* 113: 459–63
29. Cannella, V., Mydosh, J. A. 1972. *Phys. Rev. B* 6: 4220–37
30. Ruderman, M. A., Kittel, C. 1954. *Phys. Rev.* 96: 99–102
31. Kasuya, T. 1956. *Prog. Theor. Phys.* 16: 45–57
32. Yosida, K. 1957. *Phys. Rev.* 106: 893–98
33. Binder, K., Young, A. P. 1986. *Rev. Mod. Phys.* 58: 801–976
34. Brout, R. 1964. *Phys. Rev. Lett.* 14: 175–76
35. Klein, M. W., Brout, R. 1963. *Phys. Rev.* 132: 2412–26
36. Fischer, B., Klein, M. W. 1976. *Phys. Rev. Lett.* 37: 756–59
37. Fischer, B., Klein, M. W. 1979. *Phys. Rev. Lett.* 43: 289–93
38. Satija, S. K., Wang, C. H. 1978. *Solid State Commun.* 28: 617–19
39. Klenin, M. A. 1979. *Phys. Rev. Lett.* 42: 1549–52
40. Michel, K. H., Rowe, J. M. 1980. *Phys. Rev. B* 22: 1417–28
41. Edwards, S. F., Anderson, P. W. 1975. *J. Phys. F* 5: 965–74
42. DeYoreo, J. J., Meissner, M., Pohl, R. O., Rowe, J. M., Rush, J. J., Susman, S. 1983. *Phys. Rev. Lett.* 51: 1050–53
43. DeYoreo, J. J., Knaak, W., Meissner, M., Pohl, R. O. 1986. *Phys. Rev. B* 34: 8828–42
44. Moy, D., Dobbs, J. N., Anderson, A. C. 1984. *Phys. Rev. B* 29: 2160–67
45. Berret, J. F., Doussineau, P., Levelut, A., Meissner, M., Schön, W. 1985. *Phys. Rev. Lett.* 55: 2013–16
46. Loidl, A., Feile, R., Knorr, K. 1982. *Phys. Rev. Lett.* 48: 1263–66
47. Bhattacharya, S., Nagel, S. R., Fleishman, L., Susman, S. 1982. *Phys. Rev. Lett.* 48: 1267–70
48. Knorr, K., Loidl, A. 1986. *Phys. Rev. Lett.* 57: 460–62
49. Rowe, J. M., Hinks, D. G., Price, D. L., Susman, S., Rush, J. J. 1973. *J. Chem. Phys.* 58: 2039–42
50. Loidl, A., Knorr, K., Rowe, J. M., McIntyre, G. J. 1988. *Phys. Rev. B* 37: 389–98
51. Lüty, F. 1981. *Defects in Insulating Crystals*, ed. V. M. Turkevitch, K. K. Swartz, pp. 69–89. Berlin: Springer-Verlag
52. Haussühl, S. 1973. *Solid State Commun.* 13: 147–50
53. Michel, K. H., Naudts, J. 1977. *Phys. Rev. Lett.* 39: 212–15; Michel, K. H., Naudts, J. 1977. *J. Chem. Phys.* 67: 547–58
54. Rowe, J. M., Rush, J. J., Chesser, N. J., Michel, K. H., Naudts, J. 1987. *Phys. Rev. Lett.* 40: 455–58
55. Loidl, A., Schräder, T., Knorr, K., Mertz, B., Böhmer, R., et al. 1989. *Z. Phys. B.* In press
56. Bourson, P., Gorczyca, G., Durand, D. 1987. *Cryst. Lattice Defects Amorphous Mater.* 16: 311–37
57. Elschner, S., Knorr, K., Loidl, A. 1986. *Z. Phys. B* 61: 209–15
58. Lewis, L. J., Klein, M. L. 1986. *Phys. Rev. Lett.* 57: 2698–2701
59. Michel, K. H. 1989. *Phys. Rev. B.* In press

60. Schräder, T., Loidl, A., Vogt, T., Frank, V. 1989. *Physica B* 156/157: 95–97
61. Jäckle, J. 1986. *Rep. Prog. Phys.* 49: 171–231
62. Suga, H., Matsuo, T., Seki, S. 1965. *Bull. Chem. Soc. Jpn.* 38: 1115–24
63. Mertz, B., Loidl, A. 1985. *J. Phys. C* 18: 2843–48
64. Mertz, B., Loidl, A. 1987. *Europhys. Lett.* 4: 583–89
65. Moriya, K., Matsuo, T., Suga, H., Lüty, F. 1984. *Proc. Int. Conf. Defects Insulating Crystals, Utah*, pp. 239–40
66. Matsuo, T., Kishimoto, I., Suga, H., Lüty, F. 1986. *Solid State Commun.* 58: 177–79
67. Rowe, J. M., Rush, J. J., Susman, S. 1983. *Phys. Rev. B* 28: 3506–11
68a. Algra, H. A., De Jongh, L. H., Reedijk, J. 1979. *Phys. Rev. Lett.* 42: 606–9
68b. Jayaprakash, C., Riedel, E. K., Wortis, M. 1978. *Phys. Rev. B* 18: 2244–55
69. Loponen, M. T., Dynes, R. C., Narayanamurti, V., Garno, J. P. 1980. *Phys. Rev. Lett.* 45: 457–60
70. Meissner, M., Spitzmann, K. 1981. *Phys. Rev. Lett.* 46: 265–68
71. Hunklinger, S., Sussner, H., Dransfeld, K. 1976. *Adv. Solid State Phys.* 16: 267–91
72. Mertz, B., Böhmer, R., Loidl, A., Knaak, W., Meissner, M., Berret, J. F. 1989. *Phys. Rev. B*. In press
73. Mertz, B., Eisele, B., Böhmer, R., Loidl, A. 1989. *Phys. Rev. B*. In press
74. Meissner, M., Knaak, W., Sethna, J. P., Chow, K. S., DeYoreo, J. J., Pohl, R. O. 1985. *Phys. Rev. B* 32: 6091–93
75. Sethna, J. P., Chow, K. S. 1985. *Phase Trans.* 5: 317–40; Sethna, J. P. 1986. *Ann. NY Acad. Sci.* 484: 130–49
76. Birge, N. O., Jeong, Y. H., Nagel, S. R., Bhattacharya, S., Susman, S. 1984. *Phys. Rev. B* 30: 2306–8
77a. Wu, L., Ernst, R. M., Jeong, Y. H., Nagel, S. R., Susman, S. 1988. *Phys. Rev. B* 37: 10444–47
77b. Ernst, R. M., Wu, L., Nagel, S. R., Susman, S. 1988. *Phys. Rev. B* 38: 6246–56
78. Grannan, E. R., Randeria, M., Sethna, J. P. 1988. *Phys. Rev. Lett.* 60: 1402–5
79. Sahu, D., Mahanti, S. D. 1982. *Phys. Rev. B* 26: 2981–3000
80. Vogel, H. 1921. *Phys. Z.* 22: 645–46
81. Fulcher, G. S. 1925. *J. Am. Ceram. Soc.* 8: 339–55
82. Palmer, R. G., Stein, D. L., Abrahams, E., Anderson, P. W. 1984. *Phys. Rev. Lett.* 53: 958–61
83. Soulethie, J., Tholence, J. L. 1985. *Phys. Rev. B* 32: 516–19
84. Loidl, A., Knorr, K., Feile, R., Kjems, J. K. 1983. *Phys. Rev. Lett.* 51: 1054–57; Loidl, A., Feile, R., Knorr, K., Kjems, J. K. 1984. *Phys. Rev. B* 29: 6052–62
85. Loidl, A., Feile, R., Knorr, K., Renker, B., Daubert, J., et al. 1980. *Z. Phys. B* 38: 253–62
86. Loidl, A., Feile, R., Knorr, K. 1981. *Z. Phys. B* 42: 143–49
87. Feile, R., Loidl, A., Knorr, K. 1982. *Phys. Rev. B* 26: 6875–80
88. Kwiecien, J. Z., Leung, R. C., Garland, C. W. 1981. *Phys. Rev. B* 23: 4419–25
89. Garland, C. W., Kwiecien, J. Z., Damien, J. C. 1982. *Phys. Rev. B* 25: 5818–23
90. Knorr, K., Volkmann, U. G., Loidl, A. 1986. *Phys. Rev. Lett.* 57: 2544–47
91. Loidl, A., Knorr, K. 1986. *Ann. NY Acad. Sci.* 484: 121–29
92. Volkmann, U. G., Böhmer, R., Loidl, A., Knorr, K., Höchli, U. T., Haussühl, S. 1986. *Phys. Rev. Lett.* 56: 1716–19
93. Cole, K. S., Cole, R. H. 1941. *J. Chem. Phys.* 9: 341–51
94. Julian, M., Lüty, F. 1977. *Ferroelectrics* 16: 201–4
95. Kondo, Y., Schoemaker, D., Lüty, F. 1979. *Phys. Rev. B* 19: 4210–16
96. Knorr, K., Loidl, A. 1982. *Z. Physik B* 46: 219–24
97. Sethna, J. P., Nagel, S. R., Ramakrishnan, T. V. 1984. *Phys. Rev. Lett.* 53: 2489–92
98. Johari, G. P. 1985. *J. Chim. Phys.* 82: 283–91
99. Goldbart, P., Sherrington, D. 1985. *J. Phys. C* 18: 1923–40
100. Gross, D. J., Kanter, I., Sompolinsky, H. 1985. *Phys. Rev. Lett.* 55: 304–7
101. Kanter, I., Sompolinsky, H. 1986. *Phys. Rev. B* 33: 2073–76
102. Kirkpatrick, T. R., Thirumalai, D. 1987. *Phys. Rev. B* 36: 5388–97
103. Carmesin, H. O., Binder, K. 1987. *Z. Phys. B* 68: 375–90; Carmesin, H. O., Binder, K. 1987. *Europhys. Lett.* 4: 269–74
104a. Michel, K. H. 1987. *Z. Phys. B* 68: 259–64
104b. Bostoen, C., Michel, K. H. 1988. *Z. Phys. B* 71: 369–76
105. Elschner, S., Albers, J., Loidl, A., Kjems, J. K. 1987. *Europhys. Lett.* 4: 1139–44
106a. Fossum, J. O., Garland, C. W. 1988. *Phys. Rev. Lett.* 60: 592–95
106b. Fossum, J. O., Wells, A., Garland, C. W. 1988. *Phys. Rev. B* 38: 412–22

THE METAL–INSULATOR TRANSITION IN EXPANDED FLUID METALS

F. Hensel and H. Uchtmann

Institute of Physical Chemistry, University of Marburg,
D-3550 Marburg, Hans-Meerwein-Straße,
Federal Republic of Germany

INTRODUCTION

Over the past two decades, a considerable amount of effort has been centered on the experimental and theoretical investigation of liquid metals expanded by heating toward the liquid-vapor critical point. Much of the activity is motivated by the large number of current and potential applications of fluid metals as high temperature working fluids for advanced energy technologies. From the scientific point of view, the main object of this effort is to find out how the properties of metals vary with large changes in density, large enough to change the liquid metal into a nonmetal at large enough expansion. Figure 1 shows why such a metal to nonmetal transition must occur with decreasing density. The discontinuous liquid-vapor phase change of a metal at low temperatures near the triple point is obviously accompanied by a discontinuous metal(M)-insulator(I) transition. In such a situation the liquid metal is reasonably well described by the nearly-free-electron model, whereas in the dilute insulating vapor the great majority of the electrons are attached to their parent atoms occupying spatially localized atomic orbitals. It follows that the precise nature of the electronic interactions between atoms must change on going by a suitable combination of temperatures and pressures continuously along the dashed path round the critical point from the liquid-like (M) to vapor-like (I) densities. Somewhere along this line a transition range must exist where metallic properties evolve into those characteristic of non-metals.

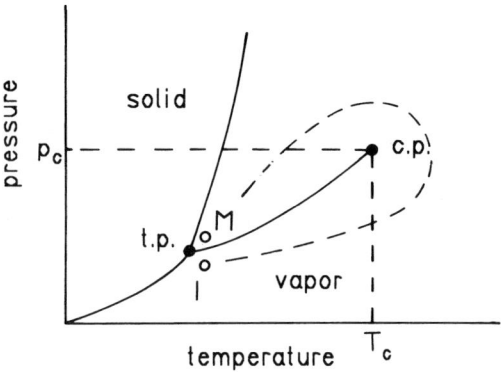

Figure 1 Schematic phase diagram in the pressure-temperature plane. M and I denote metal and insulator, respectively.

The relation of this metal-nonmetal transition to the liquid-vapor phase separation has remained a long-standing problem. This subject was first discussed by Landau & Zeldovitch (1), who suggested the possibility of separating first-order electronic and liquid-vapor transitions in fluid metals. Subsequent theoretical attempts that deal simultaneously with thermodynamic and electronic properties of fluid metals (2–8) reach similar conclusions but are still insufficient to provide a clear-cut answer to this problem. Another possibility is that the liquid-vapor and the metal-insulator transitions coincide and there is only one critical point, as first proposed by Krumhansl (9).

In the absence of an adequate theory, the obvious alternative is to explore the matter experimentally. Several attempts have been made during the past two decades to determine the basic phase behavior of the fluids, namely the equation of state, together with the basic electronic properties such as the electrical conductivity, but the subject has remained elusive until quite recently. Part of the difficulty arises because, as Table 1 shows, a combination of high temperatures and pressures is required to bring a fluid metal sample somewhere near its critical point. The critical point is at low enough temperature and pressure to be studied with conventional static techniques for only a very few metals (Hg, Cs, Rb, K, and Na). Even for these metals, the accuracy with which properties can be measured is severely limited by the highly reactive nature of metals at high temperatures and by the severe problems with the control and measurement of temperature in any high temperature–high pressure experiment. The latter becomes particularly important in studies close to the critical point where the analysis of experimental measurements can easily be hampered by the

Table 1 Critical temperatures, pressures, and densities of some metals

Metal	T_c/K	p_c/MPa	ρ_c/g cm^{-3}	References
Hg	1751	167.3	5.8	(10)
Cs	1924	9.25	0.38	(11)
Rb	2017	12.45	0.29	(11)
K	2280	161.0	0.19	(12)
Na	2485	248.0	0.30	(13)
Al	5730	182.0	0.42	(14)
Pt	9300	950.0	4.7	(14)
Mo	14300	570.0	2.9	(15)

presence of spurious effects due to temperature gradients. It is therefore not surprising that relatively accurate experimental results in the critical region are available for only the three metals with the lowest critical points: monovalent Cs and Rb and divalent Hg. The experimental information for these metals is accurate enough to permit determination of the asymptotic behavior of the properties near the critical point.

As indicated in Table 1, the estimated critical temperatures of most metals lie well above those of Hg and the alkali metals, thus making it impossible to study them in the critical region with static methods. Only less accurate transient methods such as shock waves, exploding wires, and laser heating (15–17) reach temperatures and pressures high enough to explore, for example, the critical region of molybdenum. These techniques have been applied to determine, for example, the equation of state and the velocity of sound of expanded Pb and Ta (18, 19). It has not been possible, however, to obtain measurements close to the critical point or to determine the location of the critical point (T_c, p_c, ρ_c) with sufficient certainty.

Experiment and theory up to about 1983 with all the important references can be found in various review articles (20–25) and conference proceedings (26). Therefore, an extended review of the subject is unnecessary. Instead, we have selected for attention those new developments that have not been adequately reviewed elsewhere, i.e. recent experimental and theoretical work on the vapor-liquid critical point of metals and its relation to the metal-nonmetal transition.

MONOVALENT METALS

A great deal of effort has been devoted to the experimental (11, 27–31) and theoretical (32–36) investigation of expanded fluid alkali metals because as elemental monovalent metals they closely resemble the expanded crystals with half full bands considered by Mott (37, 38) in his original discussion of the metal-nonmetal transition due to correlation. Most of this effort has focused on cesium because of its relatively low critical temperature. Measurements such as those of the electrical conductivity and the equation of state (27) clearly indicate that near the critical point cesium undergoes a metal-nonmetal transition. This may be seen at a glance from Figures 2a and 2b in which a selection of the most accurate density ρ and electrical conductivity σ results are presented in form of isotherms as a function of pressure at sub- and supercritical temperatures. Near the critical point the conductivity drops sharply, thus showing a strong effect of the phase transition on the electronic structure. However, there is no indication of a sharp (first-order) electronic transition except across the liquid-vapor phase boundary.

The close correlation between the behavior of the density and that of the conductivity convincingly shows that the variation of density is the dominant factor governing the metal-nonmetal transition. As the liquid metal is expanded, two effects are expected to occur: the average coordination number decreases and the avarage near-neighbor distance increases. In general, to describe the properties of a liquid metal at different densities, these two factors have to be taken into consideration. The central importance of these effects has stimulated neutron diffraction measurements of the structure factors of fluid Cs and Rb up to the critical points (30, 39). Figure 3a displays as a typical example the Fourier transform of the structure factor, i.e. the pair correlation function $g(R)$, of Cs at various temperatures and densities ranging from the melting point up to the critical region. The following noteworthy changes in $g(R)$ occur with decreasing density or increasing temperature. The intensities of the main peak of $g(R)$ is strongly reduced and broadened, whereas the peak position R_1 shifts only slightly toward higher R. Now the pair correlation function $g(R)$ is related to the radial distribution function $n(R) = 4\pi R^2 n g(R)$, which determines the number of neighboring atoms $n(R)\,dR$ in a spherical shell of radius R and thickness dR centered on a particular atom of interest. Detailed analysis (30) of data such as those displayed in Figure 3a shows that for Cs and Rb the average coordination number N_1, defined by the area under the first peak in $n(R)$, tends to decrease approximately linearly as the density is decreased by thermal expansion. Because there is usually considerable overlap between the first- and second-neighbor peaks, experimental values of N_1 depend sensitively on the method employed to define and integrate the first neighbor peak. For reasons of reliability and consistency we used the method of "symmetrical main maximum" (see *inset* in Figure 3b) for all data shown in Figure 3b. It is evident that the dominant effect of thermal expansion on the structure of Cs and Rb is a reduction

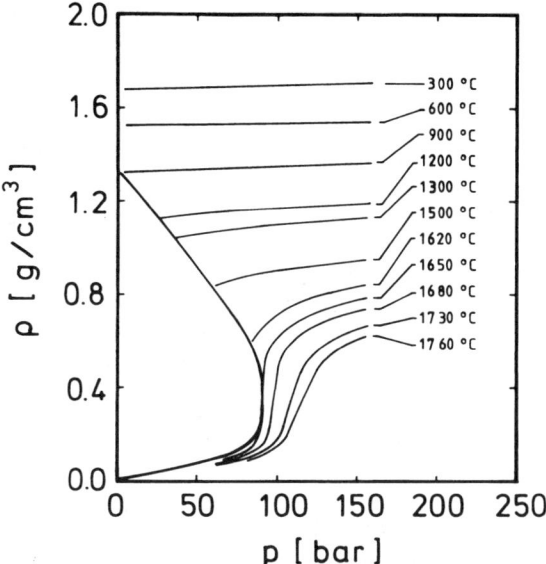

Figure 2a Equation of state data of fluid cesium at sub- and supercritical temperatures as a function of pressure.

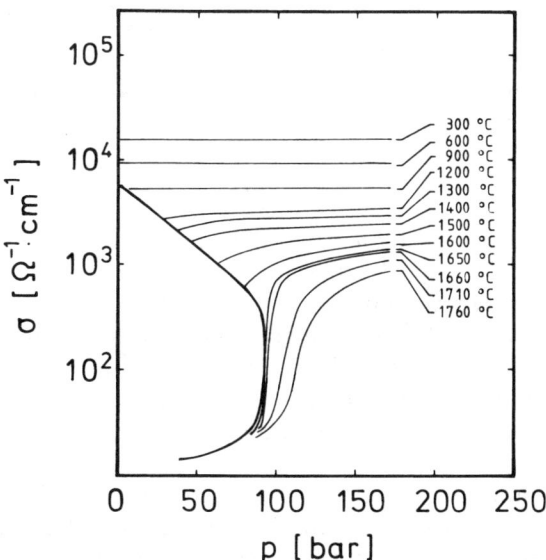

Figure 2b Electrical conductivity isotherms of fluid cesium at sub- and supercritical conditions.

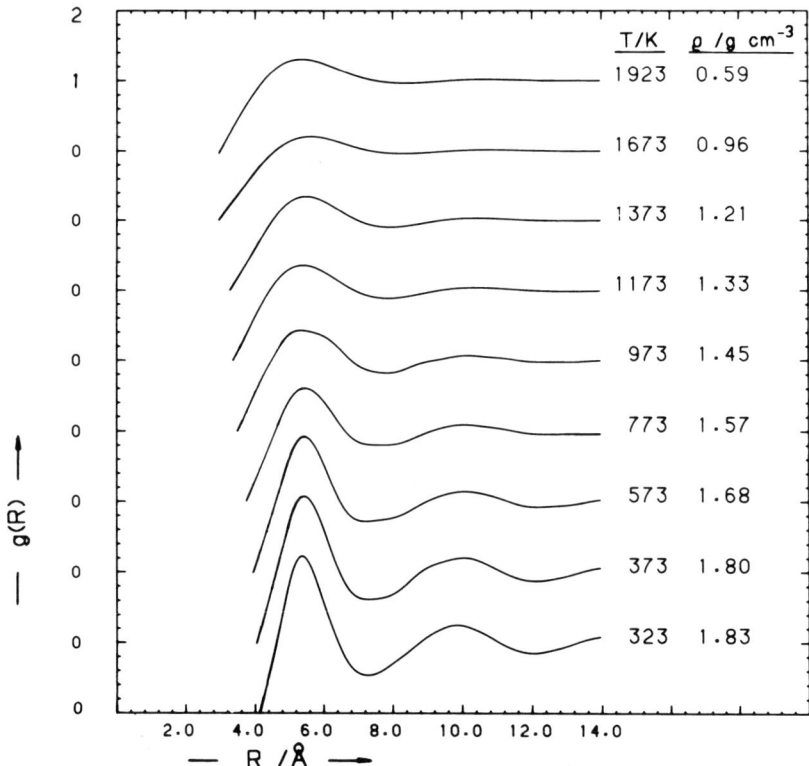

Figure 3a Pair correlation function $g(R)$ of expanded liquid cesium at conditions near the liquid-vapor coexistence.

of the average coordination number N_1 rather than an increased near neighbor distance. Data for the nonmetal argon (40) exhibit the same trend. This simply reveals the important point that thermal expansion of a liquid reorganizes the short-range order; it does not only increase the interatomic distances as it would in crystals. The knowledge of this feature has been particularly important for theoretical attempts to model the electronic structure of expanded metals by the use of band structure considerations (8, 32).

The structural data are especially suited to explore the limitations of the nearly free electron (NFE) approach for the expanded alkali metals (30) because the electrical transport is correlated with the variation of $g(R)$. In the NFE-model the electrical dc-conductivity σ is described by the Ziman formula (41):

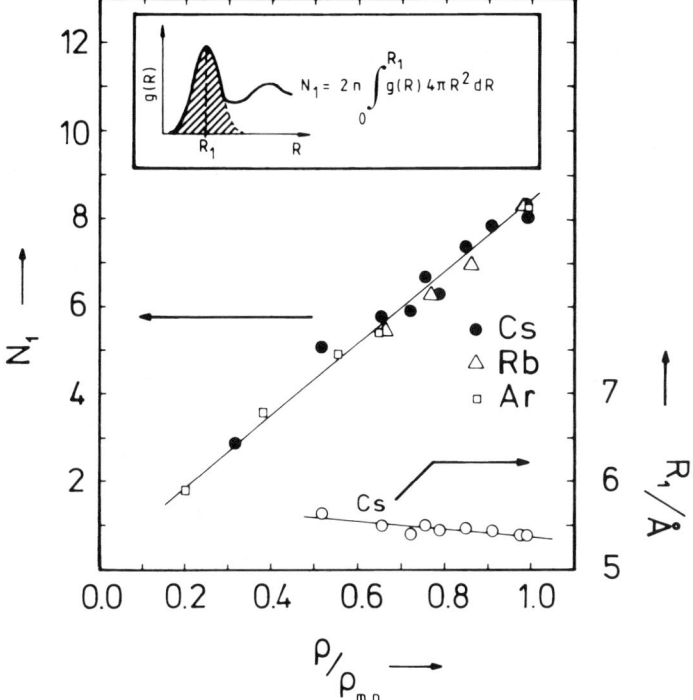

Figure 3b Average number of nearest neighbors N_1 for expanded liquid cesium, rubidium, and argon together with the average next neighbor distance R_1 of cesium as a function of reduced density (ρ_{mp} density at the melting point).

$$\sigma^{-1} = \frac{3\pi m^2 \Omega}{4h^3 e^2 k_F^6} \int_0^{2k_F} S(Q)V(Q)^2 Q^3 \, dQ \qquad 1.$$

where the structure factor $S(Q)$ is given by

$$S(Q) = 1 + \frac{4\pi n}{Q} \int_0^\infty R[g(R)-1] \sin QR \, dR \qquad 2.$$

with n the number density of atoms. $V(Q)$ denotes the screened ion pseudopotential, k_F the wavenumber of the electrons at the Fermi surface, and Ω the atomic volume. For $V(Q)$ the Ashcroft empty core potential (42) can be used combined with a density dependent dielectric function $\varepsilon(Q)$ that takes into account exchange and correlation effects (43). The only parameter in this model potential is the radius r_c, which was chosen to match the measured value of the dc-electrical conductivity near the melting point.

In Figure 4 the results of these calculations are compared with the experimental values of the dc-conductivity (27). The agreement is satisfactory for densities higher than 1.3 g cm^{-3}, where the Hall effect (44) can also be described by the NFE-theory. It is obvious from Figure 4, however, that the applied formalism breaks down for densities smaller than 1.3 g cm^{-3}, i.e. about three times the critical density ρ_c. This is very close to the range where the electron mean free path calculated from the measured conductivity σ (assuming NFE-behavior) becomes smaller and approaches a value comparable with the mean interatomic distance $a = n^{-1/3}$. In this range, the weak scattering NFE-approximation breaks down, thus indicating the onset of the strong thermodynamic state dependence of the electronic structure as the metal-nonmetal transition is approached.

The Metal-Nonmetal Transition

The problem is to understand by what mechanism the metal-nonmetal transition occurs. Mott (38) first pointed out the important fact that

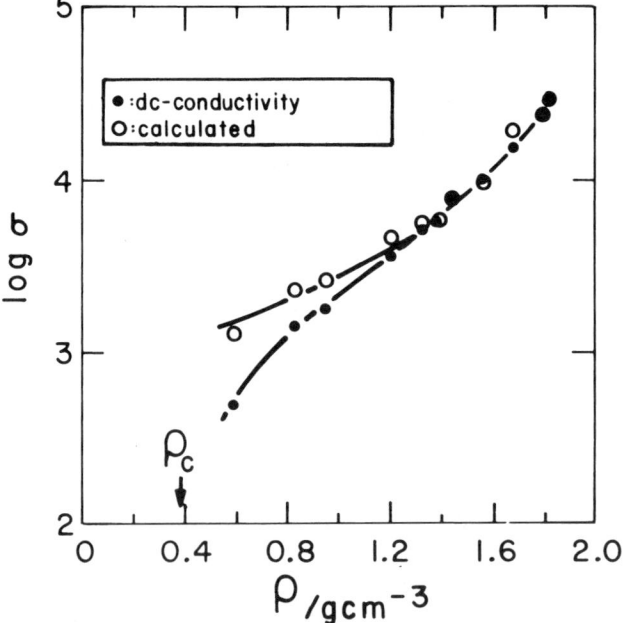

Figure 4 Measured and calculated (NFE-model) electrical conductivity σ of expanded liquid cesium as a function of density at conditions near the liquid-vapor coexistence.

electron-electron interaction must play a central role for the metal-nonmetal transition in expanded crystalline alkali metals. Without such interactions according to the Bloch-Wilson band model, these materials would always be metallic. Two effects have to be taken into account: The long-range Coulomb forces between charges and the short-range Coulomb interaction, namely the strong mutual repulsion experienced by two electrons occupying the same atomic site (37). Mott has presented convincing arguments (45) that the metal-nonmetal transition in expanded alkali metals is mainly a consequence of the short-range intra-atomic Coulomb repulsion that favors the localization of electrons to reduce double occupancy of individual sites. But, as Mott pointed out, in a fluid this mechanism can be affected in an important way by structural disorder, which can cause localization of states by the Anderson process (46) when the structural disorder is great enough.

A detailed theoretical analysis of the role of the intra-atomic repulsion in a metal (47) shows that the metallic state near the metal-nonmetal transition is highly correlated, having a low instantaneous fraction of doubly occupied sites. The correlated metal has an enhanced density of states and enhanced values for the paramagnetic susceptibility.

Consequently, magnetic properties and especially the effects of spin paramagnetism provide sensitive indications for the mechanism of the metal-nonmetal transition in expanded alkali metals. The metal has an enhanced density of states and, consequently, enhanced values for the paramagnetic susceptibility and electronic specific heat.

The possible presence of large correlation effects in the alkali metals was first shown by magnetic susceptibility measurements for expanded liquid cesium (28). An increasingly strong enhancement of the total mass susceptibility was observed with decreasing density for densities smaller than 1.3 g cm^{-3}, i.e. in the range where NFE-behavior breaks down (cf. Figure 4). Similar susceptibility enhancements have subsequently been observed in expanded liquid rubidium (48) and sodium (49). Nuclear magnetic resonance measurements of the Knight shift (50) in Cs also showed the low-density enhancement, confirming that the effect arises from the electron spin contribution to the total susceptibility. Thus the available experimental results favor a strong role for electron correlations in the metal-nonmetal transition of liquid alkali metals (32, 33, 35, 48, 51).

A distinction between different mechanisms is complicated, however, by the strong interplay between the vapor-liquid critical point density fluctuations and the rapid change in the electronic structure as the metal-nonmetal transition is approached in the critical region. Magnetic susceptibility measurements (28) for alkali metal atoms in the vapor phase show, for example, that a high concentration of molecular aggregates forms as the vapor density increases. These clusters have ionization energies substantially lower and electron affinities substantially higher than the corresponding values of the single atom (52). This effectively increases the probability of electron transfer from an atom to a larger cluster. The formation of a large portion of neutral or charged diamagnetic aggregates in the critical region of fluid alkali metals is completely consistent with the behavior of the magnetic properties (28). Thus, the conductivity observed at the critical density of Cs may be due to electrons from ionized Cs-clusters (36).

DIVALENT MERCURY

As Table 1 shows, mercury has the lowest critical temperature for any fluid metal. This property is significant experimentally in view of the severe difficulties associated with accurate measurements in the critical region. For this reason several of mercury's physical properties have been studied experimentally as a function of temperature and pressure up to and beyond the critical point, including equation of state data (53–58), electrical conductivity (10, 53–58), thermopower (10, 59–62), Hall coefficient (63), Knight shift (64, 65), optical reflectivity (66–68), optical absorption (67, 69, 70), sound velocity (71), specific heat (72), and viscosity (73). Each of these properties exhibits remarkable variations with density. For example, the dc-conductivity (Figure 5) decreases by more than seven orders of magnitude as the density is reduced from 13.6 g cm^{-3} to 2 g cm^{-3}. A comparison of the conductivity, Hall coefficient, and the Knight shift shows that for densities down to about 11 g cm^{-3} the properties of mercury can be described by the NFE-theory of metals but, with further decreasing density, a rather gradual metal-semiconductor transition occurs in the range $\rho = 9$ to 8 g cm^{-3}.

As is well known, this type of metal-semiconductor transition is predicted by the Bloch-Wilson band model to occur for an expanded divalent metal such as mercury when the 6s-valence and 6p-conduction bands no longer overlap. In a crystal, a real energy gap appears and widens as the density decreases. Mott (74) has proposed that the general features of the crystalline model survive in the liquid state with band edges smeared out by disorder. Thus, the density of states, $N(E)$, is expected to tail into the gap owing to the loss of long-range order. The tails overlap in the region of the Fermi-energy E_F, thus replacing the real energy gap of the crystal by a pseudogap or a minimum in $N(E)$ at E_F. The pseudogap depends strongly on density. When the magnitude of $N(E)$ in the pseudogap decreases with sufficient expansion to a negligibly small value, the optical properties of mercury must become compatible with the opening-up of an

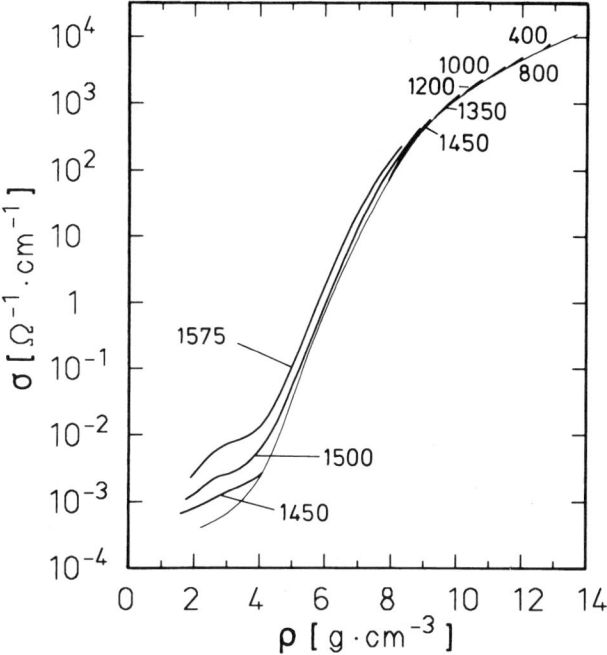

Figure 5 Electrical conductivity σ of fluid mercury at constant sub- and supercritical temperatures as a function of density ρ. The temperatures are in degrees centigrade.

energy gap. Figure 6 shows the optical conductivity $\sigma(\omega)$ measured by reflectivity (66) along the coexistence line of liquid mercury at various densities and temperatures. The change in the shape of the $\sigma(\omega)$-curves nicely illustrates the gradual diminution of metallic properties with decreasing density. In the nearly-free-electron range, ρ is larger than 11 g cm^{-3} and the low frequency optical conductivity is Drude-like. A gradual change from metallic to nonmetallic behavior occurs between 11 and 9 g cm^{-3}. For still smaller densities, the shape of the $\sigma(\omega)$-curves is characteristic of a substance with a real energy gap. A similar conclusion has been reached in a direct analysis (75) of the reflectivity data using a particular model of the density of states. In any case, the behavior of $\sigma(\omega)$ supports the view that for densities lower than 9 g cm^{-3}, either a true energy gap opens or a range of energy exists that is so thinly populated with states that their contribution to the optical properties is negligibly small. This view is completely consistent with the behavior of the electrical transport data (55) and with the sharp drop in the Knight shift (65) for densities smaller than 9 g cm^{-3}.

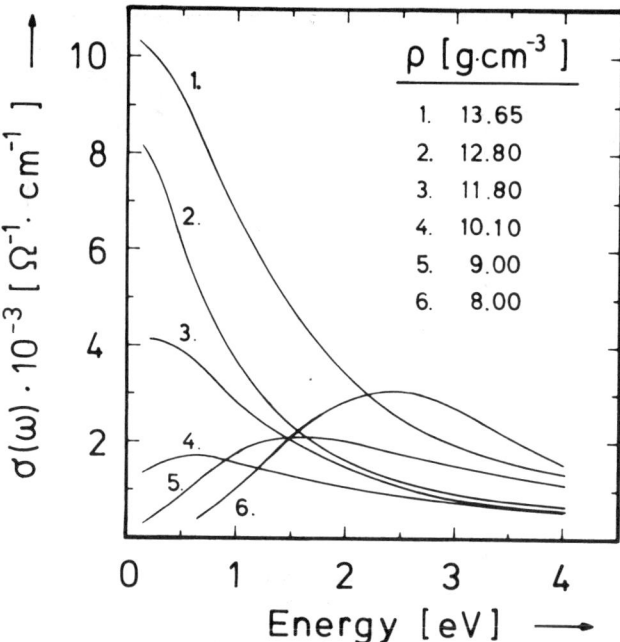

Figure 6 Optical conductivity $\sigma(\omega)$ of expanded mercury at conditions near the liquid-vapor coexistence; the corresponding densities ρ are given in the figure.

In the preceding, we have assumed that the variation of density is the most influential factor governing the behavior of $N(E)$ of expanded mercury. But when $N(E)$ is sensitive to density, density fluctuations must play a significant part in the optical properties. This becomes especially evident when the optical absorption is studied at even lower densities in the vapor (67, 69, 70). At very low densities a line spectrum is observed in which the main absorption lines at 4.89 eV and 6.7 eV correspond to transitions between the $6s$ ground state and the $6p$ triplet and singlet state of the Hg atom. As the density is increased, the sharp lines broaden due to interactions with neighboring atoms, resulting in a relatively steep absorption edge, which moves rapidly to lower energies with increasing density. Detailed analysis of these data (76, 77) shows that a uniform density increase is insufficient to explain the observed line broadening. Clusters of atoms created by density fluctuations have to be explicitly taken into account. The absorption edge is then lowered by the environment of the atom being excited, and the edge is thus explained in terms of absorption by excitonic states of large randomly distributed clusters. From the large values of the absorption coefficient it can be concluded that the singlet exciton (6^1p_1) with large oscillator strength broadens faster than the triplet exciton (6^3p_1) with small oscillator strength.

The shift of the effective gap in the absorption spectrum with increasing density can also be viewed in terms of a nonlinear enhancement of the real part of the dielectric constant ε_1. In fact, ε_1 is given by a Kramers-Kronig relation as an integral over the optical absorption coefficient (78). Data for ε_1 of Hg at the constant photon energy of 1.27 eV are shown in Figure 7 in the form of isotherms plotted versus density. There is an obvious indication for a strong interplay between the liquid vapor critical point density fluctuations and the optical properties. At the lowest, least metallic densities, ε_1 is only slightly dependent on temperature and follows the Clausius-Mosotti model of the dielectric constant of induced dipoles. At higher densities there is a strongly temperature dependent upward deviation from Clausius-Mosotti behavior. The fall to the negative values seen in the metallic state around 9 g cm^{-3} is quite gradual. The most

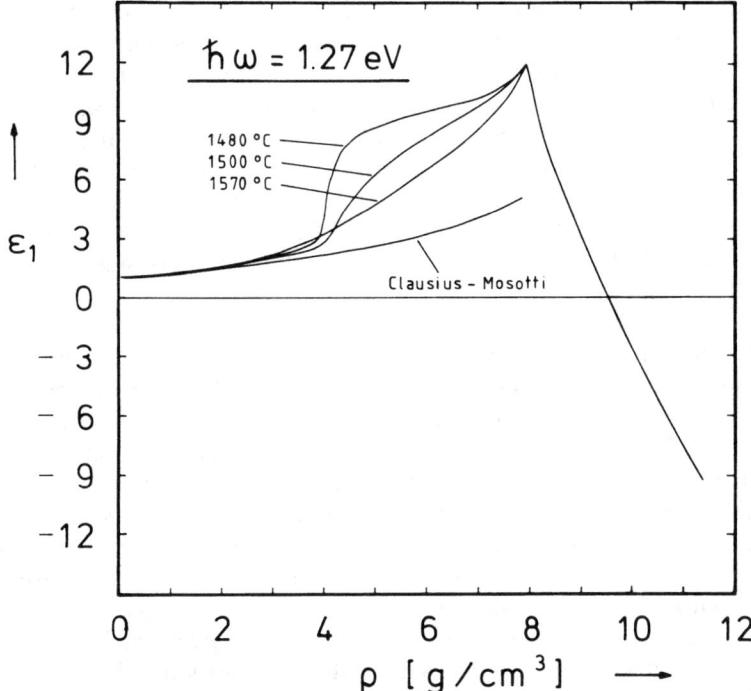

Figure 7 Real part of the dielectric constant ε_1 of fluid Hg at the constant photon energy $\hbar\omega = 1.27$ eV as a function of density at constant temperatures.

characteristic feature of the data is the strong temperature dependence of ε_1 when the critical point is approached. The analysis of the ε_1-data in the critical region close to the critical isochore as a function of temperature (70) clearly shows the presence of a large anomalous critical contribution in ε_1, reaching a maximum of about 70% of the background at a temperature about 0.1% above the critical point. By contrast, for a nonmetallic substance like CO (79), only a very weak dielectric anomaly of 0.1% is observed at a comparative distance from the critical point. We consider this contrast as evidence for a strong interplay between the vapor-liquid critical point density fluctuations and the large changes in the electronic structure as the metal-nonmetal transition is approached.

There have been several attempts to model the electronic structure of expanded liquid mercury by means of band structure calculations for appropriate crystalline structures. The justification for the application of band models to liquid metals is that the gross features of the electronic structure are determined largely by local properties such as the number and distance of nearest neighbors. The simplest approach is to assume an appropriate crystal structure with a variable lattice parameter (80). Other approaches are to assume that expansion changes the structure but not the nearest neighbor distance (69, 81–83). There is no direct structural evidence for this in expanded mercury but cesium, rubidium, and argon do behave so (Figure 3b), and their coordination numbers correspondingly decrease linearly with decrease in density between the melting and the critical point. The calculations indicate that a gap should not open until a density in the region of the critical point is reached. This is in contrast to the strong evidence from measurements of the optical reflectivity, Knight shift, and allied quantities that a real gap in the density of states opens at a density of about 9 g cm^{-3}.

The only successful approach to this problem was carried out by Franz (84), who considered the role of fluctuations in local coordination numbers caused by the disordered structure of an expanded fluid. Her model is also based on the assumption of a linear decrease in average coordination number with decreasing density, but in addition it takes into account the fact that the actual local coordination numbers will be distributed randomly over a range of values surrounding the mean value. As the density is reduced, some atoms will have such low coordination numbers that interactions with their neighbors will be insufficient to cause overlap between their s and p states; they will thus have an energy gap between their s and p states. This model leads to good agreement between the calculated density for the metal-semiconductor transition and its measured value.

METAL-NONMETAL TRANSITION IN CLUSTERS In view of the intimate link between fluctuations in local coordination numbers, i.e. microscopic clustering, and the macroscopically observed metal-nonmetal transition in fluid mercury, an understanding of the electronic structure of isolated small mercury clusters becomes essential. Mercury has a closed-shell atomic configuration, which means that the very small clusters are insulating and bonded through weak van der Waals forces, in contrast to the corresponding bulk material, which has metallic properties. Thus a size-dependent metal-nonmetal transition can be expected to occur in these clusters. Figure 8 shows the behavior of experimental ionization thresholds for mercury clusters produced in supersonic expansions. The ionization energies have been determined by energy-resolved mass spectrometry (85) and by photoelectron-photoion coincidence spectroscopy (86). They agree well with older results for smaller mercury clusters obtained by an electron impact ionization study (87). The ionization energies in Figure 8 are plotted

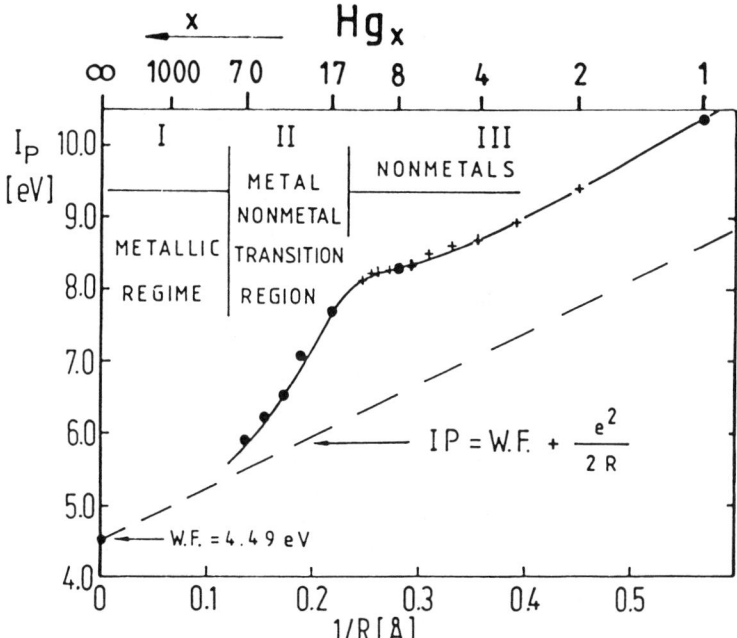

Figure 8 Comparison of the experimentally determined ionization potential data with the predictions of the classical spherical-droplet model [*circles*, this work; *crosses*, data taken from Ref. (87)]. We plot the ionization potential vs $1/R$, with R the radius of a sphere with the same volume as an x-atomic metal cluster (assuming atomic volume derived from bulk metallic density).

as a function of the reciprocal radius $1/R$, where R represents the radius of an idealized sphere that contains x Hg-atoms, each having the atomic volume v_0. The v_0-value is calculated from the molar volume of the solid. The underlying assumption is that isolated mercury clusters have a strong tendency to exist as close-packed aggregates exhibiting an overall spherical geometry.

For comparison, Figure 8 also shows the ionization energies (*dashed line*) calculated with the classical electrostatic equation, which gives the energy necessary to remove an electron from a uniformly conducting sphere having the same dimension as the cluster. The appropriate expression, which connects cluster ionization potentials with the bulk polycrystalline work function by an $1/R$ term, with clusters having higher ionization potentials for smaller particle size, is also given in Figure 8, where w represents the work function, R the radius of the equivalent sphere, and e the elementary charge. The results show that for small clusters of mercury atoms one can distinguish between two extreme situations, i.e. the metallic and the van der Waals-type modifications. Mercury atoms have an s^2-atomic configuration that is widely separated in energy from the first unoccupied atomic p-orbital, and so they give rise to small clusters with fully occupied van der Waals-type weakly binding s-bands. The change to the observed bulk metal properties of large Hg_x-clusters ($x > 70$) can only be achieved via overlap of the full s-derived valence band and the empty p-derived conduction band. The experimental results indicate that for Hg_x-clusters, the electronic properties of the bulk metal evolve gradually with increasing particle size in the size range $20 < x < 70$. The photoelectron spectra (88) clearly indicate that the evolution of the valence band structure toward the metallic state is almost fully accomplished for clusters larger than Hg_{70}. This size-dependent transformation from a nonmetallic to metallic state is again induced by the increase of the average coordination number.

The strong variation in the electronic structure in the course of this metal-nonmetal transition, which manifests itself in a correspondingly strong size-dependence of nearly all properties, also noticeably influences the kinetic features of the vapor-liquid phase transition of mercury. Measurements of the critical supersaturation for the homogeneous nucleation of mercury in the temperature range 260 to 400K (89) show that none of the current theories for homogeneous nucleation satisfactorily predicts the observed critical supersaturations. The measured values are about three orders of magnitude lower than the values predicted by the conventional Becker-Döring-Zeldovitch theory. It is noteworthy that the change in the value of the bulk liquid surface tension necessary to bring the classical nucleation theory in agreement with the experimental observation is about

40%. In contrast, most molecular liquids require only a very small adjustment of the bulk liquid surface tension to bring nucleation theory and experiment into agreement. An important difference between mercury and the molecular liquids is that in the former a size-dependent metal-nonmetal transition takes place.

The size-dependent gradual transition to metallic properties in isolated mercury clusters clearly must be part of any detailed consideration of the critical point phase transition of Hg. Because the decrease in the ionization potentials with increasing cluster-size is accompanied by a corresponding increase in the electron affinities (52), physical clustering arising from the reaction

$$l\,\mathrm{Hg}^0 \leftrightarrows \mathrm{Hg}_n^\oplus + \mathrm{Hg}_s^\ominus \quad \text{with} \quad l = n+s$$

can be viewed as an important mechanism for the occurrence of several interesting phenomena in conductivity, thermopower, and optical properties of mercury in the critical region (10, 68, 90, 91).

The most direct evidence for a strong interplay between the liquid-vapor critical point transition and the electrical transport properties of mercury stems from recent, very accurate simultaneous measurements of the electrical conductivity σ, the thermoelectric power S, and the density ρ (10). This may be seen at a glance from Figure 5, which shows the behavior of the electrical conductivity σ in the vicinity of the critical point as a function of density ρ at constant temperature T for a few selected isotherms. At the metal-nonmetal transition density of 9 g cm^{-3}, where according to optical and Knight shift measurements a real gap in the density of states opens, the conductivity reaches a value of about 200 ohm^{-1} cm^{-1}, i.e. a value that has often been cited as a characteristic indication for the limit of metallic behavior in liquids (37). As ρ falls below 9 g cm^{-3}, the conductivity falls more rapidly, but no discontinuity is observed because thermal activation of charge carriers across the gap broadens the transition range. It is clear from Figure 5 that there is a close correlation between the slope of the σ-curves and the critical point. The steepest fall in σ of Hg is observed at the critical density, where according to calculations employing optical and thermodynamic data about 1% of the valence electrons of mercury are thermally excited across the gap (10).

The patterns of the σ-curves are especially interesting for densities below the critical density, where fluid mercury forms a dense, partially ionized, gas with a relatively small degree of ionization. In this case, in addition to screened Coulomb interaction among the charges, electron-neutral interaction plays an important role for the transport (92). In this region, i.e. for densities well below the metal-nonmetal transition range, and for temperatures close to the critical temperature $T_c = 1478°C$, the thermally ionized quasi-free electrons can be trapped by density fluctuations (10). Perhaps the term "enhanced scattering" may be preferable to "trapped," because the traps are shallow (93).

Both the density dependence of σ for densities smaller than $\sigma_c = 5.8$ g cm^{-3} and the strong positive temperature dependence of σ for temperatures close to the critical temperature are completely consistent with this assumption. Heating fluid mercury to remove it from the vapor-liquid critical region decreases the magnitude of the microscopic density fluctuations and thereby diminishes the number and quality of trapping centers. The large temperature coefficient of the conductivity near the critical region (10) indicates that the entropy changes associated with the trapping of an electron is large.

As is well known, thinking of the thermoelectric power S of a conductor as a transport entropy of electric charge is valid and often helpful (94). Consequently, thermoelectric measurements are especially suited to study the interplay between the liquid vapor critical point density fluctuations and the electrical transport. From simultaneous measurement of S and density ρ it is possible to evaluate the density dependence of the thermoelectric power at constant temperature near T_c (Figure 9). A remarkably strong increase of S up to large positive values is observed in the density region where electron-neutral interaction becomes important. The energy transport by neutrals induced by the electron current gives a large contribution (a drag effect) to the thermoelectric power in the region of small degree of ionization. This finding strongly supports the suggestion that the electrons can be trapped by density fluctuations, and that there is a negative heat of transport that is present only if the density of electrons is small enough that they do not compete unduly for the surrounding atoms. A negative heat of transport can be envisaged if the trapping process involves entropy and if the fast electrons interact with the fluid less than the slow ones do and carry less heat with them.

LIQUID-VAPOR-TRANSITION

The occurrence of the metal-nonmetal transition implies a change from metallic to other interatomic interactions, and therefore changes in structural and thermodynamic properties may well appear. Neutron diffraction results (30, 39) such as those displayed in Figure 3a have provided useful information about such changes. Detailed theoretical analysis of such data (95, 96) shows that they reflect clear changes in the density dependence of the attractive part of the effective interionic interaction if screening is reduced as the metal-nonmetal transition is approached. Interestingly, the region of density in which these changes occur is the same as that for which

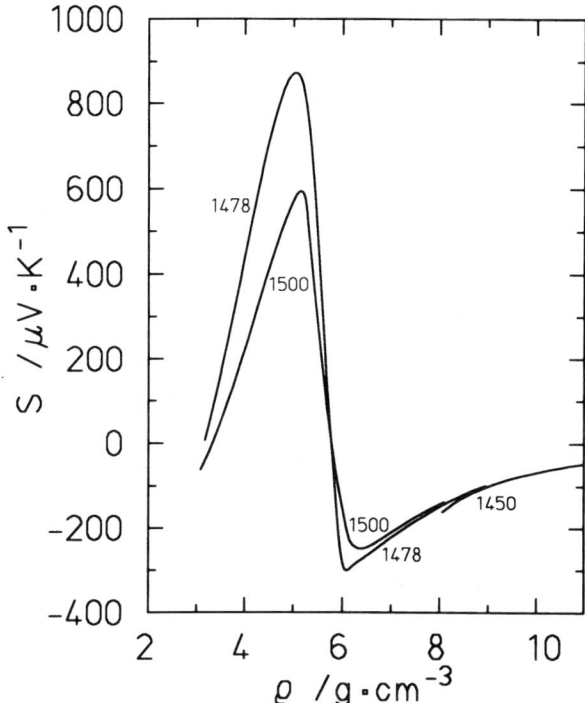

Figure 9 Thermopower S of fluid mercury at constant temperatures close to the critical as a function of density. The temperatures are in degrees centigrade.

the magnetic susceptibility data (28) indicate the presence of many-electron correlation effects.

An ultimate aim of structural measurements on expanded metals is to extract true interatomic potential functions as a function of density from the observed pair correlation functions $g(R)$. One way is (97) to invert $g(R)$-data directly to obtain a potential. This procedure requires a theory that relates the potential to $g(R)$. Whether the existing liquid state theories are accurate enough is a question of considerable controversy, but recent theoretical work (98) is leading to approximate procedures that can yield quite accurate pair potentials from structural data. The application of these methods is inextricably bound up with three-body correlations, which in turn can be related, at least in part, to the density derivatives of $S(Q)$. Data for the isothermal density derivatives of expanded cesium are now available (99).

Experimental observations for Cs, Rb, and Hg indicate that the gross change in the electronic properties in the course of the metal-nonmetal transition that manifests itself in a correspondingly strong thermodynamic-state-dependence of the effective interparticle interaction occurs in the vicinity of the critical region. It is not surprising, therefore, that the liquid-vapor phase behavior of metals differs strongly from that of normal molecular fluids, as demonstrated by the comparison of the coexisting liquid (ρ_L) and vapor (ρ_V) densities of Rb and the inert element Xe in Figure 10. The data to be compared are represented in dimensionless form by the critical constants. Poor reduced correlation between the alkali metal and the inert element is observed. Metals and nonmetals cannot be included together in one group obeying a principle of corresponding states.

The reduced diagram in Figure 10 demonstrates a second interesting consequence of the strong thermodynamic-state-dependence of the effective particle interaction in metallic fluids as the critical region is traversed. Fluid metals violate the 100 year old empirical law of rectilinear diameter (100) over a surprisingly large temperature range. By contrast, the deviations from this law are extremely (mostly immeasurably) small for the coexistence curves of essentially all nonmetallic one-component fluids (101). The law states that the locus of the tie-line mid-points $\rho_d = \frac{1}{2}(\rho_L + \rho_V)$ is a linear function of T. Since both ρ_L and ρ_V approach the limiting density ρ_c at the liquid-vapor critical point, the law can be written

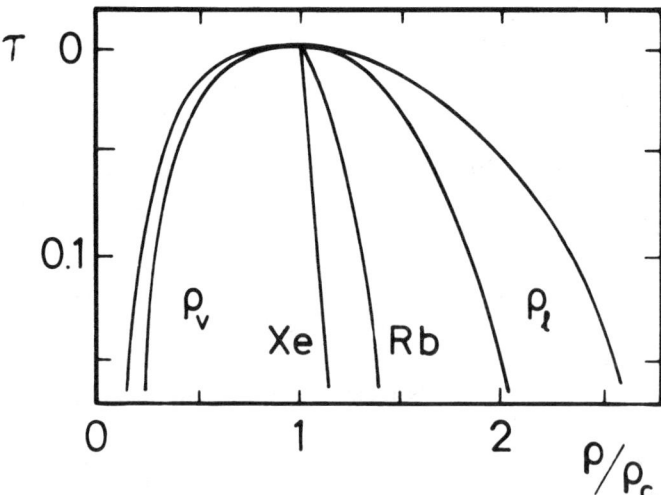

Figure 10 The reduced densities of the coexisting vapor and liquid of xenon (*inner curves*) and rubidium (*outer curves*).

$$\rho_d - \rho_c = D_1 \tau \qquad 3.$$

where $\tau = T_c - T/T_c$ and D_1 is a constant.

It has been suggested (34) that the contrast between the diameter data of metallic and nonmetallic one-component fluids arises from many-body effects whose magnitudes distinguish the particle interactions in metallic fluids from those in nonmetallic fluids. In particular, it is argued that the strong thermodynamic state dependence of the effective interactions in a metal, especially as the metal-nonmetal transition is traversed, corresponds to the mixing of thermodynamic fields present in certain solvable lattice models (102). These models, thermodynamic arguments (103), and renormalization-group studies (104) predict that the average value of the density, i.e. the diameter, will have the form

$$\rho_d - \rho_c = D_0 \tau^{(1-\alpha)} + D_1 \tau + \ldots \qquad 4.$$

where $\alpha = 0.11$ is the same exponent that describes the behavior of the constant volume specific heat c_V. Since $(1-\alpha) = 0.89$ is not very different from unity, the true singularity is difficult to separate from the analytic temperature term. The coefficient D_1 does not even have to be much larger than D_0 for the analytic term to dominate the entire range accessible to experimentation. The latter causes the difficulty in observing the $(1-\alpha)$-singularity for most nonmetallic fluids (see e.g. Xe in Figure 10).

Hitherto the most convincing experimental evidence for the existence of the $(1-\alpha)$ term for one-component systems has been the analysis of the rectilinear diameters of Cs, Rb (31), and Hg (10). The values for $(1-\alpha)$ are very close to the theoretically predicted value of 0.89. This finding strongly supports the hypothesis that the strong thermodynamic-state-dependence of the effective interparticle interactions, which is normally negligible for typical nonmetallic fluids, is responsible for the strength of the singular term in the diameters of liquid-vapor coexistence curves of metals (34).

CONCLUSIONS

Experimental results for fluid metals near the liquid-vapor critical point show profound changes in the electronic structure of fluid metals in that region. A metal-nonmetal transition occurs with decreasing density, which manifests itself in a correspondingly strong thermodynamic-state-dependence of the effective interparticle interaction. A rigorous theory of fluid metals must take into account that the very existence of this transition noticeably influences the thermodynamic and kinetic features of the vapor liquid phase transition of metals.

Literature Cited

1. Landau, L., Zeldovitch, G. 1943. *Acta Phys. Chim. USSR* 18: 194
2. Norman, G. E. 1971. *Zh. Eksp. Teor. Fiz.* 60: 1686 (*Sov. Phys.-JETP* 33: 912)
3. Odagaki, T., Ogita, N., Matsuda, H. 1975. *J. Phys. Soc. Jpn.* 39: 618
4. Ebeling, W., Kraeft, W. D., Kremp, D. 1976. *Theory of Bound States in Plasmas and Solids.* Berlin: Academic
5. Nara, S., Ogawa, T., Matsubara, T. 1977. *Prog. Theor. Phys.* 57: 1474
6. Mott, N. F. 1978. *Philos. Mag.* 37: 377
7. Ogawa, T., Nara, S., Matsubara, T. 1980. *J. Phys. Paris Colloq.* C8 41: 77
8. Yonezawa, F., Ogawa, T. 1982. *Suppl. Prog. Theor. Phys.* 1982(72): 1
9. Krumhansl, J. A. 1965. In *Physics of Solids at High Pressures*, ed. C. T. Tomizuka, R. M. Emrick, p. 425. New York: Academic
10. Götzlaff, W., Schönherr, G., Hensel, F. 1988. *Z. Phys. Chem. N.F.* 156: 219
11. Jüngst, S., Knuth, B., Hensel, F. 1985. *Phys. Rev. Lett.* 55: 2160
12. Freyland, W., Hensel, F. 1972. *Ber. Bunsenges. Phys. Chem.* 76: 347
13. Binder, H. 1984. PhD thesis. Univ. Karlsruhe, FRG
14. Gathers, R. G. 1986. *Rep. Prog. Phys.* 49: 341
15. Seydel, U., Fucke, W. 1978. *J. Phys. F* 8: L157
16. Martynuk, M. M., Panteleichuk, O. G. 1976. *High Temp.* (*USSR*) 14: 1075
17. Shaner, J. W., Gathers, G. R., Minichino, C. 1976. *High Temp.–High Press.* 8: 125
18. Shaner, J. W., Hixson, R. S., Winkler, M. A., Brown, J. M. 1987. In *Strongly Coupled Plasma Physics*, ed. F. J. Rogers, H. E. Dewitt, p. 395. New York: Plenum
19. Hixson, R. S., Winkler, M. A., Shaner, J. W. 1986. *Physica B* 140: 893
20. Cusack, N. E. 1978. In *Metal Non-Metal Transition in Disordered Systems*, ed. L. R. Friedman, D. P. Tunstall, p. 455. Edinburgh: Scott. Univ. Summer Sch. Phys.
21. Freyland, W. 1981. *Commun. Solid State Phys.* 10: 1
22. Alekseev, V. A., Jakubov, I. T. 1983. *Phys. Rep.* 96: 1
23. Freyland, W., Hensel, F. 1985. In *The Metallic and the Nonmetallic States of Matter*, ed. P. P. Edwards, C. N. R. Rao, p. 93. London: Taylor & Francis
24. Endo, H. 1982. *Suppl. Prof. Theor. Phys.* 1982(72): 100
25. Hensel, F. 1987. In *Large Finite Systems*, ed. J. Jortner, A. Pullman, B. Pullman, p. 345. Dordrecht: Reidel
26. Gläser, W., Hensel, F., Lüscher, E., eds. 1986. *Proc. 6th Int. Conf. on Liquid and Amorphous Metals.* München: Oldenbourg
27. Hensel, F., Jüngst, S., Noll, F., Winter, R. 1985. In *Localisation and Metal Insulator Transitions*, ed. D. Adler, H. Fritsche, p. 109. New York: Plenum
28. Freyland, W. 1979. *Phys. Rev. B* 20: 5140
29. El-Hanany, U., Brennert, G. F., Warren, W. W. 1983. *Phys. Rev. Lett.* 50: 540
30. Winter, R., Bodensteiner, T., Gläser, W., Hensel, F. 1987. *Ber. Bunsenges. Phys. Chem.* 91: 1327
31. Hensel, F., Jüngst, S., Knuth, B., Uchtmann, H., Yao, M. 1986. *Physica B* 139: 90
32. Warren, W. W., Mattheis, L. F. 1984. *Phys. Rev. B* 30: 3103
33. Franz, J. R. 1984. *Phys. Rev. B* 9: 1565
34. Goldstein, R. E., Ashcroft, N. W. 1985. *Phys. Rev. Lett.* 55: 2164
35. Mott, N. F., Davis, E. A. 1979. *Electronic Processes in Noncrystalline Materials.* Oxford: Clarendon
36. Hernandez, J. P. 1986. *Phys. Rev. A* 34: 1316
37. Mott, N. F. 1974. *Metal-Insulator Transitions.* London: Taylor & Francis
38. Mott, N. F. 1949. *Proc. Phys. Soc. London Sect. A* 62: 416
39. Franz, G., Freyland, W., Gläser, W., Hensel, F., Schneider, E. 1980. *J. Phys. Paris Colloq.* C8 41: 192
40. Pings, C. J. 1968. In *Physics of Simple Liquids*, ed. H. N. V. Temperley, J. S. Rowlinson, G. S. Rushbrooke, p. 387. Amsterdam: North-Holland
41. Faber, T. E. 1972. *Introduction to the Theory of Liquid Metals.* Cambridge: Cambridge Univ. Press
42. Ashcroft, N. W. 1966. *Phys. Lett.* 23: 48
43. Heine, V., Abarenkov, I. 1964. *Philos. Mag.* 9: 451
44. Even, U., Freyland, W. 1975. *J. Phys. F* 5: L104
45. Mott, N. F. 1969. *Philos. Mag.* 19: 835
46. Anderson, P. W. 1958. *Phys. Rev.* 109: 1492
47. Brinkman, W. F., Rice, T. M. 1970. *Phys. Rev. B* 2: 4302
48. Freyland, W. 1980. *J. Phys. Paris Colloq.* C8 41: 74
49. Bottyan, L., Dupree, R., Freyland, W. 1983. *J. Phys. F* 13: L173

50. El-Hanany, U., Brennert, G. F., Warren, W. W. 1983. *Phys. Rev. Lett.* 50: 540
51. Rose, J. H. 1981. *Phys. Rev. B* 23: 552
52. Castleman, A. W., Keesee, R. G. 1988. *Science* 241: 36
53. Hensel, F., Franck, E. U. 1966. *Ber. Bunsenges. Phys. Chem.* 70: 1154
54. Kikoin, I. K., Sechenkov, A. P. 1967. *Phys. Met. Metallogr.* 24: 5
55. Schönherr, G., Schmutzler, R. W., Hensel, F. 1979. *Philos. Mag. B* 40: 411
56. Yao, M., Endo, H. 1982. *J. Phys. Soc. Jpn.* 51: 966
57. Hubbard, S. R., Ross, R. G. 1983. *J. Phys. C* 16: 6921
58. Götzlaff, W. 1988. PhD thesis. Univ. Marburg, FRG
59. Schmutzler, R., Hensel, F. 1971. *Phys. Lett.* 35: 55
60. Duckers, L. J., Ross, R. G. 1972. *Phys. Lett. A* 38: 291
61. Neale, F. E., Cusack, N. E. 1979. *J. Phys. F* 9: 85
62. Yao, M., Endo, H. 1982. *J. Phys. Soc. Jpn.* 51: 1504
63. Even, U., Jortner, J. 1972. *Phys. Rev. Lett.* 28: 31
64. El-Hanany, U., Warren, W. W. 1975. *Phys. Rev. Lett.* 34: 1276
65. Warren, W. W., Hensel, F. 1982. *Phys. Rev. B* 26: 5980
66. Hefner, W., Schmutzler, R. W., Hensel, F. 1980. *J. Phys. Paris Colloq.* C8 41: 62
67. Ikezi, H., Schwarzenegger, K., Simons, A. L., Passner, A. L., McCall, S. L. 1978. *Phys. Rev. B* 18: 2494
68. Hefner, W., Hensel, F. 1982. *Phys. Rev. Lett.* 48: 1026
69. Overhof, H., Uchtmann, H., Hensel, F. 1976. *J. Phys. F* 6: 523
70. Uchtmann, H., Brusius, U., Yao, M., Hensel, F. 1988. *Z. Phys. Chem. N.F.* 156: 151
71. Suzuki, K., Inutake, M., Fujiwaka, S., Yao, M., Endo, H. 1980. *J. Phys. Paris Colloq.* C8 41: 66
72. Levin, M., Schmutzler, R. W. 1984. *J. Non-Cryst. Solids* 61: 83
73. Tippelskirch, H. V., Franck, E. U., Hensel, F., Kestin, J. 1975. *Ber. Bunsenges. Phys. Chem.* 79: 889
74. Mott, N. F. 1966. *Philos. Mag.* 13: 989
75. Krohn, C. E., Thompson, J. C. 1980. *Phys. Rev. B* 21: 2619
76. Bhatt, R. N., Rice, T. M. 1979. *Phys. Rev. B* 20: 466
77. Uchtmann, H., Popielawski, J., Hensel, F. 1981. *Ber. Bunsenges. Phys. Chem.* 85: 555
78. Ziman, J. M. 1964. *Principles of the Theory of Solids.* Cambridge: Cambridge Univ. Press
79. Pestak, M. W., Chang, M. H. W. 1981. *Phys. Rev. Lett.* 46: 939
80. Devillers, M. A. C., Ross, R. G. 1975. *J. Phys. F* 5: 73
81. Fritzson, P., Berggren, K. F. 1976. *Solid State Commun.* 19: 385
82. Yonezawa, F., Ishida, Y., Martino, F., Asano, S. 1976. In *Liquid Metals*, ed. R. Evans, D. A. Greenwood, p. 385. Bristol: Inst. Phys.
83. Mattheis, L. F., Warren, W. W. 1977. *Phys. Rev.* 16: 624
84. Franz, J. R. 1986. *Phys. Rev. Lett.* 57: 889
85. Rademann, K., Kaiser, B., Even, U., Hensel, F. 1987. *Phys. Rev. Lett.* 59: 2319
86. Rademann, K., Kaiser, B., Rech, T., Hensel, F. 1989. *Z. Phys. D.* In press
87. Cabaud, B., Hoareau, A., Melinon, J. 1980. *J. Phys. D* 13: 1831
88. Rademann, K. 1989. *Ber. Bunsenges. Phys. Chem.* In press
89. Martens, J., Uchtmann, H., Hensel, F. 1987. *J. Phys. Chem.* 91: 2489
90. Hernandez, J. P. 1985. *Phys. Rev. A* 31: 932
91. Uchtmann, H., Hensel, F., Overhof, H. 1980. *Philos. Mag. B* 42: 583
92. Höhne, F. E., Render, R., Röpke, G., Wegner, H. 1984. *Physica A* 128: 643
93. Thirumalai, D. 1987. In *Large Finite Systems*, ed. J. Jortner, A. Pullman, B. Pullman, p. 231. Dordrecht: Reidel
94. De Groot, S. R. 1952. *Thermodynamics of Irreversible Processes.* New York: Interscience
95. McLaughlin, I. L., Young, W. H. 1984. *J. Phys. F* 14: 1
96. Kahl, G., Hafner, J. 1984. *Phys. Rev. A* 29: 3310
97. Johnson, M. D., Hutchinson, P., March, N. H. 1964. *Proc. R. Soc. London Ser. A* 282: 283
98. Reatto, L. 1988. *Philos. Mag.* 58: 37
99. Winter, R., Hensel, F., Bodensteiner, T., Gläser, W. 1988. *J. Phys. Chem.* 92: 7171
100. Cailletet, L., Mathias, E. C. 1886. *C. R. Acad. Sci.* 102: 1202
101. Goldstein, R. E., Parola, A., Ashcroft, N. W., Pestak, M. W., Chen, M. H. W., deBruyn, J. R., Balzarin, D. A. 1987. *Phys. Rev. Lett.* 58: 41
102. Rowlinson, J. S. 1970. *Adv. Chem. Phys.* 41: 1
103. Mermin, N. D. 1971. *Phys. Rev. Lett.* 26: 957
104. Nicoll, J. F. 1981. *Phys. Rev. A* 24: 2203

ELECTROLYTES DISSOLVED IN POLYMERS

J. M. G. Cowie and S. H. Cree

Department of Chemistry, Heriot-Watt University,
Edinburgh, EH14 4AS, Scotland

INTRODUCTION

The incorporation of ions in polymer matrices can be achieved in a number of ways (1, 2). Polymeric structures can be synthesized specifically contain acidic or basic groups that on subsequent neutralization produce a macromolecular matrix with covalently bound anionic or cationic sites and their associated counterions. Such structures are known as ionomers or ionenes, depending on whether the site attached to the polymer is anionic or cationic, respectively.

Recent work on ion implantation techniques has demonstrated that Li^+, F^-, and I^- can be incorporated into polymer films and that enhanced conductivity levels can then be observed (3–8). This method is still relatively new, and a much easier way by which electrolytes can be included in a bulk polymer is by a simple dissolution process. If successful, it is possible to produce a solid solution that is either in a homogeneous amorphous form or a polymer-salt complex of definable stoichiometry, or indeed a mixture of both. This latter area has seen a significantly large increase in research activity in the last ten years, stimulated mainly by the observation by Armand and co-workers (9) that such systems might prove useful as electrolytic media in high energy density dry battery manufacture. This development was based on original work by Wright et al (10–12). Although the investigation of ionic conductivities in these polymer electrolyte systems has tended to dominate interest in this area, other physical properties of the systems have been examined and are reviewed here in addition to a brief commentary on the conduction studies.

CRITERIA FOR DISSOLUTION

To achieve the dissolution of electrolytes in a polymer, thereby producing a homogeneous solution, some form of interaction between the polymer chains and the electrolyte is necessary. Interaction is most easily obtained when there is an electron donor atom in the polymer chain that can coordinate with the cation of the salt through a Lewis acid-base reaction. Typically, suitable polymers will contain oxygen, nitrogen, or sulphur, and many such polymers are now available that can dissolve certain salts with relative ease. Early work on low molecular weight polyethers showed that the addition of an electrolyte enhanced the viscosity of the solution formed, compared with that of the original polymer, thus indicating that some interaction was taking place between the two components (2). This work was extended to an examination of other physical properties by Moacanin & Cuddihy (13), who observed that when lithium perchlorate was dissolved in polypropylene glycol (PPG) it caused a volume contraction that for a 10.4 wt% solution of the salt in the polymer was equivalent to an applied pressure of 18.95×10^6 Pa at 298K. This finding was interpreted as indicative of a strong polymer-salt interaction involving the Li^+ ion and the ether oxygen in PPG. Equally significant was the noted increase in the glass transition temperature of the polymer mixture, which rose monotonically as more salt was added. The increase led to a concomitant increase in the modulus of the system. The authors suggested, in qualitative terms, that the Li^+ was poorly screened by the large perchlorate anion, thereby allowing facile interaction with the ether oxygens of the polymer and aiding dissolution of the salt. This reaction would allow interchain interactions to take place, thus involving the Li^+ and leading to physical crosslinking of the polymer chains and the observed physical changes.

It is clear from the evidence that has subsequently been gathered from a wide range of systems by Armand et al (14) and Shriver's group (15) that the ability of a polymer to interact efficiently with the cation and effect a separated ion pair in the matrix has a strong influence on the solubility characteristics. To a first approximation the solubility of a salt in a polymer can be related to the lattice energy of the salt, as can be seen in Table 1 where the solubilities of various salts in poly(ethylene oxide) (PEO) are compared. Table 1 shows that salts are more likely to be soluble in PEO when there is a combination of a low lattice energy and an associated large anion. That correlation tends to be an oversimplification, however, and solubilities can also depend strongly on the structure of the polymer.

Thus PEO, which can be thought of as a "hard" base, will complex strongly with "hard" acid cations such as Li^+, Na^+, Mg^{2+}, and Ca^{2+}, but much more weakly with softer acids such as Ag^+ and Hg^{2+}. Polythio-

Table 1 A comparison of the tendency for miscible PEO-salt mixtures to form and the lattice energies of the salts. Values in parentheses are either estimated or calculated theoretically

	Li$^+$	Na$^+$	K$^+$	Rb$^+$	Cs$^+$
F$^-$	No 1036	No 923	No 821	No 785	No 740
Cl$^-$	Yes 853	No 786	No 715	No 689	No 659
CH$_3$COO$^-$	— 881	No 763	— 682	— 656	— (682)
NO$_3^-$	— 848	No 756	— 687	— 658	No 625
NO$_2^-$	— —	No 748	— 664	— 765	— (598)
Br$^-$	Yes 807	Yes 747	No 682	No 660	No 631
N$_3^-$	— 818	No 731	— 658	— 632	— 604
BH$_4^-$	— (778)	Yes (703)	— (665)	— (648)	— (628)
I$^-$	Yes 757	Yes 704	? 644	No 630	No 604
SCN$^-$	Yes 807	Yes 682	Yes 616	Yes 619	Yes 568
ClO$_4^-$	Yes 723	Yes 648	— 602	— 582	— 542
CF$_3$SO$_3^-$	Yes ($\leq$725)	Yes ($\leq$650)	Yes ($\leq$605)	Yes ($\leq$585)	Yes ($\leq$550)
BF$_4^-$	Yes (699)	Yes 619	— 631	— 605	— (556)
BPh$_4^-$	Yes ($\leq$700)	Yes ($\leq$630)	Yes ($\leq$630)	Yes ($\leq$600)	Yes ($\leq$550)

ethers, on the other hand, which are softer bases, will tend to interact more readily with the softer acids.

These interactions must be strong enough to overcome not only the lattice energy of the salt but also the loss of conformational entropy as the polymer chain reorients to form a more ordered structure when complexing with the ion. This latter parameter may not be large in a bulk polymer matrix and tends to be ignored, although it has also been suggested that a low cohesive energy and high chain flexibility are necessary prerequisites for a good host polymer to allow facile reorientation of the chain to take place. Thus the entropy loss is not negligible.

Polymer chain geometry is also an important feature. It is well known that crown ethers and cryptands are excellent solvators for metal cations. The crown ethers are the cyclic analogues of PEO, and the repeat unit (CH_2–CH_2–O) seems to be a favorable arrangement for effective interaction with alkali metal cations in particular. The PEO chains are capable of adopting a helical conformation with an oxygen-lined cavity that presents the optimum distances for oxygen-ion interactions similar to the crown structures, and one might expect other polyethers to be equally effective. Neither poly(methylene oxide) with a (CH_2–O) repeat unit nor poly(trimethylene oxide) with a ($CH_2CH_2CH_2$–O) structure is capable of forming homogeneous polymer-salt mixtures (14). This inability has been ascribed to their unfavorable geometry and their inability to form complexing structures with the appropriate spatial placing of the oxygen environment. Poly(propylene glycol), repeat unit

$$(-CH_2-CH-O-)$$
$$\quad\quad\quad |$$
$$\quad\quad\quad CH_3$$

has the same chain geometry as PEO but is less effective in dissolving electrolytes. In this molecule the steric hindrance of the methyl unit forces the polymer chain to adopt a much more open helical structure, which, although it can produce an oxygen-lined cavity in its helical conformation, cannot bind the cations as tightly as the PEO structure. The differences can be seen quite clearly when space-filling models of both structures are examined, as shown in Figures 1 and 2. Similarly the isomeric form poly(vinyl methylether), whose repeat unit is

$$(-CH_2-CH-)$$
$$\quad\quad |$$
$$\quad\quad O-CH_3,$$

is even less effective, and recourse to molecular models shows that only short linear sequences of oxygen can be formed, which cannot be transformed into a helix with an oxygen-lined inner side (see Figure 3). Thus structure and the ability to form compatible environments for ion complexation are clearly important factors in the dissolution process.

Other oxygen-containing polymer hosts that have been shown to be capable of dissolving electrolytes include

poly(epichlorohydrin) $(-O-CH_2-CH-)$ (16),
$$\quad\quad\quad\quad\quad\quad\quad\quad\quad\quad\quad |$$
$$\quad\quad\quad\quad\quad\quad\quad\quad\quad\quad\quad CH_2Cl$$

poly(ethylene succinate) $[-O(CH_2)_2-O-C(CH_2)_2-C-]$ (17, 18),
$$\quad\quad\quad\quad\quad\quad\quad\quad\quad\quad\quad\quad\quad\quad\quad \| \quad\quad\quad \|$$
$$\quad\quad\quad\quad\quad\quad\quad\quad\quad\quad\quad\quad\quad\quad\quad O \quad\quad\quad O$$

ELECTROLYTES DISSOLVED IN POLYMERS 89

(a)

(b)

Figure 1 Molecular models of poly(ethylene oxide) showing the ability of the polymer to adopt a helical structure with an internal oxygen-lined environment.

poly(β-propiolactone) [–O(CH$_2$)$_2$C–] (19),
$\|$
O

and

poly(ethylene adipate) [–O(CH$_2$)$_2$–O–C–(CH$_2$)$_4$–C–] (20),
$|\|$
OO

all of which contain elements of the PEO repeat unit. Siloxanes are, however, quite unsuitable, partly because of the inappropriate spacing of

Figure 2 Molecular model of a section of a poly(propylene oxide) chain, showing an oxygen-lined cavity for ion complexation. Note that the size of this cavity is much larger than that formed in PEO.

the oxygen atoms in the chain, and partly because of a suppression of the oxygen donor number.

More success has been achieved with the nitrogen analogues of PEO, as the (C–N) bond length (0.147 nm) is only slightly longer than the (C–O) bond length of 0.143 nm. Thus poly(ethylene imine) (21–23) R = H and its derivatives

$$(-CH_2CH_2N-) \atop R \qquad (24)$$

can form structures with suitable donor atom environments. This is not the case for the poly(alkylene sulphide)s [$(CH_2)_nS$], where the (C–S) bond (0.182 nm) is much longer and helical structures are much less likely to be formed (25). Also the coordination can be more complex, as soft Lewis acids are now more likely to dissolve in these media and the coordination may also involve the oxygen atoms in the counterion, if present.

Hannon & Wissbrun (26, 27) examined the solubility behavior of zinc, copper, cadmium, and calcium nitrates in a series of oxygen-containing polymers and found that the trend in the ability to form homogeneous solutions decreased in the order: cellulose acetate, poly(vinyl alcohol)

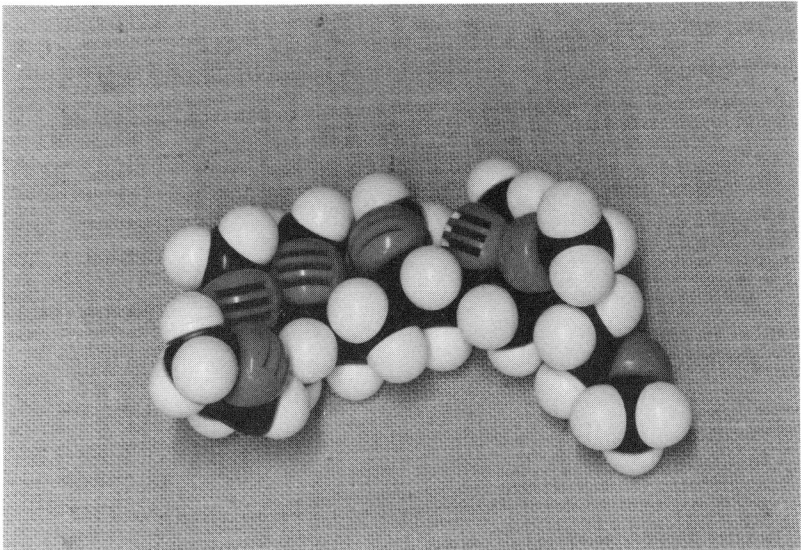

Figure 3 Molecular model of a section of a poly(vinyl methyl ether) chain that demonstrates the inability of this structure to form a crown ether-like cavity. The figure shows the longest section with an oxygen-lined environment that can be formed by this model.

> poly(vinylacetate)/poly(methylacrylate) > poly(methyl methacrylate). They also found that calcium thiocyanate was readily dissolved in a polymer formed from bisphenol A and epichlorohydrin, probably aided predominantly by the latter structure.

Recently an attempt was made to predict the ability of PEO to form complexes with sodium halides, and to relate this to the lattice energies (28). The model used was one in which three ethylene oxide units were associated with one mole of NaX and complex formation was assumed to take place if the lattice energy of the complex was less than the combined lattice energies of the salt and the crystalline ethylene oxide segment. The lattice energy of the complex was further subdivided into inter- and intracontributions, but entropy effects were neglected. A semiquantitative explanation of the solubilities of NaBr and NaI and the insolubility of NaCl in PEO was obtained, but much remains to be done in this area before a fully satisfactory theoretical prediction is obtained.

Some polymer-electrolyte mixtures have been found to be temperature sensitive, particularly when the interaction is weak. Wintersgill et al (29) have demonstrated that amorphous complexes of PPG with KSCN and NaI are unstable above 60°C and 85°C, respectively, when phase separation of the salts from the polymer was observed. This is an important consideration in ion conduction studies in which systems are often operated at temperatures as high as 120°C–140°C.

PHASE DIAGRAMS

It is obvious from the foregoing that on a macroscopic scale, a number of discrete phases may be observed in a polymer-electrolyte mixture that are dependent not only on the intrinsic properties of the polymer and the salt but also on the temperature of observation and the concentration of the added salt. In many cases a crystalline polymer-salt complex or series of complexes of specific composition can be identified and this leads to the detection of a number of eutectics in the phase diagram. In addition, a polymer that is able to solvate the added electrolyte yields an interesting amorphous phase, now universally accepted as the phase in which ion conduction takes place in these materials. Furthermore, either uncoordinated neutral salt or charged species, such as triplets or higher order anions and cations, may cluster together to yield larger aggregates, and these can lead to more elaborate phase diagrams.

A number of methods have been used in an attempt to determine the presence and amount of each of the above phases. Distinct solution and stoichiometric complexes can be identified by optical microscopy (30, 31). Due to differences in the relaxation of nuclei in an amorphous and a crystalline environment, NMR can be employed to great effect (32–34). Numerous X-ray diffraction studies have been carried out, although in general larger unit cells with low symmetry make indexing of the powder patterns difficult (30, 35, 36). These latter studies are widely complemented by differential scanning calorimetry (d.s.c.) analysis, which provides information on the presence and thermal stability ranges of the discrete phases (37–39).

On the basis of such information it is possible to construct a phase diagram for a polymer-electrolyte system that is analogous to the solubility curves and eutectic behavior for simple mixtures, if amorphous phases are equated with the liquidus and crystalline phases with the solidus phases. One should be aware, however, that polymers do not always obey the phase rule and that metastable states are often found. Slow structural reorganization may often retard crystallization, with the result that one may be dealing with a system that is only approaching chemical equilibrium. In particular phase distributions and compositions are highly dependent on the method of preparation and the thermal history (39–42). Such observations have led some workers to suggest that polymer-electrolyte phase diagrams should be referred to as "pseudophase diagrams" (37).

By far the most widely documented phase diagrams are those of PEO-alkali metal salts, as a direct result of the technological push toward the development of high energy density batteries and the use of these solutions as the solid electrolyte phase. The phase diagram for PEO-LiSO$_3$CF$_3$ has been studied by several groups (30, 34, 41–43), and NMR has been used by Minier et al (34) to probe the environment of the ^{1}H, ^{7}Li, and ^{19}F nuclei. The short relaxation times were attributed to the crystalline phase, and the longer T_2 values were taken as indicative of the amorphous phase. Deconvolution of the transverse magnetization curves allowed the determination of both the concentration and the volume fraction of the phases. Evidence for the existence of a eutectic at low salt concentrations ([O]/Li$^+$ = 100) has been presented by Sorensen & Jacobsen (44) and Robitaille & Fauteux (30). In the most recent study of this system, by Zahurak et al (37), a progressive change in the X-ray diffraction pattern was noted, with the peak due to the crystalline phase in PEO decreasing in intensity as more salt was added to the pure polymer. Eventually the crystallinity in the PEO disappeared at the composition (PEO)$_{3.5}$–LiSO$_3$CF$_3$ and the presence of another crystalline phase was identified. This phase was postulated to be a sandwich-type complex comprising two helical turns of the polymer with a single cation in the middle. The presence of free salt was also detected but only at the higher concentrations (PEO : salt) of (1:1). These data in combination with d.s.c. analyses enabled the construction of the pseudophase diagram shown in Figure 4. Also presented in this paper is the pseudophase diagram for the PEO–LiBF$_4$ system, using the same characterization procedures.

Dissolution of LiClO$_4$ in PEO seems to be a much more complex exercise. X-ray studies indicate that three distinct crystalline complexes are formed.

Robitaille & Fauteux (30) have constructed the pseudophase diagram for this system, which shows the existence of a (PEO)$_6$–LiClO$_4$ phase that melts at 65°C. A eutectic with a melting point at 50°C was also located, together with a (PEO)$_{10}$–LiClO$_4$ species. No eutectic was observed between 65°C and 160°C, the latter being the melting point of the (PEO)$_3$–LiClO$_4$ complex. These features are illustrated in Figure 5. Ferloni et al (42) have reported similar behavior for this system but found the eutectic at 30°C, thus illustrating the inherent difficulties in measuring these diagrams accurately.

Crystalline complexes have also been documented (30) for the PEO–LiAsF$_6$ system, where (3:1) and (6:1) polymer-salt species have been identified. James et al (45) and Robitaille & Prud'homme (46) have examined the interactions between PEO and sodium salts. A phase diagram for the PEO–NaSCN system has been produced in which the stoichiometry of

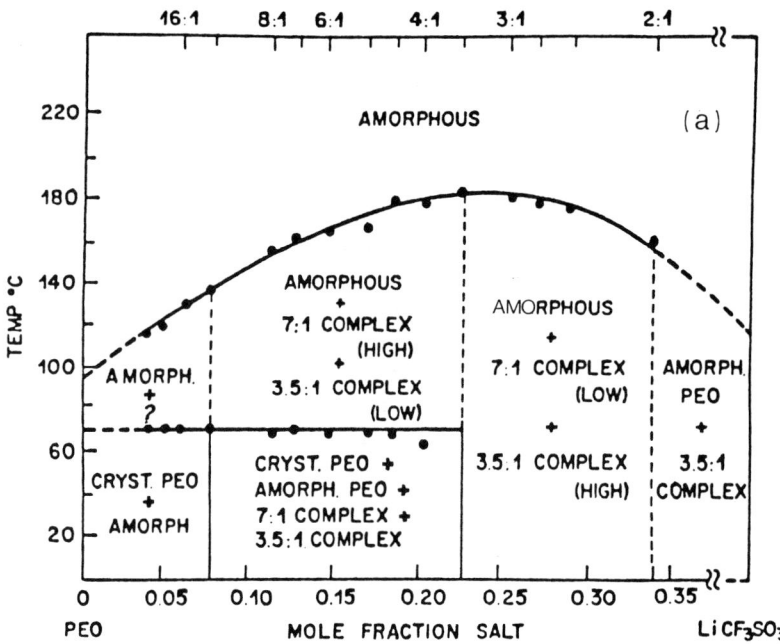

Figure 4a The pseudophase diagram for the PEO–LiCF$_3$SO$_3$ system constructed from X-ray analysis and the d.s.c. analysis shown in (*b*) and (*c*). (Reproduced from Ref. 37 with permission from the American Chemical Society.)

the complex is (PEO)$_3$–NaSCN. This has since been confirmed by Hibma (35).

Typical of the work in this area is that reported by Chiang et al (47), who used d.s.c. to show that the crystalline melting endotherm of PEO at 68°C decreased as NaI was added; at and above 10 mol% NaI a second endotherm appeared at 195°C (Figure 6). This was attributed to specific complex formation between NaI and the polymer. Extension of this work by Gorecki (48), Berthier et al (33), and Minier et al (34) has established the stoichiometry of the complex to be (PEO)$_3$–NaI. Physiochemical characterization of this system by Fauteux et al (38) in d.s.c., wide angle X-ray scattering (WAXS), polarized light optical microscopy, and a.c. conductivity studies has enabled them to establish the pseudophase diagram, which has three major features: the presence of a single salt rich intermediate complex (PEO)$_3$–NaI; a eutectic between this and pure PEO; and an inconsistent melting point for the complex at about 200°C.

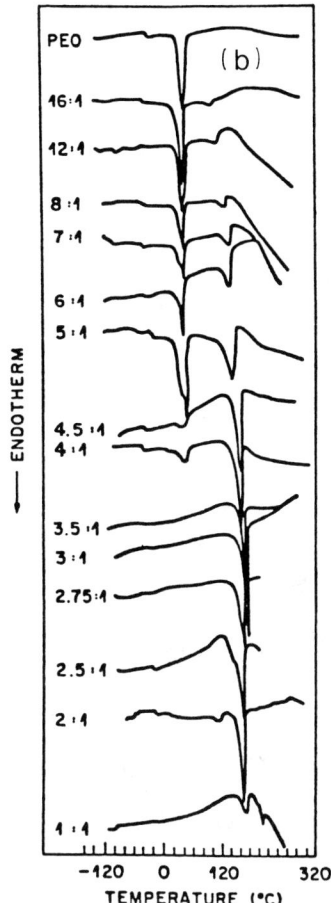

Figure 4b D.s.c. traces for PEO–LiCF$_3$SO$_3$ mixtures of different composition, used to construct the pseudophase diagram. (Reproduced from Ref. 37 with permission from the American Chemical Society.)

The phase behavior of other mixtures has also been established, e.g. PEO with NH$_4$SCN (49), NaSCN (50), LiAsF$_6$ (30) (the latter is analogous to that for PEO–LiClO$_4$), and we are aware of the imminent publication of phase diagrams for divalent electrolytes Cu(CF$_3$SO$_3$)$_2$ (51) and ZnCl$_2$ mixed with PEO (L. J. Alcacer, private communication). As yet no phase diagrams have appeared involving other polymer hosts such as polyethylene imine (PEI), and although d.s.c. and X-ray diffraction studies on the NaI (23, 53) and NaSO$_3$CF$_3$ (21) mixtures with PEI have been reported, these have not yet been subjected to a full analysis.

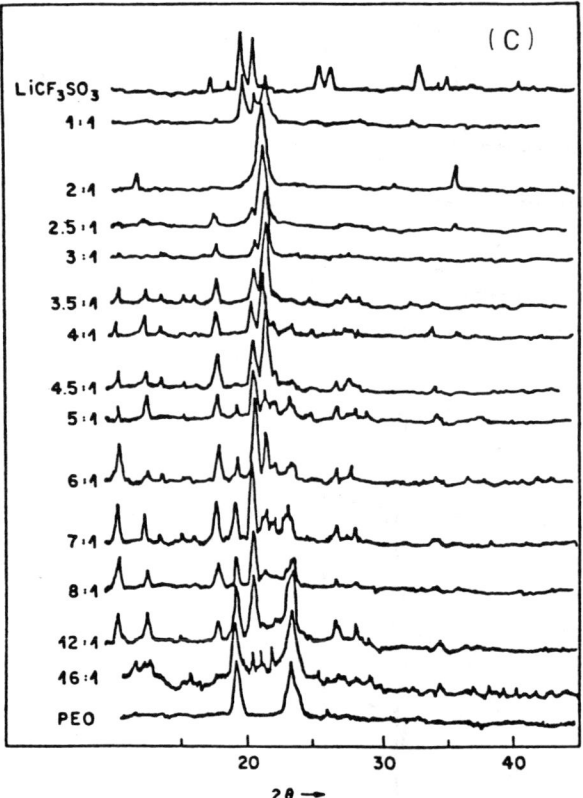

Figure 4c X-ray diffraction patterns for a range of PEO–LiCF$_3$SO$_3$ mixtures (CuKα radiation), used to construct the pseudophase diagram. (Reproduced from Ref. 37 with permission from the American Chemical Society.)

STRUCTURE DETERMINATION

X-Ray and EXAFS

In a number of cases, direct investigation of the microscopic structure of polymer-electrolyte mixtures has been hampered by the complicated phase behavior illustrated in the preceding section. Some stoichiometric complexes are highly crystalline and have been studied by X-ray diffraction techniques. Oriented films or fibers have been used to overcome the difficulty of indexing powder patterns (11, 16, 54–56). X-ray fiber photographs of PEO mixed with KSCN and NaSCN have been published by several workers, including Wright et al (36), who reported that the PEO chains

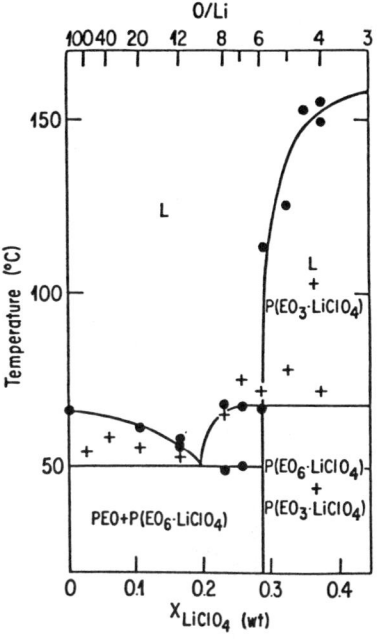

Figure 5 Pseudophase diagrams for the PEO–LiClO$_4$ system; (●) denotes transition points observed by microscopy, and (+) denotes those observed by conductivity measurements. (Reproduced from Ref. 30 with permission from the *Journal of Electrochemistry*.)

assumed not a single helix but rather a double stranded helical structure with the cations placed within the PEO cavity.

Hibma (35) has indexed the powder patterns from crystalline complexes of PEO with NaI, NaClO$_4$, NaBr, NaSCN, and KSCN. The latter two species were analyzed in greatest detail, and the PEO-metal cation ratios were determined to be 4 for KSCN and 3 for the NaSCN complexes, respectively. Interestingly, Hibma, contradicting previously reported models, stated that the K$^+$ cation resides outside the polymer helix. Complementary d.s.c. and X-ray diffraction studies on PEI complexed with NaI were carried out by Chiang et al (23). These studies revealed a decrease in PEI crystallinity with increasing addition of NaI. At a mole ratio of 0.15 NaI, no X-ray pattern due to pure PEI or NaI could be observed, and the authors concluded that the polymer had indeed dissolved the electrolyte. For mole ratios of electrolyte ≥ 0.3, diffraction maxima were observed arising from the crystalline PEI–NaI complex, but unfortunately no oriented fiber samples could be prepared that would have allowed a more detailed structural analysis. A discrete NaI phase was observed when the concentration of electrolyte exceeded a mole ratio of 0.3. Shriver et al (21) also investigated PEI as a polymer host, this time with NaSO$_3$CF$_3$ as added electrolyte. Again the system passed from an amorphous phase to a

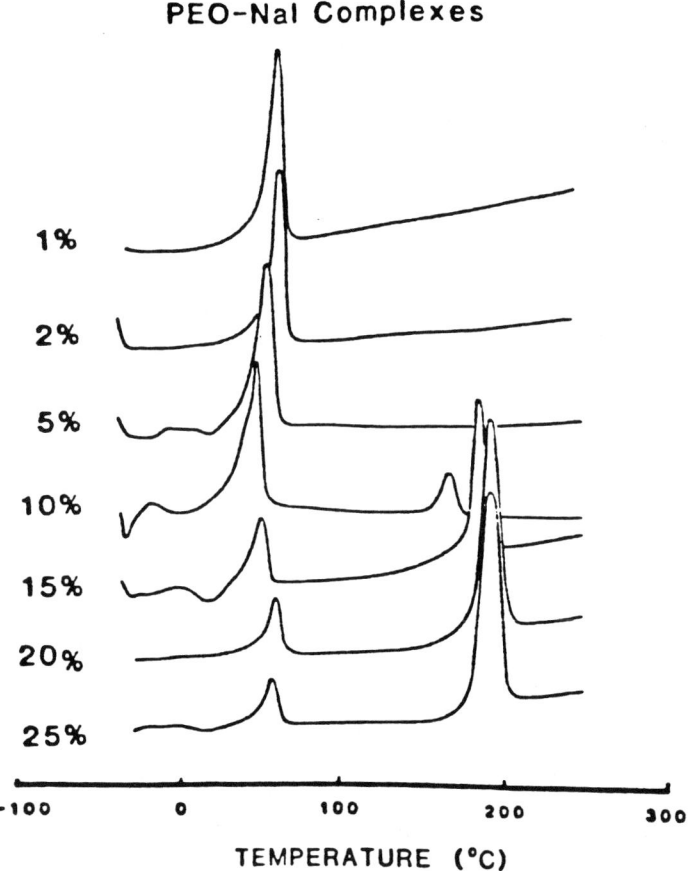

Figure 6 D.s.c. traces for mixtures of PEO and increasing mole % concentrations of NaI. All samples were heated previously to 260°C then cooled to −30°C at 20°C/min. (Reproduced from Ref. 47 with the permission of Elsevier Science Publishers.)

crystalline one as the salt concentration increased, and a partly crystalline complex was identified from X-ray diffraction patterns with a stoichiometry of $(PEI)_4$–$NaSO_3CF_3$. More recently, Chatani & Okamura (57) performed a detailed study of the structure of the PEO–NaI complex. They prepared samples over a range of mole ratios to determine the stoichiometry in the complex crystals and concluded that the structure $(PEO)_3$–NaI, as proposed by Hibma (35), was correct. The Na^+ and I^- ions were found to alternate, forming a zigzag ion chain around which PEO, in a 2/1 helix, was coiled, making a fiber period of 7.98 Å. The crystal structure of the PEO–NaI complex is shown in Figure 7.

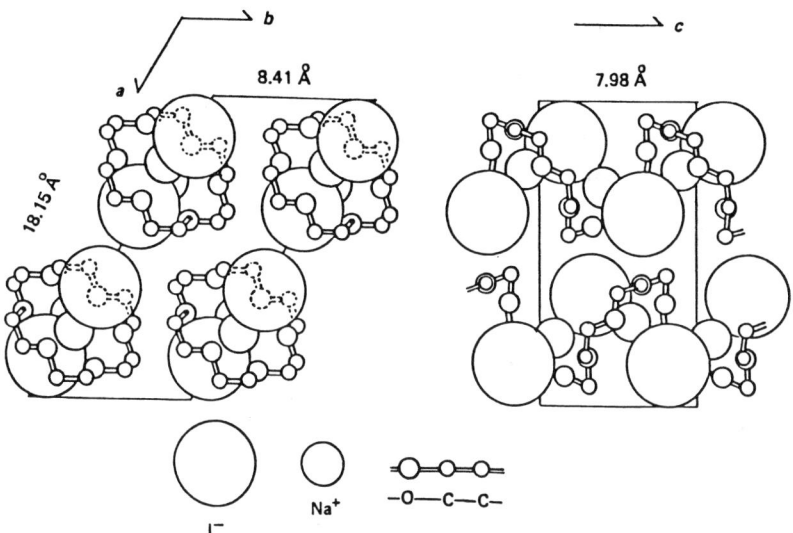

Figure 7 Crystal structure of the P(EO)$_3$–NaI complex. [Reproduced from Ref. 57 by permission of the publishers, Butterworth & Co. (Publishers) Ltd. ©.]

These X-ray diffraction studies have given valuable evidence for, and information on, the complexation of electrolytes by polymers but they are unable to provide knowledge of the precise environment of the salt in the amorphous phase. The special ability of the EXAFS (extended X-ray absorption fine structure) technique to examine local structure in amorphous polyether-electrolyte mixtures was first exploited by Catlow et al (58) in their study of rubidium salts dissolved in PEO. The technique takes advantage of the fact that oscillations observed on the high energy side of the absorption edges in the X-ray spectrum contain information about the local environment of the cation being probed. Details of the technique can be found in a recent review (59).

Catlow et al (58) performed EXAFS measurements on (PEO)$_x$–RbSCN and (PEO)$_x$–RbI complexes for $x = 4,8$ in the temperature range $-193°C$ to $180°C$. They found that the cations were incorporated in a remarkably well-defined environment; in both cases Rb$^+$ is surrounded by a coordination shell of four oxygen atoms. In the RbSCN system, the nitrogens of the anions were also included in the local environment, but the iodide ion was located outside the oxygen coordination shell in the latter system.

A word of caution was voiced by Chadwick & Worboys (59) to the effect that the EXAFS technique yields information on the *average* local structure and is, therefore, best suited to the study of systems in which the atom of interest is located predominantly in one environment. Inhomogeneities and/or thermal motion would place limitations on the interpretations derived with this method.

SPECTROSCOPIC METHODS

Evidence for specific interactions between polymer hosts and dissolved electrolytes has also come from both infrared and Raman spectroscopy (60–62). Papke et al (63) have used these techniques to study interactions of PEO with NaBr, NaI, NaBF$_4$, NaSO$_3$CF$_3$, NaSCN, KSCN, RbSCN, and CsSCN. They found that the IR (CH$_2$) rocking modes at 800–1000 cm^{-1} were particularly sensitive to conformational changes, and they used this property in conjunction with the Raman M$^+$–O symmetric stretching bands to support the presence of cation–oxygen atom interactions in the complexes. Examination of space filling models also indicated that the polyether chain was capable of wrapping itself round the K$^+$ and Na$^+$ cations to produce an oxygen-lined cavity and a PEO helix with a compressed $T_2GT_2\bar{G}$ conformation. However, the larger Rb$^+$ and Cs$^+$ cations were deemed to be too large for this model and to lie outside the helix, where they were complexed by oxygens from adjacent chains, i.e. there was both inter- and intrachain complexation, depending on the type of electrolyte used, but in all cases amorphous mixtures could be obtained.

The temperature dependence of the complexation between PPG and NaSCN was studied by Teeters & Frech (64) with Raman spectroscopy. They observed that a low frequency torsional or bending motion of the polymer backbone, which occurred at 239 cm^{-1}, was sensitive to the presence of NaSCN and that the absorption narrowed and increased to 265 cm^{-1} when the salt was dissolved in the polymer. The authors proposed that this could be caused either by complexation of the salt and the ether oxygens in the polymer backbone or by an enhancement of the rigid, more extended coiled conformations in the polymer. The frequencies of the CN and SC stretching vibrations at 2063 cm^{-1} and 755 cm^{-1}, respectively, differed from the values found for solvated or free ions in THF or DMF solutions but were much closer to the values of 2066 cm^{-1} and 754 cm^{-1} found by Chabanel et al (65) in THF and Maynard et al (66) in DMF for ion pairs in these same solvents. This would suggest incomplete solvation with strong ion pair formation in the polymer electrolyte system.

Shriver et al (67) chose to use the internal vibration modes of the BH$_4^-$, BD$_4^-$, and BF$_4^-$ anions to study interactions between PEO and the sodium salts of these anions. The high frequency B–H or B–D stretching vibrations at 2250 cm^{-1} and 1650 cm^{-1}, respectively, are readily perturbed by changes in their environment or geometry and can be used to probe the local environment of the anion. The vibrational band structure of the BH$_4^-$

anion in the polymer matrix is rather complex and has been interpreted as indicating a lowering of the anion symmetry from its original tetrahedral arrangement. This symmetry lowering was taken as evidence for extensive ion pairing in the polymer-$NaBH_4$ system in line with the ion pairing observed for thiocyanate salts.

For the $NaBF_4$ system, low frequency deformation modes and far infrared data were used to analyze the complexation, because the B–F stretches are obscured by intense C–O–C asymmetric stretching bands around 1100 cm^{-1}. The BF_4^- Raman bands, which were found to be essentially the same as those detected in aqueous or molten $NaBF_4$, were also used in the analysis. In contrast with the previously described systems, this electrolyte appears to exist as solvated ions in PEO rather than as ion pairs. Infrared spectroscopy has also been used by Shriver et al (25) to study the interaction of polyalkylene sulphides with silver nitrate. Formally the nitrate ion band at 1050 cm^{-1} is only Raman active, but it can be seen in the solid state infrared spectrum of some nitrate salts, where its appearance is attributed to distortion of the ion from D_{3h} symmetry and a shift or splitting of the band is indicative of ion pairing. Dissolution of $AgNO_3$ in poly(pentamethylene sulphide) (P5S) over a concentration range of (6:1) to (1:1) ratio of (S:Ag^+) produced splitting of the 1050 cm^{-1} band and suggested the presence of extensive ion pairing. Examination of space filling models of P5S showed that conformations that allow coordination of a cation by two or more adjacent sulphur atoms on the main chain were unfavorable. This finding, together with X-ray diffraction data on low molecular weight analogues, led the authors to predict that the Ag^+ ions were tetrahedrally coordinated with combinations of sulphur atoms from one or more chains but that the oxygens from the nitrate counterions were also incorporated in the coordination shell.

Poly(vinyl acetate) (PVAc) is also known to form complexes with inorganic electrolytes. Wintersgill et al (68) have used FTIR to examine the interactions. In spite of anhydrous working conditions, the FTIR analysis showed that traces of water were present. The water proved extremely difficult to remove and was in fact directly involved in the polymer-electrolyte complexation. The diagnostic test for complexation was taken to be the position of the ester carbonyl stretch frequency of 1737 cm^{-1}, which decreased from this original value to 1708 cm^{-1} for mixtures of polymer and electrolyte. Similar shifts in the ester stretching frequency from 1240 cm^{-1} to lower values, depending on the amount of salt added, were also observed. This behavior was considered to be consistent with the model proposed by Wissbrun & Hannon (27), shown on page 102.

Infrared spectroscopy may therefore be used to good effect to determine the presence of small amounts of water in polymer-salt systems, as its

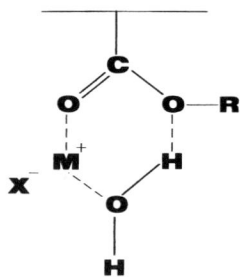

presence can have important ramifications for the analysis of ion conduction in the polymer electrolyte formed. Bruce et al (69) used infrared analysis to determine whether water of hydration had been removed from the electrolyte $Hg(ClO_4)_2$ after treatment with molecular sieves for five hours in an attempt to ensure use of totally anhydrous PEO-$Hg(ClO_4)_2$ mixtures for conductivity studies. The results indicated a 30 to 40-fold reduction in the intensity of the H_2O adsorption bands. Thus infrared spectroscopy can detect tightly bound residual water in polymer electrolytes and establish whether persisting association of anions and cations occurs in these coordinating polymer systems.

Owing to the low dielectric constant of polyethers a significant amount of free ions in the matrix is not expected. Most of the evidence so far tends to support this, although the extent of ionization is probably also a function of the salt concentration. Using $LiClO_4$ and $LiSO_3CF_3$ salts, MacCallum et al (70) and Cameron and co-workers (71–73) examined the conductivity of low molecular weight polyether-electrolyte systems and calculated the relative proportions of free ions, ion pairs, ion triplets, and higher aggregates present. Their general conclusions were that at salt concentrations normally encountered in polymer-electrolyte mixtures, the salts are present predominantly as ion pairs or triplets and that the concentration of free ions is relatively small. These conclusions were not supported, however, by the work of Cheradame et al (74) on PEO-polyurethane networks containing $LiSO_3CF_3$, as these workers reported complete dissociation into Li^+ and $CF_3SO_3^-$ ions, or by Sandahl et al (75), who found that $NaCF_3SO_3$ dissolved in PPG is also largely dissociated below a composition of (7:1). This point remains to be resolved.

Examination of these data suggests that the relative increase in the glass transition temperature T_g produced when electrolytes are dissolved in a polymer is a qualitative indicator of the solvating ability of the polymer (76–83). This conclusion is illustrated in Figure 8, which shows the variation of T_g for four electrolytes dissolved in a comb-branched polymer with PPG side chains (84). The increase ΔT_g is not uniform; it is lowest

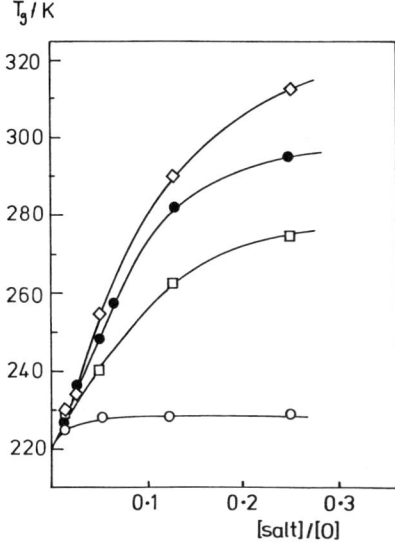

Figure 8 Variation of the glass transition temperature T_g with salt concentration for NaClO$_4$ (diamond), LiClO$_4$ (filled circle), ZnCl$_2$ (square), and LiCl (empty circle) dissolved in poly[di-poly(propyleneglycol) itaconate]. [Reproduced from Ref. 84 by the permission of the publishers, Butterworth & Co. (Publishers) Ltd. ©.]

for LiCl and highest for NaClO$_4$. With the ionic conductivity as a complementary guide to dissolution, it was found that a large ΔT_g corresponded to a high conductive response (large number of charge carriers), whereas a low ΔT_g corresponded to a low conductivity, thus suggesting that effective solvation of the ions led to an inevitable increase in T_g by virtue of the mechanism of solvation. It has also been demonstrated, by using a simple set of equilibria among the uncomplexed polymer, the electrolyte, and the polymer solvated species, that the increase in T_g can be modeled and from this an estimation of the degree of ionization can be obtained (90). The analysis suggested that as more electrolyte is added to the polymer, the percentage dissociation decreases quite sharply, leveling off at an approximate salt concentration of 1 mole per five propylene oxide units.

Gandini et al carried out an extensive study of the effect of added electrolytes on the physical properties of polymer networks, that comprised polyether chains crosslinked by multifunctional isocyanate units or polyols (74, 91–95). They found that a volume contraction occurred in the network after incorporation of the electrolyte, presumably due to the introduction of new cation-dipole crosslinks. A further consequence of this was that the T_g increased with added salt according to the relation

$$1/T_g = 1/T_g^0 - kc \qquad 1.$$

where c is the molar concentration of the salt, T_g^0 is the glass transition temperature of the network in the absence of salt, and k has a value of 0.27 ± 0.1 cm^3 mol^{-1} K^{-1} for PEO-salt systems. This constant k was independent of the crosslinking agent used in network formation and the nature of the dissolved salt, although only monovalent ions were studied; however, k was a function of the solvating polymer and increased when PPG replaced PEO in the network. The viscoelastic response of the systems was found to obey the Williams-Landel-Ferry (WLF) relationship for all of the networks and salts dissolved in them, and this behavior has also been confirmed by Watanabe and co-workers (96). Such behavior can be interpreted as indicating that the free volume is the most important parameter governing the response of these networks, thereby implying that any of the consequential transport properties is a function of the segmental motions in the systems. This implication has an important bearing on the interpretation of ion transport in such polymer-salt systems, a subject that has attracted much of the attention in the last decade. The proposition has found support in recent work by Sandahl et al (75) from Brillouin scattering studies on PPG–NaSO$_3$CF$_3$, which showed evidence for coupling between structural relaxation of the polymer and ionic conductivity.

A recent study of the physical behavior of linear polyurethane-electrolyte mixtures has been reported by McLennaghan & Pethrick (97).

Although ion conduction in polymers is now a large subject area and has been extensively and comprehensively reviewed, a brief summary is given here to provide a more balanced coverage.

ION CONDUCTION

The observation that significant levels of ionic conductivity can be obtained in many solid polymer electrolyte systems has resulted in an intense wave of activity directed particularly toward their application in lightweight batteries with lithium electrodes. In this respect an electrolyte dissolved in a polymer provides an attractive solid state medium for use in devices in which large currents are not required, offering many advantages over their liquid counterparts. Thus leakage and evaporation are no longer problems, and the mixtures are predominantly nontoxic and have a wide range of operating temperatures. This area has been extensively and comprehensively reviewed (15, 41, 98–100). From the multitude of publications that have appeared, the majority of which are concerned with mixtures of PEO or some modification of PEO with a range of electrolytes, a number of generalizations can be made as guidelines to the selection of, and search for, suitable systems.

1. Ion motion occurs predominantly, and most easily, in an amorphous polymer medium (33); crystallinity impairs ion transport in polymers.
2. The conductivity rises with increasing flexibility of the polymer chains; consequently, a material with a low T_g (i.e. well below ambient) is more likely to produce a high conductivity at a specified temperature than a more rigid material.
3. As conductivity is a function of the number of charge carriers in the polymer matrix, the latter should be an efficient solvent for the electrolyte, capable of ionizing it and thereby minimizing ion pair formation.

Thus although PEO has proved to be one of the best polymeric solvents for many electrolytes, it does have a significant crystalline content that restricts ion movement at temperatures below the melting point of $\sim 68°C$. The crystalline content can be eliminated by disrupting the regular structure of the PEO chains. Two major strategies have been adopted to achieve this goal.

The first is to form block copolymers or networks containing PEO segments that are too short to enter into crystalline arrays. Various methods have been used. Nagaoka et al (101) linked short PEO chains by using dimethyl siloxane units and obtained wholly amorphous copolymers if the number of ethylene oxide units n was ≤ 9. Electrolytes dissolved readily in these structures but a maximum conductivity was observed with $n \sim 8$ at 25°C. This suggests an optimum chain length for maximum electrolyte solvation, after which solvation may be impaired by incipient crystallization of the PEO. Cheradame's group (74, 91–95) chose to use polyurethane networks to obtain noncrystalline media, but they have also reported on the synthesis of block and graft copolymer structures based on PEO and PPG (93). Block copolymers of oxymethylene linked PEO segments, with n varying from 4 to 14, have been reported by Booth et al (102–104).

A second approach is to prepare regular graft copolymers or comb-branch material with PEO or PPG side chains of controlled length but always short enough to prevent crystallization. Successful structures of this type have been formed on methacrylate (105, 106), itaconate (84, 107), polyphosphazene (108, 109), polydimethyl siloxane (110, 111), and polyethylene (112) backbones. Again, best conductivity results are obtained when n is in the range 7–9. Other analogous structures have been prepared by attaching short PEO chains to polybutadiene by a solvo-mercuration-demercuration procedure (113). In all cases these modified structures are capable of dissolving the electrolytes that mix with PEO alone.

Ideally, conductivities of the order 10^{-3} S cm^{-1} at room temperature are sought, but this value has so far been difficult to achieve without the addition of a small molecule to act as a plasticizer. This has been used by Ward et al (114), who dissolved lithium salts in a series of poly(N,N-disubstituted acrylamide)s and added N,N-dimethylacetamide as a plasticizer. This gave tough films with ambient temperature conductivities of $>10^{-3}$ S cm^{-1}. However, this approach is not wholly desirable, as the plasticizer is prone to leakage. As maximum flexibility of the polymer system is a desirable feature assisting ion transport, the most flexible backbone chains known, e.g. polydimethylsiloxane, polyphosphazene, and polyethylene, have all been used and studied. The desirability of designing flexibility in these polymer-electrolyte mixtures also means that they will be predominantly viscous liquids that will be mechanically weak. They can be transformed into rubber-like films by crosslinking. Although crosslinking can restrict flexibility somewhat, no loss of conductivity occurs if crosslink densities are kept to $\leq 5\%$ in systems such as networks prepared from PEO macromers (115).

An alternative way of attaining mechanical stability, proposed by MacCallum et al (116), is to modify an elastoplastic triblock copolymer with a polybutadiene block flanked by two polystyrene blocks, by attaching PEO chains to the central butadiene chains. These structures dissolve the electrolyte in the rubber-like phase, which is held together and crosslinked by the glassy polystyrene phase. Conductivity levels achieved were comparable to other systems.

For dissolved salts LiClO$_4$, NaClO$_4$, or LiSO$_3$CF$_3$, conductivity levels of $\sim 10^{-4}$ S cm^{-1} have been obtained at ambient temperatures (112). Further improvement by lowering the T_g of the polymer alone is difficult, as the natural limit now appears to have been reached for the materials studied. This may mean that other parameters should be examined. In particular, can suitable electrolytes be designed that ionize readily and that can also plasticize the polymer medium, thereby limiting the increase in T_g normally observed when the electrolyte is dissolved? Some progress in this direction has been made, but attempts have not as yet met with outstanding success (117).

CONDUCTION MECHANISMS

Although measurement of ionic conductivity in polymer-electrolyte systems is relatively easy, obtaining a clear understanding of the mechanism of ion conduction has proved less easy. Fundamental to this understanding is the identification of the ionic species carrying the charge and the role each species plays in contributing to the total current. This identification requires an accurate measurement of the transference numbers for each ionic species, but inspection of the literature suggests that such measurement may not be a simple exercise for polymer-electrolyte mixtures. A number of possible experimental routes are available; however, a comparison of the values reported (41) shows that agreement is difficult to achieve. The situation may be complicated further if the exact nature of the numbers calculated is in some doubt, i.e. are these transport, transference, or diffusion numbers, as some of the techniques measure the ion transport from diffusional motion, which will also include ion pairs as well as free ions. The ambiguity in these measurements again raises questions regarding the microstructures of the electrolyte in the polymer, and what fraction of these are free ions, ion pairs, ion triplets, or higher aggregates? No satisfactory description has yet been put forward, and indeed it is a difficult problem to resolve. Thus, we refer here only to the reported data, and we do not attempt a value judgment on either the method or the significance of the values obtained.

Cheradame et al (95, 118, 119) used the Tubandt method with some success to study the cation transport numbers (t_+) in their highly crosslinked PEO-urethane networks, which contained various alkali metal salts. They obtained $t_+ = 0.02$–0.3 for single valence ions but a zero value of t_+ when $Mg(ClO_4)_2$ was the salt used. Their results imply that the ClO_4^- was the mobile ion in the system responsible for the charge transport and that doubly charged cations may be immobilized. Immobility is not a general phenomenon, however, as Farrington et al (86, 120) studied the PEO–$PbBr_2$ system and obtained $t_+ = 0.6$–0.7 with complex impedance spectroscopy, and other results suggest that Cd^{2+}, Pb^{2+}, Hg^{2+}, Cu^{2+} will have high mobilities whereas Mg^{2+} and Co^{2+} are essentially immobile (84–89). The lack of mobility of the latter group has been attributed to their classification as hard acids that will complex strongly with the hard PEO base, whereas the former fall in the soft acid category (41, 69).

Armand et al (121) have used a potentiometric measuring technique to examine LiI, $LiClO_4$, and $LiCF_3SO_3$ dissolved in PEO. They found t_+ values of 0.3, 0.3, and 0.7, respectively. Comparison of these values with measurements of the Li^+ diffusion coefficient, however, showed that the derived t_+ values were different. Diffusion coefficient measurements have also been reported by a number of other groups. Chadwick et al (122, 123) followed radioactive tracer diffusion in the PEO–NaSCN and concluded that both $^{22}Na^+$ and $S^{14}CN^-$ were mobile in this system. Fauteux et al (38), on the other hand, found with the same method that conductivity was predominantly anionic in the PEO–NaI system.

Values of t_+ in the range 0.34–0.41 were measured from diffusion numbers obtained with pulse field gradient NMR in the PEO–$LiCF_3SO_3$ systems (124), but even higher values of $t_+ \approx 1$ have been reported by Armand et al (9) from NMR spin-spin relaxation studies for Li^+ in PEO.

Equally high values have been obtained by Sorensen & Jacobsen (125) and Watanabe et al (126, 127) by using chronoamperometric methods to measure "the time of flight" for mobile ions. Watanabe's group reported $t_+ = 0.92$–0.99 for Li^+ in polyethylene succinate–LiSCN mixtures.

Interpretation of these data from different methods has led to some confusion, and many of the methods can be criticized. One of the more promising methods has been outlined recently by Bruce et al (128), who have developed an approach based on a combination of complex impedence and polarization studies. The a.c. impedence is determined as a function of frequency for a symmetrical cell before polarization with a small d.c. bias. When the steady state current is reached, allowance is made for electrode effects that are ignored in other methods. This method has yielded values of $t_+ = 0.46$ for Li^+ in PEO–$LiCF_3SO_3$ systems, and others are now under examination.

The area does require further urgent attention if a reliable understanding of the mechanism of ion transport is to be achieved and modeled effectively. The conductivity σ can be expressed in terms of the concentration n_i, the mobility μ_i, and the charge q_i of the ith ionic species by the summation: $\sigma = \Sigma n_i \mu_i q_i$, an expression that requires knowledge of the transference number for a specific analysis. In general terms, the conductivity in a polymer-electrolyte mixture is seen to obey, in the majority of cases, the semi-empirical Vogel, Tammann, Fulcher (VTF) equation

$$\sigma = \sigma_0 \exp[-B/(T-T_0)] \qquad 2.$$

where B and T_0 are constants whose interpretation can depend on the model applied, and σ_0 is a factor that subsumes the pre-exponential term normally believed to be proportional to the number of charge carriers in the system. Equation 2 can be interpreted further by using two approaches: the free volume and the configurational entropy models.

With the phenomenological approach of the free volume hypothesis, Eq. 2 is readily transformed into the WLF equation that is widely used to describe the relaxational behavior of polymeric materials in the vicinity of the glass transition temperature. In this model the polymer is assumed to contain empty space or "free volume," which will increase with temperature and allow greater freedom of movement for polymer chains in the bulk state. The parameter T_0 is then considered to be the temperature at which the free volume eventually vanishes. At temperatures above T_0, when substantially free volume is present, the ionic species in the polymer matrix will be able to move through the bulk polymer. This movement will be accompanied by polymer chain segmental motion and will be made easier with an increase in temperature. Cheradame's group (91, 92, 94) has shown a clear relationship between ionic motion and chain mobility and has used the WLF equation very effectively to describe the behavior of ion containing networks. In this model B is then related to the free volume of the system and can be regarded, not as an energy term, but as a function of the expansivity of the polymer-salt mixture.

The major weakness of this theory is that it ignores the obvious kinetic effects associated with long chain molecules. Adam, Gibbs & DiMarzio (129, 130) have proposed an alternative, the configurational entropy model, which attempts to overcome this deficiency. Now the mass transport mechanism is assumed to be a group co-operative rearrangement of the chain rather than a redistribution of free volume, T_0 is the temperature at which the configurational entropy is zero, and T_0 is related to T_g by ($T_0 = T_g - 50K$), whereas B is a function of the activation energy opposing the rearrangement of the polymer segmental unit. The approach appears to have merit, and analyses of several polymer-electrolyte systems (63, 112, 131–133) give reasonable values for the activation energy and values close to 50K for ($T_g - T_0$). The description is, of course, somewhat inadequate at a molecular level, and Druger, Nitzan, and Ratner (134, 135) have proposed a "percolation model" based on the theory developed by Hammersley (136) to model motion through a random medium.

Ratner proposes that an ion is capable of jumping from one site to a neighboring site in the polymer matrix if a site is available, and that the rate of this process is either 1 or 0 depending on whether a site is vacant or not. The theory manages to incorporate the essential aspects of polymer chain flexibility and the rate at which the chain segments move by recognizing that the dynamics of chain motion will represent an ever changing host environment for the ion, thereby allowing a previously inaccessible site to become available and thus a successful jump to take place. The relaxation time characterizing this event is known as the *renewal time*, and it is a fundamental parameter in the theory together with the jump rate and the probability that a jump can take place.

This theory is certainly a step in the right direction, but it requires improvement if we are to deal with complicating features such as ion-ion interactions and the influences of species other than single ions. Further progress may then depend on elucidation of the state of the electrolyte dissolved in the polymer in terms of species present, strength of polymer-ion binding, and the precise nature of the mobile species in the system.

Literature Cited

1. Holliday, L., ed. 1975. *Ionic Polymers*. London: Applied Sci. Publ.
2. Eisenberg, A., King, M. 1977. *Ion-Containing Polymers: Physical Properties and Structure*. New York/London: Academic
3. Allen, W. N., Brant, P., Carosella, C. A., DeCorpo, J. J., Ewing, C. T., Saalfeld, F. E., Weber, D. C. 1979/1980. *Synth. Met.* 1(2): 151–59
4. Hioki, T., Noda, S., Sugiura, M., Kakeno, M., Yamada, K., Kawamoto, J. 1983. *Appl. Phys. Lett.* 43(1): 30–32
5. Pehrsson, P. E., Weber, D. C., Koons, N., Campana, J. E., Rose, S. L. 1984. *Mater. Res. Soc. Symp. Proc.* 27: 429–34
6. Wasserman, B., Braunstein, G., Dresselhaus, M. S., Wnek, G. E. 1984. *Mater. Res. Soc. Symp. Proc.* 27: 423–28
7. Knott, K. F., Bello, I., Haworth, L. 1985. *Radiat. Effects* 89: 157–63
8. Schoch, K. F., Bartko, J. 1985. *Polym. Prepr.* 26(2): 166–67
9. Armand, M., Chabagno, J. M., Duclot, M. J. 1978. *2nd Int. Conf. on Solid Electrolytes*, St. Andrews paper 6.5
10. Fenton, B. E., Parker, J. M., Wright, P. V. 1973. *Polymer* 14: 589
11. Wright, P. V. 1975. *Br. Polym. J.* 7: 319–27
12. Wright, P. V. 1976. *J. Polym. Sci. Polym. Phys. Ed.* 14: 955–57
13. Moacanin, J., Cuddihy, E. F. 1966. *J. Polym. Sci. C* 14: 313–22
14. Armand, M., Chabagno, J. M., Duclot, M. J. 1979. *Fast Ion Transport in Solids*, ed. P. Vashishta, J. N. Mundy, G. K. Shenoy, p. 131. Amsterdam: North Holland
15. Ratner, M. A., Shriver, D. F. 1988. *Chem. Rev.* 88: 109–24
16. Payne, D. R., Wright, P. V. 1982. *Polymer* 23: 690–93
17. Dyson, R., Papke, B. L., Ratner, M. A., Shriver, D. F. 1984. *J. Electrochem. Soc.* 131: 586–89
18. Watanabe, W., Rikukawa, M., Sanui, K., Ogata, N., Kato, H., Kobayashi, T., Ohtaki, Z. 1984. *Macromolecules* 17: 2302–8
19. Watanabe, W., Togo, M., Sanui, K., Ogata, N., Kobayashi, T., Ohtaki, Z. 1984. *Macromolecules* 17: 2908–12
20. Armstrong, R. D., Clarke, M. D. 1984. *Electrochim. Acta* 29: 1443–46
21. Harris, C. S., Shriver, D. F., Ratner, M. A. 1986. *Macromolecules* 19: 987–89
22. Harris, C. S., Ratner, M. A., Shriver, D. F. 1987. *Macromolecules* 20: 1778–81
23. Chiang, C. K., Davis, G. T., Harding, C. A., Takahashi, T. 1985. *Macromolecules* 18: 825–27
24. Baldwin, K. R., Golder, A. J., Knight, J. 1985. *RAE Techn. Rep.* 84036
25. Clancy, S., Shriver, D. F., Ochrymowycz, L. A. 1986. *Macromolecules* 19: 606–11
26. Hannon, M. J., Wissbrun, K. F. 1975. *J. Polym. Sci. Polym. Phys. Ed.* 13: 113–26
27. Wissbrun, K. F., Hannon, M. J. 1975. *J. Polym. Sci. Polym. Phys. Ed.* 13: 223–41
28. Wright, P. V. 1989. *Polymer*. In press
29. Wintersgill, M. C., Fontanella, J. J., Greenbaum, S. G., Adamic, K. J. 1988. *Br. Polym. J.* 20: 195–98
30. Robitaille, C. D., Fauteux, D. 1986. *J. Electrochem. Soc.* 133: 315–25
31. Lee, C. C., Wright, P. V. 1982. *Polymer* 23: 681–89
32. Gorecki, W., Andreani, R., Berthier, C., Armand, M. B., Mali, M., Roos, J., Brinkmann, D. 1986. *Solid State Ionics* 18/19: 295–99
33. Berthier, C., Gorecki, W., Minier, M., Armand, M. B., Chabagno, J. M., Rigaud, D. 1983. *Solid State Ionics* 11: 91–95
34. Minier, M., Berthier, C., Gorecki, W. 1984. *J. Phys. Paris* 45: 739–44
35. Hibma, T. 1983. *Solid State Ionics* 9/10: 1101–5
36. Parker, J. M., Wright, P. V., Lee, C. C. 1981. *Polymer* 22: 1305–7
37. Zahurak, S. M., Kaplan, M. L., Rietman, E. A., Murray, D. W., Cava, R. J. 1988. *Macromolecules* 21: 654–60
38. Fauteux, D., Lupien, M. D., Robitaille, C. D. 1987. *J. Electrochem. Soc.* 134(11): 2761–67
39. Gray, F., MacCallum, J. R., Vincent, C. A. 1986. *Solid State Ionics* 18/19: 282–86
40. Minier, M., Berthier, C., Gorecki, W. 1983. *Solid State Ionics* 9/10: 1125–28
41. Vincent, C. 1987. *Prog. Solid State Chem.* 17: 145–261
42. Ferloni, P., Chiodelli, G., Magistris, A., Sanesi, M. 1986. *Solid State Ionics* 18/19: 265–70
43. Weston, J. E., Steele, B. C. H. 1982. *Solid State Ionics* 7: 81–88
44. Sorensen, P. R., Jacobsen, T. 1983. *Polym. Bull.* 9: 47–51
45. James, D. B., Stein, R. S., MacKnight,

W. J. 1979. *Bull. Am. Phys. Soc.* 24: 479
46. Robitaille, C., Prud'homme, J. 1982. *65th Can. Chem. Congr. Toronto*, Abstr. No. MA-10
47. Chiang, C. K., Davis, G. T., Harding, C. A., Aarons, J. 1983. *Solid State Ionics* 9/10: 1121–24
48. Gorecki, W. 1984. PhD thesis. Grenoble Univ., France
49. Stainer, M., Hardy, L. C., Whitmore, D. H., Shriver, D. F. 1984. *J. Electrochem. Soc.* 131: 784–90
50. Lee, Y. L., Crist, B. 1986. *J. Appl. Phys.* 60: 2683–89
51. Bonino, F., Pantaloni, S., Passerini, S., Scrosati, B. 1988. *J. Electrochem. Soc.* 135(8): 1961–65
52. Deleted in proof
53. Chiang, C., Davis, G. T., Harding, C. A., Takahashi, T. 1986. *Solid State Ionics* 18/19: 300–5
54. Shriver, D. F., Papke, B. L., Ratner, M. A., Dupon, R., Wong, T., Brodwin, M. 1981. *Solid State Ionics* 5: 83–88
55. Shriver, D. F., Dupon, R., Stainer, M. 1983. *J. Power Sources* 9: 383–88
56. Takahashi, Y., Sumita, I., Tadokoro, H. 1973. *J. Polym. Sci. A2* 11: 2113–22
57. Chatani, Y., Okamura, S. 1987. *Polymer* 28: 1815–19
58. Catlow, C. R. A., Chadwick, A. V., Greaves, G. N., Moroney, L. M., Worboys, M. R. 1983. *Solid State Ionics* 9/10: 1107–14
59. Chadwick, A. V., Worboys, M. R. 1987. In *Polymer Electrolyte Reviews*, ed. J. R. MacCallum, C. A. Vincent, 1: 275–313. London/New York: Elsevier
60. Papke, B. L., Ratner, M. A., Shriver, D. F. 1981. *J. Phys. Chem. Solids* 42: 493–500
61. Kasatani, K., Sato, H. 1986. *Chem. Soc. Jpn. Chem. Lett.* 6: 991–94
62. Toda, H., Yano, Y., Kujino, K., Kawahara, H. 1987. *J. Polym. Sci. Polym. Chem. Ed.* 25: 1745–53
63. Papke, B. L., Ratner, M. A., Shriver, D. F. 1982. *J. Electrochem. Soc.* 129: 1694–1701
64. Teeters, D., Frech, R. 1986. *Solid State Ionics* 18/19: 271–76
65. Chabanel, M., Wang, Z. 1984. *J. Phys. Chem.* 88: 1441–45
66. Maynard, K. J., Irish, D. E., Eyring, E. M., Petrucci, S. 1984. *J. Phys. Chem.* 88: 729–36
67. Dupon, R., Papke, B. L., Ratner, M. A., Whitmore, D. H., Shriver, D. F. 1982. *J. Am. Chem. Soc.* 104: 6247–51
68. Wintersgill, M. C., Fontanella, J. J., Calame, J. P., Smith, M. K., Jones, T. B., Greenbaum, S. G., Adamic, K. J., Shetty, A. N., Andeen, C. G. 1986. *Solid State Ionics* 18/19: 326–31
69. Bruce, P. G., Krok, F., Vincent, C. A. 1988. *Solid State Ionics* 27: 81–88
70. MacCallum, J. R., Tomlin, A. S., Vincent, C. A. 1986. *Eur. Polym. J.* 22(10): 787–91
71. Cameron, G. C., Ingram, M. D., Sorrie, G. A. 1986. *J. Electroanal. Chem.* 198: 205–7
72. Cameron, G. C., Ingram, M. D., Munro, B., Ross, E. 1988. *Eur. Poly. J.* 24(4): 395–97
73. Cameron, G. C., Harvie, J. L., Ingram, M. D., Sorrie, G. A. 1988. *Br. Polym. J.* 20: 199–202
74. Killis, A., LeNest, J. F., Gandini, A., Cheradame, H. 1984. *Macromolecules* 17: 63–66
75. Sandahl, J., Börjesson, L., Stevens, J. R., Torell, L. M. 1989. *Macromolecules*. In press
76. Blumberg, A. A., Pollack, S. S., Hoeve, C. A. J. 1964. *J. Polym. Sci. A* 2: 2499–2502
77. Blumberg, A. A., Wyatt, J. 1966. *J. Polym. Sci. B* 4: 653–56
78. Iwamoto, R., Saito, Y., Ishihara, H., Tadokoro, H. 1968. *J. Polym. Sci. A* 2: 1509–25
79. Yokoyama, M., Ishihara, H., Iwamoto, R., Tadokoro, H. 1969. *Macromolecules* 2: 185–92
80. Ciferri, A., Bianchi, E., Marchese, F., Tealdi, A. 1971. *Makromol. Chem.* 150: 265–70
81. Valenti, B., Bianchi, E., Greppi, G., Tealdi, A., Ciferri, A. 1973. *J. Phys. Chem.* 77: 389–95
82a. Wetton, R. E., James, D. B., Whiting, W. 1976. *J. Polym. Sci. Polym. Lett. Ed.* 14: 577–83
82b. James, D. B., Wetton, R. E., Brown, D. S. 1979. *Polymer* 20: 187–95
83. Stevens, J. R., Schantz, S. 1988. *Polym. Commun.* 29: 330–31
84. Cowie, J. M. G., Martin, A. C. S. 1987. *Polymer* 28: 627–32
85. Vassilev, K. G., Dimov, D. K., Stamenova, R. T., Boeva, R. S., Tsvetanov, Ch. B. 1986. *J. Polym. Sci. A Polym. Chem.* 24: 3541–54
86. Yang, L. L., Huq, R., Farrington, G. C., Chiodelli, G. 1986. *Solid State Ionics* 18/19: 291–94
87. Yang, L. L., McGhie, A. R., Farrington, G. C. 1986. *J. Electrochem. Soc.* 133(7): 1380–85
88. Fontanella, J. J., Wintersgill, M. C., Calame, J. P. 1985. *J. Polym. Sci. Polym. Phys. Ed.* 23: 113–20

89. Bruce, P. G., Krok, F., Evans, J., Vincent, C. A. 1988. *Br. Polym. J.* 20: 193–94
90. Cowie, J. M. G., Ferguson, R., Martin, A. C. S. 1987. *Polym. Commun.* 28: 130–32
91. Killis, A., LeNest, J. F., Gandini, A., Cheradame, H. 1981. *J. Polym. Sci. Polym. Phys. Ed.* 19: 1073–80
92. Killis, A., LeNest, J. F., Gandini, A., Cheradame, H. 1982. *Makromol. Chem.* 183: 1037–50
93. LeNest, J. F., Gandini, A., Cheradame, H. 1988. *Br. Polym. J.* 20: 253–68
94. Killis, A., LeNest, J. F., Gandini, A., Cheradame, H., Cohen-Addad, J. P. 1984. *Solid State Ionics* 14: 231–37
95. Levêque, M., LeNest, J. F., Gandini, A., Cheradame, H. 1985. *J. Power Sources* 14: 27–30
96. Watanabe, M., Sanui, K., Ogata, N., Kobayashi, T., Ohtaki, Z. 1985. *J. Appl. Phys.* 57(1): 123–28
97. McLennaghan, A. W., Pethrick, R. A. 1988. *Eur. Polym. J.* 24(11): 1063–71
98. MacCallum, J. R., Vincent C. A., eds. 1987. *Polymer Electrolyte Reviews*, Vol. 1. London/New York: Elsevier
99. Linford, R. G., ed. 1987. *Electrochemical Science and Technology of Polymers*, Vol. 1. London/New York: Elsevier
100. Armand, M. B. 1986. *Annu. Rev. Mater. Sci.* 16: 245–61
101. Nagaoka, K., Naruse, H., Shinohara, I., Watanabe, N. 1984. *J. Polym. Sci. Polym. Lett. Ed.* 22: 659–63
102. Craven, J. R., Mobbs, R. H., Booth, C., Giles, J. R. M. 1986. *Makromol. Chem. Rapid Commun.* 7: 81–84
103. Craven, J. R., Nicholas, C. V., Webster, R., Wilson, D. J., Mobbs, R. H., Morris, G. A., Heatley, F., Booth, C. 1987. *Br. Polym. J.* 19: 509–16
104. Nicholas, C. V., Wilson, D. J., Booth, C., Giles, J. R. M. 1988. *Br. Polym. J.* 20: 289–92
105. Bannister, D. J., Davies, G. R., Ward, I. M., McIntyre, J. E. 1984. *Polymer* 25: 1600–2
106. Xia, D. W., Soltz, D., Smid, J. 1984. *Solid State Ionics* 14: 221–24
107. Cowie, J. M. G., Martin, A. C. S. 1985. *Polymer Commun.* 26: 298–300
108. Blonsky, P. M., Shriver, D. F., Austin, P., Allcock, H. R. 1986. *Solid State Ionics* 18/19: 258–64
109. Ganapathiappan, S., Chen, K., Shriver, D. F. 1988. *Macromolecules* 21: 2299–2301
110. Fish, D., Khan, I. M., Smid, J. 1986. *Makromol. Chem. Rapid Commun.* 7: 115–20
111. Fish, D., Khan, I. M., Wu, E., Smid, J. 1988. *Br. Polym. J.* 20: 281–88
112. Cowie, J. M. G., Martin, A. C. S., Firth, A. M. 1988. *Br. Polym. J.* 20: 247–52
113. Bell, S. E., Bannister, D. J. 1986. *J. Polym. Sci. Polym. Lett. Ed.* 24: 165–69
114. Dobrowski, S. A., McIntyre, J. E., Ward, I. M., Davies, G. R. 1988. *Speciality Polymers '88, 3rd Int. Conf. on New Polymeric Materials*, Cambridge, U.K.
115. Cowie, J. M. G., Sadagianizadeh, K. 1989. *Polymer* 30: 509–13
116. Gray, F. M., MacCallum, J. R., Vincent, C. A., Giles, J. R. M. 1988. *Macromolecules* 21: 392–97
117. Armand, M. B. 1987. *6th Int. Conf. on Solid State Ionics*, Garmisch, Abstr. A4-6
118. Levêque, M., LeNest, J. F., Cheradame, H., Gandini, A. 1983. *Makromol. Chem. Rapid Commun.* 4(7): 497–502
119. Gandini, A., LeNest, J. F., Levêque, M., Cheradame, H. 1986. In *Integration of Fundamental Polymer Science and Technology*, ed. L. A. Kleintjens, P. J. Lemstra, 1: 250–55. London/New York: Elsevier
120. Huq, R., Chiodelli, G., Ferloni, P., Magistris, A., Farrington, G. C. 1987. *J. Electrochem. Soc.* 134(2): 364–69
121. Bouridah, A., Dalard, F., Deroo, P., Armand, M. 1986. *Solid State Ionics* 18/19: 287–90
122. Chadwick, A. V., Strange, J. H., Worboys, M. R. 1984. *Solid State Ionics* 9/10: 1155–60
123. Bridges, C., Chadwick, A. V., Worboys, M. R. 1988. *Br. Polym. J.* 20: 207–11
124. Bhattacharja, S., Smoot, S. W., Whitmore, D. H. 1986. *Solid State Ionics* 18/19: 306–14
125. Sorensen, P. R., Jacobsen, T. 1984. *Solid State Ionics* 9/10: 1147–53
126. Watanabe, M., Nagano, S., Sanui, K., Ogata, N. 1986. *Solid State Ionics* 18/19: 338–42
127. Watanabe, M., Rikukawa, M., Sanui, K., Ogata, N. 1985. *J. Appl. Phys.* 58: 736–40
128. Evans, J., Vincent, C. A., Bruce, P. G. 1987. *Polymer* 28: 2324–28
129. Gibbs, J. H., DiMarzio, E. A. 1958. *J. Chem. Phys.* 28: 373–83
130. Adam, G., Gibbs, J. H. 1965. *J. Chem. Phys.* 43: 139–46

131. Angell, C. A. 1986. *Solid State Ionics* 18/19: 72–88
132. Fontanella, J. J., Wintersgill, M. C., Calame, J. P., Smith, M. K., Andeen, C. G. 1986. *Solid State Ionics* 18/19: 253–57
133. Wintersgill, M. C., Fontanella, J. J., Smith, M. K., Greenbaum, S. G., Adamic, K. J., Andeen, C. G. 1987. *Polymer* 28(4): 633–38
134. Druger, S. D., Ratner, M. A., Nitzan, A. 1983. *Solid State Ionics* 9/10: 1115–20
135. Ratner, M. A. 1987. See Ref. 98, pp. 173–236
136. Hammersley, J. M. 1957. *Proc. Cambridge Philos. Soc.* 53: 642–45

DYNAMICS OF SOLVATION AND CHARGE TRANSFER REACTIONS IN DIPOLAR LIQUIDS

Biman Bagchi

Solid State and Structural Chemistry Unit, Indian Institute of Science, Bangalore 560 012, India

INTRODUCTION

The effects of solvent and solvent dynamics on chemical reactions, especially on charge transfer processes, have long been a subject of great importance in physical chemistry. In the past, attention was focused primarily on equilibrium solvent effects, such as the effect of solvent polarity on the reaction potential surface. In recent years it has become clear that in many fast reactions solvent dynamics can play a direct role and can affect both the rate and the outcome of a reaction profoundly. Thus, an understanding of the *time-dependent response* of a polar solvent to a changing charge distribution in a polar solute molecule is essential to understand the role of solvent in many important chemical and biological processes in liquids. Such understanding can be achieved by studying the dynamics of solvation of a newly created ion or of an instantaneously changed dipole in a polar liquid. This subject has undergone a renaissance in recent years because of the availability of ultra-short laser pulses that make it possible to study solvation dynamics directly with a time resolution hitherto impossible. An understanding of the details of solvent response to a sudden change in the charge distribution of a polar solute "probe" molecule is beginning to emerge.

Experimental studies on the dynamics of solvation are usually carried out by instantaneously creating a charged species inside a polar solvent and subsequently monitoring the emission/absorption spectrum of this species. As solvation of this charged species progresses, the solvent structure surrounding the polar solute undergoes a rearrangement, leading to a change in the energy of the solute molecule that is reflected in the change in the emission/absorption spectrum. The main idea is to create a situation in which the dynamics of solvation can be probed separately and independently of a chemical reaction. Recently, a large number of time-resolved studies of solvation of newly created dipoles, ions, and electrons in dipolar liquids have been reported, and intense activity is going on in this area. As a result, a considerable amount of quantitative information about the dynamics of solvation is now available. In some cases, such as the intramolecular electron transfer reaction (ETR) in excited dimethylaminophenyl sulfone (DMAPS) or in excited bianthryl, the rate of electron transfer and rate of solvation of the product charge transfer state can be probed independently and the role of solvent in the two processes can be compared. The new experimental results have raised many interesting theoretical questions.

Initially, continuum model theories were invoked to explain experimental results. These theories predict simple, closed-form expressions for the time dependence of the solvation energy in terms of macroscopic properties of the solvent, such as the dielectric constant and the Debye relaxation time. Thus, the continuum models provided a set of initial predictions that could be tested against experiments. However, recent experiments have shown that continuum models are inadequate to explain the results. Theoretical activity in this field was also motivated by an interesting comment of Onsager at the 1976 Banff conference on solvated electrons (1). Onsager suggested that the solvent far from the electron would relax faster than the solvent near the electron, so the solvent structure around the electron would form from outside in (1). Microscopic theories of solvation have been developed recently to check this conjecture and to explain new experimental results. These theories constitute a major improvement over the continuum models.

In this article I review experimental and theoretical studies of dynamics of polar solvation and charge transfer reactions in dipolar liquids, with emphasis on recent theoretical developments. A number of reviews have appeared on related subjects. Different aspects of solvation have been covered in two volumes edited by Dogonadze et al (2, 3). Hynes (4) has reviewed some aspects of solvent dynamic effects on charge transfer reactions. Fleming (5) and Kenny-Wallace (6) have reviewed different aspects of electron solvation. Kosower & Huppert (7) have reviewed dynamic solvent effects on electron and proton transfer reactions. Simon (8) has reviewed experimental studies on time-dependent fluorescence Stokes shifts, and Barbara & Jarzeba (9) have reviewed solvation effects on low-barrier isomerization reactions in polar liquids. Recently, a volume of *Faraday Discussions* was devoted to solvation and solvation dynamics (10). The focus of this article is rather different, with emphasis on microscopic theories of solvation and solvent dynamics and their role in charge transfer reactions.

CONTINUUM MODEL THEORIES

Homogeneous Dielectric Models

Models that invoke a homogeneous dielectric for solvent represent the simplest theory of solvation dynamics and are generalizations of the equilibrium solvation models of Born (11) and Onsager (12) to the time domain. In this model the solvent is replaced by a frequency (ω) dependent dielectric continuum, with dielectric function, $\varepsilon(\omega)$, and the polar solute by a molecular cavity of some simple shape. The solvation energy is obtained by evaluating the reaction field of the polar solute molecule inside the molecular cavity. The predictions of homogeneous continuum theories (13–15) are most simply discussed in terms of a solvation time correlation function defined by

$$C_s(t) = \frac{E_{\text{solv}}(t) - E_{\text{solv}}(\infty)}{E_{\text{solv}}(0) - E_{\text{solv}}(\infty)}, \qquad 1.$$

where $E_{\text{solv}}(t)$ is the time-dependent solvation energy. If the dipolar molecule is approximated by a spherical cavity with a point dipole at its center and the solvent by a dielectric continuum with a single relaxation time, then the continuum model predicts that the solvation time correlation function is a single exponential with a time constant given by

$$\tau_L^d = \left(\frac{2\varepsilon_\infty + \varepsilon_c}{2\varepsilon_0 + \varepsilon_c}\right)\tau_D, \qquad 2.$$

where ε_0 and ε_∞ are, respectively, the static and the infinite frequency dielectric constants of the solvent and τ_D is the Debye relaxation time of the dipolar solvent (14). ε_c is the dielectric constant of the molecular cavity. In the case of an ion, the continuum model predicts slightly faster solvation, and the time constant is given by

$$\tau_L = \left(\frac{\varepsilon_\infty}{\varepsilon_0}\right)\tau_D. \qquad 3.$$

τ_L is the longitudinal polarization relaxation time of the unperturbed solvent. For typical values of ε_0, ε_∞, and ε_c, the difference between τ_L^d and

τ_L is unimportant and we may simply state that homogeneous continuum models predict that the solvation energy correlation function should relax exponentially with a time constant given by τ_L.

The above simple expressions were derived under the assumptions that the molecular cavity is spherical and the solvent dielectric response is of simple Debye form. Both these assumptions were relaxed in a recent study (16) that investigated consequences of an ellipsoidal cavity and a non-Debye dielectric response. It was found that if the molecular cavity is ellipsoidal, the decay is bi-exponential even with a Debye form for $\varepsilon(\omega)$; the relaxation times, however, differ at most by 30 to 40% under extreme conditions. The effect of the non-Debye form was studied by using the Davidson-Cole (17) form for $\varepsilon(\omega)$. It was found that the short time dynamics were significantly affected even when the deviation from Debye behavior was small.

Inhomogeneous Dielectric Models

The homogeneous continuum model of solvation dynamics is clearly a naive theory that ignores, among other things, the details of solute–solvent interactions and the spatial and orientational order that are present in a dense dipolar liquid. Recently, an inhomogeneous dielectric continuum model was presented to rectify some of these shortcomings. In fact, inhomogeneous dielectric models have long been used in the static description of solute–solvent interactions, especially to explain the discrepancy of classical Born expression for solvation energy of an ion (18–21). The main idea behind these models is that the solvent molecules that are very close (for example, the nearest-neighbors) to the polar solute molecule are different in their polar response from those in the bulk. This difference may arise from the strong electric field of the ion or the dipole, which may produce such nonlinear effects as electric saturation, or from some specific solute–solvent interactions, such as hydrogen bonding.

Recently, Bagchi et al (22) and Castner et al (23) investigated the influence of such an inhomogeneous dielectric response on the dynamics of solvation of ions and dipoles. Two models that invoked a distant dependent dielectric constant, $\varepsilon(r)$, were studied. In the first model, $\varepsilon(r)$ was allowed to vary continuously as a function of distance (r) from the polar solute molecule. The time-dependent reaction field necessary to calculate the solvation energy was obtained by solving the modified Laplace's equation with the position- and frequency-dependent dielectric constant, $\varepsilon(r, \omega)$. The main effect of the dielectric inhomogeneity was to introduce relaxation times slower than τ_L and to make the decay non-exponential. In the second model (23), a discrete shell representation of the position-dependent dielectric function was assumed. This model is more restrictive than the first model because it imposes the quasistatic boundary conditions at each boundary of the shells. The advantage of this model is that it can be solved analytically for ionic solvation, although not for the dipolar case. The expression for the ionic case is fairly simple and intuitive. The dynamical response of a multishell solvent consists of a sum of independent responses, one from each distinct shell. The response time of a given shell is just the longitudinal relaxation time of that shell. For both the above models, the deviation of the average relaxation time from that predicted in a homogeneous continuum model increases as the dielectric constant and the length parameter, which specifies the rapidity of approach to the bulk dielectric value, increase. The continuum model result is recovered in the limit of very large solute-solvent size ratio. It is also found that for a given solvent model, the solvent response to a change in dipole moment is slower than that to a change in the charge.

The inhomogeneous dielectric models are no doubt phenomenological. The merit of these models is that they provide a simple and intuitive picture of solvation dynamics and at the same time incorporate some aspects of solute-solvent interactions. It is indeed surprising that many of the predictions of the inhomogeneous continuum models are in good agreement with experiments and with microscopic theories discussed below.

MOLECULAR THEORIES

Microscopic theories of non-equilibrium solvation dynamics have been developed only recently. Two somewhat different molecular theories have been proposed. The following outlines the basic ideas and techniques involved in these two theories.

Generalized Smoluchowski Equation Approach

The generalized Smoluchowski equation approach to solvation dynamics is based on the observation that the time-dependent solvent polarization of a dipolar liquid is related to number density of the solvent by the following expression:

$$\mathbf{P}(\mathbf{r}, t) = \mu \int d\omega \hat{\alpha}(\omega) \rho(\mathbf{r}, \omega, t), \qquad 4.$$

where $\rho(\mathbf{r}, \omega, t)$ is the position ($\mathbf{r}$), orientation (ω), and time (t) dependent density of the solvent, $\mathbf{P}(\mathbf{r}, t)$ is the polarization vector, and $\hat{\alpha}(\omega)$ is a unit vector with orientation ω and μ is the magnitude of the dipole moment of the solvent molecules. The time-dependent solvation energy may be given by the following expression:

$$E_{\text{solv}}(t) = -\frac{1}{2}\int d\mathbf{r}\, \mathbf{D}(\mathbf{r}) \cdot \mathbf{P}(\mathbf{r}, t), \qquad 5.$$

where $\mathbf{D}(r)$ is the bare electric field of the polar solute molecule whose solvation is being studied. To obtain the time-dependent polarization vector, one needs a kinetic equation for $\rho(\mathbf{r}, \omega, t)$. It is assumed that the proper hydrodynamic description for time dependence of the density field is given by the following generalized Smoluchowski equation (24–29a):

$$\frac{\partial}{\partial t}\delta\rho(\mathbf{r},\omega,t) = D_R \mathbf{V}_\omega^2\, \delta\rho(\mathbf{r},\omega,t) + D_T \nabla^2 \delta\rho(\mathbf{r},\omega,t)$$

$$- [D_R \mathbf{V}_\omega \cdot \rho(\mathbf{r},\omega,t)\mathbf{V}_\omega + D_T \nabla \cdot \rho(\mathbf{r},\omega,t)\nabla]\beta F, \qquad 6.$$

where $\delta\rho(\mathbf{r}, \omega, t)$ is the density fluctuation from the equilibrium state. The free energy term, βF, contains a mean-field contribution from intermolecular interactions and an external term,

$$\beta F = \int d\mathbf{r}'\, d\omega'\, c(\mathbf{r},\omega,\mathbf{r}',\omega')\delta\rho(\mathbf{r}',\omega',t) + \beta U_{\text{ext}}[\{\rho(\mathbf{r},\omega)\}_1 t]. \qquad 7.$$

D_R and D_T are the rotational and translational diffusion coefficients of the solvent, respectively; $\mathbf{V}_\omega$ and ∇ are the usual angular and spatial gradient operators; $c(\mathbf{r},\omega,\mathbf{r}',\omega')$ is the two-particle direct correlation function (30) of the dipolar solvent; $\beta = (k_B T)^{-1}$; k_B is Boltzmann constant; and T is the temperature. U_{ext} is the external field that comes from the solute ion or dipole whose solvation is being investigated.

Equation 6 can be derived either by using projection operator formalism (31, 24) or by using arguments similar to the ones used in time-dependent Ginzburg-Landau theory (26, 27, 32). The generalized Smoluchowski equation is expected to be reliable at intermediate wavenumbers where momentum relaxation is fast and only number conservation is important in the time scale of interest (33, 34). In addition, this equation is also valid at small wavenumbers, because in this range the translational contribution to relaxation of $\delta\rho(\mathbf{r}, \omega, t)$ is negligible, and Eq. 6 gives the correct behavior. Here and throughout this article, by small wavevector, we mean a value of k in the range $0 < k\sigma \lesssim 0.5$, where σ is the diameter of the solvent molecules. And, by intermediate wavevector, we mean $0.5 \lesssim k\sigma \lesssim 2\pi$. A generalized Smoluchowski equation (GSE) of a similar form had been used earlier in the study of the dynamics of liquid-solid transitions (33, 34). In most applications, GSE is linearized around the equilibrium solvent density, $\rho_{\text{eq}}(\mathbf{r}, \omega)$. If the local solvent distortion can be neglected, then $\rho_{\text{eq}} = \rho_0/4\pi$, where ρ_0 is the equilibrium number density of the bulk solvent.

This may not be a good approximation if solvent distortion around the polar solute molecule is significant.

POLARIZATION RELAXATION IN PURE LIQUID   From linearized GSE one can obtain an expression for polarization relaxation in a pure dipolar liquid if it is assumed that the direct correlation function is given by linearized equilibrium theories. Nichols & Calef (25) and Chandra & Bagchi (26) obtained the expression for polarization relaxation for dipolar hard spheres. Both longitudinal and tranverse components were found to relax exponentially with time constants given by

$$\tau_L(\mathbf{k}) = (2D_R)^{-1}\left[1 + p'(k\sigma)^2 - \frac{\rho_0}{3}(1 + p'(k\sigma)^2)(C_\Delta + 2C_D)\right]^{-1},$$

$$\tau_T(\mathbf{k}) = (2D_R)^{-1}\left[1 + p'(k\sigma)^2 - \frac{\rho_0}{3}(1 + p'(k\sigma)^2)(C_\Delta - C_D)\right]^{-1}, \qquad 8.$$

where $\tau_L(\mathbf{k})$ and $\tau_T(\mathbf{k})$ are the wavevector-dependent longitudinal and transverse polarization relaxation times, respectively, and $p' = (D_T/2D_R\sigma^2)$, σ is the solvent molecular diameter. p' is a measure of the relative importance of translational modes in solvent polarization relaxation. C_Δ and C_D are the anisotropic parts of the direct correlation function (30). In Figure 1 the wavevector dependence of the relaxation times is shown for both longitudinal and transverse components for several values of the parameter p'. Solvation dynamics probe the longitudinal component. If the translational contribution is neglected, the $\mathbf{k} = 0$ mode relaxes faster, in accordance with the comment of Onsager (1). If, however, the translational contribution is significant ($p' \geq 0.5$), polarization modes at intermediate wavevectors relax faster and the Onsager conjecture is no longer valid. The translational mechanism of polarization relaxation is depicted in Figure 2, where we consider relaxation of solvent structure around a positive ion. The molecules can relax either by rotating or by moving a small distance to attain the desired configuration. At small distances from the ion, which correspond, approximately, to intermediate wavevectors, the orientational relaxation is particularly slow, so in that region the translational contribution may be significant. But at large distances ($\mathbf{k} = 0$), the orientational relaxation is fast, so the translational contribution is unimportant.

To calculate time-dependent solvation energy, the values of $P_L(\mathbf{k}, t = 0)$ and $P_T(\mathbf{k}, t = 0)$ are required. The polarization fluctuation is considered to occur from the final equilibrium state, so these values are the negative of the final polarization values. These quantities can in principle be calculated with the density functional theory (35), which can give expression for $\rho_{eq}(\mathbf{r}, \omega)$ of the solvent in equilibrium with the polar solute molecule.

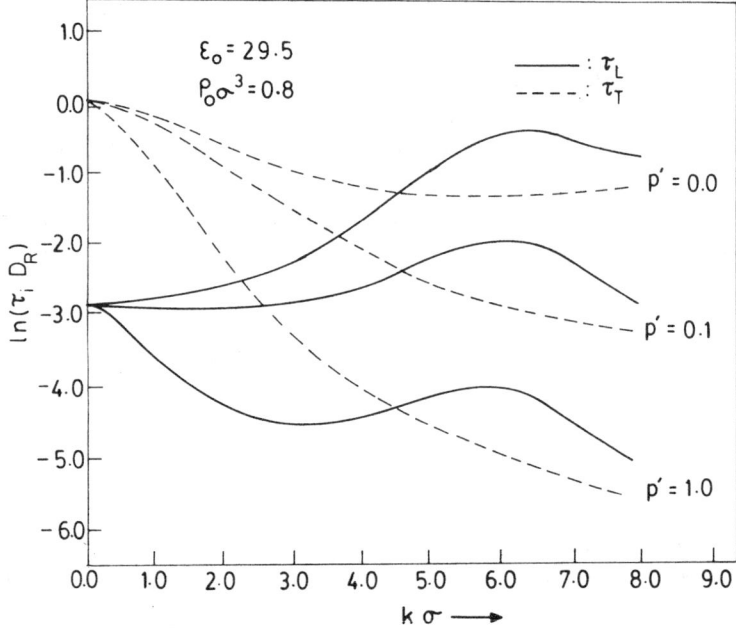

Figure 1 The dependence of longitudinal (*solid lines*) and transverse (*dashed lines*) relaxation times on wavevector **k** and translational diffusion. As the translational parameter p' ($= D_T/2D_R\sigma^2$) is increased, the relaxation becomes considerably faster at intermediate wavevectors ($k\sigma \simeq 6.3$). The values of the static dielectric constant and the reduced density are indicated on the figure (from Refs. 26–28).

This has been done recently (29a). Alternatively, one can assume a continuum model that relates the polarization vector of the final state to the displacement vector. However, this expression involves $\varepsilon(\mathbf{k})$, which is the wavevector-dependent static dielectric constant of the solvent. If one ignores the wavevector dependence of the dielectric constant, one obtains a description of solvation at the level of Born and Onsager. However, a better static description of solvation is obtained if the wavevector dependence of the dielectric constant is retained. Recently, it has been shown that these two approaches are rather similar (29a).

SOLVATION DYNAMICS OF AN ION The time-dependent solvation energy of an ion of charge Ze is given by (28)

$$E_{\text{solv}}(t) = \frac{(Ze)^2}{\pi} \int_0^\infty d\mathbf{k} \left(1 - \frac{1}{\varepsilon(\mathbf{k})}\right) e^{-t/\tau_L(\mathbf{k})} \cdot \left[\int_{k(a+\sigma/2)}^\infty dx \frac{\sin x}{x}\right]^2 \qquad 9.$$

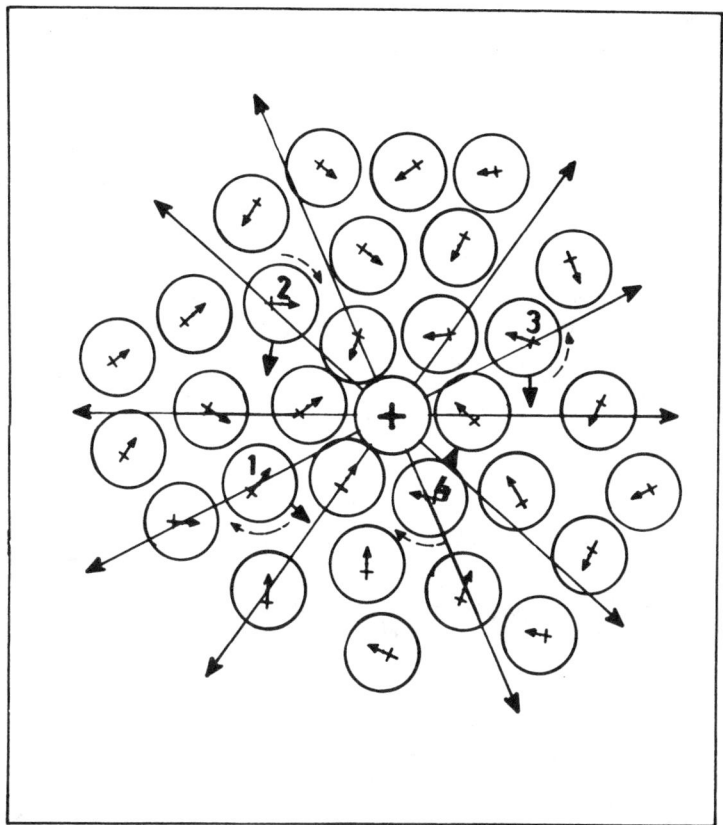

Figure 2 A pictorial description of the role of the translational diffusion mechanism in the solvation dynamics of a positive ion which is at the center of the figure. The solvent molecules, numbered 1, 2, 3, 4, can relax either by rotating (*dashed arrows*) at fixed position or by translating (*solid arrows*) a small amount in the required direction. The molecules close to the ion experience mostly the large wavevector field where orientational relaxation is slow, so the translational relaxation mechanism is important.

where a is the radius of the ion. If one makes the usual assumption $\varepsilon(\mathbf{k}) = \varepsilon_0$, the solvation time correlation function is given by

$$C_s(t) = \frac{R}{\pi} \int_0^\infty d\mathbf{q}\, e^{-t/\tau_L(\mathbf{q})} \left[\int_{q/2(R+1)}^\infty dx\, \frac{\sin x}{x} \right]^2, \text{ with } \mathbf{q} = \mathbf{k}\sigma. \qquad 10.$$

Here $R\ (= 2a/\sigma)$ is the solute-solvent size ratio. Equation 10 has several interesting predictions. It predicts that solvation of an ion is, in general, non-exponential. As R increases, we get back the continuum result of single exponential decay with the time constant given by τ_L. Equation 10 predicts that translational modes accelerate the relaxation rate. If p' is large ($p' > 0.5$), decay may again become nearly single exponential because translation will mainly affect the otherwise slow large wavevector processes that are responsible for non-exponential behavior at long times. This is seen clearly in Figure 1. However, the approximation $\varepsilon(\mathbf{k}) = \varepsilon_0$ may not be reasonable because $\varepsilon(\mathbf{k})$ has strong wavevector dependence at intermediate wavevectors.

SOLVATION DYNAMICS OF A DIPOLE   For a dipole μ, solvation energy receives contributions from both transverse and longitudinal components of the displacement vector. The longitudinal part of the solvation energy is given by (29b, Chandra & Bagchi, submitted)

$$E_L(t) = \left(\frac{4\mu^2}{3\pi a^3}\right) \int_0^\infty d\mathbf{q}' \left(1 - \frac{1}{\varepsilon(\mathbf{q}')}\right) j_1^2(\mathbf{q}') e^{-t/\tau_L(\mathbf{q}')}, \text{ with } \mathbf{q}' = \mathbf{k}a. \qquad 11.$$

where $j_1(z)$ is the spherical Bessel functon of first order of argument z. The transverse part of the solvation energy is given by

$$E_T(t) = \left(\frac{2\mu^2}{3\pi a^3}\right) \int_0^\infty d\mathbf{q}' \mathbf{q}'^2 \left(1 - \frac{1}{\varepsilon(\mathbf{q}')}\right) e^{-t/\tau_L(\mathbf{q}')} \cdot \left[\frac{\sin \mathbf{q}'}{\mathbf{q}'} + \int_{\mathbf{q}'}^\infty dx \frac{\cos x}{x}\right]^2. \qquad 12.$$

The total energy is the sum of these two parts, $E_{\text{solv}}(t) = E_L(t) + E_T(t)$. We have several interesting predictions for dipolar solvation. The zero wavevector contribution is totally absent in this case, in contrast to the ionic case, where the zero wavevector makes a significant contribution. The intermediate wavevector processes are, therefore, more important in dipolar solvation. This has the interesting consequence that in the absence of a translational contribution, dipolar solvation is slower than ionic solvation, a finding that agrees with the inhomogeneous continuum model and also with the dynamic mean spherical approximation (MSA) model, discussed below. However, translational modes become important at intermediate wavevectors. This implies that if the translational contribution is significant, dipolar solvation will benefit more than in the ionic case and dipolar solvation may proceed at a faster rate than ionic solvation. Numerical studies of the above equations confirm this (29a, Chandra & Bagchi, submitted).

Nonequilibrium Mean Spherical Approximation Model

Recently Wolynes (36) demonstrated that linearized equilibrium theories of solvation (37, 38) can be straightforwardly extended to treat dynamic situations. If only the linear response of the solvent molecules to time-varying charge is considered, the relaxation of solvation energy of an ion of charge Ze created instantaneously at time $t = 0$ is given by

$$E_{\text{solv}}(t) = (Ze)^2 \int_{-\infty}^{\infty} d\omega\, e^{i\omega t} \left[\frac{Q(\omega) - Q(0)}{2\pi i \omega} \right], \qquad 13.$$

where $Q(\omega)$ is the linear response function of the solvent. The crucial observation made by Wolynes was that the resummation of chain diagrams utilized in linear equilibrium theories of polar liquids (37, 38) in calculation of response function $Q(\omega = 0)$ can be applied to the frequency-dependent response function. By assuming Debye behavior for the frequency-dependent dielectric constant and MSA theory for dipolar hard spheres, Wolynes (36) obtained an approximate solution of the complicated equations. In this approximate solution, the solvation dynamics is bi-exponential with one time constant equal to the longitudinal time constant of the solvent. The second time constant, τ_G, gives the slow structural relaxation of the neighboring solvent molecules.

Although Wolynes' approximate solution has many appealing features, it is not quantitatively reliable. Rips et al (39, 40) solved Wolynes' MSA model exactly. The solvation energy correlation function shows a much richer behavior. The decay is, in general, non-exponential and is a sensitive function of the dielectric constant of the solvent and of the solute-solvent size ratio. The theoretical curve falls in between the continuum model prediction and the decay with time constant τ_D.

Nichols & Calef (25) and Rips et al (40) independently solved non-equilibrium MSA for solvation of a dipole. In this case the dynamics are considerably more complicated than those for an ion. The general nature of the decay is similar, but solvation of a dipole was found to occur on a time scale considerably longer than that for an ion, in agreement with the inhomogeneous dielectric model and also with the generalized Smoluchowski equation approach.

The non-equilibrium MSA model is a significant improvement over the continuum model theories. At this point, the major limitation of the dynamic MSA model is that it is not clear how to include the effect of the translational modes of the solvent. Second, the theory is to be supplemented with an externally determined frequency-dependent dielectric constant that is not satisfactory. Third, the MSA model is known to be unreliable for strongly polar liquids (41, 42). Similar calculations for more accurate integral equation theories, like linearized hypernetted chain (LHNC) (41, 42), should prove useful.

Other Theoretical Studies

Loring & Mukamel (43) have presented a lattice model calculation of the solvation of an ion. This model, which was first analyzed by Zwanzig (44) in the context of dielectric relaxation, is composed of point dipoles located at the sites of a cubic lattice. In addition to experiencing the electric field of other dipoles, each dipole undergoes rotational Brownian motion as a result of interactions with a bath. Loring & Mukamel (43) carried out a numerical evaluation of the polarization relaxation in this system following a sudden introduction of a point charge at one of the lattice sites. The distance-dependent polarization relaxation was found to occur in agreement with Onsager's conjecture of slow relaxation at the immediate vicinity of the charge and fast relaxation away from the charge. A detailed correlation function formalism based on the Liouville equation for the density matrix has been presented recently (45, 46). This theory is based on a general connection between the theory of rate processes and nonlinear optical processes in solution. The theory has been applied (46, 47) to calculate non-equilibrium solvation effects on fluorescence and hole burning lineshapes of model systems in polar solvents. Friedrich & Kivelson (47) studied the linear response of a polar liquid to a sudden imposition of an ion and concluded that the response of the solvent depends on all length scales, so time-dependent response will consist of an infinite number of time scales ranging between τ_L (the continuum limit) and τ_D (the single particle limit).

Computer Simulation Studies

Several computer simulation studies of solvation dynamics in model dipolar liquids have been carried out. Maroncelli et al (48) performed molecular dynamics simulations of electric field correlation function at a polar Lennard-Jones solute in spherical clusters of 512 ST2 waters. The calculated correlation function was non-exponential. An interesting finding of this study was that a substantial part (about 50%) of the total solvation energy for a dipolar solute comes from interactions with the water molecules in the first solvation shell only. Vijayakumar & Tembe (49) reported their results of molecular dynamics simulations of electric field correlation in a dipolar liquid near an electric charge. The correlation function was found to be non-exponential in general. A study of solvation dynamics of an instantaneously changed dipole was carried out by Karim et al (50) using the TIP S4P model of water. The conclusions of this study were similar to those of Maroncelli et al (48).

Maroncelli & Fleming (51) recently reported the results of their detailed study of the solvation energy correlation function following step function jumps in the solute's charge, dipole moment, and quadrupole moment. The solute was immersed in an ST2 water cluster of 512 molecules. The relaxation behavior was found to differ substantially depending on the type of multipole jump. Maroncelli & Fleming concluded that the Onsager picture of monotonically increasing relaxation time with distance is not correct for the systems studied, a conclusion that is in accord with the theoretical study of Chandra & Bagchi (25–29). The structure of the first solvation shell was found to play an important role in solvation dynamics, along with the translational modes of the solvent molecules. Maroncelli & Fleming also investigated the validity of the linear response approach by varying the magnitude of charge jumps and found that the relaxation was dependent on the size of jumps, a result that implies a breakdown of the linear response assumption usually made in theoretical studies.

TIME-RESOLVED EXPERIMENTAL STUDIES

Solvation Dynamics of Dipole

Initial experimental studies on time dependence of fluorescence Stokes shift (TDFSS) subsequent to electronic excitation were carried out in late 1970s and early 1980s by using picosecond techniques. In one of the early studies, Halliday & Topp (52) investigated Stokes shift relaxation in 2-amino-7-nitrofluorene in isopropanol at various wavelengths and temperatures. The relaxation was found to depend strongly on the wavelength monitored; the decay was distinctly slower in red tail of the spectrum than in the blue edge. Lessing & Reichert (53) studied the kinetics of solvation of 4-dimethyl amino 4'-nitrostilbene in polar solvents and concluded that cooperative relaxation of solvent molecules plays an important role in solvation. Okamura et al (54) studied Stokes shift dynamics in 1-naphthylamine, also in isopropanol. Identical relaxation times for the decrease in intensity at 384 nm and the increase at 460 nm were observed; both were 52 ps, a value that is in rather good agreement with the continuum model prediction of 33 ps (14).

Subpicosecond techniques, with a much better signal-to-noise ratio that those achieved in previous studies, have been used very recently to measure TDFSS after ultrafast excitation. The uncharged, rigid molecules 1-aminonaphthalene (55), 4-aminophthalamide (56, 57), Coumarins 153 (58, 59), 102 (59), 152 (60) and 311 (59) have been used as fluorescence probes. Molecules with more complicated photophysics, for example LDS-750 (61), bianthryl (57, 62), 4-(9-anthryl)-N,N-dimethylanaline (57), bis(4-aminophenyl)sulphonate sulphone (63, 64) and Nile Red (65), have also been studied. Both protic, hydrogen-bonded solvents such as n-alcohols and amides (55, 56, 61, 63–67) and aprotic solvents such as n-nitriles

(59), glycerol triacetate (57), DMSO (61), dimethyl formamide (68), and propylene carbonate (58, 59, 68) have been used in TDFSS measurements.

Very recently, Barbara and co-workers (69) measured the solvation dynamics of 7-dimethylaminocoumarin-4-acetate ion in liquid water. In these experimental studies, the solvation time correlation function, $C_s(t)$, defined in Eq. 1, was constructed by a detailed and painstaking procedure, and a systematic study of the dependence of solvation dynamics on various solvent properties and specific solute-solvent characteristics was carried out. The basic results of these experimental measurements may be summarized as follows:

1. The observed solvation times are largely probe independent and appear to reflect primarily the properties of the polar solvents studied.
2. The solvation correlation function was, in general, non-exponential. An especially interesting aspect of $C_s(t)$ is that the log $C_s(t)$ curve shows considerable curvature, even at long times ($t > \tau_L$), thus indicating the presence of a very large number of relaxation times in TDFSS. The $C_s(t)$ curve could often be well represented by a stretched exponential form—well known in studies of glassy systems as the Kohlrausch-Williams-Watts function (70, 71).
3. The average solvation time τ_s, defined by integrating over $C_s(t)$, is generally larger than the longitudinal relaxation time and usually lies between τ_L and τ_D. In some instances, solvation times were measured to be more than an order of magnitude longer than τ_L.
4. For LDS-50 in butanol and methanol, however, the solvation time was smaller than τ_L. Clearly, some mechanisms not important in other solute-solvent systems are important in these solvents. Naturally, translational diffusion has been proposed as an alternative relaxation mechanism that plays a role in these cases.
5. There appears to be a correlation between the deviation from τ_L and the static dielectric constant ε_0 (or $\varepsilon_0/\varepsilon_\infty$).

The above experimental results clearly indicate that simple continuum models are inadequate in explaining dynamics of solvation in real polar liquids. Another obvious conclusion is that molecular aspects of solvation, such as the short-range orientational correlations present in the dipolar liquid and the specific solute-solvent interactions, are in fact important in determining much of solvation dynamics. The question then naturally arises: How good are the recent microscopic theories? Although a detailed comparison between theory and experiment is yet to be carried out, some conclusions of a rather general nature may be drawn. First, the microscopic theories can explain the non-exponential nature of solvation dynamics. The non-exponentiality arises from short-range correlations, particularly from large wavevector processes, and is a reflection of short-range spatial and orientational correlations that are present in dense dipolar liquids. These theories can also predict that the relaxation times will lie between τ_L and τ_D, in agreement with experiments. The strong dependence on the static dielectric constant of the solvent can also be partly explained. Maroncelli & Fleming (72) compared MSA results for ions with experimental results on dipolar solvation; only a qualitative agreement was observed. Similar conclusions were reached by Barbara and co-workers (69). The main limitation of MSA appears to be that it predicts too slow a relaxation at longer times, perhaps because the influence of translational diffusional modes is neglected in MSA. These translational modes are expected to enhance the long time relaxation considerably and may give a better agreement between theory and experiment. An analogous comparison of experimental results with GSE is yet to be carried out.

MIXED SOLVENTS The study of solvation dynamics in mixed solvents offers scope for interesting dynamical problems. Halliday & Topp (73) investigated Stokes shift dynamics in benzene-propanol mixtures. This case is interesting because preferential solvation of the solute may occur. Robinson and co-workers (74, 75) reported their study of 8-amino-1-naphthalenesulphonic acid in water-ethanol mixture. The decay of fluorescence was found to be non-exponential in the mixed solvent, whereas the decay was single exponential in either of the pure solvents. Robinson et al pointed out that in mixed solvents, the additional process of solvent exchange around the excited, highly polar solute molecule will play an important role in Stokes dynamics. Recently Johnson & Hudson (76) studied solvation of 3-methyl indole (3MI) in a binary mixture of cyclohexane and butanol. They observed that although solvation of 3MI in pure cyclohexane was single exponential within the time resolution of the experiment, the solvation dynamics of the mixture was non-exponential. I am not aware of any theoretical study of solvation dynamics in a binary mixture. Such studies offer interesting possibilities and should be carried out.

Dynamics of Electron Solvation

When low-energy electrons are injected into polar fluids, a series of relaxation phenomena occur that eventually lead to solvation of the electron. As in the case of dipolar solvation, solvent dynamics play a crucial role in determining the rate of these relaxation phenomena. They can be studied by monitoring the absorption spectrum of this species. Experimental studies by Rentzepis et al (77), Kenney-Wallace & Jonah (78–80), Chase & Hunt (81), Wang et al (82), and Migus et al (83) have provided considerable insight into the dynamics of electron solvation in polar liquids. Much of the experimental results can be explained in terms of a two-state model (79, 84). At short times, initial localization of the quasi-free state takes place either by a configuration fluctuation or by a preexisting trap in the liquid. This is manifested in an absorption band in IR. This band decays and an analogous band grows in the visible. This latter process corresponds to structural reorganization of solvent molecules in the vicinity of the trapping site. The first stage is very fast and is usually over in much less than a picosecond. The second stage is much slower and takes place in tens of picoseconds in n-alcohols. The structural relaxation involved in the second stage was found to correlate with dielectric relaxation time in alcohols, thus indicating that orientational relaxation is the dominant mechanism. Thus, theories of solvation dynamics of an ion can be useful here.

CHARGE TRANSFER REACTIONS AND SOLVATION DYNAMICS

Polar solvents play an essential role in charge transfer reactions in solution. When the reaction rate is substantially smaller than the rate of solvent polarization relaxation, the main role of the solvent is to modify the reaction potential surface, and the time-dependent solvent relaxations are unimportant. In this limit, the traditional theories of reaction rates, such as transition state theory (TST), are adequate. If the short-range collisional forces of the solvent are important, Kramers' theory (85) or its recent modifications (86–92) may provide a sensible description. The situation changes qualitatively when solvent polarization relaxation occurs on a time scale that is the same as or slower than the charge transfer reaction. In this limit, solvent polarization relaxation can play a critical role. Recent experiments (discussed below) have revealed that several intramolecular reactions are rate-limited by polarization relaxation. In the following I briefly discuss recent theoretical and experimental work on charge transfer reactions, with emphasis on outer-sphere electron transfer reactions.

The rate and efficiency of electron transfer depends critically on the nature of interactions between the potential energy surfaces of the reactant and the product. If the two potential surfaces interact strongly so that they repel each other, then the reaction potential surface at the reactive zone is a mixture of the two potential surfaces, and the electron transfer occurs on this mixed surface. This is called adiabatic reaction. In the opposite limit of very weak interaction between the two potential surfaces, the reaction is called non-adiabatic because both the two surfaces are directly involved in electron transfer. Solvent effects are important for both adiabatic and non-adiabatic reactions. A good discussion of the relevant concepts has been provided by Frauenfelder & Wolynes (92a).

Theoretical work on ETR has come a long way since the pioneering study by Marcus (93, 94) on adiabatic electron transfer. Marcus' theory was for a reaction with a high activation barrier and based on a continuum description of the solvent polarization. Subsequently, Levich & Dogonadze (95) presented a study of nonadiabatic ETR in which the microscopic electronic process rather than polarization relaxation of the solvent controls the rate of the reaction. Tembe et al (96) and Newton (97) included molecular information by allowing for a solvent shell around the reactants in a dielectric continuum. Recently, Calef & Wolynes (98) presented a detailed analysis of the role of solvent fluctuation on an adiabatic outer sphere ETR that explored the nature of the reaction coordinate and showed that under certain circumstances the transfer of charge between two centers reduces to one-dimensional diffusion over a barrier. The rate was found to be inversely proportional to the longitudinal relaxation time, τ_L. Calef & Wolynes compared their expression with the experimental results on ETR in bivalent Ruthenium ion of Creutz et al (99). The agreement was encouraging, although caution is needed because of many uncertainties in the values of the potential parameters. Subsequently, Erfima & Bixon (100), Zusman (101), and Alexandrov (102) extended Marcus' treatment to treat dynamic solvent effects on nonadiabatic electron transfer reactions. These treatments employed a stochastic solvent model that described the dynamics of polarization fluctuation by a simple Debye model. Thus, the molecular nature of reactant-solvent interactions were ignored.

In an interesting study, Calef & Wolynes (103) investigated the solvent effects on adiabatic ETR by using a microscopic theory of solvent structure. A detailed calculation of one-dimensional reaction-free energy surface was made by using density functional theory of inhomogeneous liquid. The main conclusion was that the continuum approximation of the solvent dramatically overestimates the reaction barrier.

The influence of the adiabaticity/nonadiabaticity parameter on reaction dynamics has also been investigated (92a, 104–106). McManis et al (105) carried out a numerical study of the barrier crossing rate and found that a moderate coupling between the two participating surfaces was sufficient to give the observed inverse τ_L dependence of reaction rate. Cline & Wolynes (106) studied stochastic dynamic models to explore the influence of the adiabaticity parameter. Recently, Sparpaglione & Mukamel (107) presented a theory of electron transfer reaction in polar media that was based on an analogy with the problem of nonlinear optical lineshapes. In their approach, they used an expansion of the density matrix in Liouville space. An interesting prediction of this calculation was a nonmonotonic dependence of ET rate on the longitudinal relaxation time constant for certain values of the potential parameters.

Experimental confirmation of the theoretical predictions of the role of solvent reorganization dynamics on electron transfer reactions has been slow to come. Gennett et al (108) studied the kinetics of ETR involving metallocene complexes by varying both solvent and reactant properties. The experimental results were interpreted with the statistical theories (96–103) of outer-sphere electron transfer reactions, and good agreement with theory was claimed. Subsequently, several studies (109–112) have found dependence of ET rate on solvent polarization relaxation. McManis et al (112) found that for self-exchange reactions in metallocene complexes in associated/highly polar solvents, the observed reaction rate was faster than the rate of solvation as measured in time-dependent Stokes shift experiments.

Recently attention has been focused on low or zero barrier intramolecular ETR. Several experimental studies have indicated that intramolecular ETR in several organic molecules in their excited state proceeds without the intervention of a large activation barrier. The review by Kosower & Huppert (7) summarizes much of the recent experimental results. I shall discuss several studies that were reported even more recently.

McGuire & McLendon (113) studied non-adiabatic electron transfer reactions between $Ru(phen)_3^{2+}$ homologues and methylviologen dispersed in glycerol in which the solvent longitudinal relaxation time was varied from 10^{-8} to 10^{-1} s. A strong dependence of electron transfer rate on τ_L was observed with $k_{Et} \propto (\tau_L)^{-0.6}$. Simon & Su reported studies of dynamics of electron transfer reactions in dimethylaminobenzonitrile (114), diethylaminobenzonitrile (114), and 4,4'-dimethylaminophenyl sulphone (115) in alcohols. The ETR was found to proceed with a very low activation barrier and with a rate that was higher than τ_L^{-1}. Kahlow et al (116) studied the intramolecular ETR of electronically excited bianthryl in various polar aprotic solvents and found that the reaction time is not given accurately by τ_L. They observed however, a strong correlation between the average solvation time and the reaction time. This indicates that whereas the continuum model itself is inadequate, solvation dynamics play a critical role in electron transfer. Huppert et al (117, 118) studied intramolecular ET rate constants on TNSDMA in various alcohol solvents. The ETR was again found to be activationless and non-exponential. The rates of ETR were found to be higher than τ_L^{-1}. Similar non-exponential decay was also observed in different reaction systems by Heisel & Miehé (119, 120) and by Rettig et al (121, 122). Kang et al (123) recently reported a detailed study on ETR in excited state S_1 bianthryl and a derivative of bianthryl. The rate of ETR in this low-barrier reaction was found to correlate well with the average rate of solvation but not with the continuum model prediction.

A theory of a low-barrier reaction will be significantly different from that of a high-barrier reaction because there is no clear separation of time scale between the motion in the reaction zone and that in the rest of the potential surface. Sumi & Marcus (124) recently presented a theory of low-barrier ET reaction that ignored the molecular nature of the solvent. This theory includes an average effect of the fast intramolecular vibrational motions of the reactant and treats the diffusive orientational motion of the solvent by using the continuum model. The low barrier reaction was modeled as a diffusive process on a harmonic surface with the solvent polarization (X) as the reaction coordinate. A reaction occurs when the solvent polarization attains a certain critical value, X_c. Such a low-barrier electron transfer reaction may be studied by using a modified Smoluchowski equation of the following form for the probability distribution $P(X, t)$:

$$\frac{\partial}{\partial t} P(X, t) = \tau_L^{-1} \frac{\partial}{\partial X} \left[\frac{\partial}{\partial X} + X \right] P - k_e(X) P,\qquad 14.$$

where $k_e(X)$ is a coordinate (solvent polarization) dependent rate of electron transfer. The form of $k_e(X)$ depends on the nature of the participating electronic potential surfaces. Several limiting forms for the position dependence of $k_e(X)$ have been investigated in detail (124, 125). If the vibrational modes of the electron transfer system make a vanishing contribution to ET, $k_e(X)$ is sharply peaked around X_c and may be represented by a delta function. On the other hand, if the contribution from vibrational modes dominates the reaction, the reaction window is wide and $k_e(X)$ may be represented by a broad Gaussian around X_c. The magnitude (k_0) of the rate at the critical polarization value, X_c, is governed by the adiabaticity/nonadiabaticity of the reaction. If the zeroth order (the diabatic) potential surfaces interact strongly, ETR is adiabatic and k_0 is very large compared to τ_L^{-1}. We then essentially have a pinhole sink (126) at $X = X_c$. In this limit, Eq. 14 can be solved analytically for the time dependence of the unreacted population, $P(t)$ (127, 128). The resulting $P(t)$ is nonexponential at short times, but becomes exponential at long times with a constant τ_L^{-1}. On the other hand, if the diabatic surfaces interact weakly, we have a nonadiabatic reaction and k_0 is finite. If k_0 is much smaller than τ_L^{-1}, the reaction rate is proportional to k_0. In this case, an analytic solution is possible only for the average rate [the time integral of $P(t)$], but even then only if $k_e(X)$ is a delta function at $X = X_c$.

The theory of barrierless ETR of Sumi & Marcus (124) is identical in many respects to the theory of barrierless isomerization reaction developed earlier by Bagchi et al (126, 128–130) and the theory of rebinding of CO to iron in heme by Agmon & Hopfield (131). Poornimadevi & Bagchi (132) have also discussed the kinetics of a very low barrier reaction in which the dependence of rate of reaction was found to be markedly non-Arrhenius. Rips & Jortner (132a) recently studied activationless outer-sphere ETR by using the stochastic Liouville equation formalism. The role of initial conditions on the dynamics of ETR was investigated. No molecular theory of barrierless ETR has yet been developed.

So far I have discussed only the electron transfer reactions. It was recognized quite early that solvation dynamics can play an important role in other charge transfer reactions (133–135) and isomerizations (136), and several studies have recently explored these aspects. The main thrust has been to understand what effect nonequilibrium solvation can have on high barrier reactions. In a series of papers, van der Zwan & Hynes investigated the role of polar solvent dynamics on idealized models of classical charge transfer (137) and dipole isomerization reactions (138, 139). For broad barrier reactions in strongly polar and slowly relaxing solvents, the reaction rate was found to be limited by solvent polarization relaxation. In such situations, the reaction was found to occur from a nonequilibrium configuration of the solvent, and the transition state theory, which assumes equilibrium solvation of the reactant, failed badly. The rate was found to be inversely proportional to τ_L. In the opposite limit of reactions with a sharp barrier in weakly polar solvents, the TST description was found to be fairly accurate. Bergsma et al (140) reported molecular dynamics simulation of a model SN2 reaction $Cl^{-1} + CH_3Cl \rightarrow ClCH_3 + Cl^{-1}$ in liquid water. It was found that reaction dynamics depend strongly on the instantaneous local configurations of the solvent at the transition barrier. Neither the TST nor Kramers theory provides an adequate description of the reaction rate. A nonequilibrium solvation model (141) developed along the ideas of van der Zwan & Hynes was found to provide a fairly accurate description of the reaction rate. Recently Zichi & Hynes (142) presented a theoretical study of nonequilibrium solvation effects on unimolecular ionic dissociation reactions in polar solvents. The rate of charge variation along the reaction coordinate was found to play a critical role in this study, which was again based on a continuum model description of solvent dynamics. Clear experimental confirmation of these nonequilibrium solvation effects on charge transfer reactions is yet to come. Eisenthal and co-workers (143), Keery & Fleming (144), and Sundström & Gillbro (145) have demonstrated the strong effect of solvent polarity on isomerization reactions. It will, of course, be a nontrivial problem to disentangle the

POLAR SOLVENT DYNAMICS

The importance of the molecular nature of solvent in solvation dynamics and charge transfer reactions calls for a quantitative understanding of microscopic processes involved in orientational and polarization relaxation of a dense dipolar liquid. In the past, theoretical study of collective orientational relaxation and dielectric relaxation had mostly been done, for understandable reasons, at the continuum level. Recently some progress has been made toward a molecular theory of polar solvent dynamics. In the following I discuss some of the recent theoretical work on these relaxation processes. Most of the theoretical work has been done on spherical diffusors, and so I limit the following discussion to such systems.

Collective Orientational Relaxation

The collective orientational relaxation is best described in terms of the following autocorrelation functions:

$$C_{lm}(\mathbf{k}, t) = \langle Y_{lm}^*(\mathbf{k}, 0) Y_{lm}(-\mathbf{k}, t) \rangle, \qquad 15.$$

where $\{Y_{lm}(\mathbf{k}, t)\}$ are the Fourier transforms of the spherical harmonics of rank l and projection m,

$$Y_{lm}(\mathbf{k}, t) = \sum_{\alpha} \exp[i\mathbf{k} \cdot \mathbf{r}_\alpha(t)] Y_{lm}[\omega_\alpha(t)], \qquad 16.$$

where $\mathbf{r}_\alpha(t)$ and $\omega_\alpha(t)$ are the position and the orientation of molecule α at time t, respectively, and the sum in Eq. 15 goes over all the molecules of the system. Of particular interest is the correlation function $\phi(\mathbf{k}, t)$ defined as

$$\phi(\mathbf{k}, t) = \frac{\sum_{m=-1}^{1} C_{1m}(\mathbf{k}, t)}{\sum_{m=-1}^{1} C_{1m}(\mathbf{k}, 0)}. \qquad 17.$$

The limit $\mathbf{k} = 0$ of this correlation function is required in the study of dielectric relaxation. The small $\mathbf{k}$ limit of $C_{lm}(\mathbf{k}, t)$ is usually referred to as the collective orientational relaxation limit. In the opposite limit of large $\mathbf{k}$, single particle motions dominate $C_{lm}(\mathbf{k}, t)$. In both these two limits, the microscopic structure of the liquid plays a passive role. The situation is, however, different at intermediate wavevectors where the short-range microscopic structure of the liquid plays an important role.

Hubbard & Wolynes (146) carried out a continuum model–based study of the effect of dielectric friction on molecular orientation by using a rotational Smoluchowski equation for the orientational distribution function. Explicit expressions for $C_{lm}(\mathbf{k} = 0, t)$ were obtained. An interesting prediction of this study was that the dielectric friction should have a decreased effect for higher l correlation functions, in contrast to the previous theory of Nee & Zwanzig (147). Berne (148) investigated the effect of long-range dipolar forces and translational diffusion on $C_{lm}(\mathbf{k}, t)$ by using a Smoluchowski equation for position and orientation–dependent density distribution function. He found that dipolar forces affect only the $l = 1$ orientational correlation function, which is bi-exponential in time. The higher order correlation functions decay exponentially with a time constant given by a simple diffusion model. Subsequently, Warchol & Vaughan (149) used Berne's method to compute correlation functions for a system of spheroids. Madden & Kivelson (150, 151) also considered the effects of molecular translation on orientational relaxation.

Building on the effort of Berne, Chandra & Bagchi (152) presented a molecular theory of collective orientational relaxation. The work was based on the same generalized Smoluchowski equation that was used in the theory of solvation described above. Analytic expressions for $C_{lm}(\mathbf{k}, t)$ were presented both for one component and for binary mixtures. It was shown that details of microscopic structure are important at intermediate wavevectors and that translational diffusion significantly accelerates collective orientational relaxation at intermediate wavevectors. The dipolar correlation function $\phi(\mathbf{k}, t)$ was found to be bi-exponential with two time constants $\tau_1(\mathbf{k}) = \tau_T(\mathbf{k})$ and $\tau_2(\mathbf{k}) = \tau_L(\mathbf{k})$. $\tau_T(\mathbf{k})$ and $\tau_L(\mathbf{k})$ are the transverse and longitudinal relaxation times given by Eq. 8. Explicit expressions for the prefactors for $\phi(\mathbf{k}, t)$ in terms of the components of total pair correlation function was also obtained. It was found that the prefactor of the transverse component is much larger than that for the longitudinal component for dipolar spheres, and their ratio is not 2 as assumed by Berne (148). In the same study (152) it was found that polar forces do not affect higher order correlation functions $C_{lm}(\mathbf{k}, t)$ in the linearized theory, in partial agreement with continuum theory.

Dielectric Relaxation

The connection between dielectric function and time-dependent molecular properties is provided by linear response theory (153). If wavevector dependence of the dielectric function is neglected, this relation is given by Fatuzzo-Mason expression (154). Nee & Zwanzig (147) performed a continuum model calculation for $\varepsilon(\omega)$ and showed that long-range polar forces can give rise to deviations from Debye behavior. A molecular theory of dielectric relaxation has been developed by Madden & Kivelson (151), who derived expression for $\varepsilon(\omega)$ in terms of liquid distribution functions. Madden & Kivelson also explored the relationship between dielectric and single particle relaxation times.

Although wavevector and frequency dependence of dielectric function is important in molecular processes, surprisingly few theoretical studies of this function have been undertaken. When wavevector dependence of the dielectric function is taken into account, the degeneracy between the longitudinal and transverse components of the function is lifted. van der Zwan & Hynes (155) and Hubbard et al (156, 157) derived expressions for $\varepsilon(k, \omega)$ from the continuum model. Loring & Mukamel (158) presented general linear response expressions for frequency and wavevector dependent dielectric function and studied this function for the Zwanzig lattice model (159). Chandra & Bagchi (160) presented a microscopic calculation for $\varepsilon(k, \omega)$ of a dense dipolar liquid. The longitudinal and transverse components of the dielectric function were found to have vastly different behavior at the intermediate values of the wavevector, **k**. The continuum model was found to break down completely at intermediate **k**, where the microscopic structure of the liquid is important. Debye-like single exponential behavior was recovered at **k** = 0 with Debye relaxation time given by $\tau_T(\mathbf{k} = 0)$. Thus, a microscopic relation between dielectric and single particle relaxation times was obtained (161). This relation differs significantly from the Glarum-Powles (162) relation at large values of static dielectric constant, although the saturation at large ε_0 and other qualitative features are similar.

CONCLUDING REMARKS

Significant progress has been made in the last decade in the experimental studies of dynamics of solvation and charge transfer reactions. The advances on the theoretical side have been less impressive. The existing microscopic theories of dynamics of solvation need to be improved in several ways. First, a realistic description of the dipolar liquid is necessary. Second, a nonlinear theory of polar solvent dynamics should be developed. This implies that we must include the nonlinear terms in the appropriate kinetic equation description. This will be especially important for a giant dipole in a binary mixture. Third, distortion of solvent structure around the polar solute should be included consistently. The theories of charge transfer reactions also suffer from the neglect of the effects of short-range correlations on dynamics. There is also a need to develop a microscopic theory of polar solvent dynamics. Nevertheless, a beginning toward a proper microscopic theory of solvation has been made. Judging from the intense current activity in this field, we can look forward to an improved understanding of solvation dynamics in the near future.

ACKNOWLEDGMENTS

It is a pleasure to thank Professor Graham Fleming for introducing me to this field, for continued collaboration, and for sending many preprints prior to publication. I am indebted to Dr. E. W. Castner, Jr. and Professor M. Maroncelli for collaboration. I thank my student Amalendu Chandra for collaboration and for much help in preparing this manuscript. I thank Professor H. L. Strauss for reading an early version of the manuscript and for valuable suggestions.

Literature Cited

1. Onsager, L. 1977. *Can. J. Chem.* 55: 1819
2. Dogonadze, R. R., Kalman, E., Kornyshev, A. A., Ulstrup, J., eds. 1985. *The Chemical Physics of Solvation, Part A: Theory of Solvation.* Amsterdam: Elsevier
3. Dogonadze, R. R., Kalman, E., Kornyshev, A. A., Ulstrup, J., eds. 1985. *The Chemical Physics of Solvation, Part B: Spectroscopy of Solvation.* Amsterdam: Elsevier
4. Hynes, J. T. 1985. *Annu. Rev. Phys. Chem.* 36: 573–97
5. Fleming, G. R. 1986. *Chemical Applications of Ultrafast Spectroscopy*, pp. 166–72. New York: Oxford
6. Kenney-Wallace, G. 1981. *Adv. Chem. Phys.* 47: 535–77
7. Kosower, E. M., Huppert, D. 1986. *Annu. Rev. Phys. Chem.* 37: 127–56
8. Simon, J. D. 1988. *Acc. Chem. Res.* 21: 128–34
9. Barbara, P. F., Jarzeba, W. 1988. *Acc. Chem. Res.* 21: 195–99
10. Solvation. 1988. *Faraday Discuss. Chem. Soc.*, Vol. 85
11. Born, M. 1920. *Z. Phys.* 1: 45
12. Onsager, L. 1935. *J. Am. Chem. Soc.* 58: 1485–93
13. Mazurenko, Yu. T., Bakshiev, N. G. 1970. *Opt. Spectrosc.* 28: 490–94
14. Bagchi, B., Oxtoby, D. W., Fleming, G. R. 1984. *Chem. Phys.* 86: 257–67
15. van der Zwan, G., Hynes, J. T. 1985. *J. Phys. Chem.* 89: 4181–88
16. Castner, E. W. Jr., Fleming, G. R., Bagchi, B. 1988. *Chem. Phys. Lett.* 143: 270–76
17. Davidson, D. W., Cole, R. H. 1951. *J. Chem. Phys.* 19: 1484–90
18. Ehrenson, S. 1981. *J. Comput. Chem.* 2: 41–52
19. Block, H., Walker, S. M. 1973. *Chem. Phys. Lett.* 19: 363–64
20. Ehrenson, S. 1987. *J. Phys. Chem.* 91: 1868–73
21. Kornyshev, A. A. 1985. See Ref. 2, pp. 77–116
22. Bagchi, B., Castner, E. W. Jr., Fleming, G. R. 1989. *J. Mol. Struc. Theor. Chem.* 194: 171–81
23. Castner, E. W. Jr., Fleming, G. R., Bagchi, B., Maroncelli, M. 1988. *J. Chem. Phys.* 89: 3519–34
24. Calef, D. F., Wolynes, P. G. 1983. *J. Chem. Phys.* 78: 4145–53
25. Nichols, A. L. III, Calef, D. F. 1988. *J. Chem. Phys.* 89: 3783–88
26. Chandra, A., Bagchi, B. 1988. *Chem. Phys. Lett.* 151: 47–53
27. Bagchi, B., Chandra, A. 1988. *Proc. Indian Acad. Sci. Chem. Sci.* 100: 353–57
28. Bagchi, B. J., Chandra, A. 1989. *J. Chem. Phys.* 90: 7338–45
29. Bagchi, B. J., Chandra, A. 1989. *Chem. Phys. Lett.* 155: 533–39
29a. Chandra, A., Bagchi, B. J. 1989. *J. Phys. Chem.* In press
30. Gray, C. G., Gubbins, K. E. 1984. *Theory of Molecular Fluids*, Vol. 1. Oxford: Clarendon
31. Zwanzig, R. 1961. In *Lectures in Theoretical Physics*, ed. W. E. Britton, B. W. Downs, J. Downs, 3: 106–41. New York: Wiley Interscience
32. Ma, S.-K., Mazenko, G. 1975. *Phys. Rev. B* 11: 4077–4100
33. Bagchi, B. 1985. *J. Chem. Phys.* 82: 5677–84; 1986. *J. Chem. Phys.* 85: 4667–68

34. Bagchi, B. 1987. *Physica A* 145: 273–89
35. Lebowitz, J. L., Percus, J. K. 1963. *J. Math. Phys.* 4: 116–23
36. Wolynes, P. G. 1987. *J. Chem. Phys.* 86: 5133–36
37. Rossky, P. J. 1985. *Annu. Rev. Phys. Chem.* 36: 321–46
38. Stell, G. 1977. In *Statistical Mechanics, Part A. Equilibrium Techniques*, ed. B. Berne. New York: Plenum
39. Rips, I., Klafter, J., Jortner, J. 1988. *J. Chem. Phys.* 88: 3246–53
40. Rips, I., Klafter, J., Jortner, J. 1988. *J. Chem. Phys.* 89: 4288–99
41. Fries, P. H., Patey, G. N. 1985. *J. Chem. Phys.* 82: 429–40
42. Stell, G., Patey, G. N., Hoye, J. S. 1981. *Adv. Chem. Phys.* 48: 183–328
43. Loring, R. F., Mukamel, S. 1987. *J. Chem. Phys.* 87: 1272–83
44. Zwanzig, R. 1963. *J. Chem. Phys.* 38: 2766–72
45. Loring, R. F., Yan, Y. J., Mukamel, S. 1987. *J. Chem. Phys.* 87: 5840–57
46. Yan, Y. J., Sparaglione, M., Mukamel, S. 1988. *J. Phys. Chem.* 92: 4842–53
47. Friedrich, V., Kivelson, D. 1987. *J. Chem. Phys.* 86: 6425–31
48. Maroncelli, M., Castner, E. W., Webb, S. P., Fleming, G. R. 1987. In *Ultrafast Phenomena*, ed. G. R. Fleming, A. E. Siegman, 5: 303–7. Berlin/New York: Springer
49. Vijayakumar, P., Tembe, B. L. 1988. *Proc. Indian Acad. Sci. Chem. Sci.* 100: 305–14
50. Karim, O. A., Haymet, A. D. J., Banet, M. J., Simon, J. D. 1988. *J. Phys. Chem.* 92: 3391–94
51. Maroncelli, M., Fleming, G. R. 1989. *J. Chem. Phys.* In press
52. Halliday, L. A., Topp, M. R. 1977. *Chem. Phys. Lett.* 48: 40–45; 1978. *J. Phys. Chem.* 82: 2273–77
53. Lessing, H. E., Reichert, M. 1977. *Chem. Phys. Lett.* 46: 111–16
54. Okamura, T., Sumitani, M., Yoshihara, K. 1983. *Chem. Phys. Lett.* 94: 339–43
55. Castner, E. W., Bagchi, B., Maroncelli, M., Webb, S. P., Ruggiero, A. J., Fleming, G. R. 1988. *Ber. Bunsenges. Phys. Chem.* 92: 363–72
56. Yeh, S. W., Philips, L. A., Webb, S. P., Buhse, L. F., Clark, J. H. 1987. In *Ultrafast Phenomena*, ed. D. H. Auston, K. B. Eisonthal, 4: 359–62. Berlin/New York: Springer
57. Nagarajan, V., Brearley, A. M., Kang, T.-J., Barbara, P. F. 1987. *J. Chem. Phys.* 86: 3183–96
58. Maroncelli, M., Fleming, G. R. 1987. *J. Chem. Phys.* 86: 6221–39
59. Kahlow, M. A., Kang, T.-J., Barbara, P. F. 1988. *J. Chem. Phys.* 88: 2372–78
60. Kahlow, M. A., Jarzeba, W., Kang, T.-J., Barbara, P. F. 1989. *J. Chem. Phys.* 90: 151–58
61. Castner, E. W., Maroncelli, M., Fleming, G. R. 1987. *J. Chem. Phys.* 86: 1090–97
62. Kang, T. J., Kahlow, M. A., Giser, D., Swallen, S., Nagarajan, V., Jarzeba, W., Barbara, P. F. 1989. *J. Phys. Chem.* In press
63. Su, S.-G., Simon, J. D. 1987. *J. Phys. Chem.* 91: 2693–96
64. Simon, J. D., Su, S.-G. 1987. *J. Chem. Phys.* 87: 7016–23
65. Castner, E. W. Jr. 1988. *Dipolar solvation dynamic studies by the time dependent fluorescence Stokes shift*. PhD thesis. Univ. Chicago, Chicago, Ill. 99 pp.
66. Rulliere, C., Declemy, A., Kottis, Ph. 1987. See Ref. 48, pp. 312–14
67. Declemy, A., Rulliere, C., Kottis, Ph. 1987. *Chem. Phys. Lett.* 133: 448–54
68. Declemy, A., Rulliere, C. 1988. *Chem. Phys. Lett.* 146: 1–6
69. Jarzeba, W., Walker, G. C., Johnson, A. E., Kahlow, M. A., Barbara, P. F. 1988. *J. Phys. Chem.* 92: 7039–41
70. Kohlrausch, R. 1847. *Pogg. Ann.* (Liepzig) 12: 393
71. Williams, G., Watts, D. C. 1970. *Trans. Faraday Soc.* 66: 80–85
72. Maroncelli, M., Fleming, G. R. 1988. *J. Chem. Phys.* 89: 875–81
73. Halliday, L. A., Topp, M. R. 1978. *J. Phys. Chem.* 82: 2415–19
74. Averbach, R. A., Synowiec, J. A., Robinson, G. W. 1980. In *Picosecond Phenomena II*, ed. R. M. Hochstrasser, p. 215. Berlin/New York: Springer
75. Robinson, G. W., Robbins, R. J., Fleming, G. R., Morris, J. M., Knight, A. E. W., Morrison, R. J. S. 1978. *J. Am. Chem. Soc.* 100: 7145–50
76. Johnson, I. D., Hudson, B. 1989. *Chem. Phys. Lett.* Submitted
77. Rentzepis, P. M., Jones, R. P., Jortner, J. 1973. *J. Chem. Phys.* 59: 766–73; Huppert, D., Rentzepis, P. M. 1976. *J. Chem. Phys.* 64: 191–96
78. Kenney-Wallace, G. A., Jonah, C. D. 1976. *Chem. Phys. Lett.* 39: 596–600; 1977. *Chem. Phys. Lett.* 47: 362–66
79. Kenney-Wallace, G. A., Hall, G. E., Hunt, L. A., Sarantidis, K. 1980. *J. Phys. Chem.* 84: 1145–50
80. Kenney-Wallace, G. A., Jonah, C. D. 1982. *J. Phys. Chem.* 86: 2572–86
81. Chase, W. J., Hunt, J. W. 1975. *J. Phys. Chem.* 79: 2835–45
82. Wang, Y., Crawford, M. K., McAu-

liffe, M. J., Eisenthal, K. B. 1980. *Chem. Phys. Lett.* 74: 160–65
83. Migus, A., Gandel, Y., Martin, J. L., Antonetti, A. 1987. *Phys. Rev. Lett.* 58: 1559–62
84. Kenney-Wallace, G. A. 1980. *Philos. Trans. R. Soc. A* 298: 309–19
85. Kramers, H. A. 1940. *Physica* 7: 284–304
86. Chandler, D. 1978. *J. Chem. Phys.* 68: 2959–70
87. Skinner, J. L., Wolynes, P. G. 1978. *J. Chem. Phys.* 69: 2143–50; 1980. *J. Chem. Phys.* 72: 4913–27
88. Northrup, S. H., Hynes, J. T. 1980. *J. Chem. Phys.* 73: 2710–14
89. Grote, R. F., Hynes, J. T. 1980. *J. Chem. Phys.* 73: 2715–32
90. Bagchi, B., Oxtoby, D. W. 1983. *J. Chem. Phys.* 2735–41
91. Carmeli, B., Nitzan, A. 1982. *Phys. Rev. Lett.* 49: 423–26
92. Carmeli, B., Nitzan, A. 1984. *J. Chem. Phys.* 80: 3596–3605
92a. Frauenfelder, H., Wolynes, P. G. 1985. *Science* 229: 337–45
93. Marcus, R. A. 1956. *J. Chem. Phys.* 24: 966–78; 1957. *J. Chem. Phys.* 26: 867–77
94. Marcus, R. A. 1964. *Annu. Rev. Phys. Chem.* 15: 155–96
95. Levich, V. G., Dogonadze, R. R. 1959. *Dokl. Akad. Nauk SSSR* 124: 123
96. Tembe, B. L., Friedman, H. L., Newton, M. D. 1982. *J. Chem. Phys.* 76: 1490–1507
97. Newton, M. D., Sutin, N. 1984. *Annu. Rev. Phys. Chem.* 35: 437–80
98. Calef, D. F., Wolynes, P. G. 1983. *J. Phys. Chem.* 87: 3387–3400
99. Creutz, C., Kroger, P., Matsubara, T., Netzel, T. L., Sutin, N. 1979. *J. Am. Chem. Soc.* 101: 5442–44
100. Efrima, S., Bixon, M. 1979. *J. Chem. Phys.* 70: 3531–35
101. Zusman, L. D. 1980. *Chem. Phys.* 49: 295–304
102. Alexandrov, I. V. 1980. *Chem. Phys.* 51: 449–57
103. Calef, D. F., Wolynes, P. G. 1983. *J. Chem. Phys.* 78: 470–82
104. Bagchi, B. 1986. *Chem. Phys. Lett.* 128: 521–27
105. McManis, G. E., Mishra, A. K., Weaver, M. J. 1987. *J. Chem. Phys.* 86: 5550–56
106. Cline, R. E. Jr., Wolynes, P. G. 1987. *J. Chem. Phys.* 87: 3836–44
107. Sparpaglione, M., Mukamel, S. 1987. *J. Phys. Chem.* 91: 3938–43
108. Gennett, T., Milner, D. F., Weaver, M. J. 1985. *J. Phys. Chem.* 89: 2787–94
109. Oppallo, M. 1986. *J. Chem. Soc. Faraday Trans. 1* 82: 339–47
110. McManis, G. E., Golovin, M. N., Weaver, M. J. 1986. *J. Phys. Chem.* 90: 6563–70
111. Grampp, G., Harrer, W., Jaenicke, W. 1987. *J. Chem. Soc. Faraday Trans. 1* 83: 161–66
112. McManis, G. E., Weaver, M. J. 1988. *Chem. Phys. Lett.* 145: 55–60
113. McGuire, M., McLendon, G. 1986. *J. Phys. Chem.* 90: 2549–51
114. Su, S.-G., Simon, J. D. 1988. *J. Chem. Phys.* 89: 908–19
115. Simon, J. D., Su, S.-G. 1987. *J. Chem. Phys.* 87: 7016–23
116. Kahlow, M. A., Kang, T. J., Barbara, P. F. 1987. *J. Phys. Chem.* 91: 6452–55
117. Huppert, D., Ittah, V., Kosower, E. M. 1988. *Chem. Phys. Lett.* 144: 15–23
118. Huppert, D., Ittah, V., Masad, A., Kosower, E. M. 1988. *Chem. Phys. Lett.* 150: 349–56
119. Heisel, F., Miehé, J. A. 1986. *Chem. Phys. Lett.* 128: 323–29
120. Heisel, F., Miehé, J. A. 1985. *Chem. Phys.* 98: 233–42
121. Rettig, W., Vogel, M., Lippert, E., Otto, H. 1986. *Chem. Phys.* 108: 381–90
122. Lippert, E., Rettig, W., Bonacić-Koutecky, V., Heisel, F., Miehé, J. A. 1987. *Adv. Chem. Phys.* 58: 1–173
123. Kang, T. J., Kahlow, M. A., Giser, D., Swallen, S., Nagarajan, V., Jarzeba, W., Barbara, P. F. 1988. *J. Phys. Chem.* In press
124. Sumi, H., Marcus, R. 1986. *J. Chem. Phys.* 84: 4894–4914
125. Nadler, W., Marcus, R. A. 1987. *J. Chem. Phys.* 86: 3906–24
126. Bagchi, B., Fleming, G. R., Oxtoby, D. W. 1983. *J. Chem. Phys.* 78: 7375–85; Bagchi, B., Singer, S., Oxtoby, D. W. 1983. *Chem. Phys. Lett.* 99: 225–31
127. Schulten, K., Schulten, Z., Szabo, A. 1980. *Physica A* 100: 599–614
128. Bagchi, B. 1987. *Chem. Phys. Lett.* 135: 558–64
129. Bagchi, B. 1987. *Chem. Phys. Lett.* 138: 315–20
129a. Bagchi, B. 1985. *Chem. Phys. Lett.* 115: 209–11
130. Bagchi, B. 1987. *J. Chem. Phys.* 87: 5393–5402
131. Agmon, N., Hopfield, J. J. 1983. *J. Chem. Phys.* 78: 6947–59
132. Poornimadevi, C. S., Bagchi, B. 1988. *Chem. Phys. Lett.* 149: 411–16
132a. Rips, I., Jortner, J. 1988. *J. Chem. Phys.* 88: 818–22
133. Kurz, J. L., Kurz, L. C. 1972. *J. Am. Chem. Soc.* 94: 4451–61

134. Albery, W. J., Kreevoy, M. M. 1968. *Adv. Phys. Org. Chem.* 16: 87–157
135. Fleming, G. R. 1986. See Ref. 5, pp. 166–72
136. Leffer, J. E., Graham, W. H. 1959. *J. Am. Chem. Soc.* 63: 687–91
137. Van der Zwan, G., Hynes, J. T. 1982. *J. Chem. Phys.* 76: 2993–3001
138. Van der Zwan, G., Hynes, J. T. 1983. *J. Chem. Phys.* 78: 4174–85
139. Van der Zwan, G., Hynes, J. T. 1984. *Chem. Phys.* 90: 21–35
140. Bergsma, J. P., Gertner, B. J., Wilson, K. R., Hynes, J. T. 1987. *J. Chem. Phys.* 86: 1356–76
141. Gertner, B. J., Bergsma, J. P., Wilson, K. R., Hynes, J. T. 1987. *J. Chem. Phys.* 86: 1377–86
142. Zichi, D. A., Hynes, J. T. 1988. *J. Chem. Phys.* 88: 2513–24
143. Hicks, J., Vandersall, M., Barbargie, Z., Eisenthal, K. B. 1985. *Chem. Phys. Lett.* 116: 18–24
144. Keery, K. M., Fleming, G. R. 1982. *Chem. Phys. Lett.* 93: 322–26
145. Sundström, V., Gillbro, T. 1984. *J. Chem. Phys.* 81: 3463–74
146. Hubbard, J. B., Wolynes, P. G. 1978. *J. Chem. Phys.* 69: 998–1006
147. Nee, T. W., Zwanzig, R. W. 1970. *J. Chem. Phys.* 52: 6353–63
148. Berne, B. 1975. *J. Chem. Phys.* 62: 1154–60
149. Warchol, M. P., Vaughan, W. E. 1976. *J. Chem. Phys.* 65: 1374–77
150. Madden, P., Kivelson, D. 1975. *Mol. Phys.* 30: 1749–80
151. Madden, P., Kivelson, D. 1984. *Adv. Chem. Phys.* 56: 467–566
152. Chandra, A., Bagchi, B. 1989. *J. Chem. Phys.* 91: 1829–42
153. Titular, U. M., Deutch, J. M. 1974. *J. Chem. Phys.* 60: 1502–13
154. Fatuzzo, E., Mason, P. R. 1967. *Proc. Phys. Soc. London* 90: 741
155. Van der Zwan, G., Hynes, J. T. 1983. *Physica A* 121: 227–52
156. Hubbard, J. B., Kayser, R. F., Stiles, P. J. 1983. *Chem. Phys. Lett.* 95: 399–401
157. Stiles, P. J., Hubbard, J. B. 1984. *Chem. Phys.* 84: 431–39
158. Loring, R. F., Mukamel, S. 1987. *J. Chem. Phys.* 86: 1272–83
159. Zwanzig, R. 1964. *J. Chem. Phys.* 38: 2766–72
160. Chandra, A., Bagchi, B. 1989. *J. Chem. Phys.* 90: 1832–40
161. Chandra, A., Bagchi, B. 1988. *J. Phys. Chem.* In press
162. Madden, P., Kivelson, D. 1984. *Adv. Chem. Phys.* 56: 492 Eq. 6.10

PICOSECOND VIBRATIONAL ENERGY TRANSFER STUDIES OF SURFACE ADSORBATES[1]

E. J. Heilweil, M. P. Casassa, R. R. Cavanagh, and J. C. Stephenson

Center for Atomic, Molecular, and Optical Physics, National Institute of Standards and Technology, Gaithersburg, Maryland 20899

INTRODUCTION

Energy transport between molecules and surfaces plays an extremely important role in many chemical and physical processes. At the microscopic level, molecules may be chemisorbed or dissociate on a substrate during a chemical reaction, they may undergo diffusion and desorption, or when adsorbed may change the physical or optical properties of the surface. Over the last few decades, breakthroughs in the development of sophisticated surface-sensitive techniques have occurred, making it possible to identify and characterize very detailed properties of these types of adsorbate-surface interactions (1-17). On the whole, these methods probe the static nature of surfaces and reveal information about the system on a relatively long (> 1 sec) timescale. When chemical transformations occur at interfaces, however, vibrational energy transfer can play a key role in determining the outcome of processes, whether thermal or caused by interactions with energetic laser (2-6), atomic (7), or molecular beams (8-10). To obtain a better understanding of these types of interactions, it is advantageous to perform direct measurements of adsorbate vibrational energy transfer. This chapter reviews advances made over the last few years to directly measure mode-specific adsorbate vibrational population or energy relaxation lifetimes (T_1) with ultrashort pulsed (ca. 10^{-12} sec) laser techniques. We also refer to T_1 studies for several molecules in solution

---

[1] The US Government has the right to retain a nonexclusive, royalty-free license in and to any copyright covering this paper.

when the results illuminate the relaxation mechanisms for surface species. The general subject of vibrational relaxation in gases, liquids, and cryogenic solids is beyond the scope of this article.

During the last few decades, vibrational spectroscopies have been used to identify adsorbate vibrational transitions in high surface area and single crystal systems (11–16). In many investigations, dynamical parameters of particular adsorbate modes were inferred from frequency domain spectral bandwidths. This approach was taken in several EELS and infrared studies. To date, many of the systems investigated exhibit broad vibrational bands to which T_1 population decay mechanisms, dephasing (T_2 mechanisms), and inhomogeneous broadening make varying contributions (17). Only when direct time-resolved measurements are made on a system can the contributions from each of these band-broadening mechanisms be extracted (18–20). For none of the systems studied have vibrational T_1 processes made a significant contribution to the observed infrared bandwidths. This delineation is borne out in several time-resolved studies of molecules in liquids (19, 20), where dephasing mechanisms are generally found to be the dominant bandwidth contribution.

Some insight into the mechanisms and magnitudes for adsorbate T_1 and T_2 processes is provided by several theoretical investigations of vibrational dynamics at surfaces (21–25). These theories are often based on simple models for phonon coupling (26–29), or for metals, relaxation by the creation of electron-hole pairs in the substrate (30–32). Generally, the predictions offered by these theories should be considered to be qualitative, and direct T_1 measurements should help point the way toward more quantitative treatments.

Major advances in performing time-resolved measurements of adsorbate dynamics are clearly needed to understand even the simplest microscopic events during chemical and physical surface transformations. Such studies would reveal vibrational mode population decay, energy transfer and coupling between other adsorbate and substrate modes, and perhaps identify surface intermediates during chemical reactions. We review this new and exciting field of study by first outlining current and emerging experimental techniques followed by recent detailed results of direct T_1 measurements for diatomic and other simple adsorbates on dielectric and metal-containing substrates.

ULTRASHORT TIME-RESOLVED OPTICAL TECHNIQUES

In the condensed phase, T_1 vibrational relaxation of many polyatomic systems occurs on the hundreds of picosecond timescale or faster (18–20).

Generally, the same vibrational mode exhibits a T_2 dephasing lifetime that is shorter than that mode's corresponding T_1 lifetime (19, 20, 33). These observations dictate that in order to measure T_1 processes for adsorbates at condensed-phase interfaces, measurement methods with picosecond or femtosecond time resolution are required. At this time, only ultrashort pulsed laser spectroscopies are able to perform these types of investigations. The following sections summarize a method used by our group at the National Institute of Standards and Technology (NIST) (34, 35) and other techniques now being developed to acquire surface T_1 information.

Single-Frequency Transient Infrared Spectroscopy

Excitation of a particular infrared-active fundamental vibrational mode of a molecule in its ground electronic state is most readily achieved by direct infrared absorption. By tuning a short infrared pulse to a $v = 0 \rightarrow 1$ vibrational transition (a mode at ≥ 1800 cm^{-1} in our experiments), and for modes with appropriate transition moments, absorption of the IR pulse transiently populates the $v = 1$ level. For sufficiently high pumping pulse fluences, a significant fraction of the ground state population (several percent or more) can be moved to the excited state. This excitation process produces $v = 1$ population exclusively if the IR pulse is resonant with the fundamental transition of interest and the mode possesses sufficiently large anharmonicity that overtone ($v = 2$ and higher) state excitation is avoided.

After excitation, the excited state population recovers back to the ground state by nonradiative decay channels. Transient population of $v = 1$ implies that sample infrared transmission T *at the pump frequency* will be increased from the thermal level T_0 to an increased value. The pump pulse therefore causes transient bleaching, which can be monitored with a weaker probe pulse at the same frequency. As the population relaxes back to equilibrium, the transient bleach signal recovers to the equilibrated Beer's law (T_0) level. For a two-state model, the quantity $\ln(T/T_0)$, i.e. the pump-induced change in probe pulse absorbance, is proportional to the $v = 1$ excited state population. For systems exhibiting exponential population decay, a plot of $\ln[\ln(T/T_0)]$ versus the time delay between arrival at the sample of the excitation and weak probe pulse yields a straight line decay with slope corresponding to the population relaxation time constant T_1. Implicit in the modeling of the relaxation as a two-level system is the notion that changes in sample absorption at the pump/probe frequency associated with transient nonequilibrium population of other modes during the relaxation are not important. This is discussed in Ref. (34) and is a good assumption if the relaxation rates of lower frequency modes are as fast or faster than $(T_1)^{-1}$ for the excited vibration. Liquid phase solute studies using both single color IR bleaching and spontaneous anti-Stokes Raman scattering (i.e. a direct measure of v = 1 population lifetimes) yielded the same T_1 values (36).

The single IR frequency pump-probe approach has been applied successfully by our group to measure T_1 for adsorbates on high surface area materials and other systems in transmission mode (see below). We have employed two Nd^{+3}:YAG-based laser systems to generate high energy tunable infrared pulses: one active-passive mode-locked system (ca. 20 ps FWHM) using optical parametric amplification (OPA) (34, 35) or difference frequency generation (DFG) (37), and a second system employing two synchronously pumped dye lasers (2 ps FWHM) with the DFG approach (38). A schematic diagram of the former apparatus used to generate wavelength tunable picosecond IR pulses and measure T_1 is shown in Figure 1. This apparatus utilizes DFG in $LiIO_3$ (39–42) between the second harmonic of the Nd^{+3}:YAG laser and the tunable output of a synchronously pumped dye laser to produce IR pulses in the 5µm range (1700 cm^{-1} and higher, 15 microjoule energy, 18 ps FWHM) (37). Similar mixing schemes with two picosecond synchronously pumped dye lasers (38) or amplified femtosecond dye lasers are being used to increase the overall time resolution. The optical parametric amplifier scheme was used to generate pulses in the 2500 to 4000 cm^{-1} region (34, 35). Such nonlinear frequency shifting methods are currently the simplest laboratory-based means for obtaining single, high power tunable IR pulses to perform these types of experiments.

To accomplish a T_1 measurement, the tunable IR pulse is passed through a retardation plate and telescope system to control the beam polarization and propagation characteristics. By careful adjustment of the polarization a small fraction (<1%) of the pulse can be reflected off a transparent beamsplitting optic such as CaF_2. The intense pumping pulse is passed through a series of mirrors (fixed optical delay) and focused onto an absorbing sample of appropriate optical density and thickness. The weaker probe pulse traverses a movable mirror assembly (variable delay) and is focused to overlap with the excitation region of the sample. A matched pair of IR detectors (e.g. InSb) monitor the transmitted probe pulse energy (I) and a small fraction of the incident probe pulse (I_0) to measure the probe pulse transmission ($T = I/I_0$) at many time delay settings. Computer-interfaced sampling and averaging techniques are employed to acquire data (with ca. 0.5% accuracy) and to control the probe pulse delay time. The instrumental time resolution is *not* determined by the time response of the IR detector (typically microsecond or longer), but by the duration of the pulses. This general description of the method applies to most ultrafast optical pump-probe experiments. The single frequency infrared method can, in principle, be used whenever samples are amenable to study by conventional IR absorption spectroscopy.

ADSORBATE VIBRATIONAL ENERGY TRANSFER   147

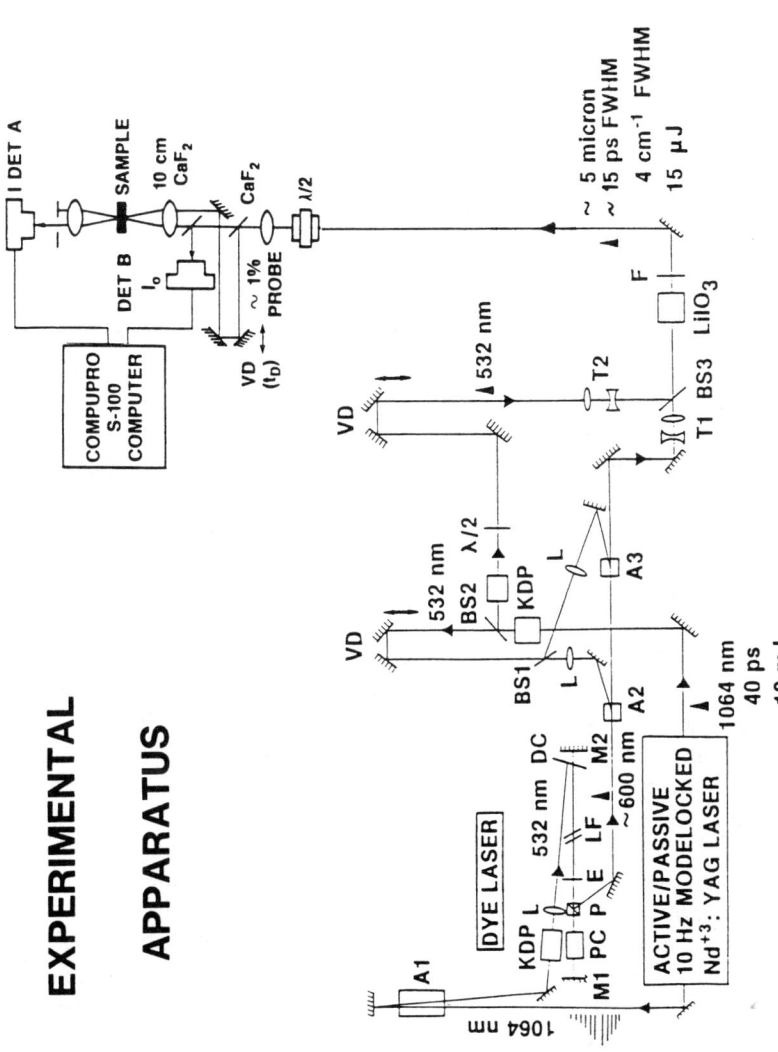

Figure 1 Picosecond apparatus for measuring CO-stretching mode T_1 lifetimes. The system produces tunable infrared 18 ps FWHM pulses by difference frequency generation in the $LiIO_3$ crystal. Components are abbreviated as follows: A, Yag or dye amplifier; KDP, doubling crystal; M, mirror; L, lens; PC, Pockels Cell; E, etalon; LF, Lyot tuning plates; DC, dye cells; BS, beam splitter; VD, variable delay; T, telescope; F, filter; Det, detector; $\lambda/2$, half-wave plate.

Infrared Pump with Narrowband Probes

According to the description in the previous section, a vibrational mode of a molecule or adsorbate is pumped by direct IR absorption, but only the T_1 relaxation time of that mode can be ascertained. In order to monitor population flow to other adsorbate internal modes or inter-adsorbate energy transfer, a second tunable IR probe pulse is needed. For systems with efficient intermode coupling and when the accepting modes are also IR active, a tunable probe pulse may be sensitive to population changes in these accepting modes. Should the population flow to IR inactive modes, alternative schemes such as anti-Stokes Raman scattering must be employed (19, 20). Campion and co-workers have obtained vibrational spectra of molecular monolayers on metal surfaces by using (without surface enhancement) spontaneous Stokes Raman scattering (43). Their work suggests that it would be conceivable, although very difficult, to detect transient excited vibrational population if a large fraction of a monolayer could be excited by short pulse excitation.

The ability to map out energy decay mechanisms and rates is ultimately determined by parameters unique to particular systems such as the magnitude of the initial excited state population, branching ratios to acceptor modes, and probing sensitivity. Only a handful of experiments have been reported that detected population flow to accepting modes of room temperature condensed-phase systems. Examples include the relaxation of the CH to the C≡C stretching modes of acetylene in CCl_4 (44), energy transfer of vibrationally excited dye molecules in C_2Cl_4 (45), and transfer from the excited NH-stretch to the C–C ring-breathing mode of pyrrole in organic solvents (46). The latter examples utilized anti-Stokes Raman probing methods to detect the ground electronic state vibrational acceptor mode transient population. Recent tunable IR probing experiments conducted by Graener, Laubereau, and co-workers have detected low energy mode transient population in CH_3Br (47, 47a) and the dissociation of ethanol oligomers in CCl_4 solution (47, 47a, 48). A transient infrared method using an upconverted tunable diode laser has successfully detected photodissociated carbon-monoxide fragments from porphyrin-containing biomolecules (49) and metal-carbonyls in solution (50). These IR absorption methods certainly could be applied to the study of T_1 for molecules on surfaces. In no instance, however, has a direct measurement been made to observe population transfer between adsorbate-surface modes.

Infrared Pump with Broadband Probe

It would be most advantageous to avoid tuning the probing pulse wavelength when searching for and measuring transient infrared spectra and dynamics. A novel method has been developed and tested in our laboratory that involves generating a broadband IR pulse by difference frequency mixing (in $LiIO_3$) between a fixed narrowband visible pulse with the output of a broadband dye laser, and then using broadband frequency upconversion in a second $LiIO_3$ crystal to shift the transient absorption spectrum back into the visible portion of the spectrum (51). This technique allows rapid acquisition of broadband IR spectra by using sensitive spectrograph-optical multichannel analyzer (OMA) detector combinations (51a). The advantage of this approach is that broadband IR spectra can be obtained for fixed time delay between the pumping pulse and broadband IR probing pulse. Information about relaxation of the excited mode and potential energy transfer to other optically active modes is monitored simultaneously by obtaining a series of "spectral snapshots" as a function of time delay between the pumping and probing pulses.

Tests were made of the approach using a modification of the apparatus shown in Figure 1. The method should be directly transferrable to the femtosecond time regime by running a femtosecond dye laser broadband (with no tuning elements) and using mixing crystals of appropriate length to avoid group velocity dispersion and pulse broadening. Transient absorption sensitivity on the order of 0.5% or better should be realized if each probe pulse is normalized to its own spectral content. This sensitivity has been achieved in visible transient absorption spectroscopic investigations (52). Such signal/noise accuracy would be obtained in the broadband IR technique by splitting the IR into two beams, only one of which would sample the "pumped" region of the sample. The two broadband IR beams would then be upconverted in different parts of the upconverting nonlinear crystal and finally dispersed by a spectrograph on two separate tracks of the two-dimensional OMA detector. The "pumped" and "unpumped" spectral tracks would then be normalized on every laser shot to increase the final sensitivity of the technique. In principle, this technique would be applicable to any system that can be studied by conventional transmission or reflection-absorption infrared approaches.

Infrared Pump with Sum-Frequency Generation Probe

Another potential approach to probing transient vibrational population at surfaces involves the interaction of two optical pulses with an adsorbate mode through second-order nonlinear processes. Since molecules at an interface exist between two phases, they are naturally in a non-centrosymmetric environment and hence $\chi^{(2)}$ mixing processes are allowed. One such interaction involves the generation of the sum frequency between a fixed-frequency field and another field tuned to an adsorbate IR-allowed vibrational transition. A more complete theoretical description of this mixing process and static spectroscopic investigations of several adsorbate-surface systems may be found in the chapter by Y. R. Shen in this volume (52a).

After tunable infrared pump pulse excitation of a surface mode, intra-adsorbate population flow can be probed by detecting the intensity of the sum-frequency signal as a function of time delay between the pump pulse and the tunable-IR/fixed-frequency probe pair. In one approach, the IR probe pulse would be tuned to the anharmonic $v = 1 \rightarrow 2$ absorption to monitor $v = 1$ population dynamics. Unlike direct infrared absorption methods, sum-frequency generation is sensitive only to modes that are IR and Raman active. So far, sum-frequency generation has been used to measure static spectra of adsorbates in a few systems. Further systems and time-resolved applications should be explored to realize the potential of this new method.

STUDIES OF ADSORBATES ON DIELECTRIC SURFACES

At the time of this review, all published reports of adsorbate time-resolved T_1 vibrational lifetimes were made on high surface area systems by using the single IR frequency pump-probe technique mentioned above. These experiments were conducted by our group at NIST over the past five years and are reviewed in the following sections. Liquid and solid phase model systems are also described that helped to identify and corroborate the characteristic T_1 lifetimes measured for the surface-adsorbate systems.

A variety of opportunities are afforded by studying high surface area materials with the techniques described above. By far the most significant benefit is that the surface area to bulk volume ratio is largest for colloidal-sized particles (ca. 10–5000 Å), thus making a large effective concentration of surface adsorbates in a small sample volume. In addition, for studies in transmission mode, samples can be made to reduce light scattering in the IR spectral region by pressing thin disks of a material or suspending it in an index matching solvent. Chemical modification of these materials is also easily accomplished, making it possible to alter the species that exist at the surface. High surface area oxides such as silica (SiO_2), alumina (Al_2O_3), titania (TiO_2), and zeolites fall into this category. For experiments in which the optical density of a dispersed sample is critical, the solid/liquid ratio can be varied. Since these types of materials have been used extensively in the field of catalysis, a rich literature devoted to their chemical and physical properties is available.

$OH(v = 1)$ Relaxation on Silica

A summary of the T_1 results for chemisorbed species on high surface area materials is given in Table 1. The first measurements of T_1 adsorbate

Table 1 T_1 decay times at room temperature for various surface vibrations

System	v (cm^{-1})	T_1 (ps)	k (10^9 s^{-1})	Notes
SiOH/vacuum	$v_{OH} = 3750$	220 ± 20	4.9	Pressed SiO$_2$ disk
SiOH/CCl$_4$	$v_{OH} = 3690$	159 ± 16	6.3	Dry SiO$_2$ dispersion
SiOH/CF$_2$Br$_2$	$v_{OH} = 3690$	140 ± 30	7.1	Dry SiO$_2$ dispersion
SiOH/CH$_2$Cl$_2$	$v_{OH} = 3660$	102 ± 20	9.8	Dry SiO$_2$ dispersion
SiOH/C$_6$H$_6$	$v_{OH} = 3625$	87 ± 30	11.0	Dry SiO$_2$ dispersion
SiOH/C$_6$D$_6$	$v_{OH} = 3625$	80 ± 30	12.0	Dry SiO$_2$ dispersion
SiOH/H$_2$O/CCl$_4$	$v_{OH} = 3690$	56 ± 10	18.0	SiOH T_1, ~ 5H$_2$O/100 Å^2 physisorbed
SiOD/vacuum	$v_{OD} = 2760$	155 ± 16	6.5	OD T_1 with 67% OH replaced by OD
BOH/vacuum BOH/CCl$_4$	$v_{OH} = 3700$	~ 70	~ 14.3	
SiNH$_2$/vacuum SiNH$_2$/CCl$_4$	$v_{NH} = 3460$ 3520	≤ 20	≥ 50	Pulsewidth limited signal, $\Delta T/T \sim 5\%$ for both stretches
SiOCH$_3$/vacuum	$v = 2860$ 3000	$< 5(?)$	$> 280(?)$	No pulse saturation observed for any CH-stretching mode
OH on zeolite ZSM-5/CCl$_4$ (Si/Al = 3600)	$v_{OH} = 3690$	140 ± 25	7.1	Pressed disk CCl$_4$ saturated
OH on ZnO	$v_{OH} = 3485$	≤ 10	≥ 100	$\Delta T/T_0 = 12\%$ pulsewidth limited
ZnO, no OH	3600–3750	≤ 10	≥ 100	Electronic excitation of free carrier absorption $\Delta T/T = 10\%$

relaxation were made on chemisorbed hydroxyl (OH) on the surface of fumed silica dispersed in CCl$_4$ at room temperature (34, 35, 53). Fumed silica, a form of amorphous silica with uniform particle size, is produced commercially and grown in a SiH$_4$ and O$_2$ flame. Depending on flame conditions, different particle sizes can be obtained. Dispersions of silica in CCl$_4$ are amenable to study because the solid can be pretreated (by heating to remove adsorbed water or by chemically modifying the surface), and the resultant sample is optically homogeneous (SiO$_2$ and CCl$_4$ are index-matched). A static infrared absorption spectrum of such a sample (Figure 2) shows a band arising from isolated surface OH groups (centered at 3690 cm^{-1}) and a very broad band arising from adsorbed water (3000 to 3600 cm^{-1}). The OH-stretching vibration is extremely anharmonic ($2\omega_e x_e$ ca. 150 cm^{-1}), thus ensuring that the v = 1 → 2 transition does not interact with the excitation and probe pulses.

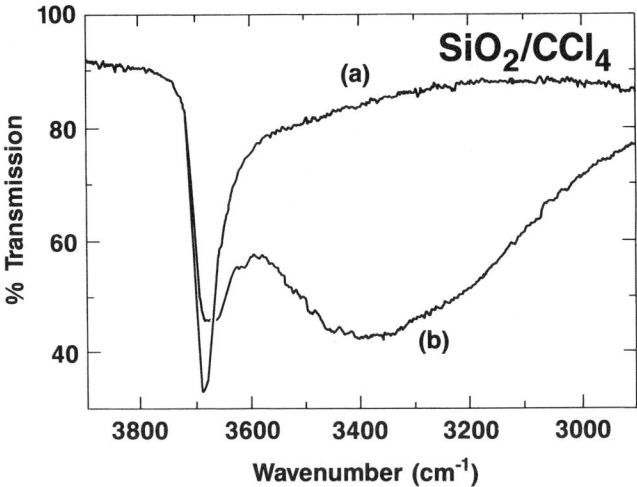

Figure 2 Absorption spectrum of SiO_2 dispersed in CCl_4: (*a*) after water removal pretreatment, showing the free hydroxyl absorption at 3690 cm^{-1}, and (*b*) with adsorbed water.

T_1 FOR OH/SiO_2 AT THE LIQUID AND VACUUM INTERFACE   The optical parametric amplifier apparatus was used for all the OH and related T_1 studies of hydrogen-containing species (see Tables 1 and 2). The T_1 measurement consists of populating OH(v = 1) with 300–400 microjoule IR pulses (tuned to the OH v = 0 → 1 transition) focused (ca. 0.2 mm beam radius) into a 1 mm liquid sample cuvette. This produces a 10–20% increase in pulse transmission, indicating that the transition was partially saturated. Following the bleaching recovery with a weak probe pulse as a function of delay time (Figure 3) yields the OH(v = 1) T_1 lifetime of 159 ± 16 ps (53). T_1 for the surface OH(v = 1) mode is independent of SiO_2 particle size (diameter > 70 Å). This first result showed that for low surface coverage (mean separation of ca. 10 Å) of a diatomic oscillator (OH) on a dielectric support (average silica particle containing 10^6 atoms at room temperature and surrounded by solvent), energy deposited in the OH oscillator remains in that mode for up to 10^4 vibrational periods. T_1 is *clearly unrelated* to the absorption bandwidth of 50–60 cm^{-1} FWHM, which, by direct Fourier transformation, would correspond to a relaxation time of about 100 fs.

Measurements were conducted to appraise the magnitude of the contribution to the OH(v = 1) vibrational relaxation rate from the CCl_4 solvent (34, 53). By pressing small quantities of fumed silica in a die, it is possible to form self-supporting disks of the material that are amenable to transmission IR spectroscopy (54–56). Placing such disks in high vacuum,

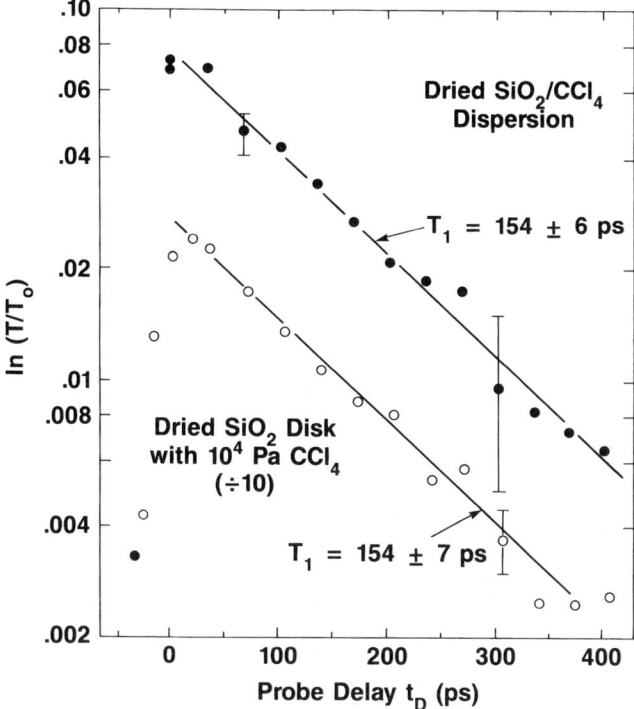

Figure 3 Examples of T_1 lifetime decays for OH(v = 1) adsorbed on silica dispersed in CCl$_4$ solvent and for a pressed silica disk saturated with CCl$_4$ vapor (at 298K).

heatable cells with IR transparent windows facilitates sample preparation and study by the transient IR method (54, 55). Removal of surface adsorbed water by heating to 625K under vacuum produces the isolated-OH absorption band at 3745 cm^{-1} with an 8 cm^{-1} FWHM bandwidth. The OH(v = 1) T_1 lifetime for this sample was found to be 220 ± 20 ps, only 40% larger than that found for SiO$_2$/CCl$_4$. Apparently, the intrinsic relaxation rate of OH(v = 1) bound to SiO$_2$ in vacuum is comparable to either the energy transfer rate from the excited OH-stretch to the low-frequency modes (<750 cm^{-1}) of CCl$_4$ or the rate of collision-induced relaxation by the solvent molecules (each surface OH experiences ca. 10^{12} collisions sec^{-1}).

A series of measurements using the same experimental approach was conducted to study the effect of different solvents on T_1 for OH(v = 1) (34). After the water removal pretreatment, silica was dispersed in solvents such as CH$_3$Cl, *n*-hexane, and benzene (see Table 1). T_1 for the OH-stretch was found to *decrease*, as the solvent molecule used was larger and its internal vibrational level structure more complex. For benzene, with the largest number of CH-stretching oscillators and greatest vibrational state density, T_1 was found to be ca. 90 ps. Despite these changes in solvent environment, it was surprising to find that the OH(v = 1) population lifetime only varied by up to a factor of two.

ISOTOPE AND TEMPERATURE DEPENDENCE In order to better understand the detailed relaxation mechanism for OH(v = 1) on silica, a series of experiments were performed on pressed silica disks in vacuum. The hydroxyl proton can be exchanged for deuterium by a controlled D_2O exposure. This "isotopic dilution" would potentially decrease hydroxyl inter-adsorbate coupling but not other OH-surface interactions (57). Changing the sample temperature, on the other hand, would affect the equilibrium population of low frequency substrate phonon modes potentially varying T_1 (58–60). Detailed results of these experiments may be found in Ref. (57).

Deuterated silica exhibits a new absorption at 2760 cm^{-1} that corresponds to the free OD-stretching vibration. This absorption has the same bandwidth and overall shape as the hydroxyl IR absorption feature. For a sample with maximum deuteration (ca. 70% OD/OH), T_1 for the residual OH-stretch population was again measured to be 220 ± 20 ps. This result implies that the free surface hydroxyls of fumed silica are sufficiently separated that an adsorbate-adsorbate coupling mechanism is ineffective. When tuned to the OD-stretching mode at 2760 cm^{-1}, T_1 [for OD(v = 1)] was found to be 135 ps at 298K. Simple considerations of reducing the energy gap between the excited hydroxyl stretching mode and the fixed internal vibrational modes of the substrate (Si–O stretches at 1100 cm^{-1} and lower) suggested that T_1 should decrease by at least an order of magnitude (58–60). The reduction of T_1 by less than a factor of two implied that the initial vibrational relaxation step probably involves multiphonon decay to modes involving the OH(D) group. The most likely energy acceptor is the Si–O–H bending vibrational mode, which drops in frequency by 30–40% upon deuteration. This significant result indicates that not only can the excited adsorbate mode have an extremely long lifetime, but the initial relaxation step probably involves only vibrational modes *spatially adjacent* to the relaxing mode.

The temperature dependence of T_1 for multiphonon relaxation of a vibrational mode embedded in a bath of other lower frequency modes has been derived by several authors (58–60). These Golden Rule formulations yield a simple analytical expression for evaluating $(T_1)^{-1}$ as a function of temperature if the frequencies of the vibrational modes that accept the energy in the initial relaxation step are known. The model is also sensitive to the number n of quanta initially created during the deactivation of the excited vibration. A plot of the temperature dependence of $(T_1)^{-1}$, which is generally expected to increase monotonically with increasing temperature, enables one to extract a value for n. By using the earlier assumption that OH(v = 1) initially decays into excited quanta of Si–O stretches (ca. 1000 cm^{-1}) or Si–O–H bends (around 900 cm^{-1}), theoretical predictions for the temperature dependence as a function of n could be made.

The experiment was conducted on a dry silica disk mounted in an evacuated UHV-compatible transmission cell that could be cooled or heated over the temperature range 100–800K. T_1 for OH and OD was measured throughout this range and was found to change by roughly a factor of two (57). The experimental results agreed very well with the theoretical prediction of $n = 4$, that is, the OH(D)(v = 1) quantum decays into four quanta of Si–O stretching or Si–O–H(D) bending motions. This high order process ensures that the initial relaxation step takes a relatively long time. However, the same theory and parameters that fit the temperature dependence would predict a larger increase in relaxation rate upon deuteration (OD versus OH) than observed, thus indicating that these models are too simple to determine vibrational energy relaxation rates quantitatively. As is seen in all following sections, the evidence that a high frequency excited vibrational mode relaxes by the simultaneous multiquantum deposition of energy into spatially adjacent lower frequency accepting modes (despite the fact that the system can have an enormous total density of states at the excitation energy) is quite general.

Model Systems for Adsorbates on Silica

Based on the hypothesis that the initial relaxation step of OH(v = 1) population (for OH on silica) involves only adjacent vibrational modes (i.e. Si–O–H bend or Si–OH stretch), an effort was undertaken to confirm this relaxation pathway by studying model systems. Molecules containing the SiOH moiety such as tertiary silanols (61), amorphous silica containing trace concentrations of hydroxyl ions (62), and layer-structured crystalline micas (63, 64) fall into this category. A brief description of the T_1 results for these systems follows.

SiOH-CONTAINING MOLECULES IN SOLUTION Silicon-containing molecules with the generic formula R$_3$SiOH (R=CH$_3$, CH$_2$CH$_3$, and C$_6$H$_5$) are commercially available and soluble in CCl$_4$ solvent at room temperature. From previous vibrational spectroscopy and structural determinations (61, 65, 66), it was felt that the energy transfer properties of these molecules might bear a strong resemblance to the silica-OH surface species. It should be noted that the vibrational modes of CCl_4 are all relatively low in frequency ($v \leq 750$ cm^{-1}), similar to the fundamental phonon spectrum of bulk SiO_2.

Dilute CCl_4 solutions of these molecules were prepared (mole fraction less than 7×10^{-3}) such that solute-solute interactions were unimportant but the OH-stretching peak intensity was large enough (ca. 50% transmission) to perform the T_1 experiment in 1 mm pathlength cells. The T_1 results for studies of these molecules as well as their deuterated analogs are summarized in Table 2. The striking similarity between T_1 measured for all of the molecular SiOH-containing analogs and that for silica dispersed in CCl_4 ($T_1 = 159 \pm 16$ ps) lends credence to the notion that the SiOH part of the molecule plays a primary role in the initial vibrational relaxation process. Additionally, the fact that the three silanol molecules, each with different R groups and therefore different vibrational density of states at 3600 cm^{-1}, yield nearly the same T_1 lifetime supports the above hypothesis. For the deuterated analogs, T_1 changes *only slightly* compared to the OH-containing molecules. This again points to the possibility that the vibrational modes of the SiOH(D) moiety play a key role in the initial decay (57).

One other important aspect of this study deserves mention here. Tertiary alcohols with molecular formula $R_3COH(D)$ (R=CH_2CH_3 and C_6H_5 as

Table 2 T_1 lifetimes for the OH and OD(v = 1) stretching vibration in alcohols and silanols in dilute CCl_4 solution (≤ 0.007 mole fraction) at 298K. Error limits are $\pm \sigma$, or 10% of T_1, whichever is greater (see Ref. 61)

Silanol[a]	$v_{OH,OD}$ (cm^{-1})	T_1 (ps)	Alcohol[a]	$v_{OH,OD}$ (cm^{-1})	T_1 (ps)
Me$_3$SiOH	3690	205 ± 21	Me$_3$COH	3614	<6[b]
Me$_3$SiOD	2722	245 ± 25	Me$_3$COD	—	—
Et$_3$SiOH	3689	185 ± 19	Et$_3$COH	3622	<6[b]
Et$_3$SiOD	2722	224 ± 22	Et$_3$COD	2673	<20[b]
φ_3SiOH	3675	206 ± 21	φ_3COH	3609	<15[b]
φ_3SiOD	2712	292 ± 29	φ_3COD	—	—
φ_2Si(OH)$_2$	3610, 3679	80 ± 15	CH$_3$OH	3641	$15-30$[b]
φ_2Si(OD)$_2$	2665, 2710	134 ± 14	CH$_3$OD	2685	52 ± 17
			CD$_3$OH	3642	73 ± 7
			CD$_3$OD	2690	79 ± 17
			EtOH	3625	70 ± 10
			φOH	3610	$5-20$[b]
			φOD	2665	$15-25$[b]
			$\varphi F_5 OD$	2640	<15[b]

[a] Me = –CH$_3$, Et = –CH$_2$CH$_3$, φ = –C$_6$H$_5$.
[b] Limits on T_1 were deduced from computer simulation.

well as methanol, ethanol, and phenol) were examined because they possess the same structure as the tertiary silanols discussed above. Vibrational spectra have also been obtained to identify many of the normal modes of these systems. Table 2 includes the T_1 results for the hydroxyl stretching mode of these molecules, where one finds a drastic reduction in relaxation lifetime for the alcohols as a group compared to the corresponding silanols. By merely changing the mass of the central atom (Si to C) but retaining the molecular structure, a reduction of T_1 by a factor of 10 or more occurs. Consideration of all possible combinations of accepting normal modes within the molecule showed that an increase in corresponding mode frequencies for the alcohols reduces the number of quanta that must be created in spatially adjacent, lower frequency accepting modes from four (for the silanols) to three or less (for the alcohols) (61). Again, the idea of direct relaxation of OH(v = 1) into SiOH-bending or SiO-stretching motions was supported by these solution-phase studies.

OH IMPURITIES EMBEDDED IN AMORPHOUS QUARTZ   Quartz and optical glass naturally contain trace amounts of hydroxyl groups in their internal amorphous structures. These impurities occur because of the decomposition of trace water at the high temperatures used to melt and form the material. The levels of OH impurity can be controlled by exclusion of water during the melting and forming process and various commercially available optical window materials contain different hydroxyl concentrations. These glasses were used for temperature-dependent studies of T_1 lifetimes for the internal OH(v = 1) (62).

Briefly, it was found that independent of glass type (OH concentration < 120 ppm), the excitation of the OH-stretch mode (ca. 60 cm^{-1} FWHM bandwidth at 295K) yielded a T_1 value of ca. 109 ps at room temperature (62). This value, again remarkably similar to that observed for free OH on the silica surface, decreased slightly as the IR pulse was tuned to the lower frequency side of the absorption band. Such an effect probably arises because the experiment samples hydroxyl groups with stronger hydrogen-bonding interactions to the surroundings as one tunes to the red. The temperature study for $100 < T < 1450K$ revealed the same $n = 4$ multiphonon decay dependence as was observed for the surface OH on SiO_2 system (57).

CRYSTALLINE MICAS   Naturally occurring alumina-silicates such as the two-dimensional sheets of the mica family offer an excellent opportunity to study hydroxyl ion vibrational relaxation in a well-ordered crystalline environment (64). Because the composition and structure of these materials are well known, and since the arrangement of the hydroxyl ions in the lattice bears an extremely close resemblance to an ordered array of adsorbates on a dielectric surface, T_1 measurements for OH(v = 1) relaxation in a series of mica families were undertaken.

For Muscovite micas from a large selection of samples, a broad OH-stretching absorption (ca. 60 cm^{-1} FWHM, room temperature) at 3630 cm^{-1} is found. Although the OH$^-$ ion orientation is well-defined in the lattice and the number density is on the order of 10^{21} cm^{-3} (close packed in any one plane), six samples yielded a relaxation time of 92 ± 13 ps. Biotite micas exhibit different OH sites with varying dipole orientations, depending on cation impurities and concentrations. High frequency OH-stretches yielded longer T_1 values ($T_1 = 221 \pm 23$ ps) than for the Muscovites, but lower frequency bands (more strongly coupled to the matrix) exhibited reduced T_1 lifetimes the same as for OH in Muscovites (64). Despite the increased probability of strong hydroxyl interactions in these systems, the qualitative similarity between the mica OH(v = 1) lifetimes and the systems described above help confirm the apparent general picture that vibrational energy relaxation in these systems occurs through a structurally localized anharmonic coupling mechanism.

Other Adsorbates Chemisorbed on Silica

Fumed silica also affords the opportunity to study other adsorbates chemically bound to the SiO$_2$ support. Reactions of the surface hydroxyls with a variety of reactants yield one-to-one replacement of the hydroxyls with such groups as –NH$_2$, –OCH$_3$, and others (54, 55). The modified material was either dispersed in CCl$_4$ solvent or pressed into disks and T_1 studied at room temperature in either liquid or UHV transmission cells (54, 55).

T_1 was found to depend strongly on the vibrational modes of a particular adsorbate. For example, the –BOH group, which is attached to the surface by a pair of bridging oxygens, gave T_1 ca. 70 ps for the OH(v = 1) relaxation. BOH groups may relax faster than the SiOH rate because the local BOH bending and stretching modes are higher in frequency than for SiOH and hence better able to accept the 3700 cm^{-1} OH(v = 1) quantum. Both the symmetric and antisymmetric NH-stretching modes of SiO$_2$ bound –NH$_2$ groups produced the same transient bleaching signal magnitude, but were difficult to time resolve ($T_1 \leq 20$ ps). No bleaching signal was observed for the methoxy group CH-stretching mode, thus indicating T_1 was only a few picoseconds or less for this system. Structural and vibrational mode arguments and comparison to T_1 measurements for these groups in condensed-phase systems (CH$_3$-stretch for CH$_3$OH in CCl$_4$) (67) pointed again to the likelihood that for adsorbates on dielectric supports, the relaxation mechanism depends strongly on the vibrational level structure of the adsorbate and that the initial relaxation step depends on structurally adjacent modes but not the total density of states at the excitation energy.

Several theoretical models and estimates of adsorbate mode relaxation have been proposed for molecules on non-metallic surfaces. A classical trajectory simulation (for the SiH-stretching mode of hydrogen on silicon, T_1 ca. 1 ns) has estimated the T_1 relaxation lifetime for a surface adsorbate mode (68). Several calculations using Fermi Golden Rule approaches have also estimated T_1 for adsorbate vibrations coupled to surface and bulk phonon modes (25–32). Measurements that directly probe the accepting mode population are clearly needed to confirm the qualitative conclusions reached about vibrational energy relaxation on dielectric surfaces.

CARBON MONOXIDE ADSORBED ON TRANSITION METALS

While many chemical processes occur in homogeneous liquid media or on dielectric supports, other important chemical reactions take place on metal and metal particle surfaces. By directly measuring the vibrational relaxation lifetime of a chemically bound adsorbate (e.g. CO) on the surface of a metal, it may be possible to determine which relaxation mechanisms are associated with these reactions. Many of the physical and chemical properties of CO-transition metal systems have been well characterized over the years by the surface science community (11–16), thus making these systems particularly attractive for T_1 measurements.

One issue that could be addressed by such experiments is to determine whether vibrational excitation of an adsorbate decays into phonon modes or rapidly couples to the free electron states of the metal (electron-hole pair excitations) (30–32). Phonon relaxation theories have been developed that treat the relaxation of small surface adsorbates (diatomics or metal-carbon stretch for CO) on metals as though the metal were composed solely of lower frequency metal-metal and adsorbate-metal vibrational oscillators (26–28). In many of these calculations, vibrational T_1 relaxation is estimated to be relatively long (as is measured for adsorbates on dielectrics). However, since transition metals are capable of supporting electron-hole pairs, vibrationally excited adsorbate modes could rapidly deposit energy into these states. Several authors have also postulated that this rapid electron-hole pair relaxation mechanism (25, 30–32) or dephasing (69, 70) is responsible for the broad absorption bandwidths observed for many adsorbate-metal systems (30–32). With Golden Rule formalisms, estimates for the electron-hole T_1 relaxation time for CO on transition metals such as platinum are found to be on the order of several picoseconds (21–27). This mechanism is also believed to be sensitive to the electronic state density at the Fermi level and metal particle size (E. Blaisten-Barojas, J. W. Gadzuk, private communication) but insensitive to temperature, since only small changes in electron state population occur upon heating.

In order to assess the applicability of these mechanisms, a series of experiments were first conducted to understand the fundamental relaxation mechanisms for the CO-stretching vibration of carbon monoxide as a ligand attached to isolated metal atoms and transition metal clusters. These measurements helped develop a solid framework for understanding the details of $CO(v = 1)$ relaxation for CO attached to metal atoms before measurements were attempted for CO adsorbed on supported platinum and rhodium particles, discussed at the end of this section.

CO Ligands of Single Metal Atom Complexes in Solution

The first set of experiments were performed on single metal atom carbonyl complexes dissolved in room temperature solutions of hydrocarbon and halogenated methane solvents at extremely high dilution (mole fraction $<10^{-4}$) (37). These included commercially synthesized monocarbonyl complexes of Rh and Ir containing triphenylphosphine (TPP), chlorine, and hydrogen ligands. Other experiments were conducted on the F_{1u} CO-stretching mode of the hexacarbonyl species $Cr(CO)_6$ and $W(CO)_6$, which have octahedral symmetry (37, 71–73). The CO-stretching vibrations of metal-carbonyls are extremely strong absorbers (typically 0.5 Debye transition moment per CO), and their vibrational frequencies normally fall in the 5 μm range (1800 to 2100 cm^{-1}) (71–73). Room temperature solutions have absorption bandwidths in the range 3–25 cm^{-1} FWHM that depend on solvent (74, 75). The apparatus shown in Figure 1 was used to generate tunable IR 18 picosecond pulses with 4 cm^{-1} FWHM bandwidth to perform the single frequency T_1 experiments.

The results for T_1 measurements made on the above systems are summarized in Table 3. For these complicated metal-carbonyl systems, the T_1 lifetime for high frequency $CO(v = 1)$ modes ranges from about 30 up to 800 ps. As for $OH(v = 1)$, the hexacarbonyl systems were also found to be sensitive to the solvent, experiencing more rapid vibrational energy relaxation in the more complex hydrocarbon solvents. The monocarbonyls with complex TPP ligands exhibit $T_1 \leq 75$ ps. These rather short lifetimes were attributed to steric interactions between the surrounding TPP ligands, which create a "benzene-like" environment around the CO ligand. This model was proposed because the hexacarbonyls were found to relax on this timescale when they are dissolved in benzene (see Table 3) (37).

As was found for OH-containing systems, the lifetime of $CO(v = 1)$ for these metal carbonyl complexes bears no relationship to the IR spectral bandwidth of the $v = 0 \rightarrow 1$ transition. The observed T_1 corresponds to

Table 3 Vibrational T_1 lifetimes for the CO($v = 1$) stretching modes of mono and hexacarbonyl metal complexes in solution at 295K (1×10^{-5} to 1×10^{-4} mole fraction). Errors in T_1 are $\pm 1\sigma$ or $\pm 10\%$, whichever is greater,[a] and v is in wavenumber (cm^{-1})

Monocarbonyl[b]	Solvent	v_{CO}	Δv (FWHM)	T_1 (ps)	Hexacarbonyl[c]	Solvent	v_{CO}	Δv (FWHM)	T_1 (ps)
Rh(CO)Cl(TPP)$_2$	CHCl$_3$	1975	26	71 ± 12	Cr(CO)$_6$	CCl$_4$	1980	9	440 ± 70
Rh(CO)H(TPP)$_3$	CH$_2$Cl$_2$	1915	25	≤20[d]	Cr(CO)$_6$	CHCl$_3$	1980	15	295 ± 30
					Cr(CO)$_6$	n-Hexane	1982	3	145 ± 25
Ir(CO)Cl(TPP)$_2$	CHCl$_3$	1961	24	50 ± 13	Cr(CO)$_6$	C$_6$H$_6$	1978	—	59 ± 6
Ir(CO)H(TPP)$_3$	CH$_2$Cl$_2$	1922	21	37 ± 20	W(CO)$_6$	CCl$_4$	1978	13	800 ± 200
Ir(CO)H(TPP)$_3$	CHCl$_3$	1922	25	29 ± 6	W(CO)$_6$	CHCl$_3$	1973	20	480 ± 50
					W(CO)$_6$	n-Hexane	1981	3	140 ± 15
					W(CO)$_6$	C$_6$H$_6$	1974	—	60 ± 6

[a] Averaged T_1 lifetimes are reported for several runs with pump energy of ca. 3 μJ.
[b] TPP = triphenylphosphine ligand.
[c] IR active v_6 or F_{1u} mode.
[d] Decay followed pulse autocorrelation, implying an upper limit of T_1 only.

population decay occurring in 10^4 vibrational periods. Just as was found for OH, discussed above, since the vibrational modes involving atoms adjacent to the carbonyl ligand are relatively low in frequency, we can account for the long T_1 lifetimes by using a multiquantum decay model. The M–C–O bend and metal-carbon stretching frequencies are about 550 cm^{-1} and 450 cm^{-1}, respectively. Four or more quanta are required to be transferred simultaneously to these modes to accept the relaxing CO-stretching quantum (at 2000 cm^{-1}) during the initial step of the relaxation process. Such $n = 4$ multiquantum processes are expected to take hundreds of picoseconds to occur.

T_1 was also measured for $Cr(CO)_6$ in the gas phase, where the lifetime was found to be a factor of two longer than for the corresponding molecule in CCl_4 solvent (76). Again, as was found for $OH(v = 1)$ on SiO_2, solvents such as CCl_4 appear to open solute-solvent energy transfer or collisionally induced relaxation channels with rates comparable to those for the isolated molecule in the absence of solvent. At this time, it is difficult to ascribe this rate enhancement as arising from collision-induced intramolecular effects or energy transfer to the solvent itself. More detailed measurements are needed to see whether energy transfers to solvent modes or the solvent collision rate is the pertinent factor.

CO Relaxation in Transition Metal Carbonyl Cluster Compounds

T_1 lifetimes were measured for CO-stretching modes of transition metal cluster compounds in room temperature solution (76) and for the same molecules supported on SiO_2 (77). These studies were designed to explore the effects on the $CO(v = 1)$ T_1 lifetime of increasing metal cluster size and to see whether energy transfer occurs to vibrational modes of the SiO_2 support (78).

Experiments were first conducted on the cluster series $Rh_2(CO)_4Cl_2$, $Rh_4(CO)_{12}$, $Rh_6(CO)_{16}$, and $Co_4(CO)_{12}$ in dilute chloroform solution (76). The spectroscopy and chemical properties of these cluster compounds have been studied for many years (79–84). In addition to their inherent interest to inorganic chemists (83, 84), these systems serve as models for CO adsorbed on more complex supported metal particle catalysts (79–82) and single crystal systems (11–16). For the Rh–CO homologous series, T_1 was found to lie in the 600–800 ps range. It was also discovered that anomalously rapid signal decays and transient *absorption* could be obtained for $Rh_6(CO)_{16}$ when the IR pulse was tuned to the peak or red edge of the CO-stretching absorption (77, 78, 85, 86). These effects were concurrent with multiple photon absorption (estimated to be up to 30 adsorbed photons per molecule) and could be eliminated by deliberately tuning the IR frequency to the high frequency absorption edge and reducing the pump pulse fluence. All three Rh systems exhibited single-exponential decays (as did the systems in the previous section) but $Co_4(CO)_{12}$ was found to have an early 50 ps component in its relaxation transient followed by a population decay with a characteristic lifetime of 350 ps. The T_1 results for these cluster compounds dissolved in $CHCl_3$ solution are summarized in Table 4.

As was found for the single metal atom carbonyl compounds, T_1 is still surprisingly long for even these large polyatomic systems. Apparently no new mechanisms need to be invoked in order to explain the relaxation timescale. That is, anharmonic coupling of the CO-stretch to the M–CO stretching or M–C–O bending vibrations could readily dominate the initial

Table 4 Compilation of T_1 lifetimes (ps) for CO(v = 1) stretching vibrations of metal-carbonyl systems in dilute $CHCl_3$ solution at 295K[a]

System	v_0	Δv	σ (Å^2)	v_{IR}	I[b]	n	T_1 (ps)	Recommended T_1
$Rh(CO)_2(C_5H_7O_2)$	2084	10	0.085	2084	16	0.1	90	90 ± 11
	2014	15	0.076	2014	23	—	87	87 ± 10
$Rh_2(CO)_4Cl_2$	2092	8	0.198	2094	3.7	0.6	687	708 ± 130
	2036	13	0.170	2043	4.0	0.4	690	750 ± 86
				2036	13	—	733	
$Rh_4(CO)_{12}$	2075	c	0.493	2080	2.4	0.5	657	612 ± 63
				2074	6.0	1.8	387	
	2043	d	0.177	2047	43	8.0	1000	—
				2042	13	—	840	
$Co_4(CO)_{12}$	2065	≈8	0.504	2068	1.3	0.3	460	396 ± 67
				2066	6.5	0.3	373	
				2065	6.0	2.0	467	
	2055	≈12	0.564	2056	8.0	2.0	263	
				2056	13	—	140	—
$Rh_6(CO)_{16}$	2077	8	1.453	2085	10	2.0	613	698 ± 98
				2082	0.8	0.6	720	
				2076	240	21	240	
				2074	10	6.0	e	

[a] Peak mode frequencies (v_0), absorption bandwidths (Δv), and IR pulse frequencies (v_{IR}) are in wavenumbers (cm^{-1}). The peak absorption cross-section σ is defined as $\ln(T) = -\rho \sigma l$. n is the average number of pump laser photons absorbed per molecule. The recommended T_1 values are for conditions that the pump and probe pulses interacted only with the indicated fundamental.
[b] I pump-pulse fluence in units of mJ/cm^2.
[c] Overlapping band composed of v_2 and v_4 modes.
[d] Overlapping band composed of v_3 and v_5 modes.
[e] 100 ps bleach recovery followed by long time absorption.

relaxation step. The long T_1 relaxation times are again independent of the vibrational state density (at 2000 cm^{-1}) for these large polyatomic systems (reaching 10^{36} states/cm^{-1} for the Rh$_6$ compound).

For the transition metal carbonyl systems with multiple CO-stretching vibrational modes, it is anticipated that intermode coupling can occur. Evidence for intermode coupling may have been observed in Co$_4$(CO)$_{12}$. In this case, an initially rapid polarization-insensitive decay is observed, which suggests that equilibration between CO-stretching modes occurs *before* the coupled population ensemble relaxes on the hundreds of picoseconds timescale. Should this be true for the metal cluster-CO systems, it should be possible to observe initially rapid population flow into the unpumped CO-stretching modes with a tunable IR probe method. These modes, in turn, would be expected to relax with the same long time constant as the initially pumped mode.

The same molecules [except Rh$_4$(CO)$_{12}$] were also adsorbed onto the surface of a pressed silica disk and T_1 was measured (77). T_1 was found to decrease by exactly a factor of four (compared to the solution results above) for each molecule upon interacting with the SiO$_2$ support. The consistency between the two sets of solution and surface T_1 data suggest that coupling of the CO-stretching mode to the resonant SiO-stretching overtone of the support (at ca. 2000 cm^{-1}) is involved in enhancing the relaxation process (77).

CO(v = 1) Relaxation for CO Adsorbed on Silica-Supported Rh and Pt Particles

The single frequency pump-probe technique was also used to determine the T_1 timescale for the terminal CO-stretching mode of CO adsorbed on Rh and Pt particles supported on silica (85–87). These complex systems, although inhomogeneous in particle size distribution and surface morphology, have been extensively studied and reported in the literature (79–81, 88, 89). Our goal was to take advantage of many previous spectroscopic studies of these supported systems (88, 89) to determine whether new mechanisms for vibrational relaxation occur when the metal cluster size approaches that necessary to support electron-hole pairs (21–27).

Samples were prepared by impregnating the same silica used in the OH(v = 1) studies with solutions of appropriate metal salts, removing the solvent by evaporation, and then reducing the dried metal salt at elevated temperature with hydrogen gas. As in previous transmission infrared studies, slurries of silica with known concentrations of aqueous salt solutions were made, followed by drying, pressing the silica into disks, and reducing with hydrogen (79–82, 88, 89). Optically transparent samples were made by impregnating a pressed SiO$_2$ disk with water-acetone metal salt solutions, allowing the solvent to evaporate, followed by evacuation and subsequent reduction of the disk under H_2 at 700K. After cooling the disks to room temperature, large metal particles (ca. 30 Å diameter) were exposed to saturation coverage of CO gas.

Transmission FTIR spectra confirmed the presence of adsorbed CO from the appearance of strong CO-stretching bands in the 2000–2100 cm^{-1} region (85–87 and references therein). The absorptions near 2070 cm^{-1} in Figure 4 are associated with CO bound to large supported Rh and Pt particles. The Pt sample exhibits only one strong asymmetric absorption band, which can be analyzed in terms of CO adsorbed on various sites of the metal particles (85, 86, 86a). Supported Rh, however, yields three new absorption bands near 2045, 2100, and 2110 cm^{-1} (other than the central CO-particle absorption), which are attributed to an isolated $Rh_2(CO)_4$ species (87). This assignment was confirmed by comparing the spectrum of Figure 4 (silica absorption subtracted) to a similar spectrum obtained for $Rh_2(CO)_4Cl_2$ directly deposited on silica. Transmission electron microscopy indicated that the Pt particles fall in the 10–40 Å diameter range, whereas the average particle diameter for supported Rh was 36 ± 6 Å.

Initial T_1 measurements were made by using the single frequency

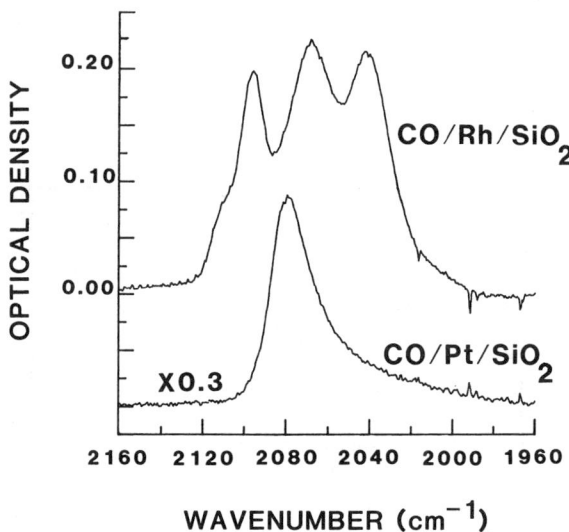

Figure 4 FTIR absorption spectra for CO/Pt/SiO$_2$ and CO/Rh/SiO$_2$ disk samples after subtraction of the silica support background spectrum. CO adsorbed on Pt and Rh particles produces the broad absorptions that peak at 2080 and 2070 cm^{-1}, respectively. The Rh sample also exhibits peaks at 2045 and 2100 with a shoulder at 2110 cm^{-1} that arises from isolated Rh$_2$(CO)$_4$ sites (see text).

approach with the apparatus shown in Figure 1. The measurements revealed that the CO(v = 1) mode for Pt particles relaxes extremely rapidly ($T_1 < 20$ ps), and when tuned to different IR frequencies, the transient response exhibited rapid bleaching recovery, *absorption*, or a mixture of these effects (85, 86). The transient absorption signals were attributed to overtone absorption phenomena, as was observed for $Rh_6(CO)_{16}$ (see above). Comparison of T_1 measured for CO adsorbed on the metal particles to that obtained for the metal-carbonyl complex and cluster compounds clearly shows that more rapid vibrational damping (by at least a factor of 20) occurs for CO on the metal particles.

Further tests were performed on the $CO/Rh/SiO_2$ system (see Figure 4) (87). When the IR pump-probe frequency was tuned to the $Rh_2(CO)_4$ absorptions, T_1 lifetimes of 140 ± 20 ps were measured. This T_1 result was consistent with $T_1 = 180 \pm 40$ ps measured for $Rh_2(CO)_4Cl_2$ deposited on silica. The band attributed to CO adsorbed on large Rh particles yielded very rapid decays, thus indicating that T_1 was again less than 20 ps (87). Apparently, the Rh/SiO_2 sample contains Rh sites that range from nearly isolated Rh_2 species to large (ca. 30–40 Å) particles that exhibit disparate CO(v = 1) relaxation times.

More recent measurements on the same CO-metal particle samples have been performed by using a CW modelocked Nd^{3+} : YAG synchronously pumped dual dye laser system with ca. 2 ps temporal resolution (87a). A bleaching recovery with 20:1 signal to noise (Figure 5) reveals an initial exponential relaxation of 7 ± 2 ps followed by a slower nonexponential tail. The decays for both Rh and Pt samples are consistent with the previous time-resolved data obtained with the longer IR pulses. Pt samples were most extensively studied with the 2 ps IR pulses. For Pt, T_1 was independent of pump-probe pulse polarization and pump fluence (2–12 mJ/cm^2) and sample temperature (100–400K). The same CO(v = 1) T_1 decay was measured for Pt particles saturated with pure $^{12}C^{16}O$ carbon monoxide or with an isotopic ratio $^{12}C^{16}O/^{12}C^{18}O = 1/3$.

For high surface coverages, rapid population exchange likely occurs between adsorbates with nearly degenerate vibrational mode energies. Strong dipolar coupling exists among CO oscillators adsorbed on metal particles or single crystal surfaces. For instance, a simple Förster dipole-dipole theory calculation (90–92) gives a rate of 2×10^{13} sec^{-1} for vibrational energy transfer between two CO adsorbates separated by a metal-metal bond distance of 3.0 Å. Unambiguous measurement of such exchange phenomena by the single frequency pump-probe technique is impossible. A narrowband tunable probe or broadband IR probing method (see TECHNIQUES section) may be able to detect population flow within an inhomogeneously broadened absorption band or inter-

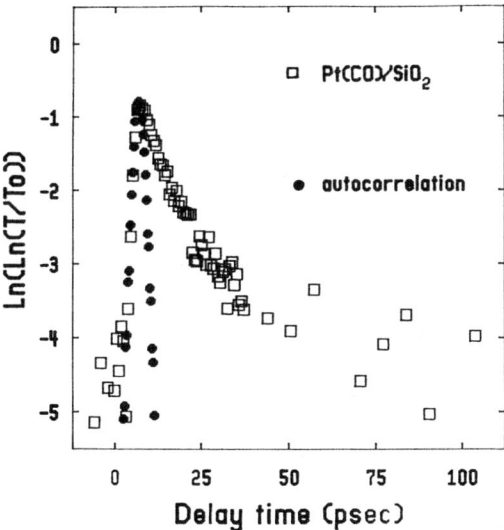

Figure 5 CO/Pt/SiO$_2$ high resolution (2 ps FWHM IR pulse duration from SHG autocorrelation) T_1 decay obtained for the IR tuned to 2090 cm^{-1} (see Figure 4). This nonexponential response is characterized by an initial decay constant of 7 ± 2 ps.

adsorbate energy flow by measuring changes in the transient infrared absorption spectrum.

These extremely fast T_1 results indicate that a new mechanism dominates the relaxation of CO adsorbed on 30 Å metal particles. Such observations are consistent with models for vibrational relaxation to electron-hole pair excitations of the metal; that is, the 2000 cm^{-1} CO(v = 1) quantum promotes a metal electron from below the Fermi level to an unoccupied state above the Fermi level (21–24, 30–32). This finding is in contrast to the model compounds capable only of supporting phonon mediated relaxation mechanisms because their HOMO-LUMO gap is much greater than the CO(v = 1) quantum. Further measurements are clearly required to identify whether the electron-hole pair or interadsorbate coupling mechanisms are responsible for the measured decay rates.

FUTURE DIRECTIONS AND SUMMARY

The measurements for surface adsorbate vibrational relaxation described in this review represent only the beginning of possible studies of vibrational energy transfer at surfaces. For the most part, the measurements made to date have focused on the relaxation of high frequency oscillators on supports with relatively low frequency fundamental modes. For these systems, the result that T_1 can be hundreds of picoseconds long and unrelated to frequency domain absorption bandwidths is important information not fully appreciated before these experiments were performed.

Comparison of surface-adsorbate systems to model compounds has been useful and necessary. For the OH and CO systems investigated, model systems have revealed fundamental mechanistic concepts and have provided insight to the measurements made for the more complicated adsorbate-surface samples. In the same sense that making analogies to model compounds has facilitated the assignment of vibrational modes of adsorbates (93–96), this notion pertains to vibrational dynamics studies as well. The interplay between model and surface systems will probably remain an important facet of future vibrational dynamics measurements.

Studies of vibrational relaxation at surfaces have been relatively rudimentary, using only single frequency infrared techniques. As mentioned above, checks of these measurements can be made by using tunable or broadband infrared probing methods. From such studies it will be possible to monitor directly energy flow out of the initially excited surface-adsorbate mode into other infrared active modes of the system. The ability to measure entire transient infrared spectra with picosecond and femtosecond time resolution should help identify complex vibrational energy transfer mechanisms. Initial studies of model compounds should shed insight into the important characteristics of intramode coupling. Eventually such techniques will be applied to adsorbate systems.

There seems to be a clear distinction between the vibration-phonon relaxation mechanism of adsorbates on dielectrics and the damping of adsorbate vibrations on metals. Even for the simple diatomic adsorbate CO on metal particles, the vastly different timescale for T_1 indicates that the relaxation mechanism for a molecule on a metal is substantially different from that for dielectrics. More detailed studies of phonon versus vibration-to-electronic relaxation are needed to clarify the importance of these mechanisms in adsorbate-metal systems (97, 98). An important step in this direction would be to examine adsorbates on metal single crystals. In these cases, as opposed to the ill-defined metal particle systems, the adsorbate coverage, orientation, and location can be better controlled and characterized for surfaces of demonstrable topography and purity. T_1 measurements utilizing new techniques for generating tunable infrared femtosecond pulses would help reveal the important relaxation pathways for a variety of adsorbates and for specific adsorbate modes as a function of coverage and temperature.

With key measurements of T_1 now available such as those described in the previous sections, it should be possible to assess the accuracy of theoretical models of vibrational relaxation and energy transfer phenomena at interfaces. We hope that these new measurements will inspire theorists to consider novel analytical approaches or computer molecular dynamics simulations for these and other complex adsorbate-surface systems.

Lastly, it is anticipated that as technical obstacles for making energy transfer measurements are eliminated, more researchers will become involved in these types of studies. As different systems and methods are explored, new concepts and detailed information about surface chemistry and related phenomena must ensue. It is hoped that as the field matures, direct time-resolved measurements of surface mediated chemistry and transient intermediates will reveal the details of complex molecular interactions and chemical transformations at solid interfaces.

ACKNOWLEDGMENTS

We are indebted to the Air Force Office of Scientific Research for providing partial financial support of this research and to all our colleagues for their continued interest and encouragement to perform this work. Special thanks are extended to Mr. Lyman Elwell for his development of the data acquisition facilities and to Dr. John Beckerle for the latest measurements on supported particle systems.

Literature Cited

1. Chabal, Y. 1988. *Surf. Sci. Rep.* 8: 211
2. Chuang, T. J. 1983. *Surf. Sci. Rep.* 3: 1 and references therein
3. Mak, C. H., Koehler, B. G., Brand, J. L., George, S. M. 1987. *J. Chem. Phys.* 87: 2340
4. Seebauer, E. G., Schmidt, L. D. 1986. *Chem. Phys. Lett.* 123: 129
5. Ready, J. F., ed. 1971. *Effect of High Power Laser Radiation*. New York: Academic
6. Eesley, G. L. 1983. *Phys. Rev. Lett.* 51: 2140
7. Beckerle, J. D., Yang, Q. Y., Johnson, A. D., Ceyer, S. T. 1987. *J. Chem. Phys.* 86: 7236
8. Ceyer, S. T., Beckerle, J. D., Lee, M. B., Tang, S. L., Yang, Q. Y., Hines, M. A. 1987. *J. Vac. Sci. Technol.* A5: 501
9. Beckerle, J. D., Johnson, A. D., Yang, Q. Y., Ceyer, S. T. 1988. *J. Vac. Sci. Technol.* A6: 903
10. Ceyer, S. T. 1988. *Annu. Rev. Phys. Chem.* 39: 479–510
11. Bradshaw, A. M., Conrad, H., eds. 1987. *Vibrations at Surfaces 1987*. New York: Elsevier
12. King, D. A., Richardson, N. V., Holloway, S., eds. 1986. *Vibrations at Surfaces 1985*. New York: Elsevier
13. Brundle, C. R., Morawitz, H., eds. 1983. *Vibrations at Surfaces*. New York: Elsevier
14. Bell, A. T., Hair, M. L., eds. 1980. *Vibrational Spectroscopies for Adsorbed Species*, ACS Symp. Ser. 137. Washington, D.C.: Am. Chem. Soc.
15. Yates, J. T. Jr., Madey, T. E., eds. 1987. *Vibrational Spectroscopy of Molecules on Surfaces*, Vol. 1. New York: Plenum
16. Sheppard, N. 1988. *Annu. Rev. Phys. Chem.* 39: 589–644
17. Tobin, R. G. 1987. *Surf. Sci.* 183: L226
18. Penzkofer, A., Laubereau, A., Kaiser, W. 1979. *Prog. Quan. Electr.* 6: 55
19. Laubereau, A., Kaiser, W. 1978. *Rev. Mod. Phys.* 50: 607
20. Seilmeier, A., Kaiser, W. 1988. *Ultrashort Laser Pulses Applic.* 60: 270–318
21. Gadzuk, J. W., Luntz, A. C. 1984. *Surf. Sci.* 144: 429
22. Persson, B. N. J. 1984. *J. Phys. C* 17: 4741
23. Persson, B. N. J. 1978. *J. Phys. C* 11: 4251

24. Chance, R. R., Prock, A., Silbey, R. 1978. In *Adv. Chem. Phys.*, ed. I. Prigogine, S. A. Rice, 37: 1. New York: Wiley. 70 pp.
25. Eguiluz, A. G. 1984. *Phys. Rev. B* 30: 4366
26. Zhdanov, V. P. 1988. *Surf. Sci.* 197: 35
27. van Smaalen, S., George, T. F. 1987. *J. Chem. Phys.* 87: 5504
28. Ariyasu, J. C., Mills, D. L., Lloyd, K. G., Hemminger, J. C. 1984. *Phys. Rev. B* 30: 507
29. Hutchinson, M., George, T. F. 1986. *Chem. Phys. Lett.* 124: 211
30. Gadzuk, J. W. 1981. *Chem. Phys. Lett.* 80: 5
31. Langreth, D. C. 1985. *Phys. Rev. Lett.* 54: 126
32. Zhdanov, V. P. 1988. *Surf. Sci.* 201: 461
33. Zinth, W., Holzapfel, W., Leonhardt, R. 1988. *Ultrafast Phenomena VI* 48: 461
34. Heilweil, E. J., Casassa, M. P., Cavanagh, R. R., Stephenson, J. C. 1985. *J. Chem. Phys.* 82: 5216
35. Heilweil, E. J., Casassa, M. P., Cavanagh, R. R., Stephenson, J. C. 1985. *Ultrashort Pulse Spectrosc. Applic.* 533: 15
36. Graener, H., Dohlus, R., Laubereau, A. 1987. *Proc. Int. Conf. Time-Resolved Vibrational Spectrosc.*, TRVS 1987. Amersfoort: The Netherlands. 24 pp. (Abstr.)
37. Heilweil, E. J., Cavanagh, R. R., Stephenson, J. C. 1987. *Chem. Phys. Lett.* 134: 181
38. Casassa, M. P., Stephenson, J. C., King, D. S. 1988. *J. Chem. Phys.* 89: 1966
39. Moore, C. A., Goldberg, L. S. 1976. *Opt. Comm.* 16: 21
40. Goldberg, L. S. 1975. *Appl. Opt.* 14: 653
41. Cotter, D., White, K. I. 1984. *Opt. Comm.* 49: 205
42. Koenig, R., Rosenfeld, A., Tam, N.-H., Mory, S. 1978. *Opt. Comm.* 24: 190
43. Campion, A. 1987. In *Vibrational Spectroscopy of Molecules on Surfaces, Methods of Surface Characterization*, 1: 345–415. New York: Plenum. 468 pp.
44. Zinth, W., Kolmeder, C., Benna, B., Irgens-Defregger, A., Fischer, S. F., Kaiser, W. 1983. *J. Chem. Phys.* 78: 3916
45. Kaiser, W., Bauerle, R. J., Elaesser, T., Hubner, H.-J., Seilmeier, A. 1988. *Ultrafast Phenomena VI* 48: 452
46. Ambroseo, J. R., Hochstrasser, R. M. 1988. *J. Chem. Phys.* 89: 5956
47. Graener, H., Laubereau, A. 1987. *Chem. Phys. Lett.* 133: 378
47a. Graener, H., Ye, T.-Q., Laubereau, A. 1989. *J. Chem. Phys.* 90: 3413
48. Graener, H., Ye, T.-Q., Dohlus, R., Laubereau, A. 1988. *Ultrafast Phenomena VI* 48: 458
49. Moore, J. N., Hansen, P. A., Hochstrasser, R. M. 1987. *Chem. Phys. Lett.* 138: 110
50. Hansen, P. A., Moore, J. N., Hochstrasser, R. M. 1988. In *Proc. Symp. Nonlinear Opt. Ultrafast Phenom., 174th Meet. Electrochem. Soc.* (Abstr.)
51. Heilweil, E. J. 1989. *Opt. Lett.* 14: In press
51a. Glownia, J. H., Misewich, J., Sorokin, P. P. 1987. *Opt. Lett.* 12: 19
52. Greene, B. I., Hochstrasser, R. M., Weisman, R. B. 1979. *J. Chem. Phys.* 70: 1247
52a. Shen, Y. R. 1989. *Annu. Rev. Phys. Chem.* 40: 327–50
53. Heilweil, E. J., Casassa, M. P., Cavanagh, R. R., Stephenson, J. C. 1984. *J. Chem. Phys.* 81: 2856
54. Heilweil, E. J., Casassa, M. P., Cavanagh, R. R., Stephenson, J. C. 1985. *J. Vac. Sci. Technol. B* 3: 1471
55. Casassa, M. P., Heilweil, E. J., Stephenson, J. C., Cavanagh, R. R. 1985. *J. Vac. Sci. Technol. A* 3: 1655
56. McDonald, R. S. 1958. *J. Phys. Chem.* 62: 1168
57. Casassa, M. P., Heilweil, E. J., Stephenson, J. C., Cavanagh, R. R. 1986. *J. Chem. Phys.* 84: 2361
58. Risenberg, L. A., Moos, H. W. 1968. *Phys. Rev.* 174: 429
59. Englman, R. 1979. In *Nonradiative Decay of Ions and Molecules in Solids*, Ch. 5–7. Amsterdam: North-Holland
60. Nitzan, A., Mukamel, S., Jortner, J. 1974. *J. Chem. Phys.* 60: 3929
61. Heilweil, E. J., Casassa, M. P., Cavanagh, R. R., Stephenson, J. C. 1986. *J. Chem. Phys.* 85: 5004
62. Heilweil, E. J., Casassa, M. P., Cavanagh, R. R., Stephenson, J. C. 1985. *Chem. Phys. Lett.* 117: 185
63. Heilweil, E. J., Casassa, M. P., Cavanagh, R. R., Stephenson, J. C. 1986. *Ultrafast Phenomena V* 46: 465
64. Heilweil, E. J. 1986. *Chem. Phys. Lett.* 129: 48
65. Landolt, H. H., Bornstein, R. 1976. *Struct. Data Free Polyatomic Mol.*, Vol. 7. (New Ser.) New York: Springer-Verlag
66. Korppi-Tommola, J. 1978. *Spectrochim. Acta A* 34: 1077
67. Fendt, A., Fischer, S. F., Kaiser, W. 1981. *Chem. Phys.* 57: 55
68. Tully, J. C., Chabal, Y. J., Raghavachari, K., Bowman, J. M., Lucchese, R. R. 1985. *Phys. Rev. B* 31: 1184
69. Trenary, M., Uram, K. J., Bozso, F.,

Yates, J. T. Jr. 1984. *Surf. Sci.* 147: 269
70. Persson, B. N. J., Ryberg, R. 1985. *Phys. Rev. Lett.* 54: 2119
71. Smith, J. M., Jones, L. H. 1966. *J. Mol. Spectrosc.* 20: 248
72. Jones, L. H., McDowell, R. S., Goldblatt, M., Swanson, B. I. 1972. *J. Chem. Phys.* 57: 2050
73. Jones, L. H., McDowell, R. S., Goldblatt, M. 1969. *Inorg. Chem.* 8: 2349
74. Bor, G., Sbrignadello, G., Noack, K. 1975. *Helv. Chim. Acta* 58: 815
75. Bor, G., Sbrignadello, G., Marcati, F. 1972. *J. Organomet. Chem.* 46: 257
76. Heilweil, E. J., Cavanagh, R. R., Stephenson, J. C. 1988. *J. Chem. Phys.* 89: 230
77. Heilweil, E. J., Stephenson, J. C., Cavanagh, R. R. 1988. *J. Phys. Chem.* 92: 6099
78. Heilweil, E. J., Cavanagh, R. R., Stephenson, J. C. 1988. *Ultrafast Phenomena VI* 48: 447
79. Imelick, B., Naccache, C., Courdurier, G., Praliaud, H., Meriaudeau, P., Gallezot, P., Martin, G. A., Verdine, J. C., eds. 1982. In *Metal-Support and Metal-Additive Effects in Catalysis*. New York: Elsevier
80. Stevenson, S. A., Dumesic, J. A., Baker, R. T. K., Ruckenstein, E., eds. 1987. In *Metal-Support Interactions in Catalysis, Sintering and Redispersion.* New York: Van Nostrand Rheinhold. 315 pp.
81. Geus, J. W., Wells, P. B. 1985. *Appl. Catal.* 18: 231
82. Wells, P. B. 1985. *Appl. Catal.* 18: 259
83. Theolier, A., Smith, A. K., Leconte, M., Basset, J. M., Zanderighi, G. M., Psaro, R., Ugo, R. 1980. *J. Organomet. Chem.* 191: 415
84. Smith, G. C., Chojnacki, T. P., Dasgupta, S. R., Iwatate, K., Watters, K. L. 1975. *Inorg. Chem.* 14: 1419
85. Cavanagh, R. R., Heilweil, E. J., Stephenson, J. C. 1987. *J. Electron Spectrosc. Relat. Phenom.* 45: 31
86. Cavanagh, R. R., Casassa, M. P., Heilweil, E. J., Stephenson, J. C. 1987. *J. Vac. Sci. Tech. A* 5: 469
86a. Hayden, B. E. 1987. See Ref. 15, p. 267
87. Heilweil, E. J., Cavanagh, R. R., Stephenson, J. C. 1988. *J. Chem. Phys.* 89: 5342
87a. Beckerle, J. D., Casassa, M. P. Cavanagh, R. R., Heilweil, E. J., Stephenson, J. C. 1989. *J. Chem. Phys.* 90: 4619
88. Yates, J. T. Jr., Duncan, T. M., Worley, S. D., Vaughan, R. W. 1979. *J. Chem. Phys.* 70: 1219
89. Robbins, J. L. 1986. *J. Phys. Chem.* 90: 3381
90. Förster, T. 1965. *Mod. Quantum Chem.* 3: 93
91. Förster, T. 1969. *Discuss. Faraday Soc.* 27: 7
92. Yardley, J. T. 1980. *Introduction to Molecular Energy Transfer.* New York: Academic. 260 pp.
93. Muetterties, E. L., Rhodin, T. N., Band, E., Brucker, C. F., Pretzer, W. R. 1979. *Chem. Rev.* 79: 91
94. Whyman, R. 1980. In *Transition Metal Clusters,* ed. B. F. G. Johnson. New York: Wiley. 545 pp.
95. Greenler, R. G., Burch, K. D., Kretzschmar, K., Klauser, R., Bradshaw, A. M. 1985. *Surf. Sci.* 152/153: 338
96. Sheppard, N. 1986. *J. Elect. Spectrosc. Relat. Phenom.* 38: 175
97. Wood, D. M., Ashcroft, N. W. 1982. *Phys. Rev. B* 25: 6255
98. Ekardt, W., Penzar, Z. 1986. *Phys. Rev. B* 34: 8444

FUNDAMENTAL MECHANISMS OF DESORPTION AND FRAGMENTATION INDUCED BY ELECTRONIC TRANSITIONS AT SURFACES

Phaedon Avouris and Robert E. Walkup

IBM Research Division, IBM Thomas J. Watson Research Center, Yorktown Heights, New York 10598

Introduction

Photons or energetic particles (e^-, H^+, He^+) can cause desorption and fragmentation at surfaces by inducing electronic transitions to dissociative states. The study of such processes is an important area in surface chemistry and physics, with many implications in basic science and technology. The scientific issues include the nature of chemical bonding at surfaces in both the ground and excited states, surface dynamical processes involving charge or energy transfer, interactions among adsorbates, and the conversion of electronic potential energy into nuclear motion. Desorption induced by electronic transitions (DIET) is widely encountered in nature and in the laboratory. For example, the surfaces of materials in the solar system and the interstellar media are exposed to energetic photons and particles that stimulate desorption processes (1). In the laboratory, DIET processes occur in almost every system involving the impact of energetic photons or charged particles on solid surfaces. By monitoring the desorbed particles, one can obtain insight into the initial binding geometry and detailed information about the relevant dynamical processes. The geometric information is reflected in the angular distributions of the desorbed particles. The experimental technique of ESDIAD (electron-stimulated-desorption-ion-angular-distribution), developed by Madey, Yates, and coworkers (2, 3), has been successfully used to examine the geometry of a variety of molecular adsorbates. Further information on this interesting subject is discussed in the literature (4–6). The processes of stimulated desorption and fragmentation have some unique characteristics that make them valuable for a variety of applications. For example, one can provide nonthermal energy to control surface reactions with very high spatial resolution by using focused electron or photon beams. Materials growth, modification, and patterning with such methods is a very active research area (7–9). DIET processes can also be a factor in electron microscopy and in surface analytical techniques such as photo-emission, Auger, and electron-energy-loss spectroscopies, low-energy electron diffraction, etc (10). In this case, desorption or fragmentation may be an unwanted side-effect caused by probing the surface with photons or electrons. On a more macroscopic scale, DIET processes play a role in plasma-wall interactions (11), in the operation of synchrotrons (12), and more generally in the stability of materials subjected to various forms of radiation.

In this review we focus on basic mechanisms for desorption and molecular fragmentation at surfaces. First we discuss the general model introduced by Menzel, Gomer (13), and Redhead (14). Several specific examples that illustrate key ingredients of this model are considered in detail. Then we discuss the competing processes of quenching at surfaces and nuclear motion. Next we discuss in more detail desorption/fragmentation due to several classes of electronic transitions including core-hole formation followed by Auger decay, negative ion formation, and above-gap excitation in semiconductors and insulators. The examples that we have chosen are illustrative, but they are by no means exhaustive. Fortunately, the area of desorption induced by electronic transitions is well-served by a number of recent books (5, 15, 16) and review articles (17–21). We refer the reader to these sources for additional information on this subject.

The Menzel-Gomer-Redhead Model

Menzel, Gomer (13), and Redhead (14) proposed a general model of desorption via a sudden (Franck-Condon) electronic transition to a repulsive state. In this respect, the Menzel-Gomer-Redhead (MGR) model is analogous to direct photo-dissociation of free molecules. However, in the case of desorption, there are new and efficient decay channels provided by the surface. After the initial electronic excitation, the adsorbed species undergoes nuclear motion on the repulsive excited-state potential energy (PE) surface. Quenching of the excitation brings the adsorbate back to the ground state, and the electronic energy is converted into substrate excitation. In the MGR model it is assumed that this substrate excitation does not influence the substrate-adsorbate bonding. The MGR model is schematically illustrated in Figure 1. Quenching of the excitation will lead

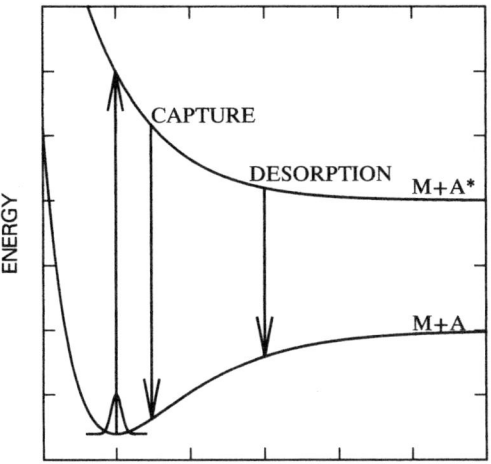

Figure 1 Schematic potential energy curves illustrating stimulated desorption by the Menzel-Gomer-Redhead model. A Franck-Condon transition from the ground state takes the system to a repulsive excited state. Quenching of the adsorbate excitation is assumed to return the system to a replica of the ground-state curve. Quenching can lead to either capture or desorption, depending on where the quenching transition occurs. The potential curves approach asymptotes labeled by the state of the metal (M) and the state of the adsorbate (A or A*).

to recapture of the adsorbate, unless it has already acquired an amount of kinetic energy sufficient to result in desorption on the ground-state PE curve. The ground-state desorption yield is thus determined by the cross section for the initial electronic excitation times the probability that sufficient kinetic energy will be gained on the excited-state curve before a quenching transition takes place. Desorption along the excited-state curve is possible only if no quenching transitions occur. Since quenching processes at surfaces can be very efficient, the net cross section for desorption will generally be substantially smaller than the analogous gas-phase cross section. The MGR model for desorption can be broadly interpreted as a general scenario for desorption induced by electronic transitions. The specific nature of the repulsive states, and the relevant quenching processes, are not specified. Because of its general nature, the MGR model provides a good starting point for a discussion of a wide range of desorption processes, from the "traditional" electron-stimulated desorption to the more recent laser-induced desorption studies. For this reason, we examine in more detail the ingredients of the MGR model, including (*a*) the nature of the electronic excitations that lead to repulsive states, (b) the nature of the decay channels provided by the surface, and (c) motion of the adsorbate-substrate system.

Desorption by Valence Excitation

The simplest electronic transitions that may result in stimulated desorption are excitations of the valence electrons, in particular, bonding to antibonding transitions. Despite the fact that bonding → antibonding transitions are often invoked in general discussions of stimulated desorption, there are relatively few specific examples where such transitions are effective. One notable example is the electron-stimulated desorption of CO from W (22, 23), which shows a desorption threshold at ~ 5 eV. At these relatively low energies, only valence excitations (or possibly valence ionization) could occur. The interaction of CO with a metal surface is generally discussed from a molecular-orbital point of view, in which the behavior of the lowest unoccupied orbital of the isolated molecule—the $2\pi^*$ orbital—plays a crucial role. The interaction of this orbital with the metal results in two new broadened levels with $2\pi^*$ character. As far as the metal-CO bond is concerned, the lower energy one, $2\pi_b^*$, has bonding character, whereas the higher energy one, $2\pi_a^*$, is antibonding. The energies of the $2\pi_b^*$ and $2\pi_a^*$ orbitals have been determined in several metal-CO chemisorption systems by using metastable de-excitation and photoemission spectroscopies for $2\pi_b^*$, and inverse-photoemission for $2\pi_a^*$. With the energies obtained from these measurements, the ~ 5 eV threshold can be assigned to the $2\pi_b^* \rightarrow 2\pi_a^*$ bonding to anti-bonding transition (24). Note that simple valence excitations, i.e. bonding to anti-bonding transitions, are often not effective at inducing desorption of particles from metal surfaces because the quenching processes can be so rapid that essentially no nuclear motion occurs before the excited state is quenched. We discuss excited-state lifetimes and quenching processes in a later section.

Desorption by Valence Ionization

Ionization of valence electrons can, in principle, result in the desorption of either positive ions or neutral particles. In fact, a wide variety of desorption mechanisms have been identified, depending on the specific nature of the relevant potential energy surfaces. For some chemisorbed species, valence ionization directly leads to repulsive states (25, 26), but for physisorbed species, the ionized adsorbate is attracted toward the surface (27) by the electrostatic image potential that dominates the interaction. For molecular adsorbates, the orientation of the molecule provides an additional degree of freedom which can play an important role (28). In the following section, we illustrate this variety of desorption processes following valence ionization, using examples from the recent literature.

A useful model system for strongly bound adsorbates was provided by recent density functional calculations of fluorine adsorbed on aluminum by Avouris et al (25, 26). The relevant potential energy curves are shown in Figure 2. The fluorine atom adsorbs on the metal as a negative ion, with an effective electonic configuration of $F^-(2p^6)$. The bonding in the ground state (curve A) can be described as a balance between an image-like attraction and short-range electron-electron repulsion. Ionization of the fluorine

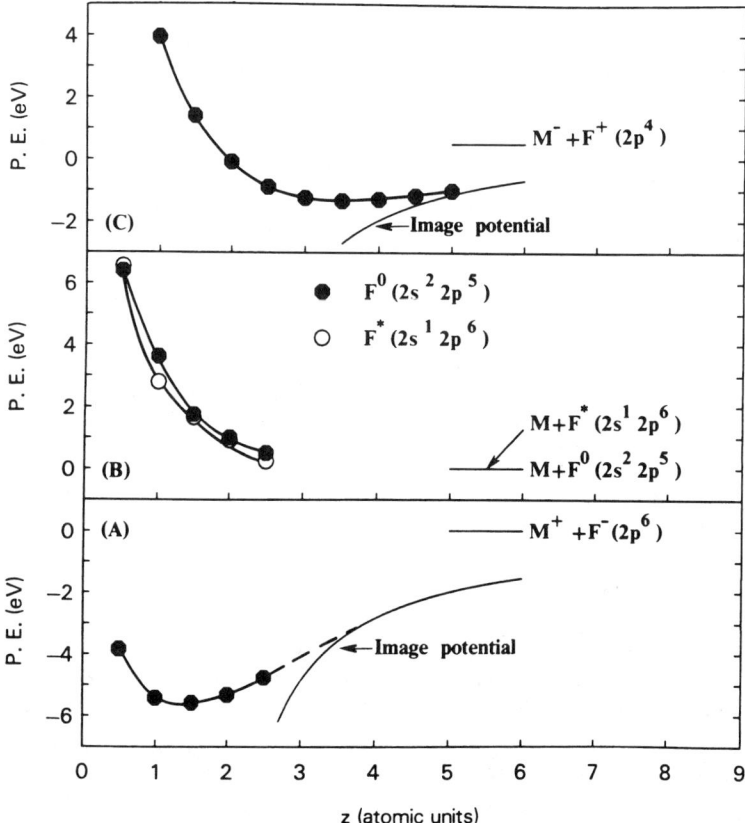

Figure 2 Local-density-functional calculations of the interaction of atomic fluorine with an aluminum surface. (A) The ground-state of chemisorbed fluorine $F^-(2p^6)$ on Al. (B) The neutral states resulting from valence ionization $F(2p^5)$ (*filled circles*) and core ionization $F(2s^1 2p^6)$ (*open circles*). (C) The positive ion state $F^+(2p^4)$. Adapted from Ref. (25).

(2p) level results in a repulsive neutral state (curve B) because the image attraction has been removed. For the neutral state, the fluorine (2p) level is located below the aluminum valence band, thus resonant electron tunnelling is blocked. Only the less efficient Auger neutralization process (29) can return the fluorine atom to the negative ion state. As a result, the desorption yield for neutral fluorine should be substantial. Note also that since the bonding in the ground state is due to the negative charge on fluorine, ionization of any level (not just the 2p valence level) will remove the negative charge and lead to a repulsive state. Figure 2(B) shows the repulsive PE curve resulting from F(2s) ionization.

The potential energy surface for F^+ interacting with the metal was also investigated (25, 26), and it illustrates the very important role of charge-transfer between the metal and the adsorbate. To use very rough arguments, one may expect F^+ to be more strongly bound than F^- and to have a smaller equilibrium separation from the surface; there is an image-like attraction in both cases, and the F^+ ion is expected to be "smaller", thus suggesting less short-range repulsion. However, the density functional calculations show just the opposite result (Figure 2 curve C). The $F^+(2p^4)$ ion is ~2.5 times less strongly bound than F^-, and the minimum in the atom-surface potential occurs at greater distances from the surface for F^+. For the negative ion state, the screening is image-like, and the electronic configuration is nominally $F^-(2p^6)$. However, for the positive ions state, the low-lying fluorine 3s and 3p levels are pulled down in energy by the presence of the 2p hole. These broadened 3s and 3p levels now extend partly below the Fermi level, and are thus partially occupied by electrons from the metal. Charge-density difference plots illustrating this behavior are shown in Figure 3. This charge-transfer screening has two very important implications for positive ion desorption: (a) the net charge on the adsorbate is reduced, thus reducing the image-like attraction, and (b) the partial occupation of the fluorine 3s and 3p levels results in increased repulsion between the adsorbate and the metal. The repulsion is increased over that for a negative ion for two reasons. First, repulsion arises because of substantial overlap between the large radius 3s and 3p orbitals and high electron density regions of the metal. The charge-transfer screened positive ion is actually "larger" than a negative fluorine ion—a reversal of the situation for free ions. Second, the image charge that screens a positive ion corresponds to a local increase in metal electron density, while the opposite is true for negative ions. This enhances the short-range repulsion for positive ions. Because of the increased repulsion and decreased image attraction, a Franck-Condon transition from the ground state (F^-) populates overall repulsive regions of the charge-transfer screened F^+ PE curve, thus providing the driving force for F^+ desorption.

An additional example of desorption by valence ionization of strongly

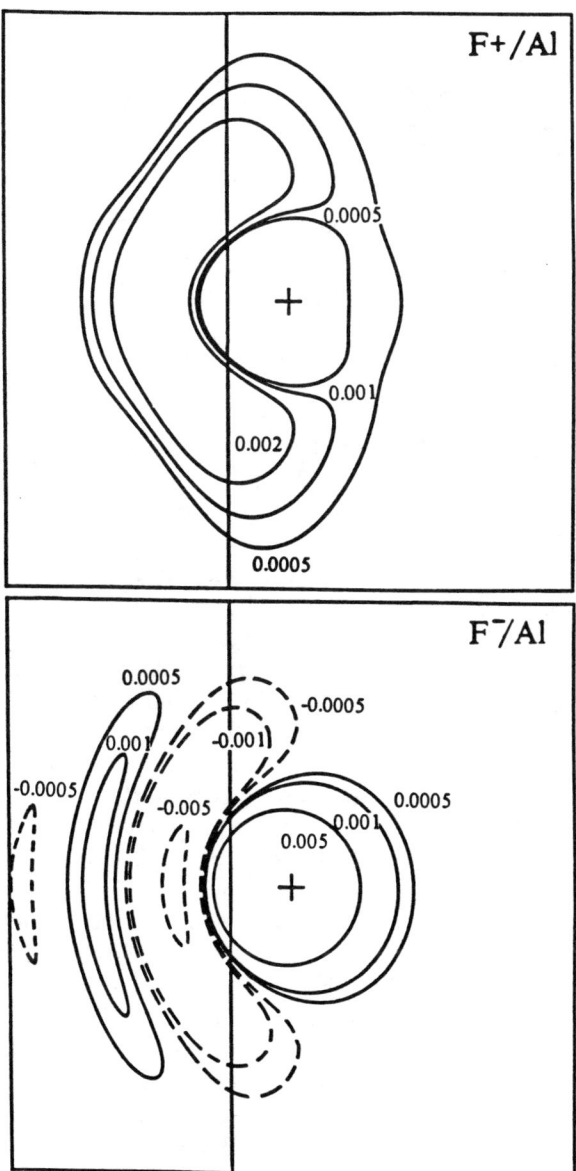

Figure 3 The *lower panel* shows the change in electron density resulting from chemisorption of fluorine on aluminum $\Delta\rho = \rho(F/Al) - \rho(F) - \rho(Al)$. *Solid lines* indicate increased and *dotted lines* indicate decreased electron density. The *vertical line* indicates the location of the jellium positive background edge. The increased electron density around the fluorine atom and the decreased electron density on the metal illustrates the formation of an ionic bond. The *upper panel* shows the electron density difference for the positive ion state $\Delta\rho = \rho(F^+/Al) - \rho(F^+) - \rho(Al)$ at the equilibrium distance for the ground state. The F^+ ion acquires screening charge from the metal in large radius $(3s, 3p)$ orbitals. The density is in atomic units. Adapted from Ref. (25).

bound species is provided by CO chemisorbed on metals. Ionization of the 5σ level of chemisorbed CO leads to both neutral CO and CO^+ desorption (30), and experimental evidence indicates that there is a substantial yield of neutral CO resulting from neutralization of CO^+ (30). The 5σ level is located below the bottom of the metal valence band, thus resonance tunnelling is blocked, and the lifetime of the $5\sigma^{-1}$ state should be reasonably long, limited by Auger neutralization. As in the case of fluorine on aluminum discussed above, charge-transfer (CT) screening is expected to play an important role. Upon 5σ ionization of CO, charge transfer from the metal into the $2\pi^*$ orbital is expected. Such CT screening of the $5\sigma^{-1}$ state is indicated by the fact that the 5σ binding energy and the $5\sigma \to 2\pi^*$ excitation energies are essentially the same (24). CT-screening reduces the image attraction, and by populating the metal-CO antibonding level, increases the short-range repulsion. Thus 5σ ionization of the ground state is expected to populate overall repulsive regions of the positive ion PE curve. Evidence that the $5\sigma \to 2\pi^*$ transition leads to a repulsive state from which neutral CO desorbs has been presented by Rubio et al (31). Core-hole states are also generally accepted to be CT-screened (32–37). The CT-screening process basically implies a mixing between ionic and excited neutral states. In principle, ions and excited-state neutral molecules can directly desorb as the result of valence ionization, with a branching ratio that depends on the details of the excited-state and positive ion PE curves. In fact, long-lived electronically excited neutral molecules have been detected in electron-stimulated desorption of chemisorbed CO (38–40), and assigned as most likely arising from the triplet-coupled state ($^3\Pi$) formed by the $5\sigma \to 2\pi^*$ transition. Ground-state neutral molecules can desorb after Auger neutralization of the original hole, provided sufficient kinetic energy has been gained. Molecules produced by distinct desorption pathways are expected to have different kinetic/internal energy and angular distributions. Such effects have been observed for CO, CO*, and CO^+ (39, 40).

A different situation prevails for physisorbed species. In this case, the ground state is essentially neutral, with a shallow van der Waals potential well located at relatively large distances from the surface. Upon valence ionization, the adsorbate will be attracted *toward* the surface by image-like interactions. This attraction leads to a mechanism for neutral atom desorption, which was first pointed out by Antoniewicz (27). In this mechanism, if sufficient kinetic energy is gained by motion on the *attractive* positive ion PE surface, then upon neutralization of the positive ion, the neutral atom is free to desorb after reflection from the surface. Note that single valence ionization cannot lead to positive ion desorption by the Antoniewicz mechanism, because the positive ion is bound by the image-like potential. There is strong evidence in favor of the Antoniewicz mechanism for desorption of noble gas atoms from metal surfaces (41, 42) and for desorption of weakly bound molecules on metals such as N_2O on Ru (43, 44).

Recent density functional calculations (P. Avouris et al, unpublished) show that charge-transfer screening is also very important in the case of physisorbed noble gas atoms. The effect of this screening is to substantially modify the positive ion PE curve compared to the "bare" image potential, $V_{\text{Image}} \simeq -e^2/[4(z-z_0)]$, where $z-z_0$ is the distance from the image plane (45). Calculations for $Ar(3p^6)$ and $Ar^+(3p^5)$ on Ag show that for Ar^+, the 4s orbital is pulled down in energy by the presence of the 3p hole, and becomes partially occupied by metal electrons. The 3p hole is located below the bottom of the Ag valence band, thus this hole cannot be filled by resonant tunnelling, and only the less efficient Auger neutralization process can occur. The partial occupation of the Ar^+ 4s level results in a weakened image attraction and additional short-range repulsion. As a result, the Ar^+ potential curve remains attractive, but it is much less attractive than a bare image potential, as shown in Figure 4. The precise

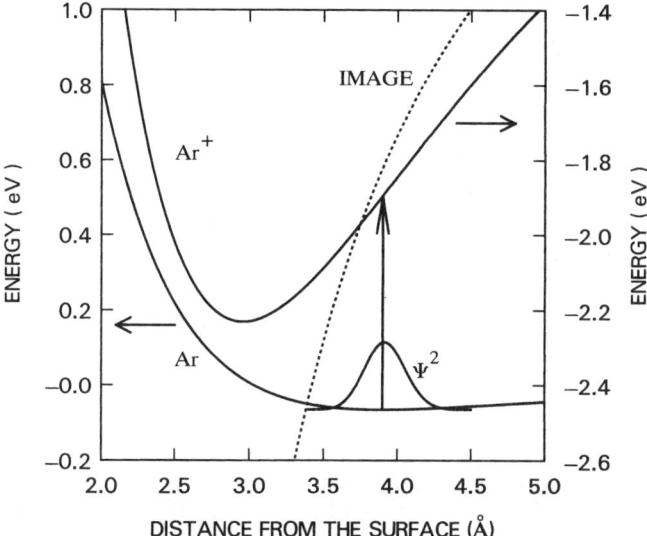

Figure 4 Potential energy curves for Ar and Ar^+ on a silver surface. The *solid curves* are the results of local density functional calculations. An image potential (*dotted line*) is also shown for comparison. For convenience, the Ar^+ and image potentials are shifted down in energy by the Ar ionization potential minus the metal work function. The probability distribution for Ar in the shallow ground-state physisorption well is indicated. Ionization of Ar results in motion toward the surface, which leads to neutral atom desorption via the Antoniewicz mechanism.

form of the positive ion PE curve is a key ingredient in theoretical models of the yield and kinetic energy distribution of neutral atoms desorbed by the Antoniewicz mechanism (44). Previous descriptions have generally used a bare image potential added to the ground-state curve for the positive ion PE curve. Recent work (P. Avouris et al, unpublished) indicates that the use of more realistic CT-screened positive ion PE curves obtained by density functional theory results in an improved description of the desorption process.

A recent example of desorption due to valence ionization involves an additional degree of freedom—the orientation of a molecule. Burns et al (28, 46, 47) have carried out detailed electron-stimulated desorption experiments on NO adsorbed on Pt(111) by using quantum-level specific laser detection of the desorbed molecules. A desorption threshold of ~ 6 eV was experimentally observed, and it was shown that only neutral NO desorbs in this low electron-energy region. The proposed mechanism involves ionization of the NO 5σ level. The resulting positive ion is CT screened to yield an effectively neutral NO molecule with the electronic configuration of NO*$(5\sigma^1 2\pi^{*2})$. This state is expected to be reasonably long-lived because the 5σ level is located below the bottom of the metal valence band, thus the 5σ hole can only be filled by the Auger neutralization mechanism. In the "equivalent binding" picture proposed by Burns et al, this CT-screened state should behave similarly to O_2 on Pt(111). O_2 is relatively weakly bound to Pt(111) (~ 0.3 eV binding energy) and the preferred orientation is lying down, in contrast to the upright, nitrogen-end-down orientation of ground-state NO. Thus upon 5σ ionization, the CT-screened positive ion state is expected to remain weakly bound to the surface, and the orientation of the molecule will change as time evolves. Substantial rotational energy is provided by the zero-point energy of the initially hindered ground-state rotor. If the Auger neutralization rate is sufficiently small, the molecule can rotate substantially (~ 1 radian) before neutralization occurs. Upon neutralization, the rotated ground-state molecule may be in an overall repulsive configuration, leading to desorption.

Support for this rotationally mediated desorption mechanism was provided by comparison between theoretical predictions and the experimental rotational and kinetic energy distributions (28, 46, 48). The calculations used a two-dimensional wavepacket propagation method, where the two relevant dimensions are the distance of the center-of-mass of the molecule from the surface, z, and the polar angle indicating the orientation of the molecule, θ. A simple model potential was assumed based on the "equivalent binding" idea. This potential was taken to be weakly bound in the z direction, in analogy to O_2 on Pt(111), and was assumed to be independent of the polar angle, θ. Using this model PE surface, the calculated rotational distributions were found to be in good agreement with experiment. There is preferential desorption of the large angular momentum components. This leads to an effective rotational temperature of the desorbed molecules of ~500K, which is larger than the zero-point energy ($\simeq 330$K). The predicted kinetic energy distribution peaks at relatively low energies (~600K), in reasonable agreement with experiment.

It would be interesting for future studies to examine several aspects of NO desorption from metal surfaces in more detail. Avouris et al (49) have proposed an alternative to the "equivalent binding" picture, in which electron correlation in the $2\pi^*$ level plays an important role in localizing the desorption-active NO*($5\sigma^1 2\pi^2$) state. For the NO/Pt(111) system, the experimental data indicate that more than one desorption process is active. For example, the vibrationally excited NO molecules (v = 1, 2, 3) have substantially higher kinetic energy components than the v = 0 molecules (47), and these fast molecules appear to have a different dependence of yield on electron energy than that observed for the slow component. This has been attributed to a possible role of multiple-hole states (47). Laser-induced desorption of NO from an oxygen-covered Ni(100) surface was reported by Budde et al (50). The laser wavelength was 193 nm (i.e. 6.4 eV photons), and a quantum level-specific detection method was employed. The authors observed both a thermal and a non-thermal desorption channel. An interesting characteristic of the nonthermal channel was the pronounced selectivity in spin-orbit states—there was a strong underpopulation of the $^2\Pi_{3/2}$ state for low rotational levels. A detailed desorption mechanism was not determined, but it should be noted that the laser could resonantly excite a specific electronic transition observed by an EELS study of NO on Ni(100) (51), probably a $2\pi_b^* \rightarrow 2\pi_a^*$ transition. In other recent studies (52), NO has been desorbed from metals by using lasers with a variety of photon energies below the electronic absorption spectrum of chemisorbed NO. Hot electrons have been invoked in this case, and the mechanism is discussed in more detail in a later section. Given the wealth of experimental data available, NO desorption from metals should continue to be an active testing ground for desorption mechanisms.

Fragmentation at Surfaces

In addition to excitations involving the substrate-adsorbate bond, interest has recently been focused on intra-adsorbate excitations that lead to fragmentation of an adsorbed molecule. The role of the substrate in this process is of particular interest. Domen & Chuang (53) have studied the photolysis of CH_2I_2 on surfaces of Ag and Al_2O_3. For the UV wavelengths employed (308 nm), the gas-phase photolysis $\hbar\omega + CH_2I_2 \rightarrow CH_2I + I$ has a quantum yield of unity. For sub-monolayer coverages of CH_2I_2 on Al_2O_3, the quantum yield was measured to be ~ 0.01–0.1, indicating the presence of a fast nonradiative decay channel that can compete with the direct photodissociation process. On Ag surfaces, no fragmentation was detected at sub-monolayer coverages, thus indicating a very efficient quenching process, as might be expected for metals. However, photolysis was possible from adsorbed multilayers on both substrates. For coverages above ~ 15 layers, a new "explosive" desorption process was observed in which large amounts of CH_2I_2 desorb in addition to the direct photolysis products. This "explosion" is due to the high density of excitation that can be produced by high fluence laser irradiation. The photolysis of CH_3Br has also been studied on both insulator (54, 55) and metal (56) surfaces. For CH_3Br on LiF, Bourdon et al (54, 55) observed photolysis products and also the desorption of parent molecules. Parent molecule desorption was attributed to the formation of a shock wave due to light absorption in the crystal. Costello et al (56) observed photo-induced dissociation of CH_3Br on Pt(111), even for sub-monolayer coverages. They interpreted their results as indicating direct photodissociation at the metal surface. Marsh et al (57) reported analogous measurements for CH_3Cl adsorbed on Ni(111). These authors presented evidence for a dissociative electron attachment process involving photo-excited metal electrons. This process is discussed in more detail in a later section.

Quenching Processes and the Lifetime of Excited States

The lifetime of excited or ionized states, and the mechanisms for quenching of the excitation by the substrate, are extremely important issues in desorption and fragmentation induced by electronic transitions. The quenching mechanisms of primary concern are resonant electron tunnelling, Auger neutralization (or deexcitation), and the generation of electron-hole pairs by field coupling. In the following section we discuss these processes, with an emphasis on the issues relevant to desorption. Additional information is available in a number of discussions in the literature (29, 58–61).

A good starting point for the description of the adsorbate-metal interactions that determine the lifetime of valence excited states is the *resonance* picture. The interaction of the discrete adsorbate valence levels with the continuum states of the metal broadens and shifts (or splits) the adsorbate levels, which are now referred to as "resonances." An electron excited into an adsorbate resonance located above the Fermi level of the metal will tunnel into unoccupied states of the metal in a characteristic time, τ, determined by the width of the resonance $\Gamma = \hbar\tau^{-1}$. A number of theoretical efforts have concentrated on calculations of the resonance widths. Following Gurney (62), early theoretical work by Gadzuk (63) and Remy (64) used a free electron description of the metal and perturbation theory to treat the atom-metal interaction. The widths calculated for s-states of alkali atoms were found to be of order 1 eV for atom-surface distances of $\simeq 3$ Å. Grozdanov & Janev (65) used a nonperturbative approach to calculate the adsorbate level width in the presence of an idealized surface potential. For Cs at 3 Å from a tungsten surface they obtained a width for the 6s level of ~ 3 eV. Local density functional (LDF) calculations by Lang et al (66) used an alkali atom analogy for the excited states of rare gas atoms at metals, and obtained a width of ~ 1 eV for Ar* on jellium at the equilibrium distance for ground-state Ar. Very recently, Nordlander & Tully (67) used the "weighted" density functional surface potential of Ossicini et al (68) to describe the metal, and the technique of "complex scaling" to calculate the resonance widths of alkali atoms and excited hydrogen atoms. For alkalis at ~ 3 Å from a model aluminum surface (jellium with $r_s = 2$), they find widths of ~ 0.3 eV. In the case of excited H, their widths are much smaller than previous calculations, and they also show a significant m-state dependence (see Figure 5). This dependence arises from the fact that degenerate states such as 2s and 2p can mix under the influence of the surface potential. As shown in Figure 5, at $\simeq 3$ Å from the surface the width of the $2s - 2p_z$ state is ~ 0.3 eV, whereas the width of the $2s + 2p_z$ state is ~ 3 eV. There is reason to believe that a better description of electron correlation on the adsorbate is needed and may lead to a further reduction in the calculated widths. An example is provided by the study of Geerlings et al (69) of electron capture by H from a cesiated

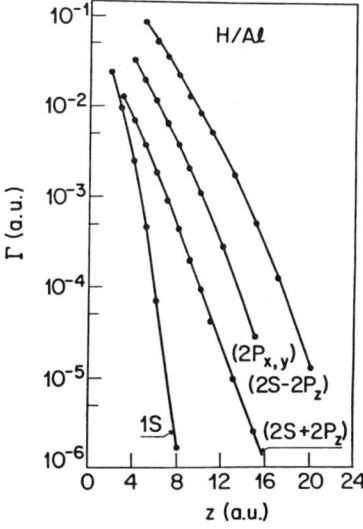

Figure 5 The width Γ of the $n = 1$ and $n = 2$ states of hydrogen on aluminum are shown as a function of distance from the positive background edge of the jellium model ($r_s = 2$). The H 2s and 2p levels mix to form hybrid states with significantly different widths. Widths and distances are in atomic units. Adapted from Ref. (67).

tungsten surface to yield H^-. If H^- is described in terms of a simple $1s^2$ configuration, a very large width is obtained (69). However, by using a better wavefunction for the negative ion—a $1s1s'$ configuration with two distinct spatial orbitals—a dramatic reduction in the calculated width was obtained. The overall trend in the calculations has been toward smaller widths as improvements in the description are made.

Experimental evidence indicates that the level widths are, in some cases, considerably smaller than the theoretical estimates. There are no direct measurements of the lifetime of excited-state at the relevant small distances from the surface, so all of the experimental information comes from linewidth or scattering measurements. In linewidth measurements, care must be taken to consider additional broadening mechanisms such as inhomogeneous broadening and vibronic broadening (70). The experimental linewidth provides only an upper limit to the lifetime broadening. With photoemission, one measures the width of the energy distribution of photo-ejected electrons, which may be (incorrectly) interpreted as the width of the initial level. For Xe on metals, the width of the $5p$ level inferred in this way is ~ 0.5 eV (at 100K) (70). Similarly, the width of the $6s$ level is inferred to be ~ 1 eV by inverse photoemission (71, 72). Thus one might expect a combined width of ~ 1.5 eV for the $5p \rightarrow 6s$ transition. In contrast, electron-energy-loss measurements of the electronic excitations of Xe on metals show $5p \rightarrow 6s$ linewidths of <0.2 eV (73). More recently, Schönhense et al (74), using a novel optical technique, reported a $5p \rightarrow 6s$ linewidth for Xe on Pd of only 0.08 eV. The much larger widths inferred from photoemission and inverse photoemission are primarily due to a wide Franck-Condon envelope. This is illustrated in Figure 6, which shows the Franck-Condon envelope for $3p$ ionization of Ar/Ag calculated with both the density functional potential curves of P. Avouris et al (unpublished) and a simple image potential. There may also be some Franck-Condon broadening of neutral excitations such as the $5p \rightarrow 6s$ electronic transition of Xe, but this is expected to be considerably smaller than for transitions involving positive ions (photoemission) or negative ions (inverse photoemission). For molecular adsorbates, there should also be substantial Franck-Condon effects that broaden the experimentally observed transitions. These effects are somewhat more difficult to evaluate because of the increased number of degrees of freedom. In addition, the resonance width also depends on lateral interactions among the adsorbates. Simulations by R. Kawai et al (unpublished) show a "motional narrowing" effect with increasing surface coverage.

Additional experimental information on level widths can be obtained from measurements of the branching ratios into different charge states by atom-surface scattering and sputtering experiments. In the case where the

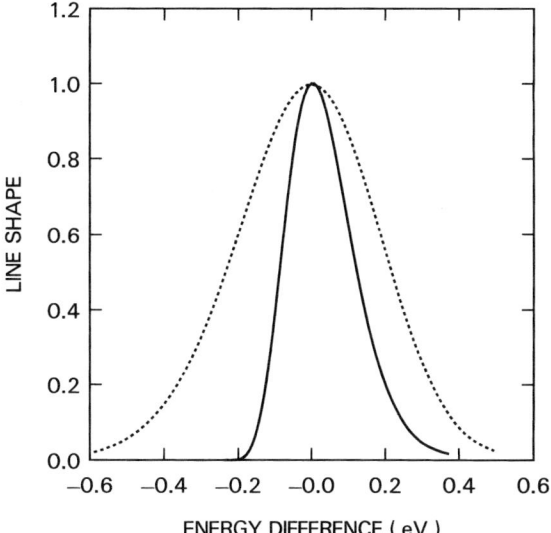

Figure 6 Franck-Condon broadening in photoionization is illustrated for Ar on a silver surface at 50K. A broad line shape results from the variation of the positive ion PE curve over the spatial extent of the initial state (see Figure 4). Two choices of positive ion PE curves are shown—the local density functional (LDF) potential (*solid line*) and an image potential (*dotted line*). The FWHM is $\simeq 0.22$ eV for the LDF potential, and $\simeq 0.45$ eV for the image potential. The maxima in both curves are shifted to zero energy. The intrinsic lifetime-broadened width is estimated to be an order of magnitude smaller than the Franck-Condon width.

affinity level crosses the Fermi level, the probability that the atom leaves as a negative ion has been shown (75) to be approximately given by: prob $\simeq \exp[-\Gamma(z_c)d/v_n]$, where $\Gamma(z_c)$ is the width (FWHM) of the level at the crossing point, d is the scale-length for the variation of level width with distance from the surface, and v_n is the normal velocity component. Yu & Lang reported Cs^+ (76) and O^- (77, 78) yields in sputtering and obtained widths of ~ 1 eV for Cs(6s) and ~ 2–3 eV for O(2p) at the equilibrium atom-surface distances. Geerlings et al (79) studied the formation of Li^- by collisions of Li^+ with a tungsten surface. They obtained a width of ~ 0.2 eV for $z \simeq 2.5$ Å.

Altogether, the experimental widths for excited-states that decay by resonant tunnelling are typically ~ 0.1–1 eV at the equilibrium atom-surface distances. This corresponds to lifetimes in the 10^{-14}–10^{-15} s range. Since the survival probability depends on the exponential of the excited-state decay rate (integrated over the trajectory), this range of lifetimes encompasses a wide range of possible desorption yields. For certain systems, desorption may proceed with substantial probability even when the excited state can decay by resonant electron tunnelling. However, for other systems, where the resonance widths at the relevant distances are ~1 eV, quenching will be very effective, implying very low yields. A clear experimental illustration of resonant tunnelling is provided in Figure 7, which shows the quenching of excited H-atoms at TiO_2 and SrF_2 surfaces (80). Excited-states that cannot be quenched by resonant tunnelling have substantially larger desorption yields.

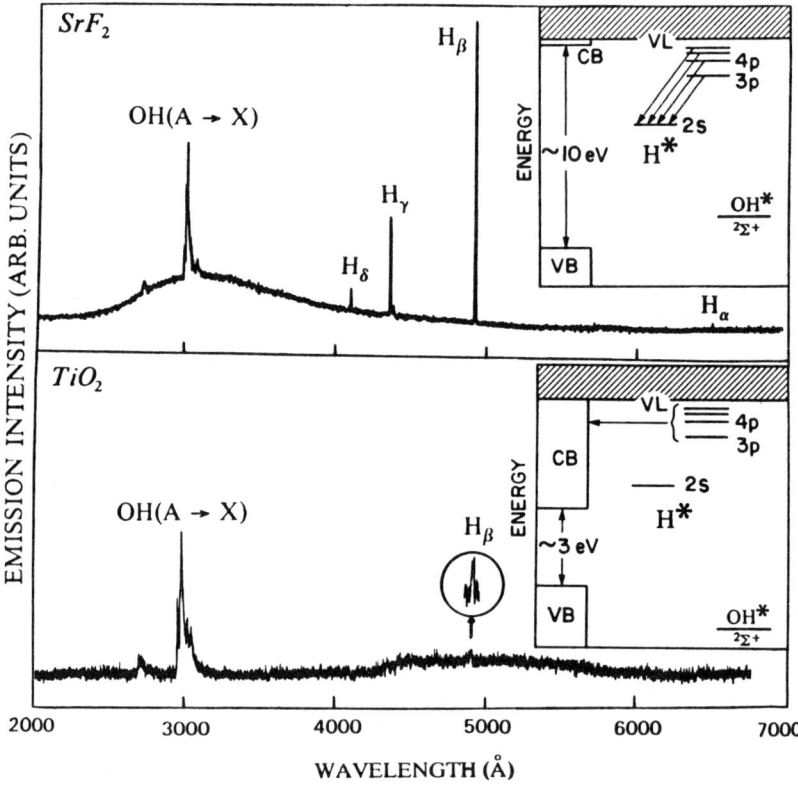

Figure 7 The importance of excited-state quenching by resonant tunnelling is illustrated for H* and OH* produced by electron impact on H_2O-exposed surfaces of SrF_2 (*top*) and TiO_2 (*bottom*). Fluorescence spectra are shown (*solid lines*) along with the band structure and the location of the excited states. For SrF_2, the H* levels are located within the bandgap, thus resonant tunnelling is blocked and the H* emission intensities are strong. For TiO_2, the H* levels fall within the conduction band, thus resonant tunnelling is allowed and the emission intensities are very weak. The OH* state is located either in the gap (SrF_2) or within the occupied valence band (TiO_2) so that OH* emission is strong in both cases. Adapted from Ref. (80).

For the case of positive ion states, resonant electron transfer from the substrate will quickly neutralize the ion if the level in which the hole is placed is in resonance with the substrate valence band. The time-scale for this one-hole hopping process is $\tau(1h) \sim \hbar/W$, where W is the width of the valence band. This time is typically $\tau(1h) \sim 10^{-15}$–10^{-16} s, which implies efficient delocalization on the time-scale for nuclear motion (10^{-14} s). However, if a two-hole state is prepared, by a multi-electron transition or by the Auger decay of a core-hole, this two-hole state can be localized provided that the hole-hole repulsion, U, is larger than the valence bandwidth, W (81, 82). For $U > W$, one-hole hopping is blocked because the valence bandwidth is too narrow to allow conversion of the repulsive potential energy, U, into kinetic energy of holes in the valence band. In this case, Ramaker et al (83, 84) have shown that only the slower two-hole hopping process is allowed, thus the localization time can be sufficient to allow substantial nuclear motion to occur. Moreover, Ramaker et al (83, 84) suggested that a class of m-hole, n-electron states may be particularly long-lived and thus important in stimulated desorption processes. The importance of multi-electron excitations in ion desorption has been demonstrated by Madey et al (85), Jaeger et al (86), Treichler et al (87), and others. Although multi-electron excitations have relatively low oscillator strengths, they appear prominently in synchrotron radiation desorption spectra because of the large survival probability of the resulting ions. The accurate assignment of such multi-electron excitations is not a simple task, but detailed experiments such as the polarization-dependent measurements by Treichler et al (87) provide valuable insight. One type of measurement that is currently lacking in this area is the analysis of the kinetic energy distributions of the ions resulting from multi-electron excitations. Such measurements could provide additional information on the nature of the final state leading to ion desorption.

If the electronic transition places a hole in an adsorbate level located below the substrate valence band, then resonant electron tunnelling is blocked. In this case, the most effective quenching process is Auger neutralization (29). Rates for Auger neutralization have been extracted from measurements of neutralization probabilities in ion-surface scattering, and from an analysis of charged or excited particles produced by sputtering. The Auger neutralization rate, Γ, is often approximated by an exponential function of distance from the surface, $\Gamma \simeq A \exp(-\alpha z)$. The exponential dependence arises because the Auger rate is related to the square of substrate wavefunctions evaluated at the adsorbate position. A rough estimate of the decay parameter, α, is given by $\alpha \simeq 2[2mE_b/\hbar^2]^{1/2}$, where E_b is a characteristic binding energy for electrons in the valence band. For metals, E_b is typically $\simeq 5$ eV, thus one has $\alpha \simeq 2$ Å^{-1}. Using this value for α, experiments on Auger neutralization of noble gas ions (29) indicate values of $A \sim 10^{17}$ s^{-1}. Thus at distances of $\simeq 3$ Å from the outermost lattice plane, typical Auger neutralization rates are $\sim 2*10^{14}$ s^{-1}. In other experiments, a wide range of parameters have been reported, indicating ratios of $A/\alpha \sim 10^6$–10^9 cm s^{-1}. Despite the wide variation of the decay rates, the overall order-of-magnitude is such that one can expect substantial nuclear motion to occur within the lifetime of an ionized (or excited) state that decays by Auger neutralization (or de-excitation). Hence such states are often desorption-active.

In some cases, studies of ESD/PSD can provide information on Auger neutralization rates. For example, P. Avouris and co-workers (unpublished) have examined Ar desorption from metal surfaces via the Antoniewicz mechanism, by using ground-state (88) and positive ion potentials obtained by local density functional calculations. Preliminary work indicates that reasonable agreement with the experimental results of P. Feulner and co-workers (unpublished) can be obtained with an exponential form of the neutralization rate by using parameters comparable to those obtained from ion neutralization measurements.

At large distances from the surface, where there is no hybridization between the adsorbate and substrate levels, an excited atom or molecule is still coupled to the substrate through the electromagnetic field. This coupling can lead to energy transfer to substrate excitations, such as electron-hole pairs or plasmons, resulting in quenching of the adsorbate excitation. For excited states with a dipole-allowed transition to the ground state, the nonradiative decay rate is given by (59, 60): $\tau_{NR}^{-1} = (\mu^2/4\hbar d^3) Im\{[\varepsilon(\omega)-1]/[\varepsilon(\omega)+1]\}$. The excited state is damped with a rate that depends on the transition dipole moment, μ, the bulk dielectric function of the substrate, $\varepsilon(\omega)$, and the inverse cube of the distance from the surface. When the distance from the surface is less than the electron inelastic mean free path, damping due to surface (rather than bulk) excitations becomes important, and the damping rate takes on a d^{-4} dependence. The relative efficiency of bulk vs. surface process is generally determined by the availability of momentum for intraband transitions in the bulk, which is reflected in the value of the inelastic mean free path. For transition metals, bulk damping can dominate even at the closest distances, but for free-electron metals and noble metals before the onset of d-band transitions, surface excitations become important. Campion et al (89) have shown that the decay of electronically excited pyrazine near a Ni surface follows the d^{-3} law at distances as close as 10 Å. In the case of excited biacetyl near an Ag surface, Alivisatos et al (90) showed that a d^{-4} dependence is followed from 30–100 Å. These large-distance formulas are not applicable at the distances important in chemisorption. However, for physisorbed systems, the decay of excited states by field coupling could be roughly valid (91). As an example, for the Xe*($5p \to 6s$) transition on Ag, classical electromagnetic damping results in a lifetime of $3*10^{-14}$ s, which is equivalent to a width of 0.02 eV. Since resonant tunnelling from the $6s$ level is possible in this case, the field coupling mechanism does not provide the major channel for decay of the excited state. However, when resonant tunnelling is not possible, field coupling could be important, and it should dominate at large distances due to the power-law dependence.

Nuclear Motion

The comparison of experimental data with calculated desorption yields and energy and/or angular distributions can provide a critical test of proposed desorption mechanisms. Such a comparison is particularly useful when the relevant potential energy (PE) surfaces are known. Nuclear motion on the PE surfaces can be described by either quantal or classical methods. For particles heavier than He, the spatial variation of the potentials is usually such that a WKB, or quasiclassical, approach would be expected to be valid. However, Schuck & Brenig (92) have shown that the presence of a spatially varying damping rate can result in a breakdown of the classical description. This occurs when the damping rate varies substantially over the wavepacket describing the state of the system. In this case, the portion of the wavepacket closest to the surface decays more rapidly than the more distant portions, thus leading to a nonclassical motion. This effect is illustrated in Figure 8. When the number of important degrees of freedom is small, a wavepacket propagation approach is preferred, particularly since there are computationally efficient methods that retain the essential quantal character (61). However, when the damping rate and its spatial gradient are not too large, a classical description is expected to be sufficiently accurate, except for hydrogen and helium. Gortel et al (44, 93) have compared classical and quantal calculations for desorption of He from tungsten, and N_2O from ruthenium. For the case of N_2O, there is little difference between the predictions until the gradient in the damping rate becomes sufficiently large. For desorption from bulk insulators (ionic solids, condensed noble gases, etc) the damping rates are expected to be considerably smaller than for adsorbates on metals, thus a classical description should be good. Indeed, when the motion of a large number of particles is important, as in the case of ion desorption from ionic solids (94), a classical description is the only practical approach.

A number of recent studies have explored different aspects of nuclear motion in stimulated desorption processes. Madey and co-workers (95, 96) have examined the effects of image forces and neutralization on the trajectories of ions. Only those ions with a sufficiently large normal velocity

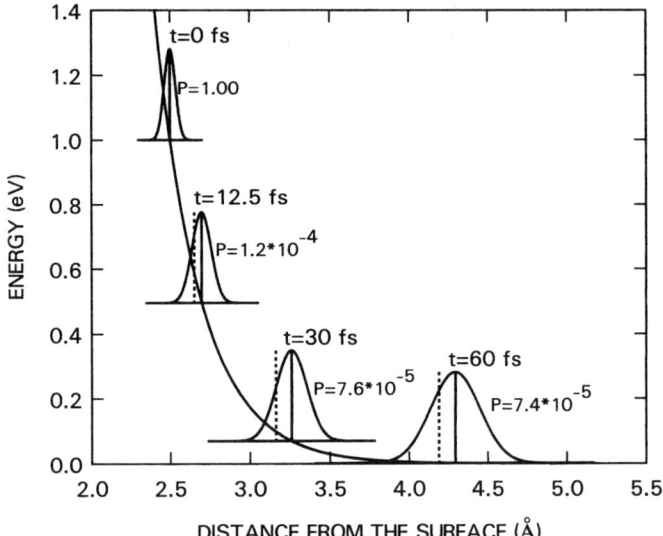

Figure 8 The quantum-mechanical motion of a wavepacket on a repulsive potential energy curve is illustrated. The spreading of the wavepacket is apparent, and the net survival probability (P) is indicated for several propagation times. For comparison, the position of a classical particle is shown as a *dotted line*. Due to the spatial variation of the damping process, the wavepacket center moves faster than the position of a classical particle.

component can overcome the long-range image attraction; ions emitted at large angles will bend back toward the surface and be recaptured. This would modify the angular distribution of desorbed ions, distorting the trajectories toward larger polar angles. Neutralization tends to compensate for this effect, because ions emitted at large polar angles have an increased residence time in the near-surface region and are thus more effectively neutralized. We (97) used a classical trajectory approach to examine the effects of lattice recoil and the origins of site-specificity in desorption from metals via the MGR model. Neutral fluorine desorption from Al(111) was considered, using an embedded-atom description of the metal (98, 99). The energy transferred to the metal due to the recoil effect was 20–30% of the initial repulsive energy, depending on the initial binding site. Recoil of the metal atoms bonded to the adsorbate produces a transient change in the metal charge density profile, which results in a site-dependent survival probability. Effects due to step edges were also considered, and the calculations (97) showed that for pure metals the neutralization behavior at step edges and flat terraces is expected to be nearly the same. For systems such as O^+ desorption from W, experimental evidence shows an important role of steps (100, 101); however, this appears to be linked to the formation of local oxide structures at steps rather than to properties of a pure metal surface. For the case of ion desorption from ionic solids, descriptions of the PE surfaces are available, and several classical trajectory studies have been carried out (94, 102, 103). These calculations show that the dynamic response of the lattice can have a dramatic effect: Some desorption channels are blocked by motion of the surrounding ions (94, 103), and in other cases (102) a strong site-specificity is predicted. These effects are discussed in more detail in a later section. For condensed noble gases, the decay of excimers can result in ejection of ground-state atoms from either the surface layer or from deeper layers (104). In this case the relevant potentials are known, and classical trajectory calculations explain the observed kinetic energy distributions rather well. Two-dimensional wavepacket propagation calculations by Stechel et al (28, 48) have been carried out for NO desorption from metal surfaces. These calculations agree well with the experimental measurements of Burns et al (28, 46), thus providing support for a rotationally mediated desorption mechanism.

Desorption Initiated by Auger Decay

A general mechanism was proposed by Knotek & Feibelman (105–107) to explain positive ion desorption from ionic solids. In the Knotek-Feibelman (K-F) mechanism, the initial step is the formation of a core-hole by the incident radiation. This core-hole can be filled by an Auger decay process, resulting in a multihole state that may be desorption active. Originally this mechanism was invoked to explain O^+ desorption from maximal valency metal oxides such as TiO_2, V_2O_3, and WO_3. Experiments (105) showed a pronounced onset in O^+ desorption associated with core-hole formation on the *metal* cation. In these oxides, the metal ion is essentially in a closed-shell configuration such as $Ti^{+4}[\ldots 3p^6]$, and the oxygen ion is also closed shell, i.e. $O^{-2}[\ldots 2p^6]$. Upon ionization of the Ti(3p) level, the 3p core-hole cannot be filled by an *intra-atomic* Auger decay process because there are no electrons occupying higher Ti levels. However, an *inter-atomic* Auger process can occur, in which the Ti(3p) hole is filled by an O(2p) electron, and one or two additional O(2p) electrons are ejected. In this way an O^+ ion can be formed, and repulsive Coulomb interactions provide the driving force for expulsion of O^+ from the surface. Auger decay of a core-hole can effectively populate desorption-active states in a wide variety of materials from ionic solids to covalently bonded adsorbates (e.g. 17, 18). This generalization of the Knotek-Feibelman mechanism is often referred to as Auger-initiated or Auger-stimulated desorption.

The rates for Auger decay into various final states are of primary importance in ion desorption by the K-F mechanism. Green, Jennison, and co-workers (108, 109) have recently reported inter-atomic Auger rates for NaF, which can be regarded as a model maximal valency ionic solid. If a hole is created in the outermost ($2p$) level of the Na^+ cation, interatomic Auger decay results in the formation of two $F(2p)$ holes on adjacent fluorine atoms. In this case it is energetically forbidden to localize two $F(2p)$ holes on one fluorine site. However, if the initial hole is formed in the $Na(1s)$ level, a much more severe Auger cascade follows. The net result of this Auger cascade is the formation of 4 to 6 individual $F(2p)$ holes, or the formation of F^+ and an additional 2 to 4 $F(2p)$ holes on adjacent sites (109). The rate for the inter-atomic Auger decay which produces F^+ is estimated to be $\sim 4*10^{13}$ s^{-1}. One might expect that formation of F^+ on the surface would lead to F^+ desorption; however, classical trajectory calculations (see the following discussion) indicate that this is not the case (except for extremely unlikely circumstances).

Perhaps the most surprising development was the prediction that because of recoil motion of the lattice, the reversal of the Madelung potential may *not* provide sufficient repulsive forces to cause ion desorption from certain ionic solids (94). For example, it had been assumed that upon localization of two holes at a fluorine site on the surface of NaF, the repulsive Coulomb forces would act to expel the F^+ ion provided that the two holes remained localized for a sufficient time. However, trajectory calculations showed that the F^+ ion does not desorb, but instead, a surface defect is formed. These calculations have recently been corroborated and extended by Green and co-workers (103). Recently, experiments have shown (110) that the positive ions that are produced by electron bombardment of alkali-halides are formed by gas-phase ionization of desorbed neutral atoms and molecules; and it appears that this mechanism can also account for the positive ions observed in PSD from alkali halides (111). The experimental results are thus in accord with the theoretical predictions.

An analysis of the natural cleavage planes of maximal valency metalhalides and oxides shows that the local surface structure plays a very important role in the dynamics of ion desorption from ionic solids (R. E. Walkup et al, unpublished). The K-F mechanism is expected to be *inoperative* for F^+ or Cl^+ desorption from halides with the rocksalt structure or the fluorite structure. However, F^+ desorption is allowed for halides with the rutile structure, such as MgF_2. For oxides, O^+ desorption is predicted to be marginally allowed (but time-delayed) for materials with the rocksalt structure (MgO), and allowed for materials with the fluorite or rutile structures. The structural factors can readily be understood by considering the forces arising from the formation of a positive ion at an anion site, as shown in Figure 9. For the rocksalt structure, there is a rather small repulsive force on the newly formed positive ion but much

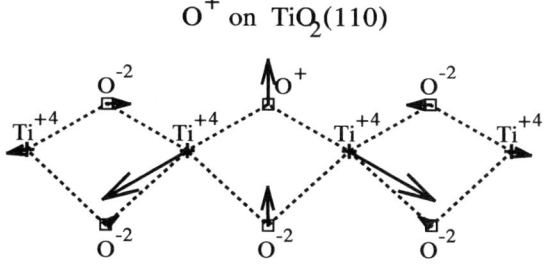

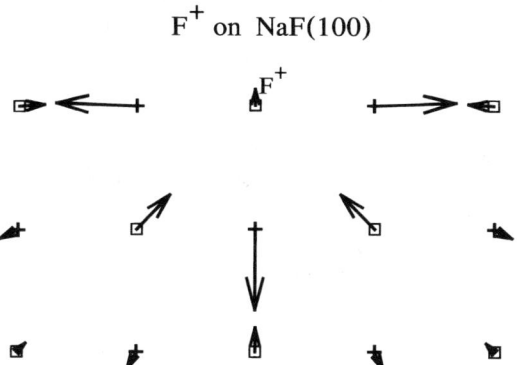

Figure 9 The forces arising from the formation of a positive ion at an anion site are illustrated for O^+ on $TiO_2(110)$ (a bridging site) and F^+ on NaF(100). For O^+ the local geometry results in a large force in the direction of the surface normal, thus prompt ion desorption occurs. In contrast, the force on F^+ is much smaller than the forces on the neighboring ions, and the resulting lattice distortions prevent F^+ desorption.

larger forces on the neighboring ions. As a result, the lattice distorts, leading to self-trapping of the F^+ ion (94). In contrast, for the rutile structure, there is a large force in the direction of the surface normal for ions in favorable sites (see Figure 9). Thus prompt ion desorption is predicted (102). Even in this case, however, the majority of the initial repulsive potential energy goes into lattice recoil motion rather than kinetic energy of the desorbed O^+ ion (102).

The most thoroughly studied maximal valence metal oxide is probably TiO_2. Recent experiments of photon-stimulated desorption of O^+ from TiO_2 (112, 113) have provided detailed information on the ion angular and energy distributions, and the dependence of ion yield on surface preparation. As in the original experiments of Knotek et al (114), a dramatic increase in O^+ yield is observed upon Ti($3p$) ionization. A remarkably complex ion desorption behavior was reported, including a progression of angular distribution patterns, depending on the sample annealing temperature. For well-annealed TiO_2(110) surfaces, the angular distribution of O^+ ions is a single spot in the direction of the surface normal, and the ion energy distribution shows a single peak at 4–5 eV. The experimental data in this case are in reasonable agreement with trajectory calculations (102), which show that only the oxygen ions in the bridging sites contribute substantially to the O^+ yield. This agreement provides a confirmation of the K-F mechanism for O^+ desorption from TiO_2.

A number of experiments illustrate extensions of the K-F mechanism. Kurtz et al (115) examined O^+ desorption from MgO(100) for photon energies near the O($1s$) and Mg($1s$) absorption edges. In addition to O($1s$) ionization, O($1s$) → core-exciton transitions were effective in stimulating O^+ desorption. For the core-exciton states, the core-hole can decay by an *intra-atomic* Auger process, which ultimately results in O^+ formation. Interestingly, no O^+ desorption was observed for photon energies near the Mg($1s$) ionization threshold, where an *inter-atomic* Auger process would be required. Bertel et al (116) investigated photon-stimulated desorption of H^+ from OH adsorbed on Ti and Cr. In this work, the OH adsorbate was observed to dissociate upon Ti($3p$) ionization, yielding H^+. Due to mixing of the Ti and OH orbitals, Auger decay of the Ti($3p$) core-hole can result in localization of two holes on the OH bond, thus leading to H^+ desorption.

The specific core-level onsets observed in Auger-initiated desorption provide an opportunity for chemically specific X-ray photochemistry. This was recently demonstrated by Yarmoff et al (117). In these experiments, F^+ desorption from partially fluorinated Si(111) was examined for photon energies near the Si($2p$) absorption edge. The fluorinated surface contains SiF, SiF_2, and SiF_3 groups. Increasing the number of fluorine atoms attached to a given Si atom results in a shift of the Si($2p$) core-level to higher binding energy. This shift is approximately 1 eV per fluorine atom. By using tunable radiation from a synchrotron source, the $2p$ levels of Si atoms in different SiF_x groups could be selectively excited, resulting in desorption of F^+ from the selected SiF_x group.

For Auger-initiated desorption, one needs to consider the possibility that secondary electrons produced by the Auger cascade could create further electronic excitation, resulting in an indirect desorption mechanism. This process was suggested by Jaeger et al (118, 119) to account for observations of H^+ desorption from NH_3 adsorbed on Ni. This effect results in a spectroscopic signature for the desorption yield that reflects the net secondary electron production. Ramaker and co-workers (120) have investigated these effects in detail. Their results indicate that when a direct Auger-initiated desorption process is active, this direct mechanism generally dominates over desorption stimulated by secondary electrons. In this case, the secondary electron contribution is typically ~35% or less. However, when core-hole decay is not effective in directly stimulating desorption, the secondary electron mechanism can dominate. A different role of secondary electrons was reported by Walkup et al (110, 121) for electron bombardment of alkali-halides. For these materials, direct desorption of positive ions is blocked by dynamic lattice distortions, but there are extremely efficient mechanisms for neutral atom desorption, and the secondary electron yields are large. Because of these conditions, the yield of positive ions and excited-state neutral atoms is dominated by *gas-phase* electron impact processes involving secondary and primary electrons and the desorbed neutral particles. Such secondary ionization and excitation processes are always present to some degree; they can dominate only when direct ion or excited neutral atom desorption is extremely unlikely compared to the desorption of ground-state neutral particles.

Dissociative Electron Attachment and Negative Ion Desorption

In the dissociative electron attachment (DEA) process, a molecule A–B captures an electron to form a transient negative ion $(A-B)^-$ which then dissociates to give $A + B^-$. This proces is applicable to adsorbed molecules, in which case desorption of either A or B^- (or perhaps both) can occur, depending on the nature of the bonding to the substrate. The electron that attaches to the molecule can be supplied either by an external electron source as in the case of ESD, or it can be a secondary electron, or a photo-excited or photo-emitted substrate electron. In contrast to direct dissociation by electron impact, the cross section for dissociative electron attachment generally peaks at rather low electron energies, i.e. < 10 eV. Since adsorption shifts and broadens the molecular levels, the influence of the substrate will modify the nature of the DEA process compared to that for isolated molecules.

The most detailed studies of DEA processes in adsorbates have been performed by Sanche and co-workers on condensed and physisorbed molecular layers (122–125). The effects of the image interaction have been investigated for O^- desorption from O_2 condensed either directly on Pt or on top of intervening layers of Ar, which serve as a spacer (125). By varying the thickness of the spacer layer, the strength of the image interaction could be controlled. Two mechanisms for O^- desorption were identified:

(a) dissociative electron attachment (DEA), $e^- + O_2(^3\Sigma_g^-) \to O_2^-(^2\Pi_u) \to O^-(^2P) + O(^3P)$, and (b) ion pair formation (IPF), $e^- + O_2 \to O_2^* \to O^- + O^+$. Ion pair formation dominates at higher electron energies (>17 eV), whereas dissociative electron attachment dominates for low electron energies (<10 eV). The image interaction affects the DEA and IPF processes in different ways. The final state for the DEA process in which the negative ion is bound to the surface is energetically favored over the state in which O^- desorbs. Also, the image attraction can result in motion of the transient O_2^- ion toward the surface. This can result in resonant electron tunnelling, a result that would effectively quench the DEA process. Thus the DEA process becomes less effective as the image interaction increases, as observed experimentally by Sambe et al (125). In contrast, image interactions increase the number of states that can contribute to O^- desorption by the IPF process, thus this process remains quite effective even in the presence of a strong image interaction (125).

For chemisorbed species, negative ion desorption has received considerably less attention than positive ion desorption; only a few reports of negative ion ESD have appeared (e.g. see 126–129). In some cases the evidence suggests a dissociative electron attachment mechanism. For example, Lichtman and co-workers (126, 127) have studied both O^- and O^+ desorption from oxidized Mo and W. They found dramatically different thresholds for negative and positive ions, ~4 eV for O^- as opposed to ~30 eV for O^+. For both Mo and W the O^- yield showed a distinct peak in the low energy region, as would be expected for a DEA process. Yu (128) examined O^- and O^+ desorption from Mo(100) exposed to O_2. The isotope effect for the O^- and O^+ yield was consistent with a direct MGR desorption mechanism for both species. Recently, negative ion (F^-) desorption from fluorine-containing molecules, PF_3, NF_3, and $(CF_3)_2CO$, adsorbed on Ru(001) was reported by Johnson et al (129). These authors measured ion angular distributions for both F^- and F^+ originating from the adsorbed parent molecules. However, from fragments of these molecules, such as F atoms directly bonded to the metal surface (which are produced by electron bombardment of the parent molecules), they could only detect F^+ ions. Some insight into the differences between positive and negative ion desorption can be obtained from basic considerations of the electronic structure, as we have discussed for fluorine on aluminum (25, 26). A negative ion such as F^- will be strongly attracted to the surface by the image interaction; however, the positive ion, F^+, is screened by charge transfer from the metal, and this leads to repulsion at small distances from the surface (see Figure 2). Thus for chemisorbed F atoms, one can expect F^+ but not F^- desorption to occur. If the F^- ion starts at a substantial distance from the surface, as in the case of the adsorbed parent molecules, then repulsion due to the dissociative electronic transition can overcome the image attraction, and F^- ions can desorb.

Another major difference between positive and negative ions involves neutralization. It is often the case that positive ion states are located below the bottom of the metal valence band. Thus resonant electron tunnelling is blocked, and Auger neutralization dominates. In this case the desorption yield can be substantial. In contrast, resonant tunnelling is usually possible for neutralization of negative ions. For most atoms, the electron affinities are small, thus resonant tunnelling of an electron from the negative ion to the metal occurs at small distances from the surface where the level widths are large. This results in very small negative ion yields. The different neutralization processes provide a qualitative explanation for the observation that negative ion yields are typically smaller than positive ion yields.

Band-gap Excitation of Semiconductors and Insulators

Desorption resulting from band-gap excitation has been observed in both semiconductors and insulators (for early reviews see 130–132). It was realized from the outset (133) that the active ingredients in these processes are the photo-generated charge carriers—electrons and holes. Which carrier is effective will be determined by the nature of the adsorbate-substrate bonding. For example, an electronegative atom or molecule, A, will be ionically bonded to a semiconductor surface, essentially as A^-. If the negative charge is captured by a hole, an essentially neutral adsorbate would result that may desorb, as in the case of fluorine on aluminum discussed above. The schematic diagram in Figure 10 illustrates this mechanism. A desorption mechanism due to photo-generated charge carriers has certain distinct experimental signatures: (a) the photo-desorption yield should follow the absorption characteristics of the substrate rather than the adsorbate, (*b*) there should be a dependence on processes that affect the carriers and/or carrier transport—such as doping, band-bending, heating, etc, and (*c*) the desorption yield may show a nonlinear dependence on photon intensity, even for very low intensities (134). These effects have been observed in several recent studies (135–139). For example, Ying & Ho (137) have studied the desorption of NO from Si(111) (7 × 7). They found that the NO desorption signal followed the Si absorption coefficient until the wavelength became smaller than a certain value, i.e. until the absorption depth was less than a distance $\delta \sim 400$ Å. For smaller absorption depths (shorter wavelength photons) the yield saturates to a value independent of the absorption coefficient. This behavior is consistent with a picture in which only those carriers generated within a distance δ from the surface are effective in stimulating desorption. Ying & Ho (137) suggested that hot holes may be the active species. Work function measure-

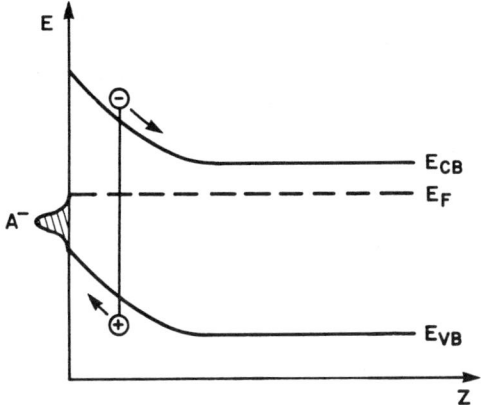

Figure 10 Desorption due to band-gap excitation of a semiconductor. An electronegative species A is adsorbed on the surface of an n-type semiconductor as A^-, thus resulting in an upward bending of the semiconductor bands. Electron-hole pairs are formed by photon excitation. Due to the band bending, the holes move toward the surface, where they can capture an electron from A^-, thus resulting in neutral A desorption.

ments indicate that chemisorption involves the formation of an NO^--like species (137). Photo-generated holes could capture the negative charge, leading to neutral NO desorption. Hot holes are implicated because the experimental distance, $\delta \sim 400$ Å is comparable to the mean free path for hole thermalization (137). In principle, hot holes may be needed to overcome an energy barrier to the carrier-adsorbate interaction. Further support for the role of holes was provided by Cs co-adsorption studies and comparisons between n- and p-type silicon substrates. Band-gap excitation of silicon by visible photons was shown by Houle (140) to enhance the evolution of SiF_3 during etching of Si by XeF_2. In this case, the photo-induced yield of SiF_3 exhibited an $I^{1.6}$ dependence on photon intensity. This suggests an overall reaction that is second order in the photo-generated carrier concentration (140).

Alkali-halides provide an interesting case in which there are very efficient mechanisms for neutral atom desorption initiated by electron-hole generation. Incident radiation with an energy above the band-gap will generate e-h pairs in the bulk. This leads to the formation of relatively stable defects, such as H centers and F centers. The bulk defect formation processes have recently been reviewed by Itoh (141) and Williams and co-workers (142). The H centers can be regarded as an extra halogen atom incorporated along the halogen chains. Diffusion of the H centers from their point of origin provides a means of transporting neutral halogen atoms to the surface, where thermal desorption of the halogen atom can occur. Also,

H centers formed near the surface can result in the ejection of energetic halogen atoms along the $\langle 110 \rangle$ directions. These processes are discussed in detail by Szymonski (143) and references therein. The fact that defects generated in the bulk result in neutral atom desorption from the surface makes the overall desorption process very efficient. For example, irradiation of NaCl by 1 keV electrons at 300°C results in desorption of ~ 5 halogen atoms per incident electron (144). Neutral alkali atoms can also desorb by a defect-mediated thermal process, but additional factors are involved. In particular it is necessary to form the neutral alkali atoms, most likely by recombination of alkali ions with electrons at defect sites. Recent experiments by Loubriel et al (145) suggest that the diffusion of F centers to the surface is the rate-limiting step. If the surface temperature is low, alkali atoms accumulate in the surface region, resulting in an alkali-enriched stoichiometry; but at high surface temperatures, the net alkali and halogen evaporation rates are equal and the material remains stoichiometric (144). In addition to ground-state neutral species, electronically excited alkali atoms (146, 147) and ionic species have also been studied. Recent experiments by Walkup et al (110, 121) have shown that for the case of electron bombardment, both the excited-state atoms and the positive ions observed experimentally are produced by gas-phase electron impact processes involving the desorbed ground-state neutral particles and secondary and primary electrons. Experimental work of Betz & Husinsky (148), Szymonski and co-workers (149), and Klekamp et al (150) have corroborated these conclusions. Excitation in the bulk is also important in rare-gas solids. For recent work on this subject see Coletti et al (151), Reimann et al (152, 153), Feulner et al (154), O'Shaughnessy et al (104), and references therein. For a discussion of analogous photo-processes in icy materials containing NH_3 and H_2O see Nishi et al (155).

Hot Electron Effects in Metals

Adsorbate fragmentation and desorption due to an interaction with photo-excited metal electrons have been suggested in recent laser-induced desorption experiments. Marsh et al (57) studied CH_3Cl on Ni(111) with excimer laser radiation. They found that 6.4 eV photons (193 nm) can dissociate adsorbed CH_3Cl by a process analogous to gas-phase photolysis. Desorption of CH_3 was detected for coverages of one monolayer up to at least 15 layers. They also observed that CH_3 radicals could be produced by 5eV photons (248 nm), which cannot dissociate gas-phase CH_3Cl. In this case, desorption could only be observed for coverages less than 8 layers. Marsh et al (57) proposed that for the 5 eV photons (and also to some extent for 6.4 eV photons), photo-excited metal electrons attach to the adsorbed CH_3Cl molecules, which dissociate producing CH_3 and Cl^- fragments.

This mechanism is consistent with the observed coverage dependence, since hot electrons from the metal cannot penetrate through a thick adsorbate film. Optically excited (hot) electrons were also implicated in a recent study of laser-induced desorption of NO from metals. Buntin and coworkers (52, 156) observed neutral NO desorption from Pt(111) upon irradiation with 1064 nm, 532 nm, and 355 nm photons from a nanosecond time-scale laser pulse. Using quantum-level specific detection, they established that the NO rotational distributions were non-Boltzmann, the spin-orbit populations were inverted, and the translational and vibrational energy distributions depended on the photon energy but were not correlated with the laser-induced surface-temperature jump. The data provided strong evidence for a nonthermal stimulated desorption process. The authors suggested that desorption could result from the interaction of optically excited (hot) electrons with adsorbed NO via its unoccupied $2\pi^*$ level. Metastable de-excitation spectroscopy of NO on metals (157) places the peak of the occupied $2\pi^*$ level at ~ 2 eV below the Fermi level, with a tail extending to E_F. Thus other effects, such as capture of an electron from the occupied $2\pi^*$ level by photo-generated holes, or NO → metal charge transfer from this occupied $2\pi^*$ level, could play an important role. In the experiments of Buntin and co-workers (52, 156), the net desorption yield was estimated to be $\sim 10^{-8}$ per photon. Thus one cannot easily rule out processes that may ordinarily be considered as unlikely.

Additional Mechanisms for Desorption by Substrate Excitation

A different type of desorption mechanism involving surface plasmon excitation was recently proposed by Träger and co-workers (158), who observed that Na atoms with high kinetic energies (~ 1.6 eV) were ejected upon visible laser illumination of Na particles deposited on LiF at low temperatures. The desorption rate was proportional to the laser power and exhibited a resonant behavior with respect to laser wavelength, which, in turn, depended on the Na particle size. For ~ 100 nm particles, the maximum yield was observed for $\simeq 400$ nm photons. The authors pointed out a correlation between the observed dependence on wavelength and particle size with the excitation of surface plasmons. The observed kinetic energy was approximately equal to the difference between the incident photon energy and the surface binding energy, thus suggesting complete conversion of the available photon energy into the desorption process. The microscopic desorption mechanism is not yet understood.

A mechanism involving a photo-acoustic shock generated by electronic excitation of the substrate was recently proposed by Polanyi and co-workers (54). These authors studied the excimer-laser-induced desorption of a variety of species from a LiF surface, including CH_3Br, H_2S, H_2O, HBr, and Xe. For the wavelengths employed, the LiF substrate is essentially transparent; however, there is some absorption of the ultraviolet laser light, which results in color center formation. The substrate temperature increase due to the laser pulse was estimated to be $<0.05K$, ruling out desorption via a simple substrate heating effect. It was proposed that nonradiative decay of the color centers generated by the laser pulse results in the production of a photo-acoustic shock that desorbs molecules by a momentum transfer process.

Concluding Remarks

Fundamental mechanisms of desorption and fragmentation induced by electronic transitions at surfaces have been extensively studied for ~ 25 years. Much progress has been made; however, a great deal of work remains, as evidenced by the intense activity in this area in the past few years. Areas for future developments include: (a) basic studies of electronically excited species interacting with surfaces, (b) real-time measurements of excited-state lifetimes and desorption dynamics, (c) energy-dependent photo-desorption spectra in the valence excitation region, (d) increased application of laser techniques both for inducing desorption and detecting the desorbed species, and (e) more complete theoretical treatments, including improvements in the relevant potential energy surfaces. The improvements in both experimental and theoretical methods coupled with the importance of this field in basic science and technology ensure that studies of stimulated desorption and fragmentation will be a very active area of chemical physics.

Literature Cited

1. Brown, W. L., Lanzerotti, L. J., Johnson, R. E. 1980. *Science* 218: 525
2. Czyzewski, J. J., Madey, T. E., Yates, J. T. Jr. 1974. *Phys. Rev. Lett.* 32: 777
3. Madey, T. E., Czyzewski, J. J., Yates, J. T. Jr. 1975. *Surf. Sci.* 49: 465
4. Madey, T. E. 1986. *Science* 234: 316
5. Stulen, R. H., Knotek, M. L., eds. 1988. *Desorption Induced By Electronic Transitions, DIET III.* Springer Ser. Surf. Sci., Vol. 13. Berlin: Springer-Verlag
6. Alvey, M. D., Yates, J. T. Jr. 1988. *J. Am. Chem. Soc.* 110: 1782
7. Osgood, R. M., Gilgen, H. H. 1985. *Annu. Rev. Mater. Sci.* 15: 549
8. Ehrlich, D. J., Higashi, G. S., Oprysko, M. M., eds. 1988. *Laser and Particle Beam Chemical Processing for Microelectronics*, Mater. Res. Soc. Symp. Proc. Vol. 101. Pittsburgh: Mater. Res. Soc.
9. Kern, D. P., Kuech, T. F., Oprysko, M. M., Wagner, A., Eastman, D. E. 1988. *Science* 241: 936
10. Madey, T. E. 1987. In *Analytical Electron Microscopy*, ed. D. C. Joy. San Francisco: San Francisco Press
11. Cohen, S. A., Cecchi, J. L., Dylla, H. F., eds. 1987. *J. Nucl. Mater.*, Vols. 145–147
12. Williams, E. M. 1988. See Ref. 5, p. 242
13. Menzel, D., Gomer, R. 1964. *J. Chem. Phys.* 41: 3311
14. Redhead, P. A. 1964. *Can. J. Phys.* 42: 886
15. Tolk, N. H., Traum, M. M., Tully, J. C., Madey, T. E., eds. 1983. *Desorption Induced By Electronic Transitions.*

Springer Ser. Chem. Phys., Vol. 24. Berlin: Springer-Verlag
16. Brenig, W., Menzel, D., eds. 1985. *Desorption Induced By Electronic Transitions, DIET II*, Springer Ser. in Surf. Sci., Vol. 4. Berlin: Springer-Verlag
17. Madey, T. E., Ramaker, D. E., Stockbauer, R. L. 1984. *Annu. Rev. Phys. Chem.* 35: 215
18. Knotek, M. L. 1984. *Rep. Prog. Phys.* 47: 1499
19. Menzel, D. 1986. *Nucl. Instrum. Methods B* 13: 507
20. Chuang, T. J. 1983. *Surf. Sci. Rep.* 3: 1
21. Ho, W. 1988. *Comments Condens. Matter Phys.* 13: 293
22. Menzel, D. 1986. *Ber. Bunsenges. Phys. Chem.* 72: 591
23. Menzel, D. 1982. *J. Vac. Sci. Technol.* 20: 538
24. Avouris, P., Bagus, P. S., Rossi, A. R. 1985. *J. Vac. Sci. Technol. B* 3: 1484
25. Avouris, P., Kawai, R., Lang, N. D., Newns, D. M. 1987. *Phys. Rev. Lett.* 59: 2215
26. Avouris, P., Kawai, R., Lang, N. D., Newns, D. M. 1988. *J. Chem. Phys.* 89: 2388
27. Antoniewicz, P. R. 1980. *Phys. Rev. B* 21: 3811
28. Burns, A. R., Stechel, E. B., Jennison, D. R. 1987. *Phys. Rev. Lett.* 58: 250
29. Hagstrum, H. D. 1954. *Phys. Rev.* 96: 336
30. Feulner, P., Treichler, R., Menzel, D. 1981. *Phys. Rev. B* 24: 7427
31. Rubio, J., López-Sancho, J. M., López-Sancho, M. P. 1982. *J. Vac. Sci. Technol.* 20: 217
32. Fuggle, J. C., Umbach, E., Menzel, D., Wandelt, K., Brudle, C. R. 1978. *Solid State Commun.* 27: 65
33. Schönhammer, K., Gunnarson, O. 1977. *Solid State Commun.* 23: 691
34. Schönhammer, K., Gunnarson, O. 1978. *Phys. Rev. Lett.* 41: 1608
35. Lang, N. D., Williams, A. R. 1977. *Phys. Rev. B* 16: 2408
36. Gadzuk, J. W., Doniach, S. 1978. *Surf. Sci.* 77: 427
37. Gumhalter, B., Wandelt, K., Avouris, P. 1988. *Phys. Rev. B* 37: 8048
38. Newsham, I. G., Hogue, J. V., Sandstrom, D. R. 1972. *J. Vac. Sci. Technol.* 9: 596
39. Alvey, M. D., Dresser, M. J., Yates, J. T. Jr. 1986. *Phys. Rev. Lett.* 56: 367
40. Feulner, P., Reidl, W., Menzel, D. 1983. *Phys. Rev. Lett.* 50: 986
41. Zhang, Q. J., Gomer, R., Bowman, D. R. 1983. *Surf. Sci.* 129: 535
42. Moog, E. R., Unguris, J., Webb, M. B. 1983. *Surf. Sci.* 134: 849
43. Feulner, P., Menzel, D., Kreuzer, H. J., Gortel, Z. W. 1984. *Phys. Rev. Lett.* 53: 671
44. Gortel, Z. W., Kreuzer, H. J., Feulner, P., Menzel, D. 1987. *Phys. Rev. B* 35: 8951
45. Lang, N. D., Kohn, W. 1973. *Phys. Rev. B* 7: 3541
46. Burns, A. R., Stechel, E. B., Jennison, D. R. 1989. *J. Vac. Sci. Technol. A* 6: 895
47. Burns, A. R., Stechel, E. B., Jennison, D. R. 1988. See Ref. 5, p. 67
48. Stechel, E. B., Jennison, D. R., Burns, A. R. 1988. See Ref. 5, p. 136
49. Avouris, P., Walkup, R. E., Kawai, R., Newns, D. M., Lang, N. D. 1988. See Ref. 5, p. 144
50. Budde, F., Hamza, A. V., Ferm, P., Ertl, G., Weide, D., Andresen, P., Freund, H.-J. 1988. *Phys. Rev. Lett.* 60: 1518
51. Avouris, P., DiNardo, N. J., Demuth, J. E. 1984. *J. Chem. Phys.* 80: 491
52. Buntin, S. A., Richter, L. J., Cavanagh, R. R., King, D. S. 1988. *Phys. Rev. Lett.* 61: 1321
53. Domen, K., Chuang, T. J. 1987. *Phys. Rev. Lett.* 59: 1484
54. Bourdon, E. B. D., Cowin, J. P., Harrison, I., Polanyi, J. C., Segner, J., Stanners, C. D., Young, P. A. 1984. *J. Phys. Chem.* 88: 6100
55. Bourdon, E. B. D., Das, P., Harrison, I., Polanyi, J. C., Segner, J., Stanners, C. D., Williams, R. J., Young, P. A. 1986. *Faraday Discuss. Chem. Soc.* 82: 343
56. Costello, S. A., Roop, B., Liu, Z.-M., White, J. M. 1988. *J. Phys. Chem.* 92: 1019
57. Marsh, E. P., Gilton, T. L., Meir, W., Schneider, M. P., Cowin, J. P. 1988. *Phys. Rev. Lett.* 61: 2725
58. Hagstrum, H. D. 1977. In *Inelastic Ion-Surface Collisions*, ed. N. H. Tolk, W. Heiland, C. W. White, pp. 1–25. New York: Academic
59. Chance, R. R., Prock, A., Silbey, R. 1978. In *Advances in Chemical Physics*, ed. I. Prigogine, S. A. Rice, 37: 1–65. New York: Wiley-Interscience
60. Avouris, P., Persson, B. N. J. 1984. *J. Phys. Chem.* 88: 837
61. Gadzuk, J. W. 1988. *Annu. Rev. Phys. Chem.* 39: 395
62. Gurney, R. W. 1935. *Phys. Rev.* 47: 479
63. Gadzuk, J. W. 1967. *Surf. Sci.* 6: 133
64. Remy, M. 1970. *J. Chem. Phys.* 53: 2487
65. Grozdanov, T. P., Janev, R. K. 1978. *Phys. Lett. A* 65: 396
66. Lang, N. D., Williams, A. R., Himpsel,

F. J., Reil, B., Eastman, D. E. 1982. *Phys. Rev. B* 26: 1728
67. Nordlander, P., Tully, J. C. 1988. *Phys. Rev. Lett.* 61: 990; 1989. *Surf. Sci.* In press
68. Ossicini, S., Bertoni, C. M., Giess, P. 1986. *Europhys. Lett.* 1: 661
69. Geerlings, J. J. C., van Amersfoot, P. W., Kwakman, L. F. T., Granneman, E. H. A., Los, J. 1985. *Surf. Sci.* 157: 151
70. Gadzuk, J. W., Holloway, S., Mariani, C., Horn, K. 1982. *Phys. Rev. Lett.* 48: 1288
71. Wandelt, K., Jacob, W., Memmel, N., Dose, V. 1986. *Phys. Rev. Lett.* 57: 1643
72. Horn, K., Frank, K. H., Wilder, J. A., Reil, B. 1986. *Phys. Rev. Lett.* 57: 1064
73. Demuth, J. E., Avouris, P., Schmeisser, D. 1983. *Phys. Rev. Lett.* 50: 600
74. Schönhense, G., Eyers, A., Heinzmann, U. 1986. *Phys. Rev. Lett.* 56: 512
75. Brako, R., Newns, D. M. 1981. *Surf. Sci.* 108: 253
76. Yu, M. L., Lang, N. D. 1983. *Phys. Rev. Lett.* 50: 127
77. Yu, M. L. 1981. *Phys. Rev. Lett.* 47: 1325
78. Lang, N. D. 1983. *Phys. Rev. B* 27: 2019
79. Geerlings, J. J. C., Rodnik, R., Los, J., Gauyacq, J. P. 1989. *Surf. Sci.* In press
80. Avouris, P., Beigang, R., Bozso, F., Walkup, R. E. 1986. *Chem. Phys. Lett.* 129: 505
81. Cini, M. 1977. *Solid State Commun.* 24: 681
82. Antonides, E., Janse, E. C., Sawatzky, G. A. 1977. *Phys. Rev. B* 15: 1669
83. Ramaker, D. E., White, C. T., Murday, J. C. 1982. *Phys. Lett. A* 89: 211
84. Ramaker, D. E. 1983. *J. Chem. Phys.* 78: 2998
85. Madey, T. E., Stockbauer, R., Flodström, S. A., van der Veen, J. F., Himpsel, F. J., Eastman, D. E. 1981. *Phys. Rev. B* 23: 6847
86. Jaeger, R., Treichler, R., Stöhr, J. 1982. *Surf. Sci.* 117: 533
87. Treichler, R., Riedl, W., Wurth, W., Feulner, P., Menzel, D. 1985. *Phys. Rev. Lett.* 54: 462
88. Lang, N. D. 1981. *Phys. Rev. Lett.* 46: 842
89. Campion, A., Gallo, A. R., Harris, C. B., Robota, H. J., Whitmore, P. M. 1980. *Chem. Phys. Lett.* 73: 447
90. Alivisatos, A. P., Waldeck, D. H., Harris, C. B. 1985. *J. Chem. Phys.* 82: 541
91. Avouris, P., Schmeisser, D., Demuth, J. E. 1983. *J. Chem. Phys.* 79: 488
92. Schuck, P., Brenig, W. 1982. *Z. Phys. B* 46: 137
93. Gortel, Z. W., Kreuzer, H. J., Feulner, P., Menzel, D. 1988. See Ref. 5, p. 173
94. Walkup, R. E., Avouris, P. 1986. *Phys. Rev. Lett.* 56: 524
95. Miskovic, Z., Vukanic, J., Madey, T. E. 1984. *Surf. Sci.* 141: 285
96. Miskovic, Z., Vukanic, J., Madey, T. E. 1986. *Surf. Sci.* 169: 405
97. Walkup, R. E., Avouris, P. 1989. *Phys. Rev. B* 39: 554
98. Daw, M. S., Baskes, M. I. 1984. *Phys. Rev. B* 29: 6443
99. Voter, A. F., Chen, S. P. 1987. In *Mat. Res. Soc. Symp. Proc.*, Vol. 82, ed. R. W. Siegel, R. Sinclair, J. R. Weertman. Pittsburgh: Mater. Res. Soc.
100. Madey, T. E. 1980. *Surf. Sci.* 94: 483
101. Bauer, E. 1983. See Ref. 15, p. 104
102. Walkup, R. E., Kurtz, R. L. 1988. See Ref. 5, p. 160
103. Green, T. A., Riley, M. E., Coltrin, M. E. 1989. *Phys. Rev. B* 39: 5397
104. O'Shaughnessy, D. J., Boring, J. W., Cui, S., Johnson, R. E. 1988. *Phys. Rev. Lett.* 61: 1635
105. Knotek, M. L., Feibelman, P. J. 1978. *Phys. Rev. Lett.* 40: 964
106. Feibelman, P. J., Knotek, M. L. 1978. *Phys. Rev. B* 18: 6531
107. Knotek, M. L., Feibelman, P. J. 1979. *Surf. Sci.* 90: 78
108. Green, T. A., Jennison, D. R. 1987. *Phys. Rev. B* 36: 6112
109. Green, T. A., Riley, M. E., Richards, P. M., Loubriel, G. M., Jennison, D. R., Williams, R. T. 1989. *Phys. Rev. B* 39: 5407
110. Walkup, R. E., Avouris, P., Ghosh, A. P. 1987. *Phys. Rev. B* 36: 4577
111. Parks, C. C., Hussain, Z., Shirley, D. A., Knotek, M. L., Loubriel, G., Rosenberg, R. A. 1983. *Phys. Rev. B* 28: 4793
112. Kurtz, R. L. 1986. *Surf. Sci.* 177: 526
113. Kurtz, R. L., Stockbauer, R., Madey, T. E. 1986. *Nucl. Instrum. Methods B* 13: 518
114. Knotek, M. L., Jones, V. O., Rehm, V. 1979. *Phys. Rev. Lett.* 43: 300
115. Kurtz, R. L., Stockbauer, R., Nyholm, R., Flodström, S. A., Senf, F. 1987. *Phys. Rev. B* 35: 7794
116. Bertel, E., Ramaker, D. E., Kurtz, R. L., Stockbauer, R., Madey, T. E. 1985. *Phys. Rev. B* 31: 6840
117. Yarmoff, J. A., Taleb-Ibrahimi, A., McFeely, F. R., Avouris, P. 1988. *Phys. Rev. Lett.* 60: 960
118. Jaeger, R., Stöhr, J., Kendelewicz, T. 1983. *Phys. Rev. B* 28: 1145

119. Jaeger, R., Stöhr, J., Kendelewicz, T. 1983. *Surf. Sci.* 134: 547
120. Ramaker, D. E., Madey, T. E., Kurtz, R. L., Sambe, H. 1988. *Phys. Rev. B* 38: 2099
121. Walkup, R. E., Avouris, P., Ghosh, A. P. 1986. *Phys. Rev. Lett.* 57: 2227
122. Sanche, L. 1984. *Phys. Rev. Lett.* 53: 1638
123. Sanche, L., Parenteau, L. 1986. *J. Vac. Sci. Technol. A* 4: 1240
124. Azria, R., Parenteau, L., Sanche, L. 1988. *J. Chem. Phys.* 88: 5166
125. Sambe, H., Ramaker, D. E., Parenteau, L., Sanche, L. 1987. *Phys. Rev. Lett.* 59: 236
126. Hock, J. L., Craig, J. H., Lichtman, D. 1979. *Surf. Sci.* 85: 101
127. Liu, Z. X., Lichtman, D. 1982. *Surf. Sci.* 114: 287
128. Yu, M. L. 1979. *Phys. Rev. B* 19: 5995
129. Johnson, A. L., Joyce, S. A., Madey, T. E. 1988. *Phys. Rev. Lett.* 61: 2578
130. Lichtman, D., Shapira, Y. 1978. *CRC Crit. Rev. Solid State Mater. Sci.* 8: 93
131. Wol'kenstein, T. 1975. *Prog. Surf. Sci.* 6: 213
132. Many, A. 1974. *CRC Crit. Rev. Solid State Mater. Sci.* 4: 515
133. Medved, D. B. 1958. *J. Chem. Phys.* 28: 870
134. Shapira, Y., McQuistan, R. B., Lichtman, D. 1977. *Phys. Rev. B* 15: 2163
135. Ekwelundu, E., Ignatiev, A. 1987. *Surf. Sci.* 179: 119
136. Kornbilt, L., Ignatiev, A. 1984. *Surf. Sci.* 136: L57
137. Ying, Z., Ho, W. 1988. *Phys. Rev. Lett.* 60: 57
138. So, S. K., Kao, F. J., Ho, W. 1988. *J. Vac. Sci. Technol. A* 6: 1435
139. Bartosch, C. E., Gluck, N. S., Ho, W., Ying, Z. 1986. *Phys. Rev. Lett.* 57: 1425
140. Houle, F. A. 1988. *Phys. Rev. Lett.* 61: 1871
141. Itoh, N. 1985. *Crys. Lattice Defects Amorphous Mater.* 12: 103
142. Williams, R. T., Faust, W. L., Craig, B. B. 1985. *Crys. Lattice Defects Amorphous Mater.* 12: 127
143. Szymonski, M. 1980. *Radiat. Eff.* 52: 9
144. Szymonski, M., Rutkowski, J., Poradzisz, A., Postawa, Z., Jorgensen, B. 1985. See Ref. 16, p. 160
145. Loubriel, G. M., Green, T. A., Richards, P. M., Albridge, R. G., Cherry, D. W., et al. 1986. *Phys. Rev. Lett.* 57: 1781
146. Tolk, N. H., Feldman, L. C., Kraus, J. S., Morris, R. J., Traum, M. M., Tully, J. C. 1981. *Phys. Rev. Lett.* 46: 134
147. Tolk, N. H., Traum, M. M., Kraus, J. S., Pian, T. R., Collins, W. E., et al. 1982. *Phys. Rev. Lett.* 49: 812
148. Betz, G., Husinsky, W. 1988. *Nucl. Instrum. Methods B* 32: 331–40
149. Postawa, Z., Rutkowski, J., Poradzisz, A., Czuba, P., Szymonski, M. 1987. *Nucl. Instrum. Methods B* 18: 574
150. Klekamp, A., Snowdon, K. J., Heiland, W. 1989. *Radiat. Eff.* In press
151. Coletti, F., Debever, J. M., Zimmerer, G. 1984. *J. Phys. Lett.* 45: L467
152. Reimann, C. T., Johnson, R. E., Brown, W. L. 1984. *Phys. Rev. Lett.* 53: 600
153. Reimann, C. T., Brown, W. L., Johnson, R. E. 1988. *Phys. Rev. B* 37: 1455
154. Feulner, P., Müller, T., Puschmann, A., Menzel, D. 1987. *Phys. Rev. Lett.* 59: 791
155. Nishi, N., Shinohara, H., Okuyama, T. 1984. *J. Chem. Phys.* 80: 3898
156. Richter, L. J., Buntin, S. A., Cavanagh, R. R., King, D. S. 1988. *J. Chem. Phys.* 89: 5344
157. Bozso, F., Arias, J., Hanrahan, C. P., Yates, J. T. Jr., Martin, R. M., Metiu, H. 1984. *Surf. Sci.* 141: 591
158. Hoheisel, W., Jungmann, K., Vollmer, M., Weidenauer, R., Träger, F. 1988. *Phys. Rev. Lett.* 60: 1649

COMPUTER SIMULATIONS OF GLOBULAR PROTEIN FOLDING AND TERTIARY STRUCTURE

*Jeffrey Skolnick**

Institute of Macromolecular Chemistry, Department of Chemistry, Washington University, St. Louis, Missouri 63130

Andrzej Kolinski

Department of Chemistry, University of Warsaw, 02-093 Warsaw, Poland

INTRODUCTION

Since Kendrew et al (1) obtained the first low resolution X-ray structure of a globular protein in 1958, the ability to predict the three dimensional (tertiary) structure of the biologically active, "native" state from the sequence of amino acids that comprise it has been a long sought objective of theoretical biophysical chemistry (2-8). This problem is of intense interest because its soluion would allow the rational design of new and novel proteins by either genetic engineering or synthetic techniques (8). On a more modest level, a partial solution would allow the prediction of the effects of amino acid substitution on conformation (site-directed mutagenesis) and thermodynamic stability (8). The prediction of tertiary structure is in principle no different from conformational analysis in small molecules; however, unlike the small molecule case, the number of possible conformations a protein can adopt is astronomical, e.g. a protein of 100 residues can have some 10^{100} possible conformations (9-12). Thus, it is an extreme example of a multiple free energy minimum problem. Systems having so many degrees of freedom are a natural candidate for computer simulations, and the present review examines the results from a number of simulation techniques that have been applied to address the problem of folding and tertiary structure prediction.

Before delving into a discussion of the various computational meth-

* Present address: Department of Molecular Biology, Research Institute of Scripps Clinic, La Jolla, California 92037.

odologies that have been employed, it is appropriate to review the salient experimental facts that any successful theory must ultimately rationalize and encompass. At high temperatures, the global conformation of a globular protein is a random coil having low segment density. The denatured form of the protein may be obtained in many cases by raising the temperature or by changing the solvent conditions (12). Both the early work of Tanford (12) and recent NMR studies (13a,b) indicate that at high temperature there is fluctuating secondary structure (α-helices, β-turns, etc) of marginal stability in the denatured state.

On cooling from the denatured state in vitro, a number of small, single-domain proteins undergo reversible renaturation to a dense, collapsed conformation having well-defined regions of secondary structure (14). In the native conformation, the hydrophobic, nonpolar residues tend to be located in the interior and the hydrophilic, polar residues tend to be located on the surface (8, 15). Moreover, the native conformation is structurally unique, and the same global conformation is obtained on successive refolding (14), although there are, of course, small fluctuations about the native state (16). Furthermore, the conformational transition appears to be well approximated by a two-state model with the equilibrium folding intermediates sparsely populated (8, 14, 17–20); i.e. either a given molecule is in the denatured state or it is in the native conformation. [There have been reports in the literature that under certain solvent conditions, a collapsed molten globule form can also exist (21)]. This is a surprising result in that the free energy difference between a residue in the native and denatured state is small (typically on the order of several hundred calories per mole of residues or so) (12, 14), and naive statistical mechanical considerations would lead one to expect that a whole manifold of conformational states should be populated.

A number of questions about the nature of the protein-folding process immediately arise. First and foremost, is the native conformation at the global free energy minimum or does it result from kinetic trapping (2)? This has long been a topic of debate, but if the proteins are sufficiently small, it appears that equilibrium thermodynamics holds (2, 14), and the majority of the approaches described below assume that the thermodynamic hypothesis is correct. Secondly, what is the mechanism by which a protein partitions phase space (9–11)? Experimentally, proteins fold from the denatured state on the order of seconds or minutes (8, 22, 23). A random search, of all the conformations of a protein (9–12) having even 10^{23} conformations would take about 10^5 years, assuming a new conformation is generated every 10 ps and every conformation is sampled just once. Thus, whatever the folding mechanism is, it is not a random search.

Turning to the nature of the interactions that are required to predict the native structure of the protein, the following points must be addressed:

1. What is the level of detail required to fold a protein? Must every interaction be known very precisely, without which the native state cannot be obtained, or are there general rules that determine a given folding motif with site specific interactions mainly involved in structural fine tuning?
2. What is the relative importance of the intrinsic local preference of an amino acid for a particular conformation (24) (short-range interactions) compared to conformational preferences that result from tertiary interactions inducing secondary structure (i.e. interactions between residues that are spatially close but far apart down the chain contour; these are also known as long-range interactions) (25)?
3. What is the relative importance of hydrogen bonding, hydrophobic, hydrophilic, van der Waals, and electrostatic interactions in determining the stability of the native state?
4. Why is the conformational transition all-or-none?

The approaches described below address various aspects of the above. However, it is not unreasonable to inquire whether the protein folding problem can be solved by brute force. For example, why not simply enumerate all the conformations and rank them in order of increasing energy? Based on a highly efficient, exhaustive conformational tree search algorithm, Lipton & Still (26) estimate that it would take over 10^{40} years of supercomputer time to search the conformational space of a 50 residue protein, assuming one structure per second is energy minimized. In other words, the problem is noncomputable. Similarly, as pointed out by numerous workers (8, 27–30), the direct application of molecular dynamics, where one solves Newton's equations of motion, is impractical. While the fundamental integration step is typically 10^{-15} seconds, one requires simulation times on the order of seconds to fold a protein (22, 23). Consequently, brute force alone is unlikely to solve the globular protein folding problem; techniques that effectively surmount the multiple minima and multiple time scale problems are required.

The methods described below and applied to protein folding are representative of the different philosophical approaches to the use of computers. One viewpoint holds that computer simulations should aim to reproduce real experiments. Hence, the most accurate available potential energy surfaces should be employed to make the calculation as realistic and quantitative as possible. When practical, this is a powerful method for determining the validity of the potential energy surface, and by agreeing in detail with real experiments, it is hoped that quantities that are inaccessible by real experiments are adequately described as well. There is another, more phenomenological viewpoint, which holds that computer simulations can provide qualitative insight into the behavior of simplified model systems that have the general features demanded by real experiments, but not the intimate details. The simulation is treated as an experiment in which one has, in contrast to the real situation, an overabundance of information about the system rather than a paucity—one is then forced to ask the right questions that convert the mass of data into qualitative insights. Clearly, the problem of protein folding is sufficiently complicated that both viewpoints have their place.

Over the years, a number of excellent reviews (27–30) and an excellent book by McCammon & Harvey (31) have appeared describing molecular dynamics and Brownian dynamics simulations (where Newton's equations of motion are modified to include random forces to mimic the solvent and thereby allow simulation of longer time processes) of the relatively short-time equilibrium and dynamic processes in proteins. These are very powerful techniques, but as they have been extensively reviewed elsewhere, we refer the reader to Refs. (27–31) for a more detailed and complete discussion.

The present review focuses on the current state of simulations designed to predict tertiary structure in globular proteins. The review is organized in order of increasing conformational complexity and, concomitantly, increasing simplicity in the level of the description of the system. We begin by describing approaches that have been applied to predict the conformations of various small constrained systems such as hypervariable loops in antibodies. Here, the approaches are rather detailed. We then summarize developments for the folding of unconstrained small polypeptides and protein fragments. Finally, highly simplified approaches to tertiary structure prediction in globular proteins are described.

STRUCTURE PREDICTION IN SMALL, CONSTRAINED PEPTIDES

There are two interrelated problems faced by the theorist in predicting the tertiary structure of a protein. It is unclear whether existing potential energy functions are adequate to predict tertiary structure, and given that suitable potential energy functions exist, one then requires practical algorithms for structure prediction. Thus, as a starting point, a number of workers have opted to study the folding of peptide fragments with constraints that greatly reduce the expanse of configuration space that must be explored. These include the grid and tree searches on small cyclic peptides (32–34). Then there are the tree searches on excised protein fragments by Moult & James (35) and the more sophisticated tree search techniques of Bruccoleri & Karplus (36). The latter approach has subsequently been applied (37) to the variable loops in the immunoglobulins, MCPC603 and HyHEL5. Of particular interest is a highly efficient random "tweak" algorithm applied to the immunoglobulins by Shenkin et al (38). In general, these approaches have been rather successful at elucidating the lowest energy conformations in these small systems, where essentially a full search of conformational space is possible. Typically this limits the methods to peptides containing no more than about 20 residues (38), and for some, to about 10 residues or less (35–37).

STRUCTURE PREDICTION IN SMALL UNCONSTRAINED POLYPEPTIDES

Build-up Procedure

A methodology related in spirit to a full conformational search is the build-up procedure developed by Scheraga and co-workers (39–41). The basic assumption is that the local intraresidue interactions are the dominant factor in determining the conformation of a protein. The procedure therefore starts with low-energy conformations of dipeptides, combines them to find low-energy conformations of tripeptides, etc. Obviously, the number of conformations will grow geometrically, and the only way to keep the problem tractable is to prune higher energy conformations (herein, the assumption of the dominance of short-range interactions enters). The key to the success of the method is choosing an energy window that allows tractability but does not eliminate the desired lowest energy structure of the full polypeptide. Vasquez & Scheraga (41) have applied the method to predict the minimum energy backbone conformation of the pentapeptide Met-enkephalin, H-Tyr-Gly-Gly-Phe-Met-OH. They find that the low-energy conformations are quite close to those found theoretically by Isogai et al (42), but that two lower energy structures were found as well. Finally, the three lowest energy conformations agree almost exactly with the three most probable conformations obtained by an adaptive importance sampling Monte Carlo algorithm (43) (see below). However, to date, it has not, to the best of our knowledge, been applied beyond a pentapeptide. Although it is probably an efficient method for small peptides, it is likely to experience severe computational difficulties due to the multiplicative nature of the number of conformations that must be generated and, more generally, due to its implicit assumption of the dominance of short-range interactions.

Monte Carlo Methods

Although the explicit search algorithms outlined above are conceptually appealing in that all conformations are delineated and their importance is assigned based on their relative energies, they are unfortunately incapable of treating the protein-folding problem. A particularly powerful method for exploring conformational space is the Monte Carlo (MC) method (44). Despite the fact that a number of variants of the Monte Carlo method have been extensively applied to study both simple liquids (44) and even to polymeric systems (45), its application to proteins has been surprisingly limited. Perhaps this was due to the belief that molecular dynamics is an inherently more efficient technique to study complex molecules than is the MC method (27, 46). Although this is probably true for the original algorithm, a number of new techniques that allow one to take larger steps in phase space have been implemented (44, 46). Moreover, methods exist in principle for the calculation of the total free energy—not just the free energy change due to a small perturbation (27–29, 31). Finally, dynamic MC algorithms permit one to simulate systems for substantially longer times than are currently possible with molecular dynamics (47). Both methods have their strengths and weaknesses, however MC is particularly appropriate for exploring the multiple conformational minima that are inherent in the globular protein folding problem.

A technique useful for relatively small polypeptides has been developed by Paine & Scheraga (43, 48, 49). The method calculates the probable conformation of a polypeptide chain by using equilibrium statistical mechanics in conjunction with an adaptive importance sampling technique. The method allows estimation of the relative free energies of any conformation and the calculation of the average conformation of the molecule. The object of the calculation is the Boltzmann probability density for a particular configuration. The SMAPPS algorithm (Statistical Mechanical Algorithm for Predicting Protein Structure) works as follows: Based on an initial guess of their probability, M conformations are selected from the whole backbone dihedral angle space. This allows for the first (crude) estimate of the partition function Z. After the initial value of Z is calculated, the probability density is modified to sample the regions of higher probability. The procedure is repeated until the probability obtained from the tentative partition function converges to the Boltzmann distribution function. A limitation of the method is the necessity to sample the backbone conformational space via a grid search. However, by emphasizing only important regions, it should allow larger systems to be studied.

The SMAPPS method employing the ECEPP/2 algorithm for the interaction energies has been applied to predict the native backbone conformation of Met-enkephalin in vacuum (49). It gave a structure (energy of −9.92 kcal/mole) in agreement with the build-up procedure described above. A lower energy structure (−13.63 kcal/mole) was obtained when side chain and peptide bond dihedral angles were allowed to relax (43). This corresponds to a type II′ β-turn involving residues 2–5. In this case, the conformation of each residue is assumed to be statistically independent of the others; this is a severe approximation. The procedure generated a hierarchy of structures in order of increasing relative free energy. The resulting lowest free energy structure contains a γ-turn involving residues Gly^2-Gly^3-Phe^4. The structure is not the lowest energy structure (which contains a β-bend) but is a lower free energy structure by about 9 kcal/mole due to the contribution of entropy. This points out the importance of including entropy in the conformational analysis. However, solution NMR indicates that a type I′ β-bend with Gly^2-Gly^3 at the corners of the bend may be present (50); the predicted γ-turn is inconsistent with the solution NMR data. Thus, something is clearly wrong—whether it is the energy functions employed (excluding solvent), the sampling procedure, or the assumed independence of the conformations of the residues is unclear.

To surmount the multiple minima problem, Li & Scheraga (51) have proposed a hybrid Monte Carlo/energy minimization procedure that works as follows. First, one employs a standard MC sampling process that generates a trial configuration by making a completely random change in a randomly selected dihedral angle. Then, the randomly chosen trial configuration is subjected to energy minimization employing the ECEPP/2 energy function. This drives the system to the nearest local energy minimum. Finally, the new local minimum is subjected to the standard Metropolis criterion (52). As is typical, the energy minimization step is the most CPU intensive and, for the case of Met-enkephalin, took 95% of the total CPU time. The advantage of the method is that it relatively rapidly sorts the local minima according to the Boltzmann criterion, and therefore in the limit of a long run should generate a Boltzmann distribution of conformational states.

In the application of the method to Met-enkephalin in the absence of water, the apparent global energy minimum conformation (whose energy is −16.02 kcal/mole) has a type II′ β-bend involving the residues Gly^2-Gly^3-Phe^4-Met^5 and was obtained from 12 out of 17 randomly generated initial conformations. Unfortunately, 5 out of 17 conformations did not converge after 10,000 iterations and had structures that differed substantially from the global minimum energy structure. Therefore, trapping in local minima is evidently a problem. In the presence of water at 25°C, no unique global minimum is found. At this point it is unclear whether the algorithm is too slow to lead to convergence, whether the potential energy itself is in error, or whether the NMR data (50), which appear to be consistent with a type I' β-bend, need to be re-examined.

To summarize the conformational predictions of the above methods as applied to Met-enkephalin in a vacuum; the SMAPPS procedure with sidechains predicts a type II' β-bend with an energy of -13.63 kcal/mole as the lowest energy state; the SMAPPS procedure predicts that a γ-turn is the lowest free energy state; Monte Carlo/energy minimization predicts that a type II' β-bend is the global energy minimum at -16.02 kcal/mole; experiment indicates that a type I' β-bend exists in solution.

Methods for Free Energy Calculation

As illustrated by the Paine & Scheraga calculation on Met-enkephalin (43), the global energy minimum is not necessarily the global free energy minimum, and thus methods for obtaining the free energy are required. The problem is simplified if the system of interest undergoes relatively small fluctuations about a local minimum. Then, procedures based on normal coordinate analysis can be employed to calculate the entropy. Gō & Scheraga (53, 54) (who assume that the dihedral angles are the only important variables) have developed a variant of the approach in which only harmonic fluctuations are included, and they have employed the method for a number of polypeptides (55–58). Subsequently, Karplus & Kushick (59) and co-workers (60) generalized the Gō & Scheraga approach and assumed that the probability distribution for the important coordinates is a normalized, multivariate distribution about the average value of the coordinates. In the harmonic approximation, only quadratic terms in the potential energy are employed. In the quasiharmonic approximation, the simulation (e.g. Monte Carlo or molecular dynamics) provides both the average value and the temperature-dependent variances and covariances of the internal coordinates. Application to n-butane and decaglycine in the harmonic approximation (59) (i.e. using a strictly harmonic potential) demonstrated that bond angle bending contributes significantly to the entropy of dense systems that exhibit strongly correlated motions. Levy et al (60) examined the role of anharmonicity by calculating (for an anharmonic potential) the entropy difference between the harmonic and quasiharmonic approximations to the entropy in α-helical decaglycine at 5, 100 and 300K. They conclude that at room temperature anharmonic corrections to the entropy are important and that the harmonic approximation serves to underestimate the entropy severely.

In contrast to the native conformation of a globular protein, which explores a relatively small region of phase space (although local multiple minima exist) (16), the denatured state of the protein, being a statistical random coil, explores a very large region of phase space. This case is identical to the problem of estimating the configurational entropy of a random coil polymer. Flory (61) and Gō & Scheraga (53, 56–58) have developed methods for estimating the configurational entropy under circumstances in which the excluded volume effect is basically ignored. More generally, Meirovitch (62–64) has developed a Monte Carlo scanning method and applied it to both self-avoiding and self-attracting random walks on lattices as well as to a continuum (65). However, the technique is unsuitable for the study of systems having a highly stable conformation, because transitions are allowed that permit all the systems to drift away from the stable conformation.

Unlike a standard MC procedure, which begins with a fully constructed chain and then allows its conformation to evolve, a scanning method is a sequential, atom-by-atom construction process of the chain, beginning at the N-terminus and finishing at the C-terminus (62–65). The key to the method is the ability to calculate the transition probability for obtaining a particular dihedral angle at the kth step in the construction process independent of the possible future outcomes, given that a specific conformation of the chain up to the kth step already exists (66). The construction of this probability requires one to integrate over all future outcomes for the remainder of the chain. The transition probabilities are of interest because the product over all the chain's dihedral angles is the Boltzmann probability distribution function for the conformation, P. If P is known, then the entropy, energy, and free energy follow from elementary statistical mechanics. In practice, an approximate probability density is constructed that depends on a small number of recent previous steps.

In a series of papers (66, 67), Meirovitch et al have developed a version of the scanning method that in principle is capable of calculating the configurational entropy for a chain of arbitrary flexibility. The method was initially applied to the hairpin and α-helical conformation of decaglycine (66) and yields the prediction that the α-helix is lower in free energy by 0.4 kcal/mole per residue. The entropy of the helix is larger due to rotation of the four terminal single bonds ($T\Delta S \simeq 0.1$ kcal/mole/residue) and the energy is lower (0.3 kcal/mole/residue). In a follow-up calculation (66), they extended the scanning procedure to the full range of phase space rather than as above, where the sampling was limited to the helical and hairpin conformations. The basic methodology is similar to the previous work, with the exception that the scanning procedure itself is used to generate the ensemble rather than a standard Metropolis scheme from which transition probabilities are extracted. At 100K, the α-helix is predicted to be the most stable conformer, whereas at 300K the statistical coil becomes the favored state. Thus for this very simple case, at least, the methodology is capable of calculating the entropy over a broad range of chain flexibility. Moreover, it should be able to treat 20 to 30 residue peptides. This is a promising technique that should be more fully exploited in the future.

TERTIARY STRUCTURE PREDICTION IN GLOBULAR PROTEINS

Until this juncture, a number of methods have been described whose application thus far has been limited to relatively small molecules. Recently, there have begun to emerge a number of techniques that may be able to treat larger systems. These are reviewed below in order of decreasing level of molecular detail.

Self-Consistent Electrostatic Field Methods

A novel idea has been proposed by Piela & Scheraga (68), which has as its basis the assumption that electrostatic interactions dominate protein folding (69–72). If so, then to find the native conformation one need only optimize the electrostatic interactions, although in subsequent work the method has been coupled to a hybrid electrostatic/Monte Carlo energy minimization approach (73). The procedure self-consistently optimizes the peptide dipole moment in the direction of the electric field and has been designated the self-consistent electric field method. The SCEF method was applied to a terminally blocked 19-residue polyalanine chain. A number of test cases were examined in which the starting conformation was placed further and further from the global energy minimum α-helix conformation; all runs successfully located the α-helix conformation.

There are, however, a number of problems with this procedure. It cannot differentiate a given topology from its mirror image, and it is not known how many distinct conformations can be generated by the SCEF procedure. Furthermore, the SCEF procedure can make large dihedral angle variations, thereby permitting the rapid sampling of many local energy minima; this is both a good and bad feature. Because it identifies defects as measured by the relative peptide dipole orientation with respect to the electric field and rapidly removes them, without the inclusion of additional criteria, it can unravel correctly folded local regions of the native state that have electrostatic defects. Finally, no estimate of the total energy of the molecule is made.

To alleviate a number of the above difficulties, Ripoll & Scheraga (73) have developed a combined SCEF and MC/energy minimization method; the former strives to obtain an optimal charge distribution, whereas the latter tries to minimize the total energy of the system. They assume that at room temperature electrostatic interactions dominate. Of particular interest is the inclusion of back-track mechanisms that temporarily suspend the Metropolis criterion when the system is caught in a deep local minimum.

From 11 starting conformations, including the random coil and the left-handed α-helix, the procedure always found the global minimum right-handed α-helix conformation of the terminally blocked 19 residue poly(L-alanine). In general, although a greater number of conformations are generated by the MC method, the acceptance ratio for the electrostatically generated conformations is larger. Furthermore, the conversion from the left-handed into the right-handed α-helix is an impressive result. Electrostatically, the backbone dipoles are well aligned with respect to the electric field, and, based on the electrostatic criterion alone, nothing should happen. However, the total energy of this conformation is substantially higher than the global energy minimum. The system waited a considerable time until a right-handed α-helix nucleated, after which it slowly grew to the full right-handed helix, which seemed stable to further global conformational modification.

Given the utility of this technique, it is reasonable to inquire whether it will be applicable to proteins. Ripoll & Scheraga report that their energy minimization step scales as the cube of the number of atoms. Therefore, at present, applications to molecules having more than 40 residues will have to be restricted to examining conformations in the vicinity of predetermined regions of phase space. Thus, the method does not appear to be able to address the full globular protein folding problem.

Simplified Folding Models

Up to now, we have described a number of approaches that treat small peptides and peptide fragments at a high level of detail. Main-chain and sometimes side-chain atoms are included, and the most detailed and realistic potential energy surfaces that are extant are employed. Recognizing that the number of degrees of freedom and the complexity of the potential energy surface must be reduced in order to treat the folding of globular proteins, a number of authors have developed simplified globular protein models designed to provide qualitative insights.

Levitt & Warshel (74, 75) made a very early attempt to simplify the model of a globular protein and used an α-carbon representation for each residue in pancreatic trypsin inhibitor (PTI); i.e. every backbone residue is a sphere to which are attached rigid side chains, and successive residues are linked by virtual bonds. Each backbone residue has a single torsional degree of freedom. The energy function consists of van der Waals interactions between residues, peptide hydrogen-bond interactions, side chain–

solvent interactions based on accessible surface area, and nonbonded near neighbor interactions. The system is started out in an arbitrary conformation that is then subject to energy minimization. After the system finds a local energy minimum, random displacements in the directions of the normal mode vibrations are made. These moves work well when the chains are unfolded but do not work as well for compact, folded conformations. In the latter case, normal mode thermalization can lead to large displacements from the local minimum, so large in fact that subsequent energy minimization may place the system in another energy minimum. To avoid the problem of large initial forces, holding potentials were employed to keep the system close to the starting conformation until the large initial forces had decayed. Furthermore, they found it necessary to introduce pushing potentials to allow the system to surmount steep, highly nonquadratic local minima where normal mode thermalization was ineffective.

For a range of parameters and conditions, native-like PTI conformations were generated. The importance of this early simulation lies in the demonstration that folded conformations can be obtained by using a simplified coordinate representation. Subsequently, these simulations were subject to intense criticism by Nemethy & Scheraga (76) and Hagler & Hönig (77), who argued that the use of pulling and pushing potentials is arbitrary—when and when not to invoke them is not clear *a priori*. Furthermore, they argued that since extended structures having a helix at one end or a fully extended state constitute the initial states, the correctly folded structure is built into the initial configuration. A somewhat similar procedure that used pushing and pulling potentials was subsequently employed by Kuntz et al (78) and by Robson & Osguthorpe (79). The former employ a detailed potential energy surface for PTI and rubredoxin that includes a repulsive potential, a potential related to the virtual bond length, hydrophobic interactions, a centrosymmetric potential that assumes a target radius for the distance of the hydrophobic and hydrophilic residues from the center of mass of the molecule, and a forcing function for the disulfide bonds in PTI and the sulfur/iron bonds in rubredoxin. Starting from extended conformations, they successfully folded compact structures that resemble the desired native conformations of these molecules. Robson & Osguthorpe initiate folding in a PTI model from a conformation having the predicted secondary structure and a disulfide closing potential as well as from a nearly extended chain, and have emphasized the importance of turn regions for native state structure formation.

Gō and co-workers have employed a series of highly simplified two- and three-dimensional lattice models to represent globular proteins (80–86). Typically, the chain is constructed as a consecutive sequence of square (cubic) lattice sites for the two- (three-) dimensional case. Side chains are represented as nearest neighbor lattice points. In most cases, they arbitrarily restrict tertiary interactions to those residues that have contacts in the native state. We return to this point below. In many cases, all the long-range interactions are assigned a uniform stabilization parameter. For short-range interactions, they assume that the energetically preferred local conformation of each residue is the same as in the native state. They also include the possibility of hydrophobic interactions by classifying each unit as polar or nonpolar. The entropy of the system is assumed to decrease in proportion to the number of vacant lattice sites next to nonpolar units. The conformation of the chain is modified by a dynamic Monte Carlo procedure in which one starts out in an initial configuration and then subjects the chain to local micro modifications, e.g. a local translation or rotation. This method can simulate the time behavior of a stochastic system if the elementary time unit is defined as the time required for each of the beads on average to be subject to all the possible local modifications. Basically, this entails a Monte Carlo solution of a master equation for the time evolution of the system (44). Whether the solution is physical depends on the nature of the local modifications employed. If dynamic information is desired in addition to their equilibrium properties, then care must be taken to ensure that the time scale is not distorted.

Based on the study of two-dimensional lattice models (80, 82–86), Gō and co-workers conclude that:

1. The all-or-none character of the unfolding and folding transition arises from the specificity of the long-range interactions.
2. The nonspecific component of the long-range interactions decreases the cooperativity and the transition rate by stabilizing incorrectly folded intermediates.
3. Short-range interactions reduce the cooperativity of the transition but accelerate the folding rate.
4. Medium-range interactions (e.g. loop formation) can counteract the kinetic slowdown resulting from the nonspecific component of the long-range interactions.

Gō and co-workers successfully folded a three-dimensional cubic lattic model of bovine pancreatic trypsin inhibitor (BPTI) from the denatured state (85), but were unable to do so for a model of lysozyme that included only native long-range interactions (81). This failure is ascribed to the presence of mixtures of mirror image intermediate states. Because the potentials employed possess spherical symmetry, a given conformation and its mirror image are isoenergetic. The failure to fold lysozyme to the native conformation (or its mirror image) indicates that the MC jump algorithm that they employed is inefficient at surmounting deep local minima. Gō and co-workers were, however, sometimes successful in refolding lysozyme from partially unfolded conformations.

More recently, Taketomi, Kano & Gō (86) examined a two-dimensional 49-residue cubic lattice polymer, where each residue is assigned an energy that favors the "native conformation" and where the only allowed long-range interactions are between residues that are nonbonded nearest neighbors in the native state. The model protein consists of two "α-helices" at right angles, which are then followed by five antiparallel strands of β-sheet. In addition to the native protein, two substitutions of a pair of residues, in one case in the middle of the β-sheet and in the other case at the α-helix/β-sheet junction, were made. These substitutions are assumed to affect only the short-range interactions. In all three cases, transitions that are thermodynamically and kinetically all-or-none are observed. Not surprisingly, these amino acid substitutions, modeled solely as having unfavorable short-range interactions, slightly decrease the stability of the native state and slow down the folding transition. Unfortunately, a detailed analysis of the intermediate states was not carried out, but based on the temperature dependence of the free energy of activation for folding with the two kinds of defects, Taketomi et al conclude that the β-sheet assembles prior to its attachment to the α-helix. They are vague about whether the α-helices have already assembled. Furthermore, they conclude that the process of assembly occurs nearly at the activated state of the transition. This bears a certain similarity to the "cardboard box" model of Goldenberg & Creighton (87), who conjecture that the activated state is a high energy distorted form of the native conformation. Here the activated state is also dominated by the enthalpy, but it is a partially assembled native state, having approximately half the number of native contacts. The entropy of the transition state and the native state are quite close. Whether the discrepancy with the "cardboard box" model is due to the fact that these are lattice proteins incapable of having high energy distorted forms or whether the difference is more fundamental is unclear. Furthermore, unlike the classical nucleation model of protein folding (10), which assumes that once a critical size nucleus near the denatured state forms, the free energy is downhill in the direction of the reaction coordinate, here substantial assembly has occurred prior to the transition state. Thus, the qualitative mechanism of assembly in the simulations is in agreement with experiment (88).

Krigbaum & Lin (89) have used a bcc lattice model of PTI and investigated the relative folding efficacy of a centrosymmetric vs. a local interaction potential. Residues are divided into the categories of hydrophobic, hydrophilic, and indifferent, and two kinds of centrosymmetric potentials are employed. The first type assumes that the potential of a given kind of residue is a quadratic function of the distance of the residue from a target radius, and the second type further biases the hydrophobic residues to lie near the center of mass of the protein. The local potential model is based on a quasichemical approximation to the interaction free energy between pairs of residues. The initial conformations consist of the best lattice representation of the native molecule, a random coil, a random coil with intact disulfides, and one having an intact helix. Both kinds of potentials work equally well at forming compact structures having radii of gyration quite close to BPTI. Nevertheless, in no case is the native structure recovered, but the presence of the native disulfides produces collapsed structures having the largest number of native hydrophobic contacts. The authors conclude that use of a lattice sufficiently restricts conformational space so that the folded structures are closer to the native PTI than those obtained by Levitt & Warshel (74, 75) and Kuntz et al (78).

Miyazawa & Jernigan (90) have performed an off-lattice MC simulation of the folding and unfolding pathways of BPTI. The model protein consists of the backbone C^α and side chain C^β atoms. The side chains are fixed, and the backbone dihedral angles are allowed to vary in ten degree increments. A hard core repulsion at short distances between all residues is implemented. The short-range part of the potential consists of an empirical energy obtained from the backbone dihedral angle distributions compiled by Nemethy & Scheraga (76) and an interaction term that favors the native conformation of a given residue by -1 kcal/mole. Long-range interactions are accounted for by assigning an interresidue contact energy of -2 kcal/mole, if nonbonded residues i and j lie within a certain distance range in both the native and the given conformation. The presence of disulfide crosslinks is ignored, a severe approximation. By employing a biased Monte Carlo sampling technique due to Wall et al (91) and others (92, 93), Miyazawa & Jernigan (90) generated a total of 230,000 conformations spanning the completely denatured to the completely folded state, and they classify each conformation according to its conformational energy; a conformational entropy, $S(E)$, is assigned to each energy E.

Miyazawa & Jernigan (90) have examined the origin of the concavity in the $S(E)$ vs E plot that is responsible for the maximum in the free energy function $[F = E - TS(E, T)]$ between the native and denatured states that must be present if a two state model is valid and conclude, in agreement with Gō (80–86, 94, 95), that the all-or-none transition is due to the site-specific long-range interactions.

Miyazawa & Jernigan have also examined the order of appearance of tertiary structure and conclude that folding initiates at a β-turn and in the α-helix. Tertiary contacts then grow to include the native pair of interacting β-sheets. Subsequently, the folded structure then grows outward to the N and C terminii, until the fully assembled structure results. The order of assembly in this equilibrium-folding simulation is in strong disagreement with the experimental studies of Creighton (8, 96). The discrepancy with experiment that occurs for this folding model most likely arises from its neglect of topological constraints exerted by the crosslinks, as BPTI, in fact, will denature if the crosslinks are reduced. This implies that the effect of loop entropy (17, 97, 98) (the reduction in configurational entropy when random coil conformations are constrained relative to when they are free) may be rather important.

A major problem with the above approaches is that they ignore interactions between any spatially close pair of residues that are not in contact in the native state. A consequence of such an approximation is a reduction in the number of allowed tertiary interactions and thus, it is not at all surprising that the calculated conformational transitions are all-or-none. To some extent it is built in a priori, i.e. a whole spectrum of possible intermediates do not interact and therefore will contribute marginally to conformational averages. Furthermore, the conclusion that short-range interactions cannot produce an all-or-none transition is well known. Single-chain synthetic helix coil transitions are continuous, i.e. a given molecule is likely to occupy the full spectrum of states from the unfolded to the folded conformation (99). Moreover, the ability to refold a protein when only native-like contacts are allowed is really a test of the algorithm and nothing more; it has no predictive value. If a structure is specified in advance, then successful folding indicates that the algorithm is sufficiently efficient to find the target structure. Failure indicates that the algorithm is incapable of folding even this simplified test case, and it will surely fail if the full range of interactions are allowed. Success in folding means the algorithm should be tried for the more general case, but there is no guarantee of success in the latter. Although these approaches represent reasonable first steps, algorithms that can successfully refold to the native conformation from the denatured state without restricting interactions to purely native contacts are required, and the thermodynamics of the globular protein conformational transition must emerge naturally without ad hoc restrictions.

Over the past several years, Skolnick, Kolinski (100–107) and co-workers have developed a series of lattice methods designed to do precisely this. A tetrahedral lattice, α-carbon representation of a protein is employed. Each individual site represents an amino acid residue, and all nonbonded residues that are nearest neighbors are allowed to interact. This interaction may be attractive, repulsive, or neutral, but it depends only on the residue types and not on whether the contact occurs in the lowest energy, native state. This approach is very much in the minimalist spirit of Gō and coworkers (80–86); the objective is to develop a series of idealized models that reproduce the general features of globular protein folding. For small proteins, the thermodynamics of the conformation transition must be all-or-none, and the system must reproducibly fold to and unfold from the unique native state (8, 14, 17–20). The models are only made more complicated when the simpler realization is unable to reproduce a desired feature of the physics. The basic approach has its philosophical origins in polymer physics, from which the methodology of many of the earlier simulations are derived (108).

In the earliest simulations (100, 101), the *trans* (t) state is assumed to be energetically favored over the *gauche* plus (g^+) and *gauche* minus (g^-) states, nonspecific attractive interactions between any pair of nonbonded nearest neighbors are allowed, and various realizations of dynamic Monte Carlo algorithms are employed. Two kinds of collapse transitions are seen. If attractive interactions dominate over the local stiffness (the intrinsic preference for a t over a g^+ or g^- state), then collapse to a dense random globule is observed. The collapse transition is continuous, with no evidence for a bimodal population of expanded conformations in equilibrium with dense random globules. If, however, the chain possesses substantial local stiffness before collapse (although it is not a rod, but is still a random coil), then collapse to a highly ordered, β-barrel-like conformation in equilibrium with an unfolded, random coil conformation results. The conformational transition is two-state in this sense. Furthermore, the fraction of β-states (*trans*) is increased on collapse. That is, secondary structure is induced by tertiary interactions. The mean length of a β-stretch is approximately the same on successive refolding, but the turn locations are variable. Thus, a manifold of β-barrel-like collapsed conformations exists, each in all-or-none equilibrium with the unfolded state. Although promising, these models did not yet reproduce the qualitative behavior of real proteins in two essential respects. There is too much secondary structure in the denatured state, and there is a manifold of compact β-barrel conformations that is populated.

Since the previous model had too much secondary structure in the denatured state, perhaps by making the denatured state devoid of any preference for secondary structure, the conformational transition would more closely resemble that seen in real proteins. Thus, in a subsequent work (102), *trans* and *gauche* states are isoenergetic, but if any nonbonded pair of nearest neighbor states are in a t conformation, they are energetically stabilized by a cooperative interaction embodied in a parameter ε_c. The extent of stabilization depends on the number of bonded nearest neighbor *trans* states. This model undergoes a very sharp transition from a voluminous random coil to a dense, nonunique random globule having stretches of local secondary structure. The reason the collapsed states did not form highly ordered, extended β-sheet structures is as follows. Unlike a β-barrel configuration whose configurational entropy is low, dense random globules have a relatively high configurational entropy, i.e. there are a very large number of ways of arranging the short β-strands. If, however, there is a local energetic preference for β-states, then the favorable energetic term can successfully compete with the unfavorable entropic term, and the resulting lowest free energy structures are highly extended β-barrels. This conclusion is borne out in a model that includes the ε_c cooperativity term plus a marginal local preference for *trans* states and that produces a two-state transition (without any site specificity whatsoever) to a highly extended, but nonunique β-barrel-like structure. Inclusion of uniform attractive interactions merely shifts the transition temperature. Thus, a local energetic preference seems to be required to shift the equilibrium from a dense random glubule state to ordered native-like structures, and an inevitable conseqence of these models is that there must be some fluctuating, marginal secondary structure in the denatured state, a fact in agreement with experiment (12, 13a,b). Unfortunately, this model fails to reproduce an essential feature of globular proteins (at least for small proteins); the ability to recover a unique tertiary structure on successive refolding (14).

Among the problems encountered in the above simulations was the very poor sampling efficiency of dense conformations. Thus, a new MC sampling scheme was proposed that allows rapid relaxation of conformational defects (104). Furthermore, although collapse to extended structures can be obtained with uniform interactions, generalization to real proteins should include the presence of hydrophobic and hydrophilic residues (8, 15). Pairs of hydrophobic residues are assumed to interact with an attractive potential of mean force, hydrophilic/hydrophobic pairs interact with a repulsive potential of mean force, and hydrophilic/hydrophilic pairs can be indifferent, attractive, or repulsive. Qualitatively identical behavior is found. Finally, we incorporated into the primary sequence a statistical tendency to form turns, the importance of which has been suggested previously (13a,b, 25, 30, 79, 109–112). (On a diamond lattice, a turn consists of three residues in a $g^{\pm}g^{\mp}g^{\pm}$ conformation.) In the turn-neutral case, based on short-range interactions, the t and g states are isoenergetic, and thus turns are induced only by tertiary interactions. In the "strong-turn" case, g^+ and g^- states are isoenergetic, and, based on short-range interactions, a native turn is but one of eight equally likely conformations.

Incorporating the above, Skolnick et al (104) subsequently examined the nature of the β-hairpin/four-member β-barrel equilibrium; we refer the reader to the literature for the details. Of greater interest here (103, 104) is the identification of sufficient conditions required to obtain the unique (and hypothetical) four-member β-barrel structure (N of Figure 1) in all-or-none equilibrium with the random coil state (D of Figure 1). These are an alternating sequence of hydrophobic and hydrophilic residues appropriately punctuated by turn-neutral regions at the desired location of the native turns. Furthermore, loop entropy (17, 97, 98), when coupled with the hydrophobic/hydrophilic pattern, keeps the turn conformations tight. The preference due to tertiary interactions for a bend at a turn-neutral location decreases the free energy of the in-register conformation sufficiently from the out-of-register states so that the latter do not contribute to the population. Juxtaposition of all the above features (without site-specificity—all the hydrophobes are identical) is what is required to produce an all-or-none conformational transition to a unique state. The models of Gō et al (80–86) and Miyazawa & Jernigan (90), with their site specific, native-like interaction potentials, had all these essential elements as well, and merely represent a far more restrictive example.

Skolnick et al (106) subsequently treated the folding of the six-member, β-barrel Greek key motif (113), which has a topology close to that in the real protein plastocyanin (114). Figure 2 displays a random coil conformation (A) in equilibrium with the folded native conformation shown in a side (B) and top view (C) of Figure 2. The latter shows the numbering of the β-strands used below. Unlike the simple topology of a β-barrel, there is a reversal of strand direction in order that strand 6 be able to interact with strands 1 and 3. Thus, there is a large entropic barrier that the loop between strands 5 and 6 must surmount before folding to the native conformation. A primary sequence constructed from an alternating hydrophobic/hydrophilic pattern of residues consistent with β-sheet formation in the native structure, bend-neutral regions consistent with the

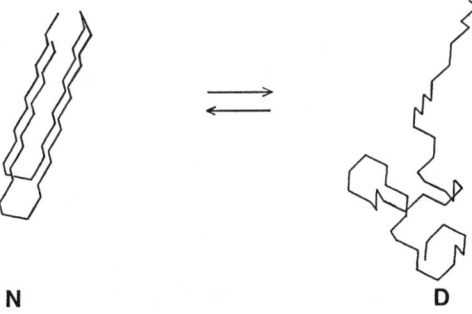

N   **D**

Figure 1 N is the native conformation of the four-member, β-barrel in equilibrium with a representative random coil conformation D.

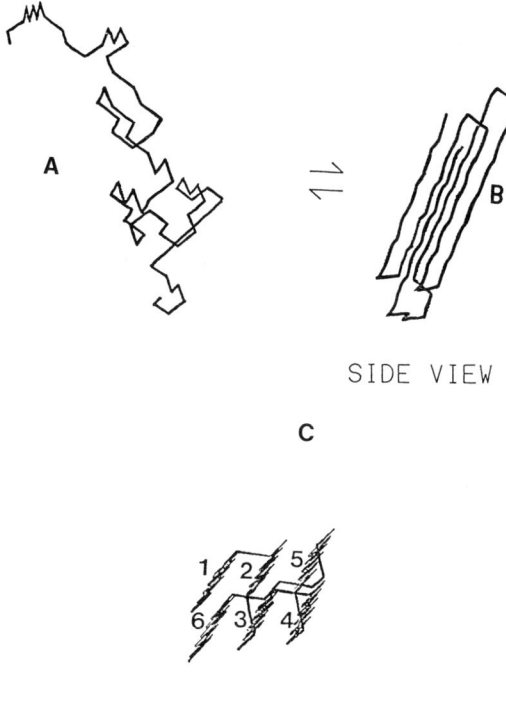

Figure 2 Representative random coil conformation (*A*) in equilibrium with the native six-member, β-barrel Greek key shown in a side view (*B*) and top view (*C*). The latter indicates the numbering scheme for the β-strands.

location of the tight turns in the native state, and a loop that has a statistical preference for a small manifold of states including the native conformation are sufficient to produce an all-or-none transition to the unique Greek-key native state. There is a marginally populated, folding intermediate involving a β-barrel composed of strands 2–5, perhaps with β-strand 1 attached. These intermediate states become more populated if the strength of interaction of the loop for the turn between β-strands 3 and 4 is diminished. This can transform the two-state model into a three-state model with the four-member β-barrel as the equilibrium intermediate.

The range of thermal stability of the six member β-barrel can be augmented by increasing the preference of the residues in the putative tight turn regions for *gauche* states. Conversely, the uniqueness of the conformational transition can be destroyed by making the loop based on short-range interactions indifferent to the native state conformation. In this case, while at very low temperature, the unique six member β-barrel is observed. In the transition region not only is the four member β-barrel intermediate substantially populated, but so are out-of-register conformations of strand 6. Due to the effect of entropy, the loop literally tugs at β-strand 6, pulling it into out-of-register conformations. Thus, loops play an important role in determining the conformational stability of proteins.

If the general methodology is correct, it should also be able to reproduce the α-helical protein motif. Thus, the same approach was taken by Sikorski & Skolnick (105, 107) to fold the three variants of left-handed, four-helix bundles (Figure 3). The first of these variants (see 0 of Figure 3) has three

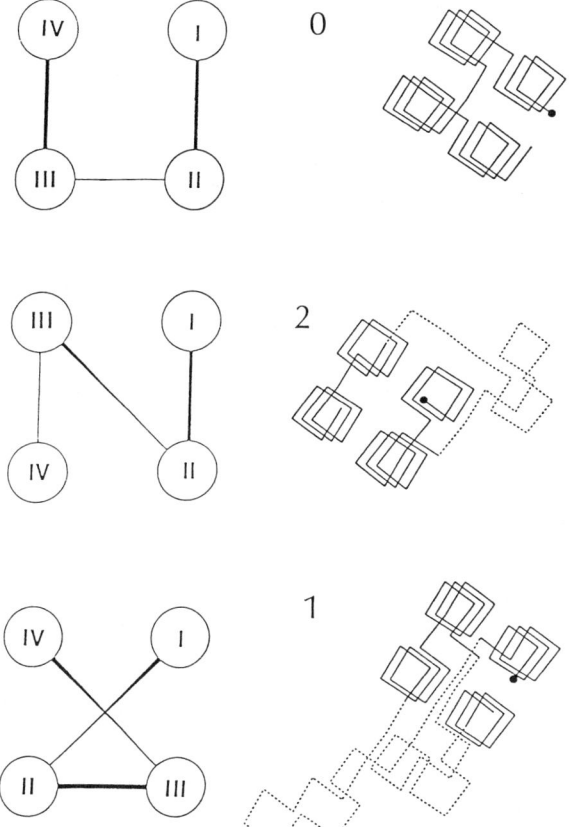

Figure 3 Representative native state topologies of four-helix bundles, side by side with their diamond lattice representations obtained on folding from the denatured state. Shown are topologies with three tight bends (0), one long loop (1), and two long loops (2).

tight bends as is found in cytochrome c′ (113, 115), myohemerythrin (113, 116), and putatively in the synthetic globular protein of Regan & DeGrado (117); the second variant (1 of Figure 3) has two tight bends and one long loop as in apoferritin (118); and the final variant (2 of Figure 3) has two long loops and a tight bend and occurs in porcine growth hormone (119). Although these topologies could be obtained by assuming a local preference for the g^- state (a sequence of g^- states produces the diamond lattice realization of an α-helix), Sikorski & Skolnick (105, 107) opted for the introduction of helical wheel interactions (120). That is, if residues i to $i+4$ are in an α-helix, then this conformation is energetically favored. Residues involved in the four-helix bundle motif are assumed to have an amphipathic primary sequence; those in the putative loop are hydrophilic and attractive with all the other hydrophilic residues; and tight bend residues may be either bend-neutral or the native bend is one of four degenerate lowest energy states. Sufficient conditions to fold the four-helix bundle with tight bends (Structure 0 of Figure 3) are a central turn-neutral region flanked by two amphipathic tails (the out-of-register conformations in the α-helix are sufficiently higher in free energy that they are not appreciably populated). For topologies with loops, only the statistical loop regions plus amphipathic helices need be included in the primary sequence; i.e. the tight bend regions need not be specified at all. They are localized because of the interaction of the loop with the helix that stabilizes the native conformation and the energetic and loop entropic cost of fraying α-helices or shifting their registration. In all cases, to a very good approximation, an all-or-none transition to the desired native conformation is obtained.

Thus, this series of simulations indicates that the general rules of protein folding are rather robust and are to a large degree insensitive to details. What appears to be required is a general pattern of hydrophobic and hydrophilic residues consistent with the native topology and the presence of turn or loop-type residues at essential locations that produce the uniqueness of the tertiary structure. These studies further suggest that site-specific interactions are mainly involved in structural fine tuning of a given topology.

Skolnick & Kolinski (125) undertook a series of dynamic Monte Carlo simulations in which only local moves that reproduce the correct Rouse-like dynamics for a random coil polymer are allowed (47). Therefore it was hoped that such a simulation would provide insight into the protein folding pathways. For the case of the Greek key, folds initiate at or near one of the β-turns in the four-member barrel (strands 2–5). For the case shown in Figure 4, the initiation site involves strands 2–3 followed by the fairly rapid assembly of a β-hairpin. Of course, the structure might dissolve;

SIMULATION OF PROTEIN FOLDING 229

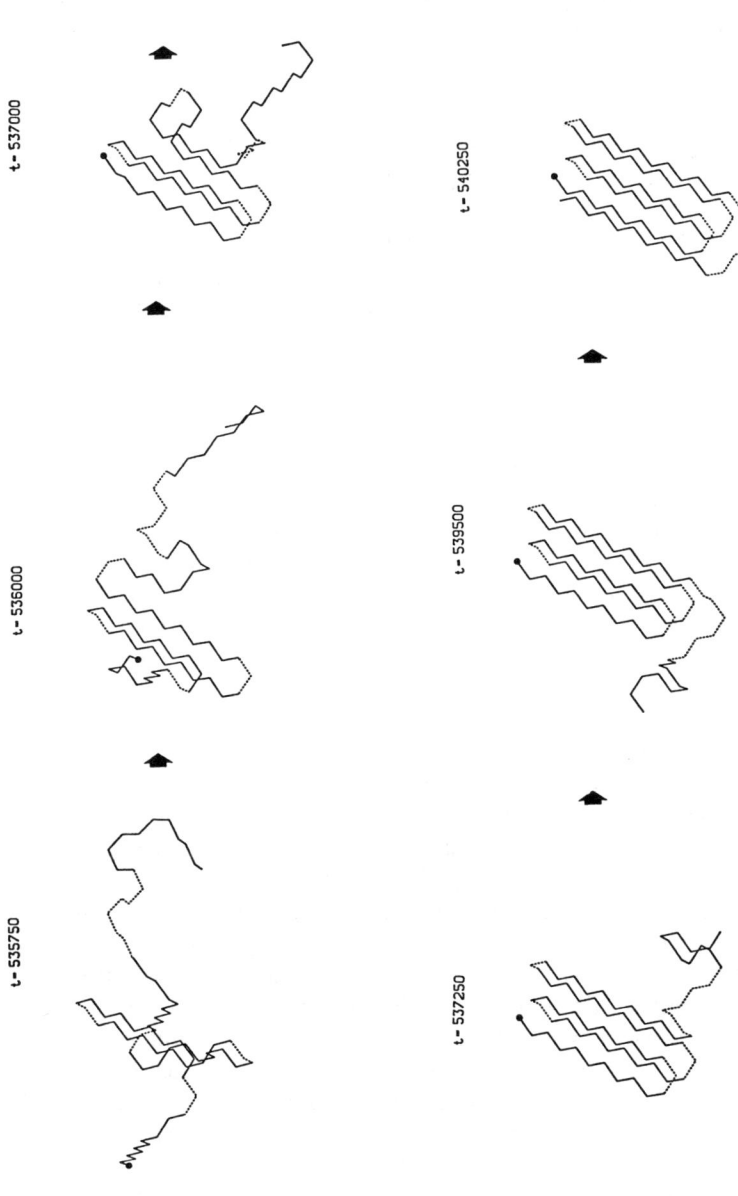

Figure 4 Representative folding pathway of the six-member β-barrel Greek key. The sections of the chain that in the native state form the loop and the turns (β-strands) are indicated by the *dashed* (*solid*) lines. The *solid circle* labels the N terminus. The time, t, is from the start of the run from which the trajectory is obtained.

there is the constant competition between structure dissolution and formation throughout the entire course of assembly. All β-turns, 2–3, 3–4, and 4–5 occur as folding initiation sites, but the turn between strands 2–3 is somewhat more preferred because, due to the excluded volume effect, closed loops are more likely the closer one gets to the end of the chain (121, 122). After initiation, the remaining strands of the β-barrel (strands 2–5) zip-up via an on-site construction mechanism; that is, the existing β-strand tertiary structure acts as scaffolding onto which subsequent β-strands assemble. Assembly of the four-member β-barrel intermediate is relatively rapid, typically taking about 10–20% of the total time to fold. The folding process is then punctuated by a very long pausing time as the remaining random coil tail thrashes about as it hunts for the narrow entropic pass to the native state. Now, the partially assembled structure hinders assembly. For example, the tail might be on the same side as strands 1–2 and must rattle about until it can go under or around the assembled portions of the protein. Eventually, the loop and strand 6 find themselves near strands 1 and 3; assembly is then very rapid. Unfolding proceeds essentially in reverse.

Figure 5 shows a similar series of snapshots of a representative folding trajectory of the four-helix bundle with two long loops obtained by Sikorski & Skolnick (126). A similar overall mechanism is observed. Folding initiates at or very close to a helix turn, and then a substantial fraction of the native structure rapidly assembles (three out of the four helices), using existing secondary structure onto which subsequent secondary structures assemble. Then a pause occurs, which is associated with the random search of phase space for the narrow entropic pass, before the final elements of secondary structure rapidly assemble.

The overall picture of protein folding that emerges from these simulations is a mechanism of punctuated "on-site construction." The secondary structure (e.g. β-sheets or α-helix) found in the native conformation is assembled on site; it does not form first and then diffuse into contact with another element of preformed secondary structure. In the case of β-strands, the algorithms have a kind of defect move that does allow performed β-strands to diffuse into contact. This does not happen often, because the β-strands are marginally stable and are constantly forming and dissolving. It is faster to initiate near a turn and assemble the conformation on site by dragging the pieces from the random coil state than it is to form the marginally stable β-strands in isolation and wait for these relatively large elements of secondary structure to diffuse together. The latter mechanism is assumed in the diffusion collision adhesion models that rely upon "prefabricated" construction (11, 123, 124) (see 124 for a representative simul-

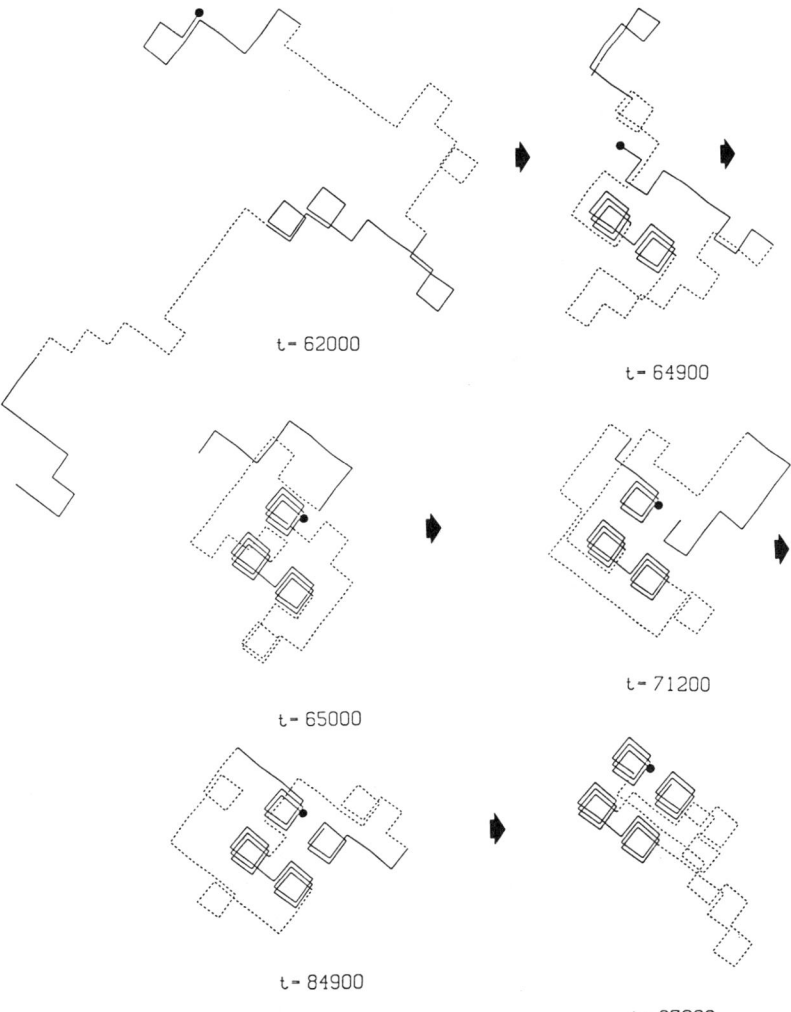

Figure 5 Representative folding pathway of the four-helix bundle having two long loops. The sections of the chain that in the native state form the loops (α-helices) are indicated by *dashed* (*solid*) lines. The *solid circle* labels the N terminus. The time, t, is from the start of the run from which the trajectory is obtained. One time unit corresponds on average to that when each bead in the chain is subject to each of the elementary micro modifications.

ation assuming preformed helices). Finally, on-lattice folding of α-helical hairpins indicates that the diffusion-collision mechanism of preformed helices does not successfully compete with the on-site construction mechanism, for the same reasons described above for the β-protein folding pathway.

It should be pointed out that there are, however, a number of severe problems with a diamond lattice representation of a protein. For example, it is impossible for a β-sheet to be parallel to an α-helix [which is the preferred geometric arrangement in real mixed motif α/β proteins (113)]. Furthermore, the β-strands cannot have a twist; the turn conformations are limited in number and are not truly representative of native-like turns. Thus, A. Kolinski, J. Skolnick and A. Sikorski (unpublished) have developed an alternative, 24 nearest neighbor lattice representation of a protein that is a more faithful model of the allowed topologies in globular proteins. In the context of a more realistic potential, including side chains, unique four-member β-barrel structures, α-helical hairpins, and mixed α/β motifs have already been successfully folded. The general folding rules seem to be the same as for the diamond lattice. Thus, some of the general conclusions about protein folding derived from the very simple diamond lattice models appear to carry over to far more complicated lattice descriptions of proteins as well.

SUMMARY

In summary, although a large number of disparate techniques have been applied to predict the tertiary structure of globular proteins from their amino acid sequence, the solution is not yet at hand. Methodologies for predicting the conformation of constrained, small protein fragments appear to be successful. As the size of the system increases, the level of detail of the treatment decreases; approaches that employ very detailed potentials appear to be limited to about 30–40 residues. Although this is a major advance, methods that reduce the effective number of degrees of freedom are clearly required. Lattice representations coupled to highly efficient Monte Carlo procedures appear to be one such approach. Thus, although a number of theoretical advances in the computer simulation of globular protein structure have been made, much work remains to be done before the globular protein folding problem is solved.

Acknowledgments

This work was supported in part by National Institute of Health grant GM-37408 from the Division of General Medical Sciences, United States Public Health Service.

Literature Cited

1. Kendrew, J. C., Bodo, G., Dintzis, H. M., Parrish, R. G., Wyckoff, H., et al. 1958. *Nature* 81: 662–66
2. Anfinsen, C. B. 1973. *Science* 181: 223–30
3. Jaenicke, R., ed. 1980. Protein Folding, *Proc. 28th Conf. German Biochem. Soc.* Amsterdam: Elsevier/North Holland. 587 pp
4. Ptitsyn, O. B., Finkelstein, V. A. 1980. *Q. Rev. Biophys.* 13: 339–86
5. Ghelis, C., Yon, J. 1982. *Protein Folding.* New York: Academic
6. Gō, N. 1983. *Annu. Rev. Biophys. Bioeng.* 12: 183–210
7. Wetlaufer, D., ed. 1984. *The Protein Folding Problem.* Boulder, Colo.: Westview
8. Creighton, T. E. 1985. *J. Phys. Chem.* 89: 2452–59
9. Levinthal, C. 1968. *J. Chim. Phys.* 65: 44–45
10. Wetlaufer, D. B. 1973. *Proc. Natl. Acad. Sci. USA* 70: 697–701
11. Karplus, M., Weaver, D. L. 1976. *Nature* 160: 404–6
12. Tanford, C. 1968. *Adv. Protein. Chem.* 23: 121–282
13a. Dyson, H. J., Rance, M., Houghten, R. A., Lerner, R. A., Wright, P. E. 1988. *J. Mol. Biol.* 201: 161–200
13b. Wright, P. E., Dyson, H. J., Lerner, R. A. 1988. *Biochemistry* 27: 7167–75
14. Privalov, P. 1979. *Adv. Protein Chem.* 33: 167–241
15. Kauzmann, W. 1959. *Adv. Protein Chem.* 14: 1–63
16. Elber, R., Karplus, M. 1987. *Science* 255: 318–21
17. Schellman, J. A. 1952. *CR Trav. Lab. Carlsberg Ser. Chim.* 29: 230–59
18. Tanford, C. 1962. *J. Am. Chem. Soc.* 84: 4240–47
19. Brandts, J. F., Lumry, R. 1963. *J. Phys. Chem.* 67: 1484–94
20. Wetlaufer, D. B., Malik, S. K., Stoller, L., Coffin, R. L. 1964. *J. Am. Chem. Soc.* 86: 508–14
21. Ptitsyn, O. B. 1987. *J. Protein Chem.* 6: 273–93
22. Garel, J. R., Baldwin, R. L. 1973. *Proc. Natl. Acad. Sci. USA* 70: 3347–51
23. Kawajima, K., Hiraoka, Y., Ikeguchi, M., Sugai, S. 1985. *Biochemistry* 24: 874–81
24. Fasman, G. 1987. *Biopolymers* 26: S59–79
25. Lewis, P. N., Gō, N., Gō, M., Kotelchuck, D., Scheraga, H. A. 1970. *Proc. Natl. Acad. Sci. USA* 65: 810–15
26. Lipton, M., Still, W. C. 1988. *J. Comput. Chem.* 9: 343–55
27. Howard, A. E., Kollman, P. A. 1988. *J. Med. Chem.* 31: 1669–75
28. Kollman, P. A., van Gunsteren, W. F. 1987. *Methods Enzymol.* 154: 430–49
29. Berendsen, H. J. C. 1987. *Comp. Phys. Commun.* 44: 233–42
30. McCammon, J. A. 1984. *Rep. Prog. Phys.* 47: 1–46
31. McCammon, J. A., Harvey, S. C. 1987. *Dynamics of Proteins and Nucleic Acids.* Cambridge: Cambridge Univ. Press
32. Madison, V. 1985. *Biopolymers* 24: 97–103
33. Hall, D., Pavitt, N. 1985. *Biopolymers* 24: 935–45
34. White, D. N. J., Kitson, D. H. 1986. *J. Mol. Graph.* 4: 112–19
35. Moult, J., James, M. N. G. 1986. *Proteins* 1: 146–63
36. Bruccoleri, R. E., Karplus, M. 1987. *Biopolymers* 26: 137–68
37. Bruccoleri, R. E., Haber, E., Novotny, J. 1988. *Nature* 335: 564–68
38. Shenkin, P. S., Yarmush, D. L., Fine, R. M., Wang, H., Levinthal, C. 1987. *Biopolymers* 26: 2053–85
39. Scheraga, H. A. 1973. *Current Topics in Biochemistry*, ed. C. B. Anfinsen, A. N. Schechter, pp. 1–42. New York: Academic
40. Scheraga, H. A. 1983. *Biopolymers* 22: 1–14
41. Vasquez, M., Scheraga, H. A. 1985. *Biopolymers* 24: 1437–47
42. Isogai, Y., Nemethy, G., Scheraga, H. A. 1977. *Proc. Natl. Acad. Sci. USA* 74: 414–18
43. Paine, G. H., Scheraga, H. A. 1987. *Biopolymers* 26: 1125–62
44. Binder, K. 1986. *Monte Carlo Methods in Statistical Physics.* Berlin: Springer-Verlag. 411 pp.
45. Baumgartner, A. 1984. *Annu. Rev. Phys. Chem.* 35: 419–35
46. Kollman, P. 1987. *Annu. Rev. Phys. Chem.* 38: 303–16
47. Kolinski, A., Skolnick, J., Yaris, R. 1987. *J. Chem. Phys.* 86: 1567–85
48. Paine, G. H., Scheraga, H. A. 1985. *Biopolymers* 24: 1391–1436
49. Paine, G. H., Scheraga, H. A. 1986. *Biopolymers* 25: 1547–63
50. Gupta, G., Sarma, M. H., Sarma, R. H., Dhingra, M. M. 1986. *Fed. Eur. Biochem. Soc.* 198: 245–50
51. Li, Z., Scheraga, H. A. 1987. *Proc Natl. Acad. Sci. USA* 84: 6611–15
52. Metropolis, N. A., Rosenbluth, A. W.,

52. Rosenbluth, M. N., Teller, A. H., Teller, E. 1953. *J. Chem. Phys.* 21: 1087–92
53. Gō, N., Scheraga, H. A. 1969. *J. Chem. Phys.* 51: 4751–67
54. Gō, N., Scheraga, H. A. 1976. *Macromolecules* 9: 535–42
55. Gō, N., Lewis, P. N., Scheraga, H. A. 1970. *Macromolecules* 3: 628–34
56. Gō, M., Gō, N., Scheraga, H. A. 1971. *J. Chem. Phys.* 54: 4489–4503
57. Gō, N., Gō, M., Scheraga, H. A. 1974. *Macromolecules* 7: 137–39
58. Gō, M., Scheraga, H. A. 1984. *Biopolymers* 23: 1961–77
59. Karplus, M., Kushick, J. N. 1981. *Macromolecules* 14: 325–32
60. Levy, R. M., Karplus, M., Kushick, J., Perahia, D. 1984. *Macromolecules* 17: 1370–74
61. Flory, P. J. 1969. *Statistical Mechanics of Chain Molecules*, pp. 49–93. New York: Wiley-Interscience. 432 pp.
62. Meirovitch, H. 1983. *Macromolecules* 16: 249–52
63. Meirovitch, H. 1985. *Macromolecules* 18: 569–73
64. Meirovitch, H. 1985. *Phys. Rev. A* 32: 3709–15
65. Meirovitch, H., Scheraga, H. A. 1986. *J. Chem. Phys.* 84: 6369–75
66. Meirovitch, H., Vasquez, M., Scheraga, H. A. 1987. *Biopolymers* 26: 651–71
67. Meirovitch, H., Vasquez, M., Scheraga, H. A. 1988. *Biopolymers* 27: 1189–1204
68. Piela, L., Scheraga, H. A. 1987. *Biopolymers* 26: S33–58
69. Levitt, M., Chothia, C. 1976. *Nature* 261: 552–58
70. Wada, A. 1976. *Adv. Biophys.* 9: 1–63
71. Perutz, M. F. 1978. *Science* 201: 1187–91
72. Jernigan, R. L., Miyazawa, S., Szu, S. C. 1980. *Macromolecules* 13: 518–25
73. Ripoll, D. R., Scheraga, H. A. 1988. *Biopolymers* 27: 1283–1303
74. Levitt, M., Warshel, A. 1975. *Nature* 253: 694–98
75. Levitt, M. 1976. *J. Mol. Biol.* 104: 59–107
76. Nemethy, G., Scheraga, H. A. 1977. *Q. Rev. Biophys.* 10: 239–352
77. Hagler, A. T., Hönig, B. 1978. *Proc. Natl. Acad. Sci. USA* 75: 554–58
78. Kuntz, I. D., Crippen, G. M., Kollman, P. A., Kimelman, D. 1976. *J. Mol. Biol.* 106: 983–94
79. Robson, B., Osguthorpe, D. J. 1979. *J. Mol. Biol.* 132: 19–51
80. Taketomi, H., Ueda, Y., Gō, N. 1975. *Int. J. Pept. Protein Res.* 7: 445–49
81. Ueda, Y., Taketomi, H., Gō, N. 1978. *Biopolymers* 17: 1531–48
82. Gō, N., Taketomi, H. 1978. *Proc. Natl. Acad. Sci. USA* 75: 559–63
83. Gō, N., Taketomi, H. 1979. *Int. J. Pept. Protein Res.* 13: 235–52
84. Gō, N., Taketomi, H. 1979. *Int. J. Pept. Protein Res.* 13: 447–61
85. Gō, N., Abe, H., Mizuno, H., Taketomi, H. 1980. *Protein Folding*, ed. N. Jaenicke, pp. 167–81. Amsterdam: Elsevier/North Holland. 587 pp.
86. Taketomi, H., Kano, F., Gō, N. 1988. *Biopolymers* 27: 527–59
87. Goldenberg, D. P., Creighton, T. E. 1985. *Biopolymers* 24: 167–82
88. Creighton, T. E. 1988. *Proc. Natl. Acad. Sci. USA* 85: 5082–86
89. Krigbaum, W. R., Lin, S. F. 1982. *Macromolecules* 15: 1135–45
90. Miyazawa, S., Jernigan, R. L. 1982. *Biopolymers* 21: 1333–63
91. Wall, F. T., Windwer, S., Gans, P. J. 1962. *J. Chem. Phys.* 37: 1461–65
92. Premilat, S., Hermans, J. Jr. 1973. *J. Chem. Phys.* 59: 2602–12
93. Premilat, S., Maigret, B. 1977. *J. Chem. Phys.* 66: 3418–25
94. Gō, N. 1976. *Adv. Biophys.* 9: 65–113
95. Gō, N. 1975. *Int. J. Pept. Protein Res.* 7: 313–23
96. Creighton, T. E. 1978. *Prog. Biophys. Mol. Biol.* 33: 231–97
97. Jacobson, H., Stockmayer, W. H. 1950. *J. Chem. Phys.* 18: 1600–6.
98. Flory, P. J. 1956. *J. Am. Chem. Soc.* 78: 5222–35
99. Poland, D., Scheraga, H. A. 1970. *Theory of Helix-Coil Transitions in Biopolymers*, pp. 1–105. New York: Academic. 597 pp.
100. Kolinski, A., Skolnick, J., Yaris, R. 1986. *J. Chem. Phys.* 85: 3585–97
101. Kolinski, A., Skolnick, J., Yaris, R. 1986. *Proc. Natl. Acad. Sci. USA* 83: 7267–71
102. Kolinski, A., Skolnick, J., Yaris, R. 1987. *Biopolymers* 26: 937–62
103. Skolnick, J., Kolinski, A., Yaris, R. 1988. *Proc. Natl. Acad. Sci. USA* 85: 5057–61
104. Skolnick, J., Kolinski, A., Yaris, R. 1989. *Biopolymers.* In press
105. Sikorski, A., Skolnick, J. 1989. *Biopolymers.* In press
106. Skolnick, J., Kolinski, A., Yaris, R. 1989. *Proc. Natl. Acad. Sci. USA* 86: 1229–1233
107. Sikorski, A., Skolnick, J. 1989. *Proc. Natl. Acad. Sci. USA* 86: 2668–72
108. Baumgartner, A. 1984. *Application of the Monte Carlo Method in Statistical*

Physics, ed. K. Binder, Chapter 5. Berlin: Springer-Verlag
109. Venkatachalam, C. M. 1968. *Biopolymers* 6: 1425–36
110. Rose, G. D., Winters, R. H., Wetlaufer, D. B. 1976. *FEBS Lett.* 63: 10–16
111. van Gunsteren, W. F., Berendsen, H. J. C. 1982. *Biochem. Soc. Trans.* 10: 301–9
112. Karplus, M., McCammon, J. A. 1981. *CRC Crit. Rev. Biochem.* 9: 293–315
113. Richardson, J. S. 1981. *Adv. Protein Chem.* 34: 167–339
114. Guss, J. M., Freeman, H. C. 1983. *J. Mol. Biol.* 169: 521–63
115. Weber, P. C., Bartsch, R. G., Cusanovich, M. A., Hamlin, R. C., Howard, A., et al. 1980. *Nature* 286: 302–4
116. Hendrickson, W. A., Ward, K. B. 1977. *J. Biol. Chem.* 252: 3012–18
117. Regan, L., DeGrado, W. F. 1988. *Science* 241: 976–78
118. Ford, G. C., Harrison, P. M., Rice, D. W., Smith, J. M. A., Treffry, A., et al. 1984. *Philos. Trans. R. Soc. London Ser. B* 304: 551–65
119. Abdel-Meguid, S. S., Shieh, H.-S., Smith, W. W., Dayringer, H. E., Violand, N. V., et al. 1987. *Proc. Natl. Acad. Sci. USA* 84: 6434–37
120. Schiffer, M. R., Edmundson, A. 1967. *Biophys. J.* 1: 121–35
121. Matsushita, Y., Noda, I., Nagasawa, M., Lodge, T. P., Amis, E. J., et al. 1984. *Macromolecules* 17: 1785–89
122. Teramoto, E., Kurata, M., Yamakawa, H. J. 1958. *J. Chem. Phys.* 28: 785–91
123. Karplus, M., Weaver, D. 1979. *Biopolymers* 18: 1421–27
124. Lee, S., Karplus, M., Bashford, D., Weaver, D. 1987. *Biopolymers* 26: 481–506
125. Skolnick, J., Kolinski, A. 1989. *J. Mol. Biol.* Submitted
126. Sikorski, A., Skolnick, J. 1989. *J. Mol. Biol.* Submitted

PHYSICAL CHEMISTRY AT ULTRAHIGH PRESSURES AND TEMPERATURES

Raymond Jeanloz

Department of Geology and Geophysics, University of California, Berkeley, California 94720

INTRODUCTION

High-pressure research has undergone a revolution in the past 15 years due to a greater than 50-fold increase in the combined pressure-temperature conditions that can be achieved experimentally (Figure 1). This has been accomplished by a breakthrough in instrumentation, most notably in the development of the diamond-anvil cell (1–3). Thus, 1975 marked the first time that pressures in excess of 10^6 atmospheres (100 GPa) could be sustained in the laboratory under controlled, equilibrium conditions (4).

The significance of being able to reach ultrahigh pressures, of the order of 10^{11} Pascal and above, is that the energy density achieved on compression is comparable to bonding energies. Therefore significant changes are expected, and indeed found, in the electronic states, chemical bonding, and atomic packing of condensed matter. Perhaps the best known example is the metallization of hydrogen that is currently expected to occur at pressures of ~ 200–500 GPa (5–12). As hydrogen is the most abundant element in the universe, its metallization is not only a dramatic illustration of pressure-induced changes in bonding character, but also plays an important role in determining the internal state and evolution of the giant planets Jupiter, Saturn, Uranus, and Neptune (13, 14). Although there is every reason to believe that the metallization of hydrogen will be confirmed by experiment, it should be noted that this has not yet been accomplished: At the time of writing this article, metallic hydrogen remains a theoretically predicted and not an experimentally observed phase.*

*Note added in proof: H. K. Mao and R. J. Hemley (1989. *Science*. In press) have just reported the first observation of metallic hydrogen at pressures of 200–250 GPa, in good agreement with the prior estimates (Figure 1).

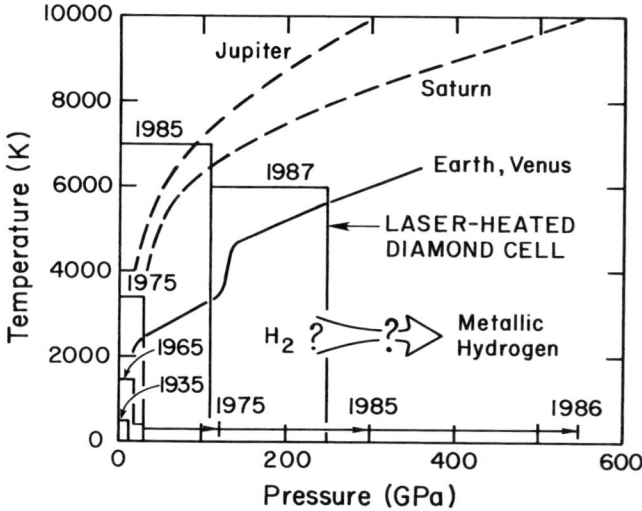

Figure 1 Summary of sustained pressure-temperature conditions that have been achieved in the laboratory (*arrows* at 300K and fields with dates) as compared with the pressures above which hydrogen is expected to metallize (*open arrow*). All of the advances shown after 1965 are based on the diamond cell, laser heating having been used to generate peak temperatures of ~3000–7000K. Average temperature-pressure profiles through the entire depth range of Earth and Venus, and as modeled for the interiors of Jupiter and Saturn, are shown as *solid* and *dashed curves* (the *curves* shown here represent conditions that include over 55% and 95% of the volumes of Jupiter and Saturn, respectively) (2, 3, 5–12, 20, 23, 26, 28, 32, 34).

With pressure it is possible to access much more extreme states of condensed matter than is usually the case, for example by changing temperature alone. This is illustrated by considering the thermodynamic perturbation, the change in free energy, induced either by pressure or by temperature. The comparison for a simple salt, CsI, clearly shows that pressures on the order of 100 GPa affect the thermodynamic state far more than heating and melting the crystal (Figure 2) (15, 16). In fact, for CsI such ultrahigh pressures even exceed vaporization in the absolute change of free energy that is involved (~2.0 eV = 0.4 MJ/mol for vaporization at zero pressure) (16). Thus, it is not surprising that the initially highly ionic bonding in CsI is completely altered at pressures above 50–100 GPa, as described below. With the range of conditions now available in the laboratory, new forms of condensed matter are created at high pressure, either alone or in combination with high temperature.

The purpose of this article is to summarize the recent advances that have been made in experimental research at ultrahigh pressures and temperatures. The emphasis is on static techniques, in which the conditions

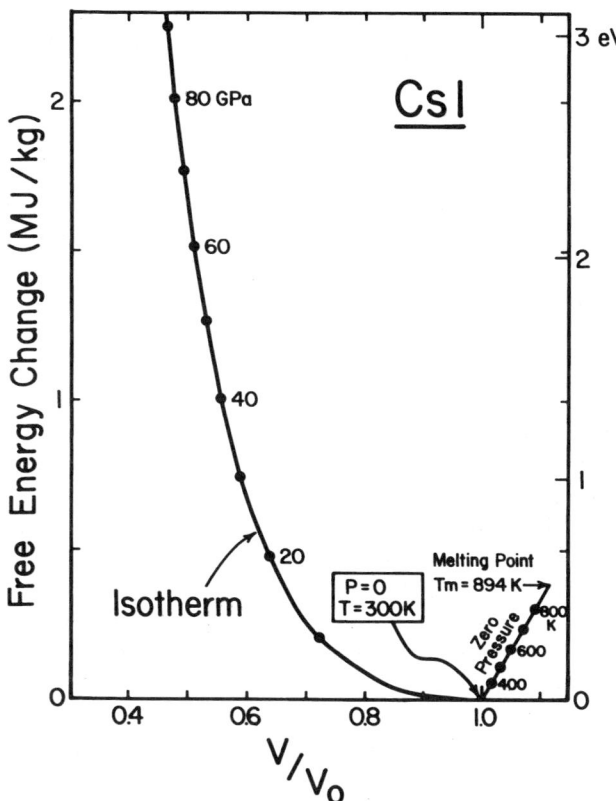

Figure 2 Absolute value of the Helmholtz free energy of CsI as a function of volume, either upon compression at 300K (isotherm curve) or upon heating at zero pressure. All values are shown relative to the free energy at ambient conditions (volume V_0), and pressures and temperatures are shown at 10 GPa and 100K intervals along the respective curves (15, 16, 69).

can be sustained, rather than on the complementary dynamic techniques involving shock-wave compression (17–19). This choice reflects the fact that the static techniques, and particularly the use of the laser-heated diamond cell, have developed much more recently than the shock-wave techniques. In addition, the ability to sustain pressure and temperature virtually indefinitely ensures that thermodynamic equilibrium is achieved in the static experiments, at least at elevated temperatures. Equilibrium is not so readily ensured in shock experiments, yet these continue to play a central role in high-pressure research; calibration of the diamond-cell experiments, for example, relies heavily on shock-wave measurements.

After summarizing the essential features of the laser-heated diamond cell, I describe three recent topics of high-pressure research. These illustrate the changes in bonding character, magnetic order, crystal structure and liquid structure that are achieved in the 10^{10}–10^{11} Pascal and 10^3K range of pressure and temperature; they confirm that the conditions shown in Figure 1 induce profound changes in the chemical properties of condensed matter. Of course, much more extreme conditions are required to alter the nuclear states and even the deep (core) electrons in matter. As such conditions—which are typical of stellar interiors—are far beyond what can be sustained in the laboratory, they are not included in the present discussion.

LASER-HEATED DIAMOND CELL

The concept of the high-pressure diamond cell is simple, involving two diamond crystals acting as ultrahard opposed anvils (Figure 3). Although opposed-anvil designs have been extensively used in high-pressure research throughout this century, it was only around 1970 that the diamond cell was developed to its current capability (1, 2, 20). Prior to that time, from the late-1950s onward, the main emphasis in using diamond anvils was as windows, thus providing direct optical access to the sample under pressure (20–22). Subsequently, a great effort was made to extend the pressure and combined pressure-temperature (P-T) range of the diamond cell. This was largely motivated by researchers in geophysics and planetary physics wanting to reproduce the P-T conditions deep inside planets (4, 23).

The breakthrough in achieving ultrahigh pressures came in appreciating the importance of aligning the diamond anvils (maintaining closely parallel culet faces even at high pressure) and in using a gasket, a metal foil to contain the sample between the diamonds (Figure 3) (20, 24–26). In both cases, the idea is to eliminate stress concentrations and to minimize stress gradients across the anvils: Though very hard and strong, diamond is also brittle and shatters readily once a fracture is initiated. Thus, further increases in pressure since the mid 1970s have dominantly been accomplished by reducing the effective area of highest pressure between the culets (26–29). This is generally a better way of achieving ultrahigh pressures than simply to increase the force on the diamonds, because it is difficult to maintain culet alignment if the anvils are put under extremely high loads.

In comparison with other techniques, the highest pressures are achieved in the diamond cell at the expense of sample volume, which is typically in the picoliter to femtoliter range. Inside this volume, a pressure medium and calibration standard are enclosed along with the sample. The purpose

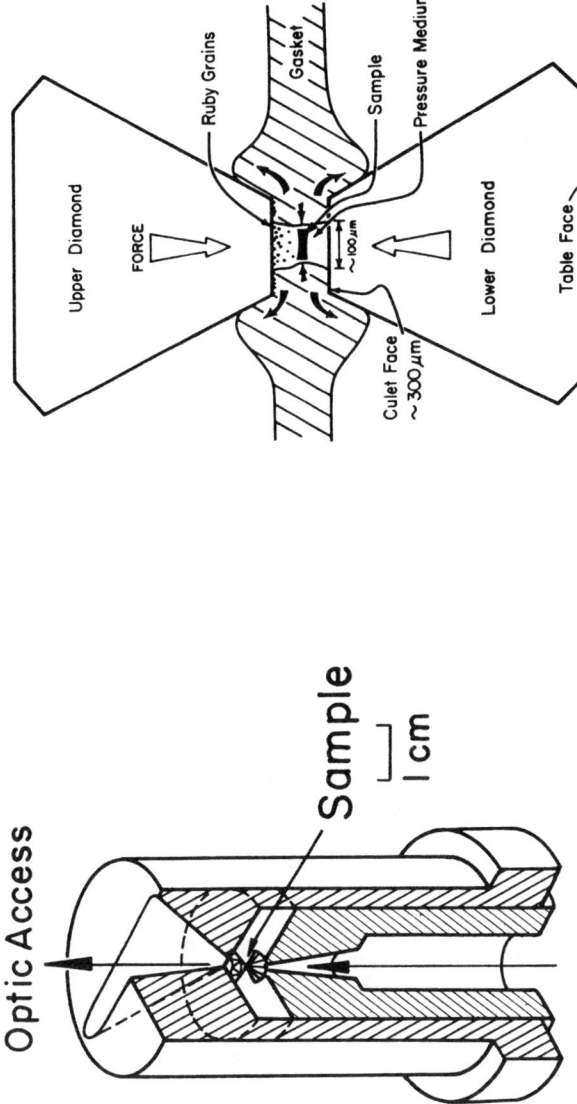

Figure 3 Schematic diagrams illustrating the diamond cell. *Left*: Two single-crystal gem quality diamonds are compressed in a piston-cylinder assembly. The sample is placed between the opposed points of the diamonds and can be viewed through the diamonds along the axis of force. Except for energies that are blocked by two-phonon transitions (~ 0.2–0.5 eV) and interband electronic transitions (~ 5.5 eV–10 keV), the diamonds act as windows to electromagnetic radiation extending from far-infrared to hard X-ray and γ-ray energies; thus, a wide variety of spectroscopic and diffraction techniques can be used to examine the sample under pressure (1, 2, 122, 123). *Right*: Details of the sample and pressure medium contained inside a gasketted hole that is centered between the culets of the diamond anvils. Typical dimensions are shown; additionally, each diamond is about 2.5 mm thick, the gasket is a foil ~ 200–300 μm in initial thickness, and the sample is usually ~ 5–20 μm thick at high pressures. Upon compression, the gasket squeezes inward on the sample and pressure medium, and also pushes around the tips of the diamonds as shown by the *arrows*. A small amount (<5% by volume) of fine-grained ruby powder (<5 μm grain size) is included in the sample volume for pressure calibration (from Ref. 124).

of the medium, ideally a fluid or an extremely soft solid, is to maintain as nearly hydrostatic a pressure as possible. The highest-pressure fluid known at 300K (a methanol-ethanol-water mixture) solidifies above about 14 GPa (1, 30); at higher pressures, solidified He, Ar, or Xe provide almost perfectly hydrostatic conditions (2, 31). Alternatively, shear stresses are diminished or eliminated altogether if the sample is heated at high pressures, especially if it is melted (32). For such high-temperature experiments, reactivity of the sample with the pressure medium is a much more serious problem than are deviations from hydrostaticity. Therefore, a highly refractory compound (e.g. Al_2O_3) or powder made from the sample material itself is often used as a pressure medium in ultrahigh P-T experiments (it is advantageous to use a medium of low thermal conductivity in such experiments) (33, 34).

In order to reach and sustain the highest temperatures inside the diamond cell, a continuous wave (cw) Nd:YAG laser beam is focused into the sample area (Figure 4) (35, 36). With a power of up to 20–25 Watts in

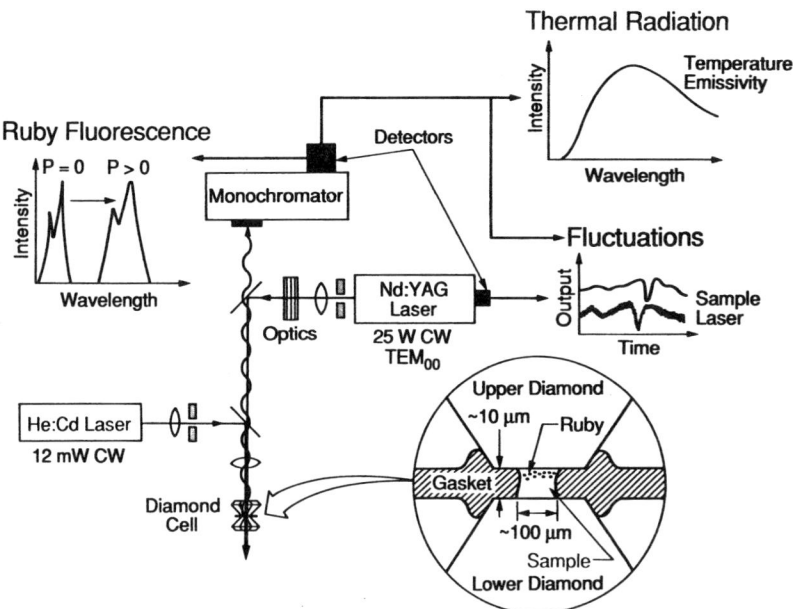

Figure 4 Illustration of ruby-fluorescence pressure calibration (*left*) and spectroradiometrically calibrated laser heating (*right*) inside the diamond cell. The inset (*lower right*) shows a close-up view of the sample between the diamond anvils (cf. Figure 3). Details of the pressure-calibration and laser-heating techniques are given elsewhere (29, 36, 38–41).

the TEM_{00} mode at 1064 nm wavelength, temperatures of $\sim 1000-7000$ Kelvin are generated within an ellipsoidal volume of sample material ~ 20 μm × 10 μm in diameter and thickness. Rather than being uniform, the temperature decreases by as much as $\sim 10^3 K/\mu m$ away from the hottest zone at the center of the sample (32, 37). This is because the diamond anvils act as nearly infinite heat sinks around the tiny focal volume of the laser (recall that diamond has the highest thermal conductivity known for any material at ambient conditions). As the hot and therefore highly reactive portion of the sample is surrounded by colder sample material (or a refractory pressure medium), which is itself in contact with the nearly unheated diamond anvils, the temperature gradients are actually beneficial in physically containing and both chemically and thermally isolating the sample that is being heated. The benefit of having strong temperature gradients within the laser-heated diamond cell is somewhat diminished, however, by the fact that the gradients must be measured in order to obtain quantitative results.

Temperature is measured spectroradiometrically by observing the blackbody-like thermal emission from the hot spot in the sample. The thermal radiation at visible wavelengths is collected through the same microscope optics that are used to focus the heating laser, and its spectrum is measured by way of a low-resolution monochromator (Figure 4) (36, 38). The spatial variation of temperature is obtained by a tomographic approach that involves observing the image of the hot spot through a movable slit. In this manner the temperature distribution inside the diamond cell and the uncertainty in that distribution (caused largely by fluctuations in laser output) are quantitatively determined. It is worth noting that more uniform temperatures can be obtained with an external resistance heater than with laser heating; but the peak temperatures achieved inside the diamond cell are much lower than those considered in this article (131).

The pressure is most readily calibrated by way of the ruby-fluorescence technique (39, 40). This involves exciting the R_1, R_2 fluorescence doublet of ruby, usually with a low-power He:Cd laser focused into the sample volume between the diamonds: By including a small amount of finely ground ruby into the sample, the fluorescence can be observed directly through the diamond anvil. The fluorescence shifts to lower energy with pressure, and the pressure dependence of wavelength has been accurately calibrated to at least 200 GPa ($d\lambda/dP \sim 0.3$ nm/GPa; 27, 31, 41–45). Therefore, the pressure distribution across the sample can be mapped out. Ideally, for a hydrostatically compressed sample, the pressure is found to be uniform; in less ideal circumstances, deviations from hydrostaticity can be quantitatively determined via ruby fluorescence (119).

The ruby fluorescence technique is usually not applied at high temperatures because the fluorescence intensity decreases with temperature, becoming difficult to observe above about 800K (25, 46, 132). Instead, it is usually adequate to measure the pressure at room temperature, both before and after laser heating (32–34). In this manner the pressure relaxation caused by heating the sample is determined, and this relaxation can be used to infer the pressure inside the laser-heated zone. Alternatively, pressure can be monitored by X-ray diffraction of a calibration standard mixed in with the sample (47). Indeed, it is by way of X-ray diffraction of materials with well-known pressure-volume equations of state that the ruby fluorescence "pressure scale" has been calibrated (41). With the bright X-ray sources offered by synchrotrons, it is possible to map the detailed pressure distribution across a sample at ultrahigh pressures (48), and this may also turn out to be an excellent way of calibrating pressures during laser heating.

As an example of current experimental capabilities, Figure 5 shows the highest-pressure melting curve measured to date for any material (34). It also illustrates the consistency of results obtained by two entirely different techniques, shock-compression and use of the laser-heated diamond cell. The shock experiments intersect the melting curve only at one point, yet because no calibrations are needed in these experiments (pressure and temperature are directly measured) and because very high P-T conditions are achieved, they provide a significant verification of the diamond-cell experiments.

CsI: METALLIZATION AND DISPROPORTIONATION

Insulator-Metal Transition

The expectation that the bonding in CsI changes at pressures in the 10^{10}–10^{11} Pascal range (Figure 2) is confirmed by direct observation. At 300K the initially transparent salt is seen to become colored (yellow to orange to red with increasing pressure), finally becoming opaque as pressure is raised above 55 GPa (49). This effect is completely reversible on decompression, and is simply due to the fundamental energy gap between valence and conduction bands decreasing with pressure (Figure 6): CsI transforms from an insulator to a semiconductor and then to a metal by 110 GPa (50–54).

Reflectivity measurements at optical energies between 0.5 and 1.5 eV and at pressures between 90 and 170 GPa confirm that CsI becomes metallic at 110 GPa (55). The 1.0 eV reflectivity of CsI against the diamond anvil is only 4% at 170 GPa, but is calculated to yield a 22% reflectivity were the sample in contact with a medium of unit refractive index ("vacuum" at 1.7×10^{11} Pascal!). The relatively small value of optical reflectivity

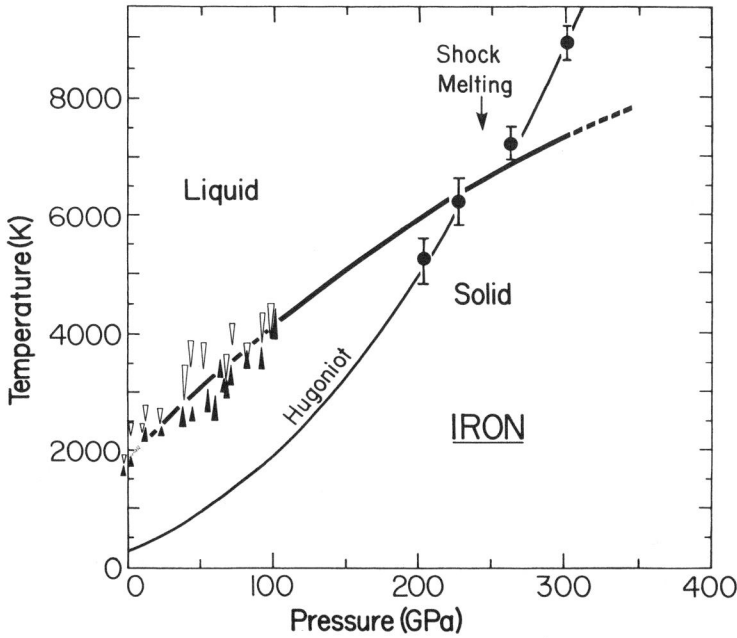

Figure 5 Melting temperature of Fe as a function of pressure. The melting curve is bracketed from above and below by experiments in the laser-heated diamond cell up to 100 GPa (*open* and *closed triangles*, respectively; the size of each symbol indicates the uncertainty in temperature). An extrapolation of this curve to 250 GPa is in excellent agreement with the temperature achieved under shock-loading at this pressure (the curve labeled *Hugoniot*). Iron is known to melt at 250 GPa under shock compression based on sound-velocity measurements, and the Hugoniot temperature is determined by spectroradiometry (*closed circles* with error bars). The *stippled curve* at low pressures indicates the pressure-dependence of the melting temperature as measured in previous experiments (from Ref. 34).

indicates that high-pressure CsI is not a simple metal: Interband transitions may influence the optical properties at visible wavelengths. Still, this is the first example of an alkali halide being converted from an insulator to a metal.

Cesium iodide is readily metallized under pressure because of its small initial band gap ($E_g = 6.3$ eV) and its large compressibility (zero-pressure bulk modulus or incompressibility $K_0 = 11.9$ GPa) (15, 50, 51, 56, 57, 65). The volume compression is $V/V_0 = 0.44$ at metallization. For comparison, RbI, KI, and CsBr are expected to be the next most easily metallized alkali halides, based on their low bulk moduli ($K_0 = 11.0, 12.0, 15.7$ GPa), small initial band gaps ($E_g = 6.1, 6.2, 7.3$ eV), and rapid band-gap closures observed under pressure (52–61). Note that whereas CsI and CsBr are

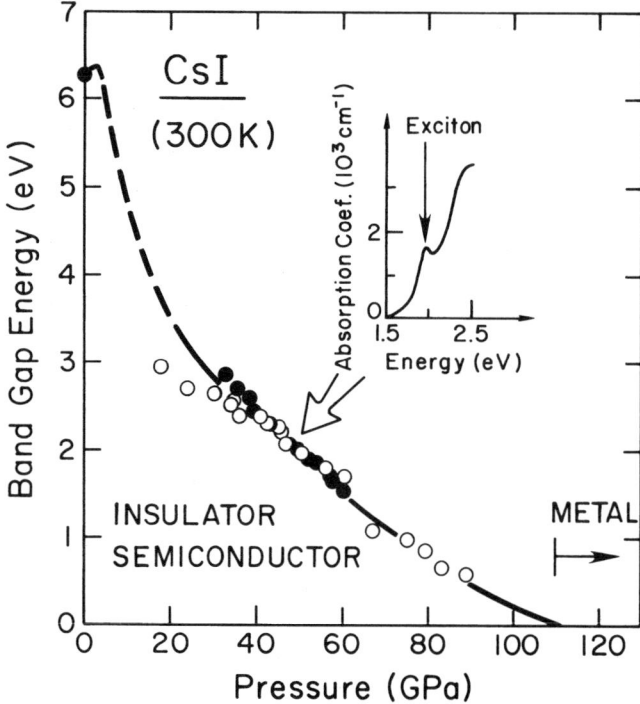

Figure 6 Energy across the band gap of CsI as measured under pressure at 300K. The results of optical absorption measurements on single crystals (*closed symbols*; 53) and polycrystalline samples (*open symbols*; 49–52, 54, 69) are shown along with an example of a single-crystal spectrum collected at 50 GPa (*inset*; 53). The band gap initially increases with pressure, when the lower-most conduction band is *s*-like, and then decreases once the lower-most conduction band becomes *d*-like.

both in the B2 (CsCl-type) structure at zero pressure, RbI and KI transform from the B1 (NaCl-type) to the B2 structure at pressures of 0.4 and 1.8 GPa: The transformation involves an increase in nearest neighbor (cation-anion) distance, though an overall decrease in volume, and therefore is likely to increase the band gap slightly. Extrapolations of the currently available spectroscopic and X-ray diffraction measurements suggest that the volume compressions and pressures at which RbI, KI, and CsBr metallize are $V/V_0 = 0.34$–0.35 and $P \sim 130$, 170, 250 GPa, respectively (62). Among compounds isoelectronic with CsI, BaTe metallizes at 25 GPa (63, 64); and most recently, two groups have reported that Xe, which is also isoelectronic (but not isostructural) with CsI, metallizes at 130–170 GPa (133, 134). This is apparently the first observation of an "inert" gas being transformed to a metal.

Several theoretical calculations (65–68) demonstrate that over the entire pressure range of Figure 6, the band gap of CsI is at the Brillouin zone center and involves transitions out of I 5 p-like states at the top of the valence band (Γ_{15} and Γ_8^- symmetries in nonrelativistic and relativistic schemes, respectively). The bottom of the conduction band changes with pressure, however. Above a few GPa, the Cs 6 s-like Γ_1 (or Γ_6^+ relativistic) state initially at the bottom of the conduction band rises above the Γ_{12} (or Γ_8^+ relativistic) state that is mainly of Cs 5-d character, but also contains I 5-d character. Thus, metallization occurs when the energy difference between I 5-p and Cs 5-d states, which are both increasing in energy with pressure, vanishes. The charge transfer from Cs$^+$ to I$^-$ ions decreases with pressure and the cesium takes on increasing d character, becoming more transition-metal-like under pressure.

The change in the Cs$^+$ ion under pressure is apparently the cause of a structural distortion that is observed in CsI (50, 51, 60, 69–72). At about 40 GPa, the initially cubic B2 structure transforms spontaneously (and reversibly) to the tetragonal CuAuI structure; no volume change is resolved across the transition. Evidently the distortion is triggered by the first-neighbor repulsion becoming significantly reduced at high compression, but it is currently uncertain whether or not this is a first-order transition (69–72). We speculate that the increased d character of the cesium induces a static Jahn-Teller-like distortion of the structure, whereby the cost in lattice energy is made up in the energy gained by splitting the occupied bands. The structural distortion also helps to precipitate closure of the band gap and hence metallization (67, 72). This idea does not explain the further tetragonal → orthorhombic distortion that has been reported for CsI at about 60 GPa (50, 73). The orthorhombic distortion is expected to increase slightly the rate of metallization, relative to the tetragonal form (67), but it is not clear how or why the orthorhombic structure is stabilized (69–72) (n.b. shear stresses in the high-pressure experiments may also be significant in stabilizing the tetragonal and orthorhombic structures, or at least in determining the pressures at which these appear; 50). Nevertheless, given that CsI, CsBr, and CsCl all transform to the tetragonal structure at a compression $V/V_0 \sim 0.54$ (50, 51, 60, 61, 74), whereas neither KI nor RbI show any deviations from cubic symmetry to $V/V_0 < 0.5$ (referenced to the B2 structure; 59, 73), it is clear that pressure-induced changes in the cesium ion play an important role in causing the cubic structure to distort.

Pressure-Induced Decomposition

The volume of elemental cesium decreases rapidly with pressure, both because Cs is compressible and because it undergoes structural (polymorphic) and isostructural electronic transitions (75–77). By 10 GPa, for example, the relative volume $V/V_0 = 0.27$. Although more complex in detail, the cesium atom can be viewed as undergoing a volume collapse as its outermost electron shell changes from the $6s$ to the $5d$ level (78–80). Qualitatively, this is explained by the fully symmetric s level sensing the pressure more readily—increasing in energy more rapidly—than the d level. Similarly, I_2 goes through structural and electronic transitions under pressure, becoming a molecular metal by 18 GPa (81–84) (the transition from diatomic I_2 to monatomic I may occur in the metallic state at 21 GPa, but the evidence is inconclusive so far; 83–86).

The result of both Cs and I_2 volumes decreasing so rapidly with pressure is that the elements have a smaller volume than the compound CsI at pressures above 5.5 GPa (Figure 7). A calculation of the Gibbs free energy difference between CsI and $Cs + 1/2\, I_2$ demonstrates that the compound is unstable relative to the elements at pressures above 65 GPa at 300K (77). This is not to say that the ions themselves are undergoing an electronic

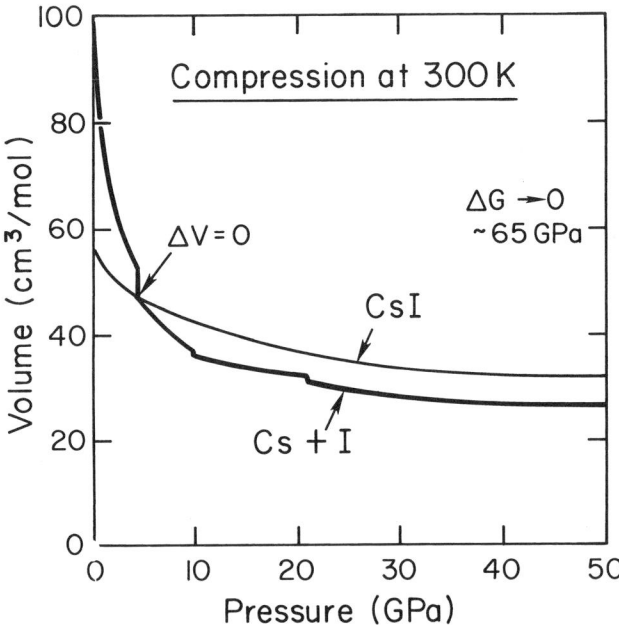

Figure 7 Comparison of the volumes of CsI and its constituents, Cs and $1/2\, I_2$, as a function of pressure at 300K. The volume difference vanishes below 5.5 GPa, and the difference in Gibbs free energy between the compound and its elemental constituents is expected to vanish at 65 GPa. At higher pressures, $Cs + I$ (or $1/2\, I_2$) is more stable than CsI (after Ref. 77).

transition within the compound, but that the compressed elements are more stable than the CsI. As the entropy of the compound is likely to be significantly lower than that of the assemblage Cs + I (or 1/2 I_2), increasing temperature would lead to CsI becoming unstable at a lower pressure than 65 GPa.

These thermodynamic conclusions have been confirmed by showing that Cs and I_2 are formed from CsI at pressures between 40 and 60 GPa. In order to overcome the kinetic hindrances that prevent the decomposition of the sample at 300K, the CsI was laser-heated to a temperature exceeding 6000K while at a peak pressure of 60 GPa. This is well above the melting point of CsI at high pressures (87). The formation of Cs upon laser heating was unequivocally confirmed by in-situ X-ray diffraction at high pressures and by compositional analyses of the quenched sample, and the presence of I_2 was shown indirectly by optical and electrical resistivity measurements at pressure (77). Aside from this direct evidence that CsI decomposes to Cs and I_2 at ultrahigh P-T conditions, the occurrence of decomposition explains the time dependence and magnitude of electrical conductivity measured under shock: e.g. a metallic conductivity of $\sim 10^4$ S/m at 40 GPa, a pressure at which the band gap is 2.5 eV (88). Finally, partial decomposition of CsI is in accord with the existing shock-wave (Hugoniot) equation of state, sound velocity, and temperature measurements (87, 89, 90).

That CsI is metastable as a compound above $\sim$ 40–60 GPa, because of compression and electronic transitions occurring in its elemental constituents, implies that the metallic form of CsI is also metastable. More interesting is the possibility that this is a general phenomenon: The stability of many compounds may change drastically under pressure because of the constituent elements undergoing electronic transitions (cf. the case of CeY; 130). Among the alkali halides, the next most likely candidate for pressure-induced dissociation is RbI, at an estimated 90 GPa, followed by CsBr and KI at somewhat higher pressures (77). Whether any alkali halide forms a stable metallic compound under pressure is therefore open to question. The answer depends on the stabilities of the elements, as enhanced by electronic transitions, relative to the rate of band-gap closure as functions of pressure.

FeO: A HIGH-PRESSURE ALLOY

An example of more complex phase relations at ultrahigh P-T conditions is offered by FeO, wüstite. This system reflects the interplay among electronic, magnetic, and structural transitions, including melting. At zero pressure, FeO is ideally an insulator, in accord with expectations from band theory that includes crystal-field splitting (91). Therefore it is not surprising that the melting relations for the Fe-FeO system are typical of those for metal-salt systems, including the presence of immiscible metallic+ionic melts over a wide range of temperatures and compositions (Figure 8) (92–94). In reality, however, the iron oxide is invariably nonstoichiometric (Fe-deficient), and its electrical conductivity at ambient conditions is that of a semiconductor (95, 126).

Wüstite is antiferromagnetic below 180K at $P = 0$, and the Néel temperature increases with pressure, exceeding 300K by about 18 GPa (96, 97). Thus, FeO remains antiferromagnetic and nonmetallic to pressures well in excess of 100 GPa (97, 98). Experiments at simultaneously high temperatures and pressures reveal that FeO becomes a metal above 70 GPa and ~ 1000K, however. In particular, shock-wave experiments prove that the conductivity is metallic, both in its magnitude (comparable to Fe) and in its characteristic decrease with increasing shock pressure and temperature (99). As summarized in Figure 8, diamond-cell experiments illustrate completely different temperature dependencies for the electrical conductivity just below and above the 70 GPa transition pressure. The jump in conductivity as the sample is laser-heated at 73 GPa is entirely reversible (*double arrow*), and the conductivity remains constant within the $\sim \pm 10\%$ resolution of the data as temperature is further increased. In contrast, the 66 GPa sample exhibits semiconducting behavior below the melting temperature, and the activation energy describing the temperature dependence of the conductivity is comparable to that at zero pressure, ~ 0.06 eV (95, 100).

The FeO phase diagram can be explained in terms of a Mott-transition model that characteristically includes transitions from a magnetically ordered phase at low temperatures to a paramagnetic nonmetallic phase at high-T low-P and a metallic phase at high-T high-P conditions (101). That is, the spin ordering and electron delocalization counteract each other, the high temperatures and pressures suppressing the former and enhancing the latter. Unfortunately, the crystal structure of the metallic phase is currently unknown, and its determination requires techniques that are barely at the current state of the art. Nevertheless, there is a strong indication that the solid-melt equilibria change drastically because of the pressure-induced electronic transition in FeO (100). First, the melt is known to be metallic at 66 GPa from the conductivity measurements in the laser-heated diamond cell (Figure 8). Second, determinations of melting temperatures for the Fe–FeO system at 80 GPa are most readily explained in terms of complete solution across the metallic binary system. As only three compositions have been examined, it is possible that stable crystalline phases remain to be discovered at intermediate compositions, yet it is clear

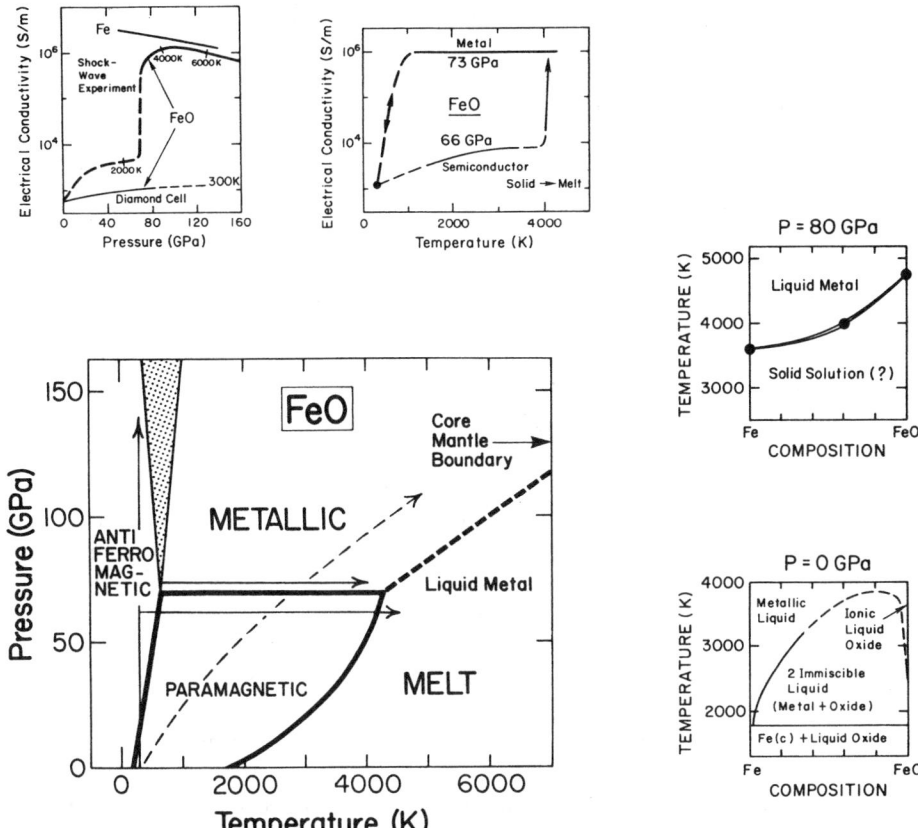

Figure 8 Summary of FeO phase diagram. The *upper panels* show the electrical conductivity measurements that help to locate the phase boundaries. Measurements were taken along paths indicated by *arrows* in the phase diagram: Results under shock-wave loading (*dashed arrow*) and at 300K in the diamond cell are in the *left upper panel*, and results as functions of temperature at 66 GPa and 73 GPa in the laser-heated diamond cell are in the *right upper panel*. The boundary between antiferromagnetic and metallic phases is not well constrained by these data, so it is indicated with shading in the P-T diagram. The panels on the *right* illustrate the melting relations for the Fe–FeO system as measured at zero pressure (*bottom right*) and as inferred from melting determinations for three compositions at 80 GPa (closed symbols: *top right*) (after Refs. 98, 100).

that the solid-melt equilibria at 80 GPa are entirely different from the metal-salt type of phase diagram at zero pressure. The FeO has been converted from an ionic ceramic at ambient conditions to a metallic alloy at ultrahigh pressures and temperatures.

STRUCTURAL TRANSFORMATIONS IN MELTS

Structural transformations among crystalline polymorphs are well known to occur as functions of either temperature or pressure. Less well documented are the transformations expected in liquid structures. Yet analyses of melting curves long ago suggested that melt structure changes with pressure (102). For example, the decrease in the melting slopes dT_m/dP, of alkali halides at pressures below the B1–B2 transition is ascribed to the change in nearest-neighbor coordination being spread over a broader pressure range in the liquid than in the solid. Specifically, it is thought that a fraction of the liquid is already in eightfold coordination below the pressure of the B1–B2 transition, at which the solid changes from sixfold to eightfold coordination. Monte Carlo simulations provide a quantitative model of this change in liquid structure, showing that the cation-anion distance increases in the liquid as the coordination number is increased under compression (103, 104). Compression alone, without a coordination change, would cause the cation-anion distance to decrease.

Silicate melts are more highly structured than alkali halide or metallic liquids (105). They include small polymeric units, such as chains and rings of SiO_4 tetrahedra, but because the Si–O bond is much stronger than the bonding in conventional polymers, silicate melts can be profitably examined at ultrahigh pressures (106, 107). One consequence of the strong bonding, however, is that high temperatures are required to study silicate melts.

Recent experiments demonstrate that silicate liquids undergo large volume compressions at ultrahigh pressures (Figure 9). This has been shown by two kinds of measurements. Shock-wave equation-of-state (Hugoniot) measurements on molten silicate samples directly reveal that volume compressions $V/V_0 \sim 0.65$ are achieved by 40 GPa (108, 109). The samples are initially at 1700K, and they reach temperatures close to 4000K at these peak pressures. A second approach involves combining diamond-cell measurements of the melting curve with measurements of the compression and the thermal expansion of the crystalline polymorph at high P-T conditions (110). Modeling of the entropy change on fusion leads (typically) to a maximum estimate of the melt density by way of the Clausius-Clapeyron equation. From this analysis the liquid equation of state is obtained along the melting curve at high pressures.

Neither the shock-wave data nor the analysis of the diamond-cell experiments yield isothermal equations of state for the melt, yet because the thermal expansion corrections are relatively small at high pressures, the two sets of results can be compared with little difficulty. As shown in Figure 9, $(Ca,Mg,Fe)SiO_3$ compositions achieve a volume of about 24–25 cm^3/mol at 50 GPa and 3000–4000K. This is clearly smaller than the

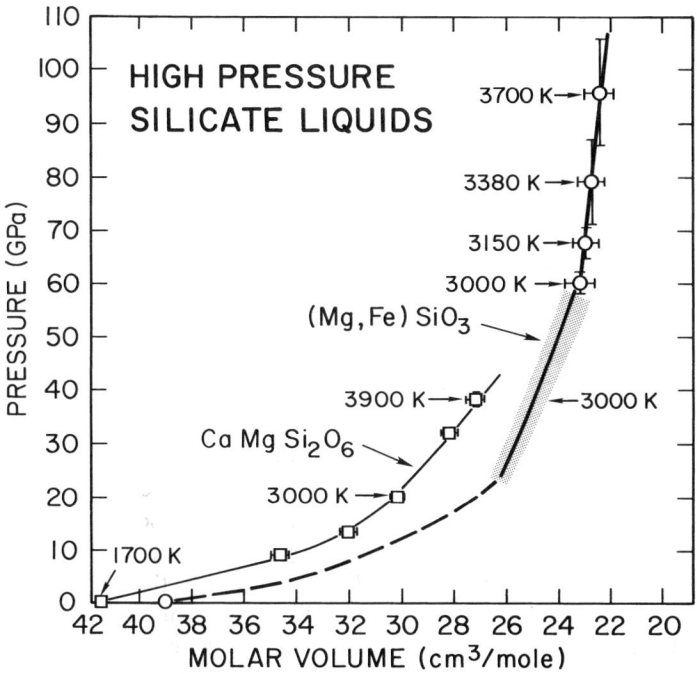

Figure 9 High-pressure equations of state of (Ca,Mg,Fe)SiO$_3$ melts. The compression of liquid CaMgSi$_2$O$_6$ is directly obtained from the Hugoniot measurements, whereas the (Mg,Fe)SiO$_3$ melt compression is derived by combining measurements of the melting curve and crystal equation of state from diamond-cell experiments (*error bars* and *stippling* indicate the estimated uncertainties). The difference between the two curves above 30 GPa can be entirely accounted for by thermal expansion. For comparison, the molar volumes of the pyroxene (low-pressure) and perovskite (high-pressure) crystalline polymorphs are 31.5–33.0 cm^3/mol and 24.4–26.7 cm^3/mol, respectively, for the same range of composition (111, 124, 125). These values are for ambient conditions, but the effects of pressure and temperature very nearly cancel out at ∼30–50 GPa and 3000–4000K (from Ref. 110).

volume of the low-pressure crystalline polymorph (pyroxene) extrapolated to the same P-T conditions (111). In fact, the liquid volume at this pressure is virtually indistinguishable from the volume expected for the high-pressure, perovskite-structured polymorph at equivalent conditions (32, 33, 112). As the main structural difference between the low and high-pressure crystalline polymorphs is that silicon changes from a tetrahedral to an octahedral coordination with oxygen, the same coordination change is inferred to occur in the melt at pressures of about 20–50 GPa.

The result of the coordination change is that the melt density closely approaches the density of the coexisting crystals at these pressures. By the Clausius-Clapeyron equation, the melting temperature is therefore expected to be nearly independent of pressure. A melting slope close to zero is what is observed experimentally at 30–50 GPa, and the pressure-induced decrease in the melting slope is completely analogous to that observed at lower pressures among the alkali halides (32, 110). As with the alkali halide melts, computer simulations of silicate liquids also show the coordination increasing with pressure. The molecular dynamics results suggest that pressures of tens of GPa are required for the silicon coordination to change, in accord with the experimental data (106, 107).

That changes in silicon coordination, from tetrahedral to octahedral and distorted octahedral, are largely responsible for the melt compression at ~ 10 to 50 GPa is supported by infrared absorption spectra obtained on glasses at high pressures (113). Normally, because it is a kinetically frozen form of the liquid, glass does not yield reliable information on melt structure. The present case is different, however, because of the remarkable observation that (a) the glass coordination does change with pressure and (b) the change is entirely reversible (though with some hysteresis) as pressure is cycled (Figure 10). That is, the glass structure is apparently not frozen with respect to pressure-induced changes, although some kinetic hindrances are undoubtedly present (113, 129). Hence, the kinetically unrestricted liquid at high temperatures must undergo similar coordination changes at pressures less than or equal to those required for the glass (ignoring the effect of temperature on the pressure range over which the coordination changes in the melt). Compared with the large activation energies that characterize high-pressure structural transformations among crystalline silicates (~ 60 to 300 kJ/mol; 114), the glasses exhibit a vanishingly small activation energy for such transformations at pressures above 10 GPa. Thus, low temperatures are sufficient for the amorphous state despite high temperatures being required for the crystalline form to transform. Although this result is not fully understood, it is entirely compatible with the pressure dependence of the crystallization rate measured in SiO_2 glass up to about 2 GPa (115, 116).

One consequence (or cause?) of the coordination change is that the nature of the Si–O bond in the melt and glass also changes with pressure (113). The SiO_4 tetrahedron is characterized by strong, covalent sp^3 bonding whereas the higher coordination is expected to involve a weaker, more ionic sp^3d^2 hybridized bond (117). Experimental support for this bond weakening has been obtained for silica glass under pressure. At 300K, the static yield strength of amorphous SiO_2 decreases by more than one order of magnitude between 27 and 65 GPa, the same pressure range over which the spectroscopic measurements indicate that the silicon coordination is

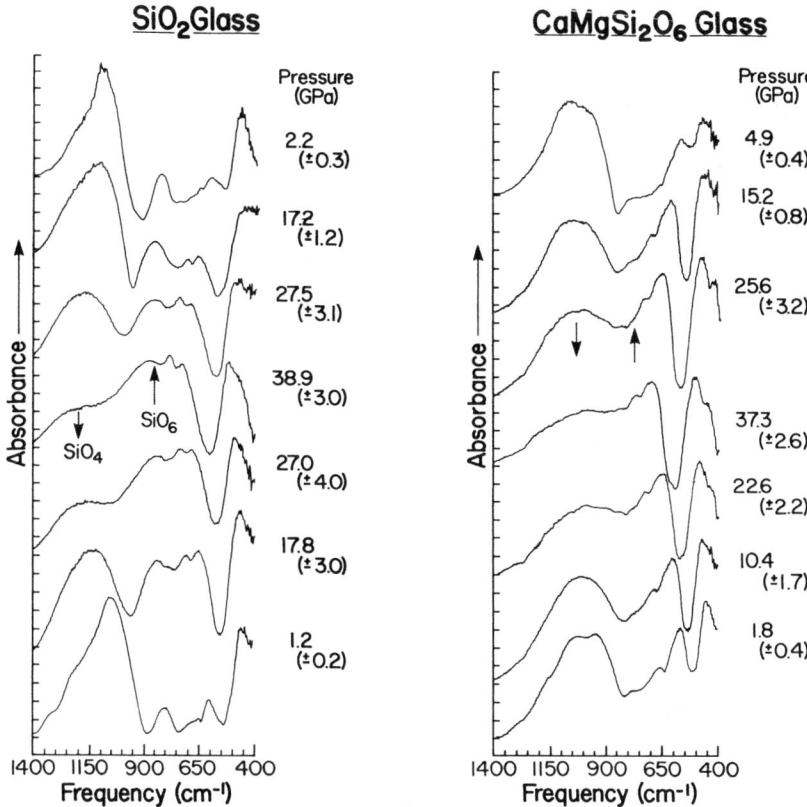

Figure 10 Representative mid-infrared absorption spectra of SiO_2 glass (*left*) and $CaMgSi_2O_6$ glass (*right*) as functions of pressure at 300K. As pressure is increased (*top* to *bottom*), the tetrahedral stretching mode characteristic of SiO_4 units decreases and the amplitude at frequencies characteristic of octahedral (SiO_6) modes increases, as shown by the *arrows*. On decreasing pressure, the spectral changes are almost completely reversed. Thus, coordination increases with pressure in these glasses, changing reversibly despite the low temperatures: e.g. the ratio of temperature to the melting temperature $T/T_m < 0.1$ (from Ref. 113).

increasing (118). Were it not for the change in bonding and atomic-packing geometry, such a decrease in strength would be highly anomalous because strength invariably increases on compression, and the SiO_2 glass is greatly densified at pressures above 10–20 GPa (119, 127, 128). Hence these experiments illustrate the wide range of structure and bonding that are achieved in amorphous phases at ultrahigh pressures, even if the temperature is low.

CONCLUSIONS

The effect of ultrahigh pressures and temperatures is to influence profoundly the physical chemistry of condensed matter. The outermost (valence) electron states can be completely altered, and the structural, magnetic, and electronic changes that result are intimately interrelated. All three cases discussed here, for example, illustrate how changes in bonding character cause and are simultaneously caused by changes in atomic packing at ultrahigh pressures.

In addition, several of the results that have been described are of direct interest to planetary geochemists. As the terrestrial planets consist of a rocky or ceramic shell (mantle and crust) surrounding an iron alloy core that is largely molten, at least for Earth and Venus, the high-pressure melting behavior of iron (Figure 5) offers the primary constraint on the temperature at the planetary center (34). The abundance of oxygen (about 60 atomic percent in the silicates and oxides of the mantle and crust) and the change in its bonding character at high pressures, make it likely that the Earth's core is primarily an iron oxide alloy (98, 100). Indeed, the boundary between the crystalline-silicate mantle and the liquid alloy core is thought to involve intense chemical reactions, and oxygen is dissolved into the core over geological time (100, 120). It is in this liquid oxide alloy of the core that the main magnetic field of the Earth is produced.

Finally, the separation of the planet into compositionally distinct zones, that is, the geochemical differentiation into crust, mantle, and core, is largely brought about through partial melting and separation of the melt from the region where it forms (leaving behind unmelted crystals). At shallow depths, silicate melts tend to be much less dense than the coexisting crystals; hence the melts rise buoyantly to the surface, emerging as volcanic eruptions. At the high pressures of the deep interior, however, melts tend to be of a density comparable to the surrounding rock because of the pressure-induced changes in melt structure (32, 108–110, 113). Therefore, melts tend not to separate readily from the crystals throughout most of the Earth's mantle: Geochemical differentiation and heat loss from the interior by migration of melts toward the surface are inhibited at depth, thus helping to maintain our planet as a geologically active body (121). In short, changes in material properties brought on by ultrahigh pressures are of interest for reasons extending from basic studies in physical chemistry to understanding the pheomena that have controlled the evolution of the Earth's interior.

Acknowledgments

I have benefitted from numerous discussions with Q. Williams, E. Knittle, and C. Meade, as well as the comments of R. J. Hemley, A. Jayaraman, H. K. Mao, and B. O'Neill. This work was supported by the National Science Foundation and the National Aeronautics and Space Administration.

Literature Cited

1. Jayaraman, A. 1983. *Rev. Modern Physics* 55: 65–108
2. Jayaraman, A. 1986. *Rev. Sci. Instrum.* 57: 1013–31
3. Hemley, R. J., Bell, P. M., Mao, H. K. 1987. *Science* 237: 605–12
4. Mao, H. K., Bell, P. M. 1976. *Science* 191: 851–52
5. Wigner, E. P., Huntington, J. B. 1935. *J. Chem. Phys.* 3: 764–70
6. Friedli, C., Ashcroft, N. W. 1977. *Phys. Rev. B* 16: 662–72
7. Min, B. I., Jansen, H. J. F., Freeman, A. J. 1986. *Phys. Rev. B* 33: 6383–90
8. Ceperley, D. M., Alder, B. J. 1987. *Phys. Rev. B* 36: 2092–2106
9. Mao, H. K., Bell, P. M., Hemley, R. J. 1985. *Phys. Rev. Lett.* 55: 99–102
10. Bell, P. M., Mao, H. K., Hemley, R. J. 1986. *Physica* 139/140B: 16–20
11. Hemley, R. J., Mao, H. K. 1988. *Phys. Rev. Lett.* 61: 857–60
12. Barbee, T. W. III, Garcia, A., Cohen, M. L., Martins, J. L. 1989. *Phys. Rev. Lett.* 62: 1150–53
13. Stevenson, D. J. 1982. *Annu. Rev. Earth Planet. Sci.* 10: 257–95
14. Hubbard, W. B. 1984. *Planetary Interiors*. New York: Van Nostrand Reinhold. 334 pp.
15. Barsch, G. R., Chang, Z. P. 1971. In *Accurate Characterization of the High Pressure Environment*, ed. E. C. Lloyd, pp. 173–89. Washington, D.C.: Natl. Bur. Standards Special Publ. 326
16. Kubaschewski, O., Alcock, C. B. 1979. *Metallurgical Thermochemistry*. New York: Pergamon. 449 pp. 5th ed.
17. Rice, M. H., McQueen, R. G., Walsh, J. M. 1958. *Solid State Phys.* 6: 1–63
18. Al'tshuler, L. V. 1965. *Soviet Physics Uspekhi* 8: 52–91
19. Ahrens, T. J. 1987. *Methods Exp. Phys.* 24A: 185–235
20. Block, S., Piermarini, G. 1976. *Phys. Today* 29: 44–55
21. Weir, C. E., Lippincott, E. R., Van Valkenburg, A., Bunting, E. N. 1959. *J. Res. Natl. Bur. Stand. Sect. A* 63: 55–62
22. Van Valkenburg, A. 1963. In *High Pressure Measurement*, ed. A. Giardini, E. C. Lloyd, pp. 87–94. Washington, D.C.: Butterworth
23. Bassett, W. A. 1979. *Annu. Rev. Earth Planet. Sci.* 7: 357–84
24. Bassett, W. A., Takahashi, T., Stook, P. W. 1967. *Rev. Sci. Instrum.* 38: 37–42
25. Piermarini, G. J., Block, S. 1975. *Rev. Sci. Instrum.* 46: 973–79
26. Moss, W. C., Halquist, J. O., Reichlin, R., Goettel, K. A., Martin, S. 1986. *Appl. Phys. Lett.* 48: 1258–60
27. Mao, H. K., Bell, P. M., Dunn, K. J., Chrenko, R. M., DeVries, R. C. 1979. *Rev. Sci. Instrum.* 50: 1002–9
28. Xu, J., Mao, H. K., Bell, P. M. 1986. *Science* 232: 1401–6
29. Jephcoat, A. P., Mao, H. K., Bell, P. M. 1987. In *Hydrothermal Experimental Techniques*, ed. G. C. Ulmer, H. L. Barnes, pp. 469–506. New York: Wiley
30. Fujishiro, I., Piermarini, G. J., Block, S., Munro, R. G. 1981. *Proc. VIII AIRAPT Conf.*, ed. S. Bergman, 2: 608
31. Mao, H. K., Xu, J., Bell, P. M. 1986. *J. Geophys. Res.* 91: 4673–76
32. Heinz, D. L., Jeanloz, R. 1987. *J. Geophys. Res.* 92: 11437–44
33. Knittle, E., Jeanloz, R. 1987. *Science* 235: 668–70
34. Williams, Q., Jeanloz, R., Bass, J., Svendsen, B., Ahrens, T. J. 1987. *Science* 236: 181–82
35. Ming, L. C., Bassett, W. A. 1974. *Rev. Sci. Instrum.* 45: 1115–18
36. Jeanloz, R., Heinz, D. L. 1984. *J. Physique Paris* 45(C8): 83–92
37. Bodea, S., Jeanloz, R. 1989. *J. Appl. Phys.* In press
38. Heinz, D. L., Jeanloz, R. 1987. In *High-Pressure Research in Mineral Physics*, ed. M. H. Manghnani, Y. Syono, pp.

113–27. Washington, D.C.: Am. Geophys. Union
39. Forman, R. A., Piermarini, G. J., Barnett, J. D., Block, S. 1972. *Science* 176: 284–85
40. Piermarini, G. J., Block, S., Barnett, J. D., Forman, R. A. 1975. *J. Appl. Phys.* 46: 2774–80
41. Mao, H. K., Bell, P. M., Shaner, J. W., Steinberg, D. J. 1978. *J. Appl. Phys.* 49: 3276–83
42. Heinz, D.L., Jeanloz, R. 1984. *J. Appl. Phys.* 55: 885–93
43. Bell, P. M., Xu, J., Mao, H. K. 1986. In *Shock Waves in Condensed Matter*, ed. Y. M. Gupta, pp. 125–30. New York: Plenum
44. Nellis, W. J., Moriarty, J. A., Mitchell, A. C., Ross, M., Dandrea, R. G., Ashcroft, N. W. 1988. *Phys. Rev. Lett.* 60: 1414–17
45. Knittle, E., Wentzcovitch, R. M., Jeanloz, R., Cohen, M. L. 1989. *Nature* 337: 349–52
46. Shimomura, O., Yamaoka, S., Nakazawa, H., Fukunaga, O. 1982. In *High-Pressure Research in Geophysics*, ed. S. Akimoto, M. H. Manghnani, pp. 49–60. Tokyo: Center Acad. Pub. Japan
47. Jamieson, J. C., Fritz, J. N., Manghnani, M. H. 1982. See Ref. 46, pp. 27–48
48. Brister, K. E., Vohra, Y. K., Ruoff, A. L. 1988. *Rev. Sci. Instrum.* 59: 318–21
49. Asaumi, K., Kondo, Y. 1981. *Solid State Commun.* 40: 715–18
50. Asaumi, K. 1984. *Phys. Rev. B* 29: 1118–20
51. Knittle, E., Jeanloz, R. 1984. *Science* 223: 53–56
52. Makarenko, I. N., Goncharov, A. F., Stishov, S. M. 1984. *Phys. Rev. B* 29: 6018–19
53. Itie, J. P., Polian, A., Besson, J. M. 1984. *Phys. Rev. B* 30: 2309–11; *J. Physique Paris* 45(C8): 47–51
54. Williams, Q., Jeanloz, R. 1986. *Phys. Rev. Lett.* 56: 163–64
55. Reichlin, R., Ross, M., Martin, S., Goettel, K. A. 1986. *Phys. Rev. Lett.* 56: 2858–60
56. Eby, J. E., Teegarden, K. J., Dutton, D. B. 1959. *Phys. Rev.* 116: 1099–1105
57. Teegarden, K., Baldini, G. 1967. *Phys. Rev.* 155: 896–907
58. Simmons, G., Wang, H. 1971. *Single Crystal Elastic Constants and Calculated Aggregate Properties: A Handbook*, Cambridge: MIT Press. 370 pp. 2nd ed.
59. Asaumi, K., Suzuki, T., Mori, T. 1983. *Phys. Rev. B* 28: 3529–33
60. Huang, T. L., Brister, K. E., Ruoff, A. L. 1984. *Phys. Rev. B* 30: 2968–69
61. Knittle, E., Rudy, A., Jeanloz, R. 1985. *Phys. Rev. B* 30: 588–90
62. Jeon, S. J., Porter, R. F., Vohra, Y. K., Ruoff, A. L. 1987. *Phys. Rev. B* 35: 4954–58
63. Grzybowski, T. A., Ruoff, A. L. 1984. *Phys. Rev. Lett.* 53: 489–92
64. Syassen, K., Christensen, N. E., Winzen, H., Fischer, K., Evers, J. 1987. *Phys. Rev. B* 35: 4052–59
65. Onodera, Y. 1968. *J. Phys. Soc. Jpn.* 25: 469–80
66. Aidun, J., Bukowinski, M. S. T., Ross, M. 1984. *Phys. Rev. B* 29: 2611–21
67. Satpathy, S., Christensen, N. E., Jepsen, O. 1985. *Phys. Rev. B* 32: 6793–99
68. Shindo, K., Nishikawa, A., Hasegawa, A. 1986. *J. Phys. Soc. Jpn.* 55: 3283–84
69. Knittle, E., Jeanloz, R. 1985. *J. Phys. Chem. Solids* 46: 1179–84
70. Vohra, Y. K., Duclos, S. J., Ruoff, A. L. 1985. *Phys. Rev. Lett.* 54: 570–73
71. Christensen, N. E., Satpathy, S. 1985. *Phys. Rev. Lett.* 55: 600–3
72. Baroni, S., Gianozzi, P. 1987. *Phys. Rev. B* 35: 765–69
73. Vohra, Y. K., Brister, K. E., Weir, S. T., Duclos, S. J., Ruoff, A. L. 1986. *Science* 231: 1136–38
74. Brister, K. E., Vohra, Y. K., Ruoff, A. L. 1984. *Phys. Rev. B* 30: 2968–69
75. Takemura, K., Minomura, S., Shimomura, O. 1982. *Phys. Rev. Lett.* 49: 1772–75
76. Takemura, K., Syassen, K. 1985. *Phys. Rev. B* 32: 2213–17
77. Williams, Q., Jeanloz, R. 1987. *Phys. Rev. Lett.* 59: 1132–35
78. Louie, S. G., Cohen, M. L. 1974. *Phys. Rev. B* 10: 3237–45
79. Glötzel, D., McMahan, A. K. 1979. *Phys. Rev. B* 20: 3210–16
80. Tups, H., Takemura, K., Syassen, K. 1982. *Phys. Rev. Lett.* 49: 1776–79
81. Balchan, A. S., Drickamer, H. G. 1961. *J. Chem. Phys.* 34: 1948–49
82. Riggleman, B. M., Drickamer, H. G. 1963. *J. Chem. Phys.* 38: 2721–24
83. Takemura, K., Fujii, Y., Minomura, S., Shimomura, O. 1979. *Solid State Commun.* 30: 137–39; 1980. *Phys. Rev. Lett.* 45: 1881–84
84. Fujii, Y., Hase, K., Ohishi, Y., Hamaya, N., Onodera, A. 1986. *Solid State Commun.* 59: 85–89
85. Shimomura, O., Takemura, K., Aoki, K. 1982. In *High Pressure in Research and Industry*, ed. C. M. Backman, T. Johannisson, L. Tegner, 1: 272. Uppsala: Arkitektkopia

86. Pasternak, M., Farrell, J. N., Taylor, R. D. 1987. *Phys. Rev. Lett.* 58: 575–78
87. Radousky, H. B., Ross, M., Mitchell, A. C., Nellis, W. J. 1985. *Phys. Rev. B* 31: 1457–62
88. Gatilov, L. A., Kuleshova, L. V. 1981. *Sov. Phys. Solid State* 23: 1663–65
89. Swenson, C. A., Shaner, J. W., Brown, J. M. 1986. *Phys. Rev. B* 34: 7924–35
90. Williams, Q., Jeanloz, R. 1987. In *Shock Waves in Condensed Matter*, ed. S. C. Schmidt, N. C. Holmes, pp. 73–75. New York: Elsevier
91. Adler, D. 1968. *Rev. Modern Physics* 40: 714–36
92. Darken, L. S., Gurry, R. W. 1946. *J. Am. Chem. Soc.* 68: 798–816
93. Distin, P. A., Whiteway, S., Masson, C. 1971. *Can. Metallurg. Q.* 10: 13–18
94. Bredig, M. A. 1964. In *Molten Salt Chemistry*, ed. M. Blander, p. 367. New York: Wiley
95. Bowen, H. K., Adler, D., Auker, B. H. 1975. *J. Solid State Chem.* 12: 355–59
96. Okamoto, T., Fujii, H., Hidaka, Y., Tatsumoto, E. 1967. *J. Phys. Soc. Jpn.* 23: 1174
97. Yagi, T., Suzuki, T., Akimoto, S. 1985. *J. Geophys. Res.* 10: 8784–88
98. Knittle, E., Jeanloz, R. 1986. *Geophys. Res. Lett.* 13: 1541–44
99. Knittle, E., Jeanloz, R., Mitchell, A. C., Nellis, W. J. 1986. *Solid State Commun.* 59: 513–15
100. Knittle, E., Jeanloz, R. 1989. *J. Geophys. Res.* In press
101. Brandow, B. H. 1977. *Adv. Phys.* 26: 651–808
102. Stishov, S. M. 1969. *Sov. Phys. Uspekhi* 11: 816–29
103. Ross, M., Rogers, F. J. 1985. *Phys. Rev. B* 31: 1463–68
104. Ross, M., Wolf, G. 1986. *Phys. Rev. Lett.* 57: 214–17
105. Waseda, Y. 1980. *The Structure of Non-Crystalline Materials*. New York: McGraw-Hill. 326 pp.
106. Matsui, Y., Kawamura, K., Syono, Y. 1982. See Ref. 42, pp. 511–24
107. Angell, C. A., Cheeseman, P. A., Tammadon, S. 1982. *Science* 218: 885–87
108. Rigden, S. M., Ahrens, T. J., Stolper, E. M. 1984. *Science* 226: 1071–74
109. Rigden, S. M., Ahrens, T. J., Stolper, E. M. 1989. *J. Geophys. Res.* In press
110. Knittle, E., Jeanloz, R. 1989. *Geophys. Res. Lett.* 16: 421–24
111. Jeanloz, R., Thompson, A. B. 1983. *Rev. Geophys. Space Phys.* 21: 51–74
112. Jeanloz, R., Knittle, E. 1989. *Philos. Trans. R. Soc. London Ser. A.* In press
113. Williams, Q., Jeanloz, R. 1988. *Science* 239: 902–5
114. Knittle, E., Jeanloz, R. 1987. See Ref. 38, pp. 243–50
115. Fratello, V. J., Hays, J. F., Spaepen, F., Turnbull, D. 1980. *J. Appl. Phys.* 51: 6160–64
116. Jeanloz, R. 1988. *Nature* 332: 207
117. Spackman, M. A., Hill, R. J., Gibbs, G. V. 1987. *Phys. Chem. Minerals* 14: 139–50
118. Meade, C., Jeanloz, R. 1988. *Science* 241: 1072–74
119. Meade, C., Jeanloz, R. 1988. *J. Geophys. Res.* 93: 3261–69
120. Knittle, E., Jeanloz, R. 1989. *Geophys. Res. Lett.* In press
121. Jeanloz, R., Morris, S. 1986. *Annu. Rev. Earth Planet. Sci.* 14: 377–415
122. Edwards, D. F., Philipp, H. R. 1985. In *Handbook of Optical Constants of Solids*, ed. E. D. Palik, pp. 665–73. New York: Academic
123. Waseda, Y. 1984. *Novel Application of Anomalous (Resonance) X-Ray Scattering for Structural Characterization of Disordered Materials*. New York: Springer-Verlag. 183 pp.
124. Jeanloz, R. 1989. In *Mantle Convection*, ed. W. R. Peltier. New York: Gordon & Breach. In press
125. Clark, S. P. Jr., ed. 1966. *Handbook of Physical Constants*. New York: Geol. Soc. Am. 587 pp.
126. Hazen, R. M., Jeanloz, R. 1984. *Rev. Geophys. Space Phys.* 22: 37–46
127. Meade, C., Jeanloz, R. 1987. *Phys. Rev. B* 35: 236–44
128. Hemley, R. J., Jephcoat, A. P., Mao, H. K., Ming, L. C., Manghnani, M. H. 1988. *Nature* 334: 52–54
129. Hemley, R. J., Mao, H. K., Bell, P. M., Mysen, B. O. 1986. *Phys. Rev. Lett.* 57: 747–50
130. Jayaraman, A., Sherwood, R. C., Williams, H. J., Corenzwit, E. 1966. *Phys. Rev.* 148: 502–8
131. Schiferl, D., Fritz, J. N., Katz, A. I., Schaefer, M., Skelton, E. F., Qadri, S. B., Ming, L. C., Manghnani, M. H. 1987. See Ref. 38, pp. 75–83
132. Sato-Sorensen, Y. 1987. See Ref. 38, pp. 53–59
133. Goettel, K. A., Eggert, J. H., Silvera, I. F., Moss, W. C. 1989. *Phys. Rev. Lett.* 62: 665–68
134. Reichlin, R., Brister, K. E., McMahan, A. K., Ross, M., Martin, S., Vohra, Y. K., Ruoff, A. L. 1989. *Phys. Rev. Lett.* 62: 669–72

THEORY OF ADSORBATE INTERACTIONS

Peter J. Feibelman

Sandia National Laboratories, Albuquerque, New Mexico 87185

INTRODUCTION

For at least 20 years, surface scientists have been promising that their work would lead to a materials science of surface chemistry. But it is only recently, as a result of the development of a variety of new experimental tools, that the focus of surface science has turned from "the surface structure problem," i.e. what atoms are on a static surface and where they are relative to one another, to the mechanisms of the fundamental processes of surface chemistry, such as energy transfer between reactants and surfaces, sticking, dissociation, diffusion, reaction, and desorption. The solution of the structure problem means that it is now possible to study these processes on a variety of interesting, *well-defined*, single-crystal surfaces. In parallel, it implies that surface theorists need to turn their attention from what have become "conventional" chemisorption problems to situations relevant to dynamical phenomena on single crystals. The most significant missing ingredient in our ability to simulate surface processes, at present, is a predictive understanding of the energy of an adsorption system as a function of the locations of the surface atoms that are involved. This article is a discussion of the status of chemisorption theory, aimed at assessing available and developing techniques that might provide such an understanding. Reflecting the nature of most of the relevant work in this area, the focus is almost exclusively on chemisorption on metals.

Interactions among adsorbates and between adsorbates and surfaces are obviously at the heart of surface physics and chemistry. Surface microscopies (1–2) as well as other surface-sensitive experimental techniques (3–

5) have made it possible to measure certain aspects of these interactions. At the same time, numerical simulation methods have been perfected that make it possible to calculate how a wide range of surface processes would occur if one knew the strengths of adsorbate-surface interactions as functions of the positions of adsorbates and their surface neighbors (6). Early studies (7–8) of such "energy-hypersurfaces" were based on rather unrealistic, Anderson model (9) calculations, and although they succeeded in explaining general features of adatom-adatom interactions, notably oscillatory long-range effects, contact with experiment was not achieved even in their most elaborate versions (10a,b). There followed a lengthy hiatus, during which surface electronic structure work focused on improved realism in the description of clean surfaces and highly ordered, high-coverage, static adsorption systems, in support of efforts to determine atomic geometries (11a,b). The result of these efforts has been to show that calculations based on local density functional (LDF) theory (12) can accurately predict not only structural, but also elastic properties of high-symmetry surfaces (13a–d). Since similar methods also describe gas-phase molecules (14), it has become both reasonable and desirable to develop LDF-based methods to treat the case of a few atoms in proximity to an otherwise perfectly ordered single-crystal surface, which encompasses the atomic arrangements pertinent to all the basic surface phenomena. A small body of such efforts has now been reported. A variety of competing methods for the calculation of surface atom energetics has also been set forth. In what follows, the new methodologies, as well as the significance of their initial output, are discussed.

Delineating the steps in a surface chemical reaction is not yet possible in most cases. This is why current theory and experiment are aimed at the *elementary* processes that must contribute: energy transfer to a surface and sticking of reactants, diffusion along or into a solid of whatever degree of perfection, dissociation, association, and desorption. The prime theoretical goals are to make contact with measured values of energies characterizing these phenomena, for example, adsorption, diffusion, and dissociation barrier heights, vibration frequencies of adsorbed fragments, etc, and their dependence on local crystal perfection or the presence of defects of various kinds. Since these energies correspond to low partial surfaces coverages and/or atomic arrangements of low symmetry, their evaluation poses a major challenge. The essential difficulty is that as a rule, the computational effort required to solve an electronic structure problem grows rapidly with a number of inequivalent atoms. For example, the usual methods of solid state physics, based on translational periodicity, grow as the cube of the number of atoms in the repeated cell. Quantum chemistry approaches to surface energetics, where a surface is modeled as a small cluster of surface atoms, typically grow even faster than the cube of the number of cluster atoms.[1] This is why the "supercell" in solid-state physics type calculations has generally been so small that atoms in neighboring cells must interact, whereas the cluster in quantum chemical calculations generally contains few if any "substrate atoms", whose coordination is that of an atom in an actual substrate. Because of these difficulties, the conventional methods of electronic structure theory have made very limited contributions to our understanding of adsorbate interactions, and there is a strong impetus to develop new methods for the low symmetry situations of interest. This is what underlies continuing efforts to develop a reasonably accurate semi-empirical theory of adsorption energetics (16a–d) and what makes the newly introduced "first-principles scattering-theory method" (17a–c) seem particularly exciting.

In the first part of this article, I briefly discuss the experimental data base regarding adsorbate interactions, and methods used to extract microscopic interaction energies from it. Then I describe various theoretical approaches to the problem of the energy hypersurface for low-coverage adsorption. These include methods based on the Anderson model such as that of Grimley (7a,b) and of Einstein & Schrieffer (8), attempts to relate surface chemical behavior to properties of the surface with no reactants present ("response theories") (18, 19a,b), quantum chemical and superlattice "first-principles" total energy calculations, semi-empirical methods such as the "effective medium" and "embedded atom" approaches (16a–d), and finally the first-principles scattering theory method (17a–c). My emphasis is on the advantages and disadvantages of the various methodologies, and on highlighting contributions that they have made to our understanding of surface chemical phenomena.

Before proceeding, it is worth noting that although the focus of this article is adsorbate interactions, theoretical methods for their treatment are not different in principle from what one would use to study the electronic structure of any kind of surface "point-defect," e.g. a single chemisorbed atom or a surface vacancy. Thus much of what follows in the theoretical sections of this article might as well be entitled: "Theory of Surface Point Defects."

[1] In local density functional calculations, no worse than three-center sums need to be evaluated to obtain the Hamiltonian, while matrix diagonalization demands grow as the cube of the matrix dimension. In Hartree-Fock calculations, the exchange integrals lead to four-center sums. When many-body effects are included, sums involving higher numbers of centers are necessary. This can clearly cause problem sizes for modest numbers of atoms to become impossibly large.

EXPERIMENTAL VALUES OF ADSORBATE-ADSORBATE INTERACTION ENERGIES

Generally, measurements of adsorbate-adsorbate and adsorbate-surface interaction energies either stem from structural observations, or involve dynamical phenomena such as atomic or molecular beam scattering, or surface kinetics. Two complementary structural techniques have been used to deduce effective interactions between adsorbates. The analysis of low energy electron diffraction (LEED) angular beam profiles as a function of surface temperature, pioneered by Lagally and collaborators, gives information concerning the sizes and shapes of islands in the course of adlayer growth (3). Simulation of the results using Ising lattice gas models can then be used in an attempt to determine the energy parameters of a lattice gas Hamiltonian (20). This technique is applicable for a wide class of adsorption systems, i.e. not only adsorbed metal atoms but also gas species such as O or CO, and even H, if its presence induces local substrate reconstruction. The statistics of the LEED method are intrinsically good because the diffraction data represent averages over a large number of islands. On the other hand, since sensitivity to adsorbate dimers and trimers is low, direct information concerning pair and three-body interaction energies is unavailable. At the same time, values of these fundamental parameters obtained via the lattice gas simulations are generally ambiguous (20).

The classic example of this sort of work concerns the growth of an O monolayer on W(110) (21). For coverages below 1/2 ml, and at sufficiently low temperature, O-atoms form $p(2 \times 1)$ islands by occupying nearest-neighbor sites in every other (111) row of W(110) hollows. At higher coverages, $p(2 \times 2)$ and $p(1 \times 1)$ phases are observed. Monte Carlo lattice gas simulations of the temperature-coverage phase diagram of this system have permitted one to draw conclusions regarding the signs of first and second neighbor O pair interactions, and to establish rough inequalities bounding the strengths of the shortest-ranged O triplet interactions (20). But the number of O-interaction parameters in the lattice gas Hamiltonian is too large to obtain their values uniquely via the simulations. Of course, if one were able to predict the interactions of O's on W(110) theoretically, it would be straightforward to determine whether the predictions were in or out of agreement with experiment. Passing this sort of test is a major aim of current efforts in chemisorption theory.

An alternate way to measure adsorbate-adsorbate interaction energies is to use a microscopy, either field ion microscopy (FIM) (1a-e) or perhaps scanning tunneling microscopy (STM) (22). One uses the microscope to compile a histogram of adsorbate geometrical configurations over many observations, holding the substrate temperature fixed. By comparing the actual statistical distribution to what one would expect if there were no interaction, one extracts the desired energy vs adatom geometry. This method is complementary to the LEED technique in that (*a*) it is a real space approach, providing actual images of the adsorbate locations; (*b*) it is best adapted to low surface coverages, (*c*) the gathering of statistics is arduous. It is of course only a useful method for adsorbates that do not desorb in the course of an experiment—this is why FIM studies of adatom interactions have been limited to transition metal atoms adsorbed on transition metal tips. FIM studies of adatom clusters by Tsong, Kellogg, Fink, Ehrlich, Bassett and their collaborators have established the surprising fact that the interactions of adsorbed atoms have little in common with the interactions of the same species of atoms in a homogeneous situation (1a–e). Re, for example, is a refractory metal, i.e. it has a high cohesive energy (8.10 eV/atom). On the other hand, two Re atoms adsorbed on W(110) in nearest neighbor sites interact *repulsively*, even though the metallic radii of W and Re are virtually identical (23a,b). Two W adatoms in neighboring adsorption sites on W(110) do bind to one another (24, 25); however, it costs only 285 meV to separate them, in comparison to a cohesive energy per bond of bulk W equal to 2.1 eV (24). Explaining this effect has been one of the first successes (26) of the recently introduced "self-consistent scattering theory of chemisorption" (17a–c), as is reviewed below (see APPLICATIONS OF THE SCATTERING THEORY METHOD TO ADATOM INTERACTIONS).

Adsorption energy hypersurfaces, of course, contain much more information than the equilibrium sites and binding energies of the surface atoms. To extract this additional information experimentally, it is necessary to study dynamical phenomena such as vibration, surface diffusion, scattering, and desorption. Vibration frequencies of adsorbates, commonly measured by using IR absorption (27), or inelastic electron (28) or He atom scattering (29), depend on the presence and species of neighboring adsorbates or defects, and thus can be used as a diagnostic of inter-adatom or adatom-defect interactions (5). An important example involves the "promotion" of surface catalytic reactions in which the rate-limiting step is the breaking of a strong bond. Co-adsorption of a surface modifier species such as K may speed up bond breaking and thereby enhance the efficiency of a catalyst (30a–d; however, see also 30e). It is therefore of great interest to monitor the vibration frequencies of adsorbed fragments or molecules versus K-coverage (31). Softening of the vibrations provides a quantitative measure of K-induced bond weakening.

Field ion as well as field emission microscopy (FEM) can be used to extract information concerning surface diffusion barriers. In the FIM technique (1a–e), one observes the location of an adsorbate or adsorbate-cluster at fixed time intervals. From its random walk, observed as a function of temperature, one extracts a diffusion constant vs temperature, and if the temperature dependence is exponential, one obtains the height of the surface diffusion barrier. The FEM technique, due to DiFoggio & Gomer (2), involves the observation of temporal fluctuations of field-emitted current from a small region of a crystal plane, corresponding to deviations from a homogeneous adsorbate distribution. This amounts to a direct measurement of the adsorbate density autocorrelation function, from which a diffusion constant can be extracted.

Diffusion constants, and particularly their dependence on the presence of co-adsorbed species, have recently been the subject of a number of laser-induced desorption (LID) studies (4a,b). A sharp laser pulse evaporates adsorbates from a disc-shaped region of a previously uniformly covered surface. Thereafter, the remaining adsorbates diffuse back into the disc. After a fixed time has elapsed, the laser is fired onto the same region of the sample once again. The number of adsorbates that evaporate is evidently a measure of the diffusion rate.

Many other techniques can be used to derive experimental information concerning energy hypersurfaces. For example, spectroscopic measurements can be used to obtain structural information, e.g., whether a co-adsorption system exists as a mixed or segregated phase as a function of temperature and partial coverages. This sort of information can be used to construct a phase diagram and thereby help determine inter-adpecies interaction energies. A recent review by White and Akhter provides many references of this nature (5).

THEORETICAL APPROACHES TO ADSORBATE-ADSORBATE INTERACTIONS

Anderson-model-based Calculations

The first theoretical issue in the study of adsorbate interactions was to explain why adsorbate superlattice structures form at relatively low coverages—numerous low energy electron diffraction (LEED) experiments show that they do. The very existence of a superlattice whose repeat distance is greater than an atomic diameter implies that adsorbates can find it more favorable to sit farther apart than closer together. This implies that adatom interactions must be relatively long-ranged and oscillatory. It is not hard to understand how such behavior can come to be, at least for a metal surface. Each adsorbate perturbs the substrate electron gas. Because of the sharpness of the Fermi surface, the adsorbate-induced, substrate charge disturbance has a long-ranged oscillatory tail (a manifestation of Friedel oscillations). The interaction of the oscillatory tails associated with two adsorbates can be expected to be oscillatory and long-ranged.

Koutecky was the first to propose a through-metal interaction mediated via the coupling of two adsorbates' orbitals to the wave functions of a substrate conduction band (32). Grimley (7a,b) embodied this idea in a "two-adsorbate Anderson model" (9) and, within the Hartree-Fock approximation, showed that the long-range behavior of the interaction is oscillatory. Einstein & Schrieffer (8) considerably embellished Grimley's work, first by treating two atoms adsorbed on the surface of a semi-infinite model crystal ["s-band cubium(100)"] rather than on an infinite one-dimensional chain, and then by considering and estimating the relative importance of multi-adsorbate (e.g. trio) interactions (10a,b).

Although this approach to the adsorbate interaction problem permits one to understand how nonintuitive, low-coverage LEED structures can come to be, it has been singularly unsuccessful in making quantitative contact with experiment—this despite efforts at incorporating "self-consistency" (largely charge neutrality), correlation and band degenacy effects (10a,b, 33). Recent work (26) explains the quantitative failure of the Anderson-model approach:

1. In actual adsorption systems, bonding sites are not discrete. In the presence of neighboring adsorbates, a particular adsorbate will move to a more favorable location, e.g. higher above the surface. As it does so, the overlaps of its orbitals with those of the substrate and with the neighboring adsorbates will change.
2. When adsorbates interact with each other, their interactions with the surface are affected. This consequence of the fact that each adsorbate only has a finite number of valence electrons is hardly a new idea—it is at the heart of bond-order bond-length correlations. In principle, this phenomenon is included in the Anderson-model. In practice it is necessary to give the adsorbates and substrate enough orbital flexibility for the rehybridizations to occur that embody the bond-order bond-length correlation. Extensive reviews of Anderson-model type calculations of adatom interactions are given by Einstein (10a) and by Muscat & Newns (10b).

Response Theory

A major goal of surface science is to identify properties of chemically active surfaces that characterize their activity. For example, the work function of a given crystal surface is related to its electronegativity, and thus should be useful in predicting the ionicity of its bonds with various adatom species. This, in turn, can help in predicting two adatoms' mutual electrostatic interaction. Evidently, the work function of a single crystal surface is much easier to calculate than the energetics of the same surface with adatoms on it. Thus it would be very important if we could correlate work functions with adatom interactions, rather than having to calculate the interactions directly.

In general, what we are after are properties of a high-symmetry surface (the work function is only one example) that describe how it will *respond* to the presence of adsorbates. (If the response is linear, so much the better, but linearity is not necessary for the idea of a response theory to be a useful one.) This pursuit is certainly not a new one. In the realm of homogeneous phase chemistry, it is well-established that what happens when two molecules interact is governed by the nature of their individual "frontier orbitals" (34). Recently, in what might be viewed as a paradigm of the response theory approach, a similar idea was the essence of an attempt by myself and D. R. Hamann to explain how electronegative surface modifiers (P, S, and Cl) can be effective poisons of the CO methanation reaction even when they are present only at very low coverages on a catalytic surface (19a,b). The experiments that led to this work, by Goodman and collaborators (35a–e), showed: (*a*) that the methanation reaction on Ni is "structure-insensitive" and therefore not defect-driven, i.e. its rate per surface atom is virtually the same on a single crystal as on fine, supported Ni particles; (*b*) that one S atom effectively poisons about 10 Ni sites, a fact established by a measurement of turnover number on Ni(111) vs S-coverage; and (*c*) that the effect is electronic rather than steric, since the inhibition of the reaction by the roughly equal radius atoms, P, S, and Cl increases in parallel with their electronegativities. One can think of several ways that electronegative poisons might affect CO methanation. The challenge is to find one in which a single S atom can poison the reaction over as many as 10 neighboring Ni's.

The work in Refs. (19a,b) is based on the premise that since the screening length in or near a metal is quite short, any *long-ranged* electronic effect of adsorbed electronegative modifiers must be a consequence of an *unscreened* surface property. Accordingly, instead of trying to correlate the poisoning effect of P, S, and Cl with screened quantities such as the charge densities or electric fields that they induce, we focused on their effect on the local density of electronic states at the Fermi energy (the E_F-LDOS). This quantity, which can be thought of as a position-dependent measure of the number of metal frontier orbitals, determines the energy cost of polarizing the electron gas. Since electrons must be transferred into the region between the surface and an adspecies in order to form a bond, the "stiffer" the substrate electron gas is, the more difficult it will be to form a bond. Thus if the presence of an atom like S modifies the surface electronic structure in the direction of opening up a gap, the cost of transfering electrons into the bonding region will increase, and chemical activity can be expected to slow down. At the same time, the E_F-LDOS is an unscreened quantity. The phenomenon of screening is a consequence of a system's minimizing its electrostatic energy by being as locally neutral as is possible within the constraints of quantum mechanics. But the E_F-LDOS represents only the charge density associated with a narrow electron energy window. It need not be screened, because any decrease in E_F-LDOS at a particular location can be compensated by an increase in LDOS at other energies, such that overall charge neutrality remains optimized. The perturbation of the E_F-LDOS by a P, S, or Cl atom might therefore be expected to persist over considerably larger distances than the induced charge density or electrostatic fields.

The results of self-consistent linearized augmented plane wave (LAPW) calculations, for hypothetical 1/3 and 1/4 ml modifier adsorption structures on Rh(100), bear out this expectation. For reasonable heights above a Rh(001) surface, the charge density induced by a Cl atom is screened within a single lattice spacing, while the disturbance it causes in the E_F-LDOS is large at a distance of 1.6 lattice spacings along the surface [i.e. as far as possible from a Cl in the 1/4 ml structure of Refs. (19a,b)], which is farther than the distance to the second-nearest fourfold hollow. This contrast is illustrated in Figures 1 and 2. It is also interesting that the reduction in the E_F-LDOS is predicted to be largest for Cl and smallest for P, a trend that parallels the observed poisoning efficiency.

One can hardly conclude from these results that we have uncovered *the* mechanism by which the electronegative poisons act. We have found, however, a property of the Rh(100) surface with no reactants present that certainly affects various steps in the methanation reaction by making the metal electron gas less polarizable near the surface, and that is affected over substantially larger distances than a lattice parameter from modifier atoms. One would hope that further work will provide a more detailed understanding of the relation between modifier-induced reductions of the E_F-LDOS and chemical activity. This seems a distinct possibility, as a result of the development of the scanning tunneling microscope (22)—the signal from such an instrument is a direct measure of the local density of states close to the sample surface (36).

At the same time, we need to learn how other modifier effects on surface electronic structure might affect reactivity. Nørskov et al (18), for example, point to the possible effects of the electric field induced by the poison

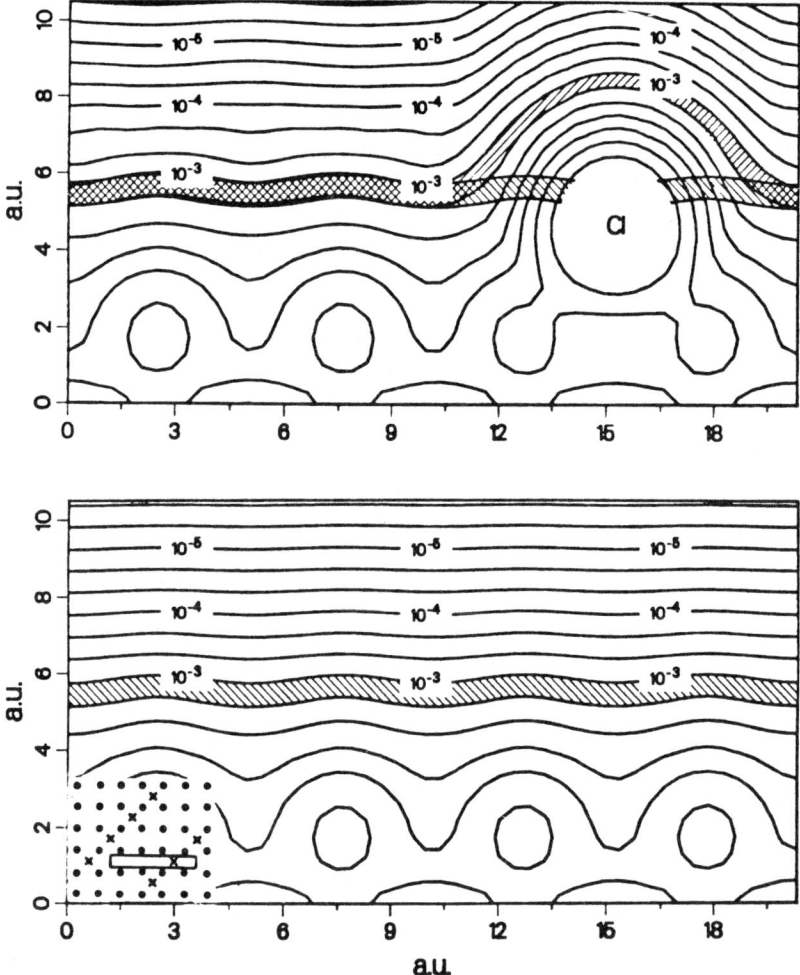

Figure 1 After the Linearized Augmented Plane Wave (LAPW) results of Ref. (19b), valence charge densities for (*A*) a two-layer Rh(001) film covered on both sides with 1/4 monolayer of Cl, and (*B*) the clean two-layer Rh(001) film. The plotting plane is perpendicular to the surface and intersects the adatom site (x in the *inset*) and two empty four-fold hollow sites, as are shown enclosed in the *rectangle* in the *inset*. The *inset* represents a top view of the surface, in which Rh atoms are shown as *solid circles* and Cl's as *crosses*. To facilitate comparison of the plots, the region between the contours of charge density $\sim 10^{-3}$ electrons/bohr^3, about 4 bohr above the Rh nuclei, has been hatched, and the *hatched region* from panel (*B*) has been transcribed onto panel (*A*). Note that the effect of the Cl atom dies off very fast as a function of distance from the Cl.

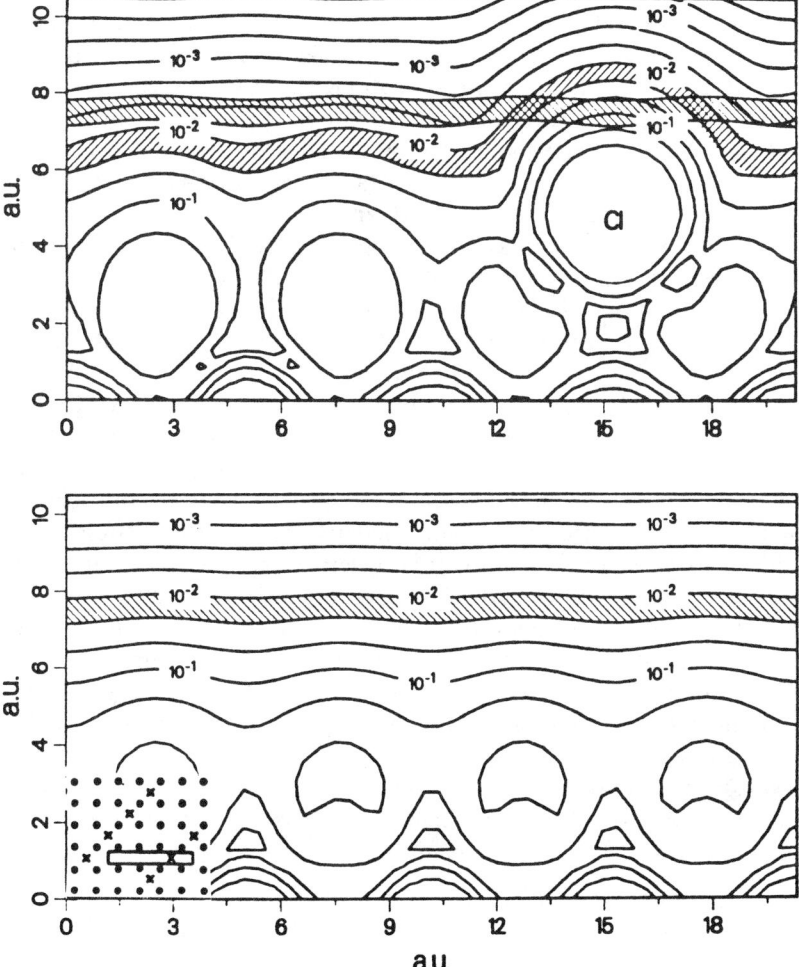

Figure 2 Same as Figure 1, except that the quantity being plotted is the Fermi-level local density of states (E_F-LDOS). Note that the effect of the Cl atoms is not well-screened in this case, and thus a low Cl-coverage can be expected to affect the E_F-LDOS over a large region of the Rh surface. Above the region distant from the Cl atom in panel (*A*), the *hatched region* transcribed from the clean Rh plot lies farther out into the vacuum than the corresponding hatched region for Cl/Rh. This means that with Cl present, one must move closer to the surface to find the same E_F-LDOS. Thus Cl *reduces* the E_F-LDOS at a fixed height. Note that the E_F-LDOS changes by $10^{1/3}$ from each contour to the next in these plots. Thus in the present case, the Cl has reduced the E_F-LDOS by a factor of ~ 2 over much of the surface. With Cl's adsorbed on only one side of the film, the reduction would be considerably smaller, perhaps 30%.

atoms. They note that if an adatom is negatively charged, the electric field in its neighborhood opposes charge transfer from the substrate outward. Just such charge transfer into the ad-CO ($2\pi^*$) anti-bonding orbital is thought to be necessary for CO-dissociation. Thus CO-dissociation should be inhibited near an electronegative adatom. To explore this picture, Nørskov et al calculate the electric potential near an isolated Cl-atom adsorbed on "Jellium," and study its falloff with distance. The idea, a response theory idea in fact, is that the nature of the Cl-covered surface without CO present can be used to predict what would happen if CO were there. The strength of the induced potential is found to fall to 0.01 Ry within ~ 3 Å parallel to the surface at the height of the Cl. This rapid falloff reflects good screening by the Jellium substrate. Higher above the surface, where there are fewer substrate electrons, screening is poorer, and the induced fields extend farther, implying a possibility of long-range effects on co-adsorbed species that are not tightly bound to the surface. As in the case of Ref. (19a,b), however, the poison-induced-electric-field model has not led to quantitative conclusions concerning CO dissociation (or, it hardly seems necessary to add, methanation). This is because (a) the Jellium model of Ref. (18) leads to exaggerated field strengths (its properties approximate those of Al, which is considerably more electropositive than the transition metals on which methanation occurs), (b) the CO dissociation pathway is unknown, and therefore the significance of the field strength as a function of height above the surface is unclear, and (c) no total-energy calculation has ever been performed in which CO and S are co-adsorbed.

Thus, what has as yet emerged from the response theories of Refs. (18) and (19a,b) are hypothetical mechanisms of poisoning by electronegative surface modifiers, but nothing quantitative.

Conventional First-principles Methods and Their Elaborations

As noted in the introduction, the low symmetry of the problems of interest makes developing a theory of adsorbate interactions on crystal surfaces difficult. Indeed, the corresponding electronic structure calculations might appear to be insoluble, because they correspond to situations in which there are formally an infinite number of inequivalent atoms. (Atoms in shell n of an adsorbate's neighbors see a different environment than those in shell $n+1$, at least in principle, for all n.) The reason that this formal inequivalence does not render it impossible to develop a practical theory of adsorbate interactions is the phenomenon of screening. Essentially all efforts to develop a first-principles theory of adsorbate-adsorbate interactions start from the fact that because of screening, an adsorbate's effects are spatially localized. Thus only atoms in its immediate neighborhood are substantially affected by its presence. Different approaches to the calculation of adsorbate interactions correspond to different ways of "embedding" this group of atoms in the remainder of the crystal, or not embedding it at all.

CLUSTER CALCULATIONS   The most straightforward way to try to take advantage of the screening of adsorbate effects is to replace the substrate crystal by those of its atoms which are affected by adsorbates, above some tolerance level. This means that the electronic structure of one or more adsorbates on a crystal is viewed as an ordinary molecular-structure problem, including, in principle, the most sophisticated treatment of many-electron correlation effects. The trouble with this simplest "cluster" approach is that, in general, one cannot expect rapid convergence with the number of atoms representing the substrate (37).[2] This is already a serious issue when considering the chemisorption of a single atom or molecule. The problem is worse if one wishes to study two or more adsorbates together, which perturb an even larger region of a surface. It should come as no surprise, then, that the literature contains few cluster calculations of adsorbate interactions.

The reason for slow convergence with the number of "substrate atoms" is that atoms at a cluster edge have the wrong coordination. The charge density within a screening radius around each edge atom therefore differs from what it would be in an actual crystal substrate, and if the screening spheres around the edge atoms intersect those of the adsorbates, then the electronic structure of the adsorbates is affected by cluster truncation. It should be emphasized that this is not "just a math problem." It represents real physics. The reduction of the coordination numbers of cluster edge atoms implies that the bonds to their remaining neighbors will be strengthened via bond-order bond-strength correlations. The adatoms must not sense this effect if their energetics is to represent that of adatoms on an extended crystal. Unfortunately, the number of atoms in a three-dimensional cluster grows rapidly with the number of atomic shells that are

---

[2] Experience shows that the possibility of performing meaningful cluster calculations must be evaluated on a case-by-case basis. The sorts of distinctions that are important are discussed by Upton et al (37). These authors show that whereas chemisorption of electronegative adspecies can be fruitfully studied on Ni clusters, that of electropositive ones cannot, because of very slow convergence of the electron affinity with cluster size. They also point out that different adsorbate properties tend to converge at different rates. For example, bond lengths often converge rapidly with cluster size, whereas densities of states converge slowly. Cleverness in the choice of cluster symmetry also plays a role in convergence rates. If cluster symmetry is too high, level degeneracies distort densities of states substantially and retard convergence.

included, and beyond that, the computational cost of molecular structure codes grows as a power of 3 or worse with the number of atoms. These facts imply that more or less serious compromises are necessary if the results of a cluster calculation are to be compared to chemisorption data from single-crystal studies.

Less serious compromises presumably apply to problems such as the rate-determining steps of "structure-insensitive" surface catalytic reactions, e.g. the methanation of CO on Ni. The fact that the per-surface-atom rates of such reactions are roughly the same on various single-crystal faces and on finely-divided small catalyst particles (35a–e) implies that the important energies in the problem will be correctly represented by cluster results. This does not mean that all aspects of the cluster electronic structure should be close to what obtains on a single crystal, only those that determine the reaction rate.

There are also situations for which the "back-bonding" that leads to bond-order bond-length correlations is minimized. If strong back-bonding always occurred, one would expect the separation of the outer two layers of any crystal surface to be contracted relative to that in the bulk. There are, however, many surfaces for which the outer layers' spacing is virtually "ideal" (i.e. unchanged relative to the bulk). In these cases, the electrons in the bonds that are cut to form the surface manifestly do not move into the region below the first layer. Analogously, certain cluster facets can be expected to be ideal, and thus involve only weak perturbations of the electron density inside the cluster. Such facets mitigate the cluster convergence problem and suggest optimal choices of cluster geometry (37).

An example of a more serious compromise, entailed by convergence requirements and used in numerous cluster calculations of chemisorption on Ni (38), is the treatment of the $3d$s as core-orbitals, via the introduction of a $3d$-pseudopotential (39a,b). Eliminating $3d$ valence-orbitals considerably reduces the size of the electronic structure problem for a cluster containing a given number of Ni atoms. On the other hand, if $4s$-$3d$ polarization is impossible, the physics of adatom interactions on Ni is forced to be similar to that on Cu (40a,b). How serious an approximation this is depends on the problem under consideration. The adsorption energetics for single H-atoms on Cu and Ni may be very similar, whereas the probability of H_2-dissociation on Cu and Ni surfaces are rather different. It is worth remembering that one of the main arguments in favor of the quantum-chemical cluster-approach to adsorption is that it permits one systematically to improve the description of electronic correlations, in contrast to calculations for extended systems, which can only be carried out with the introduction of the effective, but apparently not improvable, local density functional method. The significance of this argument is in question if in order to implement the more sophisticated treatment of correlations it is necessary to eliminate the distinction between Ni and Cu.

The case of O adsorption on Ni(100) provides an important example of how efforts to reduce the size of a sophisticated cluster calculation can lead to an incorrect understanding of chemisorption physics. Specular electron energy loss spectroscopy (EELS) data (41a,b) indicate a softening of the O-surface symmetric-stretch vibration as O coverage of Ni(100) increases from a $p(2 \times 2)$ quarter monolayer to a $c(2 \times 2)$ half monolayer. In an effort to explain this fact, Upton & Goddard (42) carried out correlated calculations of a single O chemisorbing in a four-fold hollow, on a 20 Ni-atom cluster. To make the calculation feasible, they replaced the Ni $3d$-orbitals with a pseudopotential (39a,b), and represented the valence state of each Ni via a single $4s$ orbital. With these approximations, they found two minima in adsorption energy versus O atom height above the surface (at heights of 0.26 and 0.90 Å), whose energies were rather close. The state in which the O lies close to the surface was identified with $c(2 \times 2)$O/Ni(100). A simple spring model as well as a complete dynamical calculation using the Upton & Goddard energetics confirmed that if O–Ni springs are almost in the surface plane, the symmetric-stretch mode will be soft (43). Early LEED analysis had not been extended to heights as small as 0.26 Å and therefore could not confirm or refute Upton & Goddard's idea (42, 44). However, subsequent off-specular EELS (45), extended X-ray absorption fine structure (EXAFS) (46a,b), and other data (47a–c) showed that the O atoms in both the $c(2 \times 2)$ and $p(2 \times 2)$ structures sit 0.9 Å above the Ni(100) surface. This led to a reinvestigation of the Upton & Goddard cluster calculation, culminating in the discoveries by Bauschlicher et al that the Upton & Goddard double minimum is an artifact of neglecting Ni $4p$ orbitals (48), and that a soft symmetric-stretch vibration with O residing 0.72 Å above the surface is predicted by a 5 O-atom, 25 Ni-atom cluster calculation in which the O atoms are arranged in a $c(2 \times 2)$-like structure (49). The moral of this cautionary tale is that compromises aimed at improving the convergence of cluster calculations are not to be taken lightly.

IMPROVED CLUSTER TERMINATION   A general approach to improving convergence of cluster calculations is to alter the way in which truncation is effected. For example; instead of simply removing the substrate atoms that are deemed to be unaffected by the adsorbates, one can terminate the cluster with a shell of H-atoms or some sort of "pseudoatoms" (50a,b). The idea is to restore the correct bond order to the outermost shell of "substrate atoms." This is particularly helpful for substrates such as tetrahedral semiconductors, whose atoms are attached to one another by single directed bonds. It is less clear that it can help in the case of metals.

Whitten and collaborators (51–56) have taken the idea of improved cluster truncation a good deal farther, embedding a small cluster containing the adsorbates and perhaps 25 substrate neighbors in a larger one, whose electronic structure is treated more approximately. The hope is that a simplified treatment of the larger cluster may be adequate to render truncation errors in the adsorption region negligible. The large cluster's electronic structure is treated at the Hartree-Fock level, in a minimal orbital basis [e.g. only $4s$ electrons, in a calculation of H-adsorption on Ni(111) (52)]. The atoms near the adsorbate are given more variational freedom (i.e. $3d$, $4s$, and $4p$ orbitals in the H/Ni(111) calculation). In addition, electronic correlations in the adsorbate-metal bond are accounted for via a configuration-interaction calculation.

This method has been applied to H_2-dissociation on Ti(0001) (53), Cu(100) (54), and over a Ti substitutional atom in a Cu(100) surface (55), to CO adsorption and dissociation on Ti(0001) (56), and to adsorption of an H atom on Ni(111) (53). In the recent H/Ni(111) calculation (52), the H-atom equilibrium position is found above several symmetry sites. Predicted bond lengths, binding energies, and vibration frequencies for H in hcp and fcc three-fold hollows are found to be in quite reasonable agreement with experimental results for H/Ni(111) (57a–c). The earlier calculations, for diatomics, involved only rather crude geometry optimization. Reasonable predictions did emerge from this work, e.g. that dissociation of H_2 on Cu(100) is activated whereas on Ti(0001) it is not, and that two H atoms on Ti(0001) prefer not to occupy immediately adjacent three-fold hollows. But detailed contact with experiment was not made.

Despite its successes, important questions remain to be answered concerning the embedding scheme of Whitten et al:

1. How stable are the results with respect to the choice of the simplified basis, and to other seemingly *ad hoc* aspects of the description of the larger cluster?
2. In an adsorption calculation in which electronic correlations are only allowed in the neighborhood of the adsorbates, how do we assess the extent to which apparent "adsorbate binding energy" is actually energy gain associated with *substrate* correlations permitted in the adsorbate region?
3. Why is it legitimate to describe the clean substrate in the Hartree-Fock approximation? This approximation generally yields a poor description of the electronic energy level dispersions of solids. Does it describe their structure and energetics accurately?
4. Finally, what are the effects of the mismatch of the minimal basis,

Hartree-Fock states of the larger cluster and the energy-levels appropriate to the more complete basis, correlated inner cluster? Do not electron wave-functions of the interior cluster reflect at the boundary with the outer cluster because of the different one-electron potentials in the two regions? To what extent might such reflection affect the calculated energetics?

LOCAL DENSITY FUNCTIONAL CLUSTER CALCULATIONS   Not all cluster total energy calculations are based on the use of "quantum-chemistry methods." An important advantage of the local density functional treatment of electronic exchange and correlation (12) is that it makes possible calculations of the electronic structure of condensed matter systems that "only" grow as the third power of the number of inequivalent orbitals (see footnote[1] above). This is slower than what is required by Hartree-Fock plus configuration-interaction methods, and means that the LDF method permits calculations for somewhat larger clusters than straightforward quantum chemistry techniques, without the compromises described above. A review of LDF and Hartree-Fock cluster studies of the adsorption of the small molecules, H_2O, NH_3, and CO on various metals, including predictions of bonding site preference energies, admolecule orientations, vibration frequencies, etc., has recently been published by Müller (58). Its main conclusions are that where electrostatic effects are important, or if results are desired concerning unscreened quantities such as densities of states, clusters must be impractically large to give converged answers (cf. 37). On the other hand, unpublished tests are claimed to show that LDF calculations of the geometric structures and the energetics of single molecules adsorbed on clusters containing as few as ten substrate atoms are accurate to ~ 0.04 bohr and ~ 0.1 eV, respectively. This claim is somewhat problematic—an accuracy level of 0.1 eV in the adsorption energy of H_2O/Al means a 20% systematic error due to cluster truncation.

LDF cluster calculations of adsorbate-adsorbate interaction energies are not yet available, to my knowledge.

SUPERLATTICE CALCULATIONS   A different way to take advantage of the screening of adsorbate effects is the superlattice method, which in principle permits the direct application of conventional surface band structure methods to adsorbate interaction problems. Here a large unit cell containing the adsorption complex of interest is repeated periodically in two dimensions. By virtue of the periodic repetition of the cell, Bloch's theorem is applicable, and there is no impediment to carrying out a band structure calculation other than its size. However, as in the case of the cluster method, the mathematical artifice of periodic repetition leads one to a computational problem that involves significantly more atoms than are actually affected by the screened potential of the adsorbates. Specifically, if we do not wish to study the effects of adsorbates in one cell interacting with those in the next, the cells must contain at least two shells of the adsorbates' surface neighbors. But then all the deeper layers of the cell (in what is typically taken to be a several layer film) must contain equal numbers of atoms—and many of these lie outside the adsorbate screening spheres. Generally, self-consistent band-structure calculations are carried out within the context of the local density functional theory of exchange and correlation. This means that, at best, they generally grow in computational demands as the third power of the volume of the unit cell. As a result, superlattice calculations of adsorbate energetics that are converged with respect to the size of the unit cell are too expensive to perform, and have not been.

Immersion Energy-based Methods

One of the main motivations for computing few-adatom energy hypersurfaces is to make dynamical simulations of fundamental surface phenomena reliable. There are, however, two aspects to "reliability." Not only must each energy on a calculated hypersurface be accurate, but many adatom geometries must be considered, to minimize errors in interpolating between them. In this respect, full first-principles calculations seem less than ideal. One day, if highly parallel computer architectures fulfill their promise, it may be practical to perform many first-principles energy and force evaluations for interesting systems. But currently, many scientists who aim to simulate surface dynamics have abandoned computationally demanding ab initio approaches in favor of much simpler and very much less costly methods based on the idea that, to leading order, the energy of an atom in or on a solid equals the energy of the same atom immersed in a homogeneous electron gas of the "local" electron density. There are several incarnations of this approach: the quasiatom (16a) or effective medium theory (EMT) (16b), the embedded atom method (EAM) (16c), and improved versions of these (59a–c). Generally, their use is restricted to problems in which the substrate is a simple metal or a filled or nearly filled *d*-band metal. Adsorbates of this nature have been considered, as well as H, O, and C atoms. Little has been published relating to the adsorption of a few atoms, e.g. two of them, because of difficulties in the methodology (see e.g. 60). Clearly, progress will have to be made in generalizing the applicability of the immersion energy–based methods if they are to be very useful in understanding surface chemistry.

Generally, "immersion-energy" calculations start from work such as that of Puska et al (61), who obtain the binding energies of single atoms in a homogeneous electron gas as a function of electron density. For all but the inert gases, the binding energy vs density starts from zero at zero density, achieves a maximum at a species-dependent optimum density, and decreases monotonically thereafter. This behavior results from the competition between the attractive electrostatic and exchange-correlation energies, which are dominant at low densities, and the repulsive kinetic energy, which dominates when the electron density is high. In the EMT of Nørskov and collaborators, the leading contribution to the energy of an atom near a surface is the LDF immersion energy evaluated at electron density, $\bar{n}(\mathbf{r})$, which is an average of the substrate density near the adatom site, $\mathbf{r}$ (16b, 59a). In the EAM of Daw & Baskes, it is a semi-empirical immersion energy evaluated at a density equal to the sum of the substrate atom densities at $\mathbf{r}$ (16c).

If immersion energy were the only contribution to the binding of an adatom, unphysical consequences would clearly ensue. In particular, defining n_{opt} to be the optimum density for the adatom, any $\mathbf{R}$ on the sheet $n(\mathbf{R}) = n_{\text{opt}}$ would be an energetically equivalent adsorption site. That is, there would generally be an infinite set of binding sites in the surface unit cell, and the restoring force for a displacement along the sheet, $n(\mathbf{R}) = n_{\text{opt}}$, would be zero. To arrive at a realistic chemisorption theory, the immersion energy must therefore be supplemented by additional contributions. These represent "core repulsion" and "hybridization" (59a). The core repulsion contribution corrects the energy of immersion in the homogeneous electron gas for the fact that near the cores of an adatom's neighbors, the substrate charge density is *very* inhomogeneous. This correction is assumed to be a sum of short-ranged pairwise interactions in the EAM (16c); this assumption is justified in the most recent formulation of the EMT, by Jacobsen et al (59a), aimed at problems involving more than one atom in, on, or of a host. In the EAM, the form of the core-interactions is determined empirically, via fits to cohesive and elastic properties. In the EMT, it is fitted to LDF results. The "hybridization" correction, in the EMT, is attributed to rapid variation in the local density of states at an adatom site, due to the substrate, as compared to the smooth variation when the host is a homogeneous electron gas. This is expected to be important for d-band substrates but not for simple metals, and is included in the former case, via a simple perturbation theory expression based on an Anderson model (9). In the EAM, no hybridization energy is included explicitly; empirical fitting of the immersion and core repulsion energies is assumed to incorporate any such energy. On the other hand, EAM calculations have till now been limited to simple, filled, and nearly filled d-band metals, for which hybridization contributions are expected to be small (59a).

A big advantage of the immersion-energy-based methods is the transparent picture of chemisorption to which they lead. For example, the height at which an adatom resides above a surface is dominated by its desire to be at a host electron density equal to its n_{opt}, while the restoring force that determines the adatom's vibration frequency along the surface normal is directly related to the gradient of the substrate density at the binding site (62). This transparency also applies to adatom-adatom interactions. Thus, at a given height above a surface, two neighboring adatoms will each sense a local electron density that is greater than if the adatoms were isolated. This means that in general, neighboring adatoms will sit higher above a surface than isolated ones, in agreement with what one would expect on the basis of bond-order bond-length ideas (63). Similar arguments apply to relaxation of the positions of the adatoms' neighbors.

On the other hand, whether immersion energy ideas "explain" the physics underlying surface phenomena can be a question of taste. For example, Ho & Bohnen have shown via detailed first-principles calculations how differences in d-band positions and d-orbital sizes explain why the Au(110) surface reconstructs while the Ag(110) surface does not (64a). In the immersion energy–based studies of the same problem, the different reconstruction behavior is simply attributed to the different elastic constants of bulk Au and Ag (64b). Another recent paper shows that the surface phonon spectrum of Cu(001) can be accurately calculated from an EAM energy functional that contains no empirical surface information (65). In both of these instances, the EAM or EMT result is interesting and informative but does not provide any microscopic detail.

A more serious issue concerning the immersion energy methods is their general applicability to problems involving more than one adatom. Improvements in the EAM have made it possible to study migration of adsorbed noble metal clusters on noble metal surfaces (66). However, admolecules of interest in surface chemistry often involve directional bonds, whereas the treatment of "covalency" or "hybridization" effects is the weakest aspect of the EMT and EAM schemes. Even to study H_2 dissociation on Ni one must apparently introduce an ad hoc reformulation of the EAM (see e.g. 60). To take account of the imperfect screening of the electrostatic potentials of ionically bonded adspecies, one also needs to go beyond the approximation schemes currently in use. For example, consider the coadsorption of a Na and a F atom on a metal surface, in non-nearest neighbor sites. The electrostatic attraction of these adatoms together with their screening charges is clearly not included in the sum of the Na and F immersion energies plus a core repulsion. Thus, it seems fair to say that it is unclear whether the appealing simplicity and transparency of the immersion energy methods will be available for studies of many of the important phenomena of surface chemistry.

First-principles Scattering Theory Calculations

In addressing the fundamental phenomena of surface chemistry, conventional electronic structure calculations are computationally expensive to converge because they require one to determine the electronic structure of the substrate and the adsorption complex *at the same time*. In the new "scattering theory of adsorption" (17a–c), the perfect substrate, and adsorption on it, are treated separately. Thus one very difficult problem is replaced by two easier ones. The first "easier problem" is to determine the electronic structure of the unperturbed substrate (henceforth referred to as the "host"). Its size is governed by the number of atoms in the host's unit cell, and there are often only a few if the substrate is represented as a thin slab. The second easier problem is to determine the wave-functions of the adsorption system as a whole, which are Bloch waves of the host scattered by the *screened* adsorbate-induced potential. The size of a scattering theory problem is determined by the volume of space in which the scattering potential is nonzero, or, in an orbital basis, by the number of orbitals that overlap the scattering potential. By virtue of screening, this volume is small [under most conditions],[3] and so, therefore, is the size of the scattering problem.

In the mid-1950s, when the idea of using scattering theory to study point-defect problems was first introduced by Koster & Slater (68), large-scale computing was not a possibility. Therefore the scattering theory method was used to solve empirical tight-binding Hamiltonians representing point-defect problems, and to use the defect-induced change in the sum of single-electron energies to monitor system energetics. Twenty years later, with supercomputers on the horizon, one could contemplate self-consistent Koster-Slater type calculations, in which the scattering potential is determined by the condition that the system energy be minimized. Such calculations were first performed by Baraff & Schlüter (69a) and by Bernholc et al (69b) in studies of the spectroscopy of point defects in bulk semiconductors. An important simplification of the self-consistent scattering theory method was proposed by Williams, Feibelman & Lang (WFL) (17a), in which the electronic structure problem is cast into a basis of localized orbitals at the very beginning. The resulting *matrix* Kohn-Sham equation (70) for the LDF defect problem can then be treated via a *matrix* version of scattering theory. This is an especially useful step if one contemplates calculations in which the host lattice relaxes near an adsorbate—it eliminates the sharply varying potentials and complicated

---

[3] If the adsorbate is charged, and the charge resides substantially outside the surface imge planes then the adsorbate-induced electrostatic potential falls off slowly.

basis projections (72a,b) that beset earlier scattering-theory based techniques.[4] It also avoids problems associated with projecting a singular, coordinate space, crystal Green's function into a local basis (17a). What follows is a brief overview of the self-consistent, matrix scattering theory method and a description of the first results obtained through its use.

SCATTERING THEORY FORMALISM Self-consistent scattering theory is aimed at solving the Kohn-Sham (70) energy minimization problem of local density functional theory (12) for point-defect problems. In the matrix Green's function version of the method (17a), one takes advantage of the screening of the defect-induced charge and potential by immediately introducing a basis of local orbitals, $\{\tilde{\phi}_i\}$, to represent the electronic structure of the perfect crystal, and an additional set of local functions, $\{a_i\}$, for any adatoms. (Here and below, a tilde over a symbol refers to a perfect crystal property.) In this basis, a zeroth order Kohn-Sham equation whose solution embodies the properties of the perfect host, and which allows for the subsequent inclusion of adsorbate-host interactions, is

$$\sum_j (H_{ij}^0 - E^{0(n)} S_{ij}^0) c_j^{0(n)} = 0, \qquad 1.$$

where

$$H_{ij}^0 - E S_{ij}^0 = \left(\begin{array}{c|c} (\varepsilon_i - E)\delta_{ij} & 0 \\ \hline 0 & \tilde{H}_{ij} - E\tilde{S}_{ij} \end{array} \right). \qquad 2.$$

In Eq. 2, the ε_i represent adatom energy levels and $\tilde{H}_{ij}$ and $\tilde{S}_{ij}$ are the Hamiltonian and overlap matrices of the perfectly periodic host. Since the matrix of Eq. 2 is block diagonal, and $\tilde{H}_{ij}$ and $\tilde{S}_{ij}$ correspond to a periodic solid (a few-layer film in surface applications to date), Eq. 1 can be solved with conventional band-structure methods. (Of course, $\tilde{H}_{ij}$ depends on the $c_j^{0(n)}$ implicitly, via its dependence on the electrostatic and the LDF exchange-correlation potential. Thus Eq. 1 represents generalized Hartree equations, a non-linear matrix problem.)

By virtue of the zeroes in the off-diagonal blocks in Eq. 2, the adatoms do not interact with each other or with the host. We now remove this restriction, and also allow host atoms to move from their original locations. Choosing a new basis, $\{\phi_i\}$, for the host atoms in their new positions, the Kohn-Sham equation for the adatoms interacting with the perturbed lattice is

---

[4] When lattice motion is considered in a coordinate space representation, one subtracts the unperturbed from the perturbed Hamiltonian, leading to a sharply varying potential near any nucleus. This is true even in a pseudopotential based scheme, if the pseudopotentials are strong.

$$\sum_j (H_{ij} - E^{(n)} S_{ij}) c_j^{(n)} = 0. \qquad 3.$$

Since there are an infinite number of orbitals in an infinitely extended host, and since neither H_{ij} nor S_{ij} is block diagonal, Eq. 3 cannot be viewed as anything other than an infinite set of simultaneous equations. Thus, unlike Eq. 1, which is a computationally tractable $N \times N$ problem, where N is the number of orbitals centered in the perfect host's repeated cell, the direct numerical solution of Eq. 3 is not possible.

The scattering theory method breaks the intractable Eq. 3 into two problems, each of which can be solved. To begin, one requires the two bases $\{\phi_i\}$ and $\{\tilde{\phi}_i\}$ to coincide for atoms of the host that are outside the region where adatoms reside or where host atoms have been moved. Since all orbitals in the zeroth order and interacting system are spatially localized, and since the defect potential is screened, it follows that δH_{ij} and δS_{ij} defined by

$$\delta H_{ij} = H_{ij} - H_{ij}^0, \quad \delta S_{ij} = S_{ij} - S_{ij}^0 \qquad 4.$$

vanish except when the orbitals i and j correspond to atoms near the defect.

One now identifies solving Eq. 3 with evaluating the Green's function *defined by*

$$\sum_k (Z S_{ik} - H_{ik}) G_{kj}(Z) = \delta_{ij}. \qquad 5.$$

This identification stems from the realization that if $G_{kj}(Z)$ is known, then the one-electron density matrix can be evaluated via contour integration in the Z-plane, and all other observables can be calculated. Equation 5 is solved by recasting it as a Dyson's (or scattering theory) equation. A zeroth order Green's function is defined by

$$\sum_k (Z S_{ik}^0 - H_{ik}^0) G_{kj}^0(Z) = \delta_{ij}. \qquad 6.$$

Simple algebra then leads to

$$G_{ij}(Z) = G_{ij}^0(Z) + \sum_{kl} G_{ik}^0(Z)(\delta H_{kl} - Z \delta S_{kl}) G_{lj}(Z). \qquad 7.$$

This scattering theory equation is a soluble version of the apparently intractable Eq. 3 because δH_{kl} and δS_{kl} vanish when their indices correspond to any but the M orbitals that appreciably overlap the unscreened defect potential. This means that Eq. 7 is an $M \times M$ set of simultaneous equations, where M is a finite number (typically on the order of a few hundred in applications to date). The quantity $G_{ij}^0(Z)$, which is evidently required in Eq. 7, can be computed via the usual spectral representation.

A key feature of the matrix method is that it makes no difference whether the subscripts of G^0 and G refer to the same orbitals or different ones, corresponding to the case that one or more host atoms has been moved. The fact that Eq. 7 yields a solution of Eq. 5 can be shown by simple algebra that is independent of the nature of the orbitals to which the indices refer.

The self-consistent aspect of the scattering theory method remains to be discussed. "Self-consistency" is achieved when the electron density, $\rho(r)$, used in the evaluation of H_{ij} is the same as that which emerges from the formula

$$\rho(r) = \oint dZ/2\pi i\, G_{ij}(Z)\phi_i^*(r)\phi_j(r), \qquad 8.$$

where the Z-integration runs over a contour surrounding the occupied state energies on the real Z-axis. In actual calculations, the contour integrals are done on a mesh of between 20 and 40 Z-points in the upper half Z-plane, while the agreement of input and output charge densities is accomplished by an accelerated relaxation-iteration scheme (73a,b).

Once self-consistency has been achieved to an acceptable tolerance, observables can be computed. The most important of these are the total energy of the system and the gradient of the energy with respect to nuclear positions [which can be directly evaluated using a Hellmann-Feynman type formula (17c)]. Accuracy of the numerical procedures is tested by checking that energies at a sample of neighboring adsorption geometries agree with predictions based on using the calculated forces. Once accuracy of the numerics has been verified, calculated forces can be used to relax an adsorption geometry to equilibrium, to determine transition geometries and barrier energies for diffusion, to calculate spring constants, or whatever the problem at hand may dictate. Some results of such calculations are reviewed in the following section. They show that realistic calculations of adsorption energetics can provide the answers to important questions and expand our conceptual understanding of adatom interaction phenomena.

APPLICATIONS OF THE SCATTERING THEORY METHOD TO ADATOM INTERACTIONS A computer program embodying the self-consistent scattering theory of adsorption has not been in existence for very long. So only a limited set of applications to adatom interactions have been completed. These are reviewed in this section.

As described in the second section above, the field ion microscope has for some time been used to measure the energy of pairs of transition metal atoms adsorbed on various d-band metal crystal surfaces, as a function of relative adatom positions. Such studies show that the interaction between two adatoms is generally much smaller than what one would guess on the basis of crystalline cohesion on a per bond basis. In fact, in the case of two Re atoms adsorbed on W(110), the attractive component of the interaction is apparently too weak to bind the adatom pair (23a,b). The FIM has also been used to measure single and adatom cluster diffusion constants and barrier energies. Remarkably, one often finds that adatom-dimer diffusion barriers are lower than for monomers of the same species. Explaining these results qualitatively is one of the first successes of the scattering-theory method.

Although the FIM studies of Refs. (23–25) (also see 1a–e) were carried out for d-band metal adatoms and substrates, there is no reason to expect that similar adatom energetics will not be found for simple metals. In any case, for a first study of adatom interactions using the newly developed scattering-theory computer code, it seemed reasonable to investigate a simple metal substrate and sp-bonding adsorbates. Accordingly, I began by calculating the energetics of one and then a pair of Al atoms self-adsorbed on a two-layer Al(001) film (26). The single Al is found to reside at a height of 3.30 bohr above the upper plane of Al nuclei, in a four-fold hollow site. Two Al's in neighboring four-fold hollows, however, are attracted to one another, each one moving 0.09 bohr towards its partner but at the same time also moving 0.3 bohr higher up off the surface. The binding energy of the Al adatom pair in neighboring hollows is 0.07 eV relative to separated adatoms, while the cohesive energy per bond of bulk Al is known to be 0.557 eV.

This latter result is interesting. It says that the separation energy of an Al adatom pair on Al(001) is roughly 1/8 of the cohesion per bond of bulk Al. This ratio is very similar to that which is observed, for example, for a pair of W's adsorbed on W(110) (74). The key to understanding the calculated results for the adsorbed Al dimer is that the pair resides higher off the surface than each Al would if it were isolated. Why is the dimer repelled from the surface? Evidently, two Al's in neighboring hollows share a pair of substrate neighbors and must compete for their valence electrons. This competition means that each Al's bond to the surface is weaker than it would be if it were isolated from its adatom neighbor. At the same time the two Al's are large enough to form a direct bond with each other. The fact that they do is indicated by their motion in each other's direction. This discussion makes it clear that the energy gain associated with the formation of a bond between the two ad-Al's comes at the cost of Al-surface binding energy. The partial cancellation between adatom-adatom attraction and effective adatom-surface repulsion explains why the energy to separate an ad-Al pair is small. If the pair is pulled apart, the price of rupturing the direct adatom-adatom bond must be paid, but it is compensated by a strengthening of each ad-Al's bond to the substrate. Thus, the bulk Al cohesive energy per bond is large when one removes an atom from a crystal to "infinity," because that atom's electrons go from bonding to nonbonding states. On the other hand, when an ad-Al dimer separates, the valence electrons move from the most favorable binding state to one that is somewhat less favorable. This involves a much smaller energy change.

Further calculations were done to determine the energy of the transition state in which an Al adatom traverses a bridge to a neighboring four-fold hollow, and to see how this barrier energy is changed if there is a second ad-Al nearby. Again the results agree qualitatively with FIM observations. For a single ad-Al, the energy required to move it to a bridge site saddle-point from its equilibrium site in a four-fold hollow is 0.80 eV. On the other hand, if initially there was a second ad-Al in the neighboring hollow, then the energy required to move to the saddle point is only 0.66 eV. This reduction in energy implies a lower diffusion barrier for an adatom pair than for a single Al. The explanation of this result is again related to bond-order bond-length ideas. The adatom pair resides higher above the surface than either Al would if isolated. But higher above a surface, the adatom-surface potential is generally less corrugated, which means that the energetic distance from the "valleys" to the "passes" is reduced. In addition, as one atom of a pair pays the price of mounting the saddle, the second atom can gain energy by returning toward its isolated atom bonding configuration, thus recouping some of the energy expense of its partner. These two effects account for most of the reduction found in the dimer diffusion barrier (26).

A more recent study of adatom interactions, using the scattering theory method, was aimed at testing the widely accepted idea that impurity atoms on a surface tend to migrate to surface defects (75). This idea underlies one of the common explanations of how small coverages of surface modifiers can drastically reduce surface catalytic activity. In particular, if the catalytically active sites are a small number of defects, then only a small number of modifier atoms are required to block them and turn off a reaction.

To test the generality of the idea that "modifier atoms" migrate to defects, I considered the co-adsorption of a chalcogen atom, either S or Te, and an Al on a five-layer Al(001) film. The self-adsorbed Al atom represents a model defect, and the S or Te atom a model "poison." To begin, I computed the equilibrium heights above the surface of the ad-Al and, separately, of the S or Te. Then I placed the Al and chalcogen atoms at these heights in neighboring four-fold hollows. The result, for both S

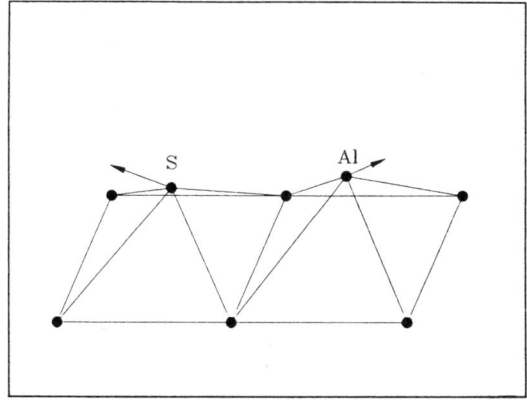

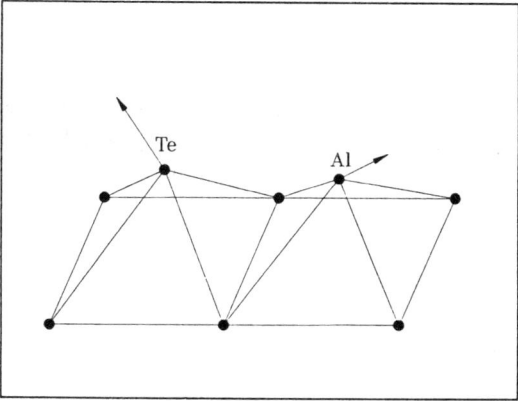

Figure 3 After Ref. (75), diagrams indicating relative locations and forces on S-Al and Te-Al dimers adsorbed on Al(001). Each adatom is at the height where it would reside if isolated from its partner. *Arrows* indicate relative directions and magnitudes of calculated forces for the dimer geometry shown.

and Te, was that the adatoms repel. In fact, if a S and an Al are forced to find their metastable equilibrium locations in neighboring hollows on Al(001), the total system binding energy is predicted to be about 0.25 eV higher than if they are isolated from one another. For a Te and an Al, the reduction in total binding energy is only 0.22 eV.

This result is not completely surprising. After all, two Re atoms, whose bond might be expected to be very strong, are known not to bind to each other at all on the W(110) surface. Still, some interpretation is in order. The first question is: What drives the chalcogen-Al repulsion? Figure 3 provides an important clue. As in the case of the adsorbed Al dimer, when the Al and chalcogen atoms are placed in their isolated-atom equilibrium positions in neighboring four-fold hollows, the forces on both of them are found to point away from the surface. Again the reason is that in forming bonds to the surface, atoms adsorbed in neighboring hollows compete for valence electrons from the pair of substrate neighbors that they have in common. This means that each atom bonds to the surface less strongly than it would if isolated. Given this fact, each atom optimizes the total energy by finding a more favorable position. The difference between the Al dimer and the Al-chalcogen pairs is that chalcogens have valence two, and are satisfied by moving toward their two Al unshared substrate neighbors. Since Al atoms have valence three, a pair of ad-Al's wants to form three strong bonds and accomplishes this by shortening the interadatom bond somewhat as both atoms move away from the surface.

Obviously, a great deal of work lies ahead in the computation and understanding of adatom interactions via the scattering theory method. Among problems of current interest are the effects of local lattice relaxations around the adsorbate complex on both interaction energies and diffusion barriers. Work is also in progress concerning adsorbates and substrates of actual interest in surface chemistry, e.g. H/Pt, and on effects of substitutional adsorbates, as might be found at the surface of an alloy catalyst.

ACKNOWLEDGMENT

This work supported by the US Department of Energy under Contract No. DE-AC04-76DP00789.

Literature Cited

1a. Tsong, T. T. 1988. *Surf. Sci. Rep.* 8: 127; 1983. *Phys. Scr.* T4: 17
1b. Bassett, D. W. 1983. In *Surface Mobilities on Solid Materials*, ed. V. T. Binh, pp. 83–108. New York: Plenum
1c. Bassett, D. W., Tice, D. R. 1975. In *The Physical Basis for Heterogeneous Catalysis*, ed. E. Drauglis, R. I. Jaffee. New York: Plenum
1d. Ehrlich, G. 1984. In *Chemistry and Physics of Solid Surfaces*, ed. R. Vanselav, R. Howe, 5: 283. Berlin: Springer-Verlag
1e. Kellogg, G. L., Tsong, T. T., Cowan, P. 1978. *Surf. Sci.* 70: 485
2. DiFoggio, R., Gomer, R. 1982. *Phys. Rev. B* 25: 3490
3. Lagally, M. G., Wang, G. C., Lu, T. M. 1978. *CRC Crit. Rev. Solid State Mater. Sci.* 7: 233
4a. Brand, J. L., Deckert, A. A., George, S. M. 1988. *Surf. Sci.* 194: 457
4b. Mak, C. H., Koehler, B. G., Brand, J. L., George, S. M. 1987. *J. Chem. Phys.* 87: 2340
5. White, J. M., Akhter, S. 1988. *CRC Crit. Rev. Solid State Meter. Sci.* 14: 131
6. Tully, J. C. 1984. In *Many Body Phenomena at Surfaces*, ed. D. Langreth, H. Suhl, p. 377. New York: Academic
7a. Grimley, T. B. 1967. *Proc. Phys. Soc. London* 90: 751
7b. Grimley, T. B., Walker, S. M. 1969. *Surf. Sci.* 14: 395
8. Einstein, T. L., Schrieffer, J. R. 1973. *Phys. Rev. B* 7: 3629
9. Anderson, P. W. 1961. *Phys. Rev.* 124: 41
10a. Einstein, T. L. 1978. *CRC Crit. Rev. Solid State Mater. Sci.* 7: 261

10b. Muscat, J., Newns, D. M. 1978. *Prog. Surf. Sci.* 9: 1
11a. Feibelman, P. J., Hamann, D. R. 1980. *Phys. Rev. B* 21: 1385
11b. Feibelman, P. J., Hamann, D. R., Himpsel, F. J. 1980. *Phys. Rev. B* 22: 1734
12. Lundqvist, S., March, N. H., eds. 1983. *The Theory of the Inhomogeneous Electron Gas.* New York: Plenum
13a. Fu, C. L., Freeman, A. J., Wimmer, E., Weinert, M. 1985. *Phys. Rev. Lett.* 54: 2261
13b. Ho, K. M., Bohnen, K. P. 1985. *Phys. Rev. B* 32: 3446; 1987. *J. Vac. Sci. Technol. A* 5: 462; 1987. *Europhys. Lett.* 4: 345
13c. Feibelman, P. J., Hamann, D. R. 1987. *J. Vac. Sci. Technol. A* 5: 424
13d. Hamann, D. R. 1987. *J. Electron Spectrosc. Relat. Phenom.* 44: 1
14. Jones, R. O. 1984. See Ref. 6, p. 175
15. Deleted in proof
16a. Stott, M., Zaremba, E. 1980. *Phys. Rev. B* 22: 1564
16b. Nørskov, J. K., Lang, N. D. 1980. *Phys. Rev. B* 21: 2136
16c. Daw, M. S., Baskes, M. I. 1984. *Phys. Rev. B* 29: 6443
16d. Lundqvist, B. I., Fonden, T., Idiodi, J., Johnsson, P., Mällo, A., Papadia, S. 1987. *Prog. Surf. Sci.* 25: 191
17a. Williams, A. R., Feibelman, P. J., Lang, N. D. 1982. *Phys. Rev. B* 26: 5433
17b. Feibelman, P. J. 1985. *Phys. Rev. Lett.* 54: 2627
17c. Feibelman, P. J. 1987. *Phys. Rev. B* 35: 2626
18. Nørskov, J. K., Holloway, S., Lang, N. D. 1984. *Surf. Sci.* 137: 65
19a. Feibelman, P. J., Hamann, D. R. 1984. *Phys. Rev. Lett.* 52: 61
19b. Feibelman, P. J., Hamann, D. R. 1985. *Surf. Sci.* 149: 48
20. Ching, W. Y., Huber, D. L., Lagally, M. G., Wang, G. C. 1978. *Surf. Sci.* 77: 550
21. Wang, G. C., Lu, T. M., Lagally, M. G. 1978. *J. Chem. Phys.* 69: 479
22. Binnig, G., Rohrer, H., Gerber, Ch., Weibel, E. 1982. *Appl. Phys. Lett.* 40: 178; 1982. *Phys. Rev. Lett.* 49: 57; 1983. *Phys. Rev. Lett.* 50: 120
23a. Tsong, T. T., Casanova, R. 1981. *Phys. Rev. B* 24: 3063
23b. Fink, H. W., Ehrlich, G. 1984. *J. Chem. Phys.* 81: 4657
24. Tsong, T. T., Casanova, R. 1980. *Phys. Rev. B* 22: 4632
25. Bassett, D. W., Tice, D. R. 1973. *Surf. Sci.* 40: 499
26. Feibelman, P. J. 1987. *Phys. Rev. Lett.* 58: 2766
27. Chabal, Y. J. 1988. *Surf. Sci. Rep.* 8: 211; and references therein
28. Ibach, H. 1987. *J. Vac. Sci. Technol. A* 5: 419
29. Toennies, J. P. 1987. *J. Vac. Sci. Technol. A* 5: 440
30a. Bonzel, H. P. 1988. *Surf. Sci. Rep.* 8: 43
30b. Bugyi, L., Solymosi, F. 1987. *Surf. Sci.* 188: 475
30c. Kiskinova, M., Pirug, G., Bonzel, H. P. 1984. *Surf. Sci.* 140: 1
30d. Ertl, G., Lee, S. B., Weiss, M. 1982. *Surf. Sci.* 114: 529
30e. Whitman, L. J., Ho, W. 1988. *Surf. Sci.* 204: L725
31. Pirug, G., Bonzel, H. P. 1988. *Surf. Sci.* 199: 371; and references therein
32. Koutecky, J. 1958. *Trans. Faraday Soc.* 54: 1038
33. Burke, N. R. 1976. *Surf. Sci.* 58: 349
34. Fulkui, K. 1982. *Science* 218: 747
35a. Kiskinova, M., Goodman, D. W. 1981. *Surf. Sci.* 108: 64
35b. Goodman, D. W., Kiskinova, M. 1981. *Surf. Sci.* 105: L265
35c. Goodman, D. W., Kelley, R. D., Madey, T. E., Yates, J. T. Jr. 1980. *J. Catal.* 63: 226
35d. Goodman, D. W. 1986. *Annu. Rev. Phys. Chem.* 37: 425
35e. Kiskinova, M. P. 1988. *Surf. Sci. Rep.* 8: 359
36. Tersoff, J., Hamann, D. R. 1983. *Phys. Rev. Lett.* 50: 1998; 1985. *Phys. Rev. B* 31: 805
37. Upton, T. H., Goddard, W. A. III, Melius, C. F. 1979. *J. Vac. Sci. Technol.* 16: 531
38. Upton, T. H., Goddard, W. A. III. 1981. *CRC Crit. Rev. Solid State Mater. Sci.* 10: 261
39a. Melius, C. F., Olafson, B. D., Goddard, W. A. III. 1974. *Chem. Phys. Lett.* 28: 457
39b. Melius, C. F., Bisson, C. L., Wilson, W. D. 1978. *Phys. Rev. B* 18: 1647
40a. Harris, J., Andersson, S. 1985. *Phys. Rev. Lett.* 55: 1583
40b. Harris, J., Andersson, S., Holmberg, C., Nordlander, P. 1986. *Phys. Scr.* T13: 155
41a. Andersson, S. 1979. *Surf. Sci.* 79: 385; 1976. *Solid State Commun.* 20: 229
41b. Lehwald, S., Ibach, H. 982. In *Vibrations at Surfaces*, ed. R. Caudano, J. M. Guilles, A. A. Lucas. New York: Plenum
42. Upton, T. H., Goddard, W. A. III. 1981. *Phys. Rev. Lett.* 46: 1635
43. Rahman, T. S., Black, J., Mills, D. L. 1981. *Phys. Rev. Lett.* 46: 1469
44. Demuth, J., Rhodin, T. 1974. *Surf. Sci.* 43: 249

45. Szeftel, J. M., Lehwald, S., Ibach, H., Rahman, T. S., Black, J., Mills, D. L. 1983. *Phys. Rev. Lett.* 51: 268
46a. Stöhr, J., Jaeger, R., Kendelewicz, T. 1982. *Phys. Rev. Lett.* 49: 142
46b. Norman, D. 1986. *J. Phys. C* 19: 3273
47a. Rosenblatt, D. H., Tobin, J. G., Manson, M. G., Davis, R. F., Kevan, S. D., et al. 1981. *Phys. Rev. B* 23: 3828
47b. Rieder, K. H. 1983. *Phys. Rev. B* 27: 6978
47c. Barker, J. A., Batra, I. P. 1983. *Phys. Rev. B* 27: 3138
48. Bauschlicher, C. W. Jr., Walsh, S. P., Bagus, P. S., Brundle, C. R. 1983. *Phys. Rev. Lett.* 50: 864
49. Bauschlicher, C. W. Jr., Bagus, P. S. 1984. *Phys. Rev. Lett.* 52: 200
50a. Redondo, A., Goddard, W. A. III., McGill, T. C., Surratt, G. T. 1977. *Solid State Commun.* 21: 991
50b. Redondo, A., Goddard, W. A. III., Swarts, C. A., McGill, T. C. 1981. *J. Vac. Sci. Technol.* 18: 498
51. Whitten, J. L., Pakkanen, T. A. 1980. *Phys. Rev. B* 21: 4357
52. Yang, H., Whitten, J. L. 1988. *J. Chem. Phys.* 89: 5329
53. Cremaschi, P., Whitten, J. L. 1981. *Surf. Sci.* 112: 343
54. Madhavan, P., Whitten, J. L. 1982. *J. Chem. Phys.* 77: 2673
55. Fischer, C. R., Whitten, J. L. 1982. *Phys. Rev. Lett.* 49: 344
56. Fischer, C. R., Whitten, J. L. 1984. *Phys. Rev. B* 30: 6821
57a. Christmann, K., Schober, O., Ertl, G., Neumann, M. 1974. *J. Chem. Phys.* 60: 4528
57b. Christmann, K., Behm, R. J., Ertl, G., Van Hove, M. A., Weinberg, W. H. 1979. *J. Chem. Phys.* 70: 4168
57c. Ho, W., DiNardo, N. J., Plummer, E. W. 1980. *J. Vac. Sci. Technol.* 17: 314
58. Müller, J. 1986. *Surf. Sci.* 178: 589
59a. Jacobsen, K. W., Nørskov, J. K., Puska, M. J. 1988. *Phys. Rev. B* 35: 7423
59b. Kress, J. D., De Pristo, A. E. 1987. *J. Chem. Phys.* 87: 4700; 1988. *J. Chem. Phys.* 88: 2596
59c. Voter, A. F., Chen, S. P. 1986. *Mater. Res. Soc. Symp. Proc.* 82: 2596
60. Foiles, S. M., Baskes, M. I., Melius, C. F., Daw, M. S. 1987. *J. Less-Common Met.* 130: 465
61. Puska, M. J., Nieminen, R. M., Manninen, M. 1980. *Phys. Rev. B* 24: 3037
62. Nordlander, P., Holloway, S., Nørskov, J. K. 1981. *Surf. Sci.* 136: 59
63. Pauling, L. 1960. *The Nature of the Chemical Bond.* Ithaca, NY: Cornell Univ. Press. 3rd ed.
64a. Ho, K. M., Bohnen, K. P. 1987. *Phys. Rev. Lett.* 59: 1833
64b. Foiles, S. M. 1987. *Surf. Sci.* 191: L779
65. Nelson, J. S., Sowa, E. C., Daw, M. S. 1988. *Phys. Rev. Lett.* 61: 1977
66. Voter, A. F. 1988. Modeling of optical thin films. *SPIE* 821: 214–26
67. Deleted in proof
68. Koster, G. F., Slater, J. C. 1954. *Phys. Rev.* 95: 1167
69a. Baraff, G. A., Schlüter, M. 1979. *Phys. Rev. B* 19: 4965
69b. Bernholc, J., Lipari, N. O., Pantelides, S. T. 1980. *Phys. Rev. B* 21: 3545
70. Kohn, W., Sham, L. J. 1965. *Phys. Rev. A* 140: 1133
71. Deleted in proof
72a. Baraff, G. A., Kane, E. O., Schlüter, M. 1980. *Phys. Rev. B* 21: 5662
72b. Lindefelt, U., Zunger, A. 1981. *Phys. Rev. B* 24: 5913
73a. Broyden, C. G. 1965. *Math. Comput.* 19: 577
73b. Srivastava, G. P. 1984. *J. Phys. A* 17: L317
74. Tsong, T. T., Casanova, R. 1980. *Phys. Rev. B* 21: 4564
75. Feibelman, P. J. 1988. *Phys. Rev. B* 38: 12133

TRANSITION METAL OXIDES[1]

C. N. R. Rao

Solid State and Structural Chemistry Unit, Indian Institute of Science, Bangalore 560012, India

INTRODUCTION

Transition metal oxides constitute probably one of the most interesting classes of solids, exhibiting a variety of structures and properties (1–3). The nature of metal–oxygen bonding can vary between nearly ionic to highly covalent or metallic. The unusual properties of transition metal oxides are clearly due to the unique nature of the outer d-electrons. The phenomenal range of electronic and magnetic properties exhibited by transition metal oxides is especially noteworthy. Thus, we find oxides with metallic properties (e.g. RuO_2, ReO_3, $LaNiO_3$) at one end of the range and oxides with highly insulating behavior (e.g. $BaTiO_3$) at the other. There are also oxides that traverse both these regimes with change of temperature, pressure, or composition (e.g. V_2O_3, $La_{1-x}Sr_xVO_3$). Interesting electronic properties also arise from charge density waves (e.g. $K_{0.3}MoO_3$), charge ordering (e.g. Fe_3O_4), and defect ordering (e.g. $Ca_2Mn_2O_5$, $Ca_2Fe_2O_5$). Oxides with diverse magnetic properties anywhere from ferromagnetism (e.g. CrO_2, $La_{0.5}Sr_{0.5}MnO_3$) to antiferromagnetism (e.g. NiO, $LaCrO_3$) are known. Many oxides possess switchable orientation states as in ferroelectric (e.g. $BaTiO_3$, $KNbO_3$) and ferroelastic [e.g. $Gd_2(MoO_4)_3$] materials. No discovery in solid state science has created as much sensation, however, as that of high-temperature superconductivity in cuprates (4). Although superconductivity in transition metal oxides has been known for some time, the highest T_c reached was around 13K; we now have oxides with T_cs in the region of 130K. The discovery of high T_c oxides has focused worldwide scientific attention on the chemistry of metal oxides and at the same time revealed how inadequate is our understanding of these fascinating materials.

---

[1] Contribution No. 595 from the Solid State and Structural Chemistry Unit.

In this article, I discuss some of the important aspects of the physical chemistry of transition metal oxides of current interest. In so doing, I had to be necessarily selective, as this is the first time that this vast topic has been reviewed here. I survey the electronic and magnetic properties as well as the structure of defect oxides and point out salient features of the different types of metal-insulator transition exhibited by metal oxides. The superconductivity of cuprates and other oxides is discussed at length in view of its timeliness. I briefly touch on some aspects related to synthesis and characterization before concluding the review with a look at future possibilities.

ELECTRONIC, MAGNETIC, AND RELATED PROPERTIES: AN OVERVIEW

In order to understand the relation between the structure and properties of oxides, it is necessary to have a proper description of the valence electrons. The two limiting descriptions of outer electrons in solids are the band theory and the ligand-field theory. In the band model, applicable to collective electron systems or systems in which the overlap between the orbitals of neighboring atoms is large, the energy U required to transfer a valence electron from one orbital to an other singly occupied orbital on an equivalent site is small compared to the bandwidth, W. In the ligand-field theory, applicable to localized electron situations, as in coordination compounds, U is large compared to W. When $U \approx W$, we have strongly correlated electrons in solids. Whereas outer s and p electrons interact strongly with neighboring atoms and are described by a collective-electron model, outer f electrons, which are tightly bound to the nuclei and well screened from the neighboring atoms, are best described by the localized electron model. Outer d electrons have an intermediate character, as they are not screened from the neighboring atoms by outer core electrons. Because of this property, d electrons exhibit itinerant electron properties as well as localized electron properties in transition metal oxides. Electronic properties of even simple oxides such as CoO and NiO do not conform to the predictions of the elementary band theory; these monoxides, which should be metals because of the partially occupied bands, are actually insulators.

The unusual properties of transition metal oxides that distinguish them from the metallic elements and alloys, covalent semiconductors, and ionic insulators are due to several factors:

1. Oxides of d-block transition elements have narrow electronic bands, because of the small overlap between the metal d and the oxygen p

orbitals. The bandwidths are typically of the order of 1 or 2 eV (rather than 5 to 15 eV as in most metals).
2. Electron correlation effects play an important role, as expected because of the narrow electronic bands. The local electronic structure can be described in terms of atomic-like states [e.g. $Cu^{1+}(d^{10})$, $Cu^{2+}(d^9)$ and $Cu^{3+}(d^8)$ for Cu in CuO] as in the Heitler-London limit.
3. The polarizability of oxygen is also of importance. The divalent oxide ion, O^{2-}, does not exactly describe the state of oxygen, and configurations such as O^{1-} have to be included, especially in the solid state. This gives rise to polaronic and bipolaronic effects. Species such as O^{1-}, which are oxygen holes with a p^5 configuration instead of the filled p^6 configuration of O^{2-}, can be mobile and correlated.
4. Many transition metal oxides are not truly three-dimensional, but have low-dimensional features. For example, La_2CuO_4 and La_2NiO_4 with the K_2NiF_4 structure are two-dimensional compared to $LaCuO_3$ and $LaNiO_3$, which are three-dimensional perovskites. Because of the varied features of individual oxides, it has not been possible to establish satisfactory theoretical models for complex transition metal oxides.

Empirical approaches have been found to be convenient to describe the electronic structures and properties of transition metal oxides. Based on empirically derived criteria for cation-cation and cation-anion-cation overlaps, Goodenough (2) has attempted to rationalize the nature of d-electrons in metal oxides. In this approach, conceptual phase diagrams are constructed in terms of the transfer energy, b_{ij}, which is related to cation-cation separation or the covalent mixing parameter of the cation-anion orbitals. Simple rules have also been proposed based on considerations of cation-oxygen-cation overlap and cation-cation separation. The following is one such rule:

$$R_c^{3d} = 3.2 - 0.05m - 0.03(Z - Z_{Ti}) - 0.04S_i(S_i + 1) \text{ Å}.$$

Here, R_c is the critical cation-cation separation, m the formal charge on the cation, Z the atomic number, and S_i the net atomic spin. Since the radial extension of the $4d$ and $5d$ orbitals is larger than that of the $3d$ orbitals, $R_c(5d) > R_c(4d) > R_c(3d)$. The covalent mixing parameter increases with m and shows a minimum where S is maximum; the cation-oxygen overlap integral is higher for an anion sp_σ orbital than for a p_Π one. One-electron energy-level diagrams that take into account the most probable hybridization between the cationic and anionic orbitals are quite useful in understanding and predicting electronic properties. Such a diagram for ReO_3 would show that this oxide is metallic because of the partly filled π^* band.

In simple transition metal monoxides possessing the NaCl structure, we find a 180° cation-oxygen-cation interaction. Those monoxides with a cation-cation separation higher than a critical value are insulators. Thus, TiO, with a short cation-cation distance ($R < R_c$), is metallic and VO ($R \sim R_c$) is a semimetal. TiO is Pauli-paramagnetic while VO shows temperature-dependent susceptibility at low temperatures. The Neèl temperature increases in the order MnO, FeO, CoO, NiO, accompanying the increase in the cation-anion overlap. Non-stoichiometric $Mn_{1-x}O$ samples show spin-glass behavior. Oxygen-deficient EuO shows a sharp drop in resistivity and becomes metallic at around 50K. Stoichiometric EuO (which is ferromagnetic) shows a transition from an insulating state to a metallic state upon application of pressure because of the promotion of a $4f$ electron to the $5d$ conduction band.

In dioxides of transition metals possessing the rutile structure, 135° cation-oxygen-cation interaction is possible between corner-shared octahedra, and 90° cation-anion-cation interaction is possible between edge-shared octahedra; a cation-cation interaction can also occur in the c-direction. These oxides can therefore become metallic through cation-cation or cation-oxygen-cation interaction. Metal-metal bonding occurs in these oxides depending on the c/a ratio, and such oxides show monoclinic distortion (e.g. VO_2). Tetragonal VO_2 ($R < R_c$) is, however, metallic; WO_2 and MoO_2 ($R < R_c$) are also metallic. CrO_2, with the longest c/a ratio, is a metallic ferromagnet, since one of the d-electrons is in the π^* band formed through cation-anion-cation interaction.

Sesquioxides of the first row transition metals possessing the corundum structure exhibit interesting properties, e.g. Ti_2O_3 and V_2O_3 undergoing temperature-induced transitions from an insulating state to a metallic state. Cation-cation interactions in the basal plane, as well as cation-oxygen-cation (135° and 90°) interactions, play a role in bestowing such properties to these oxides. Fe_2O_3 exhibits the well-known first-order spin-flip transition (Morin transition).

ABO_3 perovskites ideally have 180° cation-oxygen-cation interactions of the B-site cation; cation-cation interaction is remote because of the large distance associated with the cube-face diagonal. The influence of the A cations on the B-O covalency is indirect. Figure 1 lists some important perovskites. Those with the same d-electron configurations are grouped together in the columns. In each column the entries are arranged in the order of decreasing cation-anion transfer energy, b. The *dotted lines* in Figure 1 representing $b_\Pi = b_m$ (b_m is the critical value for spontaneous magnetism), $b_\Pi = b_c$, and $b_\sigma = b_c$ (b_c is the critical value of the transfer energy) separate oxides exhibiting localized electron behavior from those with collective electron properties. Compounds in column 1 are insulators because the B

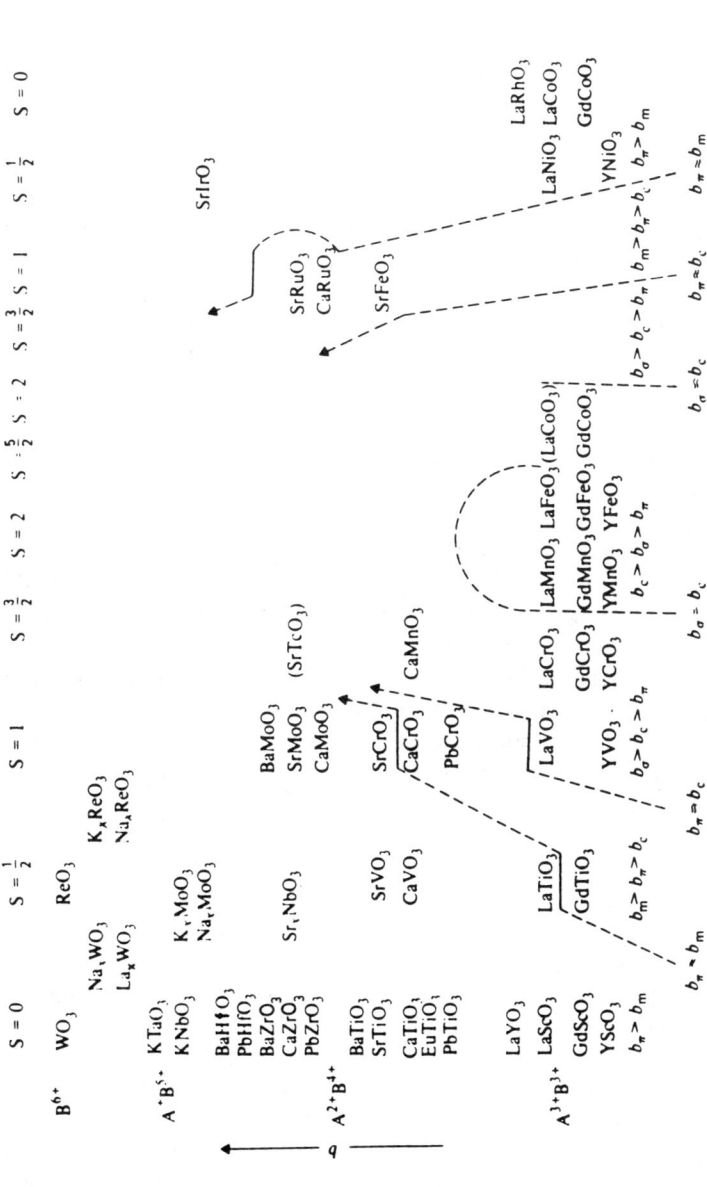

Figure 1 Perovskite oxides containing transition-metal ions in different spin configurations. Oxides are grouped into regions based on the transfer energy, b [following Goodenough (2)].

cations are of d^0 electron configuration. Most of the compounds in column 2 (spin $S = 1/2$) are metallic and Pauli paramagnetic; the line $b_\Pi = b_m$ separates LaTiO$_3$ from GdTiO$_3$ because GdTiO$_3$ is a semiconductor with a ferromagnetic Curie temperature (T_c) of 21K. AMoO$_3$ (A = Ca, Sr, Ba) and SrCrO$_3$ in the third column ($S = 1$) are metallic and Pauli paramagnetic. Other compounds in this column are semiconducting and antiferromagnetic. The line $b_\Pi = b_m$ separates the metallic, Pauli paramagnetic SrCrO$_3$ from CaCrO$_3$, which is an antiferromagnetic semimetal. The line $b_\Pi = b_c$ separates PbCrO$_3$ from LaVO$_3$ because the latter exhibits a crystallographic transition at a temperature lower than the Neèl temperature (T_N) characteristic of localized electrons. The region $b_m > b_\Pi > b_c$ appears to be narrow, as revealed by electrical, magnetic, and associated properties. Pressure experiments are valuable in the study of this region; thus, $dT_N/dP < 0$ in CaCrO$_3$ whereas $dT_N/dP > 0$ in YCrO$_3$ and CaMnO$_3$. Since increasing pressure increases b_Π (by decreasing lattice dimensions), $dT_N/dP > 0$ for $b_\Pi < b_c$ (localized behavior) and $dT_N/dP < 0$ for $b_m > b_x > b_c$ (collective behavior). Compounds in columns 4, 5, and 6 are antiferromagnetic insulators. Since intra-atomic exchange, given by $S(S+1)$, decreases the covalent mixing, the maxima in the curves $b_\Pi = b_c$ and $b_\sigma = b_c$ corresponding to smallest values of b_Π and b_σ occur in the middle of the columns with $S = 5/2$. LaFeO$_3$ has a higher T_N than LaCrO$_3$ because of greater superexchange through σ bonding. The rare earth orthoferrites, which are antiferromagnetic insulators, exhibit parasitic ferromagnetism. The important contributions here are: (a) the Fe^{3+} spins canted in a common direction either by cooperative buckling of the oxygen octahedra or by anisotropic superexchange, and (b) canting of antiferromagnetic rare earth sublattice because of the interaction between the two sublattices.

LaCoO$_3$ is shown twice in Figure 1, both in $S = 2$ and $S = 0$ columns because Co^{3+} in this solid can have either the low-spin or the high-spin configuration. The compound exhibits a transition from a localized electron state to a collective electron state (metal-insulator transition) at ~1200K. In the ninth column of Figure 1, perovskites containing d^4 cations are placed. Of the three compounds in this column, SrRuO$_3$ is a ferromagnetic metal ($T_c = 160$K) and CaRuO$_3$ is antiferromagnetic ($T_N = 110$K) with a weak ferromagnetism. Since both the compounds have the same RuO$_3$ array, the change from ferromagnetic to antiferromagnetic coupling is of significance. SrFeO$_3$ is placed in the same column on the assumption that Fe^{4+} ($3d^4$) is in the low-spin state, but there is reason to believe that Fe^{4+} in this oxide is in the high-spin state down to 4K. CaFeO$_3$, on the other hand, seems to undergo disproportionation of Fe^{4+} to Fe^{3+} and Fe^{5+} below 290K. In the next to the last column containing $S = 1/2$

B cations, metallic and Pauli paramagnetic $LaNiO_3$ should be separated from antiferromagnetic $YNiO_3$ to show that in $LaNiO_3$, $b_\sigma > b_m$, whereas in $YNiO_3$, $b_\sigma < b_m$. Similarly, in the last column, $LaCoO_3$ should be separated from $LaRhO_3$ because the latter is a narrow gap semiconductor with a filled $t_{2g}(\pi^*)$ band and an empty $e_g(\sigma^*)$ band.

Oxides such as $LaNi_{1-x}Mn_xO_3$, in which Mn^{2+} ions are present in a metallic oxide host, show spin-glass behavior (5); and these materials become insulators with an increase in x (6a,b). In $La_{1-x}Sr_xCoO_3$ and $La_{1-x}Sr_xMnO_3$, the material becomes metallic and ferromagnetic with an increase in x; the latter system is a well-known example of the Zener double exchange mechanism (6a,b).

Properties of a large number of perovskites have been compiled by Goodenough & Longo (7) and Nomura (8). Some of the oxide perovskites show superconductivity. One of the first oxides found to exhibit a reasonably high T_c (13K) was $BaPb_{1-x}Bi_xO_3$ (9). All the new high T_c cuprates possess perovskite-related structures. The perovskite motif occurs in many interesting classes of oxides; the K_2NiF_4 structure involving KF and $KNiF_3$ layers is the foremost. Properties of the oxides of K_2NiF_4 structure have been reviewed by Ganguly & Rao (10). In this structure, there is 180° B–O–B interaction in the basal plane and B–O–O–B interaction perpendicular to it. A tolerance factor can be defined for this structure similar to that in the perovskites. These two-dimensional oxides have electronic and magnetic properties that are distinctly different from those of the corresponding three-dimensional perovskites. Accordingly, $LaNiO_3$ is metallic and Pauli-paramagnetic whereas La_2NiO_4 exhibits two-dimensional antiferromagnetic ordering around 200K and a semiconductor-metal transition around 600K. While $LaCoO_3$ shows a spin-state transition of Co^{3+} and associated paramagnetism, La_2CoO_4 seems to exhibit antiferromagnetic ordering at fairly high temperatures (11). La_2CuO_4 is a low-resistivity oxide but not metallic, whereas $LaCuO_3$ is a metal; in La_2CuO_4 and other rare earth cuprates of this structure, copper has no magnetic moment. It has been suggested that La_2CuO_4 (orthorhombic-tetragonal transition $\sim$505K, $T_N \sim$ 290K) is in a quantum fluid state wherein the spins are ordered over long distances instantaneously, but no measurable time-averaged moment has been detected (12a). $La_2NiO_{4+\delta}$ (orthorhombic-tetragonal transition $\sim$240K, $T_N \sim$ 70K) shows strong two-dimensional magnetic correlations and also large in-plane spin velocities and in some ways has a behavior similar to that of La_2CuO_4 (12b). There are indications that $La_2NiO_{4+\delta}$ becomes superconducting at low temperatures just as $La_2CuO_{4+\delta}$ (12c,d).

In the $LaO(LaNiO_3)_n$ family, ($n = 1 = La_2NiO_4$, $n = \infty = LaNiO_3$), the electrical conductivity increases with increase in n (three-dimensional character), becoming essentially metallic when $n = 3$ (13a). $SrRuO_3$ is a metallic ferromagnet, but Sr_2RuO_4 is a paramagnetic insulator.

A strict comparison of the properties of three- and two-dimensional oxides can be made only when the d-electron configuration of the transition metal ion, B, is the same. A comparative study of two such systems has been made with respect to their electrical and magnetic properties (13a,b). For example, members of the $La_{1-x}Sr_{1+x}CoO_4$ system are all paramagnetic semiconductors with a high activation energy for conduction, unlike $La_{1-x}Sr_xCoO_3$ ($x \geq 0.3$), which is metallic and ferromagnetic (6a,b). $La_{0.5}Sr_{1.5}CoO_4$ shows a magnetization of $0.5\mu_B$ at 0K (compared to $1.5\mu_B$ of $La_{0.5}Sr_{0.5}CoO_3$), but the high-temperature susceptibilities of the two systems are comparable. In $SrO(La_{0.5}Sr_{0.5}MnO_3)_n$, both magnetization and electrical conductivity increase with increase in n, approaching the value of the perovskite, $La_{0.5}Sr_{0.5}MnO_3$. $LaSrMn_{0.5}Ni_{0.5}(Co_{0.5})O_4$ shows no evidence of long-range ferromagnetic ordering, unlike the perovskite $LaMn_{0.5}Ni_{0.5}(Co_{0.5})O_3$; high-temperature susceptibility behavior of these two insulating systems is similar. $LaSr_{1-x}Ba_xNiO_4$ exhibits high electrical resistivity with the resistivity increasing proportionately with the magnetic susceptibility. High-temperature susceptibility of $LaSrNiO_4$ and $LaNiO_3$ are comparable. Susceptibility measurements show no evidence for long-range ordering in $LaSrFe_{1-x}Ni_xO_4$, unlike in $LaFe_{1-x}Ni_xO_3$ ($x \leq 0.35$), and the electrical resistivity of the former system is considerably higher.

Among the other interesting oxide families with the perovskite motif, mention should be made of the oxides of the Aurivillius family (14a,b), which possess the formula $(Bi_2O_2)^{2+}$ $(A_{n-1}B_nO_{3n+1})^{2-}$; typical members are $Bi_4Ti_3O_{12}$ ($n = 3$) and $BaBi_4Ti_4O_{15}$ ($n = 4$). The $A_nB_nO_{3n+2}$ formed by the Na–Ca–Nb–O system and the $A_{n+1}B_nO_{3n+1}$ family formed by the Sr–Ti–O and La–Ni–O systems (13a, 15) are of interest. Polytypic structures of perovskites wherein the AO_3 layer can be cubic or hexagonal with respect to the adjacent layers show large periodicities (e.g. $BaCrO_3$, $BaRuO_3$).

Oxide spinels AB_2O_4 are well-known magnetic systems exhibiting a variety of interesting properties (2, 3). Thus, ferrimagnetic $CoCr_2O_4$ has a conical spiral configuration. The cooperative Jahn-Teller effect shown by some of the spinels (e.g. $FeCr_2O_4$) is of considerable interest (16). Other oxides showing this effect are rare earth zircons (e.g. $TbVO_4$, $DyVO_4$) and $PrAlO_3$ (17, 18). In vandate spinels, $AV_2^{3+}O_4$, the d-electrons are localized when $2.88 \text{ Å} < R_{V-V} < 2.97 \text{ Å}$. Fe_3O_4, which is an inverse spinel, has been of much interest in the past several decades, and I discuss some aspects of this oxide further on in the review. It is noteworthy that the spinels $Li_{1-x}M_x^{2+}Ti_2O_4$ (M = Mg, Mn) and $Li_{1+x}Ti_{2-x}O_4$ show superconductivity (19a,b). In certain spinels such as $Ga_{0.8}Fe_{0.2}NiCrO_4$, spin-glass ordering with randomly frozen clusters has been noticed (20).

Oxide pyrochlores of the general formula $A_2B_2O_7$ show interesting electronic properties [see (21) for a review]. Ferromagnetic pyrochlores of rare earths have been described recently (22). A composition dependent metal-semiconductor transition has been found in $A_2(Ru_{2-x}A_x)O_{7-y}$ where A = Bi or Pb (23).

Transitions from the low-spin to the high-spin state of transition metal ions in oxide systems have been documented and models developed to explain the transitions (24). In $LaCoO_3$, low-spin Co^{3+} ions transform to the high-spin state upon increasing the temperature. Spin-state transitions have been found in niobium compounds as well (25). Quasi-two-dimensional oxides of K_2NiF_4 structures also exhibit such transitions (13a,b).

Hexagonal, cubic, and intergrowth bronzes formed by WO_3 with alkali, hydrogen, and other metals have been well-documented in the literature (26a–c). Of these, the intergrowth bronzes, in which strips of the hexagonal bronze intergrow with strips of WO_3, sometimes recurrently, are especially interesting. Electrical transport and other properties of WO_3 bronzes have been reviewed in the literature (27a,b). MoO_3 forms different varieties of bronzes (28): Blue bronzes of the type $A_{0.3}MoO_3$ (A = K, Tl, Rb), which are quasi-one-dimensional metals with charge density wave (CDW) instability; purple bronzes, $A_{0.9}Mo_6O_{17}$ (A = Na, K), which are quasi-two-dimensional metals with CDW instability; $Li_{0.9}Mo_6O_{17}$, which is one-dimensional and superconducting ($T_c \approx 2K$); red bronzes $A_{0.33}MoO_3$ (A = K, Tl, Rb), which are semiconducting; and $Li_{0.33}MoO_3$, which is violet and three-dimensional with low resistivity (29a,b). Hydrogen molybdenum bronzes, H_xMoO_3, of different compositions ($0 < x \leq 2.0$) with structures related to MoO_3, have been characterized (28). Conductivity measurements have been made on some of these hydrogen bronzes (30). Di- and mono-phosphate tungsten bronzes of the type $A_x(P_2O_4)_2(WO_3)_{2m}$ and $A_x(PO_2)_4(WO_3)_{2m}$, with A = Na, K, Rb, or Ba and possessing hexagonal tunnels, have been studied (31a–c). The tunnels may be empty, as in the monophosphate bronze $P_4W_8O_{32}$ ($m = 4$), or occupied, as in the diphosphate tungsten bronzes. Presence of defects and microstructures related to the adaptability of the phosphate groups to the WO_3 matrix have been examined. Anisotropic electronic properties of $CsP_8W_8O_{40}$, which has a unique structure, have been measured (32). Recently, layered alkali metal-MoO_3 bronzes as well as hexagonal bronzes of the type $K_yW_{1-x}Mo_xO_3$ have been prepared by a novel low-temperature reaction of the alkali metal iodide with the parent oxide (33).

Ferroics

Materials possessing two or more orientation states or domains that can be switched from one to another through the application of one or more appropriate forces belong to a general class called ferroics (34). In a ferromagnet, the orientation state of magnetization in domains is switched by the application of a magnetic field. In a ferroelastic, the direction of spontaneous strain in a domain is switched by the application of mechanical stress. In a ferroelectric, spontaneous electric polarization is altered by the application of an electric field. These three ferroics are primary ferroics, because they are governed by switchability of the properties. Metal oxides provide many examples of ferroics. $BaTiO_3$, $KNbO_3$ and the $Bi_2A_{n-1}B_nO_{3n+3}$ family of oxides are ferroelectric, whereas $PbZrO_3$ and $NaNbO_3$ are antiferroelectric.

Secondary ferroic properties occur as induced quantities, and the orientation states in these solids differ in the derivative quantities that characterize the induced effects (e.g. induced electric polarization characterized by dielectric susceptibility). Thus, $SrTiO_3$ is a secondary ferroic showing ferrobielectricity. NiO is ferrobimagnetic, whereas Cr_2O_3 is ferromagnetoelectric; Cr_2BeO_4 exhibits magnetoferroelectricity. Oxides such as $Pb(Mg_{1/3}Nb_{2/3})O_3$ with 10% $PbTiO_3$ are relaxor ferroelectrics, and $PbZr_{1-x}Ti_xO_3$ is a well-known electro-optic material.

Oxides exhibiting certain paired properties are especially interesting:

Ferroelectric-ferroelastic: $Gd_2(MoO_4)_3$, $KNbO_3$
Ferroelectric-ferromagnetic: $Bi_9Ti_3Fe_5O_{27}$
Ferroelectric-antiferromagnetic: $YMnO_3$, $HoMnO_3$
Antiferroelectric-antiferromagnetic: $BiFeO_3$
Ferroelectric-semiconducting: $SrTiO_3$, $YMnO_3$
Ferroelectric-superconducting: $SrTiO_3$

DEFECT OXIDES

It has been known since the 1920s that stoichiometric FeO does not fall in the stability range of iron (II) oxide. In fact, a large variety of oxides exhibit nonstoichiometry and wide homogeneity ranges. Nonstoichiometric oxides are mixed valent with nonintegral electron/atom ratios. Very few nonstoichiometric oxides, have, however point defects in high concentrations, and their structures can generally be understood in terms of defect ordering or complexation. New structural principles such as crystallographic shear and block structures have emerged in our efforts to understand defect oxides. $NbO_{2.4906}$, $NbO_{2.4167}$, $PrO_{1.714}$ and such solids are not to be merely considered as oxides with irrational ratios of the constituent atoms, but as crystallographically well-defined $Nb_{53}O_{132}$, $Nb_{12}O_{29}$, and Pr_7O_{12}, etc. I briefly examine the kinds of defect situations encountered in trasition metal oxides (3, 18).

There is considerable evidence for superlattice ordering of point defects in many oxides (35, 36). Thus, TiO and VO, with $\sim 20\%$ vacancies, both have ordered defect structures. The high-temperature form of TiO with an averaged NaCl structure spans a wide range of compositions between $TiO_{0.65}$ and $TiO_{1.25}$ at 1770K (37). The low-temperature monoclinic form ($T < 1270K$) has a narrow composition range. The high-temperature form seems to have short-range order of its vacancies (38). The defect ordering in VO is quite different from that in TiO; VO and TiO are therefore immiscible. The superstructure in VO is formed at the oxygen-rich end, where the oxygen sublattice is filled and some V atoms occupy tetrahedral sites (e.g. $V_{52}O_{64}$, $V_{244}O_{320}$). Each tetrahedral cation has four vacant octahedral sites as nearest neighbors. The cluster so formed is topologically similar to the NaCl structure and gives rise to a $2\sqrt{2} \times 2\sqrt{2} \times 2$ superstructure. In the $V_{52}O_{64}$ superlattice, all the vacant octahedral sites are in the clusters, and there are no free vacancies (39).

Wüstite, which is always cation deficient, with the range $Fe_{0.85}O–Fe_{0.95}O$, is understood in terms of Koch-Cohen clusters, with the ratio of vacancies to tetrahedral Fe^{3+} of 3.25. Clusters with a ratio of ~ 2.5 are also found (40). Calculations show such clusters to be formed with a net lowering of free energy; binding energy of the 13:4 cluster is 2.1 eV (41). Starting from the rock salt structure, oxidation through the nonstoichiometric range may involve the following stages (36): isolated vacancies → dipolar associates → 4:1 clusters → 6:2, 8:3, 13:4 and other similar complex defect clusters → corner-shared 16:5 clusters → Fe_3O_4. Such clusters could be present in $Mn_{1-x}O$ as well. Although careful studies of Wüstite have been carried out by means of satellites in diffraction patterns and diffuse scattering, its constitution at high temperatures is still unclear, and no single cluster species seems to be able to explain the structure (42, 43). The thermodynamics of defects in $Mn_{1-x}O$ and $Fe_{1-x}O$ have been investigated in detail (44), but a real understanding will emerge only when we know the exact nature of defects. Energies of defect clusters in $M_{1-x}O$ (M = Mn, Fe, Co, or Ni) have also been calculated by using molecular orbital theory; extended defects are shown to be stable in MnO and FeO (45). A cluster component method (46) has been employed to understand defect ordering in $Fe_{1-x}O$ and $Mn_{1-x}O$.

In oxides with a fluorite-related structure, anion-deficient as well as anion-excess stoichiometries are known. In MO_{2-x} (M = Pr or Y-doped ZrO_2), Bevan clusters (47) are known to occur with tightly bound vacancies along $\langle 111 \rangle$ and 6-coordinated central ions with six 7-coordinated cations surrounding them. Oxygen excess in UO_2 (e.g. U_4O_9) is understood in terms of Willis clusters of different types (2:1:2, 2:2:2, 3:4:2). Defect energy calculations have been performed (48), but we do not yet fully understand the nature of anion-excess fluorites. The thermodynamics of defect formation in UO_2 and CeO_2 have been examined (44).

Point defects in oxides are eliminated by crystallographic shear (cs), and different types of cs planes are found in WO_3, MoO_3, and TiO_2 type structures, which give rise to homologous series of oxides (35, 49). An isolated cs plane or a random array of cs planes is referred to as the *Wadsley defect*. Crystallographic shear places can be regarded as translation modulations of the parent structure; the translation boundaries are cs planes. We have just begun to understand the mechanism of formation and ordering of cs planes. Elastic strain appears to play an important role in ordering. Both continuum and atom site models have been useful in illuminating this problem (49, 50).

Even slightly reduced rutile ($TiO_{1.997}$) shows the presence of cs planes. In slowly cooled samples of TiO_{2-x} ($0.0 < x \leq 0.01$), pairs of cs planes precipitate and separate subsequently (51). Novel $\{100\}$ platelet defects occur along with cs planes when $0 < x \leq 0.0035$, and this can be understood in terms of cationic interstitial defects (51). The extrinsic and/or intrinsic nature of extended defects has been analyzed in TiO_{2-x} and WO_{3-x} by drawing Burgers circuits directly onto electron micrographs (52). The defect structure of TiO_{2-x} has been discussed at some length in the light of experimental and theoretical results (53).

The structures of many oxides, especially those of Nb (e.g. $Nb_{12}O_{29}$, $Nb_{25}O_{62}$, Nb_2O_5), are best described as block structures resulting from the operation of two sets of nearly orthogonal crystallographic shear. These structures have been studied extensively by electron microscopy (54, 55). Point defects as well as Wadsley defects can occur in block structures.

A remarkable feature of the Ti_nO_{2n-1} ($n = 10-14$) system is that the ($\bar{1}21$) cs plane swings continuously through all possible orientations to ($\bar{1}32$), thus giving rise to ordered phases for any composition between $TiO_{1.900}$ and $TiO_{1.937}$. Such a continuous series of ordered structures, called *infinitely adaptive structures* (56), has been found in other systems (e.g. Ta_2O_5–WO_3). Elastic strain energy plays a crucial role in forming such structures (57). Another impressive solid state phenomenon found in oxides is that of recurrent intergrowth of two structurally related units, which gives rise to a new homologous series of materials. Some examples are $Bi_4A_{m+n-2}B_{m+n}O_{3(m+n)+6}$, formed by the intergrowth of two oxides of the formula $Bi_2A_{n-1}B_nO_{3n+3}$; intergrowth bronzes of the formula A_xWO_3; and $A_{n+1}B_nO_{3n+1}$ (A = Sr or La, B = Ti or Ni), formed by the intergrowth of AO, with different number of layers of ABO_3. Hexagonal barium ferrites, $Ba_{2n+p}M_{2n}Fe_{12(n+p)}O_{22n+19p}$, where $M = Zn$, Ni, etc and with $n = 1$–47, are also examples of recurrent intergrowths structures. The subject of intergrowth structures has been discussed at some length recently (58a,b). It becomes difficult to distinguish recurrent intergrowths from infinitely adaptive structures in some instances. Disordered intergrowth of related members of a family of oxides is of common occurrence.

Perovskites form A-site vacancies commonly; the bronzes are well-known examples. B-site vacancies are not favored but are found in certain oxides with highly covalent B–O bonds and strong B–B interaction. Anion-deficient perovskites with vacancy ordering are of common occurrence (e.g. $Ca_2Fe_2O_5$, $CaMnO_{2.667}$). In $Ca_2Mn_2O_5$, the Mn ions are square-pyramidally coordinated, whereas in $Ca_2Fe_2O_5$, the Fe ions are alternately present in octahedral and tetrahedral sites. In $Ca_2Fe_{2-x}Mn_2O_5$, ordering of transition metal ions with octahedral, tetrahedral, and square-pyramidal coordinations has been found (59). Magnetic and crystal structures of such oxides have been examined, as typified by a recent study of Sr_2CoFeO_5 (60). Phases of the type $Ca_xLa_{1-x}FeO_{3-y}$, $CaTi_{1-x}Fe_xO_{3-y}$ and $CaFe_xMn_{1-x}O_{3-y}$ have been studied extensively (61a,b, 62). The defect structure of a novel oxygen-deficient 6H polytypic oxide, $BaMn_{1-x}Fe_xO_{3-d}$ has been studied (63). Structural features of $CaMnO_{3-x}$ over the $0 < x \leq 0.5$ range have also been studied (64), as has the dependence of the structure and the electronic state of $SrFeO_3$ on composition and temperature (65). Anion-excess stoichiometry in perovskite oxides is accommodated by A and B-site vacancies. The available experimental results on defect perovskites have been reviewed adequately (3, 66). Molecular orbital calculations by the method of moments have been carried out to understand the defect patterns in perovskites (67).

Oxide pyrochlores, $A_2B_2O_6O'$, can tolerate vacancies at the A and O' sites, thus giving phases of the type $A_2B_2O_6\square$ or (ABO_3) and $A\square B_2O_6\square$ or AB_2O_6 ($\square$ = vacancy). Typical examples are $Tl_2Nb_2O_6$ and $Tl_2U_2O_6$ (21). Novel defect pyrochlores of the type $ABi_2B_5O_{16}$, where $A = Cs$, Rb and $B = Ta$, Nb have been described (68).

Oxides of the K_2NiF_4 structure such as La_2MO_4 ($M = Co$, Ni, Cu) generally seem to possess La-deficiency. Oxygen-deficient Ca_2MnO_4 can be topotactically reduced to $Ca_2MnO_{3.5}$ (69). $Ca_2FeO_{3.5}$ seems to have a different type of defect ordering compared to the manganese analogue (70). Oxygen-excess $La_2NiO_{4+\delta}$ has been examined recently (12b, 71).

MIXED VALENCE

Many transition metal compounds exhibit the phenomenon of mixed valence, wherein the metal is present in more than one oxidation state.

Properties of such compounds are generally determined by the rate of electron transfer between the different oxidation states. A classification of these compounds has been made on the basis of the valence delocalization coefficient, and their properties have been reviewed (3, 72). This type of mixed valence is different from that prevalent in rare-earth and actinide materials in which valence fluctuation, heavy fermion behavior, and superconductivity are found. Depending on the relative energies of the f^n configuration and the Fermi-level due to non-f electrons, three electronic regimes are distinguished: (a) the magnetic regime, (b) the Kondo regime, and (c) the fluctuating valence regime (73). EuO and SmS show valence fluctuation under pressure due to the promotion of an f-electron to the conduction band.

In mixed valent rare earth oxides of the type Pr_6O_{11}, Pr_7O_{12}, and Tb_4O_7 related to the fluorite structure, the electronic conductivity is proportional to the product $[M^{3+}][M^{4+}]$, since it is controlled by the hopping (diffusion) mechanism. The conductivity reaches a maximum when this product becomes a maximum, the point at which the sign of the charge carriers also changes from n- to p-type (74). Oxides of the type Ti_3O_5 and V_3O_5 undergo metal-insulator transitions whereas oxides of the type Co_3O_4 are insulators. The presence of more than one oxidation state is readily recognizable from the formula in certain oxides such as Fe_3O_4, Pb_3O_4, V_nO_{2n-1}, Ti_nO_{2n-1}, and Pr_6O_{11}, but this is not so in oxides such as $BaBiO_3$ and Sb_2O_4. The last two have Bi/Sb ions in 3+ and 5+ states. Even in oxides of the type $BaPbO_3$, mixed valence has been invoked (75). Metal ions in the lower oxidation state (2+ or 3+) can be leached out by acid dissolution from Pb_3O_4, Pr_6O_{11}, Tb_4O_7 and similar insulating oxides, thus leaving only the dioxides in the solid state.

Fe_3O_4, with the inverse spinel structure, undergoes a ferrimagnetic-paramagnetic transition around 850K and a transition associated with charge ordering (*Verwey transition*) around 123K. The latter transition and the electronic properties of the oxide through the transition have been a subject of much study. An entire issue of *Philosophical Magazine* (Vol. B 42, No. 10, 1980) was devoted to this topic. Yet there is considerable uncertainty about the transition and the mechanism of conduction (76a,b, 77). The transition is markedly dependent on stoichiometry, and becomes second-order at large stoichiometric deviations. Most of the data on transport properties seem to suggest a small polaron model. The observed entropy change $[(R \ln 2)/\text{mol } Fe_3O_4]$ suggests the presence of dimer units below the transition temperature. Randomization of the Fe^{3+} and Fe^{2+} ions from an ordered state seems to be too naive a description of this fascinating transition.

$Ln_{1-x}Sr_xCoO_3$ (Ln = La, Nd, etc) becomes metallic when $x \geq 0.3$, and the itinerancy of the *d*-electrons is associated with ferromagnetism. Mössbauer studies clearly show that cobalt has an average oxidation state between 3+ and 4+ (78). The rate of electron transfer obviously determines the nature of mixed valency in such oxides. $MoFe_2O_4$ is another oxide with fast electron transfer, and hence it has an average oxidation state of +2.5 for iron (79). Many other oxides show such behavior (80–82).

METAL–NONMETAL TRANSITIONS

The band structure of a crystalline solid made up of an even number of electrons can be made to change over to a structure wherein the empty and filled bands cross or overlap due to a change in pressure or temperature or by suitable doping. Such band-overlap or crossover transitions are generally accompanied by change in crystal structure and, in some instances, magnetic ordering as well. The celebrated Mott transition can occur from a metallic to a nonmetallic state when the band-width decreases sufficiently that it becomes smaller than the intrasite electron-electron energy because of localization induced by electron correlation. Localization can also occur because of disorder, as in amorphous materials, thus giving rise to a M–NM transition (*Anderson transition*); in such a transition, the band-width becomes less than the width of the distribution of random site energies. In spite of voluminous work and a general agreement on model patterns, we have yet to understand fully this fascinating phenomenon, which occurs in many transition metal oxides.

In this section, I examine a few illustrative cases and models. Greater details are available in recent reviews of the subject (83–86). The different types of metal-nonmetal (M–NM) transitions found in metal oxides are the following:

1. Pressure-induced transitions, as in NiO, in which the pressure increases the wavefunction overlap between neighbors to induce a change from localized to itinerant behavior of electrons.
2. Transitions as in Fe_3O_4 involving charge-ordering.
3. Transitions as in $LaCoO_3$ that are initially induced because of the different spin configurations of the transition metal ion; electron transfer between the two spin states initiates a process that eventually renders the oxide metallic around 1200K.
4. Transitions as in EuO arising from the disappearance of spin polarization band-splitting effects when the ferromagnetic Curie temperature is reached.
5. Compositionally induced transitions, as in VO_x, $La_{1-x}Sr_xCoO_3$, and

$LaNi_{1-x}Mn_xO_3$, in which changes of band structure in the vicinity of the Fermi level are brought about by a change in composition or are due to disorder-induced localization.

6. Transitions in two-dimensional systems, such as La_2NiO_4, in which Ni–O–Ni interaction can only occur in the *ab* plane (unlike in the three-dimensional analogue in $LaNiO_3$).

7. Temperature-induced transitions in a large class of oxides such as Ti_2O_3, VO_2, and V_2O_3.

The last category, involving temperature-induced M–NM transitions, deserves some elaboration. In Ti_2O_3, a second-order transition occurs around 410K, accompanied by a gradual change in the rhombohedral c/a ratio and a 100-fold jump in conductivity; the oxide remains paramagnetic throughout. A simple band-crossing mechanism occurring with the change in the c/a ratio explains this transition. Accordingly, substitution of Ti by V up to 10% in Ti_2O_3 makes the system metallic; the c/a ratio of this metallic solid solution and the high-temperature phase of TiO_3 are similar. In VO_2, a first-order transition occurs around 340K, accompanied by a change in structure (monoclinic to tetragonal) and a 10,000-fold jump in conductivity; the material remains paramagnetic throughout. A crystal distortion model wherein a gap opens up in the low-temperature low-symmetry structure adequately explains the transition. Substitution of trivalent ions such as Cr^{3+} and Al^{3+} for vanadium in VO_2 leads to a complex phase diagram with at least two insulating phases whose properties are significantly different from those of the insulating phase of pure VO_2. These phases are now fairly well understood (87).

The N–NM transition in V_2O_3 and its alloys has been a subject of a large number of publications. Pure V_2O_3 undergoes a first-order transition (monoclinic-rhombohedral) at 150K accompanied by a 10 million-fold jump in conductivity and an antiferromagnetic-paramagnetic transition. Application of pressure makes V_2O_3 increasingly metallic, thus suggesting that it is near a critical region; accordingly, doping with Ti or Cr has a marked effect on the transition; the former has a positive pressure effect and the latter a negative pressure effect. V_2O_3 also shows a second-order transition around 400K with a small conductivity anomaly. Mere crystal distortion or magnetic ordering cannot explain the large conductivity jump at 150K. The current status of the V_2O_3 transition is best presented in terms of Figure 2.

The schematic diagrams in Figure 2 show several interesting features. In V_2O_3 (98.5%)–Cr_2O_3 (1.5%), there are three transitions as shown in Figure 2(*a*). In the 0–150K range, the alloy is an antiferromagnetic insulator (AFI). There is a sharp transformation to the paramagnetic metallic

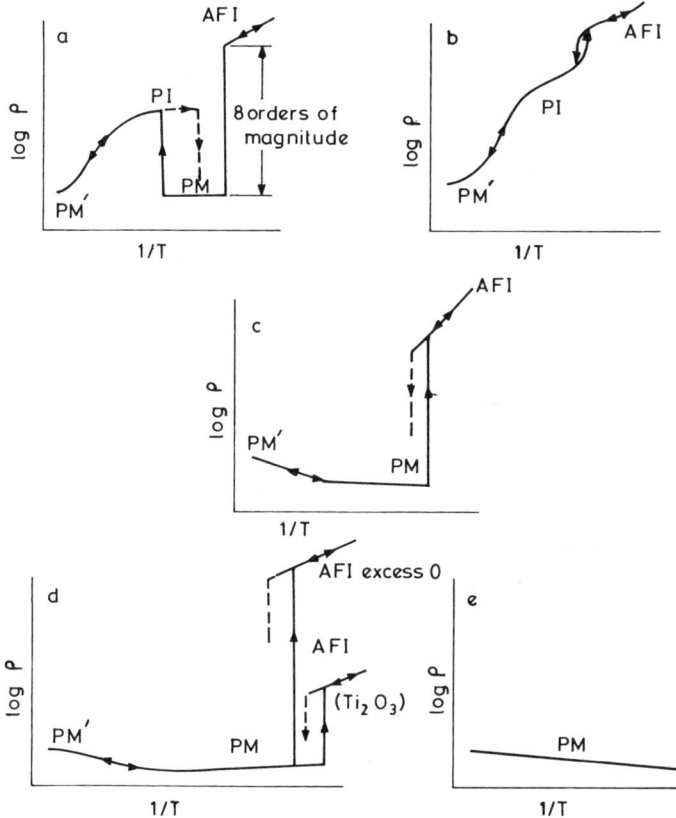

Figure 2 Schematic diagram depicting the changes of resistivity, *P*, with temperature, *T*, in the V_2O_3 alloy system as a function of dopants [from Honig & Spalek (85b)]. See text for a discussion of the different graphs.

state (PM), which prevails between 150 and 300K. At that point, another first-order transition transforms the oxide to a paramagnetic insulating state (PI). In the 300–1000K range, the resistivity gradually diminishes with increasing temperature, and beyond 650K, the alloy is in another paramagnetic metallic phase (PM′), which resembles the PM state. The three transitions are altered by minor changes through alloying. Electrical properties of the V_2O_3 (97%)–Cr_2O_3 (3%) are given in Figure 2(*b*). The PM phase is eliminated altogether, and the alloy goes directly from the AFI to the PI state via a sharp ($x > 0.03$) transition; this is followed by a gradual transition to the PM′ state. In pure V_2O_3, the resistivity exhibits a different pattern, as shown in Figure 2(*c*); there is a first-order transition near 150K that links the AFI and PM phases. With increasing temperature in the 350–650K range, an anomalous rise in resistivity occurs and the undoped or lightly Cr_2O_3-doped V_2O_3 passes continuously from the PM to the PM' state. Figure 2(d) gives the situation for V_2O_3 (99%)–Ti_2O_3 (1%) and for nonstoichiometric V_2O_3. The AFI–PM transition is shifted to lower temperature, and the resistivity discontinuity diminishes with increasing Ti/cation vacancy concentration. When the Ti content is increased beyond 5.5 or the vacancy concentration passes beyond 0.9, all transitions are eliminated and the PM phase is retained as shown in Figure 2(e). Below approximately 10–15K, however, the metallic phase transforms to an antiferromagnetic metal (AFM).

A variety of theoretical models have been proposed to explain M–NM transitions in metal oxides (83–85, 88). Here I briefly examine the mechanisms involving electron correlation and disorder. The phenomenon of correlation driven M–NM transition (*Mott transition*) was first pointed out by Mott (88). The simplest model for correlation effects in solids is that due to Hubbard (89). The possible phases of this model have been investigated in the one-body or Hartree-Fock approximation (90). The one electron density of states develops a gap when $(U/zt) > U_c \approx 1$. Here, z is number of near neighbors, t is the amplitude, and (U/zt) is dimensionless. The Hubbard gap increases with U, and at large U the material is an AFI whereas at small U it is a PM. The Hartree-Fock approximation is not satisfactory for M–NM transitions in which charge fluctuations, spin fluctuations, temperature, and mean-field have a comparable energy scale. Gutzwiller (91) introduced a different kind of approach that emphasizes local correlation effects and the relevance of this model to M–NM transitions due to correlation was pointed out by Brinkman & Rice (92). This model is satisfactory near the transition, but properties at $T \neq 0$ and of the insulating phase are not described, as spatial correlations have been completely ignored. Features such as orbital effects, electron-lattice interaction, Coulomb interaction, and disorder have also been ignored in this treatment.

Anderson (93) showed that when randomness exceeds a critical value, an electron locates itself around an appropriate potential fluctuation so that the state is spatially localized rather than extended, as in the case of weak disorder. The ratio W/Zt, where W is the spread of bandwidth, depends on the energy of the electron and on the dimensionality and connectivity of the lattice. Consequences of localization were clearly enunciated by Mott (94). A system becomes metallic or insulating depending on whether the states near the Fermi energy are extended or localized. With increasing disorder or decreasing Fermi energy, the mobility edge crosses the Fermi energy and the system becomes insulating. From this model, one obtains the minimum metallic conductivity in three and two dimensions; the latter is a universal constant. The conductivity of a metal drops from σ_{min} to zero at the transition. There is considerable evidence for the change of the transport regime in disordered systems at σ_{min}. However, at very low temperatures close to critical disorder, it has been found that conductivity goes continuously to zero at the localization transition. A scaling theory of localization has been proposed to circumvent this difficulty (95). There are also theories that take into account interaction effects in disorder-driven M–NM transitions (96), but as yet no theories treat interaction or correlation effects sufficiently accurately to lead to the Mott transition and also include disorder effects.

The well-known criterion for the M–NM transition is that due to Mott. It states,

$$n^{1/3} a_H^* \approx 0.25$$

where a_H^* is the shallow state radius or atomic orbital size and n is the carrier density. This criterion is spectacularly successful (97) over a wide density range, although the transition is not discontinuous as predicted. It can be shown that the Anderson localization criterion, $\xi t = W$ (where ξ is the localization length), and the Hubbard criterion, $U \approx Zt$, are similar to Mott's criterion. In the Hubbard model, the incommensurate system is always metallic whereas the commensurate electronic system exhibits a M–NM transition. The effect of disorder on such a model has yet to be explored. A criterion due to Herzfeld (98) states that for a metal, the ratio of molar refractivity and molar volume (R/V) is greater than or equal to unity. This criterion holds for all elemental metals; a thermodynamic criterion based on latent heat of evaporation is found to be equally satisfactory (99).

In the absence of exact models, M–NM transitions in real systems have been explained qualitatively in terms of the available models. For example, for the V_2O_3 transition, Kuwamoto et al (100) proposed a simple model wherein the density of states curve for the d-band has a set of high peaks and deep valleys in alternation. The Fermi level in close proximity to one of the minima can be replaced by a band gap that is opened by Cr or Al doping. Change in oxygen stoichiometry or Ti doping shifts the Fermi level so as to render the material metallic. Honig & Spalek (85b) have worked out a thermodynamic model for V_2O_3 that uses different free energy expressions for electrons in the localized and itinerant regimes. Disorder has been invoked to explain transitions in oxide systems such as $La_{1-x}Sr_xVO_3$ (101) and $La_{1-x}Sr_xCoO_3$ (102). Studies of systematics of M–NM transitions across a related series of oxides have yielded valuable results, as discussed in a preceding section. It is noteworthy that in binary transition metal compounds, the transfer from insulating behavior to metallic behavior occurs in the vicinity of oxides (103).

Before I close this discussion, I briefly present some of the complex oxide systems exhibiting compositionally controlled M–NM transitions that may be especially appealing to chemists. $Ln_{1-x}Sr_xMO_3$ (Ln = La, Pr, Nd; M = V, Mn, Co) show M–NM transitions with increase in x (6a,b, 80). Thus, $La_{1-x}Sr_xCoO_3$ becomes metallic when $x = 0.3–0.5$. Metallicity in the M = Mn and Co systems is accompanied by ferromagnetism. In $LaNi_{1-x}M_xO_3$ (M = Cr, Mn, Fe, or Co), the system goes from having metallic to insulator behavior above a critical value x; at the crossover, the system shows Mott's σ_{min} value. $La_{4-x}Ba_{1+x}Cu_5O_{13+\delta}$ is metallic when $x \approx 0$, but as x increases, it becomes insulating (104). The pyrochlore system $Bi_{2-x}Gd_xRu_2O_7$ exhibits an M–NM transition with increase in x (105).

SUPERCONDUCTIVITY

Although a variety of inorganic and organic solids had been investigated for superconductivity in recent decades, the highest transition temperature attained was around 23K in Nb alloys of the A15 family. Metal oxides had also been explored earlier, and the highest superconducting transition temperature found in them was around 13K in $Ba(Pb, Bi)O_3$ (9) and $Li_{1+x}Ti_{2-x}O_4$ (19a,b). The new generation of oxide superconductors discovered since the first announcement of 30K superconductivity in $La_{2-x}Ba_xCuO_4$ by Bednorz & Müller (4) has pushed the transition temperature up to 130K. Structure-property relations in the high-temperature superconducting cuprates have been reviewed in some detail very recently (106–110a,b), and references to the original literature on all the aspects discussed here may be found in these and other references cited. I examine here the salient features of these high-temperature superconductors that have caused unprecedented excitement. High T_c superconductivity in cuprates and other oxides has revealed our incomplete knowledge of the electronic structure of transition metal oxide systems.

Properties of Cuprate Superconductors

High-temperature superconducting cuprates discovered in the last two years belong to the following families: (a) $La_{2-x}M_xCuO_4$ (M = Ca, Sr, or Ba) of the K_2NiF_4 structure (111–113); (b) the $LnBa_2Cu_3O_7$ (123) system (107, 114), where Ln = Y, La, Nd, Sm, Eu, Gd, Dy, Ho, Er, Tm, or Yb; (c) the $Bi_2(Ca, Sr)_{n+1}Cu_nO_{2n+4}$ system, with $n = 1, 2, 3, 4$ (110a,b, 115–117); (d) the $Tl_2Ca_{n-1}Ba_2Cu_nO_{2n+4}$ system, with $n = 1, 2, 3, 4$ (110a,b, 118–

121); (e) the $TlCa_{n-1}Ba_2Cu_nO_{2n+3}$ system, with $n = 1, 2, 3, 4$ (122–124); and (f) $Pb_2Sr_2ACu_3O_8$, with $A =$ Ln or Ln+Sr or Ca (125). A three-dimensional oxide without Cu, $Ba_{1-x}K_xBiO_3$ has been found to exhibit (126) a T_c of $\sim 30K$. Figure 3 shows the structures of the first three families of cuprates in order to illustrate certain commonalities. The structures of the Tl cuprates are similar to those of the Bi cuprate in Figure 3. All the cuprates, (a) to (f), possess defect perovskite layers and all but the 123 compounds also contain rock-salt type M–O layers. All of them contain two-dimensional Cu–O sheets, and the 123 compounds have one-dimensional Cu–O chains in addition. The coordination of Cu is essentially square-planar, and the Cu–O bond distance is around 1.9 Å, indicative of high covalency. Oxides of the $La_{2-x}M_xCuO_4$ family are ordinarily tetragonal and become orthorhombic around 180K (127). The T_cs are in the 25–40K range (at an optimal value of x), depending on the M ion. Substitution of La by other rare earth ions or of Cu by Ni, Zn, and similar ions adversely affects the superconducting transition temperature. In the 123 compounds, the Ln ion has little effect on the T_c, but the T_c is markedly dependent on the oxygen stoichiometry, δ. In the case of $YBa_2Cu_3O_{7-\delta}$, T_c is nearly constant ($\sim 90K$) up to $\delta = 0.2$, but drops to a constant value of $\sim 55K$ between δ of 0.2 and 0.4; further increase in δ lowers the T_c until it becomes nonsuperconducting when $\delta \approx 0.6$ (128). The structure is orthorhombic over the entire δ range of 0.0–0.60 but becomes tetragonal when $\delta \geq 0.60$. It is not clear whether any special structural feature is associated with the 55K T_c plateau, although the orthorhombic lattice parameters are not related in this region, unlike in the high T_c region ($a \neq b \approx c/3$). Gd and Dy cuprates of this family also show this behavior.

The formal mixed valence of Cu is considered to be essential for the superconductivity in these cuprates. Yet we find superconductivity in $YBa_2Cu_3O_{6.5}$ ($T_c \sim 45K$), which should contain Cu only in the 2+ state. Intergrowth of the O_7 and the O_6 phases is probably responsible for this observation. The presence of Cu^{2+} in $YBa_2Cu_3O_{6.5}$ is evidenced from EPR spectroscopy, in contrast to well-annealed samples with $\delta < 0.2$.

In the 123 compounds, ordered orthorhombic structures with 90K T_c are found only when the 01 oxygen in the Cu–O chains are fully populated and ordered; distribution of the chain oxygen between the 01 and the 05 sites gives rise to disordered orthorhombic structures with low T_cs (129a,b). Equal population of the 01 and 05 sites, just as complete depletion of the 01 oxygen gives rise to tetragonal structures. $YBa_2Cu_3O_{7-\delta}$ samples with high δ can be oxidized to the $\delta \approx 0.0$ composition, but the diffusion of oxygen is a highly activated process.

$LnBa_2Cu_3O_7$ may be considered to be the $x = 1$ member of the more

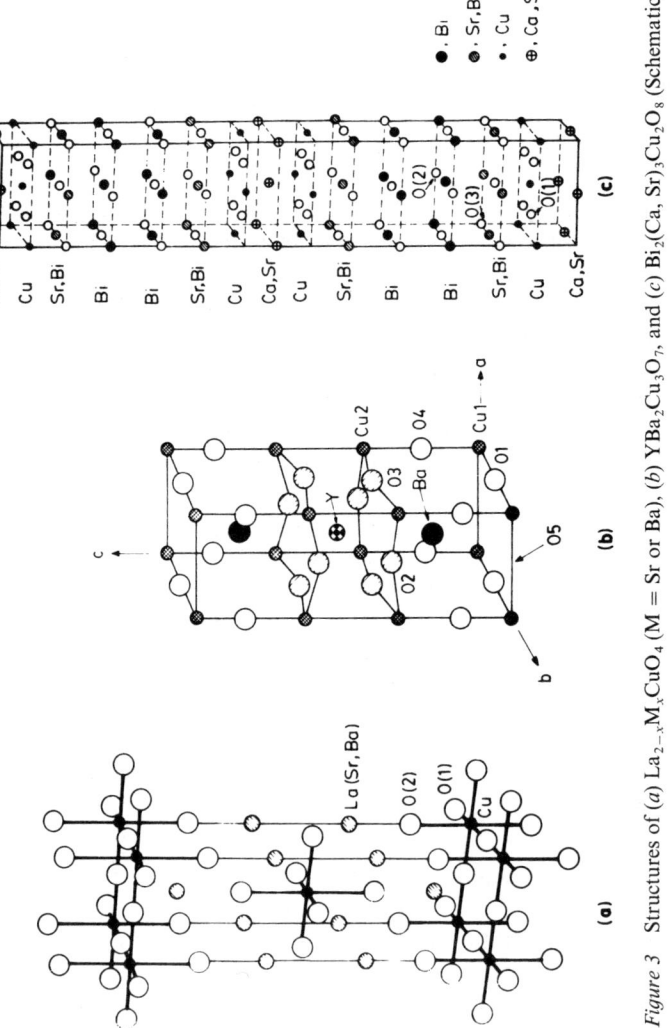

Figure 3 Structures of (a) $La_{2-x}M_xCuO_4$ (M = Sr or Ba), (b) $YBa_2Cu_3O_7$, and (c) $Bi_2(Ca, Sr)_3Cu_2O_8$ (Schematic).

general $Ln_{3-x}Ba_{3+x}Cu_6O_{14+\delta}$ family. A common occurrence in the 336 and 123 compounds is the exchange between the Ln and Ba sites, especially when Ln is a large rare earth ion such as La. Such an exchange does not occur in the $YBa_2Cu_3O_7$ because of the small size of yttrium. The Ba ion in $YBa_2Cu_3O_7$ can be replaced by La, but the T_c is lowered and there is also oxygen excess. The smallest Ln ion tolerated by the 123 structure is Yb.

Orthorhombic 123 compounds show extensive twinning occurring during their formation from the high-temperature tetragonal structures. Across the twin boundaries there is a 90° rotation of the a and b axes. Although twins may play a crucial role in determining properties such as the critical current density of the 123 compounds, they are not the cause of superconductivity. Thus, orthorhombic $PrBa_2Cu_3O_7$, which is not superconducting, shows twins. The reason for the absence of superconductivity in the 123 compounds of Ce, Pr, and Tb has been suspected to lie in the bivalency (3+ and 4+) of these lanthanide ions, but the exact cause is not yet understood. Orthorhombicity was considered to be a necessary criterion for high T_c in the 123 compounds for some time, but tetragonal $YBa_2Cu_3O_7$, where Co, Fe, or Ga partly substitute for Cu, has since been found to show high T_c behavior; the compounds, however, seem to possess some orthorhombic distortion. Substitution of Cu by Ni and Zn adversely affects the oxygen stoichiometry and lowers the T_c.

Members of the $Bi_2(Ca, Sr)_{n+1}Cu_nO_{2n+4}$ and $Tl_2Ca_{n-1}Ba_2Cu_nO_{2n+4}$ series have similar structures and contain two Bi, Tl–O type rock-salt layers. The bismuth cuprates show modulation in the structure. Members of the $TlCa_{n-1}Ba_2Cu_nO_{2n+3}$ have a single Tl–O type rock-salt layer. The Bi, Ca, and Sr sites in the Bi cuprates are interchangeable, and the compositions are never exactly 2122 or 2223 as described by the general formula. Bi can be partly substituted by Pb (up to ~25%) and this generally favors the formation of better monophasic compositions with slightly enhanced T_cs. In the $Bi_2(Ca, Sr)_{n+1}Cu_nO_{2n+4}$ series, the first three members with c-parameters of ~25, 31, and 38 Å have been characterized; the T_cs are 60 ± 20, 85 ± 5, and 107 ± 3K, respectively; the $n = 4$ member also seems to have a T_c close to 110K (117).

In the $Tl_2Ca_{n-1}Ba_2Cu_nO_{2n+4}$ series, the $n = 1$, 2, and 3 members (c-parameters 23, 29, and 36 Å) show T_cs of 80, 110, and 125K, respectively. The $n = 2$ and 3 members of the $TlCa_{n-1}Ba_2Cu_nO_{2n+3}$ series show T_cs of 90 and 115K, respectively, which are lower than those of the corresponding members of the Tl_2 series. In the Tl cuprates, just as in the Bi cuprates, we see a progressive increase in T_c as well as in the c-parameter with the number of Cu–O sheets only up to $n = 3$ (Figure 4); when $n > 3$, the thallium cuprates do not seem to show a further increase in T_c (121). A

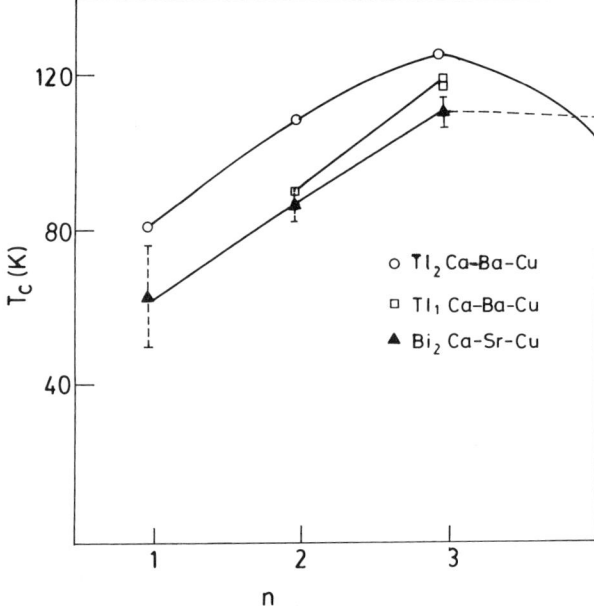

Figure 4 Variation of superconducting transition temperature in thallium and bismuth cuprate systems with the number of Cu–O sheets.

feature common to the Bi and Tl cuprates is the presence of disordered intergrowths (58). It is noteworthy that $YBa_2Cu_4O_8$ with two Cu–O chains (edge-shared square planar units) has only a T_c of ~80K compared to ~90K of $YBa_2Cu_3O_7$ (130).

It has not been possible to prepare well-defined superconducting compositions of the Tl–Ca–Sr–Cu–O system with the general formula $Tl_{1,2}(Ca, Sr)_{n+1}Cu_nO_{2n+3,4}$. These members are, however, stabilized by partial substitution of Tl by Pb (131a,b). Thus, progressive increase in x in $Tl_{1-x}Pb_xCaSr_2Cu_2O_y$ increases the T_c up to 90K when $x = 0.5$; $Tl_{0.5}Pb_{0.5}CaSr_2Cu_3O_7$ has a T_c of 120K. Progressive substitution of Ca by Y in $TlCa_{n-1}Ba_2Cu_nO_{2n+3}$ and $Bi_2(Ca, Sr)_{n+1}Cu_nO_{2n+4}$ lowers the T_c, until superconductivity is lost upon complete replacement (132a). The layered Pb cuprates (125) contain Cu mainly in the 1+ state (132b) and show T_c around 70K. Whereas Pb in $Pb_2Sr_2Ca_{1-x}ln_xCu_3O_8$ is mainly in the 2+ state, in $Tl_{0.5}Pb_{0.5}CaSr_2Cu_3O_7$ and $Bi_{1+\delta}Pb_{0.5}(Ca, Sr)_{n+1}Cu_nO_{2n+4}$ it is in the 4+ state (C. N. R. Rao, unpublished results). Recently, a series of cuprates of the formula $TlCa_{1-x}ln_xSr_2Cu_2O_{7+\delta}$ with T_c in the 60–90K range has been discovered (132c). There is a possibility of competition between electron and hole superconductivity in this system.

Properties of Superconducting Cuprates

Some of the relevant properties of the superconducting state are the critical field, the critical current, the magnetic penetration depth, and the coherence length. The $H_{c1}(O)$ and $H_{c2}(O)$ of $YBa_2Cu_3O_7$ parallel to the c-axis are around 1 and 120T, respectively; the magnetic penetration depth is ~ 900 Å. The coherence length (size of the Cooper pair) is 10–30 Å in the ab-plane of $YBa_2Cu_3O_7$ as well as other cuprates, but only about 3 Å perpendicular to the plane (133). The meaning of such a small coherence length is not clear. Anisotropy is also found in the magnetic and electrical properties of $YBa_2Cu_3O_7$ and the other cuprates. The critical current of a superconductor should be at least $\sim 10^5$ amp cm^{-2} for magnetic and other applications at the operating temperatures. Although values of 10^5 amps cm^{-2} or higher have been reported in films and single grain materials of $YBa_2Cu_3O_7$, it has not been possible to obtain good, reproducible samples because of the presence of grain boundaries and weak flux pinning. The situation is the same with Ti and Bi cuprates.

No measurable ^{18}O isotope effect has been observed in $YBa_2Cu_3O_7$ (134). Infrared absorption, point-contact tunnelling spectroscopy, and other measurements (135) suggest the superconducting gap, 2Δ, to be around 3–4 (k_BT_c). It is not yet clear whether there is a specific heat discontinuity at T_c in the oxide superconductors.

The Cu ions in the superconducting cuprates are EPR-silent, but the cuprates show intense nonresonant absorption of microwaves (136a,b). The presence of Cu^{2+} in $YBa_2Cu_3O_{7-\delta}$ is seen from magnetic measurements in the non-superconducting $YBa_2Cu_3O_6$, which is antiferromagnetic ($T_N = 450K$). $LnBa_2Cu_3O_7$ compounds with Ln = Gd, Dy, etc show magnetism at low temperatures due to the Ln ion (137).

The d–d correlation energy is the largest relevant energy in these oxides. The density of states near the Fermi energy in La_2CuO_4 and $La_{2-x}M_xCuO_4$ is small as revealed by photoemission spectra. An angle-resolved photoemission study of $Bi_2CaSr_2Cu_2O_8$ has shown that there are p-like bands near and below E_F, with a very small width (~ 0.5 eV). It appears that $(\varepsilon_p - \varepsilon_d)$ is around 1–3 eV, and the pd mixing integral t_0 is in the range of 1–2 eV (138).

Nature of Copper and Oxygen

The state of copper in the cuprate superconductors is of seminal importance to the mechanism of superconductivity in these materials. In stoichiometric La_2CuO_4, Cu is present in the 2+ state; this would not be the case in $La_{2-x}M_xCuO_4$ or $YBa_2Cu_3O_7$. It has been generally believed that the presence of Cu in the mixed valent (2+, 3+) state, in such doped oxides, is essential to explain the magnetic and superconducting properties. Oxygen-excess in these materials is expected to create holes on Cu in the form of Cu^{3+}. However, X-ray absorption and photoemission measurements show no evidence for the presence of Cu^{3+} in $YBa_2Cu_3O_7$. Instead, they show the presence of copper in the 1+ state and holes on oxygen, O^{1-} (139–141). This is corroborated by electron energy loss spectroscopy, which shows an $s-p$ transition of oxygen (142). X-ray photoelectron spectroscopy and Auger spectroscopy also show that for all practical purposes, there is no Cu^{3+} in $YBa_2Cu_3O_7$, $Bi_2(Ca, Sr)_3Cu_2O_{8+\delta}$, and $Tl_2Ca Ba_2Cu_2O_8$, but there is evidence for the presence of the Cu^{1+} state. Evidence for the dimerization of oxygen holes giving rise to peroxo-type species, O_2^{2-}, has also been presented based on $O(1s)$ spectra. It is believed by many workers today that mobile oxygen holes are responsible for the superconductivity of the cuprates. In the Bi and Tl cuprates, the Bi–O and Tl–O layers seem to donate such holes.

Chakraverty et al (143) have proposed that the $d_{x^2-y^2}$ orbital of copper (in the Cu–O sheets) overlaps with the p_σ orbital of oxygen (formed by a combination of p_x and p_y orbitals), forming a broad Cu–O band consistent with high covalency of the Cu–O bands. They also propose that the holes are in the p_σ state (within the Cu–O band) and that they are more favored by the d^{10} state Cu^{1+} ions than by $Cu^{2+}(d^9)$ ions. They suggest that a O^{1-}–Cu^+–O^{1-} state (designated as a peroxiton) is energetically favored compared to a hole bipolaron, O^{1-}–Cu^{2+}–O^{1-}. The presence of holes in p_σ also explains the absence of anti-ferromagnetism in the cuprates. Guo, Langlois & Goddard (144) suggest, based on cluster calculations, that oxidation of Cu beyond Cu^{2+} creates oxygen p_Π holes bridging two Cu^{2+} sites. The p_Π holes are ferromagnetically coupled to adjacent Cu^{2+} d-electrons, and hopping of the p_Π holes in the Cu–O sheets from site to site is responsible for the conductivity. The p_Π of these workers is not the p_z orbital but a $p_{x,y}$ orbital of oxygen. It is not clear that these will form narrow bands distinctly separated from the broad Cu–O band involving $d_{x^2-y^2}$ (Cu) and oxygen p_σ orbitals. High energy spectroscopy experiments (145a,b) clearly show that the holes are in the CuO_2 planes and have essentially $p_{x,y}$ character. Cluster calculations support existence of the copper states found from core level spectroscopies (146). Oxygen hole-pairing has been suggested by a few theoreticians (147, 148). We can relate the presence of O^{1-} holes to the average charge, p, of $(Cu-O)^p$, since p is known to be related to the superconducting properties of the cuprates (149a). A finite positive value of p can only result from $(Cu^{2+}-O^{1-})^{1+}$ in combination with $(Cu^{2+}-O^{2-})^0$ and $(Cu^{1+}-O^{1-})^0$ and the relative importance of the last two may determine whether a particular cuprate is a superconductor or an insulator. The importance of oxygen holes is also underscored by the discovery of relatively high T_c in $Ba_{1-x}K_xBiO_3$ (126). In the Pb cuprates (125), there is no possibility of Cu^{3+} occurring, even according to the chemical formula, and this would suggest a role for Cu^{1+} (132b), proposed earlier for 123 oxides and other systems (140a–c). The discovery of the so-called electron superconductor $Nd_{2-x}Ce_xCuO_4$ (149b) is interesting, but it is not yet clear that the electron donated by Ce goes to the Cu $3d$ orbital (149c).

Theoretical Approaches

Models for the electronic structures of superconducting cuprates range from the strong crystal field limit to the correlated weak crystal field limit. A one-band Hubbard model with a novel ground state involving low-spin Cu has been proposed (150). A simple two-band model with a strongly correlated d-type band and weakly correlated p-type band hybridizing with it has been examined (147, 151). There are, however, many real complications with these models (152). Electronic structure calculations are believed to be inadequate in highly correlated systems. Useful information has been obtained, however, from self-consistent one-electron theory as well as from numerical calculations on Cu–O clusters (153, 154). Calculations on the problem of two holes hybridizing with the oxygen p-band have also been performed (155).

Any theoretical model has not only to explain the varied properties of the superconducting state, such as the short coherence length, non-exponential dependence of NMR relaxation rate on temperature, the energy gap, resistivity and heat capacity behavior near T_c, etc, but also the optical, magnetic, transport, NMR relaxation, and other properties of the normal state (152). A number of models have been proposed. They make use of both phonon and electronic mechanisms. A spin-bag theory has been discussed (156). The most original idea is the resonating valence band theory of Anderson and co-workers (150, 157). In this theory, the spin 1/2 of the d^9 configuration and two dimensionality of the Cu–O layers make for a non-Neèl magnetic ground state (the RVB state) with characteristic excitations. Singlet pairs present in this ground state are preformed Cooper pairs with a binding energy close to J and of the size of interatomic spacing. Upon introduction of holes, this system transports charge and becomes a superconductor. Hole-pairing via interlayer hopping has also been considered (158). Correlated electron models have been used to examine the problem of hole mobility or the attraction of charge carriers (152). The nature of oxygen defects and the Fermi level location in $YBa_2Cu_3O_{7-\delta}$ and $La_2CuO_{4-\delta}$ have been examined in the light of tight-binding calculations (159).

SYNTHESIS AND CHARACTERIZATION

A wide range of conditions have been employed to synthesize transition metal oxides. These include high temperatures and pressures, carefully controlled oxidizing or reducing atmospheres, hydrothermal conditions, skull and arc melting, and so on (3, 160, 161). Oxides are most commonly prepared by the ceramic method, involving repeated grinding and heating of the reactant powders of oxides, carbonates, etc (often in pellet form) or sealed tube reactions. There has been increasing interest in preparing oxides under milder, less energy-consuming conditions. The precursor method has been employed effectively to achieve homogeneous mixing of reactant species on an atomic scale (162). This method reduces the diffusion distances to 100 Å or so instead of 10,000 Å or more in the ceramic method. Besides precursor compounds, solid solutions of carbonates, nitrates, hydroxides, and cyanides have been employed for the purpose. The method also enables the synthesis of novel oxides, which are otherwise difficult to prepare. Topochemical reactions similarly yield unusual oxides; e.g. the synthesis of MoO_3 of ReO_3 structure by the topochemical dehydration of $MoO_3 \cdot H_2O$ is a good example (163a,b). $Mo_{1-x}W_xO_3$ has also been prepared by such a dehydration reaction. Many examples of topochemical reactions giving novel oxides are reported in the literature (3, 162). A topochemical reaction deserving special mention is the insertion of atomic species into oxide hosts (164). Thus, lithium and other alkali metals have been inserted into a variety of oxides such as VO_2, TiO_2, MnO_2, ReO_3, and Fe_3O_4. The subject of intercalation has been reviewed adequately in the literature (3, 164, 165). Deintercalation of lithium and other species can be carried out readily by employing mild oxidizing conditions. Many new examples of intercalation and deintercalation are constantly being reported. Recently, topochemical reactions of Li_xNbO_2 (166) and lithium insertion to $W_{19}O_{55}$ (167) have been reported.

Ion exchange has been carried out in oxides having layered, tunnel, or close-packed structures. Such reactions are also topochemical and are carried in aqueous solutions or molten media (3, 162). Conversion of $LiNbO_3$ to $HNbO_3$ by treatment with hot aqueous acid is an example (168). The mechanism of this reaction appears to be reverse of the transformation of ReO_3 to rhombohedral $LiReO_3$ (169). Hydrogen can also be inserted into oxide holes in the presence of a Pt catalyst (162). Potentialities of exchange reactions for synthetic purposes are immense. Many interesting exchange reactions have been reported in the literature; two recent examples are the effect of intercalated alkylammonium ion on the cation exchange properties of $H_2Ti_3O_7$ and the exchange properties of

$Na_4Ti_9O_{20} \cdot x \cdot H_2O$ (170a,b). Synthesis of metastable TiO_2 and layered $K_2Ti_4O_9$ by a topotactic dehydroxylation is another interesting example (171).

Among the other chemical methods of synthesis, the chemical vapor deposition technique is well known. Fused salt electrolysis has been employed to synthesise oxides such as complex oxides of Mo and Mo bronzes (28, 172a,b). The pyrochlores $Pb_2[Ru_{2-x}Pb_x^{4+}]O_{7-y}$ and $Bi[Ru_{2-x}Bi_x^{5+}]O_{7-y}$ have been prepared from a strongly alkaline medium under oxidizing conditions (173). The sol-gel route has proved to be extremely successful in synthesizing a variety of oxides (3, 162, 174), including the superconducting cuprates (175).

Many oxides have been prepared by arc melting. A novel method of preparing some of the transition metal oxides is by the crucible-free method (176). Single crystals of complex oxides such as La_2NiO_4 and Fe_3O_4 have been prepared by such skull melting. High pressure methods in synthesis have been reviewed (3, 177, 178). Use of high pressure enables stabilization of unusual oxidation states (e.g. $CaCrO_3$, $La_2Pd_2O_7$, $GdNiO_3$). Fe(V) has been stabilized in La_2LiFeO_6; high-spin Fe(IV), low-spin Ni(III), and Co(IV) are the other states so stabilized under high oxygen pressure. $YBa_2Cu_4O_8$ has been prepared recently under high oxygen pressure (130).

Techniques of characterization of oxide materials have advanced rapidly in the last decade (3, 179). Although single crystal X-ray crystallography continues to be a useful technique, profile analysis of X-ray and neutron diffraction patterns of powders by the Rietveld method has emerged to become a powerful tool (179, 180). A new generation of ultra-high resolution powder X-ray and neutron diffractometers (181, 182) allows *ab initio* determination of oxide structures by using synchrotron X-rays and high-intensity (spallation) neutron sources (183, 184). Epithermal neutrons are useful to study crystal field transitions in oxides such as PrO_2 and $BaPrO_3$ (185).

High-resolution electron microscopy has been used routinely to study local structure of oxides at atomic and unit cell levels. Composition characterization by X-ray emission, electron energy loss spectroscopy, etc carried out in the electron microscope is becoming more useful for heavy elements and also for oxygen (186a–c). Electron energy loss spectroscopy is also useful in characterizing oxidation states of metals (187), and the technique has been reviewed recently (188).

High-resolution NMR spectroscopy has been of great value in the study of zeolites and other oxidic materials. The technique has also been useful in the study of phase transitions and other phenomena, but a wider use employing transition metal nuclei is likely in the future. EXAFS, XANES, and related X-ray absorption techniques are useful in determining the oxidation state and coordination of the transition metal ion in complex oxide materials (including catalysts), as illustrated by the recent studies on cuprate superconductors (139–141). Electron spectroscopies (XPS, UPS, Auger, EELS, etc) have been used widely to investigate transition metal oxides; a noteworthy technique to study electron states of the metal is the one based on Auger intensity ratios (189a–c). These high-energy spectroscopies have played an important role in determining the states of copper and oxygen in superconducting cuprates (140–141). Scanning tunneling microscopy (190a,b) is now another powerful tool in the arsenal of solid state scientists to study oxide materials.

EPILOGUE

We are yet a long way from fully understanding the electronic structures and properties of transition metal oxides. Considerable scope remains for carrying out good measurements, designing novel oxides with desired properties, and developing useful models. High T_c superconducting oxides are all quasi two-dimensional cuprates. There is a good probability that high T_c will be discovered in other oxide materials, including three-dimensional systems. In this context, the possible occurrence (12c,d) of superconductivity in layered nickelates is noteworthy. It is also interesting that the 30K superconductor $Bi_{1-x}K_xBiO_3$ has rather unusual properties, such as a large ^{18}O isotope effect (191) and the absence of static magnetic order (192), unlike in the cuprates. An important aspect that needs to be understood is the role of holes on oxygen (O^{1-}) and their possible dimerization (O_2^{2-}) in transition metal oxides. Holes on oxygen have been found in many transition metal oxides purported to contain the metals in high oxidation states, such as $La_2NiO_{4+\delta}$, $LiNiO_2$, $LaNiO_3$, and Ba_2CuO_5 (143, 193, 194; C. N. R. Rao et al, unpublished results). It is known that anion hole pairing gives rise to S–S, Se–Se, and Te–Te bonds in the chalcogenides containing $Cu^{1+}(d^{10})$ ions. There is some evidence that peroxide-like species arising from hole pairing may be present in certain transition metal oxides, including the superconducting cuprates (195). Unless the exact description of the states of oxygen and the metal ions becomes possible, understanding the properties of transition metal oxides will be difficult. In this regard, a breakthrough is needed in theoretical approaches to understand phenomena such as metal-insulator transitions and superconductivity in transition metal oxides.

ACKNOWLEDGMENTS

I thank the Department of Science and Technology, the University Grants Commission, and the US National Science Foundation for supporting my research efforts in transition metal oxides.

Literature Cited

1. Rao, C. N. R., Subbarao, G. V. 1970. *Phys. Status Solidi A* 1: 597–652; Rao, C. N. R., Subbarao, G. V. 1974. *Transition Metal Oxides*, NSRDS-NBS Monogr. 49. Washington, DC: Natl. Bur. Stand.
2. Goodenough, J. B. 1971. *Progr. Solid State Chem.* 5: 149–399; Goodenough, J. B. 1974. In *Solid State Chemistry*, ed. C. N. R. Rao. New York: Dekker; Goodenough, J. B. 1963. *Magnetism and the Chemical Bond*. New York: Wiley
3. Rao, C. N. R., Gopalakrishnan, J. 1986. *New Directions in Solid State Chemistry*. Cambridge: Cambridge Univ. Press
4. Bednorz, J. G., Müller, K. A. 1986. *Z. Phys. B* 64: 189–96
5. Vasanthacharya, N. Y., Ganguly, P., Goodenough, J. B., Rao, C. N. R. 1984. *J. Phys. C* 17: 2745–60
6a. Rao, C. N. R., Ganguly, P. 1985. In *Localization and Metal–Insulator Transitions*, ed. D. Adler, H. Fritzsche. New York: Plenum
6b. Rao, C. N. R., Prakash, O., Bahadur, D., Ganguly, P., Nagabhushana, S. 1977. *J. Solid State Chem.* 22: 353–64
7. Goodenough, J. B., Longo, J. M. 1970. *Landolt-Börnstein Tabellen*, New Ser., III/4a. Berlin: Springer-Verlag
8. Nomura, S. 1978. *Landolt-Börnstein Tabellen*, New Ser., III/12a. Berlin: Springer-Verlag
9. Sleight, A. W., Gillson, J. L., Bierstedt, F. E. 1975. *Solid State Commun.* 17: 27–31
10. Ganguly, P., Rao, C. N. R. 1984. *J. Solid State Chem.* 53: 193–216
11. Mohan Ram, R. A., Ganguly, P., Rao, C. N. R., Honig, J. M. 1988. *Mater. Res. Bull.* 23: 501–5
12a. Shirane, G., Endoh, Y., Birgneau, R. J., Katsner, M. A., Hidaka, Y., Oda, M., Suzuki, M., Murakami, T. 1987. *Phys. Rev. Lett.* 59: 1613–15
12b. Aeppli, G., Buttrey, D. J. 1988. *Phys. Rev. Lett.* 61: 203–5
12c. Rao, C. N. R., Ganguli, A. K., Nagarajan, R. 1989. *Pramana J. Phys.* 32: 177–80
12d. Spalek, J., Kakol, Z., Honig, J. M. 1989. *Phys. Rev. Lett.* To be published
13a. Mohan Ram, R. A., Ganapathi, L., Ganguly, P., Rao, C. N. R. 1986. *J. Solid State Chem.* 63: 139–47
13b. Rao, C. N. R., Ganguly, P., Singh, K. K., Mohan Ram, R. A. 1988. *J. Solid State Chem.* 72: 14–23
14a. Aurivillius, B. 1950. *Akiv Kemi* 2: 519–27
14b. Hutchison, J. L., Anderson, J. S., Rao, C. N. R. 1977. *Proc. R. Soc. London Ser. A* 355: 301–13
15. Portier, R., Carpy, A., Fayard, M., Galy, J. 1975. *Phys. Status Solidi A* 30: 683–89
16. Englman, R. 1972. *The Jahn-Teller Effect in Molecules and Crystals*. London: Wiley
17. Gehring, G. A., Gehring, K. A. 1975. *Rep. Prog. Phys.* 38: 1–89
18. Rao, C. N. R., Rao, K. J. 1978. *Phase Transitions in Solids*. New York: McGraw-Hill
19a. Johnston, D. C., Prakash, H., Zachariasen, W. H., Viswanathan, R. 1973. *Mater. Res. Bull.* 8: 777–82
19b. Lambert, P. M., Harrison, M. R., Edwards, P. P. 1988. *J. Solid State Chem.* 75: 332–46
20. Muraleedharan, K., Srivastava, J. K., Marathe, V. R., Vijayaraghavan, R., Kulkarni, J. A., Darshane, V. S. 1985. *Solid State Commun.* 55: 363–66
21. Subramanian, M. A., Aravamudan, G., Subbarao, G. V. 1983. *Progr. Solid State Chem.* 15: 55–143
22. Subramanian, M. A., Torardi, C. C., Johnson, D. C., Pannetier, J., Sleight, A. W. 1988. *J. Solid State Chem.* 72: 24–30
23. Beyerlein, R. A., Horowitz, H. S., Longo, J. M. 1988. *J. Solid State Chem.* 72: 2–13
24. Rao, C. N. R. 1985. *Int. Rev. Phys. Chem.* 4: 19–38
25. Imoto, H., Simon, A. 1982. *Inorg. Chem.* 21: 308–14
26a. Kihlborg, L. 1978. *Chem. Scr.* 14: 187–97
26b. Eckstrom, T., Tilley, R. J. D. 1980. *Chem. Scr.* 16: 1–16
26c. Ramanan, A., Gopalakrishnan, J., Uppal, M. K., Jefferson, D. A., Rao, C. N. R. 1984. *Proc. R. Soc. London Ser. A* 395: 127–40
27a. Hagenmuller, P. 1971. *Progr. Solid State Chem.* 5: 71–144
27b. Doumerc, J. P., Pouchard, M., Hagenmuller, P. 1985. See Ref. 84, pp. 287–328
28. Greenblatt, M. 1988. *Chem. Rev.* 88: 31–53
29a. Collins, B. T., Ramanujachary, K. V., Greenblatt, M., McCarroll, W. H., McNally, P., Waszczak, J. V. 1988. *J. Solid State Chem.* 76: 319–27
29b. Tsai, P. P., Potenza, J. A., Greenblatt,

M. 1987. *J. Solid State Chem.* 69: 329–35
30. Barbara, T. M., Gammie, G., Lyding, J. W., Jonas, J. 1988. *J. Solid State Chem.* 75: 183–87
31a. Domenges, B., Goreand, M., Labbe, Ph., Raveau, B. 1983. *J. Solid State Chem.* 50: 173–80
31b. Hervieu, B., Domenges, B., Raveau, B. 1985. *Chem. Scr.* 54: 10–16
31c. Domenges, B., Hervieu, M., Raveau, B., O'Keefee, M. 1988. *J. Solid State Chem.* 72: 155–72
32. Wang, E., Greenblatt, M. 1988. *J. Solid State Chem.* 76: 340–44
33. Ganguli, A. K., Ganapathi, L., Gopalakrishnan, J., Rao, C. N. R. 1988. *J. Solid State Chem.* 74: 228–31
34. Newnham, R. E., Cross, L. E. 1981. In *Preparation and Characterization of Materials*, ed. J. M. Honig, C. N. R. Rao. New York: Academic
35. Anderson, J. S., Tilley, R. J. D. 1974. In *Surface and Defect Properties of Solids*, ed. M. W. Roberts, J. M. Thomas, Vol. 3. London: Chem. Soc.
36. Anderson, J. S. 1984. *Proc. Indian Acad. Sci. (Chem. Sci.)* 93: 861–904
37. Watanabe, D., Tearesaki, O., Jostons, A., Castles, J. R. 1970. In *The Chemistry of Extended Defects in Nonmetallic Solids*, ed. L. Eyning, M. O'Keeffe. Amsterdam: North-Holland
38. Terauchi, H., Cohen, J. B. 1979. *Acta Cryst. A* 35: 646–52
39. Morinaga, M., Cohen, J. B. 1979. *Acta Cryst. A* 35: 745–54, 975–80
40. Gavarri, J. 1978. Doctoral thesis. Univ. Paris VI
41. Catlow, C. R. A., Mackrodt, W. C., eds. 1982. *Computer Simulation of Solids. Lect. Notes Phys.* Berlin: Springer-Verlag
42. Bauer, E., Pianelli, A., Aubry, A., Jeannot, F. 1980. *Mater. Res. Bull.* 15: 323–28
43. Garstein, E., Mason, T. O., Cohen, J. B. 1986. *J. Phys. Chem. Solids* 47: 759–73
44. Tetot, R., Gerdanian, P. 1985. *J. Phys. Chem. Solids* 46: 1131–39
45. Grimes, R. W., Anderson, A. B., Heuer, A. H. 1986. *J. Am. Ceram. Soc.* 69: 619–23
46. Men, A. N., Carel, C. 1985. *J. Phys. Chem. Solids* 46: 1185–93
47. Thornber, M. R., Bevan, D. J. M. 1970. *J. Solid State Chem.* 1: 536–42
48. Catlow, C. R. A., Chadwick, A. V., Corish, J. 1983. *J. Solid State Chem.* 48: 65–72
49. Tilley, R. J. D. 1980. In *Chemical Physics of Solids and Surfaces*, ed. M. W. Roberts, J. M. Thomas, Vol. 8. London: Chem. Soc.
50. Catlow, C. R. A., James, R. 1980. In *Chemical Physics of Solids and Surfaces*, ed. M. W. Roberts, J. M. Thomas, Vol. 8. London: Chem. Soc.
51. Blanchin, M. G., Bursill, L. A., Smith, D. J. 1984. *Proc. R. Soc. London Ser. A* 391: 351–63; Bursill, L. A., Blanchin, M. G., Smith, D. J. 1984. *Proc. R. Soc. London Ser. A* 391: 373–85
52. Bursill, L. A., Smith, D. J. 1984. *Nature* 309: 319–21
53. Millot, F., Blanchin, M. G., Tetot, R., Marucco, J. F., Poumellec, B., Picard, C., Touzelin, B. 1987. *Prog. Solid State Chem.* 17: 263–93
54. Wadsley, A. D., Andersson, S. 1970. In *Perspectives in Structural Chemistry*, ed. J. D. Dunitz, J. A. Ibers, Vol. 3. New York: Wiley
55. Gruehn, R., Mertin, M. 1980. *Angew. Chem. Int. Ed. Engl.* 19: 505–12
56. Anderson, J. S. 1973. *J. Chem. Soc. Dalton Trans.* 1973: 1107–15
57. Kittel, C. 1978. *Solid State Commun.* 25: 319–22
58a. Rao, C. N. R., Thomas, J. M. 1985. *Acc. Chem. Res.* 18: 113–18
58b. Rao, C. N. R. 1985. *Bull. Mater. Sci.* 7: 155–70
59. Vidyasagar, K., Ganapathi, L., Gopalakrishnan, J., Rao, C. N. R. 1986. *J. Chem. Soc. Chem. Commun.*, pp. 449–50
60. Battle, P. D., Gibb, T. C., Lightfoot, P. 1988. *J. Solid State Chem.* 76: 334–39
61a. Alario-Franco, M. A., Gonzalez, J. M. G., Valletregi, M., Grenier, J. C. 1983. *J. Solid State Chem.* 49: 219–27
61b. Gonzalez, J. M. G., Valletregi, M., Alario-Franco, M. A., Grenier, J. C. 1983. *Mater. Res. Bull.* 18: 285–90
62. Rodriguez, J., Fontcuberta, J., Longworth, G., Valletregi, M., Gonzalez, M. G. 1988. *J. Solid State Chem.* 73: 57–64
63. Caignaert, V., Hervieu, M., Domenges, B., Nguyen, N., Pannetier, J., Raveau, B. 1988. *J. Solid State Chem.* 73: 107–17
64. Reller, A., Thomas, J. M., Jefferson, D. A., Uppal, M. K. 1984. *Proc. R. Soc. London Ser. A* 394: 223–39
65. Takano, M., Okita, T., Nakayama, N., Bando, Y., Takeda, Y., Yamamoto, O., Goodenough, J. B. 1988. *J. Solid State Chem.* 73: 140–48
66. Rao, C. N. R., Gopalakrishnan, J., Vidyasagar, K. 1984. *Indian J. Chem. A* 23: 265–78
67. Burdett, J. K., Kulkarni, G. V. 1988. *J. Am. Chem. Soc.* 110: 5361–68

68. Ehlert, M. K., Greedan, J. E., Subramanian, M. A. 1988. *J. Solid State Chem.* 75: 188–96
69. Poeppelmeier, K. R., Leonowicz, M. E., Scanlon, J. C., Longo, J. M., Yelon, W. B. 1982. *J. Solid State Chem.* 45: 71–77
70. Vidyasagar, K., Gopalakrishnan, J., Rao, C. N. R. 1985. *Inorg. Chem.* 23: 1206–11
71. Buttrey, D. J., Ganguly, P., Honig, J. M., Rao, C. N. R., Schartman, R. R., Subbanna, G. N. 1988. *J. Solid State Chem.* 74: 233–37
72. Day, P. 1981. *Int. Rev. Phys. Chem.* 1: 149–94
73. Varma, C. M. 1985. *Comments Solid State Phys.* 11: 221–24
74. Subbarao, G. V., Ramdas, S., Mehrotra, P. N., Rao, C. N. R. 1970. *J. Solid State Chem.* 2: 377–84
75. Ganguly, P., Hegde, M. S. 1988. *Phys. Rev. B* 37: 1988–92
76a. Aragon, R., Buttrey, D. J., Shepherd, J. P., Honig, J. M. 1985. *Phys. Rev. B* 31: 430–38
76b. Honig, J. M. 1986. *Proc. Indian Acad. Sci. Chem. Sci.* 96: 391–410
77. Honig, J. M. 1985. See Ref. 84, pp. 261–86
78. Bhide, V. G., Rajoria, D. S., Rao, C. N. R., Ramarao, G., Jadhao, V. G. 1975. *Phys. Rev. B* 12: 2832–44
79. Ramdani, A., Gleitzer, C., Gavoille, G., Cheetham, A. K., Goodenough, J. B. 1985. *J. Solid State Chem.* 60: 269–82
80. Rao, C. N. R., Ganguly, P. 1985. See Ref. 84, pp. 329–58
81. Gibb, T. C., Greatrex, R., Greenwood, N. N., Puxley, D. C., Snowdon, K. G. 1974. *J. Solid State Chem.* 11: 17–25
82. Rao, C. N. R. 1987. In *Valence Fluctuation*. New York: Plenum
83. Mott, N. F. 1974. *Metal-Insulator Transitions*. London: Taylor & Francis; Mott, N. F. 1984. *Rep. Prog. Phys.* 47: 909–23
84. Edwards, P. P., Rao, C. N. R., eds. 1985. *The Metallic and the Nonmetallic States of Matter*. London: Taylor & Francis
85a. Honig, J. M., Van Zandt, L. L. 1975. *Annu. Rev. Mater. Sci.* 5: 225–50
85b. Honig, J. M., Spalek, J. 1986. *Proc. Indian Natl. Sci. Acad. A* 52: 232–64
86. Milligan, M. F., Thomas, G. A. 1985. *Annu. Rev. Phys. Phys. Chem.* 36: 139–58
87. Villeneuve, G., Hagenmuller, P. 1985. In *Localization and Metal-Insulator Transitions*, ed. H. Fritzsche, D. Adler. New York: Plenum
88. Ramakrishnan, T. V. 1985. See Ref. 84, pp. 23–64; Ramakrishnan, T. V. 1985. *Proc. Indian Natl. Sci. Acad. A* 52: 217–31
89. Hubbard, J. 1964. *Proc. R. Soc. London Ser. A* 281: 401–15
90. Economu, E. N. 1981. *Phys. Rev. B* 24: 5806–14
91. Gutzwiller, M. C. 1965. *Phys. Rev. A* 137: 1726–33
92. Brinkman, W., Rice, T. M. 1970. *Phys. Rev. B* 2: 4302–6
93. Anderson, P. W. 1958. *Phys. Rev.* 109: 1492–1504
94. Mott, N. F., Davis, E. A. 1979. *Electronic Processes in Noncrystalline Solids*. Oxford: Clarendon. 2nd ed.
95. Abrahams, E. A., Anderson, P. W., Licciardello, D. C., Ramakrishnan, T. V. 1979. *Phys. Rev. Lett.* 42: 673–76
96. Lee, P. A., Ramakrishnan, T. V. 1985. *Rev. Mod. Phys.* 57: 287–337
97. Edwards, P. P., Sienko, M. J. 1981. *J. Am. Chem. Soc.* 103: 2967–72
98. Herzfeld, K. F. 1927. *Phys. Rev.* 29: 701–5
99. Rao, C. N. R., Ganguly, P. 1986. *Solid State Commun.* 57: 5–6
100. Kuwamoto, H., Honig, J. M., Appel, J. 1980. *Phys. Rev. B* 22: 2626–32
101. Mott, N. F., Pepper, M., Pollitt, S., Wallis, R. H., Adkins, C. J. 1975. *Proc. R. Soc. London Ser. A* 345: 169–80
102. Rao, C. N. R., Bhide, V. G., Mott, N. F. 1975. *Philos. Mag.* 32: 1277–81
103. Wilson, J. A. 1985. See Ref. 84, pp. 215–60
104. Vijayaraghavan, R., Mohan Ram, R. A., Ganguly, P., Rao, C. N. R. 1988. *Mater. Res. Bull.* 23: 719–23
105. Goodenough, J. B., Hamnett, A., Teller, D. 1985. In *Localization and Metal-Insulator Transitions*, ed. H. Fritzsche, D. Adler. New York: Plenum
106. Nelson, D. L., Whittingham, M. S., George, T. F., eds. 1987. *Chemistry of High-Temperature Superconductors*, ACS Symp. Ser. 351. Washington, DC: Am. Chem. Soc.
107. Rao, C. N. R. 1988. *J. Solid State Chem.* 74: 147–62; Rao, C. N. R. 1988. *Mod. Phys. Lett. B* 2: 1217–21
108. Rao, C. N. R., ed. 1988. *Chemistry of Oxide Superconductors*. Oxford: IUPAC/Blackwell
109. Müller, J., Olsen, J. L., eds. 1988. *Proc. Interlaken M^2HTSC Conf. Physica C* 153–155: 1–1774
110a. Rao, C. N. R., ed. 1988. *Chemical and Structural Aspects of High-Temperature Superconductors, Progress in High-*

Temperature Superconductivity, Vol. 7. Singapore: World Scientific
110b. Rao, C. N. R., Raveau, B. 1989. *Acc. Chem. Res.* 22: 106–13
111. Uchida, S., Takagi, H., Kitazawa, K., Tanaka, S. 1987. *Jpn. J. Appl. Phys.* 26: L1–4
112. Chu, C. W., Hor, P. H., Meng, R. L., Gao, L., Huang, Z. J., Wang, Y. Q. 1987. *Phys. Rev. Lett.* 58: 405–9
113. Ganguly, P., Mohan Ram, R. A., Sreedhar, K., Rao, C. N. R. 1987. *Solid State Commun.* 62: 807–10
114. Wu, M. K., Ashburn, J. R., Torng, C. J., Hor, P. H., Meng, R. L. et al. 1987. *Phys. Rev. Lett.* 58: 908–11
115. Maeda, H., Tanaka, Y., Fukutomi, M., Asano, T. 1987. *Jpn. J. Appl. Phys.* 27: L209–13
116. Subramanian, M. A., Torardi, C. C., Calabrese, J. C., Gopalakrishnan, J., Morrissey, K. J., et al. 1988. *Science* 239: 1015–17
117. Rao, C. N. R., Ganapathi, L., Vijayaraghavan, R., Rao, G. R., Murthy, K., Mohan Ram, R. A. 1988. *Physica C* 156: 827–33; 1989. *J. Solid State Chem.* 79: 177–80
118. Sheng, Z. Z., Hermann, A. M. 1988. *Nature* 332: 55–57, 138–40
119. Torardi, C. C., Subramanian, M. A., Calabrese, J. C., Gopalakrishnan, J., Morrissey, K. J., et al. 1988. *Science* 240: 631–33
120. Ganguli, A. K., Nanjundaswamy, K. S., Subbanna, G. N., Rajumon, M. K., Sarma, D. D., Rao, C. N. R. 1988. *Mod. Phys. Lett. B* 2: 1169–76
121. Maignan, A., Michel, C., Hervieu, M., Martin, C., Groult, D., Raveau, B. 1988. *Mod. Phys. Lett. B* 2: 681–86; Harvieu, M., Maignan, A., Martin, C., Michel, C., Provost, J., Raveau, B. 1988. *Mod. Phys. Lett. B* 2: 1103–7
122. Ganguli, A. K., Subbana, G. N., Rao, C. N. R. 1988. *Physica C* 156: 181–82
123. Hervieu, M., Maignan, A., Martin, C., Michel, C., Provost, J., Raveau, B. 1988. *J. Solid State Chem.* 75: 212–16
124. Parkin, S. S. P., Lee, V. Y., Nazzal, A. I., Savoy, R., Bayers, R., La Placa, S. J. 1988. *Phys. Rev. Lett.* 61: 750–53
125. Cava, R. J., Batlogg, B., Krajewski, J. J., Rupp, L. W., Schneemeyer, L. F., et al. 1988. *Nature* 336: 221–23
126. Cava, R. J., Batlogg, B., Krajewski, J. J., Farrow, R., Rupp, L. W., et al. 1988. *Nature* 332: 814–16
127. Day, P., Resseinsky, M., Prassides, K., David, W. I. F., Moze, O., Soper, A. 1987. *J. Phys. C* 20: L429–33
128. Johnston, D. C., Jacobson, A. J., Newsam, J. M., Lewandowski, J. T.,
Goshorn, D. P., et al. 1987. *Symp. Ser.* 351: 136–51
129a. Beech, F., Miraglia, S., Sontoro, A., Roth, R. S. 1987. *Phys. Rev. B* 35: 8778–83
129b. Jorgensen, J. D., Beno, M. A., Hinks, D. G., Soderholm, L., Volin, K. J., et al. 1987. *Phys. Rev. B* 35: 7915–22
130. Karpinski, J., Kaldis, E., Jilek, E., Rusiecki, S., Bucher, B. 1988. *Nature* 336: 660–62
131a. Ganguli, A. K., Nanjundaswamy, K. S., Rao, C. N. R. 1988. *Physica C* 156: 788–90
131b. Subramanian, M. A., Torardi, C. C., Gopalakrishnan, J., Gai, P. L., Calabrese, J. C., Askew, T. R., Flippen, R. B., Sleight, A. W. 1988. *Science* 242: 249–51
132a. Ganguli, A. K., Nagarajan, R., Nanjundaswamy, K. S., Rao, C. N. R. 1989. *Mater. Res. Bull.* 24: 103–7
132b. Rao, C. N. R., Bhat, V., Nagarajan, R., Rao, G. R., Sankar, G. 1989. *Phys. Res. B.* In press
132c. Rao, C. N. R., Ganguli, A. K., Vijayaraghavan, R. 1989. *Phys. Rev. B.* In press
133. Umezawa, A., Crabtree, G. W., Liu, J. Z. 1988. *Physica C* 153–155: 1461–64
134. Batlogg, B., Cava, R. J., Jayaraman, A., van Dover, R. B., Kourouklis, G. A., et al. 1987. *Phys. Rev. Lett.* 58: 2333–36
135. Van Bentum, P. J. M. 1988. *Physica C* 153–155: 1718–25
136a. Bhat, S. V., Ganguly, P., Ramakrishnan, T. V., Rao, C. N. R. 1987. *J. Phys. C* 20: L559–61
136b. Portis, A. M., Blazey, K. W., Muller, K. A., Bednorz, J. G. 1988. *Europhys. Lett.* 5: 487–91
137. Dunlap, B. D., Slaski, M., Hinks, D. G., Soderholm, L., Beno, M., et al. 1987. *J. Magn. Magn. Mater.* 68: L139–45
138. Takahashi, T., Matsuyama, H., Yoshida, H. K., Okabe, Y., Hosoya, S., et al. 1988. *Nature* 334: 691–93
139. Bianconi, A., De Sourtis, M., Flank, A. M., Fontaine, A., Lagarde, P., et al. 1988. *Physica C* 153–155: 1760–65
140a. Sarma, D. D., Ganguly, P., Sreedhar, K., Rao, C. N. R. 1987. *Phys. Rev. B* 37: 2371–73
140b. Sarma, D. D., Rao, C. N. R. 1988. *Solid State Commun.* 65: 47–48
140c. Rao, C. N. R., Sarma, D. D., Rao, G. R. 1989. *Phase Transitions.* In press
141. Fuggle, J. C., Fink, J., Nücker, N. 1988. *Int. J. Mod. Phys. B* 1: 1185–1226
142. Fuggle, J. C., Weijs, P. J. W., Schorl,

R., Sawatzky, G. A., Fink, J., et al. 1988. *Phys. Rev. B* 37: 123–30
143. Chakraverty, B. K., Sarma, D. D., Rao, C. N. R. 1988. *Physica C* 156: 413–17
144. Guo, Y., Langlois, J., Goddard, W. A. III. 1988. *Science* 239: 896–99
145a. Fink, J., Nücker, N., Romberg, H., Fuggle, J. C. 1989. *Phys. Rev.* In press
145b. Himpsel, F. J., Chandrasekhar, G. V., McLean, A. B., Shafer, M. W. 1988. *Phys. Rev. B* 38: 11946–50
146. Sarma, D. D. 1988. *Phys. Rev. B* 37: 7948–52
147. Emery, V. J. 1987. *Phys. Rev. Lett.* 58: 2794–96
148. Hirsch, J. E. 1987. *Phys. Rev. Lett.* 59: 228–31
149a. Takura, Y., Torrance, J. B., Huang, T. C., Nazzal, A. I. 1988. *Phys. Rev. B* 38: 7156–59
149b. Takura, Y., Takagi, H., Uchida, S. 1989. *Nature* 337: 345–47
149c. Rajumon, M. K., Sarma, D. D., Vijayaraghavan, R., Rao, C. N. R. 1989. *Solid State Commun.* 70: 875–77
150. Anderson, P. W. 1987. *Science* 235: 1196–99
151. Zhang, F. C., Rice, T. M. 1987. *Phys. Rev. B* 37: 3759–65
152. Ramakrishnan, T. V., Rao, C. N. R. 1989. *J. Phys. Chem.* In press
153. Schluter, M., Hybertsen, M. S., Christensen, N. E. 1988. *Physica C* 153–155: 1217–22
154. Fujimori, A. 1987. *Phys. Rev. B* 35: 8814–18
155. Eskes, H., Sawatzky, G. A. 1988. *Phys. Rev. Lett.* 61: 1415–17
156. Schrieffer, J. R., Wen, X. G., Zhang, S. C. 1988. *Phys. Rev. Lett.* 60: 944–46
157. Baskaran, G., Zou, Z., Anderson, P. W. 1987. *Solid State Commun.* 63: 973–77
158. Wheatley, J. M., Hsu, T. C., Anderson, P. W. 1988. *Phys. Rev. B* 37: 627–32
159. Burdett, J. K., Kulkarni, G. V., Levin, K. 1987. *Inorg. Chem.* 26: 3650–52
160. Hagenmuller, P., ed. 1972. *Preparative Methods in Solid State Chemistry.* New York: Academic
161. Honig, J. M., Rao, C. N. R., eds. 1981. *Preparation and Characterization of Materials.* New York: Academic
162. Rao, C. N. R., Gopalakrishnan, J. 1987. *Acc. Chem. Res.* 20: 228–35
163a. Ganapathi, L., Ramanan, A., Gopalakrishnan, J., Rao, C. N. R. 1986. *J. Chem. Soc. Chem. Commun.*, pp. 62–63
163b. McCarron, E. W. III. 1986. *J. Chem. Soc. Chem. Commun.*, pp. 336–37
164. Schollhorn, R. 1980. *Angew. Chem. Int. Ed. Engl.* 19: 983–93
165. Whittingham, M. S., Jacobson, A. J., eds. 1982. *Intercalation Chemistry.* New York: Academic
166. Kumada, N., Muramutu, S., Muto, F., Kinomura, N., Kikkawa, S., Koizumi, M. 1988. *J. Solid State Chem.* 73: 33–39
167. Rosique, C., Gonzales, J., Valletregi, M., Alario-Franco, M. A. 1988. *J. Solid State Chem.* 76: 313–18
168. Rice, C. E., Jackel, J. L. 1982. *J. Solid State Chem.* 41: 57–61
169. Cava, R. J., Santoro, A., Murphy, D. W., Zahurak, S., Roth, S. 1982. *J. Solid State Chem.* 42: 251–56
170a. Izawa, H., Kikkawa, S., Koizumi, M. 1987. *J. Solid State Chem.* 69: 336–42
170b. Clearfield, A., Lehto, J. 1988. *J. Solid State Chem.* 73: 98–106
171. Tourmoux, M., Murchand, R., Brohan, L. 1986. *Prog. Solid State Chem.* 17: 33–52
172a. Banks, E., Wold, A. 1974. In *Solid State Chemistry*, ed. C. N. R. Rao. New York: Dekker
172b. McCarroll, W. H., Darling, C., Jakubicki, G. 1983. *J. Solid State Chem.* 48: 189–94
173. Horowitz, H. S., Longo, J. M., Lewandowski, J. T. 1981. *Mater. Res. Bull.* 16: 489–93
174. Sen, A., Chakraverty, D. 1986. *Proc. Indian Natl. Sci. Acad. A* 52: 159–75
175. Nagano, M., Greenblatt, M. 1988. *Solid State Commun.* 67: 595–602
176. Osiko, V. V., Borik, M. A., Lomonova, E. E. 1987. *Annu. Rev. Mater. Sci.* 17: 101–60
177. Joubert, J. C., Chenavas, J. 1975. In *Treatise in Solid State Chemistry*, ed. N. B. Hannay, Vol. 5. New York: Plenum
178. Hagenmuller, P. 1986. *Proc. Indian Natl. Sci. Acad. A* 52: 102–16
179. Rao, C. N. R., Rao, K. J., Gopalakrishnan, J. 1985. *Annual Reports C.* London: Royal Soc. Chem.
180. Cheetham, A. K. 1986. *Proc. Indian Natl. Sci., Acad. A* 52: 25–33; Cheetham, A. K. 1981. In *Non-Stoichiometric Oxides*, ed. O. T. Sorenseon. New York: Academic
181. Cox, D. E., Hastings, J. B., Thomlinson, W., Prewitt, C. 1983. *Nucl. Instrum. Methods* 208: 573–78
182. Johnson, M. W., David, W. I. F. 1985. *Rutherford-Appleton Lab. Rep.* N. 85/112
183. Altfield, J. P., Sleight, A. W., Cheetham, A. K. 1986. *Nature* 322: 620–22
184. Cheetham, A. K., David, W. I. F., Eddy, M. M., Jakeman, R. J. B., Johnson, M. W., Torardi, C. C. 1986. *Nature* 320: 46–48

185. Loong, C. K. 1985. *J. Appl. Phys.* 57: 3772–78
186a. Jefferson, D. A. 1982. *Philos. Trans. R. Soc. London Ser. A* 305: 535–44
186b. Cheetham, A. K., Skarnulis, A. J. 1981. *Anal. Chem.* 53: 1060–64
186c. Egerton, R. F. 1982. *Philos. Trans. R. Soc. London Ser. A* 305: 521–30
187. Rao, C. N. R., Sparrow, T. G., Williams, B. G., Thomas, J. M. 1984. *J. Chem. Soc. Chem. Commun.*, pp. 1238–40
188. Williams, B. G. 1987. *Prog. Solid State Chem.* 17: 87–143
189a. Rao, C. N. R., Sarma, D. D. 1982. *J. Solid State Chem.* 45: 14–28
189b. Rao, C. N. R. 1986. *Philos. Trans. R. Soc. London Ser. A* 318: 37–50
189c. Reviere, J. C. 1982. *Philos. Trans. R. Soc. London Ser A* 305: 545–55
190a. Binnig, G., Rohrer, H. 1984. *Physica B* 127: 37–45
190b. Tersoff, J., Hamann, D. R. 1985. *Phys. Rev. B* 31: 805–12
191. Hinks, D. G., Richards, D. R., Dabrowski, B., Marx, D. T., Mitchell, A. W. 1988. *Nature* 335: 419–21
192. Uemura, Y. J., Sternlieb, B. J., Cox, D. E., Brewer, J. H., Kadono, R., et al. 1988. *Nature* 335: 151–53
193. Rao, C. N. R., Ganguly, P., Hegde, M. S., Sarma, D. D. 1987. *J. Am. Chem. Soc.* 109: 6893–94
194. Shafer, M. W., Penney, T., Olson, B. L. 1987. *Phys. Rev. B* 36: 4047–50
195. Dai, Y., Manthiram, A., Campion, A., Goodenough, J. B. 1988. *Phys. Rev. B* 38: 5091–93

OPTICAL SECOND HARMONIC GENERATION AT INTERFACES[1]

Y. R. Shen

Department of Physics, University of California, and
Materials and Chemical Sciences Division,
Lawrence Berkeley Laboratory, Berkeley, California 94720

INTRODUCTION

Surface characterization is necessarily the first step toward understanding a surface or interface. The development of surface characterization tools is therefore essential for the advance of surface science. Among the existing techniques (see e.g. 1), many involve emission, absorption, or scattering of massive particles; their applications are limited to samples situated in a vacuum. The optical techniques are more flexible, but they often suffer from the lack of surface specificity and sensitivity. Recently, however, optical second harmonic generation (SHG) has been proven to be a most effective and versatile probe for surface studies.

As a second-order nonlinear optical effect, SHG is forbidden under the electric-dipole approximation in a medium with inversion symmetry. At a surface or interface, the inversion symmetry is necessarily broken. For this reason, the process is highly surface specific for interfaces between two centrosymmetric media (see e.g. 2). It also has a sensitivity that allows the detection of a submonolayer of molecules at interfaces in most cases. There are a number of obvious advantages of using SHG as a surface probe: The experimental setup is simple. Because it is optical, the technique is

---

[1] The US Government has the right to retain a nonexclusive, royalty-free license in and to any copyright covering this paper.

applicable to all interfaces accessible by light. As the output is coherent and monochromatic, it can be easily separated from the input and the background luminescence or scattering by spatial and spectral filtering. The technique has inherently high spatial, spectral, and temporal resolutions, and is capable of nondestructive, in-situ remote sensing of a surface on a picosecond or subpicosecond time scale.

Surface SHG was first considered and formulated by Bloembergen & Pershan in 1962 (3). Later, the physical origins of surface nonlinearities were discussed (4, 5) and experiments of SHG from media with inversion symmetry emphasizing the surface effects were carried out (5–7). The submonolayer sensitivity of the surface SHG in detecting adsorbates on metals was actually observed in early experiments (8, 9). Unfortunately, the work went unnoticed for many subsequent years. In 1974, surface-enhanced Raman scattering on noble metals was discovered (10). It soon caught the fancy of many researchers (see e.g. 11). In the course of development, it was recognized that Raman scattering can be considered as a two-photon process, which is nonlinear, and if one nonlinear optical process shows surface enhancement because of enhanced local fields on a metal surface, other nonlinear optical processes should experience similar surface enhancement (12). This was readily demonstrated by SHG from a roughened Ag surface. Monitoring of oxidation and reduction of Ag on a Ag electrode in an electrolytical solution indicated that SHG has the sensitivity to detect a small fraction of a molecular monolayer (13). A quantitative estimate then revealed that even without surface enhancement, SHG should be capable of detecting molecular monolayers at any surface or interface (certainly not limited to metal surfaces) (14). Thus, SHG as a possible surface probe was rediscovered. Today, we have witnessed successful applications of the technique in almost all disciplines of surface science.

A number of review articles on SHG for surface studies have been published (15, 16). In this paper, I focus more on the theoretical basis of SHG. I first discuss the wave propagation problem of how SHG from an interface layer can be treated as radiation from a nonlinear polarization sheet, with the boundary effects taken into account by the Fresnel coefficients' acting as macroscopic local-field correction factors. The bulk contribution to surface SHG is also considered. I then discuss the physics behind the surface nonlinearities. I discuss how structural and field discontinuities at an interface give rise to a nonvanishing surface nonlinear susceptibility and how difficult it is to eliminate the bulk contribution in surface SHG. I then briefly describe the experimental arrangement, and then give a number of examples of the applications of surface SHG to show what types of information one can obtain from SHG measurements.

BASIC THEORY OF SURFACE SECOND HARMONIC GENERATION

The general theory of surface SHG follows the original work of Bloembergen & Pershan (3). For a more rigorous derivation of the results, the readers are referred to Refs. (17–20). Here, I present only the essential points, with emphasis on physical understanding.

First consider the simple case of radiation of a polarization sheet imbedded in a dielectric medium with a dielectric constant ε', as shown in Figure 1. This polarization sheet is described by

$$\mathbf{P}^s = \boldsymbol{\wp}^s \delta(z) \exp(i\mathbf{k}\cdot\mathbf{r}-i\omega_s t) \qquad 1.$$

assuming $\hat{x}$–$\hat{z}$ is the plane of incidence. The radiation field generated by $\mathbf{P}^s$ can be obtained from the solution of the wave equation

$$\nabla \times (\nabla \times \mathbf{E}) + (\varepsilon/c^2)(\partial^2/\partial t^2)\mathbf{E} = (-4\pi/c^2)(\partial^2/\partial t^2)\mathbf{P}^s \qquad 2.$$

with

$$\nabla \cdot \varepsilon \mathbf{E} = -4\pi \nabla \cdot \mathbf{P}^s$$

subject to the boundary conditions [a detailed derivation can be found in (21)]

$$\Delta B_x = -4\pi i(\omega_s/c)P^s_y; \qquad \Delta B_y = +4\pi i(\omega_s/c)P^s_x$$
$$\Delta E_x = -(4\pi/\varepsilon')(\partial P^s_z/\partial x); \qquad \Delta E_y = -(4\pi/\varepsilon')(\partial P^s_z/\partial y)$$
$$\Delta B_z = 0; \qquad \Delta D_z = -4\pi[(\partial P^s_x/\partial x)+(\partial P^s_y/\partial y)]. \qquad 3.$$

where Δf refers to $f(z=0^+)-f(z=0^-)$. We find, for the radiation field in medium 1,

$$E_p(\omega_s) = \frac{i4\pi k_1}{\varepsilon_2 k_{1z}+\varepsilon_1 k_{2z}}[k_{2z}\wp^s_x+(\varepsilon_2/\varepsilon')k_x\wp^s_z]\exp(i\mathbf{k}_1\cdot\mathbf{r}-i\omega_s t)$$

$$E_s(\omega_s) = \frac{i4\pi k_1}{k_{1z}+k_{2z}}[(\varepsilon_1)^{-1}k_1 \wp^s_y]\exp(i\mathbf{k}_1\cdot\mathbf{r}-i\omega_s t) \qquad 4.$$

where $\mathbf{k}_1$ and $\mathbf{k}_2$ are wavevectors of the radiation in media 1 and 2, respectively, and the subscripts p and s denote the $\hat{p}$- and $\hat{s}$-polarized components of the field. For radiation in medium 2, the fields are obtained by interchanging 1 and 2 and replacing $+\wp^s_z$ by $-\wp^s_z$ in Eq. 4. The same results can be derived by treating the polarization sheet as a collection of radiation dipoles and summing over coherently the radiation from such dipoles, with reflections at the boundaries properly taken into account.

For comparison, we also calculate radiation from the polarization sheet of Eq. 1 in the absence of boundary surfaces, i.e. $\varepsilon' = \varepsilon_1 = \varepsilon_2$. The radiation fields in Eq. 4 then reduce to

$$E_{p0}(\omega_s) = \frac{i2\pi k_1}{\varepsilon_1 k_{1z}}(k_{1z}\not{p}_x^s + k_x\not{p}_z^s)\exp(i\mathbf{k}_1\cdot\mathbf{r}-i\omega_s t)$$

$$E_{s0}(\omega_s) = \frac{i2\pi k_1^2}{\varepsilon_1 k_{1z}}\not{p}_y^s \exp(i\mathbf{k}_1\cdot\mathbf{r}-i\omega_s t). \qquad 5.$$

From Eqs. 4 and 5, we find

$$(E_p/E_{p0})_x = L_{xx} \equiv \frac{2\varepsilon_1 k_{2z}}{\varepsilon_2 k_{1z}+\varepsilon_1 k_{2z}}$$

$$(E_p/E_{p0})_z = L_{zz} \equiv \frac{2\varepsilon_1 k_{1z}(\varepsilon_2/\varepsilon')}{\varepsilon_2 k_{1z}+\varepsilon_1 k_{2z}}$$

$$(E_s/E_{s0})_y = L_{yy} \equiv \frac{2k_{1z}}{k_{1z}+k_{2z}}. \qquad 6.$$

One may recognize that these L factors are just the transmission Fresnel coefficients relating the field components in medium 1 to the corresponding ones in the polarization sheet (the middle layer in Figure 1) for fields with wavevectors $\mathbf{k}_1$ and $\mathbf{k}_2$ in media 1 and 2, respectively. Thus, physically, $\ddot{\mathbf{L}} = (L_{xx}, L_{yy}, L_{zz})$ can be considered as a macroscopic local-field correction factor on the radiation from the polarization sheet due to the presence of the boundary surfaces. We can then express Eq. 4 in the form

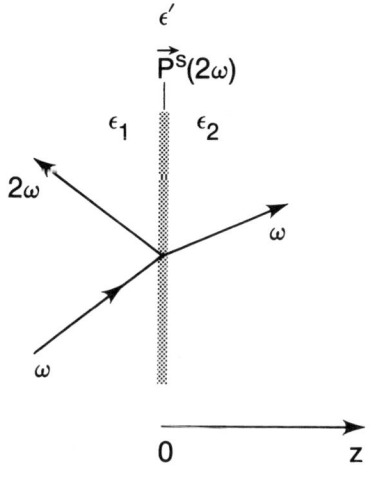

Figure 1 Geometry of second harmonic generation from an interface in the reflected direction. The polarization sheet $\mathbf{P}^s$ is imbedded in a thin layer of dielectric constant ε'.

$$\mathbf{E}(\omega_s) = \overset{\leftrightarrow}{L} \cdot \mathbf{E}_0(\omega_s) \qquad 7.$$

with the components of $\mathbf{E}_0$ given in Eq. 5.

We now consider the case where $\mathbf{P}^s(\omega_s)$ is a nonlinear polarization at $\omega_s = 2\omega$ induced by an incoming field at ω from medium 1. If the field in the layer of the polarization sheet is $\mathbf{E}_L(\omega)$, then the surface nonlinear polarization can be written as

$$\mathbf{P}^s(2\omega) = \overset{\leftrightarrow}{\chi}{}_s^{(2)} : \mathbf{E}_L(\omega)\mathbf{E}_L(\omega). \qquad 8.$$

This field $\mathbf{E}_L(\omega)$ is of course different from the incoming field $\mathbf{E}_I(\omega)$, and is related to the latter by the Fresnel coefficients given in Eq. 6. We then arrive at the final result for the SH radiation field in medium 1 generated by the surface nonlinear polarization $\mathbf{P}^s(2\omega)$

$$E_p(2\omega) = i(4\pi\omega/c)[L_{xx}(2\omega)\chi_{s,xjk}^{(2)}L_{jj}(\omega)L_{kk}(\omega) + (k_x(2\omega)/k_{1z}(2\omega))$$
$$\times L_{zz}(2\omega)\chi_{s,zjk}^{(2)}L_{jj}(\omega)L_{kk}(\omega)]E_{Ij}(\omega)E_{Ik}(\omega)$$
$$E_s(2\omega) = i(2\pi k_1^2(2\omega)/k_{1z}(2\omega)\varepsilon_1(2\omega))L_{yy}(2\omega)\chi_{s,yjk}^{(2)}L_{jj}(\omega)L_{kk}(\omega)$$
$$\times E_{Ij}(\omega)E_{Ik}(\omega). \qquad 9.$$

In the absence of the boundary surfaces, all Ls become unity, and Eq. 9 reduces to Eq. 5 with $\mathbf{P}^s = \overset{\leftrightarrow}{\chi}{}_s^{(2)} : \mathbf{E}_I(\omega)\mathbf{E}_I(\omega)$. Thus, the physical implication of the results is clear. The expressions for the radiation fields in other geometries, for example, incoming field from medium 2 or SH output radiating into medium 2, can be readily obtained by a simple modification of Eq. 9.

The SH output intensity can now be calculated from the field in Eq. 9. We find, after some manipulation,

$$I(2\omega) = c\varepsilon_1(2\omega)|\mathbf{E}(2\omega)|^2/2\pi$$
$$= \frac{32\pi^3\omega^2 \sec^2\theta_{2\omega}}{c^3\varepsilon_1^{1/2}(2\omega)\varepsilon_1(\omega)}|\mathbf{e}'(2\omega)^\dagger \cdot \overset{\leftrightarrow}{\chi}{}_s^{(2)} : \mathbf{e}'(\omega)\mathbf{e}'(\omega)|^2 I_I^2(\omega) \qquad 10.$$

where $\mathbf{e}'(\Omega) \equiv \overset{\leftrightarrow}{L}(\Omega) \cdot \hat{e}(\Omega)$, $\hat{e}(\Omega)$ is the unit vector describing the polarization of the field at Ω, and $I_I(\omega)$ is the input laser intensity. If the input laser beam is in the form of a pulse with a pulsewidth width T and a beam cross-section A, then the SH output, in terms of photons/pulse, is given by

$$S(2\omega) = \frac{32\pi^3\omega \sec^2\theta_{2\omega}}{\hbar c^3\varepsilon_1^{1/2}(2\omega)\varepsilon_1(\omega)}|\mathbf{e}'(2\omega) \cdot \overset{\leftrightarrow}{\chi}{}_s^{(2)} : \mathbf{e}'(\omega)\mathbf{e}'(\omega)|^2 I_I^2(\omega) AT. \qquad 11.$$

We have assumed so far that only a surface nonlinear polarization sheet in a boundary layer between two media contributes to the SHG. In general, however, the bulks of the media are also nonlinear. For simplicity, let us assume that only medium 2 is nonlinear, with a bulk nonlinear polarization

$$\mathbf{P}^B(2\omega) = \ddot{\chi}_B^{(2)} : \mathbf{E}_2(\omega)\mathbf{E}_2(\omega) \qquad 12.$$

induced by the field $\mathbf{E}_2(\omega)$ in medium 2. It can be shown, from the solution of the wave equation with the boundary conditions of Eq. 3, that the results in Eqs. 4 and 9–11 for SHG in the reflected direction still hold if we replace $\mathbf{P}^s$ and $\ddot{\chi}_s^{(2)}$ by $\mathbf{P}^s_{\text{eff}}$ and $\ddot{\chi}_{s,\text{eff}}^{(2)}$ with

$$P^s_{\text{eff},j} = P^s_j + iP^B_j/(k_{2z}(2\omega) + 2k_{2z}(\omega))f_j(2\omega)$$

$$(\chi^{(2)}_{s,\text{eff}})_{ijk} = (\chi^{(2)}_s)_{ijk} + (\chi^{(2)}_B)_{ijk}/(k_{2z}(2\omega) + 2k_{2z}(\omega))f_i(2\omega)$$

$$\times f_j(\omega)f_k(\omega) \qquad 13.$$

where $f_j(\Omega) = \varepsilon_2(\Omega)/\varepsilon'(\Omega)$ for $j = z$ and $f_j = 1$ for $j = x, y$.

It is clear from Eq. 13 that in order for SHG to be surface-specific, we must have $|P^s| \gtrsim |P^B/2k_2|$ or $|\chi_s^{(2)}| \gtrsim |\chi_B^{(2)}/2k_2|$. This can only be true, in general, if the medium has an inversion symmetry. In this case $\chi_B^{(2)}$ vanishes under the electric-dipole approximation. If we assume that the surface is a simple termination of the bulk structure, and that both $\chi_s^{(2)}$ and $\chi_B^{(2)}$ arise from the electric-quadrupole contribution that responds to the spatial variation of the fields, we should find $|\chi_s^{(2)}| \sim |\chi_B^{(2)}/k|$. This is because the field varies in the bulk on the scale of $1/k$ whereas it varies at the surface on the scale of surface layer thickness. More generally, $\chi_s^{(2)}$ may also have an electric-dipole contribution if the surface has the structure of a polar layer. In those cases, as discussed below, the surface contribution can dominate over that of the bulk in SHG.

To end this section, I give a crude estimate of the strength of surface SHG. The typical nonresonant value of $\chi_s^{(2)}$ for a polar monolayer of molecules adsorbed at an interface is 10^{-14}–10^{-15} esu, as one would find from a perturbation calculation. If we take $|\chi^{(2)}| = 10^{-15}$ esu, $I(\omega) \sim 10$ MW/cm^2, $A \sim 0.2$ cm^2, $T \sim 10$ nsec, and $\theta_{2\omega} \sim 45°$ in Eq. 11, the signal strength S is of the order of 10^4 photons/pulse. Such a large signal should be readily detectable, therefore pointing to the submonolayer sensitivity of surface SHG.

SURFACE NONLINEAR SUSCEPTIBILITIES

From surface SHG measurements, we can only expect to find $\ddot{\chi}_{s,\text{eff}}^{(2)}$. For surface studies, however, we are interested in $\ddot{\chi}_s^{(2)}$, which is directly related to the surface properties. Generally, it is impossible to separate the contribution of $\ddot{\chi}_{s,\text{eff}}^{(2)}$ from that of $\ddot{\chi}_B^{(2)}$ in the measurements (19), but in many cases, the contribution of $\ddot{\chi}_B^{(2)}$ can be neglected or subtracted.

For a better understanding of $\ddot{\chi}_s^{(2)}$, let us first consider the physical mechanisms responsible for $\ddot{\chi}_s^{(2)}$. In Figure 1, we treat the surface or interface as an infinitesimally thin layer with a dielectric constant ε' and a surface nonlinear polarization $\mathbf{P}^s = \ddot{\chi}_s^{(2)} : \mathbf{E}_L(\omega)\mathbf{E}_L(\omega)$. This is of course only an approximation. In reality, the interface layer has a finite thickness and a continuous variation of properties from one medium to the other. We must define (17–19)

$$P_i^s = \int_I P_i^B \, dz \qquad \text{for } i = x, y$$

$$P_z^s / \varepsilon' = \int_I [P_z^B / \varepsilon(z)] \, dz \qquad \qquad 14.$$

where I denotes the interfacial layer. Note that for the z-component, it is P_z^s/ε' instead of P_z^s that appears as the effective source term for the radiation field $\mathbf{E}_p$, as seen in Eq. 4.

In general, the bulk nonlinear polarization $\mathbf{P}^B$ can be expanded into a series of multipole terms (see e.g. 22)

$$\mathbf{P}^B = \mathbf{P}^{(2)}(2\omega) - \nabla \cdot \ddot{Q}^{(2)}(2\omega) - \frac{c}{i2\omega} \nabla \times \mathbf{M}^{(2)}(2\omega) + \ldots, \qquad 15.$$

where $\mathbf{P}$, $\ddot{Q}$, and $\mathbf{M}$ denote electric-dipole polarization, electric-quadrupole polarization, and magnetization, respectively. All terms are quadratic functions of the field and/or their derivatives. For simplicity, we neglect $\mathbf{M}$ in the following discussion. Substitution of Eq. 15 into Eq. 14 then yields (17–19)

$$P_i^s / \eta_i = \int_I s_i(z) [\mathbf{P}^{(2)}(z) - \nabla \cdot \ddot{Q}^{(2)}(z)]_i \, dz$$

$$= \int_I \left[s_i(z) P_i(z) + Q_{iz}(z) \frac{\partial}{\partial z} s_i(z) \right] dz - [s_i(z) Q_{iz}(z)]_0^{0+} \qquad 16.$$

where, $\eta_i = s_i = 1$ for $i = x, y$, and $\eta_z = \varepsilon'$ and $s_z(z) = E_z(z)/D_z$ for $i = z$, with D_z being the z-component of the displacement current. We have used 0^- and 0^+ to denote the boundaries of the interface layer with media 1 and 2, respectively. We can now define the surface nonlinear susceptibility $\ddot{\chi}_s^{(2)}$ by the relation

$$P_i^s(2\omega) = \chi_{s,ijk}^{(2)} F_j(\omega) F_k(\omega) \qquad 17.$$

with

$$F_j(\omega) = E_j(\omega) \quad \text{for } j = x, y$$
$$= D_j(\omega)/\varepsilon'(\omega) \quad \text{for } j = z$$

because E_x, E_y, and D_z are continuous across the interface layer. From Eq. 16, we can then show

$$\chi^{(2)}_{s,ijk}(2\omega) = \eta_i(2\omega) \int_I \left\{ \chi^D_{ijk} s_i(2\omega) s_j(\omega) s_k(\omega) \right.$$

$$\left. + \chi^Q_{izjk} s_j(\omega) s_k(\omega) \frac{\partial}{\partial z} s_i(2\omega) \right\} dz + \eta_i(2\omega) \left\{ [\chi^Q_{izjk} s_i(2\omega) s_j(2\omega) s_k(\omega)]_{z=0^-} \right.$$

$$\left. - [\chi^Q_{izjk} s_i(2\omega) s_j(\omega) s_k(\omega)]_{z=0^+} \right\}. \quad 18.$$

The nonlinear susceptibilities χ^D and χ^Q refer to quantities originated from the electric-dipole and electric quadrupole contributions.

We can identify the physical origins of the various terms in Eqs. 16 and 18 (17–20). In both equations, the first term in the integral is the electric-dipole term, which is nonvanishing even if the bulk has an inversion symmetry. This is because the inversion symmetry is necessarily broken at the interface. This term can be dominating if the structure of the interface layer is highly noncentrosymmetric. The second term in the integral comes from the field discontinuity, i.e. the rapid variation of the normal component of the field E_z across the interface. This is often known in the literature as the electric-quadrupole (nonlocal) contribution to the surface nonlinearity. If the dielectric constants of the two bounding media are matched, the field becomes continuous across the interface and this particular term should vanish. As mentioned above, this term gives the same order-of-magnitude contribution to the surface SHG as the bulk with inversion symmetry. The third term, outside the integral, comes from the structural disparity between the two bounding media. This is a term that has been forgotten in the earlier papers on surface SHG. It is uniquely determined by the bulk parameters, and should in fact be called a bulk contribution. The existence of this term indicates that one can never have a $\chi^{(2)}_{s,ijk}$ with only the surface contribution if z ever appears in the subindices. Thus, generally speaking, the real surface nonlinear susceptibility consists of two parts: one is the local, electric-dipole term arising from the structural discontinuity at the interface and the other is the nonlocal electric-quadrupole term arising from the field discontinuity at the interface.

In many cases, the interface has a rather well-defined structural symmetry. One can then expect $\ddot{\chi}^{(2)}_s$ to have a particular symmetric form.

Table 1 lists the independent, nonvanishing elements of $\ddot{\chi}_s^{(2)}$ for various two-dimensional symmetry classes. Different components of $\ddot{\chi}_s^{(2)}$ can be measured by using different polarization combinations, and/or by varying the orientation of the interface with respect to the incoming beam. The observed symmetry of $\ddot{\chi}_s^{(2)}$ then allows us to infer the structural symmetry of the interface layer.

From the surface science point of view, it is most important that the macroscopic $\ddot{\chi}_s^{(2)}$ can be related to the surface properties. Unfortunately, the detailed microscopic understanding of $\ddot{\chi}_s^{(2)}$ is still lacking. We can offer here only the following general picture. For a molecular monolayer adsorbed on a surface, the surface nonlinear susceptibility can be written as

$$\ddot{\chi}_s^{(2)} = \ddot{\chi}_{ss}^{(2)} + \ddot{\chi}_m^{(2)} + \ddot{\chi}_i^{(2)}. \qquad 19.$$

Here, $\ddot{\chi}_{ss}^{(2)}$ is the contribution from the bare substrate surface, $\ddot{\chi}_m^{(2)}$ is the contribution from the molecular layer isolated from the substrate, and $\ddot{\chi}_i^{(2)}$ arises from the interaction between the molecule and the substrate. Whether a particular term dominates in Eq. 19 or not depends on the

Table 1 Independent nonvanishing elements of $\ddot{\chi}_s^{(2)}(2\omega)$ for surfaces of various symmetry classes (surface is in the $\hat{x}$–$\hat{y}$ plane)

Symmetry classes	Location of mirror plane	Independent nonvanishing elements
C_1	No mirror	xxx, xxy = xyx, xyy, yxx, yxy = yyx, yyy, xxz = xzx, xyz = xzy, yxz = yzx, yyz = yzy, zxx, zxy = zyx, zyy, xzz, yzz, zxz = zzx, zyz = zzy, zzz
C_{1v}	$\hat{y}$–$\hat{z}$	xxy = xyz, yxx, yyy, xxz = xzx, yyz = yzy, zxx, zyy, yzz, zyz = zzy, zzz
C_2	No mirror	xxz = xzx, xyz = xzy, yxz = yzx, yyz = yzy, zxx, zyy, zxy = zyx, zzz
C_{2v}	$\hat{x}$–$\hat{z}$, $\hat{y}$–$\hat{z}$	xxz = xzx, yyz = yzy, zxx, zyy, zzz
C_3	No mirror	xxx = −xyy = −yxy = −yyx, yyy = −yxx = −xxy = −xyx, xxz = xzx = yyz = yzy, zxx = zyy, xyz = xzy = −yxz = −yzx, zzz
C_{3v}	$\hat{y}$–$\hat{z}$	yyy = −yxx = −xxy = −xyz, xxz = xzx = yyz = yzy, xxz = xzx = yyz = yzy, zxx = zyy, zzz
C_4, C_6	No mirror	xxz = xzx = yyz = yzy, zxx = zyy, xyz = xzy = −yxz = −yzx, zzz
C_{4v}, C_{6v}, or isotropic	$\hat{x}$–$\hat{z}$, $\hat{y}$–$\hat{z}$	xxz = xzx = yzy = yyz, zxx = zyy, zzz

material system. In some cases, asymmetric molecules with large second-order nonlinear polarizabilities can yield a dominating $\vec{\chi}_m^{(2)}$ if they are aligned on the surface. In other cases, $\vec{\chi}_m^{(2)}$ may be negligible and the only effect of the adsorbates on $\vec{\chi}_s^{(2)}$ is a modification of $\vec{\chi}_{ss}^{(2)}$ by $\vec{\chi}_i^{(2)}$. This happens, for example, with metal and semiconductor surfaces. Although it is possible to make a reasonable guess at the physical origins of the various terms in Eq. 19 for a given surface system, a more quantitative picture would require a thorough understanding of the microscopic properties of the surface system. It is possible, for example, to use the bond theory to calculate $\vec{\chi}_s^{(2)}$ for semiconductor and insulator surfaces, but one needs prior knowledge of the surface structure.

I now discuss what types of information one can deduce from measurements of $\vec{\chi}_s^{(2)}$. I mentioned above the possibility of learning about the structural symmetry of an interface from the symmetry of $\vec{\chi}_s^{(2)}$. Microscopically, $\vec{\chi}_s^{(2)}$ can be written as

$$\chi_{s,ijk}^{(2)} = \int \left\{ \sum_{g,n,n'} \langle g|r_i|n\rangle \langle n|r_j|n'\rangle \langle n'|r_k|g\rangle \rho_{gg}^0 A_{nn'} \right\} f(\Omega)\, d\Omega$$

with

$$A_{nn'} = (e^3/\hbar^2)\{[(\omega-\omega_{n'g})(2\omega-\omega_{ng})]^{-1} + [(\omega-\omega_{n'g})(\omega+\omega_{ng})]^{-1}$$
$$+ [(\omega+\omega_{n'g})(2\omega+\omega_{ng})]^{-1}\} \quad 20.$$

where ρ_{gg}^0 is the population in the $\langle g|$ to state, ω_{ng} refers to the transition frequency from $\langle g|$ to $\langle n|$, and $f(\Omega)$ is a distribution function of the parameter, or a set of parameters, Ω, with $\int f(\Omega)\,d\Omega = N_s$ = number of molecules per unit area. We have assumed here, for simplicity, that the microscopic local field correction is negligible, the damping constants in the frequency dominators are unimportant, and the interface layer is composed of a set of localized molecules. Extension to the more general case is fairly straightforward, but is not essential for our discussion.

The above equation shows that we should be able to deduce the transition frequencies of the interfacial system from the resonant dispersion of $\vec{\chi}_{s,ijk}^{(2)}$. The distribution function $f(\Omega)$ can be a function of the surface molecular density and molecular orientation. If the matrix elements in Eq. 20 depend only on the molecular orientation but not on the surface density, then measurements of $\vec{\chi}_{s,ijk}^{(2)}$ can yield information about the average surface density and the average molecular orientation separately. The distribution function can also vary with position on the surface and with time. The spatial dependence of $\vec{\chi}_s^{(2)}$ gives the average molecular distribution on the surface, and the time dependence of $\vec{\chi}_s^{(2)}$ allows in-situ monitoring of surface dynamics or reactions. I illustrate these various applications of

SHG measurements of $\bar{\chi}_s^{(2)}$ by a number of practical examples in the Applications section below.

EXPERIMENTAL CONSIDERATIONS

Before I discuss the actual SHG experiments, I first briefly describe a typical experimental setup and certain precautions. One of the advantages of SHG as a surface tool is the simplicity of the experimental arrangement. As shown in Figure 2, the basic elements in the setup are nothing but a laser and a photodetection system with proper spectral filtering. Mode-locked, Q-switched, and CW lasers can all be used. Which one is more advantageous depends on the system to be investigated. The SH signal from a smooth surface is coherent and highly directional; therefore, collection of the signal is a trivial matter.

Equation 11 shows that with a pulsed laser, the SH signal should increase with the increase of $I(\omega)$, A, and T; however, if the pump pulse energy $W(\omega) = I(\omega)AT$ is fixed, the signal should increase with the decrease of A and T. In practice, the laser fluence (energy/unit area) impinging on a surface is often limited by laser-induced breakdown. If we assume that the breakdown is due to laser-induced melting, and its threshold for a short-pulse excitation at 1.06 μm is $I^{th}T \sim 1 \text{ J/cm}^2$, then the maximum SH signal is (23)

$$S_{max} \sim 10^{27} f |\chi_s^{(2)}|^2 A/T \text{ photons/sec,}$$

where f is the pulse repetition rate. Let us consider a Q-switched pump laser with $W \sim 100$ mJ, $T \sim 10$ nsec, and $f \sim 10$/sec. Assuming that the

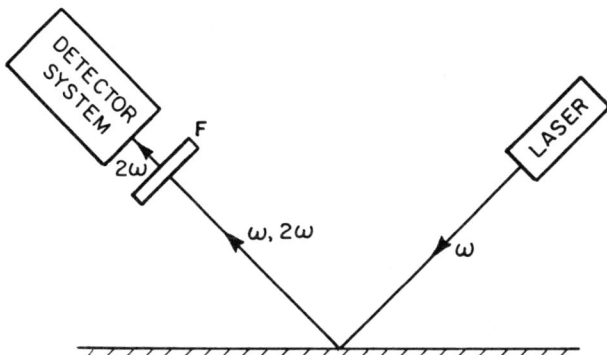

Figure 2 Experimental arrangement of second-harmonic generation from a surface. F denotes the filtering system that rejects the fundamental but passes the second harmonic.

minimum detectable signal is 10 photons/sec, we should have a minimum detectable $|\chi_s^{(2)}|$ of $\sim 3 \times 10^{-18}$ esu. Let $\chi_s^{(2)} \sim N_s \alpha^{(2)}$, where N_s is the number of surface atoms or molecules per cm², and $\alpha^{(2)}$ is the nonlinear polarizability. Then, for typical values of $\alpha^{(2)}$ in the range between 10^{-31} esu (for weakly nonlinear molecules) and 10^{-29} esu (for more strongly nonlinear molecules, metals, and semiconductors), the minimum surface density of atoms or molecules detectable is between $3 \times 10^{13}/\text{cm}^2$ and $3 \times 10^{11}/\text{cm}^2$. Since N_s for a monolayer is typically larger than $10^{14}/\text{cm}^2$, the submonolayer sensitivity of surface SHG is obvious. The detectability can be further improved by using a mode-locked laser. If we take $W \sim 10\,\mu\text{J}$, $T \sim 1$ psec, and $f \sim 10^3/\text{sec}$, we find a minimum detectable $|\chi_s^{(2)}|$ of $\sim 3 \times 10^{-19}$ esu.

An estimate for the CW case is also interesting. Because of thermal diffusion, the threshold power P_{th} for laser-induced melting is roughly proportional to $A^{1/2}$ for sufficiently small A. If we assume $P_{th}/A^{1/2}$ to be $\sim 10^4$ W/cm (as for Si), the maximum SH signal is

$$S_{max} \sim 10^{36} |\chi_s^{(2)}|^2 \text{ photons/sec,}$$

which is independent of A. With $A \sim 10^{-8}$ cm² from tight focusing, P_{th} is only about 1 W, a power readily obtainable from a CW laser. The minimum detectable $|\chi_s^{(2)}|$ is around 10^{-18} esu.

In the above estimates, we have not taken into account the macroscopic local-field factors or Fresnel coefficients, $L(\Omega)$, in Eq. 11. They can have significant effects on the minimum detectability of $\chi_{s,ijk}^{(2)}$. Consider for example, an interface layer of refractive index $\varepsilon' = 2$ sandwiched between two media $\varepsilon_1 = 1$ and $\varepsilon_2 = 10$, respectively. For this case, the local-field factors L_{ii} as functions of the incident angle are plotted in Figure 3. If the pump beam is incident from the low refractive index side, with an incident angle $\theta_1 \sim 45°$, then from Eq. 11 and Figure 3a, the minimum detectability of all $\chi_{s,ijk}^{(2)}$ is greatly reduced (by a factor of 3 to 10). If the pump beam is incident from the high refractive index side, on the other hand, the detectability of $\chi_{s,ijk}^{(2)}$ can be enhanced when the incident angle is around the critical angle.

APPLICATIONS

I now describe a few selective examples to illustrate the types of information one can obtain about an interface from surface SHG experiments. We should first remind ourselves that SHG measurements can only yield $(\chi_{s,\text{eff}}^{(2)})_{ijk}$, which in most cases, contains unavoidably a contribution from the bulk (17–19). In order to obtain $\chi_{s,ijk}^{(2)}$ from $(\chi_{s,\text{eff}}^{(2)})_{ijk}$, one must be able to modify the interface such that the bulk contribution can be identified and subtracted from $\chi_{s,\text{eff}}^{(2)}$. This can be accomplished by controlling atomic

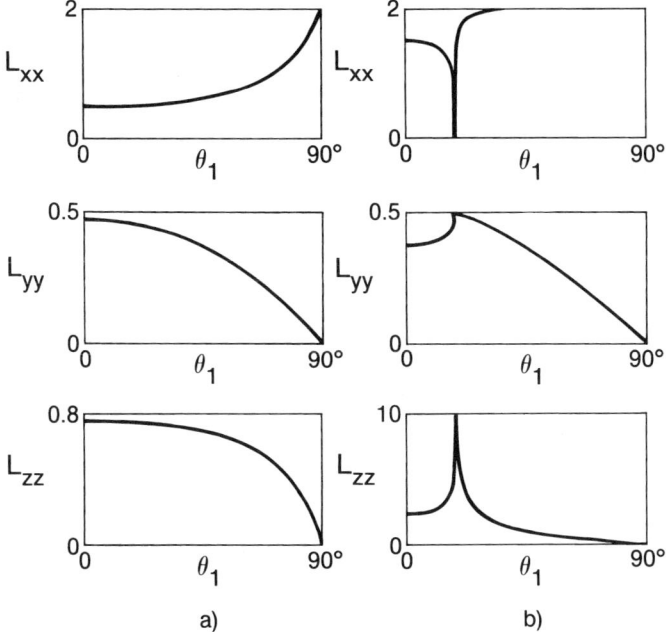

Figure 3 Variation of macroscopic local-field factors L_{ii} ($i = x, y, z$) as functions of the incident angle for a representative case of $\varepsilon_1 = 1$, $\varepsilon' = 2$, and $\varepsilon_2 = 10$. (*a*) Incoming beam from medium 1; (*b*) incoming beam from medium 2.

or molecular adsorption at the interface, as we shall discuss in the various examples below.

Study of Molecular Adsorption and Desorption at an Interface

The surface susceptibility $\ddot{\chi}_s^{(2)}$ is expected to vary with the density of adsorbed molecules at an interface. The relation should be linear if the interaction between adsorbed molecules can be neglected. We can then write

$$\chi_{s,\text{eff}}^{(2)} = (A+C)+(B-A)\theta \qquad 21.$$

where θ is the surface coverage in terms of a fraction of a full monolayer, and A, B, and C refer to constants associated with the bare surface, the covered surface, and the bulk contribution, respectively. For metals and semiconductors, A and C are often appreciable, but for insulators, A and C are usually quite small. The B constant depends on the adsorbed molecules and comes in through $\ddot{\chi}_m^{(2)}$ and $\ddot{\chi}_i^{(2)}$ in $\ddot{\chi}_s^{(2)}$ of Eq. 19. During an adsorption or desorption process, θ varies with time. The in-situ monitoring of $\chi_{s,\text{eff}}^{(2)}(t)$ allows the deduction of $\theta(t)$ via Eq. 21.

CO on Ni(111) in ultrahigh vacuum provides an example (24). The SH probing of θ can be calibrated by thermal desorption spectroscopy in a simultaneous measurement of surface SHG and thermal desorption. The surface SHG can then be used to monitor in-situ the adsorption of CO on Ni(111). The experiment was carried out with a 0.53-μm laser beam on Ni at 300K. The results are shown in Figure 4. In this case, CO adsorbs on Ni(111) only at the two-fold bridge site, and the adsorption is expected to follow the Langmuir kinetic model that predicts $\theta(t) = \theta_s(1 - e^{-\gamma t})$, where θ_s and γ are constants. Indeed, by inserting θ in Eq. 21 and using $(A+C)$, $(B-A)$, and γ as adjustable parameters to calculate the SH signal, the theoretical fit to the experimental data is remarkably good, as seen in Figure 4.

In other applications, SHG has been used to study molecular adsorption on semiconductors (25) and insulators (26), and at liquid/solid interfaces

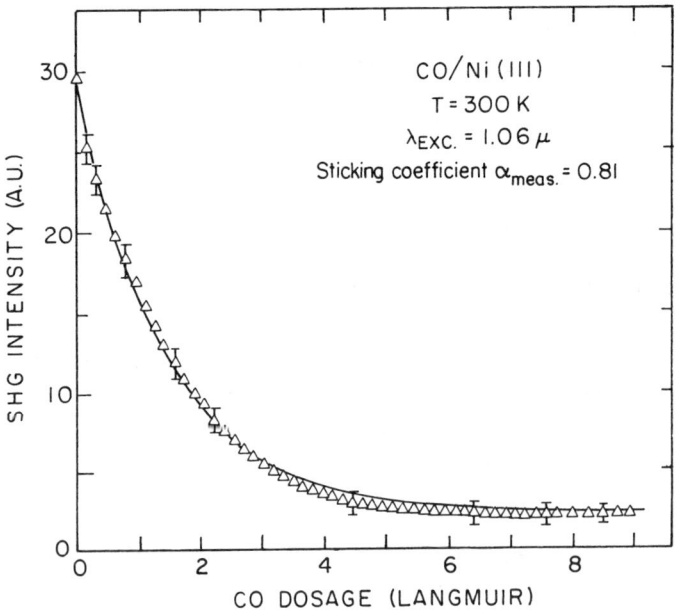

Figure 4 Second harmonic signal as a function of surface coverage of CO on Ni(111) at 300K. The solid theoretical curve derived from the simple Langmuir kinetic model is used to fit the experimental data (Δ) (after Ref. 24).

(27). It has also been employed to probe laser desorption of molecules from an interface (28, 29).

Probing Structure Symmetry of an Interface

The symmetry of $\tilde{\chi}_s^{(2)}$ provides information on the structural symmetry of an interface layer. Here, we use the Si(111) surface as an example (30). This surface has a C_3 symmetry, i.e. a three-fold symmetry without an inversion center in the plane. With a 1.06-μm laser excitation, the SHG from Si(111) is dominated by the surface nonlinearity, as concluded from the observation that a monolayer of adsorbates almost completely suppresses the SHG. If both the fundamental and the SH beams are normal to the surface, then the surface nonlinear polarization can be written as, with $\chi^{(2)}_{\xi\xi\xi} = -\chi^{(2)}_{\xi\eta\eta} = -\chi^{(2)}_{\eta\xi\eta}$,

$$P_x^s = \chi^{(2)}_{\xi\xi\xi}[-(E_x^2 - E_y^2)\cos 3\psi + 2E_x E_y \sin 3\psi]$$

$$P_y^s = \chi^{(2)}_{\xi\xi\xi}[(E_x^2 - E_y^2)\sin 3\psi + 2E_x E_y \cos 3\psi] \qquad 22.$$

where $\hat{\xi}$ is along [2$\bar{1}\bar{1}$], which is the projection of the [100] crystal axis on the (111) surface, $\hat{\eta}$ is along [01$\bar{1}$], which is orthogonal to $\hat{\xi}$ in the (111) plane, and ψ is the angle between $\hat{\xi}$ and the lab axis $\hat{x}$. The fields here are defined to be in Si. The above equation shows that if the input and output polarizations are fixed, then upon rotation of the sample about its surface normal (varying ψ), the SH signal ($\propto |P|^2$) should exhibit a six-fold symmetry. The [2$\bar{1}\bar{1}$] direction can be identified from the result. If, on the other hand, the sample is fixed (ψ unchanged) and the $\hat{x}$- and $\hat{y}$-polarized components of SHG are measured separately upon rotation of the linear polarization of the pump beam, then the SH signal should exhibit a four-fold symmetry. The latter case is illustrated in the lower two panels of Figure 5b. Again, from the patterns, the [2$\bar{1}\bar{1}$] and [01$\bar{1}$] directions can be identified.

Figure 5 is actually an example to show how SHG can be used to study surface reconstruction (30). A freshly cleaved Si(111) surface in vacuum has a (2×1) structure with a C_2 symmetry. The SHG should exhibit a two-fold symmetry. This was indeed observed experimentally, as depicted in Figure 5a along with the theoretical fit. Upon annealing, the (2×1) structure transforms into the (7×7) structure with the C_3 symmetry. The result is shown in Figure 5b. Since SHG has a nearly instantaneous response, it can be used to follow the surface structural transformation. For the $(2 \times 1) \rightarrow (7 \times 7)$ transformation of Si(111), the process has a time constant of a few tens of seconds at 275°C, which is probably dominated by surface defects.

Similar measurements have been used to probe amorphous → crystalline

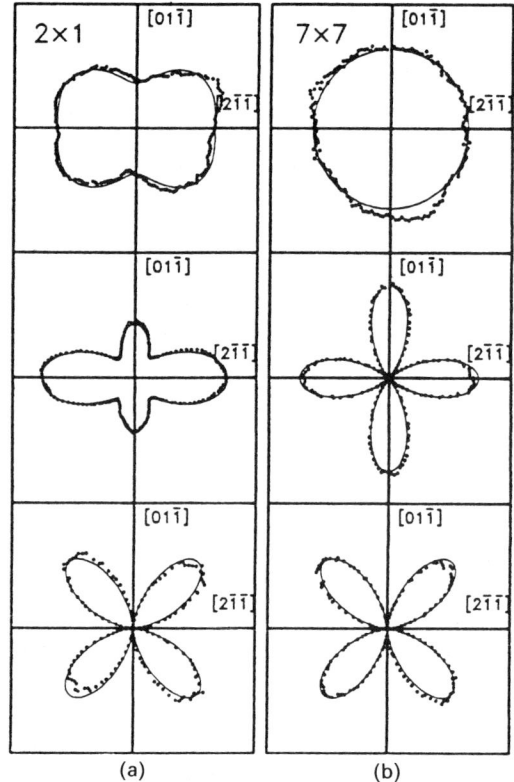

Figure 5 Second-harmonic intensity from (*a*) Si(111) (2 × 1) and (*b*) Si(111) (7 × 7) surfaces as a function of the polarization of the normally incident pump beam. The *top panels* display the total SH signal; the *middle and lower panels* show the SH signal polarized along [2$\bar{1}\bar{1}$] and [01$\bar{1}$], respectively. (Experimental = *dashed line*; theory = *solid line*.) (After Ref. 30.)

transformation of an epitaxial monolayer induced by annealing (31), surface melting (32, 33), surface disordering (31), surface reconstruction of an electrode by electrolytical process (34), and structural symmetry of a metal overlayer deposited electrochemically on an electrode (35).

Measurements of Molecular Orientations at an Interface

It is possible to use SHG to measure the average orientation of a molecular adsorbate at an interface if $\vec{\chi}^{(2)}_{s,\text{eff}}$ is dominated by the molecular susceptibility $\vec{\chi}^{(2)}_m$, so that $\vec{\chi}^{(2)}_{s,\text{eff}} \simeq \vec{\chi}^{(2)}_s \simeq \vec{\chi}^{(2)}_m$. This is the case with many asymmetric molecules possessing conjugate bonds. In more general cases, the measurements are also possible if the substrate contribution to $\bar{\chi}^{(2)}_{s,\text{eff}}$ can be subtracted off.

The principle underlying the molecular orientation measurements is easy to understand if we realize that $\chi^{(2)}_{s,ijk}$ in the lab coordinates (x, y, z) should be related to the nonlinear polarizability $\alpha^{(2)}_{\lambda\mu\nu}$ of the molecules in the molecular coordinates (ξ, η, ζ) by a simple coordinate transformation $T^{\lambda\mu\nu}_{ijk}$ that describes the molecular orientation (14, 27).

$$\chi^{(2)}_{s,ijk} = N_s \langle T^{\lambda\mu\nu}_{ijk} \rangle \alpha^{(2)}_{\lambda\mu\nu}. \qquad 23.$$

Here, the angular brackets denote an average over the orientational distribution. Measurements of $\langle T^{\lambda\mu\nu}_{ijk} \rangle$ should yield the average orientation of the molecules. Unfortunately, this is not easy unless $\alpha^{(2)}_{\lambda\mu\nu}$ is dominated by a single element, for example, $\alpha^{(2)}_{\xi\xi\xi}$, in the case of a long molecule with $\hat{\xi}$ along the molecular axis.

A study of adsorbed p-nitrobenzoic acid (PNBA) by surface SHG illustrates these points (27). PNBA forms an isotropic monolayer on glass. This can be checked by the invariance of SHG under rotation of the sample about its surface normal. The nonlinearity of PNBA is dominated by $\alpha^{(2)}_{\xi\xi\xi}$. Because the adsorbed layer is isotropic in the plane, the orientation of PNBA must have an azimuthal symmetry. It can therefore be specified by a single angle θ between $\hat{\xi}$ and the surface normal. As seen from Eq. 23, the ratio of any two independent components of $\bar{\chi}^{(2)}$ must then give a value for a weighted average of θ. For example, $\chi^{(2)}_{s,zzz}/\chi^{(2)}_{s,zxx} = 2\langle\cos^3\theta\rangle/\langle\sin^2\theta\cos\theta\rangle$. The average orientation of PNBA adsorbed on fused silica at fractional monolayer coverages has been determined in this fashion both at solid/air and at solid/ethanol interfaces. Taking the orientational distribution to be sharply peaked, θ was found to be $\sim 40°$ in ethanol and $\sim 70°$ in air. This difference might be explained by the solvation energy of PNBA in the liquid.

Orientational measurements have also been carried out with molecules at air/liquid (36, 37) and liquid/liquid interfaces (38). When molecules adsorb chemically on a substrate, the molecules form a polar monolayer. The heads-up and heads-down configurations are distinguished by a 180° phase difference in $\chi^{(2)}_s$. The phase factor of $\chi^{(2)}_s$ can be measured by a standard interference method by using a crystalline quartz plate as a reference (39, 40).

Through the orientational measurements, SHG can be used to probe the orientational phase transition of a molecular monolayer (36). A surfactant monolayer on water undergoes various two-dimensional phase transitions as the surface density is varied. For some systems, there exists a liquid-expanded → liquid-condensed transition. It was suggested that this transition might be associated with a sudden change in the molecular orientation. SHG measurements were able to confirm that this is indeed the case (36).

Surface Spectroscopy

When either ω or 2ω approaches a resonance of an interface layer, the corresponding $\chi_s^{(2)}$ is expected to exhibit a resonant enhancement. Surface SHG with a tunable laser can therefore be used as a surface spectroscopic tool to probe the resonant transitions of an interface layer. I use here dye molecules adsorbed at an air/fused quartz interface as an example (14).

Figure 6 shows the SH signal versus the SH frequency at 2ω for two samples on fused quartz, one with a half monolayer of Rhodamine 6G and the other with a half monolayer of Rhodamine 110. The observed resonant structure in SHG reflects the $S_0 \to S_2$ electronic transitions in the molecules. Because of the small difference in their molecular structures, the resonant peaks of the two dye molecules are shifted slightly from each other. The signal rise toward the low-frequency side indicates the appearance of another resonant peak as ω approaches the $S_0 \sim S_1$ transition. The SH signal from the dye molecules was exceptionally strong. With a pump laser energy of 1 mJ in a pulsewidth of 8 nsec focused to a spot of 10^{-3} cm^2, the observed signal from a half monolayer of Rhodamine 6G ($N \sim 5 \times 10^{-13}$/cm^2) was $\sim 10^4$ photons/pulse on resonance, corresponding to a nonlinear susceptibility $\chi^{(2)} \sim 7 \times 10^{-29}$ esu/molecule. The surface spectroscopic method is, of course, not limited to adsorbates on insulator surfaces. It can be used, for example, to study the charge-transfer band created by adsorption of molecules on metals and the surface or interface states of semiconductors or metals. The latter has recently been demonstrated (41).

Unfortunately, with tunable dye lasers, only the electronic transitions of adsorbates can be probed. In surface science, vibrational spectroscopy of adsorbates is of tantamount importance, since the spectra allow one to deduce the molecule-substrate interaction. One would think that this could be achieved by surface SHG using a tunable infrared laser. However, infrared detectors are generally much less sensitive than photomultipliers in the visible, so that infrared SHG is not likely to have the necessary sensitivity to detect submonolayer adsorbates. It is, however, possible to use yet another second-order process, sum-frequency generation, instead of SHG for vibrational spectroscopy. In this case, a tunable infrared laser is used to probe the vibrational transitions of the adsorbates, and the resonant spectrum can be up-converted to the visible by sum-frequency generation using a visible laser. Recent experiments show that this is indeed a viable method (42–47).

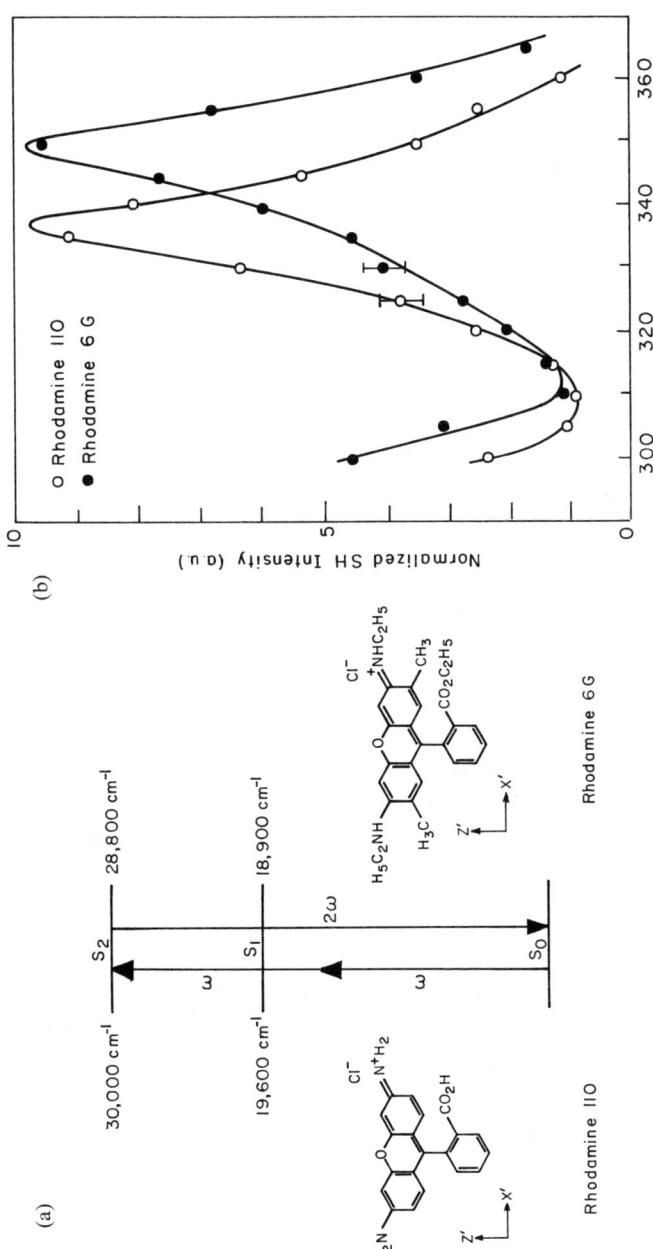

Figure 6 Resonant second-harmonic generation in Rhodamine 110 and Rhodamine 6G: (*a*) the resonant process in the two dyes with energy levels corresponding to the absorption line center for the molecules dissolved in ethanol; (*b*) the experimental SH spectrum in the region of the $S_0 \rightarrow S_2$ transition for submonolayers of the dye molecules adsorbed on fused silica. (After Ref. 14.)

Surface Monolayer Microscopy

Surface SHG measures a $\bar{\bar{\chi}}_s^{(2)}$, which spatially averaged over a surface area defined by the probing laser spot on the surface. With a laser beam properly focused, it can have a spatial resolution of the order of a wavelength. The surface specificity and monolayer sensitivity of the technique make it ideally suited for surface optical microscopy. It can actually be used to probe the morphology of a surface monolayer. Shown in Figure 7 is an example that displays an SH image of a laser-ablated hole in a Rhodamine 6G dye monolayer on fused quartz (23). As illustrated, the diameter of the hole is ~ 8 μm, which was limited by the focusing lens used in the experiment.

Studies of Surface Dynamics and Reactions

Surface SHG arises from the electronic response of the interfacial system to the incoming field. It has a response time limited only by the electronic relaxation in the system. Therefore, the technique has the capability to monitor surface dynamics and reactions with a picosecond or subpicosecond time resolution.

The use of SHG to monitor slow processes on surfaces has been demonstrated in many cases. Molecular adsorption and desorption and surface phase transformation induced by annealing, mentioned above, are just a few examples. On surface reactions, SHG has been used to probe the UV-initiated polymerization process of monolayers of monomers floating on water (48).

The possibility of using SHG to study ultrafast surface phenomena is of course most exciting. Figure 8 shows how SHG can be employed to probe the dynamics of surface melting of Si (33). In this experiment, 75-

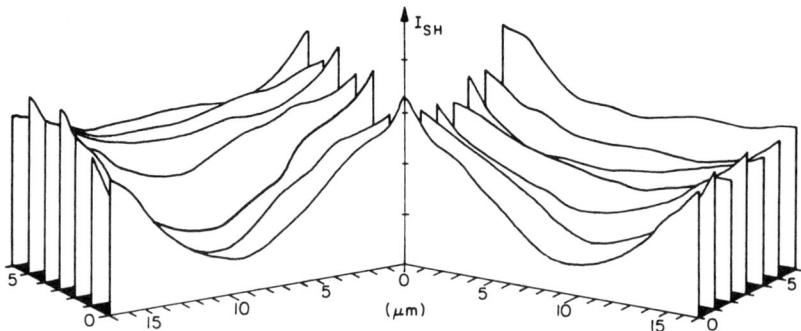

Figure 7 Second harmonic image of a laser ablated hole in a Rhodamine 6G dye monolayer on fused quartz. (Courtesy of G. T. Boyd.)

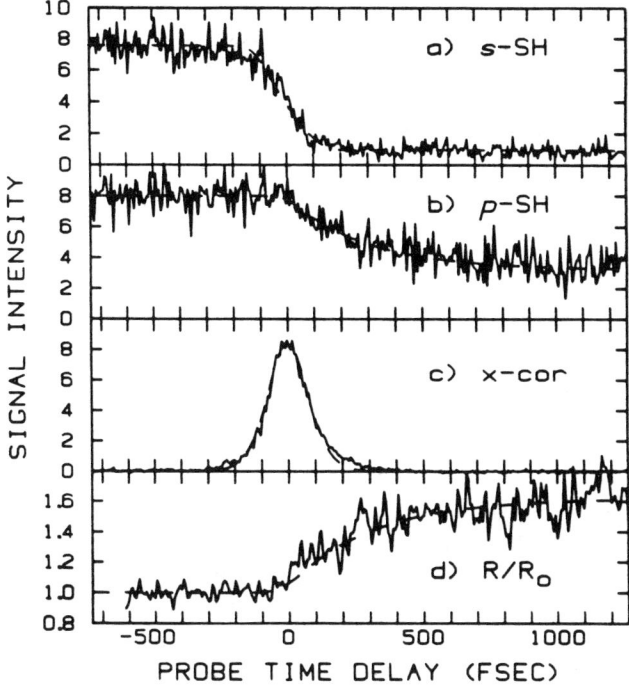

Figure 8 SHG from Si(111) versus probe time delay after the excitation of a pump pulse at two times the threshold energy for melting. Panels from top: (*a*) *s*-polarized SH, *p*-polarized fundamental, (*b*) *p*-polarized SH and fundamental, (*c*) pump-probe cross-correlation, (*d*) *p*-polarized linear reflection. *Solid line*: data; *dashed line*: theoretical fits. (After Ref. 33.)

fsec laser pulses at 610 nm were used to initiate the melting and to probe, after a time delay, via SHG from the melting surface. The probe beam was *p*-polarized, and *s*- and *p*-polarized outputs were detected separately. It can be shown that the *s*-polarized SH signal is related to $\chi^{(2)}_{s,yzz}$ and $\chi^{(2)}_{s,yxz}$, both of which should vanish if the crystalline structure of the surface layer of Si is lost. On the other hand, the *p*-polarized SH signal depends on both the isotropic and anisotropic parts of $\overleftrightarrow{\chi}^{(2)}_s$, but the isotropic part can dominate. Figures 8*a* and 8*b* show that the decays of the two polarization components are different. The result in Figure 8*a* indicates that the crystalline order is lost with a time constant of ~ 100 fsec, whereas the result in Figure 8*b* describes the change of the isotropic part in the solid → liquid transformation with a time constant of 333 fsec. The latter is consistent with the variation of linear reflection shown in Figure 8*d*.

The use of surface SHG to monitor the ultrafast dynamics of a surface chemical reaction was demonstrated in a recent experiment on the photoisomerization of DODCI (3,3′-diethyl-oxadicarbocyanine iodied) molecules at the air/water interface (43). DODCI normally exists in the *cis*-form. Photoexcitation of *cis*-DODCI from S_0 to S_1 is followed by a rapid decay to partly *cis* S_0 and partly *trans* S_0 isomers. The *trans* S_0 isomer then decays slowly back to *cis* S_0. The time development of the above isomerization process at an interface can be followed by SHG, since the nonlinearities of species differ. It was found, from SHG, that isomerization of DODCI at the air/water interface has a time constant of 220 psec. In comparison with the isomerization time of 520 psec for DODCI in water, this is significantly shorter, and is believed to be the effect of the surface environment that allows DODCI molecules to adjust to the new configuration more freely in the isomerization process.

CONCLUSION

I have discussed in this paper the theoretical basis of surface SHG for surface studies, and have shown that surface SHG can be treated as radiation from a nonlinear polarization sheet induced by an incoming wave at the surface. The boundary effects on both incoming and outgoing radiation are accounted for by the Fresnel coefficients, which appear as macroscopic local-field correction factors on the surface nonlinear susceptibility. The bulk contribution to SHG, which is not necessarily negligible even in media with inversion symmetry, can be included by defining an effective surface nonlinear susceptibility for the system. In applying SHG to surface studies, however, care must be taken to subtract off the bulk contribution or to avoid confusion led by the presence of the bulk contribution.

The surface nonlinear susceptibility $\vec{\chi}_s^{(2)}$ arises from both the structural discontinuity and the field discontinuity at an interface. As a material constant, it reflects the characteristics of the interfacial system. Thus the symmetry of $\vec{\chi}_s^{(2)}$ follows the structural symmetry of the interface layer, whereas the resonances in $\vec{\chi}_s^{(2)}$ determine the transition frequencies of the layer. Modification of the interface leads to a change of $\vec{\chi}_s^{(2)}$. Therefore, measuring $\vec{\chi}_s^{(2)}$ allows us to monitor all sorts of surface phenomena, including surface dynamic and reactions. For an adsorbed molecular layer, the ratios of various components of $\vec{\chi}_s^{(2)}$ can yield information about the orientation of the adsorbed molecules.

Surface SHG is an effective surface analytical tool. As it is forbidden in media with inversion symmetry, it is highly surface-specific. It is also sensitive enough to detect a fraction of a molecular monolayer. The technique has a number of attractive features: It is capable of in-situ, remote sensing of a surface in a hostile environment; it has high spatial, time, and spectral resolutions; and it can be applied to all interfaces accessible by light. These features have already led SHG to many useful applications. I have described a number of selected examples in the preceding section, but they were chosen to show only certain types of information that one can obtain with the technique. Many other demonstrated applications have been omitted. New applications are also being explored in many areas of science and technology, and will probably be limited only by one's imagination.

The major obstacle in the advance of the technique for surface studies is the lack of microscopic understanding of the surface nonlinear susceptibility $\vec{\chi}_s^{(2)}$. Attempts have been made, for example, to calculate $\vec{\chi}_s^{(2)}$ for simple metals from the electronic properties, but the calculations are still far from being realistic (50). No such calculation has ever been attempted for other surfaces or interfacial systems.

ACKNOWLEDGMENT

Help from Xu-dong Xiao and Wei Chen on the evaluation of the local-field factors is greatly appreciated. This work was supported by the Director, Office of Energy Research, Office of Basic Energy Sciences, Materials Sciences Division of the US Department of Energy under Contract No. DE-ACO3-76SF00098.

Literature Cited

1. Somorjai, G. 1981. *Chemistry in Two Dimensions: Surfaces*. Ithaca, NY: University Press
2. Shen, Y. R. 1984. *The Principles of Nonlinear Optics*, Chapt. 25, pp. 479–504. New York: Wiley
3. Bloembergen, N., Pershan, P. S. 1962. *Phys. Rev.* 128: 606–22
4. Bloembergen, N., Shen, Y. R. 1965. *Proc. Phys. Quantum Electron. Conf.*, Puerto Rico, pp. 119–28
5. Bloembergen, N., Chang, R. K., Lee, C. H. 1966. *Phys. Rev. Lett.* 16: 986–89
6. Lee, C. H., Chang, R. K., Bloembergen, N. 1967. *Phys. Rev. Lett.* 18: 167–70
7. Bloembergen, N., Chang, R. K., Jha, S. S., Lee, C. H. 1968. *Phys. Rev.* 174: 813–22
8. Brown, F., Matsuoka, M. 1969. *Phys. Rev.* 185: 985–87
9. Chen, J. M., Bower, J. H., Wang, C. S., Lee, C. H. 1973. *Opt. Commun.* 9: 132–34
10. Fleischmann, M., Hendra, P. J., McQuillan, A. J. 1974. *Chem. Phys. Lett.* 26: 163–66
11. Chang, R. K., Furtak, T. E., eds. 1982. *Surface Enhanced Raman Scattering*. New York: Plenum
12. Chen, C. K., de Castro, A. R. B., Shen, Y. R. 1981. *Phys. Rev. Lett.* 46: 145–48
13. Chen, C. K., Heinz, T. F., Ricard, D., Shen, Y. R. 1981. *Phys. Rev. Lett.* 46: 1010–12; 1983. *Phys. Rev. B* 27: 1965–79
14. Heinz, T. F., Chen, C. K., Ricard, D., Shen, Y. R. 1982. *Phys. Rev. Lett.* 48: 478–81
15. Shen, Y. R. 1985. *J. Vac. Sci. Technol. B* 3: 1464–66; 1986. *Annu. Rev. Mater. Sci.* 16: 69–86; 1989. *Nature* 337: 519–25
16. Richmond, G. L., Robinson, J. M., Shannon, V. L. 1988. *Prog. Surf. Sci.* 28: 1–70
17. Guyot-Sionnest, P., Chen, W., Shen, Y. R. 1986. *Phys. Rev. B* 33: 8254–63
18. Guyot-Sionnest, P., Shen, Y. R. 1987. *Phys. Rev. B* 35: 4420–26
19. Guyot-Sionnest, P., Shen, Y. R. 1989. *Phys. Rev. A* 38: 7985–89
20. Wang, C. C. 1969. *Phys. Rev.* 178: 1457–61

21. Heinz, T. F. 1982. PhD thesis. Univ. Calif., Berkeley
22. Pershan, P. S. 1963. *Phys. Rev.* 130: 919–29
23. Boyd, G. T., Shen, Y. R., Hansch, T. W. 1986. *Opt. Lett.* 11: 97–99
24. Zhu, X. D., Shen, Y. R., Carr, R. 1985. *Surf. Sci.* 163: 114–20
25. Tom, H. W. K., Zhu, X. D., Shen, Y. R., Somorjai, G. A. 1986. *Surf. Sci.* 167: 167–76
26. Mullin, C., Guyot-Sionnest, P., Shen, Y. R. 1989. *Phys. Rev. (Rapid Comm.) A* 39: 3745
27. Heinz, T. F., Tom, H. W. K., Shen, Y. R. 1983. *Phys. Rev. A* 28: 1883–85
28. Zhu, X. D., Rasing, Th., Shen, Y. R. 1989. *Chem. Phys. Lett.* 155: 459
29. Hicks, J. M., Urbach, L. E., Plummer, E. W., Dai, H. L. 1988. *Phys. Rev. Lett.* 61: 2588–91
30. Heinz, T. F., Loy, M. M. T., Thompson, W. A. 1985. *Phys. Rev. Lett.* 54: 63–66
31. Heinz, T. F., Loy, M. M. T., Thompson, W. A. 1985. *J. Vac. Sci. Technol. B* 3: 1467–70
32. Shank, C. V., Yen, R., Hirlimann, C. 1983. *Phys. Rev. Lett.* 51: 900–2
33. Tom, H. W. K., Aumiller, G. D., Brito-Cruz, C. H. 1988. *Phys. Rev. Lett.* 60: 1438–41
34. Richmond, G. L., Rojhantalab, H. M., Robinson, J. M., Shannon, V. L. 1987. *J. Opt. Soc. Am. B* 4: 228–36
35. Shannon, V. L., Koos, D. A., Richmond, G. L. 1987. *J. Chem. Phys.* 87: 1440–41; 1987. *Appl. Opt.* 26: 3579–83
36. Rasing, Th., Shen, Y. R., Kim, M. W., Grubb, S. 1985. *Phys. Rev. Lett.* 55: 2903–6
37. Rasing, Th., Berkovic, G., Shen, Y. R., Grubb, S. G., Kim, M. W. 1986. *Chem. Phys. Lett.* 130: 1–5
38. Grubb, S. G., Kim, M. W., Rasing, Th., Shen, Y. R. 1988. *Langmuir* 452–54
39. Wynne, J. J., Bloembergen, N. 1969. *Phys. Rev.* 188: 1211–20
40. Kemnitz, K., Bhattacharyya, K., Hicks, J. M., Pinto, G. R., Eisenthal, K. B., Heinz, T. F. 1986. *Chem. Phys. Lett.* 131: 285–90
41. Giesen, K., Hage, F., Riess, H. J., Steinmann, W., Haight, R., et al. 1987. *Phys. Scr.* 35: 578–81
42. Zhu, X. D., Suhr, H., Shen, Y. R. 1987. *Phys. Rev. B* 35: 3047–50
43. Hunt, J. H., Guyot-Sionnest, P., Shen, Y. R. 1987. *Chem. Phys. Lett.* 133: 189–92
44. Guyot-Sionnest, P., Hunt, J. H., Shen, Y. R. 1987. *Phys. Rev. Lett.* 59: 1597–1600
45. Guyot-Sionnest, P., Superfine, R., Hunt, J. H., Shen, Y. R. 1988. *Chem. Phys. Lett.* 144: 1–5
46. Harris, A. L., Chidsey, C. E. D., Levinos, N. J., Loiacono, D. N. 1987. *Chem. Phys. Lett.* 141: 350–56
47. Superfine, R., Guyot-Sionnest, P., Hunt, J. H., Kao, C. T., Shen, Y. R. 1988. *Surf. Sci.* 200: L445–50
48. Berkovic, G., Rasing, Th., Shen, Y. R. 1986. *J. Chem. Phys.* 85: 7374–76
49. Sitzmann, E. V., Eisenthal, K. B. 1988. *J. Chem. Phys.* 92: 4579–80
50. Rudnick, J., Stern, E. A. 1971. *Phys. Rev. B* 4: 4274–90; Liebsch, A. 1988. *Phys. Rev. Lett.* 61: 1233–36

RUBBER–LIKE ELASTICITY

B. Erman

Polymer Research Center, School of Engineering, Bogazici University, Bebek 80815, Istanbul, Turkey

J. E. Mark

Department of Chemistry and the Polymer Research Center, The University of Cincinnati, Cincinnati, Ohio 45221

INTRODUCTION

The area of rubber-like elasticity has had one of the longest and most distinguished histories in all of polymer science (1–4). For example, quantitative measurements of the mechanical and thermodynamic properties of natural rubber and other elastomers go back to 1805, and some of the earliest studies have been carried out by such luminaries as Joule and Maxwell. Also, the earliest molecular theories for polymer properties of any kind were, in fact, addressed to the phenomenon of rubber-like elasticity.

In spite of long-continuing efforts, however, much remains to be done. The main impediment to progress is the complex molecular nature of elastomeric materials. They consist of high molecular weight polymers cross linked so as to give a three-dimensional network structure (1). The very high viscosity of such a system and the copious interpenetration or entangling of the chain molecules greatly complicate the determination of physical properties at elastic equilibrium. The network structure itself, in turn, makes the material insoluble, thus making difficult the task of obtaining information on its structure for structure-property relationships. The same entangling and the usual problems of treating disordered condensed phases have also complicated the development of realistic molecular theories.

Nonetheless, there has recently been a great deal of progress in this area, particularly over the last 12–13 years. On the experimental side, new methods have been developed for the synthesis of networks of known structure, and new techniques established for network characterization. On the theoretical side, molecular theories have been put forward that are increasingly realistic in the way they treat topological features such as junction fluctuations and entangling, and the detailed structural characteristics that distinguish an elastomeric chain of one chemical type from another.

Theory and experiment up to approximately 1974 were treated in the now-classic book by Treloar (2), and the status of theory was admirably reviewed by Eichinger six years ago (5) and more recently by Heinrich et al (6). More specialized reviews, describing theoretical results obtained with particular models, have also appeared (7–9). Several recent books have reported the proceedings at symposia on elastomers and rubber-like elasticity (10–14), and two general review articles of relevance have been published in encyclopedias (15, 16). A very recent overall survey is given in the introductory book by Mark & Erman (4).

The present review briefly discusses methods for cross linking polymers into elastomeric networks and then surveys some of the standard, and also more novel, techniques for characterizing them. The theories of rubber-like elasticity are covered with the focus essentially entirely on theories that are molecular and treat elastomeric properties at equilibrium (i.e. to the exclusion of time-dependent or viscoelastic effects). Theoretical and experimental results are compared, both for the evaluation of theory and for guidance for the development of improved molecular models and more probing experimental investigations.

PREPARATION OF NETWORKS

Polymerizations with Multi-Functional Monomers

For rubber-like elasticity to be achieved, not only must the material consist of flexible polymer chains, but the chains must be joined (cross linked or end linked) into a network structure (1, 2, 4). One way of doing this is by a copolymerization in which at least one of the comonomers has a functionality ϕ of 3 or larger. This is one of the oldest ways of preparing networks but has been used mostly to prepare materials so heavily cross linked as to be relatively hard thermosets rather than highly deformable elastomers (1, 11). Recent work in this area involves the use of sol fractions and statistical arguments to obtain information on the structures of the sol phases and gel (elastomeric) phases (17–19).

Physical Aggregation

Preparation of elastomeric networks is also possible by causing physical aggregation of some of the chain segments (4, 20). Examples are the adsorption of chain segments onto filler particles, formation of polymer microcrystallites, condensation of ionic side chains onto metal ions, chelation of ligand side chains to metal ions, and microphase separation of glassy end blocks in an elastomeric triblock copolymer. These materials are now not much used in quantitative studies of rubber-like elasticity. The nature and extent of the cross linking can change with temperature, presence of diluent, and degree of deformation in an uncontrolled manner, and the "cross links" are frequently so large as to complicate greatly the theoretical analysis of any experimental results.

Random Chemical Cross Linking

Polymers can be cross linked by using chemical reactions that attack a pair of chains, at essentially random locations (1, 4). Examples are the addition of sulfur atoms to the double bonds of diene elastomers, and the attack of free radicals from peroxide thermolyses or high-energy radiation on side chains (frequently unsaturated) or on the chain backbone itself.

The covalent linkages thus formed are generally highly stable, thus avoiding some of the problems with physically cross-linked systems. These networks, however, still present the problem that the cross-linking reaction used is highly uncontrolled in that it is not known how many cross links are introduced or where they lie along the chain trajectories. It is thus very difficult to use these networks for quantitative purposes such as the development of structure-property relationships. Obtaining independent measures of their degree of cross linking (as represented directly by the number density of cross links or the number density of network chains, or inversely by the average molecular weight M_c between the cross links that mark off a network chain) is virtually impossible.

Highly Specific Chemical Cross Linking

If networks are formed by end linking functionally terminated chains instead of haphazardly joining chain segments at random, then the nature of this very specific chemical reaction provides the desired structural information (4, 20–22). Thus, the functionality of the cross links is the same as that of the end linking agent, and the molecular weight M_c between cross links and its distribution are the same as those of the starting chains prior to their being end linked.

Because of their known structures, such elastomers are now the preferred materials for the quantitative characterization of rubber-like elasticity.

Additionally, bimodal networks prepared by these end-linking techniques have very good ultimate properties (23), and there is currently much interest in preparing and characterizing such networks (4; L. K. Silva, J. E. Mark, F. J. Boerio, unpublished results) and developing theoretical interpretations for their properties (25). Such a network is shown schematically in Figure 1 (23).

Cross Linking Under Unusual Conditions

For some investigations, networks have been prepared by partially disentangling the polymer chains prior to their cross linking by either stretching or dissolution in a solvent (4, 26). Such networks are of particular interest because they come to elastic equilibrium relatively rapidly, show little stress relaxation, and yield stress-strain isotherms of an unusually simple form. Interpretation of their properties thus constitutes an important challenge to the various molecular theories of rubber-like elasticity.

In a rather novel series of experiments (4), results on the trapping of cyclic molecules during network formation were found to support the

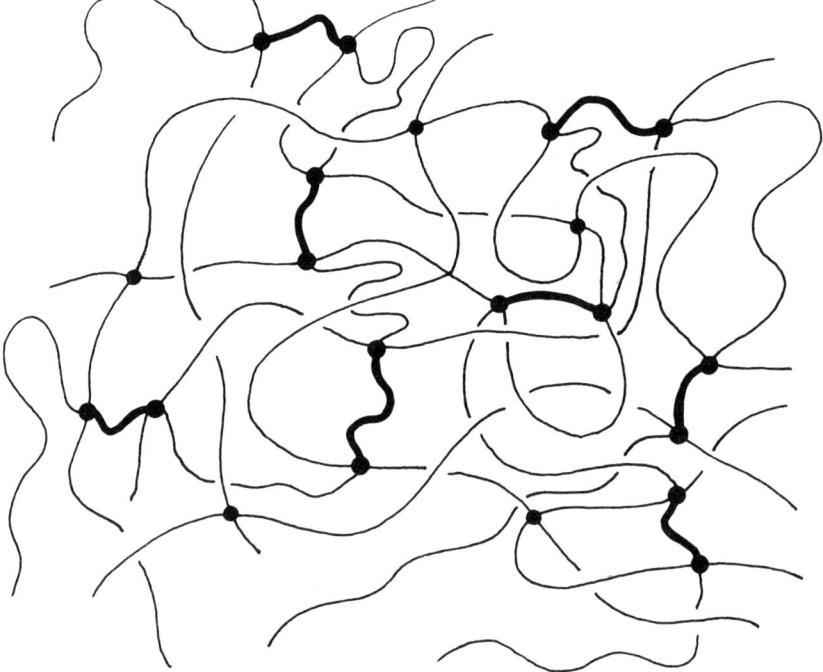

Figure 1 Sketch of a portion of a bimodal network (23). The very short polymer chains are represented by *heavy lines* and the relatively long chains by *thin lines*.

notion that "olympic" or "chain mail" networks can be formed in a similar manner. This could be done by threading functionally terminated chains through such cyclics and then difunctionally end linking them, as illustrated in Figure 2 (27).

ELASTICITY EXPERIMENTS

Mechanical Properties

These are the properties that determine the unique usefulness of elastomeric materials; not surprisingly, they are the properties that have been most extensively studied (1, 2, 4, 10–16). The relationship of primary interest is the stress-strain isotherm. Measurements to determine it are typically carried out to the rupture point of a sample, thus yielding as well the two ultimate properties of the elastomer, namely its ultimate strength and maximum extensibility. In addition to providing practical information, such results are also much used to obtain estimates of the degree of cross linking, and to test the predictions of theory. Most experimental studies have involved simple elongation, largely because of the simplicity of the techniques involved. Results obtained with other deformations such as biaxial extension, compression, shear, and torsion are, however, particularly valuable for gauging the generality of a particular theory, and more such investigations should be carried out. Some important non-mechanical properties are mentioned in the following three sections.

Swelling

In a swelling experiment the network is typically placed into an excess of solvent, which it imbibes until the dilational stretching of the chains pre-

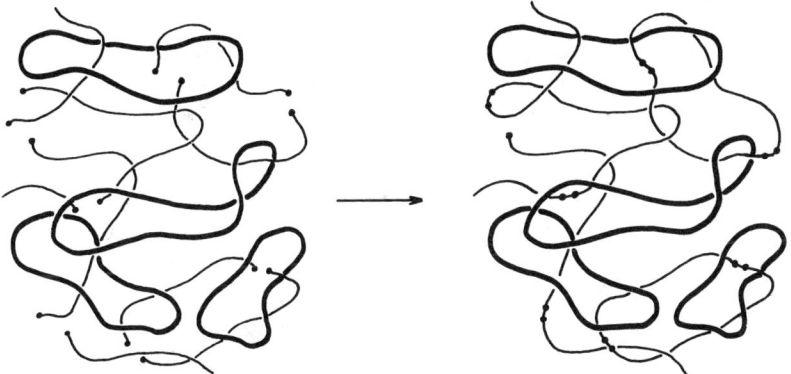

Figure 2 Preparation of an olympic or chain mail network, which has no cross links at all (27). Linear chains (*light lines*) passing through the cyclics (*heavy lines*) are difunctionally end linked to form a series of interlinked cyclics.

vents further absorption (1, 2, 4). This equilibrium extent of swelling can be interpreted to yield the degree of cross linking of the network, provided the polymer-solvent interaction parameter χ is known. Conversely, if the degree of cross linking is known from an independent experiment, then the interaction parameter can be determined. The equilibrium degree of swelling and its dependence on various parameters and conditions provide, of course, additional tests of theory.

Optical and Spectroscopic Properties

An example of a relevant optical property is the birefringence of a deformed polymer network. This strain-induced birefringence can be used to characterize chain segmental orientation, both Gaussian and non-Gaussian elasticity, crystallization and other types of chain ordering, and short-range correlations (2, 4). Other optical and spectroscopic techniques are also important, particularly with regard to segmental orientation. Some examples are fluorescence polarization, deuterium NMR, and polarized infrared spectroscopy (4, 28).

Scattering

The technique of this type of greatest utility in the study of elastomers is small-angle neutron scattering; for example, from deuterated chains in a nondeuterated host (4). One application has been the determination of the degree of randomness of the chain configurations in the undeformed state, an issue of great importance with regard to the basic postulates of elasticity theory. Of even greater importance is determination of the manner in which the dimensions of the chains follow the macroscopic dimensions of the sample. This relationship between the microscopic and macroscopic worlds is one of the central problems in rubber-like elasticity.

Some small-angle X-ray scattering techniques have also been applied to elastomers; an example is the characterization of fillers incorporated into elastomers to improve their mechanical properties (4; D. W. Schaefer, J. E. Mark, unpublished results).

ELASTICITY THEORY

Phenomenological

The phenomenological approach to elasticity theory is based on continuum mechanics and symmetry arguments rather than on molecular concepts (2, 30). It attempts to fit stress-strain data with a minimum number of parameters, which are then used to predict other mechanical properties of the same material. Its best-known result is the Mooney-Rivlin equation, which states that the modulus of an elastomer should vary linearly with reciprocal elongation.

Structureless Chains

The molecular theory of elasticity has developed along two lines. The first is based on the statistical mechanics of networks, and is mostly concentrated on network topology, determination of stress-strain behavior, number of effective cross-links, contributions from entanglements, etc. Most experimental effort is on this first aspect. In this approach, the chemical structure of the chains constituting the medium has not been considered important—only the topological nature of the network is relevant. These models are considered in the present section.

The second line of theoretical development has not received much attention (except for biregringence and thermoelasticity). In this second approach, the chemical structure and the resulting configurational characteristics of the chains are considered important. Topics are birefringence, thermoelasticity, rotational isomerization upon stretching, strain dichroism, local segmental orientation, local segmental mobility, and neutron scattering. Advances in present-day experimental techniques focus more attention on these phenomena. Additionally, changes in the swelling behavior of networks resembling first-order transitions are being studied experimentally in several laboratories. These approaches are discussed in later sections.

CLASSICAL Molecular theories of rubber-like elasticity may be conveniently discussed in two categories: (*a*) the classical theories comprising the phantom and the affine network models, introduced between 1941 and 1943, and (*b*) the modern theories introduced after 1975.

The term "classical" is used in describing the phantom and affine models because of their conceptual simplicity and their status as suitable reference systems for all later treatments of the molecular theory. Below, the characteristic features of the classical theories are outlined with extensive references to relevant literature, followed by a discussion of the modern theories and their extensions and applications to various phenomena observed by advanced experimental techniques of our day.

Phantom The theory of James & Guth, which has subsequently been termed the "phantom network theory," was first outlined in two papers (31, 32), followed by a mathematically more rigorous treatment (33–35). The theory rests on five basic assumptions:

1. The elasticity of the network resides uniquely within the individual

chains. No intermolecular effects are present, a characteristic signifying the phantom-like nature of the chains.
2. The distribution $W(\mathbf{r}_{ij})$ of the vector $\mathbf{r}_{ij}$ of a chain forming two junctions i and j in the undeformed network is equal to that of the single chain in the uncrosslinked bulk polymer.
3. $W(\mathbf{r}_{ij})$ is Gaussian.
4. All network chains are identical.
5. A small number of cross links are fixed at the boundaries of the network and their positions transform affinely with macroscopic deformation.

The configuration partition function Z_N for the network is obtained as the product of configuration functions for its v chains as

$$Z_N = Z^v \prod_{i<j} W(\mathbf{r}_{ij})$$
$$= C \prod_{i<j} \exp(-3r_{ij}^2/2\langle r_{ij}^2\rangle_0) \qquad 1.$$

where Z is the configuration integral for the free chain and the product includes all pairs ij of junctions connected by a chain. The first line of Eq. 1 results from Assumption 1 and the second line from Assumption 3. C is a constant factor and $\langle r_{ij}^2\rangle_0$ is the mean-squared end-to-end distance in the unperturbed state. Equation 1 may be written as

$$Z_N = C\exp(-\{\mathbf{R}\}^T\mathbf{\Gamma}\{\mathbf{R}\}) \qquad 2.$$

where $\{\mathbf{R}\} = \{\mathbf{R}_1, \mathbf{R}_2, \ldots, \mathbf{R}_\mu\}$ is the column vector formed by position vectors of all μ junctions, $\{\mathbf{R}\}^T$ is its transpose, and $\mathbf{\Gamma}$ is the Kirchoff matrix defined by $\Gamma_{ij} = -3/2\langle r_{ij}^2\rangle_0$ if $i \neq j$ and junctions i and j are connected by a chain (33), and $\Gamma_{ij} = 0$ if $i \neq j$ and junctions i and j are not connected by a chain. Also,

$$\Gamma_{ij} = -\sum_j \Gamma_{ij} = -\sum_j \Gamma_{ji}.$$

Following Assumption 5, the partition function of Eq. 2 may be written as

$$Z_N = C\exp[-\{\mathbf{R}_\sigma\}^T(\mathbf{\Gamma}_\sigma - \mathbf{\Gamma}_{\sigma\tau}\mathbf{\Gamma}_\tau^{-1}\mathbf{\Gamma}_{\tau\sigma})\{\mathbf{R}_\sigma\} - \{\Delta\mathbf{R}_\tau\}^T\mathbf{\Gamma}_\tau\{\Delta\mathbf{R}\}] \qquad 3.$$

where $\{\mathbf{R}_\sigma\}$ and $\{\mathbf{R}_\tau\}$ are columns of the position vectors for the fixed and free junctions, respectively, $\mathbf{\Gamma}_\sigma$ is the square matrix of rows and columns of $\mathbf{\Gamma}$ for the fixed junctions, $\mathbf{\Gamma}_\tau$ is the one for the free junctions, and $\mathbf{\Gamma}_{\sigma\tau}$ ($\mathbf{\Gamma}_{\tau\sigma}$) is the rectangular matrix having the rows (columns) for the fixed junctions and columns (rows) for the free junctions. In Eq. 3, the column vector $\{\Delta\mathbf{R}_\tau\}$ denotes the set of fluctuation vectors $\Delta\mathbf{R}_i = \mathbf{R}_i - \bar{\mathbf{R}}_i$ of instantaneous junction positions $\mathbf{R}_i$ from their time-averaged positions $\bar{\mathbf{R}}_i$. Integrating Eq. 3 over all fluctuating quantities leads to

$$Z_{N\sigma} = C\exp(-\{\mathbf{R}_\sigma\}^T(\mathbf{\Gamma}_\sigma - \mathbf{\Gamma}_{\sigma\tau}\mathbf{\Gamma}_\tau^{-1}\mathbf{\Gamma}_{\tau\sigma})\{\mathbf{R}_\sigma\}) \qquad 4.$$

from which the stored elastic free energy is obtained as

$$\Delta A_{el} = -kT\ln Z_{N\sigma} = \tfrac{1}{3}kT[\{\mathbf{R}_\sigma^0\}^T(\mathbf{\Gamma}_\sigma - \mathbf{\Gamma}_{\sigma\tau}\mathbf{\Gamma}_\tau^{-1}\mathbf{\Gamma}_{\tau\sigma})\{\mathbf{R}_\sigma^0\}]$$
$$\times (\lambda_x^2 + \lambda_y^2 + \lambda_z^2 - 3) \qquad 5.$$

where λ_t is the extension ratio along the tth principal direction and $\{\mathbf{R}_\sigma^0\}$ refers to the set of fixed position vectors in the reference state.

The term in brackets in Eq. 5 depends on the topology of the network. Rigorous work toward its derivation does not exist other than that of James (33) for a cubic lattice. Derivation by a different procedure by Flory (36) shows that for a symmetrically grown network of cycle rank ξ, the factor in brackets should equate to $3\xi/2$.

Integrating over the fixed variables in Eq. 3 gives

$$Z_{N\tau} = C\exp(\{\Delta\mathbf{R}_\tau\}^T\mathbf{\Gamma}_\tau\{\Delta\mathbf{R}_\tau\}). \qquad 6.$$

This equation leads to the calculation of mean-squared fluctuations in the network. The cross correlation of the fluctuations of two junctions, for example, is obtained according to the expression

$$\langle\Delta\mathbf{R}_i\cdot\Delta\mathbf{R}_j\rangle = \frac{\int \Delta\mathbf{R}_i\cdot\Delta\mathbf{R}_j\exp[-\{\Delta\mathbf{R}_\tau\}^T\mathbf{\Gamma}_\tau\{\Delta\mathbf{R}_\tau\}]\,d\{\Delta\mathbf{R}_\tau\}}{\int \exp[-\{\Delta\mathbf{R}_\tau\}^T\mathbf{\Gamma}_\tau\{\Delta\mathbf{R}_\tau\}]\,d\{\Delta\mathbf{R}_\tau\}}$$
$$= \partial \ln Z_{N\tau}/\partial \Gamma_{ij}, \qquad 7.$$

which simplifies to

$$\langle\Delta\mathbf{R}_i\cdot\Delta\mathbf{R}_j\rangle = \frac{3}{2}\frac{\partial}{\partial\Gamma_{ij}}\ln|\det\mathbf{\Gamma}_\tau| = \frac{3}{2}(\mathbf{\Gamma}_\tau^{-1})_{ij}. \qquad 8.$$

The characterization of fluctuations therefore requires the construction of the $\mathbf{\Gamma}_\tau$ matrix, which depends on the topology of the network, and finding its inverse as suggested by Eq. 8. Characterization of networks along these lines was made by Duiser & Staverman (37), Eichinger (38), Graessley (39, 40), Flory (36), Pearson (41), and Kloczkowski et al (42, 43) on symmetrically grown networks. It follows from the above formulation [see James (33)] that the mean positions of junctions transform affinely with macroscopic deformation, whereas the fluctuations from this mean are independent of strain. The theory also provides relationships between various molecular parameters (36), such as $\mu = 2\nu/\phi$ and $\xi = (1-2/\phi)\nu$, where μ, ν, and ϕ are the numbers of junctions, chains, and the junction functionality, respectively.

Relationships between various molecular dimensions are obtained as (4)

$$\langle \bar{r}^2 \rangle_0 = (1-2/\phi) \langle r^2 \rangle_0$$
$$\langle \Delta r^2 \rangle_0 = (2/\phi) \langle r^2 \rangle_0$$
$$\langle \Delta R^2 \rangle = [(\phi-1)/\phi(\phi-2)] \langle r^2 \rangle_0 \qquad 9.$$

where $\bar{r}$ and Δr denote the magnitude of mean chain vectors and fluctuations from them, respectively. The subscript zero denotes the unperturbed state, as usual. Equations 9 follow essentially from the application of Eq. 8 to a network in the form of a symmetrically grown tree. Work on other network topologies does not exist. Relations similar to those given by Eq. 9 have also been obtained for the phantom network in the deformed state. The mean-squared chain dimensions $\langle r^2 \rangle$ in the deformed state are related to those in the undeformed state by (4)

$$\langle r^2 \rangle = [(1-2/\phi) \sum_{i=1}^{3} \lambda_i^2/3 + 2/\phi] \langle r^2 \rangle_0. \qquad 10.$$

Similar relations between any two points on the network chains are given by Pearson (41) and recently, in a more general form, by Kloczkowski et al (42, 43).

Further references on the theory of phantom networks are given in the review article by Eichinger (5).

Affine In contrast to the phantom network model, in which only a small fraction of junctions are fixed at the surface, all junctions are fixed in the affine network. Consequently, their positions transform affinely with macroscopic strain. Between junctions, the chains are phantom-like in the sense that they may pass through each other without introducing intermolecular contributions to the elastic behavior of the network.

The idea of having the junction positions transform affinely with macroscopic strain goes back to Kuhn (44–46). Statistical treatment of the affine network is given by Flory (1, 47–49) and Wall (50–52).

The analysis is based on writing the configuration partition function as a product of two terms $Z_N = Z_1 Z_2$, where Z_1 refers to independent deformation of chains, and Z_2 introduces the correction to Z_N due to the connectivity of the network. The former is given by the expression

$$Z_1 = v! \prod_i (W(\mathbf{r}_i)\, d r_i)^{v_i}/v_i! \qquad 11.$$

where $W(\mathbf{r}_i)$ denotes the a priori probability of having the end-to-end vector of the ith chain at $\mathbf{r}_i$, and v_i is the number of chains having the end-to-end vector $\mathbf{r}_i$ after deformation. Z_2 is given as (1)

$$Z_2 = (\phi \delta V/V)^{2v(1-1/\phi)}[(2v/\phi)!]^{(\phi-1)} \qquad 12.$$

where δV is the deformation-independent volume element in which the ϕ chain ends meet and V is the total network volume. The elastic free energy of the network is obtained from the above partition functions as (1)

$$\Delta A_{\text{el}} = \frac{1}{2} v k T \left[\sum_{i=1}^{3} \lambda_i^2 - 3 - (1 - 2/\phi) \ln \lambda_1 \lambda_2 \lambda_3 \right]. \qquad 13.$$

Theoretical developments after 1975 (see below) have discredited the affine network model (36, 53–58), mainly because the assumption of embedding each junction securely into the continuum of the network volume was judged to be unrealistic. The phantom network model seems to be more appropriate in this sense.

Non-Gaussian Most molecular theories are based on the Gaussian distribution of end-to-end distances of the network chains. This distribution, however, is not suitable for very short chains, or for any chains stretched to near the limits of their extensibility (2, 4, 15, 25). In these cases the modulus shows a marked upturn at high elongations, a result that may be of practical as well as fundamental importance.

Non-Gaussian theories developed for such circumstances are either analytical, involving generalization or modification of the Gaussian limit, or simulations, typically based on rotational isomeric state models. Some of the results thus obtained are described in a later section.

MODERN The term "modern" refers to theories of rubber-like elasticity introduced after 1975 mainly to account for the disagreement between experiment and the predictions of the phantom or affine network models. All the theories in this category may essentially be regarded as corrections to the phantom network model. Corrections come in the form of contributions of entanglements to the elastic free energy.

Constrained junctions The constrained junction theory of Ronca & Allegra (53) represented the first recognition of possible correlations between the macroscopic deformation and the fluctuations of junctions. The elastic free energy of the Ronca-Allegra theory may be expressed as

$$\Delta A_{el} = \frac{\xi kT}{2}\left[I_1 - \frac{\mu}{\xi}\ln(I_3/I_1)\right] \qquad 14.$$

where $I_1 = \lambda_1^2 + \lambda_2^2 + \lambda_3^2$ and $I_3 = \lambda_1^2\lambda_2^2\lambda_3^2$ are the first and third invariants of the deformation gradient tensor, respectively. The second term in the brackets, obtained under the assumption that junction fluctuations are affine at small deformations and independent of macroscopic strain at large deformations, gives the correction term that agrees with experimental data from mechanical measurements. Ronca & Allegra concluded that whenever a nonvanishing coupling exists between the macroscopic deformation and the junction fluctuations, a logarithmic term appears in the free energy.

With physical arguments similar to those of Ronca & Allegra, Flory introduced a statistical model of the network in the presence of entanglements or constraints (54). According to the model, the constraints, which are randomly distributed around the junctions, exert additional linear spring-like forces on the fluctuating junctions. More precisely, the constraints were envisaged to form domains of constraints around each junction. The strength of the constraints were measured by a parameter $\kappa = \langle \Delta R^2 \rangle_0 / \langle \Delta s^2 \rangle_0$ in which the denominator is the mean-squared fluctuations of the junctions in the undisturbed phantom network and the numerator denotes the mean-squared dimensions of the constraint domains. The elastic free energy based on this model is derived as

$$\Delta A_{el} = \frac{\xi kT}{2}\left\{I_1 + \frac{\mu}{\xi}\sum_t [B_t + D_t - \ln(B_t+1) - \ln(D_t+1)]\right\} \qquad 15.$$

where

$$B_t = \kappa^2(\lambda_t^2 - 1)(\lambda_t^2 + \kappa)^{-2}$$

$$D_t = \lambda_t^2 \kappa^{-1} B_t. \qquad 16.$$

For $\kappa = 0$ and ∞, Eq. 15 reduces to the elastic free energy of the phantom and affine model, respectively. Predictions based on this elastic free energy are in agreement with the stress-strain data on networks. The theory was later improved (58) by considering a more detailed state of deformation of constraint domains. The resulting expression, which contained an additional parameter ζ, is shown to agree quantitatively with stress-strain swelling data (59), birefringence (60, 61), and segmental orientation (62, 63) in uniaxially stretched networks and with experimental data obtained in shear and multiaxial states of stress (64).

The Ronca & Allegra and Flory theories are both based on the idea that effects of constraints are local and decrease with increasing strain and swelling. The basic difference between the two theories is that in the Ronca-Allegra theory the fluctuations of junctions become exactly affine as the undeformed state is approached, whereas in the Flory theory they are closer to but below those of the affine state.

Attempts at greater rigor The constrained junction theory starts from a detailed molecular model of the real network. The fundamental postulate of the theory states that only strain-dependent contributions to the elastic free energy are of importance, and that these contributions vanish at infinitely large extensions or swelling. Contributions from trapped entanglements, for example, are categorically eliminated when this postulate is accepted. According to this approach, whether the trapped entanglements contribute to stress can only be determined by experiments. A complete statistical mechanical theory that does not depend on physical assumptions of the type stated above has been outlined by Freed (65), and a mathematical theory of elasticity of networks with internal constraints has been worked out by Edwards and collaborators (66–70). According to these theories, a set of internal constraints, including knots and entanglements, are assumed to be fixed during the formation of the network. These constraints are conserved under deformation, thus contributing to the elastic properties. For a canonical ensemble, for example, the thermodynamic properties are obtained from the partition function

$$Z_c(T, V) = \int_V d\Gamma \exp[-H(\Gamma)/kT]\delta[C - C(\Gamma)] \qquad 17.$$

where Γ is a point in the configurational phase space of the system, and $C(\Gamma)$ denotes the phase functions describing the constraints. The partition function given by Eq. 17 is for a particular constraint C. The free energy $A(T, V)$ for the system is obtained as the ensemble average

$$\Delta A_{el} = -kT \langle \ln Z_c(T, V) \rangle_{\{c\}}, \qquad 18.$$

which extends over the whole set of constraints $\{c\}$ operating in the network.

The statistics of the network represented by Eqs. 17 and 18, although general, present difficulties. Drastic simplifications are required to reduce the problem to a soluble form. Deam & Edwards (70) simplified the problem by treating the single entanglement represented as)(→ 8. Effects of multiple entanglements were introduced later by Ball et al (71) by the "slip-link" model. Multiple entanglements operating on a chain are represented by a slip link that is free to slide along the chains. According to the model, the presence of other entanglements restrict the extent to which the links may slide. The mathematical difficulties in averaging the logarithm of the partition function given by Eq. 18 are overcome by a method called the *replica trick*. The resulting elastic free energy is

$$\Delta A_{el} = \frac{1}{2}\mu kT\left\{I_1 + \frac{N_s}{\mu}\sum\left[\frac{(1+\eta)\lambda_i^2}{1+\eta\lambda_i^2} + \log(1+\eta\lambda_i^2)\right]\right\} \qquad 19.$$

where N_s/μ is the ratio of the number of slip links to junctions, and $\eta = 0.2343$.

The second term in brackets in Eq. 19 represents an entanglement contribution to the elastic free energy of a phantom network. Contributions to the modulus from this term have been shown to diminish with increasing deformation (72). The only term contributing to the modulus at high deformation is the one from the phantom network. In this limit, the physical pictures behind the Edwards and Flory models are in agreement. Significant controversy exists, however, as to whether at small deformations trapped entanglements, as proposed by the Edwards model, contribute significantly to the modulus or whether they are insignificant, as proposed by Flory. The reader is referred elsewhere (4, 21, 73) for details on this controversial issue.

The slip-link theory has subsequently been treated in a simplified mathematical approach by Edwards & Vilgis (74), and is discussed further in a recent review (9).

Rotational Isomeric State Chains

ANALYTICAL CALCULATIONS   As described below, the major applications of rotational isomeric state models have been to thermoelasticity (4, 15, 16, 75), strain birefringence (75–79), segmental orientation (80), strain dichroism (81), and rotational isomerization (82).

SIMULATIONS   Monte Carlo simulations have been used to calculate thermoelastic results through the temperature coefficient of the unperturbed dimensions (75). In the case of networks of the protein elastin, such results were used to evaluate alternative theories for the molecular deformation mechanism for this bioelastomer (83).

Stress-strain isotherms have also been calculated with this approach. Examples are unimodal networks of polyethylene and poly(dimethylsiloxane) (84), polymeric sulfur and selenium (85), short *n*-alkane chains (86), natural rubber (87), several polyoxides (88, 89), and elastin (90), and bimodal networks of poly(dimethylsiloxane) (91).

Descriptions of Various Phenomena

The expressions for the Helmholtz free energies of the phantom and affine network models as well as relations of chain dimensions to macroscopic deformation lead to simple expressions that may be used for interpretation of various phenomena.

1. The equation of state: The general thermodynamic relationship between the principal states of stress and deformation is (92)

$$t_i = (\lambda_i/V)(\partial \Delta A_{el}/\partial \lambda_i) = (\alpha_i/V)(\partial \Delta A_{el}/\partial \alpha_i)_V. \qquad 20.$$

Here, t_i is the true stress along the ith principal coordinate defined as force per unit deformed area, V is the final volume of network, and α_i is the deformation ratio relative to initial state defined as the ratio of final length along the ith principal direction to that in the initial state of the network. For networks formed at a polymer concentration v_{2C} and tested at a concentration of v_2, the relation of α_i to λ_i is $\alpha_i = (v_2/v_{2C})^{1/3}\lambda_i$. For uniaxial stress along the x_1 direction, the expressions for the elastic free energies and Eq. 20 lead to

$$t_1 = 2(\mathscr{F}kT/V)(v_{2C}/v_2)^{2/3}(\alpha_1^2 - \alpha_1^{-1}) \qquad 21.$$

where the value of the factor $\mathscr{F}$ depends on the nature of the deformation. For the phantom and affine network models, $\mathscr{F}$ is equal to $\xi/2$ and $\nu/2$, respectively. In deriving Eq. 21, the volume of the network is assumed to be constant. The stress-strain relations for the general case may also be obtained by an alternate method, yielding what are referred to as the Treloar relations (2). The shear modulus of the network, also identified as the reduced force $[f^*]$, is obtained from Eq. 21 as

$$[f^*] \equiv t_1 v_2^{-1/3}/(\alpha_1^2 - \alpha_1^{-1}) = 2(\mathscr{F}kT/V_d)v_{2C}^{2/3} \qquad 22.$$

where V_d is the volume of the dry network. The simple constancy expressed by the second equality in this equation is generally not observed experimentally, as is discussed below.

As pointed out above, the chains in both the phantom and affine network models are phantom-like, eliminating any possible intermolecular interaction. One type of intermolecular interaction that has attracted a great deal of attention is the interchain entanglement. Such entanglements have been proposed (93) to be of importance even at elastic equilibrium, and a number of experimental studies (21, 94) have been interpreted in support of this suggestion.

2. Thermoelasticity relations: In uniaxial tension, part of the total force f acting on a sample is used in changing the entropy of the chain and the other part, f_e, in changing the energy but not the entropy of the chains. The ratio f_e/f is obtained, by thermodynamics, as $f_e/f = -T[\partial \ln(f/T)/\partial T]_{L,V}$ and is also given by the molecular quantity $Td\ln\langle r^2\rangle_0/dT$. Here, $\langle r^2\rangle_0$ is the mean squared end-to-end distance of chains in the unperturbed state. Inasmuch as $\mathscr{F}$ has been eliminated, the equation for $d\ln\langle r^2\rangle_0/dT$

is valid for both the phantom and affine network models. It provides a valuable link between the macroscopically measured forces and the microscopic characteristics of the network chains.

3. Strain birefringence: The early treatments (2, 75, 76, 82, 95–97) of strain birefringence of elastomeric polymer networks were based on the affine network model. Analysis according to phantom networks, given much later (98), directs attention to the importance of introducing the state of microscopic deformation into the formulation. The components of the microscopic state of strain $\Lambda^2_{t,\text{ph}}$ for the phantom network are defined by using Eq. 12 as

$$\Lambda^2_{t,\text{ph}} = (1-2/\phi)\lambda_t^2 + 2/\phi, \quad t = 1, 2, 3. \qquad 23.$$

The refractive index difference Δn_{xy} along the two principal axes for a phantom network stretched along the x axis is obtained as

$$\Delta n_{12} = (2\pi/27)(v/V)[(n^2+2)^2\Gamma_2/n](\Lambda^2_{1,\text{ph}} - \Lambda^2_{2,\text{ph}})$$
$$= (2\pi v/27V)(1-2/\phi)[(n^2+2)^2\Gamma_2/n](v_{2C}/v_2)^{2/3}$$
$$\times (\alpha_1^2 - \alpha_1^{-2}) \qquad 24.$$

where n is the mean refractive index and

$$\Gamma_2 = (9/10)\sum_i \langle \mathbf{r}^T\hat{\boldsymbol{\alpha}}_i\mathbf{r}\rangle_0/\langle r^2\rangle_0. \qquad 25.$$

Here, the summation is over all structural units contributing to the birefringence, and $\hat{\alpha}_i$ is the anisotropic part of the polarizability tensor of the structural unit. Evaluation of Γ_2 based on the rotational isomeric state formalism has been given for various polymeric systems (75–79).

The second part of Eq. 24 differs from the corresponding expression for an affine network by the factor $1-2/\phi$. The stress-optical coefficient C is defined as the ratio of Δn_{12} to the stress τ_1

$$C = \frac{\Delta n_{12}}{\tau_1} = 2\pi(n^2+2)^2\Gamma_2/27nkT \qquad 26.$$

where the expression given by Eq. 21 is used in the denominator. The stress-optical coefficient is independent of macroscopic deformation. Moreover, it is given by the same expression for the affine network and therefore is independent of the molecular model used. However, the coefficient Γ_2, which can be calculated only for the unperturbed isolated chain, is severely affected in the bulk state by intermolecular correlations (99, 100).

4. Segmental orientation: Segmental orientation in uniaxially strained networks is conveniently represented by the second Legendre polynomial,

$S = (3\langle \cos^2 \theta \rangle - 1)/2$, where $\langle \cos^2 \theta \rangle$ is the mean-squared projection of the segment vectors along the direction of stretch, θ being the angle between the segment vector and the axis. Early theories of orientation (96, 101–107) are all based on the affine model of Gaussian networks where S is related to the orientation function by

$$S = (1/5N)(\lambda^2 - \lambda^{-1}) \qquad 27.$$

where N is the number of Kuhn segments in a chain. Later work (62) generalized Eq. 27 to phantom networks, according to which S may be written as

$$S = (1 - 2/\phi)D_0(\lambda^2 - \lambda^{-1}) \qquad 28.$$

where D_0 is obtained (35) according to the expression

$$D_0 = (3\langle r^2 \cos^2 \Phi \rangle_0 / \langle r^2 \rangle_0 - 1)/10, \qquad 29.$$

which is valid for sufficiently long real chains under small deformations. The averages shown in Eq. 29 may be obtained by the rotational isomeric state scheme for a given polymer (80). Such calculations would be representative of segmental orientation in networks with sufficiently low cross-link densities at small deformations. Rigorous calculations for shorter chains and for high deformations are not currently available.

Experimental work indicates that intermolecular correlations contribute significantly to segmental orientation (63). Approximate calculations describing the extent of orientational correlations have been given recently (62, 107).

5. Strain-dichroism and rotational isomerization in deformed networks: Preferential orientation of certain configurational sequences in network chains in the deformed state is referred to as *strain-dichroism*. Rigorous theoretical analysis of preferential orientation has been formulated by Flory & Abe (81) for the affine network model under small deformations, and the predictions compared with the experimental results of Read & Stein (108). Recent progress in spectroscopic techniques (28) direct attention to the measurement of strain dichroism.

Stretching a network results in the decrease of the entropy of the chains, simultaneously accompanied by the apportionment of its bonds and bond sequences among various rotational isomeric states. The latter is related to the energetic component f_e of the total force acting on the network. Changes in the average population of various rotational isomeric states may be analyzed in more detail by the theory of rotational isomerization with stretching. Earlier discussion of the subject may be found in Birshtein & Ptitsyn (95). A rigorous calculation scheme for rotational isomerization in affine networks with long chains under small deformations is given by

Abe & Flory (82). Calculations have shown that the extent of rotational isomerization in networks would be small. However, new techniques of preparing model networks that contain short chains and interest in their ultimate properties direct attention to this problem of rotational isomerization.

6. Chain dimensions in deformed networks: Application of small angle neutron scattering to deformed networks makes it possible to characterize, rigorously, the state of deformation at the molecular level. Experiments (109, 110) usually show that molecular deformation is closer to that represented by the phantom network model. The x-component of the molecular deformation tensor Λ_x^2 is obtained from Eq. 10 as (111–113)

$$\Lambda_x^2 \equiv \langle x^2 \rangle / \langle x^2 \rangle_0 = (1 - 2/\phi)\lambda_x^2 + 2/\phi, \qquad 30.$$

with similar expressions for the y and z components.

Various open questions in rubber-like elasticity pertaining to neutron-scattering are discussed by Bastide & Boue (114), Boue et al (115), and Picot (116). The most recent experiments of Ewen et al (117), performed by the neutron spin echo technique, indicate that junction fluctuations in networks are substantial, in support of Flory's earlier predictions (36).

7. Critical phenomena and transitions in swollen networks (118): Swelling equilibrium in a network exposed to a solvent obtains when the increase in the mixing entropy of the solvent-network equates to the decrease of the entropy of network chains upon dilation. The state of equilibrium is expressed by equating the solvent chemical potential

$$\Delta \mu_1 = \ln(1 - v_2) + v_2 + \chi v_2^2 + (v_{20}/2x_c)\lambda^{-1} - iv_2/x_c \qquad 31.$$

to zero, where χ is the polymer-solvent interaction parameter, x_c is the number of segments in the chain (the volume of a segment being equal to that of the solvent molecule), and i is the number of ionic groups per chain. Equation 31 is expressed for a phantom network model, following the preceding discussion that swollen networks are better represented by this model rather than by the affine one.

For theta and good solvents (1), the solution of Eq. 31 gives the equilibrium degree of swelling. Alternatively, x_c may be determined if the equilibrium degree of swelling is known (119). When the solvent medium is suitably altered, the network may exhibit contractions that resemble a first-order phase transition (118). The possibility of phase transitions in networks has previously been discussed by Dusek & Prins (120) and has been observed experimentally in polyacrylamide gels by Tanaka (121), Janas et al (122), and Ilavsky (123, 124).

THEORY VS. EXPERIMENT

Stress-Strain

Numerous investigations have been made on stress-strain relationships and have involved a variety of types of deformations (1, 2, 4, 74, 125–128). Typical experimental results for elongation and compression (biaxial extension) for unswollen and swollen networks are shown schematically in Figure 3 (4). The constrained junction theory successfully describes most of the features of these isotherms. Specifically, the decrease in modulus $[f^*]$ with increase in elongation α is viewed as the deformation's becoming more nonaffine as the stretching of the network chains decreases chain-junction entangling; this in turn increases junction fluctuations. The observation that the decrease in $[f^*]$ is less in the case of swollen networks results from the swelling diluent decreasing some of this entangling even at low elongations. The observation that the isotherms become approxi-

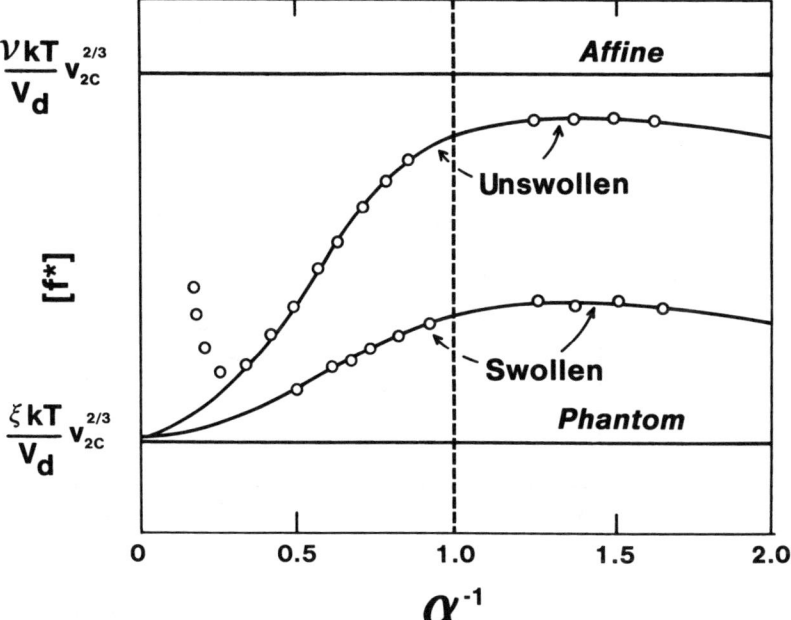

Figure 3 The reduced stress or modulus shown schematically as a function of the reciprocal of the relative length of the sample, for both elongation ($\alpha > 1$) and compression ($\alpha < 1$) (4). The *upper* and *lower horizontal lines* represent results from affine and phantom network models, respectively. *Circles* show representative data from experiments, and the *curves* are from the constrained junction theory.

mately horizontal in compression is also successfully predicted by the theory (4, 5). The upturn in [f^*] at very high elongations is a non-Gaussian effect not accounted for in the theory and must be treated separately.

Interpretation of the upturn in [f^*] in terms of the limited extensibility of the network chains first requires demonstration that reinforcement from strain-induced crystallization is not a significant part of the effect (129). Failure to test for this by swelling or increase in temperature has caused a great deal of confusion in the literature. Some upturns observed for bimodal networks of poly(dimethylsiloxane), known to be uncompromised by strain-induced crystallization, are shown by the symbols in Figure 4

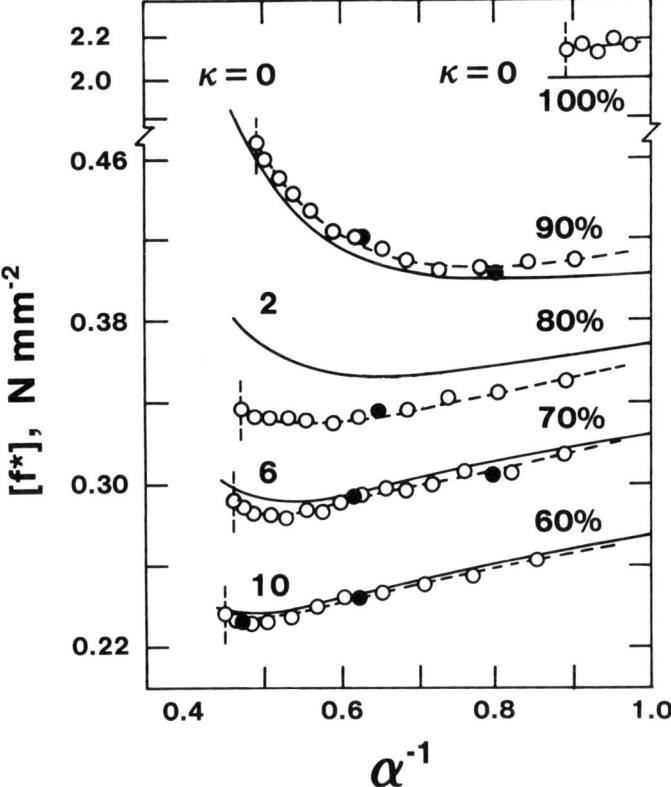

Figure 4 Mooney-Rivlin plots for some bimodal PDMS networks ($M_c = 660/18,500$ g mol^{-1}) (25). The *circles* and *dashed lines* represent experimental data; the *filled circles* locate data taken out of sequence to test for reversibility. The *solid curves* give the results of the theoretical calculations. Each set of experimental data and corresponding theoretical curve is labeled with the mol % of short chains present in the network. The values of the parameter κ specified for the theoretical curves are those required to approximate the observed decreases in reduced stress before the upturns that are modeled by the Fixman-Alben distribution.

(25). The entire isotherms, including the upturns, are seen to be well represented by the curves, which were calculated with the Fixman-Alben modification (130) of the Gaussian distribution function combined with the constrained junction theory and reasonable values of the constraint parameter κ. Experimental results such as these have also been successfully interpreted in terms of series expansions (131) or Monte Carlo simulations based on rotational isomeric state models (91).

Swelling

Experiment and theory are in at least approximate agreement regarding swelling, in that estimates of degree of cross linking from equilibrium swelling are generally in fair agreement with values obtained from mechanical properties (1, 2).

A major unresolved issue concerns the observation that the swelling or dilation modulus frequently goes through an unanticipated maximum with increase in degree of swelling (132, 133). Modifications of existing theories to reproduce this feature quantitatively have not been very successful. In a worst-case scenario, this discrepancy could undermine one of the crucial assumptions in swelling theory, namely that the mixing and elastic contributions to the swelling free energy are separable and simply additive (133).

Birefringence

The reduced birefringence shows some properties that parallel the reduced stress or modulus. In other respects, however, it can be quite different, showing for example significant *increases* with increase in degree of swelling. Nonetheless, experimental results and those from the constrained junction theory seem to be in at least fair agreement (60, 61).

Orientation

Experimental results on the elongation and swelling dependence of the reduced orientation seem to be well reproduced by theory (62, 63). Preparing networks by cross-linking a polymer in solution (134) can give very useful orientation results, but it is essential in such cases to account for changes in reference state in the interpretation of the data (135).

Scattering

The most important elasticity result of this type is obtained by small-angle neutron scattering. In particular, such studies yield information on how the radius of gyration of the network chains transforms with elongation (4). Preliminary comparisons between theory and experiment are quite encouraging (113), but many issues remain unresolved in this relatively new approach to characterizing elastomeric behavior.

SUMMARY

Although much has been accomplished in the area of rubberlike elasticity, much also remains to be done. Among the most important topics are (4) (a) improved understanding of dependence of T_g and T_m on polymer structure, (b) preparation and characterization of high-performance elastomers, (c) new cross-linking techniques, (d) improved understanding of network topology, (e) more experimental results for deformations other than elongation and swelling, (f) better characterization of segmental orientation, (g) more detailed understanding of critical phenomena and gel collapse, (h) additional molecular characterization by using NMR spectroscopy and various scattering techniques, (i) study of possibly novel properties of bioelastomers, and (j) improved molecular understanding of reinforcing effects of filler particles in elastomers.

ACKNOWLEDGMENTS

J.E.M. wishes to acknowledge the financial support generously provided by the National Science Foundation, Army Research Office, and Air Force Office of Scientific Research, and fellowship support for students by the Dow Corning Corporation and IBM-San Jose.

Literature Cited

1. Flory, P. J. 1953. *Principles of Polymer Chemistry*. Ithaca, NY: Cornell Univ. Press
2. Treloar, L. R. G. 1975. *The Physics of Rubber Elasticity*. Oxford: Clarendon. 3rd ed.
3. Morawetz, H. 1985. *Polymers: The Origins and Growth of a Science*. New York: Wiley-Interscience
4. Mark, J. E., Erman, B. 1988. *Rubberlike Elasticity. A Molecular Primer*. New York: Wiley-Interscience
5. Eichinger, B. E. 1983. *Annu. Rev. Phys. Chem.* 34: 359
6. Heinrich, G., Straube, E., Helmis, G. 1988. *Adv. Polym. Sci.* 85: 33
7. Furukawa, J. 1985. *Makromol. Chem. Suppl.* 14: 3
8. Kilian, H. G., Enderle, H. F., Unseld, K. 1986. *Colloid Polym. Sci.* 264: 866
9. Edwards, S. F., Vilgis, T. A. 1988. *Rep. Prog. Phys.* 51: 243
10. Mark, J. E., Lal, J., eds. 1982. *Elastomers and Rubber Elasticity*. Washington, DC: Am. Chem. Soc.
11. Labana, S. S., Dickie, R. A., eds. 1984. *Characterization of Highly Cross-Linked Polymers*. Washington, DC: ACS
12. Lal, J., Mark, J. E., eds. 1986. *Advances in Elastomers and Rubber Elasticity*. New York: Plenum
13. Singler, R. E., Byrne, C. A., eds. 1987. *Elastomers and Rubber Technology*. Washington, DC: US GPO
14. Kramer, O., ed. 1988. *Biological and Synthetic Polymer Networks*. London: Elsevier
15. Queslel, J. P., Mark, J. E. 1986. Elasticity. In *Wiley-Interscience Encyclopedia of Polymer Science and Engineering*. New York: Wiley-Interscience
16. Queslel, J. P., Mark, J. E. 1987. Rubberlike elasticity. In *Encyclopedia of Physical Science and Technology*. New York: Academic
17. Dusek, K. 1986. *Adv. Polym. Sci.* 78: 1

18. Stepto, R. F. T. 1986. In Ref. 12
19. Miller, D. R., Macosko, C. W. 1987. *J. Polym. Sci. Polym. Phys. Ed.* 25: 2441
20. Mark, J. E. 1981. *J. Chem. Educ.* 58: 898
21. Gottlieb, M., Macosko, C. W., Benjamin, G. S., Meyers, K. O., Merrill, E. W. 1981. *Macromolecules* 14: 1039
22. Mark, J. E. 1985. *Polym. J.* 17: 265; *Acc. Chem. Res.* 18: 202
23. Mark, J. E. 1979. *Makromol. Chem. Suppl.* 2: 87
24. Deleted in proof
25. Erman, B., Mark, J. E. 1988. *J. Chem. Phys.* 89: 3314 and relevant references cited therein
26. Mark, J. E. 1984. In *Physical Properties of Polymers*, ed. J. E. Mark, A. Eisenberg, W. W. Graessley, L. Mandelkern, J. L. Koenig. Washington, DC: Am. Chem. Soc.
27. Garrido, L., Mark, J. E., Clarson, S. J., Semlyen, J. A. 1985. *Polym. Commun.* 26: 53
28. Noda, I., Dowrey, A. E., Marcott, C. 1987. In *Fourier Transform Infrared Characterization of Polymers*, ed. H. Ishida. New York: Plenum
29. Deleted in proof
30. Ogden, R. W. 1986. *Rubber Chem. Technol.* 59: 361
31. Guth, E., James, H. M. 1941. *Ind. Eng. Chem.* 33: 624
32. James, H. M., Guth, E. 1942. *Ind. Eng. Chem.* 34: 1365
33. James, H. M. 1947. *J. Chem. Phys.* 15: 651
34. James, H. M., Guth, E. 1947. *J. Chem. Phys.* 15: 669
35. James, H. M., Guth, E. 1953. *J. Chem. Phys.* 21: 1039
36. Flory, P. J. 1976. *Proc. R. Soc. London Ser. A* 351: 351
37. Duiser, J. A., Staverman, A. J. 1965. *Physics of Non-Crystalline Solids*, ed. J. A. Prins. Amsterdam: North-Holland
38. Eichinger, B. E. 1972. *Macromolecules* 5: 496
39. Graessley, W. W. 1975. *Macromolecules* 8: 186
40. Graessley, W. W. 1975. *Macromolecules* 8: 865
41. Pearson, D. S. 1977. *Macromolecules* 10: 696
42. Kloczkowski, A., Mark, J. E., Erman, B. 1989. *Macromolecules* 22: 1423
43. Erman, B., Kloczkowski, A., Mark, J. E. 1989. *Macromolecules* 22: 1432
44. Kuhn, W. 1936. *Kolloidzeitschrifte* 76: 258
45. Kuhn, W. 1938. *Angew. Chem.* 51: 640
46. Kuhn, W. 1946. *J. Polym. Sci.* 1: 380
47. Flory, P. J., Rehner, J. 1943. *J. Chem. Phys.* 11: 512
48. Flory, P. J., Rehner, J. 1943. *J. Chem. Phys.* 11: 521
49. Flory, P. J. 1950. *J. Chem. Phys.* 18: 108
50. Wall, F. T. 1942. *J. Chem. Phys.* 10: 132
51. Wall, F. T. 1943. *J. Chem. Phys.* 11: 527
52. Wall, F. T., Flory, P. J. 1951. *J. Chem. Phys.* 19: 1435
53. Ronca, G., Allegra, G. 1975. *J. Chem. Phys.* 63: 4990
54. Flory, P. J. 1977. *J. Chem. Phys.* 66: 5720
55. Flory, P. J. 1977. In *Contemporary Topics in Polymer Science*, ed. E. M. Pearce, J. R. Schaefgen, 2: 1. New York: Plenum
56. Flory, P. J. 1979. *Polymer* 20: 1317
57. Flory, P. J. 1985. *Polym. J.* 17: 1
58. Flory, P. J., Erman, B. 1982. *Macromolecules* 15: 800
59. Erman, B., Flory, P. J. 1982. *Macromolecules* 15: 806
60. Erman, B., Flory, P. J. 1983. *Macromolecules* 16: 1601
61. Erman, B., Flory, P. J. 1983. *Macromolecules* 16: 1607
62. Erman, B., Monnerie, L. 1985. *Macromolecules* 18: 1985
63. Queslel, J. P., Erman, B., Monnerie, L. 1985. *Macromolecules* 18: 1991
64. Erman, B. 1981. *J. Polym. Sci. Polym. Phys. Ed.* 19: 829
65. Freed, K. F. 1971. *J. Chem. Phys.* 55: 5588
66. Edwards, S. F. 1968. *J. Phys. A Gen. Phys.* 1: 15
67. Edwards, S. F. 1971. In *Polymer Networks: Structural and Mechanical Properties*, ed. A. J. Chompff, S. Newman. New York/London: Plenum
68. Edwards, S. F., Freed, K. F. 1970. *J. Phys. C* 3: 739, 750, 760
69. Edwards, S. F. 1971. *Br. Polym. J.* 3: 140
70. Deam, R. T., Edwards, S. F. 1976. *Philos. Trans. R. Soc. London Ser. A* 280: 317
71. Ball, R. C., Doi, M., Edwards, S. F., Warner, M. 1981. *Polymer* 22: 1010
72. Thirion, P., Weil, T. 1984. *Polymer* 25: 609
73. Brotzman, R. W., Flory, P. J. 1987. *Macromolecules* 20: 351
74. Edwards, S. F., Vilgis, T. 1986. *Polymer* 27: 483
75. Flory, P. J. 1969. *Statistical Mechanics of Chain Molecules*. New York: Interscience
76. Nagai, K. 1964. *J. Chem. Phys.* 40: 2818

77. Liberman, M. H., Abe, Y., Flory, P. J. 1972. *Macromolecules* 5: 550
78. Ingwall, R. T., Czurylo, E. A., Flory, P. J. 1973. *Biopolymers* 12: 1137
79. Liberman, M. H., DeBolt, L. C., Flory, P. J. 1974. *J. Polym. Sci. Polym. Phys. Ed.* 12: 187
80. Erman, B., Bahar, I. 1988. *Macromolecules* 21: 452
81. Flory, P. J., Abe, Y. 1969. *Macromolecules* 2: 335
82. Abe, Y., Flory, P. J. 1970. *J. Chem. Phys.* 52: 2814
83. DeBolt, L. C., Mark, J. E. 1987. *Polymer* 28: 416
84. Mark, J. E., Curro, J. G. 1983. *J. Chem. Phys.* 79: 5705
85. Mark, J. E., Curro, J. G. 1984. *J. Chem. Phys.* 80: 5262
86. Mark, J. E., Curro, J. G. 1984. *J. Chem. Phys.* 81: 6408
87. Mark, J. E., Curro, J. G. 1985. *J. Polym. Sci. Polym. Phys. Ed.* 23: 2629
88. Curro, J. G., Mark, J. E. 1985. *J. Chem. Phys.* 82: 3820
89. Curro, J. G., Schweizer, K. S., Adolf, D., Mark, J. E. 1986. *Macromolecules* 19: 1739
90. DeBolt, L. C., Mark, J. E. 1988. *J. Polym. Sci. Polym. Phys. Ed.* 26: 865
91. Curro, J. G., Mark, J. E. 1984. *J. Chem. Phys.* 80: 4521
92. Flory, P. J. 1961. *Trans. Faraday Soc.* 57: 829
93. Langley, N. R. 1968. *Macromolecules* 1: 348
94. Oppermann, W., Rennar, N. 1987. *Prog. Colloid Polym. Sci.* 75: 49
95. Birshtein, T. M., Ptitsyn, O. B. 1966. *Conformations of Macromolecules*. New York: Interscience
96. Kuhn, W., Grün, F. 1942. *Kolloid. Zh.* 101: 248
97. Gottlieb, Yu. Ya. 1957. *Zh. Tekh. Fiz.* 27: 707
98. Erman, B., Flory, P. J. 1983. *Macromolecules* 16: 1601
99. Erman, B., Flory, P. J. 1983. *Macromolecules* 16: 1607
100. Gent, A. N. 1969. *Macromolecules* 2: 262
101. Hermans, J. J. 1946. *J. Colloid Sci.* 1: 235
102. Treloar, L. R. G. 1947. *Trans. Faraday Soc.* 43: 277
103. Ishihara, A., Hashitsume, N., Tatibana, M. 1952. *J. Appl. Phys.* 23: 508
104. Treloar, L. R. G. 1954. *Trans. Faraday Soc.* 50: 881
105. Roe, R.-J., Krigbaum, W. R. 1964. *J. Appl. Phys.* 35: 2215
106. Tanaka, T., Allen, G. 1977. *Macromolecules* 10: 426
107. Jarry, J. P., Monnerie, L. 1979. *Macromolecules* 12: 316
108. Read, B. F., Stein, R. S. 1968. *Macromolecules* 1: 116
109. Bastide, J., Duplessix, R., Picot, C. 1984. *Macromolecules* 17: 83
110. Beltzung, M., Picot, C., Herz, J. 1984. *Macromolecules* 17: 663
111. Pearson, D. S. 1977. *Macromolecules* 10: 696
112. Ullman, R. 1979. *J. Chem. Phys.* 71: 436
113. Erman, B. 1987. *Macromolecules* 20: 1917
114. Bastide, J., Boue, F. 1986. *Physica A* 140: 251
115. Boue, F., Bastide, J., Buzier, M., Collete, C., Lapp, A., Herz, J. 1987. *Prog. Colloid Polym. Sci.* 75: 152
116. Picot, C. 1987. *Prog. Colloid Polym. Sci.* 75: 83
117. Oeser, R., Ewen, B., Richter, D., Farago, B. 1988. *Phys. Rev. Lett.* 60: 1041
118. Erman, B., Flory, P. J. 1985. *Macromolecules* 19: 2342
119. Queslel, J. P., Mark, J. E. 1985. *Adv. Polym. Sci.* 71: 229
120. Dusek, K., Prins, W. 1969. *Adv. Polym. Sci.* 6: 1
121. Tanaka, T. 1978. *Phys. Rev. Lett.* 40: 820
122. Janas, V. F., Rodriguez, F., Cohen, C. 1980. *Macromolecules* 13: 978
123. Ilavsky, M. 1981. *Polymer* 22: 1687
124. Ilavsky, M., Bouchal, K. 1988. In Ref. 14
125. Pak, H., Flory, P. J. 1979. *J. Polym. Sci. Polym. Phys. Ed.* 17: 1845
126. Brotzman, R. W., Mark, J. E. 1986. *Macromolecules* 19: 667
127. Brereton, M. G., Klein, P. G. 1988. *Polymer* 29: 970
128. Gottlieb, M., Gaylord, R. J. 1983. *Polymer* 24: 1644
129. Mark, J. E. 1979. *Polym. Eng. Sci.* 19: 254, 409
130. Fixman, M., Alben, R. 1973. *J. Chem. Phys.* 58: 1553
131. Menduiña, C., Freire, J. J., Llorente, M. A., Vilgis, T. 1986. *Macromolecules* 19: 1212
132. Gottlieb, M., Gaylord, R. J. 1984. *Macromolecules* 17: 2024
133. Neuburger, N. A., Eichinger, B. E. 1988. *Macromolecules* 21: 3060
134. Dubault, A., Deloche, B., Herz, J. 1987. *Macromolecules* 20: 2096
135. Erman, B., Mark, J. E. 1989. *Macromolecules* 22: 480

VECTOR CORRELATIONS IN PHOTODISSOCIATION DYNAMICS

G. E. Hall

Department of Chemistry, Brookhaven National Laboratory, Upton, New York 11973

P. L. Houston

Department of Chemistry, Cornell University, Ithaca, New York 14853-1301

INTRODUCTION

The study of photodissociation dynamics has undergone tremendous growth since Norrish & Porter first identified the internal states of photofragments in the 1950s (1). Tunable laser sources, now available throughout the visible, ultraviolet, and vacuum ultraviolet regions of the spectrum, have largely been responsible for this revolution, since these sources make it possible not only to excite parent molecules to a well-defined internal energy but also to detect photofragments with unprecedented resolution of final state. Several recent reviews have described the new understanding of photodissociation dynamics available from such studies (2–12).

In general, the properties measured in photofragment spectroscopy can be classified into either *scalar* quantities, such as the energy in a particular degree of freedom, or *vector* quantities, such as the direction (and magnitude) of a particular type of motion. In this review, we demonstrate how correlations between the *vector* properties of photofragments can provide a better understanding of the dynamics of the dissociation process. The vector properties of most interest will be (*a*) the relative recoil velocity, **v**, of the departing photofragments, (*b*) the angular momentum vector(s), **J**, of the photofragment(s), and (*c*) the transition dipole moment, μ, of the parent compound. Note that the first two of these vectors, **v** and **J**, are properties of the fragments, whereas the third, μ, is a property of the parent compound. A fourth vector of interest is the electric field, **E**, of the dissociating light; it is defined most easily in a space-fixed frame and serves to align the transition dipole in the laboratory. The physical basis for this alignment is that in the electric dipole approximation the absorption probability is proportional to $|\mathbf{E}\cdot\mathbf{\mu}|^2$, so that immediately following absorption, the parent molecules are aligned such that the probability of finding the angle, θ, between their transition dipole moment and the electric field vector varies as $\cos^2\theta$. The consequences of this parent alignment in the laboratory frame have fascinated scientists for many years.

Recoil Anisotropy: The **E**–μ–**v** *Correlation*

Although there is some indication that scientists recognized the importance of such alignment as early as the 1930s, it was Zare & Herschbach who first drew general attention to the fact that the vector correlation between μ and **v** might lead to an anisotropic distribution of photofragments (13). For any parent molecule there is a fixed angular correlation between the direction of the dipole transition moment and the direction of fragment recoil; usually the recoil is along the bond that breaks during the photo-dissociation process. If the bond-breaking occurs on a time scale short compared to the rotational motion of the parent, the alignment of μ in the laboratory frame by the polarized photodissociation light should lead to an alignment of **v**. Zare gave the angular distribution of velocities as $I(\theta) = (4\pi)^{-1}[1+\beta P_2(\cos\theta)]$ where θ is the angle between **v** and **E** and β is a parameter that describes the degree of anisotropy ($-1 \le \beta \le 2$) (14). Thus, measurement of the anisotropy of the recoil velocity distribution can provide information about the symmetry of the transition in the parent molecule (i.e. about the alignment of the dipole moment in the molecular frame) and about the time scale of the dissociation.

The first observation of an anisotropic distribution of photofragment recoil velocities was made by Solomon, who photolyzed bromine or iodine inside a hemispheric bulb and observed that tellurium coated onto the inside surface of the bulb was etched by the recoiling halogen atoms in an anisotropic fashion (15). The next advance was made by Busch et al (16) and by Diesen, Wahr & Adler (17), who used a quadrupole mass spectrometer to detect the arrival time and angular distributions of photofragments. An alternate technique suggested by Zare & Herschbach (13, 18) is based on the Doppler effect, which provides both the direction and magnitude of the recoil velocity. Experimental confirmation of this technique was provided first by Schmiedl et al, who monitored the Doppler profile of the H atom produced in the 266-nm photodissociation of HI

(19). Two distinct magnitudes of the H atom velocity were observed, corresponding to two electronic states of the sibling I atom. Two channels displayed different angular distributions, demonstrating a further correlation between the magnitude and direction of **v**.

Rotational Alignment: The E–μ–J Correlation

A second consequence of the alignment of parent molecules by the dissociating photon is that a laboratory alignment may be induced in the rotational angular momentum of the photofragment. This alignment, like the recoil anisotropy, is strongest when the dissociation takes place in a time short compared to parent rotation. It arises because there is usually an angular correlation between the transition dipole moment and the plane of fragment rotation. Since μ is aligned in the laboratory frame by **E**, **J** will also be aligned. If the fragment is born in an excited state, its emission will be polarized, as first described by van Brunt & Zare (20), whereas if the fragment is formed in its ground state, it will preferentially absorb light of a specific polarization, as discussed by Gouedard & Lehmann (21), Fano & Macek (22), and Greene & Zare (23, 24).

The first experimental observation of rotational alignment in the photodissociation of a polyatomic molecule was reported by Chamberlain & Simons (25, 26), who saw that dissociation of H_2O, HCN, and BrCN by linearly polarized light gave rise to polarized emission from the excited photofragments. Results prior to 1984 based on detection of the polarization of photofragment emission have been summarized by Simons (27).

The v–J Correlation and the E–μ–(v–J) Correlation

It should come as no surprise that since both **J** and **v** are correlated with μ, they must also be correlated with each other. Although Case, McClelland & Herschbach suggested as early as 1978 that this correlation might have important consequences (28), it was several years before it was found nearly simultaneously and independently by four laboratories that this correlation has a profound effect on the Doppler profile of individual fragment rotational transitions (29–32). An important difference between the **v**–**J** correlation and the correlations discussed above is that, whereas the magnitude of the anisotropy and alignment both depend on the relative rates of dissociation and parent rotation, the **v**–**J** correlation does not. This is because anisotropy and alignment depend on having μ aligned in the laboratory frame; they tend to be lost if the parent rotates before it dissociates. The **v**–**J** correlation, however, is not made until the instant of dissociation; it is completely independent of the laboratory frame and so should always be informative. Of course, when the dissociation is rapid, both **v** and **J** will also display a correlation in the laboratory frame. In this case all three of the vectors μ, $\mathbf{v}$, and $\mathbf{J}$ will be aligned relative to one another and relative to a fixed axis ($\mathbf{E}$) in the laboratory frame.

Three approaches have recently been used to describe how the angular distributions between these three vectors are related to the shape of the fragment Doppler profile as measured by laser-induced fluorescence. One approach, developed by Dixon, is based on expansion of the correlated angular momentum and velocity distribution in terms of the expectation values of bipolar harmonics (33). Two other approaches, developed at Cornell, are based on averages for the absorption and fluorescence processes (29, 34). Limiting cases for specific relationships between μ, $\mathbf{v}$, and $\mathbf{J}$ can be calculated by performing the averages using classical mechanics (34), while a more general solution can be found by using the corresponding quantum mechanical density matrix method (29, 34). The results from all three methods appear to be in agreement (34). In addition, Balint-Kurti & Shapiro have recently (35) related the Doppler profile to the photodissociation amplitudes appearing in their theory of photodissociation of triatomic molecules (36). For the purposes of the discussion that follows, we summarize here the results of Dixon's analysis (33). For one-photon laser-induced fluorescence using linearly polarized light, the absorption profile can be written as

$$I(\cos \chi) = (2\Delta v_\mathrm{D})^{-1}[g_0 + g_2 P_2(\cos \chi) + g_4 P_4(\cos \chi) + g_6 P_6(\cos \chi)], \qquad 1.$$

where $\cos \chi$ is the fraction of the maximum Doppler shift, Δv_D, for a single velocity. The g_i can be expressed in terms of the renormalized moments in the bipolar spherical harmonic expansion, $\beta_0^K(k_1 k_2)$. The recoil anisotropy parameter usually called β is given by $\beta = 2\beta_0^2(20)$, where $\beta_0^2(20)$ ranges between $+1$ for a parallel transition and $-\frac{1}{2}$ for a perpendicular one. The moment $\beta_0^2(02)$ is proportional to the usual alignment parameter, $\beta_0^2(02) = \frac{5}{4} A_0^{(2)}$, and ranges from $+1$ for rotation parallel to $\mathbf{E}$ to $-\frac{1}{2}$ for rotation perpendicular to $\mathbf{E}$. The moment $\beta_0^0(22) = \langle P_2(\cos \mathbf{v} \cdot \mathbf{J}) \rangle$, sometimes called the helicity, ranges from $+1$ for $\mathbf{v}$ and $\mathbf{J}$ parallel to $-\frac{1}{2}$ for $\mathbf{v}$ and $\mathbf{J}$ perpendicular. Other bipolar moments are also available from careful analysis of the Doppler profiles.

One limitation on the quantitative analysis of the Doppler lineshapes arises from any spread in the velocity distribution of the probed photofragment. Each velocity component could in principle be characterized by a different set of vector properties, like the two components of the HI dissociation mentioned above (19). Experiments to date on the dissociation of polyatomic molecules have been limited to determining velocity-averaged values of the vector correlations from the Doppler-resolved LIF lineshapes.

Further Developments in Quantifying Vector Correlations

Several recent papers have further described new and interesting methods for determining vector correlations. Techniques for measurement of alignment and population by $1+1$ (one photon excitation, one photon fluorescence) LIF have been developed by using a variety of approaches. Case, McClelland & Herschbach (28) and McCaffery and co-workers (37–42) have used a density matrix approach, whereas Greene & Zare (23, 24) employed a tensor formalism based on work by Gouedard & Lehmann (21) and Fano & Macek (22). Kummel, Sitz & Zare have recently shown how both alignment and orientation moments can be obtained by using linearly, circularly, or elliptically polarized light for the excitation and by analyzing the emitted light with a quarter-wave plate and a linear polarizer (43a,b). Saturation effects in $1+1$ LIF measurements of populations and polarization have been reviewed by Altkorn & Zare (44).

The use of $1+1$ multiphoton ionization has also been considered. Jacobs & Zare (45) have developed both quantum and classical methods for accounting for the effects of saturation and intermediate state alignment in $1+1$ MPI measurements. Jacobs, Madix & Zare (46) then applied these methods to extract populations and alignment factors in NO. Mons & Dimicoli have discussed the use of multiphoton ionization spectroscopy to measure the state of the fragment following photodissociation with a linearly polarized source (47). As noted in earlier work (48–51), the time-of-flight (TOF) of the ionized fragments is a function of their projection along the spectrometer axis. For an MPI process that has a known sensitivity to the angular distribution of the fragment angular momentum, Mons & Dimicoli provide techniques for extracting the bipolar description of the angular correlations between μ, $\mathbf{v}$, and $\mathbf{J}$ from the TOF distribution. Winniczek et al (52) have demonstrated a novel technique in which the alignment is detected by the circular dichroism of photoelectron angular distributions.

The use of $2+n$ processes to probe photofragment population, alignment, and orientation has also been discussed (30, 53–55). The work of Dubs, Brühlman & Huber using two-photon $(2+1)$ laser induced fluorescence has been extended by Docker to include the use of arbitrary experimental geometry or polarization (53). In addition, Docker used the density matrix approach of Case, McClelland & Herschbach (28) to re-derive the equations for $1+1$ processes. Kummel, Sitz & Zare have used a tensor formalism to obtain similar information for $2+n$ processes, covering the special cases of $2+1$ LIF or $2+n$ MPI in which the ionization step is saturated (i.e. ionization is independent of orientation or alignment) (54). If the elliptically polarized light is created by passing linearly polarized light through a quarter-wave plate, the alignment and orientation moments can be independently determined by using a single experimental excitation-detection geometry.

The degree to which the alignment of photofragment angular momentum is reduced due to rotation when the parent compound has a finite predissociation lifetime has been reconsidered. The earlier work of Loge & Zare (56) for dissociation of a triatomic parent molecule has been extended and in some cases corrected by the more recent work of Nagata et al (57).

Outline of This Review

The review that follows treats the photofragmentation of neutral triatomic and polyatomic molecules. Particular attention is paid to those parent molecules in which Doppler-resolved LIF has provided vector correlations that illuminated the photofragmentation dynamics. A representative, but not encyclopedic, selection of other examples suffices to fill our space quota. This review does not discuss in detail the use of lambda doublet populations to deduce molecular frame alignments except when this technique is used in conjunction with other vector correlation methods. The reader interested in more detail on lambda doublet populations may consult the extensive references in a recent article by Alexander and a host of co-authors (58). We do not review in detail the work on polarization and anisotropy effects in the photodissociation of diatomic molecules, although we note in passing the elegant theoretical and experimental work on interference effects in such dissociations (59–70).

TRIATOMIC MOLECULES

ICN, BrCN, ClCN, and HCN

Cyanogen halides and hydrogen cyanide have played an important role in the history of vector correlation applications to molecular photodissociation dynamics. Chamberlain & Simons demonstrated in an early experiment that the $CN(B^2\Sigma^+)$ emission following photodissociation of HCN and BrCN at $\lambda \geq 125$ nm was highly polarized and inferred that the relevant excited states of these molecules are linear and of A' symmetry (26). The photodissociation of ClCN at 157 nm was later examined by Guest, O'Halloran & Zare, who found that the rotationally resolved CN emission was polarized in a manner consistent with a direct dissociation mechanism (71). After the observed alignment was corrected for the depolarizing effects of nuclear and electron spin in the photofragment, it was found that the alignment was independent of rotational state.

ICN dissociates from its first absorption continuum to yield CN in its ground electronic state and either $I\,(\equiv I\,{}^2P_{3/2})$ or $I^*\,(\equiv I\,{}^2P_{1/2})$. These two spin orbit states are separated by 7603 cm^{-1}. In a pioneering study, Nadler et al resolved the Doppler profiles of individual laser-induced fluorescence rotational lines of the CN fragment (72). From the widths of the Doppler profiles they were able to determine that the I+CN channel selectively populates high N″ levels, whereas the I*+CN channel populates low N″ levels. The shapes of the Doppler profiles indicated that both channels are the result of absorption via a parallel transition: $\beta_I = 1.3$, $\beta_{I^*} = 1.6$. Figure 1 provides sample spectra at different excitation geometries. Individual spin-rotation components were partially resolved and were found to be unequally populated.

Hall, Sivakumar & Houston examined the rotational alignment of the CN fragment following the dissociation of ICN in the 235–290 nm region and reported a strong, though not perfect, tendency for the CN rotation vector to lie perpendicular to the electric vector of the photolysis light (73). They concluded that the transition moment appeared to be a mixture of 85% parallel and 15% perpendicular near 266 nm and that it became more parallel in nature to shorter or longer wavelengths. O'Halloran, Joswig & Zare have also reported measurements of the CN alignment following dissociation of ICN at 248 nm (74). Although their measurements are in good agreement with those of Hall et al, they interpret the lack of perfect alignment to indicate the presence of strong, nonplanar interactions between the developing CN rotation and the nuclear spin and/or electronic orbital angular momentum of the iodine atom. These forces change the directions of both the velocity and angular momentum vectors, causing a decrease in the magnitudes of both β and the alignment parameter. These same out-of-plane forces have been held responsible for the population differences between the F_1 and F_2 fine structure components (75).

An exciting experiment has been recently reported by Hasselbrink, Waldeck & Zare in which ICN is dissociated with circularly polarized light at 248 nm (76). By analyzing the variation of the CN LIF signal as the probe laser is alternately left- or right-circularly polarized, the authors determined that the CN fragments were oriented, i.e. that they rotate in space with a preferred clockwise or counter-clockwise motion. Evidence was found that the direction of CN orientation correlated with the spin-orbit state of the sibling iodine atom. Black, Waldeck & Zare have recently improved the experimental apparatus to allow measurement of the orientation as a function of Doppler shift (77).

Guest & Webster have reported the alignment of the CN fragment following dissociation of ICN at 157 nm (78). The rotational alignments show a large variation with both rotational and vibrational level, indicating that several channels are present for dissociation at this wavelength.

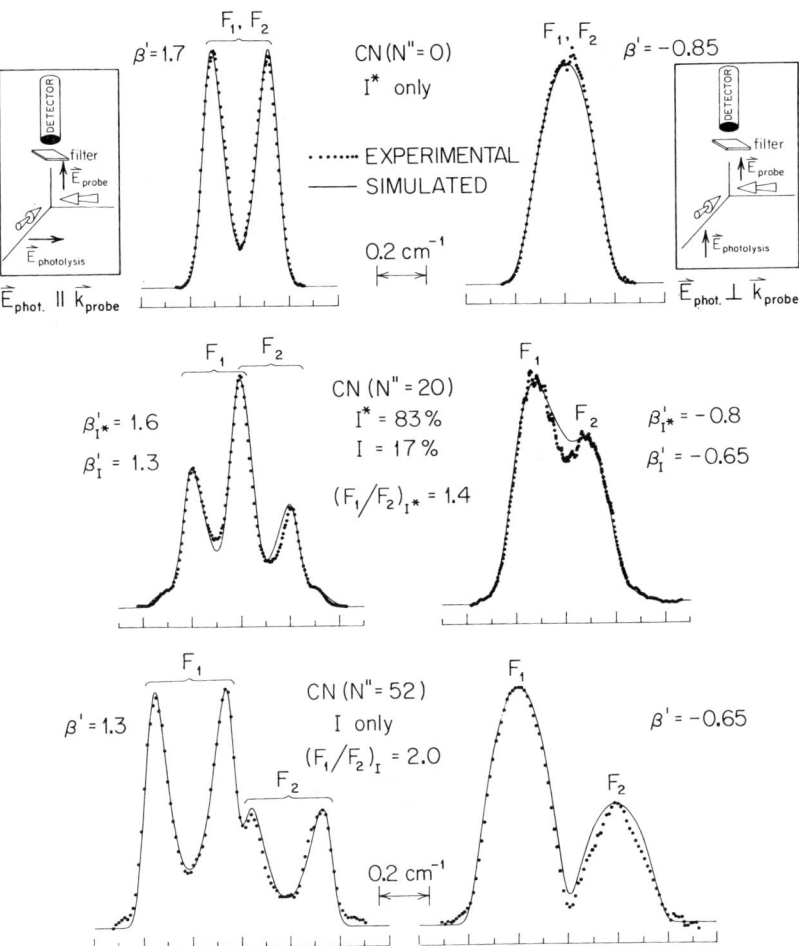

Figure 1 Sub-Doppler resolution (0.07 cm^{-1}) spectra of CN($X\,^2\Sigma$, v = 0), showing spatial anisotropies and separate F_1 and F_2 states. The *left-* and *right-hand-side entries* are associated with $\mathbf{E}_{\text{photolysis}}$ parallel and perpendicular, respectively, to $\mathbf{k}_{\text{probe}}$ (see the inserts). As N'' increases, F_1 and F_2 become apparent, and the unequal populations of these states at $N'' = 52$ is unmistakable. The *solid lines* are simulations to the data. [Reprinted with permission from Ref. (72).]

H_2O

The photodissociation of water has been examined extensively at a variety of wavelengths, and the results provide several examples of how vector correlations can be used to elucidate the dynamics of the dissociation process. We divide the discussion below into three main subsections: one-photon dissociation at wavelengths near 130 nm, two-photon dissociation near 248.5 nm, and one-photon dissociation at 157 nm.

It is first useful to summarize the excited states of water and the mechanisms by which they dissociate to give OH either in its excited ($A\,^2\Sigma^+$) or ground ($X\,^2\Pi$) states. The $C\,^1B_1$ state of water, as we see below, is predissociated by the $B\,^1A_1$ surface, which correlates to electronically excited OH. The $B\,^1A_1$ surface itself has a conical intersection with the ground $X\,^1A_1$ surface and a linear intersection with the $A\,^1B_1$ surface, both of which correlate to OH in the electronic ground state. Crossings among these states affect the branching ratio between excited and ground state OH, as well as the rotational excitation, alignment, and other properties of the OH product.

Simons and his co-workers were the first to observe that the rotationally unresolved fluorescence from OH ($A\,^2\Sigma^+$) produced in the photodissociation of water near 130 nm was polarized (79, 80). These results led to the conclusion that the rotationally excited OH fragments are formed via predissociation of the $^1A_1$ electronic state. Improvements to the experimental apparatus subsequently allowed rotational resolution of the fluorescence (81). The rotational alignments tended to be a maximum at rotational levels close to the limit imposed by energy conservation, and both the alignment and populations fell sharply for rotational levels beyond this limit, where the additional energy needed to be supplied by the parent rotation (82). The sharply decreasing alignment was taken to indicate that the parent rotational levels that contributed most to the dissociation were those in which the rotation was principally about the a axis (in the plane of the molecule), since rotation about the c axis (perpendicular to the plane) would not degrade the alignment. The authors concluded therefore that dissociation via the rotationally assisted channel was promoted when $J' = K_a$ in the parent water molecule.

Simons and his coworkers have also investigated the two-photon dissociation of water with a tunable KrF laser at wavelengths near 248.5 that excite the $C\,^1B_1$ state (83–85). Rotationally resolved fluorescence from the OH(OD) $A\,^2\Sigma^+$ state was obtained as the KrF laser was scanned across individual J_{KaKc} levels of the parent. The OH(OD) $A\,^2\Sigma^+$ fragment is formed rotationally hot as a result of the large change in bond angle in going from either the $X\,^1A_1$ or $C\,^1B_1$ state to the linear and dissociative $B\,^1A_1$ state. A potential energy diagram showing the energy as a function of bending angle is provided in Figure 2. Examination of the alignment of the OH products demonstrated that the alignment was larger for excitation of the water continuum than for excitation of the individual J_{KaKc} lines of the water $C\,^1B_1$ state. The reason for this is that predissociation of the C

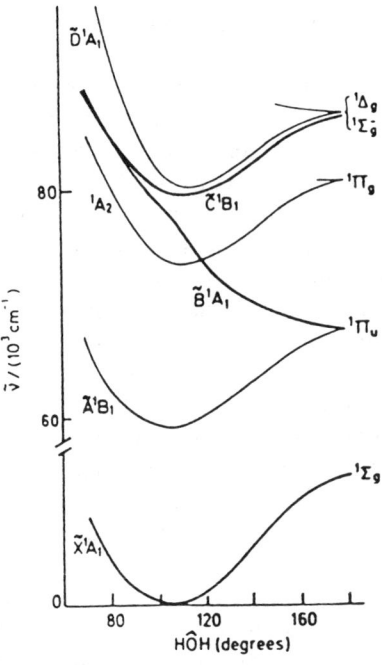

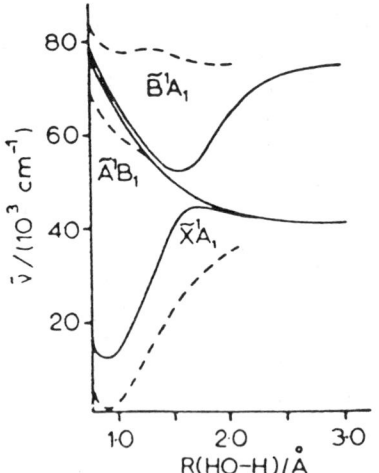

Figure 2 (*Top*) Bending potentials for the lower electronic states of H$_2$O with r(O–H) fixed at 0.96 Å. (*Bottom*) Asymmetric stretching potentials for the ground and first two singlet excited electronic states of H$_2$O. One of the O–H bond lengths is fixed at 0.96 Å. The *solid curves* show the potential for angles near linearity (170°), whereas the dotted lines are for $\theta = 104.5°$. Note the barrier to direct dissociation in bent configurations of the $B\,^1A_1$ state. [Reprinted with permission from Ref. (86).]

state requires that $J_a^2 > 0$, and rotation about the a axis (located in the plane of the water molecule) destroys the alignment. By contrast, excitation to the continuum, which the authors argue is caused by the $B\,^1A_1 \leftarrow X\,^1A_1$ transition, would lead to direct dissociation and substantially less degradation of the alignment. Water emission from the excited $C\,^1B_1$ state to the $A\,^1A_1$ state was observed and used to confirm the model for the dynamics of dissociation once the water molecule reaches the dissociative $^1A_1$ state (86). Direct trajectories, those which do not cross the linear intersection of the $B\,^1A_1$ state and the $A\,^1B_1$ state, are found to correlate to high product rotation, whereas indirect trajectories, which cross this intersection many times, lead to low product rotation, in agreement with earlier theoretical predictions (87).

Meijer et al have challenged the interpretation of one aspect of the results described above (88). These authors used a much more monochromatic tunable KrF laser and found that there was little or no nonresonant fluorescence background, thus calling into question the existence of the $B\,^1A_1$ continuum in this region. However, Simons and his co-workers have pointed out that their assignment was not based on the absorption intensity, but rather on the alignment measurements (89). At present, the extent to which there is an underlying continuum in the region of the water $C\,^1B_1$ state is certainly in question.

Atkins et al have examined the two-photon dissociation of water at 266 nm and have measured the rotational energy distribution in the OH $A\,^2\Sigma^+$ state (90). The distribution is similar to that obtained in a recent one-photon dissociation at roughly the same energy (91), but the alignment was found to be reduced.

The one-photon dissociation of water at 157 nm has been investigated in exquisite detail, both experimentally (92–96) and theoretically (95–99). This wavelength excites the $C\,^1B_1$ state, so that the transition dipole moment from the ground state ($X\,^1A_1$) is perpendicular to the plane of the water molecule. The ground-state OH ($X\,^2\Pi$) produced by the dissociation is found by using laser-induced fluorescence to be highly polarized, particularly when high rotational levels are probed (92). In a more extensive work (93), a strong J-dependent lambda doublet inversion was also found, which may provide the explanation for the astronomical OH maser. Both the Λ doublet populations and the alignment are a result of a planar dissociation, and both are substantially larger for a rotationally cooled parent molecule than for a room-temperature sample, in which out-of-plane rotations reduce both the Λ doublet selectivity and the alignment. Calculations on the surface of Staemmler & Palma (100) have been performed by Schinke et al (97) and by Balint-Kurti (98). In general, both the rotational state distributions and the Λ doublet state distributions are in excellent agreement with experiment.

A significant experimental advance was made by use of an infrared laser to prepare the water molecule in a particular rotational state of the H_2O (001) vibrational level (95, 96). Dissociation was then effected by a 193 nm photon, which excited the vibrationally excited molecules efficiently while producing little excitation of the water molecules in their ground vibrational level. A complete characterization of the rotational and electronic fine structure state distributions in the OH ($X\,^2\Pi$) product was then possible for dissociation from a well-defined state of the parent, and the agreement with theory was found to be quantitative in all six of the selected quantum numbers. In particular, the strong structure observed in the rotational distributions was matched nearly perfectly by the results of a Franck-Condon calculation. A complete ab initio treatment of the water dissociation in its first absorption band has been presented recently by Engel, Schinke & Staemmler (99). Grunewald et al have recently reported on the dynamics of the water dissociation following the weak absorption at 193 nm (101).

NO_2

The near ultraviolet photodissociation of NO_2 has been investigated at energies from 600 cm^{-1} below to 1700 cm^{-1} above the dissociation threshold for the ground-state NO and O products by Mons & Dimicoli (102). A complete analysis of the NO(v = 1) product was obtained by combining MPI of the NO fragment with TOF mass spectrometry. The negative values of the alignment parameter ($A_0^{(2)} \approx -0.2$) indicate that μ and **J** tend to be perpendicular, whereas the positive values of the anisotropy paramerter ($\beta \approx 0.5$) demonstrate that μ and **v** tend to be parallel.

Two-photon dissociation of NO_2 has also been reported by a number of authors (103–105). Kawasaki et al dissociated the NO_2 with two 308-nm photons and observed that the angular distribution of the NO fragment was described by an expansion of the form $1+\beta_2 P_2(\cos\theta)+\beta_4 P_4(\cos\theta)$, where $\beta_2 = 1.44$ and $\beta_4 = 0.44$ for the two--photon process that yields $NO(X^2\Pi)+O(^3P)$, and $\beta_2 = 0.67$ and $\beta_4 = 0.18$ for the two-photon process that yields $NO(X^2\Pi)+O(^1D)$ (104).

CS_2

Kawasaki et al have also investigated the two-photon photodissociation of CS_2 at 308 nm in a method similar to that used for NO_2 (104). The angular distribution of the sulfur photofragment was described by a value of $\beta_2 = 0.67$, but β_4 was equal to zero within experimental uncertainty because the lifetime of the intermediate CS_2 $^1B_2$ state is long compared to molecular rotation.

NOCl and NOBr

The photodissociation of nitrosyl halides has been investigated by two groups (106–109). Huber and his co-workers have used excitation of NOCl and NOBr at wavelengths near 450 or 470 nm to dissociate the parent molecules and two-photon LIF to probe the NO photofragment (106, 107). The Doppler profiles demonstrated the predictable perpendicular relationship between **v** and **J**, indicating a planar dissociation. The alignment and anisotropy measurements led to an $A'' \leftarrow A'$ assignment of the electronic transition at 470 nm in NOCl. Other interesting correlations were also noted. For example, the spin-orbit levels of the NO were formed in a 2:1 ($F_2:F_1$) ratio in the dissociation of NOCl but in a statistical ratio in the dissociation of NOBr, thus illustrating the possible randomizing effect of the larger spin-orbit coupling of the Br atom. In the dissociation of NOBr it was found that unexcited Br atoms ($^2P_{3/2}$) were always formed in coincidence with vibrationally or highly rotationally excited NO, whereas excited Br($^2P_{1/2}$) atoms were formed in coincidence with low rovibronic excitation of the NO.

Reisler and her co-workers have investigated in detail the photodissociation of NOCl at wavelengths from 180 nm to 610 nm (108–110). The recoil anisotropy, the rotational alignment, and the Λ doublet populations were used to provide the following assignments of the main absorption bands. The E band near 610 nm was assigned to an upper triplet state, $^3A'' \leftarrow ^1A'$; the D band near 480 nm and the C band near 440 nm were both assigned to $^1A'' \leftarrow ^1A'$ transitions, in agreement with the results of Huber and co-workers, with the latter band exciting the v_1 vibration of the upper state; the B band at 330 nm corresponds to excitation from $S_0(^1A')$ to $S_3(^1A')$; and the A band at 190 nm is assigned to excitation from S_0 to S_5 ($^1A'$). An interesting correlation between parent vibrational excitation and NO internal energy was noted for excitation in the lowest energy band (109). NO(v = 0 or 1) is produced when the NO stretch is unexcited or excited, respectively, and the NO rotational distributions depend strongly on the number of bending quanta excited in the parent molecule.

OCS

Photodissociation of OCS in the region from 222–248 nm has been investigated by Sivakumar et al, who monitored both the CO and S(^1D) primary photoproducts using vacuum ultraviolet LIF (29, 111). The CO fragment is produced almost exclusively in v = 0, but the rotational distribution is inverted and peaked at very high rotational levels. For dissociation at 222 nm, the rotational distribution is bimodal, with peaks at $J = 56$ and $J = 65$. Doppler profiles of the CO rotational transitions revealed (*a*) that all observed levels were produced in coincidence with $S(^1D)$, (*b*) that the fragment recoil anisotropy shifts from a distribution characterized by $\beta = 1.9$ at $J = 67$ to one characterized by $\beta = 0$ near $J = 54$, (*c*) that the CO angular momentum vector is aligned nearly perpendicular to its angular momentum vector for all rotational levels, and (*d*) that the CO angular momentum vector is also aligned parallel to the component of the transition dipole moment that lies perpendicular to the recoil velocity. These results were interpreted in terms of a model for the dissociation in which excitation takes place to two surfaces of A' and A'' symmetry derived from a bent $^1\Delta$ configuration.

The deconvolution of the rotational distribution into components corresponding to absorption via the A' and A'' states could be made only because the triple-vector (μ–v–J) correlation was extracted from the data. This correlation allows a distinction to be made between $\beta = -1$ fragmentations of a bent triatomic occurring either from an $A'' \leftarrow A'$ transition, which necessarily has an out-of-plane transition moment, or from an $A' \leftarrow A'$ transition that happens to have μ perpendicular to the recoil axis but in the molecular plane. In the first case, there is a tendency for μ and J to be parallel, and for v to be perpendicular to both μ and J, whereas in the second case, all three vectors tend to be orthogonal. These cases could be distinguished by the alignment alone, given $\beta = -1$, but a complete treatment of the other vector correlations is required even to extract a consistent value of β from the Doppler profiles.

The dissociation of OCS at 157 nm has also been investigated by Strauss et al (112, 113). Sulfur from this dissociation is produced almost entirely in the $S(^1S)$ state, whereas CO is produced in its ground electronic state and in vibrational levels from v = 0–3. The rotational distribution for each vibrational level was found to be near Boltzmann, with temperature that decreased with increasing vibrational excitation. Measurements of the CO Doppler profiles demonstrated that the dissociation takes place from a transition of predominantly parallel character. A value of $\beta = 1.8$ was derived, assuming the CO velocity and angular momentum vectors to be perpendicular.

POLYATOMIC MOLECULES

H_2O_2

The ultraviolet photodissociation of hydrogen peroxide was one of the first systems for which the influence of vector correlations on the Doppler lineshapes of photofragment laser-induced fluorescence was observed and interpreted (31, 32). Following the initial illustration of the technique, a remarkably distinctive and detailed picture of this photodissociation has emerged from the extensive and continuing studies, principally by the groups in Frankfurt and Nottingham.

Work prior to 1985 had identified the energy disposal patterns in H_2O_2 photodissociation at several convenient ultraviolet wavelengths (114–121). In the first LIF measurements (116), it was determined that, despite sufficient available energy to populate high vibrational levels of OH ($X\,^2\Pi$) and even the $A\,(^2\Sigma^+)$ state of OH (at wavelengths less than 200 nm), only vibrationless OH ($X\,^2\Pi$) was formed in detectable amounts. About 90% of the available energy is expressed as translation, with the balance in rotation. An early application of polarized, Doppler-resolved LIF to molecular photofragmentation (116) was limited by the laser linewidth, but was sufficiently well resolved to determine that the fragmentation was preferentially perpendicular to the polarization direction of the dissociation light at 248 nm, but nearly isotropic at 193 nm.

The vibrational and rotational distribution of OH ($A\,^2\Sigma^+$) was determined from dispersed fluorescence following 157 nm dissociation (118). At this wavelength, the formation of OH ($X\,^2\Pi$) + OH ($A\,^2\Sigma^+$) is a minor (7%) channel (119); the major channel leads to two OH ($X\,^2\Pi$) products. An inverted rotational distribution, peaked at $N = 21$, was observed, similar to earlier work at shorter wavelengths (114). A strong torsional angle dependence in the excited state potential was considered the likely source of the rotational excitation, based on angular momentum conservation arguments. Recent polarization analysis of the dispersed fluorescence (122) has determined a large negative alignment ($A_0^{(2)} = -0.25$) for the OH ($A\,^2\Sigma^+$) photoproducts.

The Doppler-resolved laser-induced fluorescence techniques described above have been applied to the dissociation of H_2O_2 at 266 nm (31, 123–125), 248 nm (32, 126, 127), and 193 nm (124, 128–130). Excellent summaries of this work have recently been published (127, 131). In the first application of Dixon's formulation of the theory of Doppler-resolved alignment, Gericke et al (31) extracted a set of bipolar moments from Doppler-resolved LIF excitation lineshapes of OH ($X\,^2\Pi$) photofragments in different absorption branches and in several different excitation-detection geometries. The bipolar moment $\beta_0^2(20) = \frac{1}{2}\beta$ has the strongest influence on the Doppler lineshapes. For room temperature H_2O_2 at 266 nm (31) and at 248 nm (32, 126), the recoil anisotropy, β, was found to be close to -1, the limiting value for recoil perpendicular to the transition moment of the parent molecule. This measurement, in conjunction with a quantum calculation that positions the $^1B \leftarrow {}^1A$ transition moment nearly parallel to the O–O bond (132), identifies the excited state accessed at 248 or 266 nm as ^1A.

The v–J correlation, $\beta_0^0(22)$, is found to be positive and increasing with N to a maximum value of 0.4. A positive v–J correlation is qualitatively evident in these Doppler lineshapes as a shallower dip or a sharper peak at the center of a Q line, compared with a P or R line in the same geometry. Dynamically, the positive v–J correlation of the OH fragments implies a torque directed along the recoil direction, which can arise only from motion along the torsional coordinate, or preexisting rotation of the parent on the O–O axis. Conversion of zero-point bending motion into rotation (133) or the impulsive torque of the O–O repulsion (134) would result in a negative v–J correlation. The observed value can be attributed to a mixture of these sources of rotational excitation; the helicity is a quantitative measure of their relative importance.

The alignment parameter $A_0^{(2)} = \frac{4}{5}\beta_0^2(02)$ was found to be very small, <0.1, thus indicating no strong preference for J to be parallel or perpendicular to μ. This might at first seem surprising, since the large negative value of β indicates a strong preference for v to be perpendicular to μ, and the positive value of the helicity, $\beta_0^0(22)$, indicates a tendency for J to be parallel to v, suggesting that J should be preferentially aligned perpendicular to μ. The apparent problem is resolved by recognizing that quite distinct distributions can share an average value. In particular, an alignment of $A_0^{(2)} = 0$ does not necessarily imply an absence of correlation between μ and J, but merely an average angle of 54.7°, the same as would be obtained for an uncorrelated distribution. An alignment of 0.1 corresponds to an average angle of 50° between μ and J; a helicity of 0.4 corresponds to an average angle of 40° between J and v. Reconciling these average angles with the 90° angle between v and μ requires that J be restricted near the plane of μ and v, i.e. the plane containing the C_2 symmetry axis and the O–O bond. Gericke et al (31) have recast the measured bipolar moments in terms of the expectation values of J_x^2, J_y^2, and J_z^2 in the molecular frame of the parent, which provides a simple mechanistic interpretation of the vector correlations in the case of prompt dissociation.

Dissociation of H_2O_2 at 193 nm has also been studied (124, 128–130, 135). A minor channel that produces H atoms (136) and a two-photon dissociation path that produces OH ($A\,^2\Sigma^+$) have been observed (120, 121), but the main products at 193 nm are still a pair of OH ($X\,^2\Pi$) radicals. The recoil anisotropy is near zero, a value attributed (116, 128) to a simultaneous absorption to a ^1B state, polarized along the O–O bond, and a ^1A state polarized along the C_2 axis. Shapiro has warned that observed β values may not be simple incoherent averages of the dynamics on two surfaces (137), but averages of classical trajectories on two ab initio surfaces provide remarkably good agreement with the rotational distributions and vector correlations for this dissociation (138). In contrast to the ground state, both excited state potentials show strong dependence on the torsional coordinate. The excited ^1A state has a *trans*-planar minimum and the ^1B state has a *cis*-planar minimum; each has a barrier to rotation on the O–O bond of about 3 eV. Despite the difference in excited state geometries, dissociation from either surface results in a similar rotational distribution, well-described by the rotational reflection principle, dominated by the torsional coordinate (12, 138).

The differences between H_2O_2 and D_2O_2 are quite dramatic (125, 139), both in the rotational distribution and the vector correlations. At 266 nm, the rotational distributions are approximately Gaussian, with the maximum shifting from $N_{OH} = 6$ to $N_{OD} = 11$ and the width (fwhm) remaining at about 9. This corresponds to a 30% increase in the rotational energy of photofragments in the D_2O_2 dissociation. Figure 3 shows the N dependence of four bipolar moments for photofragments of H_2O_2 and D_2O_2 at 266 nm. The positive alignment, smaller helicity, and higher rotational excitation with D_2O_2 are all consistent with a more important contribution from the impulsive torque on the OD, due to the longer lever arm from the O to the center of mass of the OD (125). An intriguing variation of the lambda doublet ratio is observed only in D_2O_2, where $\Pi(A'')$ levels are increasingly dominant in rotational levels above $N = 11$ and $\Pi(A')$ levels are dominant at lower N.

Improved experimental frequency resolution attainable with pulse-amplified continuous lasers and translationally cooled parent molecules has led to new interest in the scalar correlation of the internal energies of the two sibling fragments produced from the same parent molecule (124, 140–142). High resolution measurement of the kinetic energy distribution of a state-resolved photofragment allows the identification of the distribution of internal energy in the undetected sibling fragment (19, 50, 51, 72, 91, 111). Applying similar techniques to H_2O_2 has allowed a determination of the average internal energy of the unmeasured OH as a function of the rotational level of the measured OH fragment (124, 141). The internal energy of the OH fragments is exclusively rotational, so the measured correlation can be viewed as one between the rotational state of one OH fragment, N_a, with the average rotational quantum number of the other fragment, $\langle N_b \rangle$. A strong positive correlation of N_a with $\langle N_b \rangle$ is observed for the two OH fragments generated by 193 nm from rotationally cooled H_2O_2 in a supersonic jet (128, 135), particularly for the higher rotational states. This correlation accords well with the torsional excitation of rotation that produces a pair of counter-rotating OH radicals with

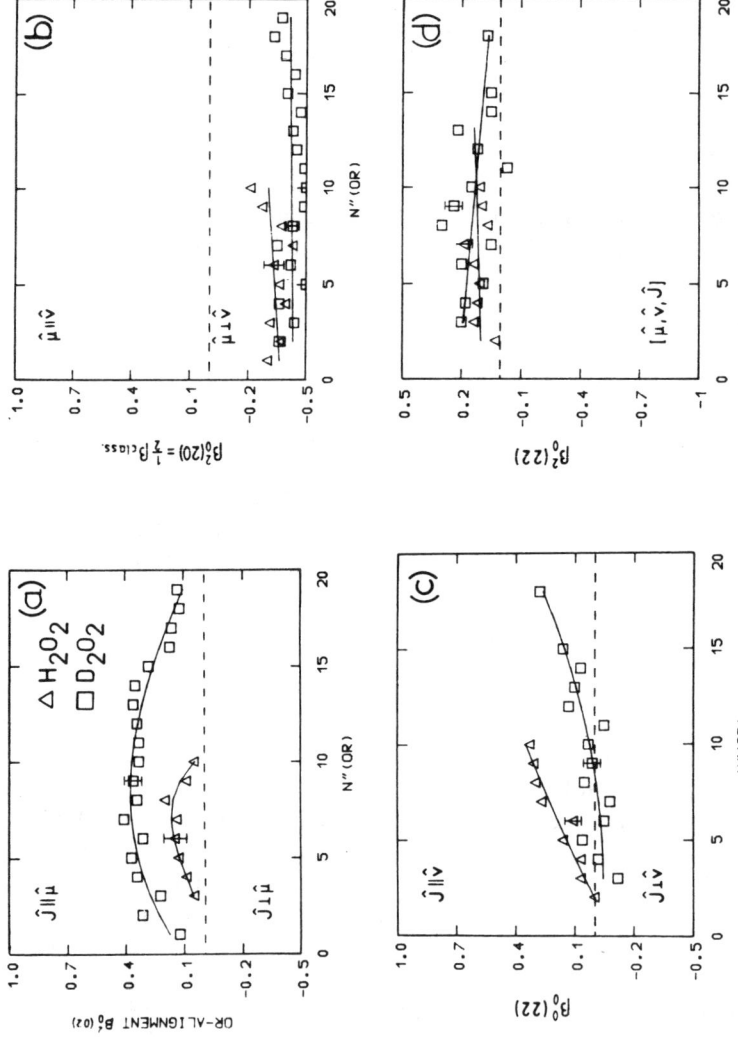

Figure 3 Bipolar moments $\beta_0^2(02)$, $\beta_0^2(20)$, $\beta_0^0(22)$, and $\beta_0^2(22)$ for the 266 nm dissociation of H_2O_2 (*triangles*) and D_2O_2 (*squares*). The moments are given as functions of the rotational quantum number N''_{OR}. (*a*) $\beta_0^2(02)$ is proportional to the rotational alignment; (*b*) $\beta_0^2(20)$ is proportional to the recoil anisotropy; (*c*) $\beta_0^0(22)$ is a measure of the **v**–**J** correlation; (*d*) $\beta_0^2(22)$ is a measure of the three-vector correlation of $\boldsymbol{\mu}$, **v**, and **J**. [Reprinted with permission from Ref. (125).]

matched rotational energies. Interestingly, at room temperature, $\langle N_a \rangle$ is nearly independent of N_a. This finding has been given an appealingly intuitive explanation (124, 142) by invoking the role of parent rotation on the O–O axis to break the symmetry of the fragmentation, adding angular momentum to one fragment and subtracting it from the other. This leads to a *negative* correlation of $\langle N_b \rangle$ with N_a that balances the positive correlation observed with rotationally cold parent molecules. Classical trajectory calculations have verified quantitatively the reasonableness of this view (142). The same argument also conforms with the observation of a *larger* helicity at room temperature than in the jet, since when the parent rotation increases the fragment rotation, it is by adding angular momentum along the velocity direction, making the **v** and **J** vectors more parallel.

Very recent efforts (143) have provided vector correlation measurements to the vibrationally mediated photodissociation of H_2O_2 (144). Sequential two-photon absorption via the $4v_{OH}$ overtone at 750 nm leads to a pair of OH ($X\,^2\Pi$) fragments (145) with an internal energy distribution that differs sharply from that obtained from one photon dissociation at 266 nm. The intermediate vibrational state allows different regions of the dissociative potential energy surfaces to be accessed by Franck-Condon transitions. Perhaps most exciting is the prospect of a penetrating look at the intramolecular dynamics of the overtone-pumped molecule, since the various vibrational modes of H_2O_2 excited at the time of the second photon absorption will leave their distinctive signatures on the photoproduct energy distributions and vector correlations.

HONO

The photochemistry, spectroscopy, and photodissociation of R–ONO and R–NO compounds has recently been reviewed (146). The R–NO compounds generally predissociate slowly and yield statistical product distributions. The R–ONO compounds generally show an intermediate behavior, neither direct nor statistical, which has led to substantial experimental and theoretical interest. In a pair of pioneering papers, Vasudev, Zare & Dixon (147, 148) applied polarized LIF and Doppler-resolved LIF to characterize the OH ($X\,^2\Pi$) product of the dissociation of HONO in the S_1 state. In addition to cold vibrational and rotational distributions, a positive alignment and a β near -1 were found. The lambda doublets and spin components of the $^2\Pi$ state were both found in nonstatistical ratios. The $S_1 \leftarrow S_0$ absorption spectrum shows a vibrational structure, assigned to a localized NO stretch. The photodissociation dynamics following selective excitation of this vibration reflect the coupling of the NO stretch in the parent molecule to the dissociation coordinate. Excitation of HONO with different amounts of energy in the NO stretch has little effect on the energy of the OH fragment (148). By cruel accident, the NO fragment cannot easily be monitored in this dissociation, despite great interest (149), due to ever-present amounts of NO and NO_2 in HONO. The alkyl nitrites are more stable and have similar absorption spectra. The details of this vibrational predissociation have been worked out in substantial detail for several of these molecules, closely related to the parent acid, HONO.

$(CH_3)_3CONO$ and $(CH_3)_2NNO$

Thorough measurements of the NO fragment from the S_1 and S_2 electronic states of t-butyl nitrite have been made by Rosenwaks and co-workers (150–152). The spectroscopy and dissociation dynamics of dimethyl nitrosamine (30, 153–155) are similar to the alkyl nitrites; a recent review has compared the photodissociation of these two molecules (156), with an emphasis on vector correlations. Huber and co-workers (30) were among the first to demonstrate the effect of **v**–**J** correlation on Doppler lineshapes, by using data from dimethyl nitrosamine. The simpler methyl nitrite has attracted even more detailed study.

CH_3ONO

A detailed understanding of the vibrational predissociation of methyl nitrite in the S_1 excited state has been attained, aided in large part by the measurement of fragment vector correlations. A recent review (157) summarizes much of this work.

$$CH_3ONO(S_0) + hv \rightarrow CH_3ONO(S_1) \rightarrow CH_3O + NO(X\,^2\Pi) \qquad 2.$$

The S_1–S_0 absorption band, like that of HONO, is dominated by a vibrational progression in the NO stretch (v_3) associated with a n-π^* transition. The vibrational energy of the NO photofragment increases as more energy is put into the v_3 mode of the parent molecule (158, 159), while the NO rotational energy is not strongly affected, thus implying dissociation on a timescale faster than intramolecular vibrational redistribution. Molecular beam photofragment mass spectral measurements on this molecule (160–162) and on the similar ethyl nitrite (134) have determined that β is negative and independent of rotational cooling, a finding that suggests a prompt dissociation on the rotational timescale. Two-photon LIF has been used (163–165) to determine a large positive NO alignment, which confirms a planar dissociation from an $A''\,S_1$ state.

An ab initio surface for the two important coordinates: r_{NO} and r_{O-NO} has been calculated (166) and used (167) to interpret the absorption spectrum and the product vibrational distributions. An adiabatic tunneling mechanism is required for fragmentation in which v_3 in the parent is exactly preserved as the vibrational state of the NO fragment, whereas the nonadiabatic mechanism of vibrational predissociation accounts for the NO fragments with lower vibrational energies. With this theoretical distinction, the observed spin selectivity (168) in some vibrational states of NO can be associated with the vibrational predissociation, while the minor, tunneling channel gives a statistical mixture of spin states. The mechanism of the spin selectivity is unclear. A recent reanalysis of Doppler lineshapes of the NO fragments (169) has used a two-photon generalization of Dixon's formalism. The 3_0^2 band of CH_3ONO was selectively excited and a high rotational level of each populated vibrational state was characterized by a set of bipolar moments, assumed to be velocity independent. The alignment and β both tend toward smaller absolute values at lower vibrational levels, whereas the **v**–**J** correlation and the lambda doublet ratio are unaffected. These trends suggest an increased lifetime for dissociation into lower vibrational levels, although comparison with improved theory may help clarify these finer points that can now be addressed experimentally.

The S_2 π–π^* state of methyl nitrite has also been studied by molecular beam photofragment TOF (160, 161) and shows a wavelength-dependent positive β. The maximum negative alignment of the NO products has been measured by the novel CDAD technique (52), which stands for circular dichroism of photoelectron angular distributions. Vector correlations were also extracted from the photoion angular distributions, following one photon dissociation and two-color, 1+1 ionization of the NO photoproduct (52).

CHOCHO

As emphasized in the Introduction, the correlation between **v** and **J** is not diminished by rotation of the parent molecule prior to dissociation, because this correlation is not made until the moment at which the molecular system crosses the transition state between parent and fragments. Glyoxal and formaldehyde illustrate that the effects of this correlation can be observed even when the parent molecule dissociates on a very slow time scale compared to rotation.

When excited to low-lying vibrational levels of S_1, glyoxal, which is *trans*-planar in the ground state, dissociates to give CO and several other possible sibling fragments:

$$CHOCHO + h\nu \rightarrow H_2CO + CO \qquad 3.$$

$$\rightarrow H_2 + 2CO \qquad 4.$$

$$\rightarrow HCOH + CO. \qquad 5.$$

Burak et al (170) have investigated this dissociation by probing the CO fragment with vacuum ultraviolet LIF. Because the dissociation is predissociative with lifetimes on the order of 1 μs, and several unresolved rotational states are excited, it is expected that any anisotropy in fragment recoil will be greatly diminished. Isotropic fragmentation, in the absence of any **v**–**J** correlation, will result in a flat-topped Doppler profile for each recoil velocity. In the presence of a correlation, however, Q-branch transitions acquire a shape different from P- or R-branch transitions, independent of the excitation-detection geometry (33, 34). If the dissociation takes place with considerable torsional motion about the C–C bond, we might expect **v** and **J** to be parallel, whereas, if the transition takes place entirely in a plane, the CO should both rotate and translate in the plane so that **v** and **J** would be expected to be perpendicular. Row (e) in Figure 4 shows the expected Doppler profiles for CO with a single speed if **v** and **J** are perpendicular. Of course, with three possible dissociation channels and with differing internal energies in the sibling fragments, the CO will have a wide distribution of speeds. Row (d) of Figure 4 shows that even after averaging over a 5600 K thermal speed distribution for the CO, the difference in Q vs P and R branch lineshapes persists. The distinctive line shapes of the Q vs P, R lines illustrated in rows (a–c) of Figure 4 are observed throughout the complete Doppler profile data set spanning $J = 20$–59. These observations are thus in accord with the planar dissociation mechanism, which also agrees with the transition state predicted from ab initio calculations (171).

H_2CO

The photofragmentation of formaldehyde to H_2 and CO products is an intensively studied case of predissociation (171–174). Following excitation in the ro-vibronically resolved S_1–S_0 absorption band, internal conversion to vibrationally excited levels of S_0 leads to unimolecular fragmentation into H_2 and CO. With ultraviolet excitation at an energy of ca. 29,500 cm^{-1}, near the ground state barrier to dissociation, the rotational energy distribution of the CO fragment (173) is inverted and highly excited, consistent with strong exit channel translation-rotation coupling (175). Vibrational excitation of the CO is minor; the v = 1 population is about 15% as large as v = 0. The H_2 product (174) is formed in a distribution of vibrational states, peaked at v = 1 and extending to v = 3. The H_2 rotational distributions are approximately Boltzmann, described by "temperatures" near 2000K. Model calculations by Schinke (176) suggest that the rotational excitation of H_2 is dominated by the Franck-Condon pro-

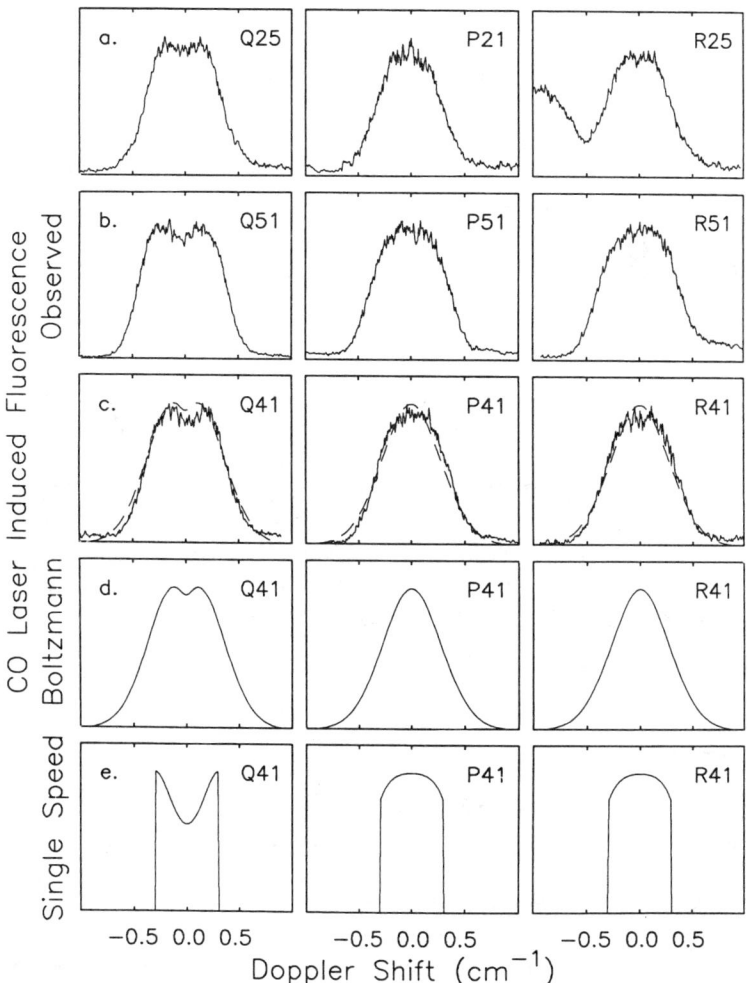

Figure 4 Calculated and experimental Doppler profiles for CO produced in the photodissociation of glyoxal. Rows (*a*)–(*c*): Representative experimental data for $Q(J'')$ (*left-hand side*), $P(J'')$ (*center*), and $R(J'')$ (*right-hand side*) transitions. The distinctive line shapes of the Q vs P, R lines persist throughout the complete Doppler profile data set spanning $J'' = 20$–59. Rows (*d*) and (*e*): Calculated Doppler profiles assuming a **v** perpendicular to **J** correlation. Row (*e*) is for a single CO recoil velocity, whereas row (*d*) is for a Boltzmann distribution of recoil speeds. The *dashed line* in row (*c*) is a superposition onto the data of the Doppler profiles shown in row (*d*). [Reprinted with permission from Ref. (170).]

jection of the transition state onto final products, and is insensitive to exit channel effects.

Recent experimental work (177) has begun to address the previously speculative topic of product correlations in this well-documented predissociation. With polarized, Doppler-resolved LIF on the H_2 photoproducts, vector correlations and state-resolved average kinetic energies have been measured by using an analysis like that applied to H_2O_2 (31, 33). The vector correlations involving the lab frame polarization could be expected to be weak, due to the long predissociative lifetime (~ 50 ns) compared to the rotational timescale. Theoretical work on the effect of parent rotation on recoil anisotropy (178) and alignment (57) guided the interpretation of β and $A_0^{(2)}$ in predissociating near-symmetric tops. The H_2 photoproducts in $v = 1$ were found to have a nearly isotropic recoil distribution in the most populated rotational states, $J = 2, 4$, and 6. The less populated $J = 0$ and $J = 8$ photofragments surprisingly show $\beta \approx 0.5$, about 5 times larger than the maximum value classically predicted possible (178) for a rotationally averaged dissociation from a symmetric top with the transition moment perpendicular to the top axis. A strong sensitivity of fragment vector correlations to the parent ro-vibronic state has been reported (177), which promises to show some influence of the parent vibrational mode selection, after accounting for the orientational distribution due to rotational state selection. The present case of a rotationally resolved excitation of the parent molecule to states of low total angular momentum calls for a quantum mechanical approach.

The correlation of **v** and **J** is rotationally invariant and might be expected to dominate the Doppler lineshapes in this predissociating molecule. The helicity is about -0.2, a value that shows a partial tendency toward perpendicularity of **v** and **J** and a non-negligible out-of-plane component to the rotational excitation.

The state-resolved average kinetic energies extracted from the H_2 Doppler profiles have allowed a direct determination of the scalar correlation of the measured H_2 internal energy state with the average rotational energy of the sibling CO molecule. The sum of total kinetic energy and H_2 vibrational energy increases with H_2 vibration, and implies that lower rotational states of CO are produced in coincidence with higher vibrational states of H_2. This result is consistent with an angular momentum conservation argument, which depends on a large, fixed value for the impact parameter of the half-collision.

These early measurements of formaldehyde product correlations show the value of the Doppler techniques even in predissociating systems, and particularly to supplement prior experimental and theoretical work.

$HONO_2$

Nitric acid is a polyatomic molecule in which a broad distribution of fragment recoil velocities prohibits an exact analysis of the Doppler lineshapes. Simons and co-workers (127, 179) have demonstrated an approximate analysis procedure that assumes no kinetic energy dependence to the vector correlations and provides a satisfactory approach in the several cases investigated. Within this approximation, the Doppler lineshapes can all be parametrized with a Gaussian distribution of kinetic energies and a set of bipolar moments for each rotational state of the probed OH fragment. For dissociation in the very weak 280 nm absorption band, about 70% of the available energy goes to the internal energy of NO_2 and most of the rest is expressed as translation, although these distributions are very broad. The measured vector correlations nonetheless provide significant constraints on the possible dissociation dynamics and excited state symmetry. A positive β implies an $A' \leftarrow A'$ transition, yet since the lowest A' excited state is much higher in energy, a vibronic transition involving an out-of-plane (A'') vibrational motion and the A'' electronic state seems to be required. A small positive alignment is attributable to a planar-to-pyramidal deformation, and a torsion along the N–OH bond in the excited state is implied by the positive correlation of **v** and **J**.

$HCOOH$ and CH_3COOH

The dissociation of formic acid (HCOOH) into OH + HCO has been studied following excitation in the n-π* absorption band at 220–250 nm (180–182). With 225 nm dissociation (180), the OH fragment is vibrationally cold, with a rotational distribution in v = 0 that approximates a 620K thermal distribution. The alignment is near zero, the lambda doublets are similarly populated, and the $F_1 : F_2$ ratio is close to statistical—all in contrast with isoelectronic HONO, but similar to the results recently found for CH_3COOH (183). The analysis of the LIF Doppler lineshapes is complicated by a very broad kinetic energy distribution, as in the case of nitric acid. The same analysis that was used for nitric acid (127) led to near zero values of the recoil anisotropy and the three-vector correlation $\beta_0^2(22)$, and a small positive value of the **v**–**J** correlation at higher rotational levels. A torsional origin of the higher rotational states is implied. The low values of β and $A_0^{(2)}$ could be accounted for by either a relatively slow predissociation or a change in geometry from planar to pyramidal at the carbon upon excitation. The formic acid absorption spectrum has resolvable rotational structure, suggesting slow dissociation, but the structured absorption may be a separate state superimposed on an underlying continuum. A weak fluorescence has been observed (180, 181) with a similarly structured excitation spectrum. It may be that the structured, fluorescent state is unrelated to a dissociative state accessed at the same wavelength, analogous to HNCO, in which a slow fluorescence and fast fragmentation from apparently separate electronic states has also been reported (184).

CH_3I

Among the polyatomic molecules, methyl iodide has become a favored test case for both experimental and theoretical studies of photodissociation. The A band of methyl iodide extends from 200–300 nm and has been attributed to three components of a n-σ^* excitation, split by spin-orbit coupling. The experimental resolution of the absorption band into components has been addressed by the wavelength dependence of magnetic circular dichroism (MCD) (185), quantum yields of spin-orbit excited iodine (I^*) (186–190), and recoil anisotropy (50, 51, 188, 191–198). Throughout the absorption band, the recoil anisotropies are close to the maximum expected for prompt dissociation following a parallel dissociation, in conflict with the interpretation of the MCD spectrum. The wavelength dependence of the I^* quantum yield is evidently dominated by nonadiabatic transitions, rather than overlapping absorption bands. The prompt, parallel dissociation of methyl iodide, along with a recoil-velocity-dependent TOF measurement, has been used as a tool to characterize the orientation of a beam of CH_3I after selection of the J, K, and M quantum numbers by a hexapole field (194).

Recent measurements of rotational distributions (50, 113, 195, 196, 199) and rotational alignments (196, 197, 199) also bear on the excited state surfaces and the dynamics associated with the crossings. An ab initio calculation (200) of the excited state surfaces as the iodine atom is displaced from the methyl C_3 axis suggests that an exit channel torque may be expected in the crossing region from the $^3Q_0$ to the $^1Q_1$ surfaces. This can account for the higher rotational excitation and negative alignment in the CH_3 produced in coincidence with I as a result of curve crossing (50, 113, 195, 196, 199).

The nascent distribution of K levels of a symmetric top photofragment can be considered a type of vector correlation. The alignment of fragment angular momentum along the recoil direction is related to the distribution of K levels for a given N in the methyl fragment, if the methyl C_3 axis is closely associated with the recoil direction. The rotational excitation in the I^* channel can come from the parent rotation (which will tend to produce $K \approx N$ rotational states of methyl) and from the transformation of zero-point bending energy into rotation (which will tend to produce $K \approx 0$ states). Preferential high K excitation is observed (199) in a 300K effusive beam and preferential low K excitation is observed (113, 196) in a supersonic cooled beam.

CONCLUSIONS

It is clear from the recent rapid increase in the use of vector correlations that this subfield of chemical physics is in a state of growth. In the future we might expect to see measurements of vector correlations become more routine, especially as more widely tunable narrow-band laser sources become available. A promising note is the possibility of using pulse amplified cw dye lasers to provide extremely high-resolution Doppler profiles. Several exciting new directions have been noted in this review. One is the use of laser preparation of the parent compound prior to dissociation, e.g. as performed in H_2O (95, 96) or H_2O_2 (143), in order to select the initial rotational level of the parent or to access different Franck-Condon regions of the dissociative state. No one has yet tested how the vector correlations are affected by systematic selection of the parent rotational state. A second new direction concerns measurements of orientation. Most studies thus far have used linearly polarized light to create and measure only alignment; the recent work from Stanford on the dissociation of ICN indicates that even more information can be obtained by using circularly polarized light (76, 77). Finally, it is now becoming possible (201) to observe the formation of vector correlations in real time with fs lasers.

ACKNOWLEDGMENTS

We would like to thank our colleagues who made pre-publication information available to us. Work at Cornell was supported by the National Science Foundation under grant CHE-8617062 and by the Air Force Office of Scientific Research under grant AFOSR-87-0017. Work carried out at Brookhaven National Laboratory was supported by the Department of Energy, Division of Chemical Sciences, Office of Basic Energy Sciences.

Literature Cited

1. Norrish, R. G. W., Porter, G. 1950. *Proc. R. Soc. London Ser. A* 200: 284
2. Simons, J. P. 1977. *Gas Kinetics and Energy Transfer*, 2: 58–95. London: Royal Soc. Chem.
3. Ashfold, M. N. R., Macpherson, M. T., Simons, J. P. 1979. *Top. Current Chem.* 86: 1–90
4. Leone, S. R. 1982. *Adv. Chem. Phys.* 50: 255–324
5. Leone, S. R. 1983. *Acc. Chem. Res.* 16: 88–95
6. Crim, F. F. 1984. *Annu. Rev. Phys. Chem.* 35: 657–91
7. Lawley, K. P., ed. 1985. *Adv. Chem. Phys.* 60
8. Jackson, W. H., Okabe, H. 1986. *Adv. Photochem.* 13: 95
9. Simons, J. P. 1987. *J. Phys. Chem.* 91: 5378–87
10. Houston, P. L. 1987. *J. Phys. Chem.* 91: 5388–97
11. Ashfold, M. N. R., Baggott, J. E., eds. 1987. *Molecular Photodissociation*

Dynamics. London: Royal Soc. Chem., 243 pp.
12. Schinke, R. 1988. *Annu. Rev. Phys. Chem.* 39: 39–68
13. Zare, R. N., Herschbach, D. R. 1963. *Proc. IEEE* 51: 173–82
14. Zare, R. N. 1972. *Mol. Photochem.* 4: 1–37
15. Solomon, J. 1967. *J. Chem. Phys.* 47: 889–95
16. Busch, G. E., Mahoney, R. T., Morse, R. I., Wilson, K. R. 1969. *J. Chem. Phys.* 51: 449; 51: 837
17. Diesen, R. W., Wahr, J. C., Adler, S. E. 1969. *J. Chem. Phys.* 50: 3635–36
18. Zare, R. N. 1964. PhD thesis. Cambridge: Harvard Univ.
19. Schmiedl, R., Dugan, H., Meier, W., Welge, K. H. 1982. *Z. Phys. A* 304: 137–42
20. Van Brunt, R. J., Zare, R. N. 1968. *J. Chem. Phys.* 48: 4304–8
21. Gouedard, G., Lehmann, J. C. 1973. *J. Phys.* 34: 693–99
22. Fano, U., Macek, J. H. 1973. *Rev. Mod. Phys.* 45: 553–73
23. Greene, C. H., Zare, R. N. 1983. *J. Chem. Phys.* 78: 6741–53
24. Greene, C. H., Zare, R. N. 1982. *Annu. Rev. Phys. Chem.* 33: 119–50
25. Chamberlain, G. A., Simons, J. P. 1975. *Chem. Phys. Lett.* 32: 355–58
26. Chamberlain, G. A., Simons, J. P. 1975. *J. Chem. Soc. Faraday Trans 2* 71: 2043–50
27. Simons, J. P. 1984. *J. Phys. Chem.* 88: 1287–93
28. Case, D. A., McClelland, G. M., Herschbach, D. R. 1978. *Mol. Phys.* 35: 541–73
29. Hall, G. E., Sivakumar, N., Houston, P. L., Burak, I. 1986. *Phys. Rev. Lett.* 56: 1671–74
30. Dubs, M., Brühlmann, U., Huber, J. R. 1986. *J. Chem. Phys.* 84: 3106–19
31. Gericke, K.-H., Klee, S., Comes, F. J., Dixon, R. N. 1986. *J. Chem. Phys.* 85: 4463–79
32. Docker, M. P., Hodgson, A., Simons, J. P. 1986. *Chem. Phys. Lett.* 128: 264–69
33. Dixon, R. N. 1986. *J. Chem. Phys.* 85: 1866–79
34. Hall, G. E., Sivakumar, N., Chawla, D., Houston, P. L., Burak, I. 1988. *J. Chem. Phys.* 88: 3682–91
35. Balint-Kurti, G. G., Shapiro, M. 1989. *J. Chem. Phys.* In press
36. Balint-Kurti, G. G., Shapiro, M. 1981. *Chem. Phys.* 61: 137–55
37. Rowe, M. D., McCaffery, A. J. 1979. *Chem. Phys.* 43: 35–54
38. Bain, A. J., McCaffery, A. J., Proctor, M. J., Whitaker, B. J. 1984. *Chem. Phys. Lett.* 110: 663–65
39. Bain, A. J., McCaffery, A. J. 1984. *Chem. Phys. Lett.* 105: 477–79
40. Bain, A. J., McCaffery, A. J. 1984. *Chem. Phys. Lett.* 108: 275–82
41. Bain, A. J., McCaffery, A. J. 1984. *J. Chem. Phys.* 80: 5883–92
42. Bain, A. J., McCaffery, A. J. 1985. *J. Chem. Phys.* 83: 2627–31; 83: 2632–40; 83: 2641–45
43a. Kummel, A. C., Sitz, G. O., Zare, R. N. 1988. *J. Chem. Phys.* 88: 7357–68
43b. Waldeck, J. R., Kummel, A. C., Sitz, G. O., Zare, R. N. 1989. *J. Chem. Phys.* 90: 4112–14
44. Altkorn, R., Zare, R. N. 1984. *Annu. Rev. Phys. Chem.* 35: 265–89
45. Jacobs, D. C., Zare, R. N. 1986. *J. Chem. Phys.* 85: 5457–68
46. Jacobs, D. C., Madix, R. J., Zare, R. N. 1986. *J. Chem. Phys.* 85: 5469–79
47. Mons, M., Dimicoli, I. 1989. *J. Chem. Phys.* 90: 4037–47
48. Mons, M., Dimicoli, I. 1986. *Chem. Phys. Lett.* 131: 298–302
49. Hall, G. E., Sivakumar, N., Ogorzalek, R., Chawla, G., Haerri, H.-P., Houston, P. L. 1986. *Faraday Discuss. Chem. Soc.* 82: 13–24
50. Ogorzalek Loo, R., Hall, G. E., Haerri, H.-P., Houston, P. L. 1988. *J. Phys. Chem.* 92: 5–8
51. Black, J. F., Powis, I. 1988. *Chem. Phys.* 125: 375–88
52. Winniczek, J. W., Dubs, R. L., Appling, J. R., McKoy, V., White, M. G. 1989. *J. Chem. Phys.* 90: 949–63
53. Docker, M. P. 1988. *Chem. Phys.* 125: 185–210
54. Kummel, A. C., Sitz, G. O., Zare, R. N. 1986. *J. Chem. Phys.* 85: 6874–97
55. Kummel, A. C., Sitz, G. O., Zare, R. N. 1988. *J. Chem. Phys.* 88: 6707–32
56. Loge, G. W., Zare, R. N. 1981. *Mol. Phys.* 43: 1419–28
57. Nagata, T., Kondow, T., Kuchitsu, K., Loge, G. W., Zare, R. N. 1983. *Mol. Phys.* 50: 49–63
58. Alexander, M. H., et al. (29 authors) 1988. *J. Chem. Phys.* 89: 1749–53 and references therein
59. Vigué, J., Grangier, P., Roger, G., Aspect, A. 1981. *J. Physique* 42: 531–35
60. Vigué, J., Beswick, J. A., Broyer, M. 1983. *J. Physique* 44: 1225–45
61. Diebold, G. J. 1983. *Phys. Rev. Lett.* 51: 1344–47
62. Diebold, G. J. 1985. *Phys. Rev. A* 32: 1458–67
63. Grangier, P., Aspect, A., Vigué, J. 1985. *Phys. Rev. Lett.* 54: 418–21

64. Kurizki, G., Ben-Reuven, A. 1985. *Phys. Rev. A* 32: 2560–63
65. Gerber, G., Möller, R. 1985. *Phys. Rev. Lett.* 55: 814–17
66. Diebold, G. J. 1986. *J. Chem. Phys.* 85: 25–33
67. Grangier, P., Vigué, J. 1987. *J. Physique* 48: 781–96
68. Baba, M., Kato, H. 1989. *J. Chem. Soc. Faraday Trans. 2* 85: In press
69. Kurizki, G., Ben-Reuven, A. 1989. *J. Chem. Soc. Faraday Trans. 2* 85: In press
70. Glass-Maujean, M., Beswick, J. A. 1989. *J. Chem. Soc. Faraday Trans. 2* 85: In press
71. Guest, J. A., O'Halloran, M. A., Zare, R. N. 1984. *Chem. Phys. Lett.* 103: 261–65
72. Nadler, I., Mahgerefteh, D., Reisler, H., Wittig, C. 1985. *J. Chem. Phys.* 82: 3885–93
73. Hall, G. E., Sivakumar, N., Houston, P. L. 1986. *J. Chem. Phys.* 84: 2120–28
74. O'Halloran, M. A., Joswig, H., Zare, R. N. 1987. *J. Chem. Phys.* 87: 303–13
75. Joswig, H., O'Halloran, M. A., Zare, R. N., Child, M. S. 1986. *Faraday Discuss. Chem. Soc.* 82: 79–88
76. Hasselbrink, E., Waldeck, J. R., Zare, R. N. 1989. *Chem. Phys.* 126: 191–200
77. Black, J. F., Waldeck, J. R., Zare, R. N. 1989. *J. Chem. Soc. Faraday Trans. 2* 85: In press
78. Guest, J. A., Webster, F. 1987. *J. Chem. Phys.* 86: 5479–90
79. Chamberlain, G. A., Simons, J. P. 1975. *Chem. Phys. Lett.* 32: 355–58
80. Macpherson, M. T., Simons, J. P. 1977. *Chem. Phys. Lett.* 51: 261–64
81. Simons, J. P., Smith, A. J. 1983. *Chem. Phys. Lett.* 97: 1–3
82. Simons, J. P., Smith, A. J., Dixon, R. N. 1984. *J. Chem. Soc. Faraday Trans. 2* 80: 1489–1501
83. Hodgson, A., Simons, J. P., Smith, A. J., Dixon, R. N. 1985. In *Photophysics and Photochemistry Above 6 eV*, ed. F. Lahmani, pp. 505–20. Amsterdam: Elsevier
84. Hodgson, A., Simons, J. P., Ashfold, M. N. R., Bayley, J. M., Dixon, R. N. 1985. *Mol. Phys.* 54: 351–68
85. Hogdson, A., Simons, J. P., Ashfold, M. N. R., Bayley, J. M., Dixon, R. N. 1985. *Ber. Bunsenges. Phys. Chem.* 89: 251–54
86. Docker, M. P., Hodgson, A., Simons, J. P. 1986. *Mol. Phys.* 57: 129–47
87. Segev, E., Shapiro, M. 1982. *J. Chem. Phys.* 77: 5604–23
88. Meijer, G., ter Meulen, J. J., Andresen, P., Bath, A. 1986. *J. Chem. Phys.* 85: 6914–22
89. Brouard, M., Docker, M. P., Hodgson, A., Simons, J. P. 1987. *J. Chem. Phys.* 86: 7246–47
90. Atkins, C. G., Briggs, R. G., Halpern, J. B., Hancock, G. 1988. *Chem. Phys. Lett.* 152: 81–86
91. Krautwald, H. J., Schnieder, L., Welge, K. H., Ashfold, M. N. R. 1986. *Faraday Disc. Chem. Soc.* 82: 99–110
92. Andresen, P., Rothe, E. W. 1983. *J. Chem. Phys.* 78: 989–90
93. Andresen, P., Ondrey, G. S., Titze, B., Rothe, E. W. 1984. *J. Chem. Phys.* 80: 2548–69
94. Andresen, P., Beushausen, V., Häusler, D., Lüft, H. W. 1985. *J. Chem. Phys.* 83: 1429–30
95. Schinke, R., Engle, V., Andresen, P., Häusler, D., Balint-Kurti, G. G. 1985. *Phys. Rev. Lett.* 55: 1180–83
96. Häusler, D., Andresen, P., Shinke, R. 1987. *J. Chem. Phys.* 87: 3949–65
97. Schinke, R., Engel, V., Staemmler, V. 1985. *J. Chem. Phys.* 83: 4522–33
98. Balint-Kurti, G. G. 1986. *J. Chem. Phys.* 84: 4443–54
99. Engel, V., Schinke, R., Staemmler, V. 1988. *J. Chem. Phys.* 88: 129–48
100. Staemmler, V., Palma, A. 1985. *Chem. Phys.* 93: 63–69
101. Grunewald, A. U., Gericke, K.-H., Comes, F. J. 1987. *Chem. Phys. Lett.* 133: 501–6
102. Mons, M., Dimicoli, I. 1988. *Chem. Phys.* 130: 307–24
103. McCoustra, M. R. S., Pfab, J. 1988. *J. Chem. Soc. Faraday Trans. 2* 84: 655–69
104. Kawasaki, M., Sato, H., Kikuchi, T., Fukuroda, A., Kobayashi, S., Arikawa, T. 1987. *J. Chem. Phys.* 86: 4425–30
105. Morrison, R. J. S., Grant, E. R. 1982. *J. Chem. Phys.* 77: 5994–6004
106. Bruno, A. E., Brühlmann, U., Huber, J. R. 1988. *Chem. Phys.* 120: 155–67
107. Ticktin, A., Bruno, A. E., Brühlmann, U., Huber, J. R. 1988. *Chem. Phys.* 125: 403–13
108. Ogai, A., Qian, C. X. W., Iwata, L., Reisler, H. 1988. *Chem. Phys. Lett.* 146: 367–74
109. Qian, C. X. W., Ogai, A., Iwata, L., Reisler, H. 1988. *J. Chem. Phys.* 89: 6547–48
110. Bai, Y. Y., Ogai, A., Qian, C. X. W., Iwata, L., Segal, G. A., Reisler, H. 1989. *J. Chem. Phys.* 90: 3903–14
111. Sivakumar, N., Hall, G. E., Houston, P. L., Hepburn, J. W., Burak, I. 1988. *J. Chem. Phys.* 88: 3692–3708

112. Strauss, C. E., McBane, G. C., Houston, P. L., Burak, I., Hepburn, J. W. 1989. *J. Chem. Phys.* 90: 5364–72
113. Ogorzalek Loo, R., Strauss, C. E., Haerri, H.-P., Hall, G. E., Houston, P. L., Burak, I., Hepburn, J. W. 1989. *J. Chem. Soc. Faraday Trans 2* 85: In press
114. Becker, K. H., Grothert, W., Kley, D. 1965. *Z. Naturforsch. A* 20: 748–49
115. Greiner, N. R. 1966. *J. Chem. Phys.* 45: 99–103
116. Ondrey, G., van Veen, N., Bersohn, R. 1983. *J. Chem. Phys.* 78: 3732–37
117. Jacobs, A., Kleinermanns, K., Kuge, H., Wolfrum, J. 1983. *J. Chem. Phys.* 79: 3162–63
118. Gölzenleuchter, H., Gericke, K.-H., Comes, F. J., Linde, P. F. 1984. *Chem. Phys.* 89: 93–102
119. Suto, M., Lee, L. C. 1983. *Chem. Phys. Lett.* 98: 152–56
120. Gölzenleuchter, H., Gericke, K.-H., Comes, F. J. 1985. *Chem. Phys. Lett.* 116: 61–65
121. McKendrick, C. B., Kerr, E. A., Wilkinson, J. P. T. 1984. *J. Phys. Chem.* 88: 3930–32
122. Gericke, K.-H., Gölzenleuchter, H., Comes, F. J. 1989. *Chem. Phys.* 127: 399–409
123. Klee, S., Gericke, K.-H., Comes, F. J. 1986. *J. Chem. Phys.* 85: 40–44
124. Gericke, K.-H., Grunewald, A. U., Klee, S., Comes, F. J. 1988. *J. Chem. Phys.* 88: 6255–59
125. Klee, S., Gericke, K.-H., Comes, F. J. 1988. *Ber. Bunsenges. Phys. Chem.* 92: 429–34
126. Docker, M. P., Hodgson, A., Simons, J. P. 1986. *Faraday Discuss. Chem. Soc.* 82: 25–36
127. August, J., Brouard, M., Docker, M. P., Hodgson, A., Milne, C. J., Simons, J. P. 1988. *Ber. Bunsenges. Phys. Chem.* 92: 264–73
128. Grunewald, A. U., Gericke, K.-H., Comes, F. J. 1986. *Chem. Phys. Lett.* 132: 121–27
129. Grunewald, A. U., Gericke, K.-H., Comes, F. J. 1987. *J. Chem. Phys.* 87: 5709–21
130. Grunewald, A. U., Gericke, K.-H., Comes, F. J. 1988. *J. Chem. Phys.* 89: 345–54
131. Comes, F. J., Gericke, K.-H., Grunewald, A. U., Klee, S. 1988. *Ber. Bunsenges. Phys. Chem.* 92: 273–81
132. Chevaldonnet, C., Cardy, H., Dargelos, A. 1986. *Chem. Phys.* 102: 55–61
133. Morse, M. D., Freed, K. F. 1983. *J. Chem. Phys.* 78: 6045–65
134. Tuck, A. F. 1977. *Trans. Faraday Soc.* 73: 689–708
135. Jacobs, A., Wahl, M., Weller, R., Wolfrum, J. 1987. *Appl. Phys. B* 42: 173–79
136. Gerlach-Meyer, U., Linnebach, E., Kleinermanns, K., Wolfrum, J. 1987. *Chem. Phys. Lett.* 133: 113–15
137. Shapiro, M. 1986. *J. Phys. Chem.* 90: 3644–53
138. Schinke, R., Staemmler, V. 1988. *Chem. Phys. Lett.* 145: 486–92
139. Gericke, K.-H., Klee, S., Comes, F. J. 1987. *Chem. Phys. Lett.* 137: 510–15
140. Gericke, K.-H. 1988. *Phys. Rev. Lett.* 60: 561–64
141. Dixon, R. N., Nightingale, J., Western, C. M., Yang, X. 1988. *Chem. Phys. Lett.* 151: 328–34
142. Schinke, R. 1988. *J. Phys. Chem.* 92: 4015–19
143. Brouard, M., Martinez, M. T., O'Mahony, J., Simons, J. P. 1989. *Chem. Phys. Lett.* 150: 6–12; 1989. *J. Chem. Soc. Faraday Trans. 2* 85: In press
144. Crim, F. F. 1987. See Ref. 11, pp. 177–210
145. Ticich, T. M., Likar, M. D., Dübal, H.-R., Butler, L. J., Crim, F. F. 1987. *J. Chem. Phys.* 87: 5820–29
146. Reisler, H., Noble, M., Wittig, C. 1987. See Ref. 11, pp. 139–76
147. Vasudev, R., Zare, R. N., Dixon, R. N. 1983. *Chem. Phys. Lett.* 96: 399–402
148. Vasudev, R., Zare, R. N., Dixon, R. N. 1984. *J. Chem. Phys.* 80: 4863–78
149. Hennig, S., Untch, A., Schinke, R., Nonella, M., Huber, J. R. 1989. *Chem. Phys.* 129: 93–107
150. Schwartz-Lavi, D., Bar, I., Rosenwaks, S. 1986. *Chem. Phys. Lett.* 128: 123–26
151. Lavi, R., Schwartz-Lavi, D., Bar, I., Rosenwaks, S. 1987. *J. Phys. Chem.* 91: 5398–5402
152. Schwartz-Lavi, D., Rosenwaks, S. 1988. *J. Chem. Phys.* 88: 6922–30
153. Dubs, M., Huber, J. R. 1984. *Chem. Phys. Lett.* 108: 123–27
154. Lavi, R., Bar, I., Rosenwaks, S. 1987. *J. Chem. Phys.* 86: 1639–40
155. Lavi, R., Rosenwaks, S. 1988. *J. Chem. Phys.* 89. 1416–26
156. Lavi, R., Schwartz-Lavi, D., Rosenwaks, S. 1989. *J. Chem. Soc. Faraday Trans. 2* 85: In press
157. Huber, J. R. 1988. *Pure and Appl. Chem.* 60: 947–52
158. Lahmani, F., Lardeux, C., Solgadi, D. 1983. *Chem. Phys. Lett.* 102: 523–28
159. Benoist d'Azy, O., Lahmani, F., Lardeux, C., Solgadi, D. 1985. *Chem. Phys.* 94: 247–56
160. Keller, B. A., Felder, P., Huber, J. R.

1986. *J. Phys. Chem.* 91: 1114–20; 1986. *Chem. Phys. Lett.* 124: 135–39
161. Felder, P., Keller, B. A., Huber, J. R. 1987. *Z. Phys. D* 6: 185–92
162. Inoue, G., Kawasaki, M., Sato, H., Kikuchi, T., Kobayashi, S., Arikawa, T. 1987. *J. Chem. Phys.* 87: 5722–27
163. Lahmani, F., Lardeux, C., Solgadi, D. 1986. *Chem. Phys. Lett.* 129: 24–30
164. Brühlmann, U., Huber, J. R. 1987. *Z. Phys. D* 7: 1–8
165. Brühlmann, U., Dubs, M., Huber, J. R. 1987. *J. Chem. Phys.* 86: 1249–57
166. Nonella, M., Huber, J. R. 1986. *Chem. Phys. Lett.* 131: 376–79
167. Hennig, S., Engel, V., Schinke, R., Nonella, M., Huber, J. R. 1987. *J. Chem. Phys.* 87: 3522–29
168. Brühlmann, U., Huber, J. R. 1988. *Chem. Phys. Lett.* 143: 199–203
169. Docker, M. P., Ticktin, A., Brühlmann, U., Huber, J. R. 1989. *J. Chem. Soc. Faraday Trans. 2* 85: In press
170. Burak, I., Hepburn, J. W., Sivakumar, N., Hall, G. E., Chawla, G., Houston, P. L. 1987. *J. Chem. Phys.* 86: 1258–68
171. Osamura, T., Schaefer, H. F., Dupuis, M., Lester, W. A. 1981. *J. Chem. Phys.* 75: 5828–36
172. Moore, C. B., Weisshaar, J. C. 1983. *Annu. Rev. Phys. Chem.* 34: 525–55
173. Bamford, D. J., Filseth, S. V., Foltz, M. F., Hepburn, J. W., Moore, C. B. 1985. *J. Chem. Phys.* 82: 3032–41
174. Debarre, D., Lefebvre, M., Péalat, M., Taran, J.-P. E., Bamford, D. J., Moore, C. B. 1985. *J. Chem. Phys.* 83: 4476–87
175. Schinke, R. 1985. *Chem. Phys. Lett.* 120: 129–34
176. Schinke, R. 1986. *J. Chem. Phys.* 84: 1487–91
177. Butenhoff, T. J., Carleton, K. L., Chuang, M.-C., Moore, C. B. 1989. *J. Chem. Soc. Faraday Trans. 2* 85: In press
178. Yang, S., Bersohn, R. 1974. *J. Chem. Phys.* 61: 4400–7
179. August, J., Brouard, M., Simons, J. P. 1988. *J. Chem. Soc. Faraday Trans. 2* 84: 587–98
180. Brouard, M., O'Mahony, J. 1988. *Chem. Phys. Lett.* 149: 45–50
181. Ebata, T., Fujii, A., Amano, T., Ito, M. 1987. *J. Phys. Chem.* 91: 6095–97
182. Ebata, T., Amano, T., Ito, M. 1989. *J. Chem. Phys.* 90: 112–17
183. Hunnicutt, S. S., Waits, L. D., Guest, J. A. 1989. *J. Phys. Chem.* In press
184. Spiglanin, T. A., Perry, R. A., Chandler, D. W. 1986. *J. Phys. Chem.* 90: 6184–89
185. Gedanken, A., Rowe, M. D. 1975. *Chem. Phys. Lett.* 34: 39–43
186. Riley, S. J., Wilson, K. R. 1972. *J. Chem. Soc. Faraday Disc.* 53: 132–46
187. Hunter, T. F., Kristjansson, K. S. 1978. *Chem. Phys. Lett.* 58: 291–94
188. van Veen, G. N. A., Baller, T., de Vries, A. E., van Veen, N. J. A. 1984. *Chem. Phys.* 87: 405–17
189. Baughcum, S. L., Leone, S. R. 1980. *J. Chem. Phys.* 72: 6531–45
190. Hess, W. P., Kohler, S. J., Haugen, H. K., Leone, S. R. 1986. *J. Chem. Phys.* 84: 2143–49
191. Dzvonik, M., Yang, S., Bersohn, R. 1974. *J. Chem. Phys.* 61: 4408–21
192. Barry, M. D., Gorry, P. O. 1984. *Mol. Phys.* 52: 461–73
193. Penn, S. M., Hayden, C. C., Carlson Muyskens, K. J., Crim, F. F. 1988. *J. Chem. Phys.* 89: 2909–17
194. Bernstein, R. B., Choi, S. E., Stolte, S. 1989. *J. Chem. Soc. Faraday Trans. 2* 85: In press
195. Black, J., Powis, I. 1988. *Laser Chem.* 9: 339–58
196. Ogorzalek Loo, R., Haerri, H.-P., Hall, G. E., Houston, P. L. 1989. *J. Chem. Phys.* 90: 4222–36
197. Thoman, J. W., Chandler, D. W., Parker, D. H., Janssen, M. H. M. 1988. *Laser Chem.* 9: 27–47
198. Chandler, D. W., Thoman, J. W., Sitz, G. O., Janssen, M. H. M., Stolte, S., Parker, D. H. 1989. *J. Chem. Soc. Faraday Trans. 2* 85: In press
199. Black, J., Powis, I. 1989. *J. Phys. Chem.* 93: 2461–70
200. Yabushita, S., Morokuma, K. 1989. *Chem. Phys. Lett.* 153: 517–21
201. Zewail, A. H. 1989. *J. Chem. Soc. Faraday Trans. 2* 85: In press

SPECTROSCOPY OF THE DIATOMIC 3d TRANSITION METAL OXIDES

A. J. Merer

Department of Chemistry, University of British Columbia, Vancouver, B.C., Canada V6T 1Y6

INTRODUCTION

The diatomic oxides of the 3d transition metals have astonishingly complicated spectra which even now are by no means fully understood. The interest in them stems from their importance in astrophysics, high temperature chemistry, and in theoretical understanding of the chemical bonding in simple metal systems.

The 3d transition metals have a special place in astrophysics because of the great stability of the nuclei. ^{56}Fe, for example, has the lowest mass per nucleon of any nucleus, so that it is the final product in the thermonuclear processes that fuel stars. The nuclei surrounding iron in the 3d transition series are almost as stable, which makes for a local maximum in the cosmic abundances of these elements (1). Because of the high cosmic abundance of oxygen and the large dissociation energies of the diatomic oxides of the earlier 3d metals, the band systems of compounds such as TiO and VO completely dominate the spectra of cooler (M-type) stars if they contain metal-rich recycled supernova material (2). Low-temperature astrophysics is high-temperature terrestrial chemistry and, not surprisingly, it is found that the thermodynamically stable high temperature forms of the metal oxides are diatomic vapors. However, by the standards of room temperature chemistry the 3d monoxides are highly refractory materials, so that the difficulty of getting them into the gas phase under nonequilibrium low temperature conditions has held back progress with their spectra.

The real problems with their spectra come from the many unpaired electrons, which produce huge numbers of low-lying electronic states, many with very high spin multiplicity. There are widespread perturbations between the electronic states, which make analysis very awkward, while the nuclei with odd atomic mass numbers usually have large nuclear spins and magnetic moments, so that hyperfine structure adds another dimension of complexity.

In the worst cases, such as ^{55}MnO, a detailed analysis is not possible without extensive wavelength-resolved fluorescence to identify the lines, and the crowding caused by the electron spin and hyperfine structure is such that the average line spacing is less than the room temperature Doppler width over perhaps half of an electronic band.

The same problems that plague analysis also apply to ab initio calculations. The difficulty lies in getting the electron correlation effects right, so that quite small changes in the model can lead to very different predictions for the energy order and properties of the low-lying electronic states (3, 4). Nevertheless, progress in understanding of the 3d oxide spectra has mainly come from combining careful experiments with sophisticated calculations, and provides a good example of how theory and experiment can work together for the benefit of both. In this article I do not attempt to review the details of the quantum chemistry, except where they bear directly on the interpretation of the spectra; my aim is to present the experimental results, pointing out what has been done and what remains to be done, as seen by someone who works with chart paper rather than print-out.

A MODEL FOR THE BONDING IN THE 3d TRANSITION MONOXIDES

Where the dipole moments of the 3d transition metal monoxides have been measured (5–7), they are in the range 3.8–4.7 D, which translates as $M^{\delta+}O^{\delta-}$ where the effective charge, δ, is 0.5–0.6. Obviously there are ionic and covalent components to the bonding, which can be represented as somewhere between M^+O^- and the covalent MO. In first approximation it is possible to treat the oxides as covalently bonded molecules, and to draw suitable molecular orbital diagrams for them. In practice, particularly toward the right-hand side of the series in the periodic table, the m.o. diagrams may look slightly unusual because there is a "hole" in a fairly deep-lying orbital, since the metal behaves more nearly as M^+. Also, the single configuration approximation for a particular state is likely to break down very severely, since it will have contributions from the ionic and the covalent configurations. All the same, molecular orbital (m.o.) diagrams such as those given in Figure 1 provide a simple starting point for understanding the electronic structure. The danger lies in interpreting them too

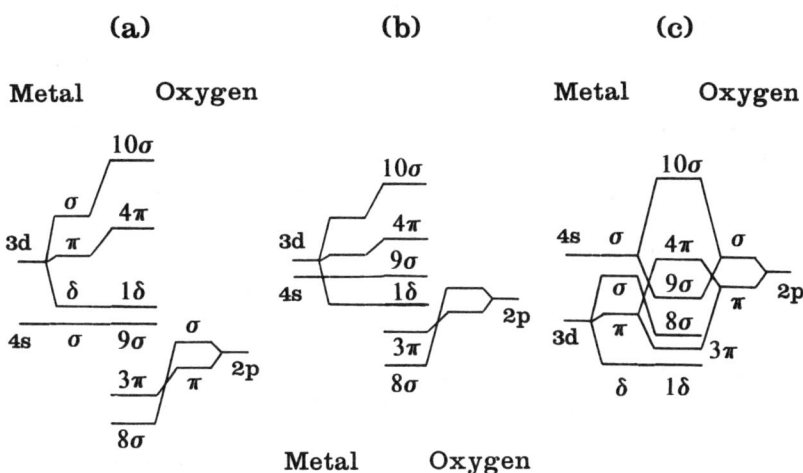

Figure 1 Molecular orbital diagrams for the 3d transition metal oxides. (a) TiO, (b) FeO, (c) CuO. With increasing atomic number the metal 3d orbital drops in energy compared to 4s, whereas the oxygen 2p rises because the metal ionization potentials increase.

literally. First, orbitals have no reality: Only the electronic states are solutions of the Schrödinger equation, and the "orbitals" change from electronic state to electronic state. Second, the Coulomb and exchange splittings between the states that arise from a particular electron configuration may be much larger than the "energy separations" of the orbitals. Finally (as indicated above), configuration interaction will play a very large part in determining the properties of a particular state in this model.

Figure 1 shows the molecular orbitals formed from the metal 4s and 3d atomic orbitals and the oxygen 2p orbital for three cases, which correspond to TiO, FeO and CuO. The atomic orbitals are drawn on a rough energy scale, in which the ionization potentials of the metal and the oxygen (off the top of the figure) are the zero, and the ionization potential of the metal has been taken as the geometric mean of the ionization potentials (I.P.s) for the neutral metal and the metal ion M^+ to allow for the partial ionic character of the bonding. The atomic orbitals are split into components of different λ (the projection of the atomic angular momentum quantum number l) by the axial electric field that exists in the molecule; finally, linear combinations of the atomic orbitals with the same λ value are formed into the molecular orbitals. There is no need to invoke the metal 4p orbitals because a glance at Condon and Shortley's "The Theory of Atomic Spectra" (8) shows that the 4p states lie 20,000–30,000 cm^{-1} above the 4s states in the first transition series. It may be argued that the higher σ orbitals are really $n = 4$ sp-hybrids (mixing the languages of m.o. and valence bond theory), but this is a matter of how one interprets an electronic wave function built from thousands of basis functions.

The trends across the $3d$ transition series are that the O $2p$ orbital catches up in energy to the metal $4s$ orbital with increasing atomic number, whereas the metal $3d$ falls rapidly, since it is a highly penetrating orbital with $l = 2$. The ground states of the oxides, as they are presently known, show these trends very well (see Table 1). The ground state of ScO is $(8\sigma)^2(3\pi)^4(9\sigma)^1$ $^2\Sigma^+$; successive electrons go into the 1δ and 4π orbitals in a high spin configuration until the multiplicity peaks at MnO, whose ground state is $(8\sigma)^2(3\pi)^4(9\sigma)^1(1\delta)^2(4\pi)^2$ $^6\Sigma^+$. At this point the 9σ orbital, which is known to be mostly metal $4s$ in character from the hyperfine structure, lies above 1δ; the next two electrons go into 1δ and then 9σ, so that the ground state of CoO is ... $(1\delta)^3(9\sigma)^2(4\pi)^2$ $^4\Delta_i$ (although the alternative ... $(1\delta)^4(9\sigma)^1(4\pi)^2$ $^4\Sigma^-$ cannot be totally ruled out yet). In NiO and CuO the unpaired electrons are in the 4π orbital. It can be seen in Table 1 how the spin-orbit coupling of the 1δ electron (that is, the metal $3d\delta$ electron) rises from 101 cm^{-1} in TiO to 190 cm^{-1} in FeO and ~ 240 cm^{-1} in CoO as the d orbitals shrink with increasing atomic number, Z; this reflects the r^{-3} dependence of the spin-orbit coupling, which is the expected behavior for an electron moving in a simple Coulomb potential.

The bond lengths in the zero-point level, r_0, indicate that the 1δ electrons behave initially as though they are slightly bonding, since r_0 shortens from ScO to VO as two δ electrons are added; there is no change in r_0 between CoO and NiO, meaning that at higher Z the 1δ electron is definitely nonbonding as it "drops out of the picture."

Table 1 Ground states of the $3d$ transition metal monoxides

	Ground state	$\Delta G_{1/2}$ (cm^{-1})	B_0 (cm^{-1})	r_0 (Å)	$A\Lambda$ or λ[a] (cm^{-1})	Electron configuration
^{45}ScO	$^2\Sigma^+$	964.65	0.51343	1.668	—	σ
^{48}TiO	$^3\Delta_r$	1000.02	0.53384	1.623	101.30	$\sigma\delta$
^{51}VO	$^4\Sigma^-$	1001.81	0.54638	1.592	2.03	$\sigma\delta^2$
^{52}CrO	$^5\Pi_r$	884.98	0.52443	1.621	63.22	$\sigma\delta^2\pi$
^{55}MnO	$^6\Sigma^+$	832.41	0.50122	1.648	0.57	$\sigma\delta^2\pi^2$
^{56}FeO	$^5\Delta_i$	871.15	0.51681	1.619	-189.89	$\sigma\delta^3\pi^2$
^{59}CoO	$^4\Delta_i$	851.7	0.5037$_0$	1.631	(-240)	$\sigma^2\delta^3\pi^2$
^{58}NiO	$^3\Sigma^-$	828.$_5$	0.505$_8$	1.631	(26)	$\sigma^2\delta^4\pi^2$
^{63}CuO	$^2\Pi_i$	629.39	0.44208	1.729	-277.04	$\sigma^2\delta^4\pi^3$

[a] The quantities $A\Lambda$ and λ are the first and second order spin-orbit splitting parameters; $A\Lambda$ is tabulated for orbitally degenerate states, λ for Σ states.

Copper oxide is anomalous because the ground state is not well described by a single configuration. Figure 1 suggests that the 4π orbital containing the unpaired electron is mainly O $2p\pi$, but the spin-orbit coupling is too big for an electron based on oxygen, and the hyperfine structure indicates that there is also a contribution from the three open shell configuration $(8\sigma)^1(9\sigma)^2(4\pi)^3(10\sigma)^1$. The main difference between CuO and NiO seems to be that there is much less $d\pi$–$p\pi$ bonding in CuO as the $3d$ orbital falls further below the $4s$. In more quantitative terms, an open d shell is needed to give such bonding; it takes 1.5 eV of energy to excite copper from its ground state, $3d^{10}4s$ ^2S to the $3d^94s^2$ ^2D level, while nickel already has an open d shell in its $3d^84s^2$ ^3F ground state.

The excited states of the oxides are fairly straightforward to interpret for the early members of the series but become progressively more obscure with increasing Z. The 8σ and 3π molecular orbitals of Figures 1a and 1b are strongly bonding orbitals formed from $3d\sigma$–$2p\sigma$ and $3d\pi$–$2p\pi$ combinations, though they are located mainly toward the oxygen atom. The 9σ m.o. (as shown by the hyperfine structure it produces) is about 80% metal $4s$ in character, which means that it points away from the oxygen atom; however, because it contains a little oxygen $2p\sigma$ and metal $3d\sigma$ (and $4p\sigma$) character, it is very slightly bonding. The 1δ orbital, essentially the metal $3d\delta$, is almost nonbonding. The 4π m.o. is slightly antibonding, and 10σ is rather more so; it is sometimes conveniently denoted σ^* (9).

Two types of excited states can arise. One comes from promotion of the electrons among the four uppermost m.o.s of Figure 1a, which are associated with the metal atom. The small increases in the bond length that occur on promotion of a 9σ or 1δ electron to 4π or 10σ can be used, together with the hyperfine parameters and the spin-orbit couplings, to give fairly specific accounts of the electron configurations. The excited states of ScO, TiO, VO and CrO that have been characterized so far are all of this type. The number of states increases rapidly with the number of available electrons: in VO, for example, there are 27 predicted low-lying electronic states for the three electrons, of which 13 are now known (9).

The other type of excited state arises by promotion of a bonding electron from the oxygen-centered 3π or 8σ to one of the metal-centered orbitals. This is equivalent to moving an electron from the oxygen to the metal in a "charge transfer transition." Not surprisingly this type of excitation causes a considerable increase in the bond length, of about 10%. Interestingly, the charge transfer states of ScO, TiO, VO and CrO are calculated to lie very high in energy. This is not what would be expected by analogy with CaO. In CaO the ground state is $(8\sigma)^2(3\pi)^4$ $^1\Sigma^+$, whereas the charge transfer states $(8\sigma)^2(3\pi)^3(9\sigma)^{1\ 1,3}\Pi$ and $(8\sigma)^1(3\pi)^4(9\sigma)^{1\ 1,3}\Sigma^+$ lie only 1–1.5 eV above the ground state (R. W. Field, private communication). The reason for the difference is that the $3d$ atomic states are about 2.5 eV lower in Sc than in Ca, so that the nature of the m.o.s changes fundamentally between CaO and ScO, with the 8σ and 3π m.o.s lying much further below 9σ in ScO. Even so, an unusually large vibrational dependence of the spin-rotation coupling in the ground state of ScO may possibly result from interactions with the charge transfer excited states (10, 11), so that these may still be low enough to have noticeable effects on the ground state. Manganese oxide is the first member of the $3d$ oxides for which a charge transfer state with a long bond length has been definitely identified; it lies near 2.2 eV. In FeO the charge transfer states drop dramatically in energy, with the $(8\sigma)^2(3\pi)^3(1\delta)^3(9\sigma)^1(4\pi)^2(10\sigma)^{1\ 5}\Phi$ and $^5\Pi$ states lying only 1.2 eV above the ground state.

It is very difficult to interpret the excited states of the later $3d$ oxides, for the reasons given above. The data for CoO and NiO are still fragmentary and, although 15 states are known for CuO, a satisfactory model for them all still eludes us. I shall not include ZnO with the $3d$ oxide series, since the Zn^+ atomic states with an open $3d$ shell lie so high that they cannot be involved in the bonding of ZnO; in any case no states of ZnO have been assigned. From Figure 1c the ground state of ZnO is expected to be $^1\Sigma^+$, and the low-lying excited states to be ...$(4\pi)^3(10\sigma)^{1\ 1,3}\Pi$ and ...$(9\sigma)^1(4\pi)^4(10\sigma)^{1\ 1,3}\Sigma^+$. Calculations of their energies (12) predict that ZnO should have band systems in the infrared but nothing in the visible region; the $^3\Sigma^+$–$^3\Pi$ transition should lie near 12,000 cm^{-1}.

It is interesting that the ligand field approach (13, 14) used so successfully by Field et al to rationalize the electronic states of the $4f$ lanthanide oxides appears to be less applicable to the $3d$ transition oxides. The reason is that the $3d$ orbitals in the transition oxides are comparatively so very much larger than the $4f$ orbitals in the lanthanide oxides that the covalent component of the bonding is correspondingly more important. For similar reasons, the ionic model of Törring et al (15), which explains the dipole moments of the alkaline earth monohalides so beautifully, may also be less applicable to the $3d$ oxides.

DESCRIPTIONS OF THE INDIVIDUAL MOLECULES

The data for the $3d$ oxides up to about 1976 have been comprehensively listed in Huber & Herzberg's tables (16). There is no need for me to go back into the early history except to say that the strongest band systems have been known since about 1910. Instead I emphasize the recent progress.

ScO

Scandium oxide is one of the 3d oxides that has been found in the spectra of M-type variable stars, where it usually accompanies TiO (17). It has two easily excited band systems in the visible region (see Table 2), which have been analyzed rotationally. The first analyses established the bond lengths, but there was great confusion over the natures of the electronic states because of the hyperfine structure. The only stable isotope, ^{45}Sc, has a nuclear spin $I = 7/2$ and a very large magnetic moment, such that the Fermi contact interaction for an unpaired electron in the 9σ ($4s\sigma$) m.o. in ScO is huge by comparison to the organic molecules that mainly had been studied up to that time. The ground state of ScO (Table 1 and Figure 1) is $(8\sigma)^2(3\pi)^4(9\sigma)^1\,{}^2\Sigma^+$, and belongs to the unusual hyperfine coupling case $b_{\beta S}$ (18, 19), where the Fermi contact parameter b is much larger than the spin-rotation coupling parameter γ.

Specifically, the relevant parts of the Hamiltonian are

$$H_{\text{spin-rotation}} = \gamma \mathbf{N} \cdot \mathbf{S}; \quad H_{\text{Fermi contact}} = b \mathbf{I} \cdot \mathbf{S}$$

where the "spin-rotation" interaction consists of the interaction between the magnetic moments of the electron spin (**S**) and the molecular rotation (**N**), together with a second-order cross-term between the spin-orbit and orbit-rotation interactions; the Fermi contact interaction results from the fact that s electron wave functions do not go to zero at the nucleus, and there is a finite probability of the electron being found inside the nucleus. In Hund's case (b) coupling (that is, where the first-order spin-orbit coupling is small or zero), the two possible situations are that the Fermi contact parameter b is larger or smaller than the parameters governing the separation of the electron spin components, such as γ. If b is larger, the vector coupling that most closely represents the physical situation is

$$\mathbf{I} + \mathbf{S} = \mathbf{G}; \quad \mathbf{G} + \mathbf{N} = \mathbf{F},$$

where the nuclear and electron spins (**I** and **S**) add first, because of the coupling term $\mathbf{I} \cdot \mathbf{S}$, to give a resultant **G**, which then adds to the rotational angular momentum (**N**) to give the total angular momentum **F**. This is known as case $b_{\beta S}$. The other extreme, known as case $b_{\beta J}$, occurs when the electron spin is coupled more strongly to **N** than to **I**; the vector coupling is then

$$\mathbf{N} + \mathbf{S} = \mathbf{J}; \quad \mathbf{J} + \mathbf{I} = \mathbf{F}.$$

This is the more usual case, and is the reason that the occurrence of $b_{\beta S}$ coupling caused such confusion in ScO. In case $b_{\beta S}$, the hyperfine energy follows from

$$\mathbf{I} \cdot \mathbf{S} = (1/2)(\mathbf{G}^2 - \mathbf{I}^2 - \mathbf{S}^2) = (1/2)[G(G+1) - I(I+1) - S(S+1)]$$

Table 2 Spectroscopic constants of the known states of the 3d transition metal oxides[a]

	T_0	$\Delta G_{1/2}$	B_0	r_0/Å	Refs.	Notes
[45]ScO						$D_0 = 7.01 \pm 0.12$ eV (32)
Systems attributed to ScO in the regions 360–391 nm and 412–430 nm (30)						
$B^2\Sigma^+$	20571.15	817.05	0.48308	1.7199	(20, 24)	
$A^2\Pi_r$	16498.13	870.0[h]	0.50277	1.6859	(25)	$\tau = 33.3 \pm 0.6$ ns (31); $\gamma = -0.0670$ cm^{-1} $(8\sigma)^2(3\pi)^4(10\sigma)^1$
	($A = 115.33$)	($^2\Pi_{3/2}$)				$\mu(v=1) = 4.2 \pm 0.2$ D (5); perturbations in $v=1$ (10, 29a); $\tau = 35.9 \pm 2.4$ ns ($^2\Pi_{1/2}$), 27.0 ± 2.5 ns ($^2\Pi_{3/2}$) (31); $p = -0.0655$ cm^{-1} ("pure precession" with $B^2\Sigma^+$). $(8\sigma)^2(3\pi)^3(4\pi)^1$
		858.9[h]				
		($^2\Pi_{1/2}$)				
$A'^2\Delta_r$	15018.9[h]	837.0[h] ($^2\Delta_{5/2}$)	0.4797	1.726	(29, 29a)	Constants deduced from perturbations in $A^2\Pi$ (29a). Appears as $A' \to X$ chemiluminescence. $(8\sigma)^2(3\pi)^4(1\delta)^1$
	($A\Lambda = 106.1$[h])	834.0[h] ($^2\Delta_{3/2}$)				
$X^2\Sigma^+$	0	964.95	0.51343	1.6683	(21–23, 25)	Spin and hyperfine constants for $v=0$ (11): $\gamma = 3.2175$, $b = 1922.534$, $c = 74.416$, $eQq = 72.240$, $c_I = 0.02181$ MHz. $(8\sigma)^2(3\pi)^4(9\sigma)^1$

[h] = from head measurements

[48]TiO

I.P. $= 6.56 \pm 0.03$ eV (48) $D_0 = 6.92 \pm 0.10$ eV (32, 61)

Absorption systems (Ne matrix) at 32,068 cm^{-1} and 32,660 cm^{-1}; MCD indicates $|\Delta\Lambda| = 1$ (47).
The 32,068 cm^{-1} system (413 nm), and one at 329 nm with $\Delta G''_{1/2} = 983$ cm^{-1}, appear in flame spectra (44).
Another emission system at 376 nm (44) with $\omega' - \omega'' = -60$ cm^{-1}, possibly singlet with $|\Delta\Lambda| = 1$.

	T_0	$\Delta G_{1/2}$	B_0	r_0/Å	Refs.	Notes
$e^1\Sigma^+$	29964.6	845.2	0.4881	1.696$_9$	(49)	$v=0$ and 1 perturbed. $(9\sigma)^1(10\sigma)^1$
$f^1\Delta$	22517.31	869.101	0.502277	1.6728	(50, 51)	$T_e(f^1\Delta) - T_e(a^1\Delta) = 19140.567$ cm^{-1}, $\omega_e = 874.104$, $\omega_e x_e = 2.501$, $\alpha_e = 0.00308$ cm^{-1} $(1\delta)^1(10\sigma)^1$. $\tau = 17.5 \pm 1$ ns (53). $(1\delta)^1(4\pi)^1$. $c-a$: β bands.
$c^1\Phi$	21288.89	909.53	0.52133	1.6420	(52)	Perturbed at $v=4$ and 5. $C-X$: α bands.
$C^3\Delta_3$	19536.63	828.36				$\tau(v=0) = 37 \pm 9$ ns (55); $\tau(v=2,$
$^3\Delta_2$	19441.47	829.01	0.48836	1.6957	(54)	$J=17) = 28.2 \pm 0.2$ ns (56).
$^3\Delta_2$	19341.68	828.64				$\alpha_e = 0.00306$ cm^{-1}. $(1\delta)^1(10\sigma)^1$.

State	T_e	ω_e		B_e	r_e	Refs	Notes
$B^3\Pi_2$	16266.797	864.334		0.50606	1.6666	(47, 54)	Λ-doubling in $^3\Pi_0$ at $J=0$ is 1.263 cm^{-1} (47).
$^3\Pi_1$	16247.951	864.207					B–X: γ' bands. $\alpha_e = 0.00318$ cm^{-1}. $(1\delta)^1(4\pi)^1$.
$^5\Pi_0$	16225.986	863.526					
$b^1\Pi$	14721.14	911.165		0.51202	1.6568	(49, 57, 59)	$T_0(b^1\Pi - d^1\Sigma^+) = 9054.039$ cm^{-1} (ϕ bands). $\alpha_e = 0.00284$ cm^{-1}; $q = 1.6 \times 10^{-4}$ cm^{-1}. $(9\sigma)^1(4\pi)^1$
$A^3\Phi_4$	14365.60	859.55					$\alpha_e = 0.00315$ cm^{-1}.
$^3\Phi_3$	14193.69	859.79		0.50571	1.6671	(54)	A–X: γ bands.
$^3\Phi_2$	14019.43	860.22					$(1\delta)^1(4\pi)^1$
$E^3\Pi_2$	12073^h						Red-degraded. Seen in matrix absorption (46) and chemiluminescence (58). No high resolution analysis. $(9\sigma)^1(4\pi)^1$. E–X: ε bands.
$^3\Pi_1$	11993^h	914^h		—	—	(46, 58)	
$^3\Pi_0$	11899^h						
$d^1\Sigma^+$	5667.10	1013.281		0.54765	1.6020	(49, 57, 59)	$\alpha_e = 0.00335$ cm^{-1}. $(9\sigma)^2$.
$a^1\Delta$	3448.32	1009.231		0.536168	1.6191	(51)	$T_0 \pm 0.04$ cm^{-1} (38) (Ar$^+$ excitation of $C^3\Delta$, $v=2$, $J=17$ at 4765 Å). $\omega_e = 1018.273$, $\omega_e x_e = 4.521$, $\alpha_e = 0.00292$ cm^{-1}. $(9\sigma)^1(1\delta)^1$. b–a: δ bands.
$X^3\Delta_3$	202.601						
$^3\Delta_2$	97.806	1000.022		0.53384	1.6226	(41, 42, 54)	$\alpha_e = 0.00304$ cm^{-1}. $(9\sigma)^1(1\delta)^1$.
$^3\Delta_1$	0						IR fundamental at low resolution (60). Radio-astrophysical search (62) unsuccessful.

I.P. = 7.25 ± 0.01 eV (80) $D_0 = 6.44 \pm 0.20$ eV (32)

Eight states between 230 and 380 nm observed in flash photolysis of VOCl$_3$ (79)

^{51}VO

State	T_e	ω_e		B_e	r_e	Refs	Notes
E	23980	833		—	—	(78)	Matrix absorption; not seen in gas by (79).
$2\,^2\Delta_r$	b+9595.85	—		0.48902	1.6828	(9)	$A \sim 50$ cm^{-1}; $\dots(9\sigma)^1(1\delta)^1(10\sigma)^1$.
$D^4\Delta_r$	19148.08	835		0.48704	1.6863	(9)	$A = 47.83$ cm^{-1}; $(9\sigma)^1(1\delta)^1(10\sigma)^1$.
$2\,^2\Pi_{3/2}$	a+8126.99	—		0.52646	1.6219	(9)	? $(9\sigma)^1(1\delta)^1(4\pi)^1$; $^2\Pi_{1/2}$ not seen.
$1\,^2\Pi$	a+7208.08	927.14		0.51987	1.6321	(9)	$A \sim 207$ cm^{-1}; $(9\sigma)^2(4\pi)^1$.
$C^4\Sigma^-$	17420.103	852.6		0.49379	1.6747	(64, 66–69, 82)	$\lambda = 0.7470$, $b = -0.0088$ cm^{-1} (68). Internal hyperfine perturbation between F_2 and F_3 at $N=5$ (68, 69). $(1\delta)^2(10\sigma)^1$. Many perturbations (64, 68, 82).
$1\,^2\Phi_r$	b	—		0.51900	1.6335	(9)	$A \sim 66$ cm^{-1}; $(9\sigma)^1(1\delta)^1(4\pi)^1$.
$B^4\Pi_r$	12605.57	901		0.5124	1.644	(66, 67, 76, 77, 86, 87)	$A \cong 63.0$ cm^{-1} (77); b is small. $(1\delta)^2(4\pi)^1$.

Table 2 (*continued*)

	T_0	$\Delta G_{1/2}$	B_0	r_0/Å	Refs.	Notes
$1^2\Sigma^+$	~12430	~1000	0.54	1.60	(9)	Perturbs $B^4\Pi$; $(9\sigma)^1(1\delta)^2$.
$1^2\Delta_r$	a	? 1019.9	0.55163	1.5845	(9)	$(9\sigma)^2(1\delta)^1$. $A \sim 158$ cm^{-1}; perturbed.
$A^4\Pi_i$	9898.878	884	0.51693	1.6368	(74)	$b = 0.026$ cm^{-1}, $A = 35.19$ cm^{-1}; $(9\sigma)^1(1\delta)^1(4\pi)^1$.
$A'^4\Phi_r$	7254.951	936.48	0.52213	1.6286	(9)	$A = 56.93$ cm^{-1}; $(9\sigma)^1(1\delta)^1(4\pi)^1$.
$X^4\Sigma^-$	0	1001.81	0.54638	1.5921	(64–69)	Shown to be the ground state by ESR (65). $b = 0.02731$, $c = -0.00413$, $\lambda = 2.0309$, $\gamma = 0.02252$ cm^{-1} (68). Internal hyperfine perturbation between F_2 and F_3 at $N = 15$ (67, 68). $(9\sigma)^1(1\delta)^2$.

The state $(1\delta)^3\,^2\Delta_i$ is calculated to lie near 22,000 cm^{-1} (28).

^{52}CrO				I.P. = 7.85 ± 0.02 eV (96)		$D_0 = 4.41 \pm 0.30$ eV (32) [4.52 eV (99)]
C	22163	585	—	—	(93)	Chemiluminescence. Presumably charge-transfer.
$B^5\Pi_r$	16502.404	732.410	0.47097	1.7108	(94, 95)	Very many small perturbations. $A = 54.93$ cm^{-1}. $(1\delta)^2(4\pi)^1(10\sigma)^1$. $\alpha_e = 0.00548$ cm^{-1}.
$A'^5\Delta_r$	~11800	820	—	—	(92, 93)	Chemiluminescence. $(9\sigma)^1(1\delta)^1(4\pi)^2$.
$A^5\Sigma^+$	8191.23	868	0.49920	1.6618	(90)	$\lambda = 0.736$, $\gamma = 0.0090$ cm^{-1}. $(1\delta)^2(4\pi)^2$.
$X^5\Pi_r$	0	884.976	0.52439	1.6213	(90,94)	$A = 63.221$, $\lambda = 1.156$ cm^{-1}, $o+p+q = 0.032$, $p+2q = 0.0069$, $10^6 q = 33$ cm^{-1}, $\omega_e x_e = 6.72$, $\alpha_e = 0.00443$ cm^{-1}. $(9\sigma)^1(1\delta)^2(4\pi)^1$. Matrix IR (100).

^{55}MnO				I.P. = 8.65 ± 0.2 eV (110)		$D_0 = 3.83 \pm 0.08$ eV (101)
$C^6\Pi$?	38950	—	—	—	(102)	Flash photolysis. Not seen in matrices (103).
$B^6\Sigma^+$	~17947	—	0.395	1.86	(70)	$\lambda \sim 3.0$ cm^{-1}. Charge transfer state; perturbs A, $v = 0$. Only one vibrational level known.
$A^6\Sigma^+$	~17894	~749	0.4633	1.714	(70, 105)	$\lambda \sim -0.13$ cm^{-1}. $(1\delta)^2(4\pi)^2(10\sigma)^1$. Internal hyperfine perturbations in $v = 1$.

$X^6\Sigma^+$	~41480	0	832.41	0.50122	1.6477	(70, 105)

$\lambda = 0.57, \gamma = -0.003, \alpha_e = 0.00406$ cm^{-1}. ESR spectrum: $b = 433, |c| = 46 \pm 12$ MHz (108). $(9\sigma)^1(1\delta)^2(4\pi)^2$. Internal hyperfine perturbations between F_3 and F_4 spin components.

$D_0 = 4.17 \pm 0.08$ eV (101)

^{56}FeO

I.P. $= 8.9 \pm 0.16$ eV (110) (102)

?	22335^h		545^h			Flash photolysis
?	19449^h		526^h			Chemiluminescence (115); exploding wire (129).
$D^5\Delta_0$	18533.9		—	0.469		Chemiluminescence (115).
$^5\Delta_1$	18366.8$_5$		670^h	0.480	1.700	(92, 112, 115, 121)
$^5\Delta_2$	18160.23		—	0.468		
$^5\Delta_4$	17901.39		671^h	0.469		

Irregular structure with many perturbations (121); $^5\Delta_3$ not found; $^5\Delta_2$ pushed down by ~50 cm^{-1}. "A_1 and A_2 systems," of (130) are $^5\Delta_1$ and $^5\Delta_4$–$X^5\Delta$. $\ldots(1\delta)^3(4\pi)^2(10\sigma)^1$ $^5\Delta_i$.

Many irregular perturbed levels with $\Omega = 0$–4 in the range 16,350–17,750 cm^{-1} seen in absorption and emission (92, 114, 115, 130). Most prominent bands in laser-induced fluorescence (121): 5791 Å ["B system" of (130)]; 5820 Å ($\Omega' = 4$–$X^5\Delta_4$); 5845 and 5975 Å ($\Omega' = 3$–$X^5\Delta_3$); 5919 Å ($\Omega' = 2$–$X^5\Delta_2$) $[\mu(\Omega' = 2) = 2.6 \pm 0.2$ D (6)]. "Orange system," possibly a separate charge transfer system from D–X. Upper state B values vary irregularly from 0.38 cm^{-1} to 0.495 cm^{-1}. Isotope shifts (121) indicate $v' > 0$. Irregular Λ-doublings (132).

$C^5\Pi_{-1}$	10808.4$_6$		581.4	0.45709		
$^5\Pi_0$	10613.5$_9$		~644	0.45325		
$^5\Pi_1$	10405.648		627.447	0.45097	1.733	(117, 131)
$^5\Pi_2$	10188.671		639.600	0.45301		
$^5\Pi_3$	9977.6		635.3	0.443		

Charge transfer complex with $B^5\Phi_i$; severely perturbed. $(8\sigma)^2(3\pi)^3(9\sigma)^1(1\delta)^3(4\pi)^2(10\sigma)^1$. Perturbing states include $a^7\Sigma^+$ (?), with $\omega \sim 650$ cm^{-1}. Λ-doublings of $^5\Pi_0$: 37.940 cm^{-1} ($v = 0$), 64.5 cm^{-1} ($v = 1$), 15.01 cm^{-1} ($v = 2$).

$B^5\Phi_1$	10446.3$_7$		615.793	0.46236		
$^5\Phi_2$	10331.4		594.3	0.447		
$^5\Phi_3$	10192.359		593.3	0.44902	1.736	(117, 131)
$^5\Phi_4$	10052.721		580.371	0.44643		
$^5\Phi_5$	9903.579		581.430	0.44460		

Electron configuration as $C^5\Pi_i$. Very heavily perturbed; perturbing states include $^3\Phi_i$. Irregular and sporadic Λ-doublings. Small perturbations between $C^5\Pi_i$ and $B^5\Phi_i$. $\ldots(9\sigma)^2(1\delta)^2(4\pi)^2$. From FeO$^-$ photoelectron spectrum

$A^5\Sigma^+$	~3990		~900	—	—	(128)
$a^7\Sigma^+$	~150?		~650	—	—	(118, 124, 125)

Only 1 level (? $v = 3$) seen in optical spectrum (118), perturbing $X^5\Delta_2$, $v = 2$. FeO$^-$ autodetachment spectrum suggests $a^7\Sigma^+$, $v = 0$ lies near $X^5\Delta_3$, $v = 0$ (124, 125). $\ldots(9\sigma)^1(1\delta)^2(4\pi)^2(10\sigma)^1$.

Table 2 (continued)

	T_0	$\Delta G_{1/2}$	B_0	$r_0/\text{Å}$	Refs.	Notes
$X^5\Delta_0$	759.4$_2$					Mm wave spectrum, $\Omega = 2$–4 (119, 120); not found in interstellar space (133). $\ldots(9\sigma)^1(1\delta)^3(4\pi)^2$. Λ-doubling of $^5\Delta_0$, $J = 0$: 0.3183 cm^{-1}. $\mu(X^5\Delta_2) = 4.7 \pm 0.2$ D (6). $\omega_e = 880.4148$, $\omega_e x_e = 4.6322$, $\omega_e y_e = 5.55 \times 10^{-4}$, $B_e = 0.518721$, $\alpha_e = 0.003825$ cm^{-1} (118).
$^5\Delta_1$	564.1$_6$					
$^5\Delta_2$	372.261	871.152	0.516809	1.6194	(6, 118–120)	
$^5\Delta_3$	184.084					
$^5\Delta_4$	0					

FeO$^-$ autodetachment spectrum (124, 125): theoretical discussion (134)

	T_0	$\Delta G_{1/2}$	B_0	$r_0/\text{Å}$	Refs.	Notes
$C\Sigma$	12225.50	—	0.4918	1.660		? Dipole bound (FeO, $a^7\Sigma^+ + e^-$); large spin-doubling.
$A^4\Delta_{5/2}$	12086.00	—	0.5045	1.639		Rapid autodetachment by spin-orbit mechanism. ? Valence.
$B^4\Delta_{7/2}$	12011.19	—	0.5046	1.639		? Dipole bound (i.e. Rydberg-like) (FeO, $X^5\Delta_4 + e^-$). $X^4\Delta_{7/2}$ lies 12,042 cm^{-1} (1.493 eV) below FeO, $X^5\Delta_4$, $v = 0$. $\ldots(9\sigma)^2(1\delta)^3(4\pi)^2$.
$X^4\Delta_{3/2}$	465.82	—	0.5018			
$^4\Delta_{5/2}$	226.123	—	0.4997	1.645		
$^4\Delta_{7/2}$	0	—	0.4972			

^{59}CoO

	T_0	$\Delta G_{1/2}$	B_0	$r_0/\text{Å}$	Refs.	Notes
				I.P. $= 8.9 \pm 0.2$ eV (110)		$D_0 = 3.94 \pm 0.14$ eV (32)
$\Omega = 7/2$, $v > 0$	15772.513	~547	0.40531	1.8183	(138)	Long progressions of red-shaded emission and absorption bands in the region 10,900–19,200 cm^{-1} (136, 137). Rotational analysis shows at least two perturbed charge transfer transitions (138).
$\Omega = 5/2$, $a + 15594.974$ $v > 0$		—	0.4250$_3$		(138)	
$\Omega = 7/2$, $v > 0$	15535.77	~550	0.422$_4$	1.77$_3$	(138)	

^{58}NiO

	T_0	$\Delta G_{1/2}$	B_0	$r_0/\text{Å}$	Refs.	Notes
$X^4\Delta_{5/2}$	a (~240)		0.5026$_6$			Matrix isolation IR spectrum (111) has $\Delta G_{1/2} = 846.4$ cm^{-1}. No ESR spectrum (127). $(9\sigma)^2(1\delta)^3(4\pi)^2$.
$^4\Delta_{7/2}$	0	851.7	0.50058	1.631	(138)	$D_0 = 3.91 \pm 0.17$ eV (32)
				I.P. $= 9.5 \pm 0.2$ eV (110)		

3d OXIDE SPECTRA 419

Long progressions of red-degraded bands, 460–920 nm, divided into six systems (16, 92, 112, 136, 137, 139); vibrational analysis very uncertain. Four upper levels of a parallel transition near 520 nm analyzed rotationally (140), three with $\Omega = 0^+$, one with $\Omega = 1$; B' values near 0.42 cm^{-1} ($r = 1.79$ Å); isotope shifts indicate $v' > 0$; perturbed.

$X^3\Sigma^-$	0	828._3	0.5059	1.631	(140)	$\lambda \sim 26$ cm^{-1} ... $(9\sigma)^2(1\delta)^4(4\pi)^2$.

^{63}CuO

$P^2\Pi_{3/2}$	25190.31	573.81	0.3881	1.845	(151)	$D_0 = 2.75 \pm 0.2$ eV (32)
$M^2\Pi_{3/2}$	23898.00	—	0.418$_8$	1.777	(149, 150)	Vibrational numbering not known; large D. 5 perturbations for $J < 67.5$; one state has $B = 0.387_6$ cm^{-1}.
$k^2\Sigma_{1/2}$	22818.18	—	0.40390	1.8091	(152)	$v = 0$ only; $p = 4B$ proves $^4\Sigma$ nature.
$I^2\Pi_{3/2}$	[22448.95	608.11	0.4164	1.782]	(151)	$v = 0$ not seen $\}$ (from isotope shifts);
$H^2\Pi_{3/2}$	[22325.80	555.90	0.4190	1.776]	(151)	$v = 0$ not seen $\}$ values quoted are for $v = 1$.
$G^2\Sigma^+$	21593.98	582.74	0.41296	1.7891	(155)	$\gamma = 0.1674$ cm^{-1}.
$F^2\Pi_i$	21221.66	600.75	0.41022	1.7951	(141, 153, 154)	$A_{v=0} = -6.24$; $A_{v=1} = -31.87$ cm^{-1}; $p(^2\Pi_{1/2}) = 0.3190$ cm^{-1}.
$E^2\Delta_{5/2}$	21103.66	722.40	0.4429	1.728	(151)	$E^2\Delta_{3/2} - X^2\Pi_{1/2}$ (0,0) band may be at 4773 Å (152).
$D^2\Delta_i$	19149.036	636.7($^2\Delta_{5/2}$)	0.42906	1.7552	(156)	$A\Lambda = -73.184$ cm^{-1}; $\tau = 1.8$ μs (157).
$C^2\Pi_i$	18890.647	631.739	0.42102	1.7719	(151, 156, 158, 159)	$A_{v=0} = -156.758$, $A_{v=1} = -127.437$; $\tau = 1.3$ μs (157); $p = -0.1480$ cm^{-1}.
$A^2\Sigma^-$	16492.37	631.02	0.43150	1.7503	(141, 153, 159)	$\tau = 0.65$ μs (157); $b = -0.031$ cm^{-1} (160); $\gamma = -0.1952$ cm^{-1}. $(3\pi)^3(4\pi)^3(10\sigma)^1$.
$\beta^2\Delta_{3/2}$	15665	604	—	—	(146, 161)	$A' \dots (3\pi)^3(4\pi)^3(10\sigma)^1$; A' and γ analyzed as an
$A'^2\Sigma^-$	15531.90	605.0	0.4382$_8$	1.737 $\}$	(146, 162)	interacting complex with $A_\perp = 7.52$ cm^{-1} (148);
$\gamma^2\Pi_{1/2}$	15470	—	0.4153	1.784	(148)	A' has $b = -0.053$ cm^{-1}.
$\alpha^4\Sigma_{(1/2)}$	15424	—	—	—	(146)	
$\beta^2\Delta_{5/2}$	15317.24	—	0.4253$_6$	1.763	(161)	$(3\pi)^3(8\sigma)^2(9\sigma)^2(4\pi)^3(10\sigma)^1$
$\gamma^2\Pi_{3/2}$	15166	—	—	—	(146)	$\dots (8\sigma)^1(9\sigma)^2(4\pi)^3(10\sigma)^1$
$\delta^2\Sigma$	12986	649.4	—	—	(146)	
$Y^2\Sigma^+$	7865	$\sim$680	—	—	(145)	$\dots (8\sigma)^2(9\sigma)^1(4\pi)^4$
$X^2\Pi_{1/2}$	277.04	627.46 $\}$	0.442085	1.7292	(142, 153)	$\mu = 4.45 \pm 0.3$ D (7) in $^2\Pi_{3/2}$. Hfs constants from microwave spectrum (142): $a - (b+c)/2 = 432$,
$^2\Pi_{3/2}$	0	631.32				$d = 143$, $b \sim -400$ MHz. No ESR (103). $\dots (8\sigma)^2(9\sigma)^2(4\pi)^3$.

^{a} Values are given in cm^{-1} units, unless otherwise indicated.

so that for $I = 7/2$ and $S = 1/2$, as in ScO, there are two hyperfine states, $G = 7/2 \pm 1/2 = 4$ or 3, separated by $4b$ for each rotational level.

If the applicability of this coupling is not recognized there appear to be twice as many electron spin components as there actually are, and the first rotational analyses (20) assigned the ground state of ScO as $^4\Sigma$. Ab initio calculations (21) gave the ground state as $^2\Sigma$, and this was immediately confirmed by molecular beam refocussing (22) and ESR experiments (23). More recent rotational analyses (24, 25) have confirmed the ESR value for the Fermi contact parameter b in the ground state. The unpaired 9σ ($4s\sigma$) electron is not present in the excited states of ScO, having been promoted to one of the three higher orbitals. As a result, the hyperfine effects in the excited states are very small and follow case $b_{\beta J}$ or $a_{\beta J}$ coupling.

The hyperfine structure in the ground state has recently been studied at ultra-high resolution by the method of laser-radio frequency double resonance in a molecular beam (10, 11). The Fermi contact parameter is found to be 1947.34 MHz, which is 83% of the value calculated for a $4s$ unpaired electron of atomic Sc (26). By comparison, the spin-rotation parameter γ is only 3.22 MHz. There is however a 30% variation of γ on vibrational excitation, which may possibly be caused by interaction with the very high-lying charge-transfer states described in the previous section. A charge-transfer state $(8\sigma)^2(3\pi)^3(9\sigma)^1(1\delta)^1\ ^4\Pi$, with a bond length near 2 Å, is calculated (27) to lie at 20,500 cm^{-1}.

Only one other low-lying state of ScO is expected (28) or known: the $(8\sigma)^2(3\pi)^4(1\delta)^1\ A'^2\Delta$ state, which has only been seen directly in chemiluminescence experiments (29) in a transition to the ground state. This transition, $A'^2\Delta - X^2\Sigma^+$, is highly forbidden and must obtain its intensity from the $A^2\Pi - X^2\Sigma^+$ system by rotational mixing of the $A'^2\Delta$ and $A^2\Pi$ states. Two band systems in the regions 360–391 nm and 412–430 nm, seen by optogalvanic spectroscopy (30), are attributed to ScO; these need to be investigated further, to see whether they involve Rydberg or charge transfer states.

TiO

Titanium oxide is the most abundant of the 3d oxides in the spectra of the cool red M-type stars. So strong are its band systems that it is used for the spectral classification of these stars on the MK system (2, 33, 34); at high resolution the rotational line strengths can even be used as a thermometer for circumstellar envelopes (35).

Because of its importance in astrophysics, considerable effort has been put into the analysis of the various TiO band systems, and TiO is probably the best understood of the 3d oxides, spectroscopically (see Table 2). The bonding in TiO is similar to that in ScO, and is best described as a double bond. The occupied bonding orbitals are 8σ and 3π; the four electrons in

3π count as only one bond because three of them come from the oxygen and only one from the Ti (36, 37). Two electrons are left over, which go into the metal-centered orbitals of Figure 1a, giving manifolds of triplet and singlet states. Sixteen low-lying states are expected, of which 11 are known.

The principal problems with the TiO spectrum have been the relative energies of the singlet and triplet manifolds, and the spin-orbit coupling intervals in the Hund's case (a) multiplet states. The ground state of TiO is now known to be $(8\sigma)^2(3\pi)^4(9\sigma)^1(1\delta)^1\ ^3\Delta_r$, with the $^1\Delta$ state from the same configuration lying 3448.32 ± 0.04 cm^{-1} above it (38). This is different from ZrO (the corresponding 4d oxide), where the ground state is $\sigma^2\ ^1\Sigma^+$, with $^3\Delta_1$ lying 1099 cm^{-1} higher (39). The states of TiO are in good case (a) or (b) coupling, so that intermultiplicity transitions are extremely weak. However, a chance coincidence between the 4765 Å line of the Ar$^+$ laser and a line of the (2, 0) band of $C^3\Delta_3$–$X^3\Delta_3$ (40) is found to give weak emission to $a^1\Delta$, as well as strong emission to $X^3\Delta$. At high resolution there is no doubt that the lower state of the weak emission has v = 0, from the rotational combination differences (38).

The spin-orbit intervals were established by the observation of spin satellite branches in the $B^3\Pi$–$X^3\Delta$ transition (41, 42). These branches arise because the ...$(1\delta)^1(4\pi)^1\ B^3\Pi$ state has quite small spin-orbit coupling, since the coupling constant is the difference between those for the 1δ and 4π electrons. As a result, the $B^3\Pi$ state uncouples fairly rapidly to case (b) with increasing rotation, so that case (a)-forbidden spin satellite branches with $\Delta\Sigma \neq 0$ become allowed.

The spin-orbit couplings and the bond lengths in the states of TiO enable the electron configurations to be assigned with some certainty (9, 36, 37). As is seen in Table 2, the Σ and Δ states with the two "nonbonding" metal electrons in the 9σ or 1δ m.o.s have $r_0 = 1.60$–1.62 Å. Excitation of one of these electrons to 4π gives Π and Φ states with $r_0 = 1.64$–1.67 Å, whereas excitation of one of these electrons to 10σ gives even-Λ states with $r_0 = 1.67$–1.70 Å. At the same time, the spin-orbit couplings are consistent with $2a_\delta \cong 95$ cm^{-1}, $a_\pi \cong 75$ cm^{-1}. For example, the ...$(1\delta)^1(4\pi)^1\ A^3\Phi$ and $B^3\Pi$ states have single configuration electronic wave functions (43) that can be written

$$|^3\Phi\rangle = |\pi^+\delta^{2+}|; \quad |^3\Pi\rangle = |\pi^-\delta^{2+}|.$$

Their spin-orbit coupling constants, $A\Lambda$, are the sum and difference of those for the individual electrons; thus

$$A\Lambda(^3\Phi) \cong 75+95 \cong 170\ \text{cm}^{-1}\ (\text{exptl} = 173.08\ \text{cm}^{-1}\ \text{from Table 2})$$

$$A\Lambda(^3\Pi) \cong -75+95 \cong 20\ \text{cm}^{-1}\ (\text{exptl} = 20.41\ \text{cm}^{-1}).$$

Similarly the spin-orbit coupling of the ...$(4\pi)^1 E^3\Pi$ state should be 75 cm^{-1} (exptl = 87 cm^{-1}), and those of the ...$(1\delta)^1 C^3\Delta$ and $X^3\Delta$ states should be 95 cm^{-1} (exptl = 97.47 and 101.30 cm^{-1}). We can say very clearly (9) that the only low-lying states of TiO not yet observed are $(1\delta)^1(1\pi)^{1\ 1}\Pi$, $(9\sigma)^1(10\sigma)^{1\ 3}\Sigma^+$, and the $^3\Sigma^-$, $^1\Gamma$, and $^1\Sigma^+$ states from $(1\delta)^2$; the first of these should lie near 23,500 cm^{-1} in the single configuration model.

Various unanalyzed transitions occur in the near UV. In TiO flames (44), systems are seen at 313, 329, and 376 nm. The 313 nm system occurs also in chemiluminescence (45) and in absorption in a neon matrix (46). The magnetic circular dichroism spectrum (47) indicates that the 313 nm system, and another system at 306 nm, are perpendicularly polarized; the vibrational structures of these systems are highly irregular, suggesting strong perturbations. The 376 nm system seen in TiO flames may be a perpendicularly polarized singlet system (44). The various upper states seen in matrix absorption (46, 47) appear to have vibrational frequencies that are higher than that of the ground state; if so, these states are presumably Rydberg states. The first I.P. of TiO is known from photoelectron spectroscopy to lie at 6.56 ± 0.03 eV (48); the ground state of TiO$^+$ is assigned as $^2\Delta$ from ab initio calculations. A second, rather stronger, photoelectron system at 8.09 ± 0.01 eV is assigned as going to the $A^2\Sigma^+$ state of TiO$^+$; four more states occur in the 10.5–12 eV range.

In view of the strength of the TiO absorption bands in cool stars (17), it is surprising that no TiO emission has been found at millimeter wavelengths in the interstellar medium (62). The fact that atomic Ti is strongly depleted in the interstellar medium, but is not present as TiO molecules, suggests that it is bound in interstellar dust grains

VO

Vanadium monoxide is another molecule that features prominently in stellar spectra. It is less abundant than TiO, and is used for the spectral classification of the coolest M-type stars (M7–M9), where the TiO bands are essentially saturated (17); a photograph of the $A^4\Pi-X^4\Sigma^-$ system of VO in late M-type stellar spectra is given by Spinrad & Wing (63).

Vanadium has two stable isotopes, ^{50}V and ^{51}V. Fortunately ^{50}V, with its gigantic nuclear spin $I=6$, is only present in 0.2% abundance: The hyperfine structure caused by the $I=7/2$ spin of ^{51}V is already sufficient to reduce most of the bands involving the ground state to blurred chaos at Doppler-limited resolution. As in the case of ScO, the hyperfine structure initially led to disputes about the nature of the ground state. For instance, Lagerqvist & Selin (64), using a high temperature arc source for their VO spectra, could not distinguish the hyperfine-broadened branches of the F_1 and F_4 ($N = J \pm 3/2$) electron spin components when they analyzed the $C^4\Sigma^- - X^4\Sigma^-$ system at 5740 Å; although their line assignments were correct, they suggested the transition was $^2\Delta - ^2\Delta$. Ten years later, the ESR spectrum (65) proved conclusively that the ground state is $^4\Sigma^-$, and shortly afterwards the hyperfine-broadened branches of the $B^4\Pi - X^4\Sigma^-$ and $C^4\Sigma^- - X^4\Sigma^-$ systems (whose linewidths range up to 0.35 cm^{-1}) were identified by Richards & Barrow (66).

Although their grating spectra could not resolve the hyperfine structure, Richards & Barrow (67) discovered a remarkable "internal" hyperfine perturbation in the $X^4\Sigma^-$ ground state. The electron spin contributions to the rotational energy in the F_2 ($N = J - 1/2$) and F_3 ($N = J + 1/2$) components are accidentally equal near $N = 15$ because of the specific values of the spin-spin and spin-rotation parameters (see Figure 2). At the crossing point, the matrix elements of the hyperfine Hamiltonian off-diagonal in J, but diagonal in N and F, become important and cause the F_2 and F_3 electron spin components to perturb each other. Seen at medium resolution, the branches involving the F_2 and F_3 electron spin components, which may be some distance apart, show small line doublings; at sub-Doppler resolution (68) there are clear avoided crossings in the hyperfine level patterns (see Figure 3). Richards & Barrow remarked that the internal

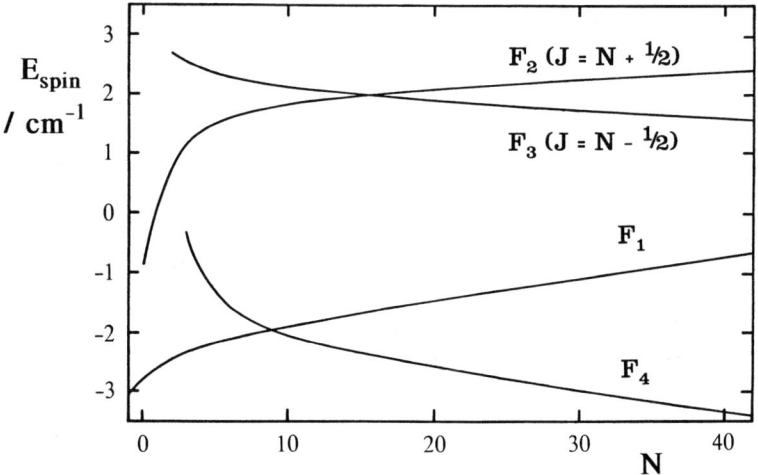

Figure 2 Energies of the four electron spin components of the $X^4\Sigma^-$ state of VO plotted as a function of the rotational quantum number N. The names F_1 to F_4 mean $J = N+3/2$ to $J = N-3/2$, where J is the total angular momentum including electron spin. (See Ref. 68.)

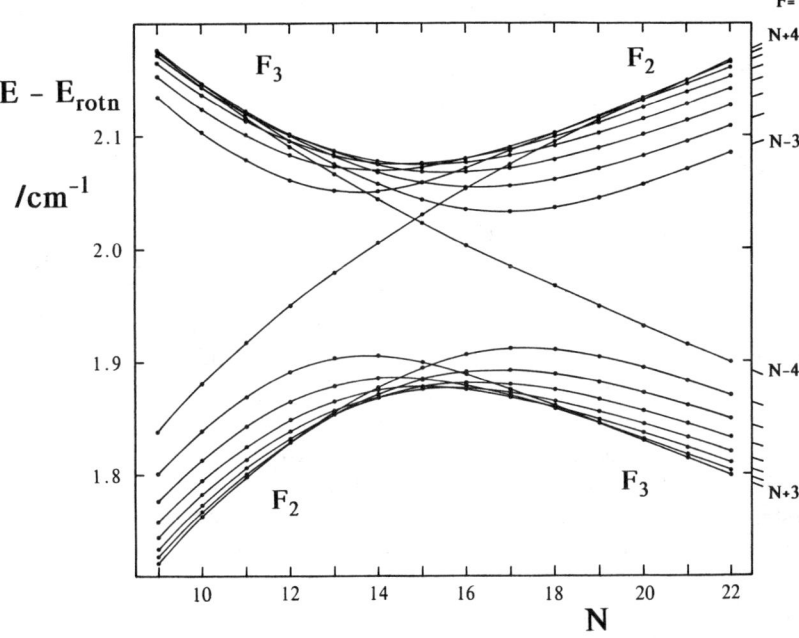

Figure 3 Calculated hyperfine energies of the F_2 and F_3 electron spin components of the $X^4\Sigma^-$ state of VO, in the region of the internal hyperfine perturbation, plotted against the rotational quantum number N. The quantum number F is the total angular momentum including electron and nuclear spin. (See Ref. 68.)

hyperfine perturbation in the $X^4\Sigma^-$ state of VO was possibly unique, but other examples are now known in the $C^4\Sigma^-$ state of VO (68, 69) and various sextet states of MnO (70) and MnS (71). In fact these perturbations will probably be the rule rather than the exception in high multiplicity Σ states in case (b) coupling.

Large hyperfine widths are found in VO whenever there is an unpaired electron in the 9σ ($4s\sigma$) m.o.; the most spectacular example is the ground state, $\ldots(9\sigma)^1(1\delta)^2 X^4\Sigma^-$, where the Fermi contact parameter of the 9σ electron (68) is 79% of that of the $4s$ state of the V atom (72). The ground state follows case $b_{\beta J}$ coupling, where the hyperfine energy expression (73) is

$$E_{\text{hfs}} \cong b(J-N)[F(F+1)-I(I+1)-J(J+1)]/2J.$$

The effect of this equation for a quartet state is that the hyperfine widths of the four electron spin components, F_1 to F_4 (that is, $N = J-3/2$ to $N = J+3/2$), are in the ratio $3:1:-1:-3$, each with its hyperfine components following a Landé-type pattern (19) as a result of the factor $F(F+1)$. The strongest hyperfine lines have the highest values of F and, in the absence of internal hyperfine perturbations, are furthest apart.

The $\ldots(9\sigma)^1(1\delta)^1(4\pi)^1 A^4\Pi$ state, which is the upper state of the 1.05 μm system illustrated in the stellar spectra of Ref. (63), has the same unpaired 9σ electron, and a virtually identical Fermi contact parameter to the ground state (74). The $A^4\Pi$ state is in case (a) coupling at low J, with a spin-orbit coupling constant $A\Lambda = 35$ cm^{-1}, though because of the magnified effects of spin-uncoupling in a high multiplicity state (9, 75), it uncouples very rapidly to case $b_{\beta J}$. As a result, the structure of the A–X system is very complex at low J, but at high J it reorganizes itself into four simple sets of main branches with narrow hyperfine linewidths, in which the upper and lower state level widths are the same (74). Even though the hyperfine structure is not resolved at Doppler-limited resolution, the evolution of the line width with J allows the Fermi contact parameter b to be determined quite accurately.

The very intense $B^4\Pi$–$X^4\Sigma^-$ system at 7900 Å (66, 76, 77) is truly awesome at high resolution because of the many branches with different line widths. The high frequency subband ($^4\Pi_{5/2}$) is fairly uncrowded, but the two central subbands are extremely difficult to analyze because the strong branches involve the hyperfine-broadened F_1 and F_4 components of the ground state. In addition, there are perturbations by a Σ state. Initially it was not possible to distinguish whether the perturbing state was $^2\Sigma^+$ or $^4\Sigma^-$, and only when all the low-lying quartet states had been accounted for (9) could the Σ state be assigned as $\ldots(9\sigma)^1(1\delta)^2\, ^2\Sigma^+$. The details of the perturbations have not been analyzed yet. It had been hoped that these intermultiplicity perturbations would give the relative locations of the doublet and quartet manifolds of VO, but so far no other transitions involving this $^2\Sigma^+$ state have been found.

Our knowledge of the states of VO has progressed considerably with the analysis of the near IR emission spectrum (9). Arguments based on the bond lengths, spin-orbit couplings and hyperfine parameters permit the electron configurations to be assigned to the states with some confidence, as described above for TiO. An energy level diagram of the states and transitions, as presently known, is given in Figure 4. Almost every single excited state has one or more rotational perturbations in each vibrational level: The Λ-doublings of the degenerate states are often anomalous, and bizarre hyperfine effects occur in which the line widths change rapidly with J.

There is still confusion about the higher excited states of VO. A band at 4185 Å (23,890 cm^{-1}) seen in an argon matrix (78) does not appear in flash photolysis experiments (79) in which eight new transitions in the

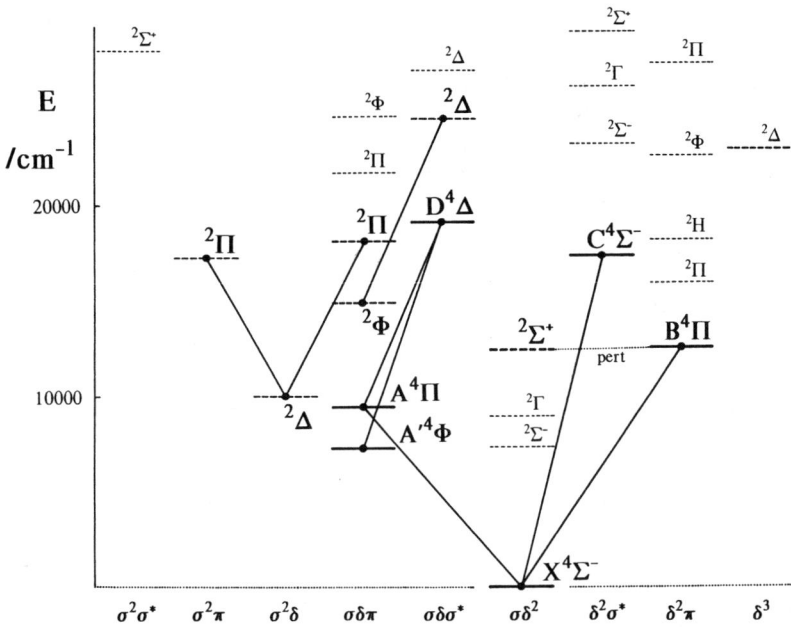

Figure 4 Electronic states and transitions of VO as presently known (see Ref. 9). Only the nine electron configurations with at least two electrons in the 9σ and 1δ m.o.s are shown. *Full lines* for the states indicate those whose absolute energies are known; *thick dashed lines* indicate states whose relative energies are known, and *thin dashed lines* show states whose energies are guessed. *Sloping lines* show rotationally analyzed transitions.

range 230–380 nm are ascribed to VO. Higher resolution is needed to answer the questions raised by these results.

The photoelectron spectrum of VO (80) gives the first two I.P.s as 7.25 ± 0.01 eV (VO$^+$, $X^3\Sigma^-$) and 8.42 ± 0.01 eV (VO$^+$, $A^3\Delta$). The energy order of these states, which differs from that in TiO, was deduced from ab initio calculations (80, 81). Two apparently singlet electronic transitions, at 840 and 920 nm, have been assigned to VO$^+$ (82); recent discovery of a third system, halfway between these two, suggests that the three may be related, possibly as components of a highly perturbed $^3\Delta$–$^3\Delta$ transition.

VO is of special interest to high resolution spectroscopists because of the new effects in its spectrum. Besides the internal hyperfine perturbations (67–69), there are good examples of $^4\Sigma$, $^4\Pi$, $^4\Delta$, and $^4\Phi$ states, all of which have been analyzed in great detail (9, 68, 74). The $X^4\Sigma^-$ and $C^4\Sigma^-$ states are beautiful examples of case $b_{\beta J}$ hyperfine coupling in quartet states, and there is an entertaining perturbation in the F_1 levels of $C^4\Sigma^-$, v = 0 in which the hyperfine structure behaves like fragments of perturbed rotational branch structure. The hyperfine splitting parameter of the perturbing state can be measured very accurately, but nothing further can be learned about it (68, 83). It has even been necessary to extend the hyperfine Hamiltonian to take account of the spin-orbit distortion of the Fermi contact interaction (84) along the lines of the corresponding distortion of the spin-rotation interaction (85, 86); both these effects appear in the $C^4\Sigma^-$ state.

CrO

Chromium oxide is the last of the 3d oxides to have been positively identified in stellar spectra. It is prominent in the "prototype" red giant star β Pegasi (88). What has been found so far in CrO is well understood, but only five quintet states are known. According to the ab initio calculations (89), the ground state should be ...$(9\sigma)^1(1\delta)^2(4\pi)^1$ $^5\Pi$, with $^7\Pi$ and $^7\Sigma^+$ charge transfer states lying within 1–1.5 eV. Nothing is known about the expected singlet and triplet manifolds.

From a spectroscopist's viewpoint CrO is extremely interesting, as it provides the first fully analyzed example of a $^5\Sigma$–$^5\Pi$ transition (90). A band of this type should have 75 branches; 51 of these were identified in the (0, 0) band of the $A^5\Sigma^+$–$X^5\Pi$ transition at 8200 cm^{-1}, thus allowing complete characterizations of the two states. The branch intensities in the 8200 cm^{-1} transition do not follow the published expressions for $^5\Sigma(b)$–$^5\Pi(a)$ transitions. The explanation (75, 91) is that spin-uncoupling effects are massively magnified in such high multiplicity states, so that the limiting coupling case formulae become inapplicable much more rapidly than they do in doublets; an interesting situation arises in which certain branches have intensities that pass through zero as the spin-uncoupling proceeds (75).

A system at 11,800 cm^{-1} reported by Gaspard and Rosen (92), which does not appear in discharge emission, has recently been found in chemi-luminescence by Devore & Gole (93); from its spin-orbit intervals it must be the transition ...$(9\sigma)^1(1\delta)^1(4\pi)^2$ $A'^5\Delta \rightarrow$ ...$(9\sigma)^1(1\delta)^2(4\pi)^1$ $X^5\Pi$. Interestingly, the analogous $E^3\Pi \rightarrow X^3\Delta$ system of TiO also appears in chemi-luminescence but not in discharge emission.

The strong $B^5\Pi$–$X^5\Pi$ system at 605 nm (94, 95) is remarkable for the density of small rotational perturbations in the upper $^5\Pi$ state; this density is far too great to be explained by the three lower quintet states and indicates that many other low-lying states must exist. A new chemi-luminescent system at 451 nm (93) must be a charge-transfer transition, since the upper state vibrational interval $\Delta G_{1/2}$ is much lower than that of the other known states. The photoelectron spectrum (96) gives the first I.P. as 7.85 ± 0.02 eV (CrO$^+$, $X^4\Sigma^-$), though the assignment of the vibrational structure in this system has been questioned following recent ab initio studies (97, 98). The $(1\delta)^2(4\pi)^1\,^4\Pi$ state appears to lie much lower in CrO$^+$ than in VO (98).

MnO

Manganese oxide has not been identified in stellar spectra though, since the strongest bandheads of its optical spectrum are not well-defined, it may be hard to confirm; the most characteristic bandhead is that of the A–X (1, 0) band at 18,652.76 cm^{-1}.

Although doublet, quartet, sextet, and octet states of MnO are expected, the spectrum is surprisingly sparse. A diffuse band near 256 nm was observed by flash photolysis (102), but this has not been confirmed by matrix work (103), and the only system in the visible region is the massively perturbed $A\,^6\Sigma^+$–$X\,^6\Sigma^+$ transition (70, 92, 102–106). The ESR spectrum (107, 108) confirms the ground state as $^6\Sigma$.

The (1, 0) band of the A–X system is not unduly perturbed, and shows the three pairs of branches, with hyperfine widths in the ratio 1:3:5, characteristic of $^6\Sigma$ states in case $b_{\beta J}$ coupling (see the discussion of VO, above). Both the $A\,^6\Sigma^+$ and $X\,^6\Sigma^+$ states have internal hyperfine perturbations between their $N = J+1/2$ and $N = J-1/2$ components (F_3 and F_4), but by accident the magnitudes of the internal hyperfine perturbations in the two electronic states are almost equal. This has the surprising result of canceling the intensities of the extra lines that would normally be expected, and produces some weird hyperfine patterns that would be unanalyzable without data from other bands involving the same levels. The $A\,^6\Sigma^+$ v = 0 level is disastrously perturbed by a charge transfer state with a much longer bond length ($B\,^6\Sigma^+$), and further unrelated rotational perturbations also occur. Because of the ^{55}Mn hyperfine structure ($I = 5/2$), the A–X (0, 0) band is certainly the most complicated band this author has ever seen.

The hyperfine structure may be a nuisance because it splits every rotational line into six components, but it helps considerably in assigning the electron spin structure. Figure 5 shows a small part of the A–X (0, 0) band at sub-Doppler resolution, where all 46 hyperfine components of the $R(26)$ line can be seen. The R_1 ($J = N+5/2$) and R_6 ($J = N-5/2$) electron spin components have the widest hyperfine structures, though their senses are opposite (as expected from the diagonal elements of the Fermi contact interaction). They can be distinguished because the F quantum numbers, which govern the line strengths, are higher for the R_1 branch; similar arguments apply to the R_2 and R_5 components, which have medium hyperfine widths. The R_3 and R_4 components can be recognized because they are distorted by ground state internal hyperfine perturbations, so that each contains five extra hyperfine lines. The reason that five extra lines

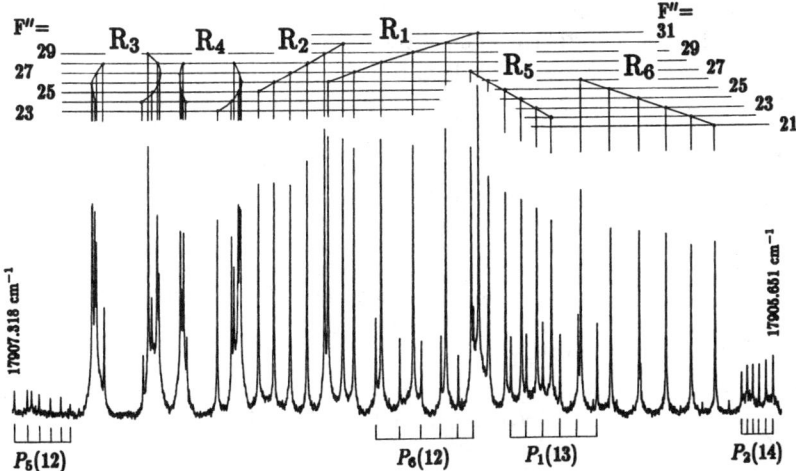

Figure 5 A small part of the $A^6\Sigma^+ - X^6\Sigma^+$ (0,0) band of MnO at sub-Doppler resolution, showing the six electron spin components of the $R(26)$ line (labeled R_1 to R_6) with their distinctive hyperfine structures. Also shown are some weak low-J P lines that have very different hyperfine widths because of perturbations (Y. Azuma, C. Thambir, and A. J. Merer, unpublished).

occur, rather than six, is that the range of F quantum numbers is different in the lower rotational levels. R_3 (26) has $J'' = 26\frac{1}{2}$, and therefore F runs from 24 to 29, whereas $R_4(26)$ has $J'' = 25\frac{1}{2}$, with F running from 23 to 28. Since the hyperfine matrix elements are diagonal in F, only the five pairs of hyperfine levels with $F = 24$ to 28 can perturb each other. (See also Figure 3 for VO.)

Charge transfer $^6\Sigma^+$ states, such as $B^6\Sigma^+$, can arise from the configurations $\ldots(3\pi)^3(9\sigma)^1(1\delta)^2(4\pi)^3$, $(8\sigma)^1(3\pi)^4(9\sigma)^2(1\delta)^2(4\pi)^2$, and $(8\sigma)^1(3\pi)^4(9\sigma)^1(1\delta)^2(4\pi)^2(10\sigma)^1$. The very great bond length change points to the third alternative (109). In this case the B–X transition should be dipole-allowed, and indeed various intensity anomalies in the spectrum may be consistent with this. More ab initio calculations are needed for MnO, since the density of perturbations in $A^6\Sigma^+$ shows that many lower-lying states must exist.

FeO

The spectrum of iron oxide is enormously complicated, and much argument has raged around it because of apparently conflicting experiments and inconsistent ab initio results. The key pieces to the puzzle have been the matrix isolation IR spectrum (111), which established that the lower state of the "orange system" (92, 112–115) is the ground state (from

its vibrational frequency), and the identification of this state as
...$(9\sigma)^1(1\delta)^3(4\pi)^2\ ^5\Delta_i$ (116). The ground state of FeO is now very well characterized, following detailed analysis of the near IR emission spectrum (117, 118) and extremely precise measurements of the millimeter wave spectrum (119, 120). Even so, only two of the four spin-orbit intervals have been measured directly, though the calculated values of the other two should be good to ~ 0.2 cm^{-1} (121, 121a).

Low-lying excited states, ...$(9\sigma)^1(1\delta)^2(4\pi)^2(10\sigma)^1\ ^7\Sigma^+$ and ...$(9\sigma)^2(1\delta)^2(4\pi)^2\ ^5\Sigma^+$, are predicted by ab initio work (3, 122, 123). At the CAS-MCSCF level (3), the $^7\Sigma^+$ state is calculated to lie below the ground $^5\Delta_i$ state, and in fact there is evidence from the autodetachment spectrum of FeO$^-$ (124, 125) for a low-lying Σ state at about the energy of $X^5\Delta_3$, v = 0, that is ~ 150 cm^{-1} above the $X^5\Delta_4$, $J = 4$ ground level of the molecule but *below the center* of the $X^5\Delta$ spin-orbit manifold. The $a^7\Sigma^+$ state perturbs the $^5\Delta_2$ component of the ground state at v = 2 (118) and is probably also the Σ state that causes the gigantic Λ-doublings in $C^5\Pi$ (see Figure 6). From the perturbations in $C^5\Pi$ we can estimate the vibrational frequency of the $^7\Sigma^+$ state as ~ 650 cm^{-1}, so that if the perturbations in $X^5\Delta_2$, v = 2, are caused by $a^7\Sigma^+$, v = 3, the optical and autodetachment spectra agree in placing $a^7\Sigma^+$, v = 0, at ~ 150 cm^{-1}. FeO has no matrix ESR spectrum, a finding that confirms the orbital degeneracy of the ground state (126, 127).

A peak in the photoelectron spectrum of FeO$^-$ at 3990 ± 100 cm^{-1} (128) appears to represent the $A^5\Sigma^+$ state; its observed vibrational frequency of 900 cm^{-1} is exactly as expected. This state has not been found in the optical spectrum of FeO, even though the dipole-allowed $C^5\Pi$–$A^5\Sigma^+$ transition should lie near 6000 cm^{-1}. Ab initio calculations (3) predict the C–A system to be 100 times weaker than the overlapping C–X system, so that it will be totally buried under it.

The $C^5\Pi$ state, and its companion $B^5\Phi$, are the upper states (117) of the very intense 1 μm system of FeO (131). They are two of the 48 electronic states that arise from the charge-transfer configuration $(8\sigma)^2(3\pi)^3(9\sigma)^1(1\delta)^3(4\pi)^2(10\sigma)^1$, of which the unseen $^7\Phi$ and $^7\Pi$ states lie lowest. The $B^5\Phi$ and $C^5\Pi$ states lie only 200 cm^{-1} apart, a distance that is much less than their spin-orbit splittings, and they are both savagely perturbed; the resulting energy level pattern (Figure 6) is truly remarkable. Even at the small scale of Figure 6, the larger perturbations show up clearly: the only spin-orbit substate that escapes unscathed is $^5\Phi_5$, v = 0. Among the perturbing states recognized are a $^3\Phi_i$ state, a Π state of uncertain multiplicity (triplet or septet), and $a^7\Sigma^+$ (see above); irregular Λ-doublings are widespread and occur even in $B^5\Phi_4$, which is an exceedingly high-order effect.

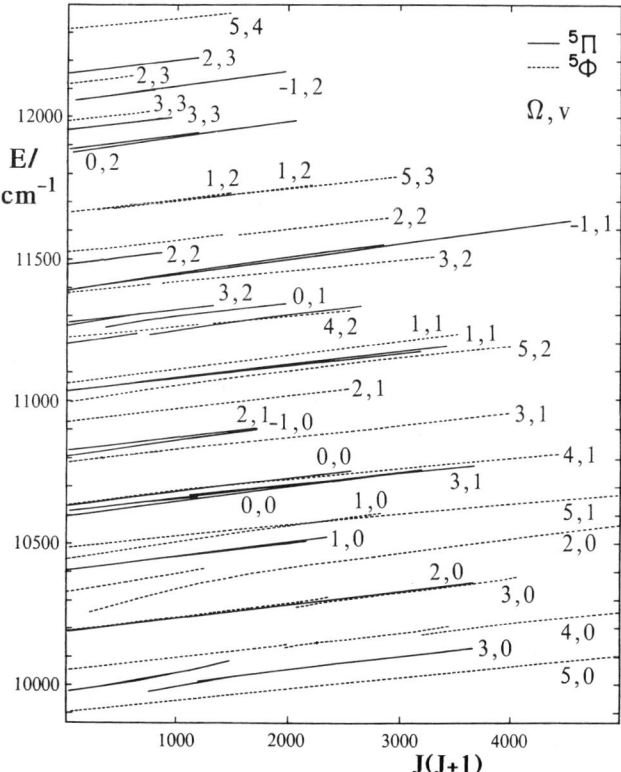

Figure 6 Energies of the assigned spin-orbit components of the $B^5\Phi$ and $C^5\Pi$ states of FeO, less $0.4\,J(J+1)$, plotted against $J(J+1)$, where J is the total angular momentum quantum number. The states are labeled by Ω and v; *dashed lines* are for the $B^5\Phi$ electronic state, *thick lines* for the $C^5\Pi$ state. Irregularities and discontinuities are rotational perturbations (A. W. Taylor, A. S.-C. Cheung, and A. J. Merer, unpublished).

With so many low-lying electronic states possible, it is not surprising that the higher electronic states are totally chaotic. The "orange system" has been studied in considerable detail by laser-induced fluorescence (121), and is found to be the shattered wreckage of a very severely perturbed $^5\Delta_i$–$^5\Delta_i$ transition, together with many "extra" levels. The upper state, $D^5\Delta_i$, is clearly $\ldots(1\delta)^3(4\pi)^2(10\sigma)^1$, which is in line with all the early 3d oxides in which the $9\sigma \to 10\sigma$ promotion produces a strong parallel band system in the 550–600 nm region. The isotope shifts of the "extra" levels show that they do not have v′ = 0; they must therefore be interlopers, belonging to lower excited states, which are obtaining their intensity from the D–X

transition. The upper state B values range randomly from 0.38 to 0.495 cm^{-1}, and no vibrational progressions can be recognized. There are myriads of rotational perturbations of all sizes, which turn the FeO orange system into a new type of chaotic diatomic spectrum in which the normal regular structure is destroyed by a dense manifold of perturbing electronic levels at lower energy.

The "extra" levels in the orange system give quite intense bands and have patterns of vibrationally resolved fluorescence that are completely at variance with their B' values and isotope shifts. To account for these anomalies and for the strength of some of the bands with $B' = 0.42$–0.44 cm^{-1}, it may be that there is a second electronic transition in this region, with a charge transfer $^5\Delta$ upper state, and that its transition moment interferes with that of $D^5\Delta_i$–$X^5\Delta_i$ for the "induced" levels; a possible electron configuration for this charge transfer state is $(8\sigma)^1(3\pi)^4(9\sigma)^1(1\delta)^3(4\pi)^2(10\sigma)^1$, analogous to the $B^6\Sigma^+$ state of MnO. Ab initio work on these states (3, 135) will be very difficult, but more is needed.

Fragments of higher excited states have been found at 19,449, 22,335, and 41,480 cm^{-1} (102, 115, 129).

CoO

The spectrum of cobalt oxide in the visible region is a long amorphous progression of red-degraded subbands extending from 900 to 520 nm (136, 137). Rotational analysis of three subbands in the middle of the system (138) shows that they are parallel-polarized and that more than one electronic transition is involved. The observed lower levels have $\Omega = 7/2$ and $5/2$ and, from the agreement of the vibrational frequency with that found in matrix IR studies (111), indicate that the ground state is $^4\Delta_i$. The absence of a matrix ESR spectrum for CoO (127) and the ab initio work by Dolg et al (122) are consistent with this assignment.

The upper states are perturbed, though not as severely as in FeO, and they have much larger hyperfine widths than the ground state. As for the electron configurations, the ground state is obviously $\ldots(9\sigma)^2(1\delta)^3(4\pi)^2$ $^4\Delta_i$. The upper states appear to be heavily-mixed combinations of the $8\sigma \to 10\sigma$ charge transfer promotion and the $9\sigma \to 10\sigma$ promotion, equivalent to the orange system of FeO. More work is needed on these systems, but the work will be very laborious because the perturbations, hyperfine structure and the extreme red-degradation of the subbands make the complete analysis of even one of them very time-consuming.

NiO

Nickel oxide has several confused band systems in the visible, extending from 430 to 920 nm (137, 139). They are strongly red-degraded, irregular, and perturbed. As with CoO, only one rotational study has been undertaken (140), which shows that the ground state is $^3\Sigma^-$. Again the agreement for the vibrational frequency between the matrix IR (111) and optical work was important in identifying the ground state. On the other hand, the absence of an ESR spectrum (127) for NiO does not imply an orbitally degenerate ground state, but rather that the $X^3\Sigma^-$ state has such a large second-order spin-orbit splitting between its $\Omega = 0^+$ and 1 components that only the nonmagnetic 0^+ component is populated in low-temperature matrices.

Four perturbed upper levels (19,270, 19,164, 19,044, 18,914 cm^{-1}) have been analyzed rotationally (140). The B' values (0.42 cm^{-1}) indicate a charge transfer upper state. Ab initio work (92, 122) agrees that the ground state should be $^3\Sigma^-$, where the dominant configurations are ...$(3\pi)^4(8\sigma)^2(9\sigma)^2(4\pi)^2$ and ...$(3\pi)^3(8\sigma)^2(9\sigma)^2(4\pi)^3$ (Ni$^+$O$^-$). For the upper states the m.o. diagram of Figure 1c is the most appropriate; there will be many possible excitations from the three nickel 3d orbitals, 1δ, 3π and 8σ, into the 4π and 10σ orbitals. Again, more work is needed.

CuO

Copper oxide is most tantalizing. Large numbers of states have been characterized, as a result of heroic analyses at Stockholm and Lille. Most of them are "only" doublets, but attempts to make sense of them have met quite limited success.

It is now well-established that CuO has a $^2\Pi_i$ ground state, from gas-phase analyses of transitions observed in a low-temperature matrix (141). As described above, the spin-orbit coupling is too large for an unpaired electron on oxygen, so the dominant configuration ...$(8\sigma)^2(9\sigma)^2(4\pi)^3$ must have some copper 3d and 4p character in the unfilled 4π m.o. ($\sim$ oxygen 2$p\pi$); the large negative Fermi contact hyperfine parameter (142) requires a sizeable contribution from ...$(8\sigma)^1(9\sigma)^2(4\pi)^3(10\sigma)^1$, where 10$\sigma$ is mostly the metal 4s orbital. The longer bond length in CuO $X^2\Pi$ compared to NiO $X^3\Sigma^-$ results because the copper 3d orbitals play a much smaller role in covalent bonding. Ab initio calculations most definitely stress the importance of Cu$^+$O$^-$ ionic terms in the wave function (143).

The low-lying excited state, $Y^2\Sigma^+$, was predicted by ab initio calculations (144) and then found in emission (145) with collisional-transfer laser pumping of the $A^2\Sigma^-$ state; the unpaired electron of the $Y^2\Sigma^+$ state is in the 9σ m.o., that is oxygen 2$p\sigma$, rather than 2$p\pi$ as in the ground state. There are several states in the 13,000–16,000 cm^{-1} region that can be observed by collisional-transfer pumping (146), some of which have been analyzed rotationally. Two very useful sets of ab initio calculations (144, 147) have clarified their identities. The "active" configurations in this energy region are ...$(3\pi)^4(8\sigma)^1(9\sigma)^2(4\pi)^3(10\sigma)^1$, best described as Cu 3$d\sigma^{-1}$ or "σ-hole," and $(3\pi)^3(8\sigma)^2(9\sigma)^2(4\pi)^3(10\sigma)^1$, or Cu $3d\pi^{-1}$ or "π-hole." The σ-hole states must be of Π symmetry, whereas the π-hole states can only be Σ or Δ. The calculations agree that the lowest of these states must be the unobserved σ-hole state $a^4\Pi$, but after this it is not always clear how to relate the observed states to the calculated states of the correct symmetry: for example, there is little doubt that $\beta^2\Delta_i$ is the lower of the two predicted π-hole $^2\Delta$ states, but it is not possible to say whether $\gamma^2\Pi_i$ is the upper or lower of the σ-hole $^2\Pi$ states.

The Kronig symmetries of the $^2\Sigma$ states, A, A', and G, make an interesting story. Originally it was thought that $A^2\Sigma$, the upper state of the very intense A–X system near 605 nm, was a Σ^+ state. However, with the discovery of $Y^2\Sigma^+$, and the realization that $Y^2\Sigma^+$ would be in "pure precession" with $X^2\Pi$, it was possible to assign the absolute parities of the Λ-doubling sublevels in $X^2\Pi$ (145), and then use the rotational selection rules to establish that the A state is $^2\Sigma^-$. The A' state is heavily perturbed by $\gamma^2\Pi_{1/2}$, so the rotational selection rules alone did not distinguish its Kronig symmetry. Instead, its large negative Fermi contact parameter, which cannot come from a single unpaired electron in a σ m.o. and is only consistent with a wave function of the type (43)

$$\Psi(^2\Sigma^-) = 6^{-1/2}[2|\pi^+\pi^-\bar{\sigma}| - |\overline{\pi^+\pi^-}\sigma| - |\pi^+\overline{\pi^-}\sigma|],$$

was used (148) to assign the A' state as $^2\Sigma^-$ also.

The higher excited states are very disorganized, and since the Φ and Π states from the "δ-hole" configuration, $(1\delta)^3(3\pi)^4(8\sigma)^2(9\sigma)^2(4\pi)^3(10\sigma)^1$, are calculated to begin near 2.2 eV (144, 147), it becomes increasingly hard to assign them. The analyses are extremely difficult in this region, and it is obvious that there is heavy mixing of states with the same Ω value. It has been pointed out (144) that more $^2\Pi_{3/2}$ states have been assigned than should exist, so that some of them must be quartet states with $\Omega = 3/2$; states such as $^4\Sigma_{3/2}$, suitably mixed by spin-orbit interaction, can be mistaken for $^2\Pi_{3/2}$ states in the absence of other evidence. The spin-orbit coupling constants of the various states change markedly with vibration, and there are very large Λ-doublings in the $^2\Pi_{1/2}$ states, thus giving clear evidence of strong interactions. The M–X transition is interesting as being the first CuO band system that was analyzed successfully (149); the $M\,^2\Pi_{3/2}$ state has five perturbations in the J range to 67.5 (150).

CONCLUSIONS

Re-reading this chapter, I realize how little we have really achieved so far. The ground states of the oxides have been characterized, and some of the low-lying states also, particularly for the early members of the series. The later members are still in considerable confusion. Part of the problem is experimental, part theoretical. It is not easy to obtain strong spectra of CoO and NiO, but this is just where the three atomic orbitals responsible for the electronic states, metal $3d$ and $4s$ and oxygen $2p$, are almost degenerate, so that the resulting states are most difficult to calculate. The principal analysis problem is that the spin-orbit coupling is large enough to produce very serious perturbations between the different spin manifolds, but small enough for the quantum number S to remain well-defined: except in a few cases, the only well-understood states have the same spin multiplicity as the ground state, whereas there may be up to four different spin manifolds present.

The keys to the interpretation of the spectra have been the spin-orbit and hyperfine parameters. With improvements in experimental techniques, the latter are becoming routinely available: There is a great need for ab initio wave functions accurate enough to reproduce the various expectation values of the electron coordinates, especially if presented in a way that allows the experimental parameters to be calculated simply. There is also a need for calculations of transition moments. Similarly, the theoreticians will need to test their models against easily calculated properties such as dipole moments and the energies of low-lying excited states.

What remains to be done is to attack the excited states, especially those with different spin multiplicity to the ground state. Real progress will come when the detailed energy orders of the excited states, and the matrix elements between them, can be calculated accurately enough, and when experimental methods are developed for routine automated assignment of wildly complicated and perturbed spectra. Almost certainly these will depend on improved computing power. The problems are worth doing, because here are simple systems that need to be understood properly, while at the same time they are among the fundamental building blocks of high temperature chemistry and cool star astrophysics.

Literature Cited

1. Landolt-Börnstein. 1965. New Series Group 6, Vol. 1, *Astronomy and Astrophysics*. Berlin: Springer-Verlag
2. White, N. M., Wing, R. F. 1978. *Astrophys. J.* 222: 209–19
3. Krauss, M., Stevens, W. J. 1985. *J. Chem. Phys.* 82: 5584–96
4. Langhoff, S. R., Bauschlicher, C. W. 1988. *Annu. Rev. Phys. Chem.* 39: 181–212
5. Rice, S. F., Field, R. W. 1986. *J. Mol. Spectrosc.* 119: 331–36
6. Steimle, T. C., Nachman, D. F., Shirley, J. E., Merer, A. J. 1989. *J. Chem. Phys.* 90: 5360–63
7. Steimle, T. C., Nachman, D. F., Fletcher, D. A. 1987. *J. Chem. Phys.* 87: 5670–73
8. Condon, E. U., Shortley, G. H. 1935. *The Theory of Atomic Spectra*. London: Cambridge Univ. Press
9. Merer, A. J., Huang, G., Cheung, A. S.-C., Taylor, A. W. 1987. *J. Mol. Spectrosc.* 125: 465–503
10. Childs, W. J. 1987. *Z. Phys. D* 7: 107–12

11. Childs, W. J., Steimle, T. C. 1988. *J. Chem. Phys.* 88: 6168–74
12. Bauschlicher, C. W., Langhoff, S. R. 1986. *Chem. Phys. Lett.* 126: 163–68
13. Rice, S. F., Martin, H., Field, R. W. 1985. *J. Chem. Phys.* 82: 5023–34
14. Schall, H., Dulick, M., Field, R. W. 1987. *J. Chem. Phys.* 87: 2898–2912
15. Törring, T., Ernst, W. E., Kindt, S. 1984. *J. Chem. Phys.* 81: 4614–19
16. Huber, K.-P., Herzberg, G. 1979. *Constants of Diatomic Molecules*. New York: Van Nostrand-Reinhold
17. Merrill, P. W., Deutsch, A. J., Keenan, P. C. 1962. *Astrophys. J.* 136: 21–34
18. Frosch, R. A., Foley, H. M. 1952. *Phys. Rev.* 88: 1337–49
19. Dunn, T. M. 1972. In *Molecular Spectroscopy, Modern Research*, ed. K. N. Rao, C. W. Mathews, Chapt. 4. New York: Academic
20. Åkerlind, L. 1962. *Ark. Fys.* 22: 41–64
21. Carlson, K. D., Ludeña, E., Moser, C. 1965. *J. Chem. Phys.* 43: 2408–15
22. Berg, R. A., Wharton, L., Klemperer, W., Büchler, A., Stauffer, J. L. 1965. *J. Chem. Phys.* 43: 2416–21
23. Kasai, P. H., Weltner, W. 1965. *J. Chem. Phys.* 43: 2553
24. Adams, A., Klemperer, W., Dunn, T. M. 1968. *Can. J. Phys.* 46: 2213–20
25. Stringat, R., Athénour, C., Féménias, J.-L. 1972. *Can. J. Phys.* 50: 395–403
26. Herman, F., Skillman, S. 1963. *Atomic Structure Calculations*. Englewood Cliffs, NJ: Prentice-Hall
27. Jeung, G. H., Koutecký, J. 1988. *J. Chem. Phys.* 88: 3747–60
28. Bauschlicher, C. W., Langhoff, S. R. 1986. *J. Chem. Phys.* 85: 5936–42
29. Chalek, C. L., Gole, J. L. 1976. *J. Chem. Phys.* 65: 2845–59
29a. Rice, S. F., Childs, W. J., Field, R. W. 1989. *J. Mol. Spectrosc.* 133: 22–35
30. Schenck, P. K., Mallard, W. G., Travis, J. C., Smyth, K. C. 1978. *J. Chem. Phys.* 69: 5147–50
31. Liu, K., Parson, J. M. 1977. *J. Chem. Phys.* 67: 1814–28
32. Pedley, J. B., Marshall, E. M. 1983. *J. Phys. Chem. Ref. Data* 12: 967–1031
33. Morgan, W. W., Keenan, P. C. 1973. *Annu. Rev. Astron. Astrophys.* 11: 29–50
34. Johnson, H. L., Morgan, W. W. 1953. *Astrophys. J.* 117: 313–52
35. Phillips, J. G., Davis, S. P. 1987. *Publ. Astron. Soc. Pacific* 99: 839–41
36. Bauschlicher, C. W., Bagus, P. S., Nelin, C. J. 1983. *Chem. Phys. Lett.* 101: 229–34
37. Sennesal, J. M., Schamps, J. 1987. *Chem. Phys.* 114: 37–42
38. Kobylyanskii, A. I., Kulikov, A. N., Gurvich, L. V. 1983. *Opt. Spectrosc.* 54: 254–55 (Engl. ed.)
39. Hammer, P. D., Davis, S. P. 1980. *Astrophys. J.* 237: L51–53
40. Linton, C., Broida, H. P. 1977. *J. Mol. Spectrosc.* 64: 389–400
41. Phillips, J. G. 1971. *Astrophys. J.* 169: 185–89
42. Hocking, W. H., Gerry, M. C. L., Merer, A. J. 1979. *Can. J. Phys.* 57: 54–68
43. Raftery, J., Scott, P. R., Richards, W. G. 1972. *J. Phys. B* 5: 1293–1301
44. Pathak, C. M., Palmer, H. B. 1970. *J. Mol. Spectrosc.* 33: 137–46
45. Dubois, L. H., Gole, J. L. 1977. *J. Chem. Phys.* 66: 779–90
46. McIntyre, K. S., Thompson, K. R., Weltner, W. 1971. *J. Phys. Chem.* 75: 3243–49
47. Powell, D., Brittain, R., Vala, M. 1981. *Chem. Phys.* 58: 355–70
48. Dyke, J. M., Gravenor, B. W. J., Josland, G. D., Lewis, R. A., Morris, A. 1984. *Mol. Phys.* 53: 465–77
49. Linton, C., Singhal, S. R. 1974. *J. Mol. Spectrosc.* 51: 194–97
50. Linton, C. 1972. *J. Mol. Spectrosc.* 50: 312–16
51. Brandes, G. R., Galehouse, D. C. 1985. *J. Mol. Spectrosc.* 109: 345–51
52. Linton, C. 1974. *J. Mol. Spectrosc.* 50: 235–45
53. Feinberg, J., Davis, S. P. 1977. *J. Mol. Spectrosc.* 65: 264–72
54. Phillips, J. G. 1973. *Astrophys. J. Suppl. Ser.* 26: 313–31
55. Steele, R. E., Linton, C. 1978. *J. Mol. Spectrosc.* 69: 66–70
56. Feinberg, J., Davis, S. P. 1978. *J. Mol. Spectrosc.* 69: 445–49
57. Pettersson, A. V., Lindgren, B. 1962. *Ark. Fys.* 22: 491–95
58. Linton, C., Broida, H. P. 1977. *J. Mol. Spectrosc.* 64: 382–88
59. Galehouse, D. C., Brault, J. W., Davis, S. P. 1980. *Astrophys. J. Suppl. Ser.* 42: 241–59
60. Gallaher, T. N., DeVore, T. C. 1979. *High Temp. Sci.* 11: 123–30
61. Hildenbrand, D. L. 1976. *Chem. Phys. Lett.* 44: 281–84
62. Churchwell, E., Hocking, W. H., Merer, A. J., Gerry, M. C. L. 1980. *Astron. J.* 85: 1382–85
63. Spinrad, H., Wing, R. F. 1969. *Annu. Rev. Astron. Astrophys.* 7: 249–302
64. Lagerqvist, A., Selin, L.-E. 1957. *Ark. Fys.* 12: 553–68
65. Kasai, P. H. 1968. *J. Chem. Phys.* 49: 4979–84
66. Richards, D., Barrow, R. F. 1968. *Nature* 217: 842

67. Richards, D., Barrow, R. F. 1968. *Nature* 219: 1244–45
68. Cheung, A. S.-C., Hansen, R. C., Merer, A. J. 1982. *J. Mol. Spectrosc.* 91: 165–208
69. Hocking, W. H., Merer, A. J., Milton, D. J. 1981. *Can. J. Phys.* 59: 266–70
70. Gordon, R. M., Merer, A. J. 1980. *Can. J. Phys.* 58: 642–56
71. Douay, M., Dufour, C., Pinchemel, B. 1988. *J. Mol. Spectrosc.* 129: 471–82
72. Childs, W. J., Poulsen, O., Goodman, L. S., Crosswhite, H. M. 1979. *Phys. Rev. A* 19: 168–76
73. Féménias, J.-L., Cheval, G., Merer, A. J., Sassenberg, U. 1987. *J. Mol. Spectrosc.* 124: 348–68
74. Cheung, A. S.-C., Taylor, A. W., Merer, A. J. 1982. *J. Mol. Spectrosc.* 92: 391–409
75. Sassenberg, U., Cheung, A. S.-C., Merer, A. J. 1984. *Can. J. Phys.* 62: 1610–15
76. Keenan, P. C., Schroeder, L. W. 1952. *Astrophys. J.* 115: 82–88
77. Veseth, L. 1975. *Phys. Scripta* 12: 125–28
78. Weltner, W. 1969. *Adv. High Temp. Chem.* 2: 85–105
79. Briggs, A. G., Kemp, R. J. 1972. *J. Chem. Soc. Dalton Trans.*, pp. 1223–26
80. Dyke, J. M., Gravenor, B. W. J., Hastings, M. P., Morris, A. 1985. *J. Phys. Chem.* 89: 4613–17
81. Carter, E. A., Goddard, W. A. 1988. *J. Phys. Chem.* 92: 2109–15
82. Merer, A. J., Cheung, A. S.-C., Taylor, A. W. 1984. *J. Mol. Spectrosc.* 108: 343–51
83. Cheung, A. S.-C., Hansen, R. C., Lyyra, A. M., Merer, A. J. 1981. *J. Mol. Spectrosc.* 86: 526–33
84. Cheung, A. S.-C., Merer, A. J. 1982. *Mol. Phys.* 46: 111–28
85. Hougen, J. T. 1962. *Can. J. Phys.* 40: 598–606
86. Brown, J. M., Milton, D. J. 1976. *Mol. Phys.* 31: 409–22
87. Grimeland, B., Veseth, L. 1979. *J. Mol. Spectrosc.* 77: 154–55
88. Davis, D. N. 1947. *Astrophys. J.* 106: 28–75
89. Bauschlicher, C. W., Nelin, C. J., Bagus, P. S. 1985. *J. Chem. Phys.* 82: 3265–76
90. Cheung, A. S.-C., Żyrnicki, W., Merer, A. J. 1984. *J. Mol. Spectrosc.* 104: 315–36
91. Kovács, I., Antal, J. 1984. *Can. J. Phys.* 62: 1603–9
92. Gatterer, A., Junkes, J., Salpeter, E.-W., Rosen, B. 1957. *Molecular Spectra of Metallic Oxides*. Vatican City: Specola Vaticana
93. Devore, T. C., Gole, J. L. 1989. *Chem. Phys.* 133: 95–102
94. Hocking, W. H., Merer, A. J., Milton, D. J., Jones, W. E., Krishnamurthy, G. 1980. *Can. J. Phys.* 58: 516–33
95. Ninomiya, M. 1955. *J. Phys. Soc. Jpn.* 10: 829–36
96. Dyke, J. M., Gravenor, B. W. J., Lewis, R. A., Morris, A. 1983. *J. Chem. Soc. Faraday Trans 2* 79: 1083–88
97. Harrison, J. F. 1986. *J. Phys. Chem.* 90: 3313–19
98. Jasien, P. G., Stevens, W. J. 1988. *Chem. Phys. Lett.* 147: 72–78
99. Balducci, G., Gigli, G., Guido, M. 1981. *J. Chem. Soc. Faraday Trans 2* 77: 1107–14
100. Serebrennikov, L. V., Mal'tsev, A. A. 1980. *Vestnik Mosk. Univ. Ser. 2 Khim.* 21: 148–51
101. Smoes, S., Drowart, J. 1984. *High Temp. Sci.* 17: 31–52
102. Callear, A. B., Norrish, R. G. W. 1960. *Proc. R. Soc. London Ser. A* 259: 304–24
103. Thompson, K. R., Easley, W. C., Knight, L. B. 1973. *J. Phys. Chem.* 77: 49–52
104. Garrett, M. B., Lee, P. S., Kay, J. G. 1966. *J. Chem. Phys.* 45: 2698–99
105. Pinchemel, B., Schamps, J. 1975. *Can. J. Phys.* 53: 431–34
106. Woodward, R., Le, P. N., Temmen, M., Gole, J. L. 1987. *J. Phys. Chem.* 91: 2637–45
107. Ferrante, R. F., Wilkerson, J. L., Graham, W. R. M., Weltner, W. 1977. *J. Chem. Phys.* 67: 5904–13
108. Baumann, C. A., Van Zee, R. J., Weltner, W. 1982. *J. Phys. Chem.* 86: 5084–93
109. Pinchemel, B., Schamps, J. 1976. *Chem. Phys.* 18: 481–89
110. Armentrout, P. B., Halle, L. F., Beauchamp, J. L. 1982. *J. Phys. Chem.* 76: 2449–57
111. Green, D. W., Reedy, G. T., Kay, J. G. 1979. *J. Mol. Spectrosc.* 78: 257–66
112. Pearse, R. W. B., Gaydon, A. G. 1965. *The Identification of Molecular Spectra*. London: Chapman & Hall. 3rd ed.
113. Barrow, R. F., Senior, M. 1969. *Nature* 223: 1359
114. Harris, S. M., Barrow, R. F. 1980. *J. Mol. Spectrosc.* 84: 334–41
115. West, J. B., Broida, H. P. 1975. *J. Chem. Phys.* 62: 2566–74
116. Cheung, A. S.-C., Gordon, R. M., Merer, A. J. 1981. *J. Mol. Spectrosc.* 87: 289–96

117. Cheung, A. S.-C., Lee, N., Lyyra, A. M., Merer, A. J., Taylor, A. W. 1982. *J. Mol. Spectrosc.* 95: 213–25
118. Taylor, A. W., Cheung, A. S.-C., Merer, A. J. 1985. *J. Mol. Spectrosc.* 113: 487–94
119. Endo, Y., Saito, S., Hirota, E. 1984. *Astrophys. J.* 278: L131–32
120. Kröckertskothen, T., Knöckel, H., Tiemann, E. 1987. *Mol. Phys.* 62: 1031–40
121. Cheung, A. S.-C., Lyyra, A. M., Merer, A. J., Taylor, A. W. 1983. *J. Mol. Spectrosc.* 102: 224–57
121a. Azuma, Y., Merer, A. J. 1989. *J. Mol. Spectrosc.* 135: 194–96
122. Dolg, M., Wedig, U., Stoll, H., Preuss, H. 1987. *J. Chem. Phys.* 86: 2123–31
123. Bagus, P. S., Preston, H. J. T. 1973. *J. Chem. Phys.* 59: 2986–3002
124. Andersen, T., Lykke, K. R., Neumark, D. M., Lineberger, W. C. 1986. *Electronic and Atomic Collisions*, ed. D. C. Lorents, W. E. Meyerhof, J. R. Peterson, pp. 791–98. Amsterdam: Elsevier
125. Andersen, T., Lykke, K. R., Neumark, D. M., Lineberger, W. C. 1987. *J. Chem. Phys.* 86: 1858–67
126. Weltner, W. 1978. *Ber. Bunsenges. Phys. Chem.* 82: 80–89
127. Van Zee, R. J., Brown, C. R., Zeringue, K. J., Weltner, W. 1980. *Acc. Chem. Res.* 13: 237–42
128. Engleking, P. C., Lineberger, W. C. 1977. *J. Chem. Phys.* 66: 5054–58
129. Malet, L., Rosen, B. 1945. *Bull. Soc. R. Sci. Liège* 14: 377–81
130. Delsemme, A., Rosen, B. 1945. *Bull Soc. R. Sci. Liège* 14: 70–80
131. Bass, A. M., Benedict, W. S. 1952. *Astrophys. J.* 116: 652–53
132. Kröckertskothen, T., Knöckel, H., Tiemann, E. 1986. *Chem. Phys.* 103: 335–43
133. Merer, A. J., Walmsley, C. M., Churchwell, E. 1982. *Astrophys. J.* 256: 151–55
134. Russek, A. 1987. *Phys. Rev. A* 36: 487–96
135. Knowles, P. J., Werner, H. J. 1985. *Chem. Phys. Lett.* 115: 259–67
136. Malet, L., Rosen,B. 1945. *Bull Soc. R. Sci. Liège* 14: 382–89
137. McQuaid, M. J., Morris, K., Gole, J. L. 1988. *J. Am. Chem. Soc.* 110: 5280–85
138. Adam, A. G., Azuma, Y., Barry, J. A., Huang, G., Lyne, M. P. J., Merer, A. J., Schröder, J. O. 1987. *J. Chem. Phys.* 86: 5231–38
139. Rosen, B. 1945. *Nature* 156: 570
140. Srdanov, V. I., Harris, D. O. 1988. *J. Chem. Phys.* 89: 2748–53
141. Shirk, J. S., Bass, A. M. 1970. *J. Chem. Phys.* 52: 1894–1901
142. Gerry, M. C. L., Merer, A. J., Sassenberg, U., Steimle, T. C. 1987. *J. Chem. Phys.* 86: 4754–61
143. Langhoff, S. R., Bauschlicher, C. W. 1986. *Chem. Phys. Lett.* 124: 241–47
144. Schamps, J., Pinchemel, B., Lefebvre, Y., Raseev, G. 1983. *J. Mol. Spectrosc.* 101: 344–57
145. Lefebvre, Y., Pinchemel, B., Delaval, J. M., Schamps, J. 1982. *Phys. Scr.* 25: 329–32
146. Appelblad, O., Lagerqvist, A., Renhorn, I., Field, R. W. 1981. *Phys. Scr.* 22: 603–8
147. Madhavan, P. V., Newton, M. D. 1985. *J. Chem. Phys.* 83: 2337–47
148. Appelblad, O., Renhorn, I., Dulick, M., Purnell, M. R., Brown, J. M. 1983. *Phys. Scr.* 28: 539–50
149. Lagerqvist, A., Uhler, U. 1967. *Z. Naturforsch.* 22B: 551–52
150. Appelblad, O., Lagerqvist, A., Lyyra, M. 1979. *Phys. Scr.* 20: 93–97
151. Appelblad, O., Lagerqvist, A. 1976. *Phys. Scr.* 13: 275–88
152. Appelblad, O., Lagerqvist, A., Lyyra, M. 1978. *Phys. Scr.* 18: 137–40
153. Appelblad, O., Lagerqvist, A. 1974. *Phys. Sci.* 10: 307–24
154. Appelblad, O., Lagerqvist, A. 1973. *J. Mol. Spectrosc.* 48: 607–8
155. Appelblad, O., Lagerqvist, A. 1975. *Can. J. Phys.* 53: 2221–31
156. Appelblad, O., Lagerqvist, A., Lefebvre, Y., Pinchemel, B., Schamps, J. 1978. *Phys. Scr.* 18: 125–36
157. Delaval, J. M., David, F., Lefebvre, Y., Bernage, P., Niay, P., Schamps, J. 1983. *J. Mol. Spectrosc.* 101: 358–68
158. Lefebvre, Y., Pinchemel, B., Bacis, R. 1976. *Can. J. Phys.* 54: 735–39
159. Antić-Jovanović, A., Pešić, D. S., Gaydon, A. G. 1968. *Proc. R. Soc. London Ser. A* 307: 399–406
160. Steimle, T. C., Azuma, Y. 1986. *J. Mol. Spectrosc.* 118: 237–47
161. Pinchemel, B., Lefebvre, Y., Schamps, J. 1977. *J. Phys. B* 10: 3215–17
162. Lefebvre, Y., Pinchemel, B., Schamps, J. 1977. *J. Mol. Spectrosc.* 68: 81–88

TRANSPORT OF ELECTRONS IN NONPOLAR FLUIDS

Richard A. Holroyd

Department of Chemistry, Brookhaven National Laboratory, Upton, New York 11973

Werner F. Schmidt

Hahn-Meitner-Institut Berlin, Bereich Strahlenchemie, D-1000 Berlin 39, West Germany

INTRODUCTION

Electron transport in nonpolar fluids has been investigated for several decades. At first the emphasis was on liquified rare gases, and later on excess electrons were also studied in molecular liquids such as alkanes. Now mobility data are available for approximately 75 nonpolar liquids (1). Progress in this field was hampered initially by impurities like O_2, CO_2, and various electrophilic compounds in the liquids. The availability of adequate methods of purification has stimulated experimental studies. Some new compounds have been studied recently, and new experimental techniques are being applied. Innovative theoretical approaches are now being tried. These efforts will eventually provide a better understanding of the basic problems of electron scattering and localization in liquids.

Excess electrons can be introduced in liquids either by ionization or by injection. Ionization may utilize single or multiphoton absorption or high energy radiation. Injection can be from a cathode immersed in the liquid (photoelectric effect) or by field emission. Most studies of electron transport require a short excitation pulse. The electrons produced thereby can be monitored by a number of techniques (2). One is that of DC conductivity. Even though ions may be concomitantly formed, electrons move much faster than molecular ions, therefore the current at short times is due to the motion of the electrons. Another technique is that of microwave conductivity, i.e. the absorption of electromagnetic radiation by free charges. A third technique is optical absorption, which is useful in certain alkane liquids with low electron mobility. Trapped electrons in such alkanes absorb in a broad band extending from the visible with a maximum in the infrared. Recent laser studies (3) show that this absorption is present in liquid hexane at 23°C within a few picoseconds after photoexcitation.

The main transport property is the drift mobility of the electron, μ_D, which is given by the ratio of v_D, the observed drift velocity to E, the applied electric field. In most cases the mobility has been studied as a function of temperature and in some cases as a function of density, pressure, and electric field. Values of μ_D for hydrocarbons at room temperature range from 0.016 cm^2 V^{-1} s^{-1} for n-tetradecane (4) to 70 cm^2 V^{-1} s^{-1} for neopentane (1). Mobilities of several hundred to several thousand have been observed for liquid methane, argon, krypton, and xenon. The lowest value of the electron mobility is still about 30 times larger than the mobility of ions in these liquids.

Several other transport properties of the electron are being measured. One is the Hall mobility (μ_H), which provides information about transport in the conduction band. The diffusion coefficient (D_e) combined with μ_D provides a measure of the temperature of electrons. The rate constant for homogeneous recombination of electrons with ions, and that of attachment to certain electrophilic molecules, give additional insight into the electron transport process. Another property intimately connected to the electron mobility is the energy of the bottom of the conduction band (V_0). At room temperature V_0 is generally between $+0.2$ and -0.7 eV for most liquids. Experimental values of V_0 for many nonpolar liquids are summarized elsewhere (2, 5–7).

Models for electron transport in liquids like argon, krypton, xenon, and methane, in which the mobility is high, consider that the electron is quasifree. Theory must deal with the enigma of a long mean free path in these liquids while the electron is surrounded by molecules, which in the gas phase are quite effective scattering centers. For liquids in which the electron mobility is low, traps are presumed to play a role. Since trapped electrons are relatively immobile, transport is envisaged to occur in one of several ways. One is a hopping model in which the electron jumps from one localized state to another nearby. Another is the two-state model, which presupposes the existence of a high mobility quasifree state to which electrons must be thermally activated from the trapped state for transport. A third possibility is diffusive motion or movement of the trapped electron (8).

In this review we concentrate on selected developments in recent years. Several reviews are available that cover the earlier studies and provide more detail on individual topics (2, 5, 9–12). The subject of mobility in liquid He is not treated in detail.

DRIFT MOBILITY STUDIES

Some recent results serve to illustrate the range of mobilities observed at room temperature in nonpolar liquids as well as the effect of molecular structure. A series of n-alkanes have been examined: μ_D was found to decrease monotonically with chain length from 0.14 cm^2 V^{-1} s^{-1} for n-pentane to 0.016 cm^2 V^{-1} s^{-1} for tetradecane (4, 13). In contrast, much higher values of the mobility are observed for compounds made up of symmetrical molecules containing many methyl groups. For example, tetrakis(dimethylamino)ethylene has an electron mobility of 2.2 cm^2 V^{-1} s^{-1} (14). The presence of the amino groups and their associated local dipoles does not markedly reduce the electron mobility. The relationship of symmetry to high mobility is further indicated by the observation that $\mu_D = 47$ cm^2 V^{-1} s^{-1} in 1,2-bis(trimethylsilyl)ethane (15). However, a lower value (20 cm^2 V^{-1} s^{-1}) was reported for hexamethyldisilane (16).

Cryogenic Liquids

Drift velocities of photoinjected electrons were measured as a function of electric field strength and temperature in liquid neon, helium, hydrogen, and deuterium (17, 18). In these liquids at low electric field strengths, the drift velocity was linearly dependent on field and low mobilities were measured. At higher field strengths a supralinear dependence of the drift velocity on field was found. These observations are characteristic of transport via localized states.

Interesting effects were found in liquid neon, wherein fast electronic signals were registered in the nanosecond time domain (Figure 1b). These electrons exhibited a lifetime that increased with applied electric field (Figure 1c). At longer times, electron signals characteristic of localized electrons were found. These effects are interpreted as being due to the process of bubble formation. In liquid neon the electrons are initially quasifree, as in solid neon; due to scattering they fall into a shallow potential well that widens into a bubble, as in the case of liquid helium.

Data on the electron drift mobility in liquid argon and xenon as a function of temperature and field strength (19, 20) were analyzed by Lekner's hot electron theory to extract information about the mean electron energy as a function of the applied electric field strength. Measurements of the thermal-electron mobility for these liquids have been extended to very low fields (19–24). In liquid argon the mobility measured at $E = 3$

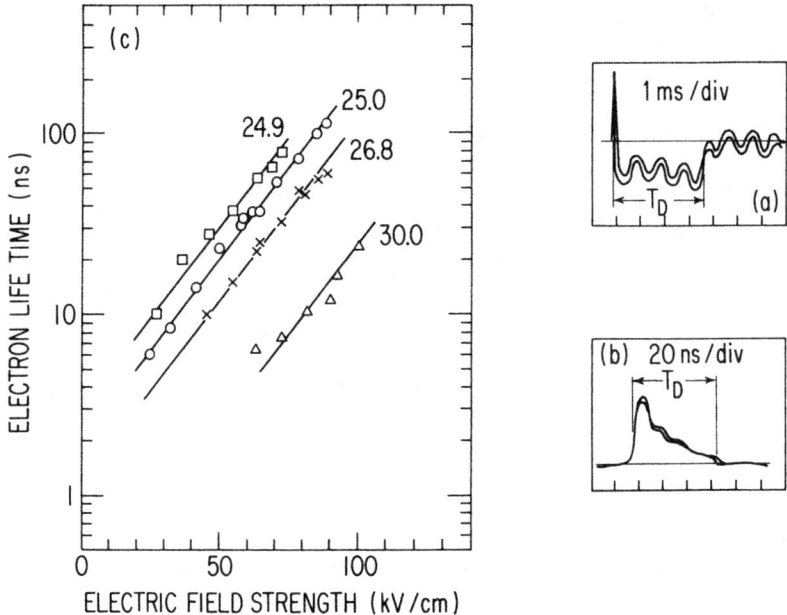

Figure 1 (a) Oscilloscopic trace of the slow electron signal due to electron bubbles in liquid neon at 25.9K, 14.3 kV cm^{-1}. (b) Oscilloscopic trace of the fast electron signal due to quasifree electrons in liquid neon, 25.9K, 52.9 kV cm^{-1}. (c) Electron lifetime in the quasifree state as a function of the electric field strength at the temperatures (in degrees Kelvin) indicated. (From Refs. 17, 18.)

V/cm was 545 cm^2 V^{-1} s^{-1} at 87K (24). The levels of electron-attaching impurities were below 1 ppb in some of these studies.

Various deuterated methanes have been investigated (25, 26). The mobility in deuterated methanes is 30 to 40% less than in CH$_4$. This isotope effect is attributed to a disruption of the sphere-like symmetry of the methane molecule (25).

Effects of Density on Mobility

Mobility data that have been measured as a function of temperature can be plotted alternatively as a function of density, as is shown in Figure 2. For symmetrical molecules like neopentane, as the density increases from the gas phase, the mobility decreases to a first minimum somewhat below the critical density, then increases to a maximum value and decreases again to a second minimum near the triple point. On freezing, the mobility increases a factor of two (1). For *n*-pentane and isopentane, the mobility decreases almost continuously as the density increases, showing, however,

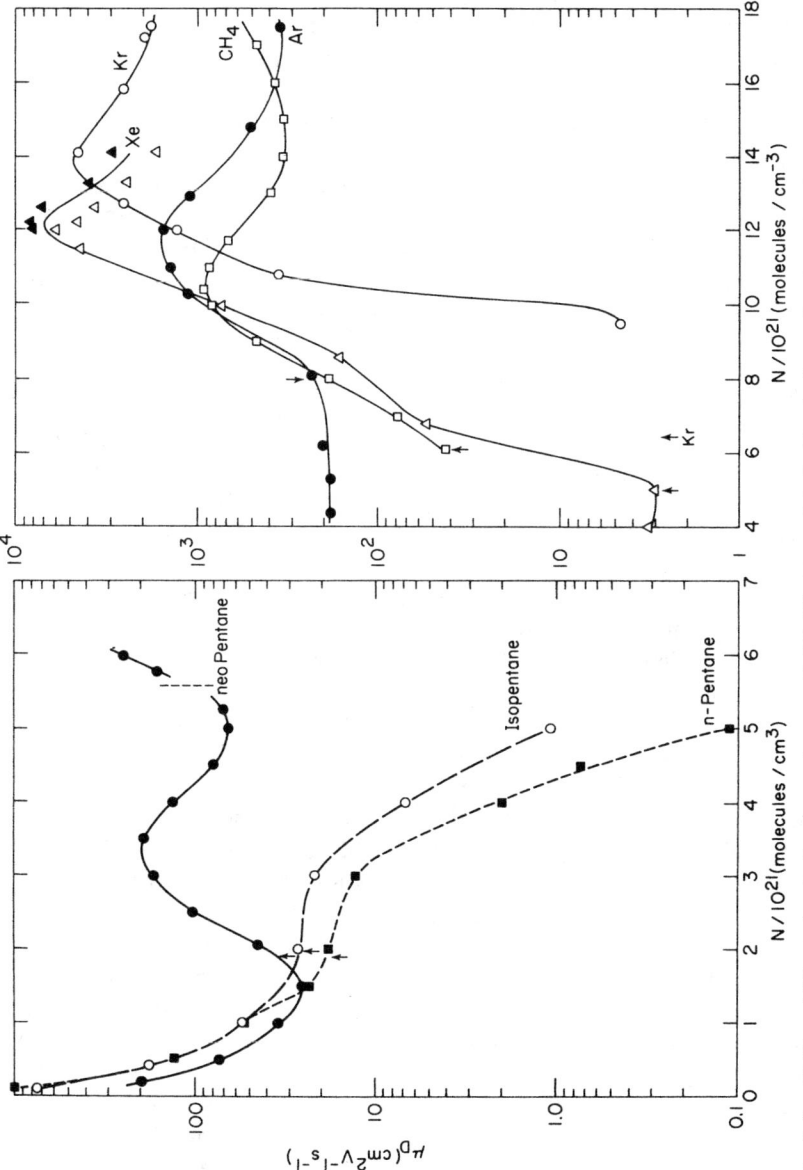

Figure 2 Left—Drift mobility vs density in pentane isomers (1, 27, 28). *Vertical dashed line* indicates freezing point for neopentane. *Right*—Drift mobility vs density in Ar (21), Kr (22), Xe (△ from 22); (▲ from 20), and methane (25). *Arrows* indicate critical densities.

some inflection point around the critical density. Most n-alkanes behave similarly. The mobility in ethane (29) exhibits a small maximum at a density of 5×10^{21} cm^{-3}.

Recently there have been several studies of hydrocarbons in which the mobility was measured isothermally at temperatures above the critical temperature. For neopentane, the isothermal data (28, 30) are in good agreement with the data of Figure 2 (*left*), which were obtained at different temperatures. This shows that the density is an important parameter for neopentane. This generalization is also approximately true for isopentane, n-pentane, and the butanes (31) for densities less than 4×10^{21} cm^{-3}. Temperature does, however, have a small effect around the critical density for n-pentane (30), n-butane (31), propane (32), and ethane (33). A satisfactory model of the mobility in this limited density region is given by the equation:

$$\mu_D = (2e/3)(2/\pi m k_B T)^{1/2}/N\sigma S(k), \qquad 1.$$

where σ is the scattering cross section and m the mass of the electron. This is similar to an equation due to Lekner (34) except the structure factor $S(0)$ has been replaced by $S(k) = RT/M\omega^2$, where M is the molecular weight and ω is the velocity of sound. However, this equation fails at higher densities.

The electron mobility in argon, krypton, and xenon exhibits large variation with density [see Figure 2 (*right*)]. In all three liquids there is a minimum at densities around the critical density, where a large variation of the mobility occurs with temperature above T_c. At higher densities, there is a maximum in the mobility for all three of these liquids.

Conduction Band Energy

An important quantity related to the mobility is the conduction band energy V_0. The energy of the bottom of the conduction band relative to vacuum is usually designated V_0. If V_0 is low enough, high mobility is likely because trapping is energetically unfavorable. On the other hand, if V_0 is high, trapping of the electron is more likely, which leads to low mobility. The density dependence of the conduction band energy (V_0) has been measured for several hydrocarbons (35–37) and for argon, krypton, and xenon (38, 39). The variation of (V_0) with density for several fluids is shown in Figure 3. As the density is increased from the gas phase, V_0 first decreases, passes through a minimum, and then increases as the density approaches the liquid phase density. For liquids composed of atoms or symmetrical molecules, like neopentane and methane, the density at which the V_0 minimum occurs is close to the density at which the electron mobility goes through a maximum (compare Figures 2 and 3).

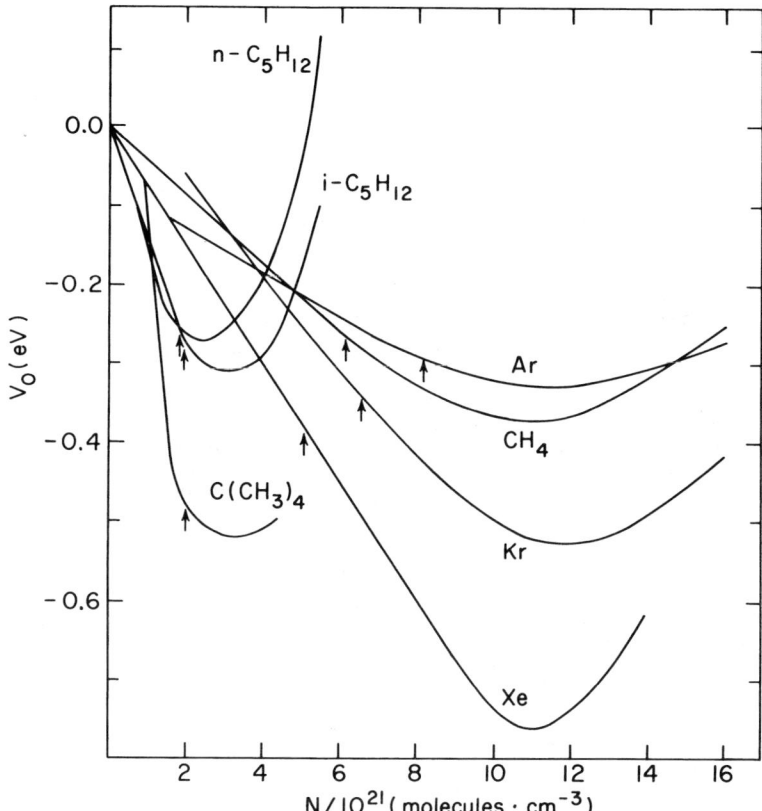

Figure 3 Conduction band energy vs density for pentanes (36), methane (37), Ar (38), Kr, and Xe (39). *Arrows* indicate critical densities.

For some time, most theoretical calculations of V_0 have been based on the Wigner-Seitz model (40). This theory has recently been updated by using first-principle pseudopotentials (41). The early theory, derived for rare gas fluids, is not totally satisfactory in describing experimental variation of V_0 as a function of density. This improvement has been shown to give a better fit to data on V_0 in argon, krypton, xenon, and methane (42). In the recent work for argon (41), the pseudopotential is optimized by fitting gas-phase scattering cross sections. The potential in the liquid is calculated within the "muffin-tin" approximation. The fit of this theory to the data for V_0 is much more satisfactory, and application to other molecular liquids would be of considerable interest.

High Pressure Effects

Electron drift mobility measurements have recently been reported for several nonpolar liquids subjected to high pressure. In these studies (43–45), static pressures as large as 3000 bar have been applied. As the pressure increases, the density of the liquid increases and the compressibility decreases.

Pressure has diverse effects, depending on the hydrocarbon. Liquids that exhibit a low mobility at 1 bar, like n-hexane, n-pentane, and 3-methylpentane, show a decrease in mobility as the pressure increases. Branched hydrocarbons in which the mobility is higher show an increase with pressure. For tetramethylsilane, however, a decrease in mobility with pressure is observed (see Figure 4).

The decreases in mobility in the n-alkanes and 3-methylpentane are interpreted to be a consequence of a shift of the equilibrium,

$$e_f \Leftrightarrow e_t, \qquad 2.$$

where e_f denotes the electron in the quasifree state and e_t the electron in the trapped state. At high pressure, the equilibrium shifts to the right. Since the electron spends a longer time in traps, the mobility, which is proportional to the fraction of electrons in the quasifree state, is reduced. The volume changes for Reaction 2 are of the order of -20 cm^3 mol^{-1} at standard conditions. These changes are attributed to electrostriction of the solvent around the electron localized in a cavity in the liquid.

The increases in mobility with pressure that are observed for branched hydrocarbons, although not totally understood, have been associated with variations in the quasifree mobility (46). The change in electron mobility in tetramethylsilane (44) with pressure provides a good test of theoretical models. In this liquid the electron remains quasifree at all times (47). Experimentally, the mobility decreases with increasing pressure at all temperatures; however, the effect is greater at higher temperatures. These relative changes in the electron mobility with temperature and pressure are predicted quite well by the deformation potential model of Basak & Cohen (48). However, the absolute magnitude of the mobility is predicted by theory to be about 2.5 times the experimental value at room temperature.

Transport via Temporary Negative Ions

Electron transport occurs in some fluids through intermediate ionic species. Some well-documented examples are solutions of small amounts of CO_2 (49) and aromatic hydrocarbons (50, 51) in nonpolar solvents. Such solutes

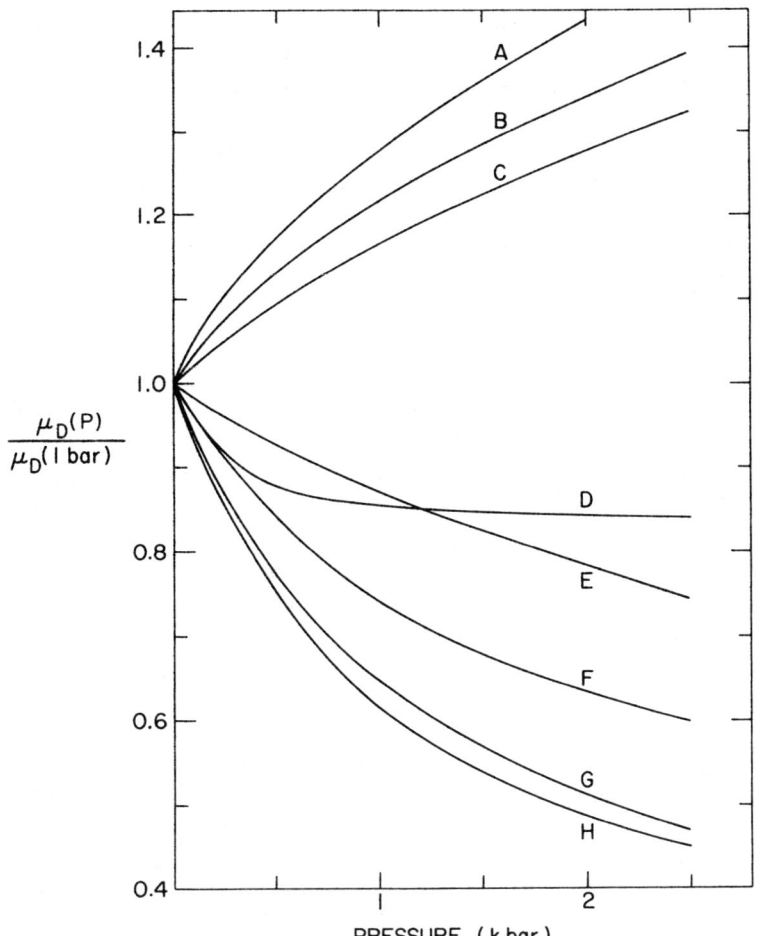

Figure 4 Relative drift mobility near room temperature as a function of pressure for (*A*) 2,2,4,4-tetramethylpentane/28, (*B*) 2,2,4-trimethylpentane/6.5, (*C*) 2,2-dimethylbutane/11, (*D*) tetramethylsilane/100, (*E*) cyclopentane/1, (*F*) 3-methylpentane/0.2, (*G*) *n*-hexane/0.07, (*H*) *n*-pentane/0.13. Numbers after each liquid indicate μ_D in cm^2 V^{-1} s^{-1} for each at 1 bar. (From Refs. 43–46.)

capture electrons but, because of their weak or negative electron affinities, the radical anions formed can thermally detach the electron.

$$A + e^- \rightleftharpoons A^-. \qquad 3.$$

Because the electron is trapped temporarily as an anion, its average mobility is reduced in the presence of such solutes. These equilibria are dependent on solvent, temperature, and pressure. An increase in temperature favors the reverse (detachment) reaction, whereas application of high pressure, at least when A is CO_2 (52), shifts the reaction to the right because of the large negative volume change, about -200 cm^3 mol^{-1}.

A related case is liquid N_2, in which the negative species mobility is low, only slightly larger than that of the cation (53). It is proposed that the electron is trapped by a rotational resonance to form N_2^-, which is then further stabilized by addition of more N_2 molecules. The overall process may be represented by:

$$e^- + mN_2 \Leftrightarrow [N_2]_m^-. \qquad 4.$$

An anion, rather than an electron in a cavity, is indicated here (53). The effect of increasing the density of N_2 in the gas phase is to shift this equilibrium to the right (54). Application of a high electric field shifts the equilibrium back to the left (18). A very similar situation appears to exist in liquid CO, where the anion and cation mobilities are again very comparable (55).

Another type of electronic transport is reported for perfluorobenzene. The mobility of the negative species in this liquid is 0.011 cm^2 V^{-1} s^{-1}, which is some 40 times the mobility of the positive ion (56). However, the electron is not free, since the electron affinity of C_6F_6 is 1.1 eV. The magnitude of the mobility is confirmed by the rates of attachment of the negative species to such solutes as CBr_4 and tetracyanoethylene, which are about 1.5×10^{11} M^{-1} s^{-1}, a value that indicates a very mobile species reacting at a diffusion-controlled rate. The negative species present in this liquid is suggested to be delocalized over a cluster of 15–20 molecules. The electron moves from one such cluster to another suitably oriented configuration of molecules. Another example of this type of transport, which has been called a "radical-ion conduction band" (57), is reported to occur in liquid CS_2 (58) and may possibly be found in other liquids.

POSITIVE HOLES Transport of charge as positive holes has been inferred in cyclohexane, *cis*-decalin, and *trans*-decalin from conductivity and chemical studies (59, 60). This conclusion has been challenged and transport via proton transfer suggested as an alternative in the case of cyclohexane (61, 62). Further experiments are needed to settle this important point.

Photoassisted Mobility of Electrons

Fast pulse techniques are contributing to the understanding of both hot and thermal electron transport. Some experiments utilize two laser pulses separated by a time delay. The first pulse ionizes the sample, producing trapped electrons (e_t); then a second laser pulse detraps the electron, producing an extra conductivity signal, while the hot electrons (e_{hot}) thermalize and retrap.

In a two-pulse experiment, Balakin & Yakovlev (63) deduced, from the integral of the current induced by the second (detrapping) pulse, that the average drift displacement ($\mu't'$) of detrapped electrons is 4×10^{-13} cm^2 V^{-1} for hexane at 185K. A similar value was found for methylcyclohexane (10, 64). For 2,2,4-trimethylpentane at 175K a much larger value of 3.8×10^{-11} cm^2 V^{-1} was observed (10). The data were analyzed according to Reaction 5.

$$e_t + h\nu \rightarrow e_{hot} \rightarrow e_f \rightarrow e_t. \qquad 5.$$

The authors conclude that the displacements observed in hexane and methylcyclohexane are due primarily to electrons while hot. In contrast, the higher value in 2,2,4-trimethylpentane indicates the electrons spend a significant time in the quasifree state (e_f) before being trapped.

Double-pulse conductivity experiments have also been done on the picosecond time scale by varying the time delay between pump and probe laser pulses and measuring the change in the integrated current observed. In liquid hexane at 23°C, the first half-life of geminate recombination is deduced thereby to be less than 9 ps (65). The results show that recombination occurs by diffusion (66) and that the mobility of the electron at picosecond times is the same as at longer times. It may be further concluded that thermalization occurs in hexane within a few picoseconds.

For short enough time delays or for fast response times, the trapped electron will still be within the coulomb field of the geminate positive ion. Drift of geminate electrons in an external electric field is in effect a change in dipole moment of the positive ion-electron pair. This polarization was observed as an extra current during a single laser pulse used to photoionize solutes in methylcyclohexane (67), and is most obvious at low temperatures. Another study showed similar results for photoionization of pyrene in 2-methylpentane (68). The effect of polarization is to add to the current signal initially but to subtract from the signal during later stages of recombination. This later negative current has been predicted for ion-pair recombination by Monte-Carlo (69) and diffusion calculations based on the Smolukowski equation (70). Calculations of the latter type are valid as long as the mean free path of the electron is short compared to the Onsager length ($e^2/\varepsilon_r k_B T$) (71). The conductivity experiments were done with a time resolution of 3–10 nsec. Better time resolution would be useful; the best that has been achieved for conductivity measurements is 100 ps (72).

HALL MOBILITY STUDIES

A very significant experimental advance in our understanding of electron transport comes from recent Hall mobility (μ_H) measurements. The Hall mobility derives from the deflection of electrons by a magnetic field **B** while they drift in an electric field **E**. The drift velocity is therefore

$$v = \mu_D \mathbf{E} + \mu_H \mu_D \mathbf{E} \cdot \mathbf{B}. \qquad 6.$$

Such deflection occurs while the electron is in the conduction band. Thus μ_H is a measure of the conduction band mobility. However, because the magnitude of the signal is quite weak, measurements so far exist mainly for high mobility liquids ($\mu_D > 10$ cm^2 V^{-1} s^{-1}).

The first results in this area showed that under most conditions there is not a large difference between the Hall and drift mobilities for neopentane and tetramethylsilane (see Figure 5) and that traps are unimportant in these liquids, at least for moderate temperatures (47, 73–76). Furthermore, in the case of neopentane, the deformation potential model (48) predicts approximately the magnitude and temperature dependence of μ_H between 20 and 140°C and also finds correctly the maximum in the mobility at 125°C. The results thus are consistent with the idea that electrons are scattered as a result of density fluctuations.

For neopentane at temperatures greater than 140°C, however, the ratio μ_H/μ_D increases and is about five near the critical temperature (75, 76). This large ratio led the authors to postulate the existence of localized states under these conditions.

This technique has recently been extended to other molecular liquids (78). The results show that the conduction band mobility, as measured by the Hall effect, is different in each liquid and varies considerably with temperature. In the case of 2,2-dimethylbutane and 2,2,4,4-tetramethylpentane, μ_H/μ_D is close to unity at moderate temperatures, thus indicating the absence of intrinsic traps in these liquids as well. Interestingly, for 2,2,4-trimethylpentane at 20°C, the Hall mobility is 22 cm^2 V^{-1} s^{-1}, which is three times the drift mobility. Thus the existence of traps in this liquid is confirmed. There is no direct measurement of the conduction band mobility in liquids of lower mobility.

HOT ELECTRONS

Hot Electron Relaxation

The energy relaxation of superthermal electrons can be followed by microwave conductivity measurements in the heavier liquified rare gases (argon, krypton, and xenon). Ionization of the liquid by high energy particles or quanta yields electrons with an initial energy of the order of 10 eV. This

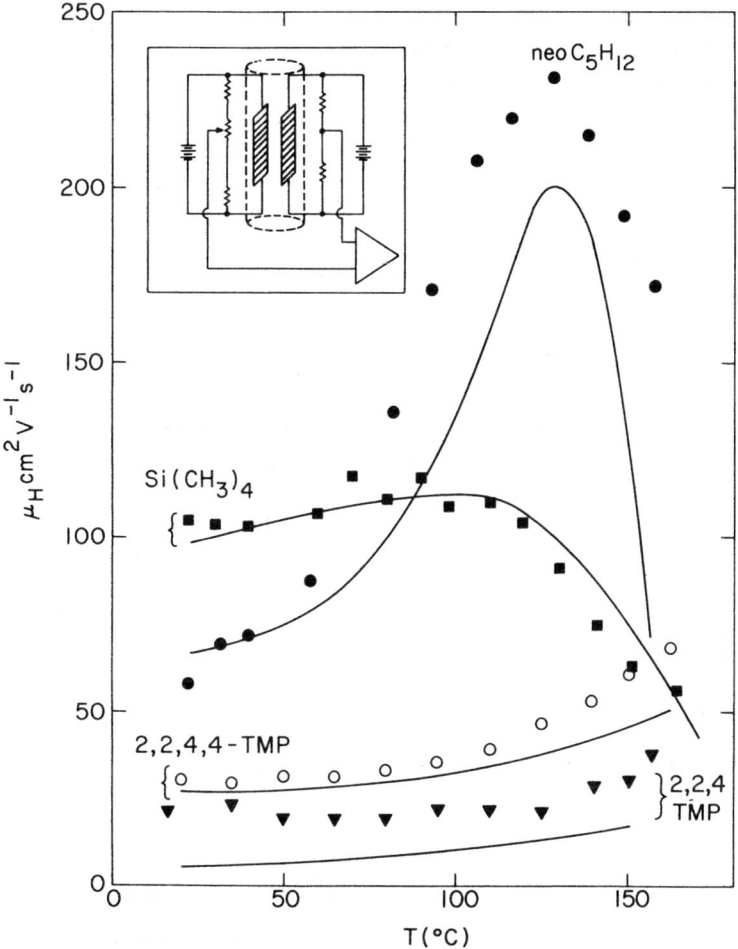

Figure 5 Hall mobilities as a function of temperature for various liquids indicated by *points*. *Line* indicate corresponding drift mobilities for each liquid. *Insert* shows schematic of Hall cell and circuit. (From Refs. 47, 76–78.)

energy is dissipated by collisions until the electron energy equals $k_B T$ for the particular liquid, which is the order of 10 meV. Measurements of the radiation-induced conductivity after a subnanosecond pulse of high energy X-rays showed a delay time for the development of the full conductivity signal because superthermal electrons exhibit a lower mobility, defined as the ratio of the drift velocity to the electric field strength, than thermal electrons. This delay time is then interpreted as the energy relaxation time or thermalization time of the electrons (79). The results obtained are compiled in Table 1. From measurements of the electron diffusion coefficient, Kubota et al (80) calculated hot electron relaxation times that were about one-half of the values given in Table 1. Their calculations pertain to electrons with at most a mean energy of 1 eV, whereas the mean kinetic energy of the electrons produced by high energy radiation in the experiments of Sowada et al is of the order of 10 eV. Estimates by Warman (81) for molecular liquids yield thermalization times smaller than 10 ps, too small to be observed directly with present-day conductivity techniques.

Electron Diffusion Coefficient

The ratio of the diffusion coefficient to the mobility was measured for electrons in liquid argon by Shibamura et al (82). The method was essentially that of Townsend, developed for gases. At the cathode plane, electrons were produced by ionization of the liquid with Po-α-particles. The electrons drifted under the influence of an applied electric field to an anode, the distance of which was set to be between 1.7 and 4.8 mm. The anode consisted of two sets of eight conducting strips, each with an effective width of 60 μm. With this experimental arrangement the diffusion-broadening of the electron cloud during its drift to the anode could be monitored. Values of the diffusion coefficient D_e as a function of the electric field strength up to 10 kV cm^{-1} were obtained. With the drift mobility μ_D, the mean electron energy could be obtained as

$$\langle \varepsilon \rangle = eD_e/\mu_D. \qquad 7.$$

The mean electron energy increased monotonically from 0.1 eV at 2 kV cm^{-1} to 0.4 eV at 10 kV cm^{-1}.

Field-Dependent Electron Attachment Reactions

The attachment of electrons to electronegative solute molecules in liquid hydrocarbons reflects the electron transport mechanism in the particular liquid and the energy dependence of the attachment cross section of the

Table 1 Thermalization times for superthermal electrons in liquefied rare gases[a]

Liquid	Temp. (K)	τ_{th} (ns)
Argon	85	0.9 ± 0.2
Krypton	117	4.4 ± 0.2
Xenon	163	6.5 ± 0.5

[a] Ref. (79).

scavenger molecule. Bakale & Schmidt (83, 84) studied electron attachment to SF_6 in liquid ethane, propane, argon, and xenon as a function of the applied electric field strength.

In ethane and propane at low temperatures, electron transport takes place via localized states. Application of a sufficiently high electric field increases the electron drift mobility but leaves the electrons in thermal equilibrium with the liquid. A concomitant increase of the rate constant is observed because the attachment rate constant is proportional to D_e, which is proportional to the mobility.

An additional effect of the electric field on the rate of diffusion-controlled reactions has been discussed by Tachiya (85). He attributed the increase of the reaction rate constant as due to an increase of the effective volume seen by the electron when it drifts in the direction of the electric field. This effect would manifest itself as a continuous increase of the rate constant with electric field. The scavenging data available so far are not adequate to verify his predictions.

In argon and xenon, liquids that show high electron mobilities, a decrease of the attachment rate constant to SF_6 is observed at field strengths at which the electron mean energy exceeds $k_B T$ for the liquid. This reflects the decrease of the cross section for electron attachment to SF_6 with increasing electron energy, as observed for the gas phase. A similar behavior was observed for electron attachment to CCl_4 in tetramethylsilane, a high mobility liquid, where above 15 kV cm^{-1} a decrease of the rate constant was observed due to the reaction of superthermal electrons. In contrast, attachment to ethylbromide increased with electric field, thus indicating that the cross section of this reaction increases with electron energy (86).

RELATED TRANSPORT PROPERTIES

Homogeneous Electron-Ion Recombination

For the homogeneous recombination of positive and negative ions in nonpolar liquids, Debye's equation has been found to hold:

$$k_r = (\mu_+ + \mu_-)e/(\varepsilon_r \varepsilon_0), \qquad 8.$$

where μ_+, μ_- represent the mobilities of the positive and negative ions, respectively, e is the electronic charge, ε_r is the relative dielectric constant of the liquid, and ε_0 is the permittivity of the vacuum. The prerequisite for the applicability of this diffusion equation is that the mean free path of the charge carrier motion be much smaller than the characteristic length that determines the actual recombination process. In nonpolar liquids the characteristic length is given by the Onsager critical distance r_c, which is determined by the condition that at such separation distance the coulomb energy of attraction equals the thermal energy $k_B T$. The mean free path Λ for ionic motion is of the same magnitude as the intermolecular separation distances (1 nm), whereas $r_c = 30$ nm for $\varepsilon_r = 2$ and $T = 300$K. If the mean free path for the charge carrier motion becomes comparable to r_c, deviations from Eq. 8 are expected. A wide variation of Λ can be found in the motion of excess electrons in neat nonpolar liquids and mixtures. Measurements of the electron positive-ion recombination rate constant k_r as a function of electron mobility followed Eq. 8 up to $\sim \mu_D = 50$ cm^2 V^{-1} s^{-1} (87). Slight deviation occurred for tetramethylsilane (88), in which $\mu_D = 100$ cm^2 V^{-1} s^{-1}. Even bigger deviations from Eq. 8 were found in liquid methane and argon (89, 90). In both liquids, the electron mobility is around 400 cm^2 V^{-1} s^{-1}, which corresponds to a mean free path of approximately 15 nm. The critical distance r_c is approximately 90 nm. Tachiya treated the electron-ion recombination process theoretically by use of a Monte Carlo technique (91). His results describe qualitatively the influence of the ratio Λ/r_c, although quantitative discrepancies still exist.

The electric field effect on k_r has also been investigated recently. In liquid methane, k_r increases with increasing field and at high field approaches the value expected from the Debye equation (92). In liquid argon, k_r initially increases with field up to a critical field, and then decreases with further increase in field (90). One theoretical study predicted that k_r would decrease at high field in methane (93); another study by Tachiya (85), who considered the fractal geometry of reactant trajectories, concluded that the reaction rate should increase with field. A disturbing feature of these investigations is that the measured electron-ion recombination rate constants in solid methane and argon, where the mobilities are approximately twice the liquid values, come close to those predicted by the Debye equation.

Electron Emission

Electron emission into the vapor phase should occur and has been observed for liquefied rare gases (94, 95) and for hydrocarbons (96). The probability of crossing the liquid/vapor barrier depends on the value of V_0, which ranges from $+1.05$ for liquid helium to -0.7 eV for tetramethyltin, and on the externally applied field strength. It is obvious that the liquid has to be sufficiently free of electron-attaching compounds in order to allow all the electrons to reach the liquid surface. Gushchin et al (97) measured the emission of hot electrons from liquid and solid argon and xenon. Electric field strengths of several kV cm^{-1} are necessary to extract 100% of the electrons. At these fields the electrons in the bulk of the material are nonthermal. By applying the Lekner theory, the following barriers (V_0-

values) were estimated: for argon (liquid) -0.065 eV, (solid) -0.02 eV; for xenon (liquid) -0.85 eV, and (solid) -0.42 eV. Although the argon values are about 0.1 eV higher than values determined by the photoelectric effect, the xenon data are 0.1 to 0.2 eV more negative.

Recently, new interest in this phenomenon has arisen from the point of view of using the emitted electrons for the reconstruction of ionizing tracks in the liquid (98). Two sets of detecting wires, multi-wire proportional counters (MWPC), above the liquid surface allow the determination of the point of emission. In another set of experiments, solutions of photoionizable compounds were used as liquid photocathodes. Photoionization produces an electron and a positive ion in the liquid layer. The electron is then extracted from the liquid and detected in the vapor phase by the MWPC (99).

THEORY OF ELECTRON TRANSPORT

The fundamental quantity describing electron transport is the drift mobility μ_D. A wide range of values has been measured as a function of temperature, electric field strength, and external pressure (1, 6, 100). The theoretical models developed for the description of the electron mobility can be classified according to the magnitude of the mobility.

First, electron mobilities <1 cm^2 V^{-1} s^{-1} are treated by transport via localized states. Here, two variants are proposed, hopping mobility or trap-modulated band mobility. Second, electron mobilities >10 cm^2 V^{-1} s^{-1} are treated as transport in extended states; the electrons are assumed to be delocalized. Third, the intermediate region from 1 to 10 cm^2 V^{-1} s^{-1} has not received a lot of attention. Classic diffusion models or percolation theory have been used to rationalize the mobility effects observed.

Difficulties exist in the explanation of electron mobilities as a function of temperature up to the critical temperature. Often several orders of magnitude are covered by the experimental data, which extend from values <0.1 to $>10^3$ cm^2 V^{-1} s^{-1}. A continuous change in the mode of transport seems to take place. The models put forward so far only address a limited variation of the electron mobility as a function of temperature, electric field strength, or external pressure. Two main approaches in the theoretical modeling of electron transport are prevalent: In one class of models concepts of solid state physics are adapted to the description of electron transport in liquids, whereas in the other, electron transport in gases serves as the starting point. Because of the relative simplicity of the system and the great amount of experimental and theoretical data available, the theory of electron transport in liquefied rare gases has engendered more intense study.

Electron Transport in Liquid Argon, Krypton, and Xenon

In the discussion of electron transport three problems have to be solved: First, the elementary process of electron-atom or electron-molecule interaction (scattering on an atom or molecule in the case of a dilute gas) has to be treated; second, the influence of spatial order or disorder of atoms or molecules has to be taken into account; and third, the correct statistics have to be applied in order to calculate macroscopically measurable quantities, as, for instance, the drift mobility. The simplest case is the motion of low-energy electrons in a dilute rare gas. The electron-atom scattering can be described in terms of a repulsive core of radius a_0, which represents the effect of the closed electron shells, and an attractive polarization potential, which acts at longer distances from the atom. As long as $N^{1/3}a_0 \ll 1$, where N is the number of atoms per cc, each scattering event is well separated in space and time from the next one. At high enough temperatures (required for an ideal gas behavior) and very low electron concentrations, the classical Maxwell-Boltzmann statistics can be applied to the electrons. The mobility is derived as

$$\mu_D = (2e/3)(2/\pi m k_B T)^{1/2}/N\sigma_0, \qquad 9.$$

where m denotes the electron mass and σ_0 denotes the cross section at zero electron mean energy. From Eq. 9 it follows that the product of μ_D and N should be constant. Any deviation from this condition indicates a departure from the initial assumption of single scattering. Thus, the primary effort in the theoretical treatments that have appeared is directed towards calculation of $1/N\sigma_0$ within the context of Maxwell-Boltzmann statistics.

In the original treatment by Cohen & Lekner (101), single scattering of the electron by an argon atom modified by a shell of next neighbor atoms was assumed. Actually, a weakening of the scattering due to the spatially correlated atoms occurs. The spatial arrangement of these next neighbors was inferred from the structure factor $S(k)$ at low energies, which is correlated to the pair correlation function. In the low energy limit, $k = 0$. The structure factor is given in terms of the isothermal compressibility χ_T as

$$S(0) = \chi_T N k_B T. \qquad 10.$$

Lekner's approach was essentially a single scattering approximation, which failed when applied to the explanation of the density variation of the mobility even though attempts had been made to rescue the original concept by modification of the scattering center (102). Multiple scattering of the electron was introduced into the picture by Polischuk (103), who was able to calculate the variation of the electron mobility with density in liquid xenon. His theoretical dependence agrees quite well with the measured data up to the mobility maximum.

Basak & Cohen (48) tried to take into account multiple scattering processes by adapting the deformation potential theory of solid state physics. The deformation potential is caused by long wavelength fluctuations of the density N. The potential was assumed to be given in terms of dV_0/dN, d^2V_0/dN^2, etc. The maximum in the mobility occurs near the density where V_0 is a minimum, i.e. when $\partial V_0/\partial N = 0$. By fitting an assumed dependence of V_0 on N, Basak & Cohen were able to reproduce the observed mobility changes. Later, Reininger et al (38) measured V_0 as a function of density for argon and xenon and applied the Basak-Cohen formalism using their V_0 data. The good agreement disappeared, but the theory reproduced the mobility maximum. Nishikawa (104) improved the situation somewhat by employing the wavelength-dependent structure factor $S(k)$ rather than the long wavelength limit $S(0)$. Disagreement between data and theory remained; at low densities the theory predicted values for the mobility too low compared to the measured data, and at high densities the predicted values were too high (see Figure 6).

Another refinement of the original Basak-Cohen model was introduced by Ascarelli (105, 106), who, in addition, took into account the scattering of electrons on phonons. The electrons absorb and emit quanta of heat, and this leads to changes in their wave vector. The agreement of the calculated dependence of mobility on density with experimental data for liquid argon and xenon is within an order of magnitude. In calculating the mobility for xenon, Ascarelli used a value of the effective mass of the electron (m^*) equal to $0.27\ m$; for argon a density-dependent effective mass was assumed.

The electron mobility in liquid argon near the critical point was analyzed by Watanabe (107), who pointed out that electron scattering occurs on critical fluctuations, with the wavelength shorter than the correlation length. Qualitatively, the drop of mobility could be explained, but for quantitative agreement improvement of the theory is indicated. Vertes (108) has calculated the electron mobility in liquid xenon by the method of partial waves in which experimental values of V_0 in xenon were used. This theory matched the experimental mobility values, with one adjustable parameter related to the average fluctuation size. He attributed the low mobility in xenon near T_c to bubble formation.

Path Integral Calculations

Path integral calculations provide a new approach to the theoretical description of excess electrons in fluids. In these studies the quantum

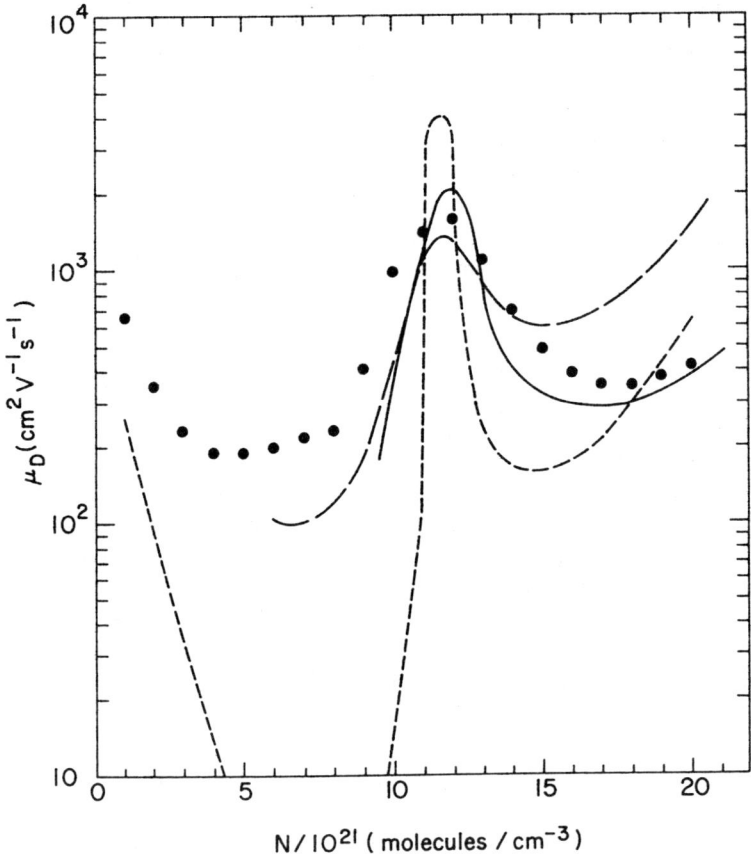

Figure 6 Electron mobility in Ar vs density. *Points* are experimental (21); *lines* are theoretical—*short dashed lines* (38), *long dashed lines* (104), *solid lines* (105).

nature of the electron is represented by a classical isomorphic ring polymer, as suggested by Feynman (109). These studies have so far focused on changes in electron mobility with density in rare gas fluids.

Chandler and collaborators (110–115) have used the path integral method to treat excess electrons in a fluid of hard spheres. The electron is excluded from the spheres, which are of diameter σ. The results show that as the hard sphere solvent density increases, the electron ring polymer is compressed, and at high enough fluid densities (ρ) the solvent will localize the electron. One advantage of this theory is that it is able to treat the electron in both the extended and localized states. Although this is an equilibrium theory, the electron mobility was calculated (111) and predicted to decrease rapidly as the electron becomes localized at densities above 0.15 σ^{-3}. Laria & Chandler (115) conclude that the electron in xenon remains in an extended state because the excluded volume is small, whereas for dense helium, because the excluded volume is large, the electron is localized.

The absorption spectrum of the excess electron in a hard sphere solvent has also been calculated for various densities (113). At a density of 0.3 σ^{-3} a maximum in the spectrum is predicted in the infrared, which shifts to higher energies as the density increases. Coker & Berne (116) calculated the absorption spectrum of the localized electron specifically for helium by averaging over a series of solvent configurations obtained from path integral Monte Carlo calculations (see below). These calculations were carried out for a supercritical helium fluid at 309K. A broad absorption band is predicted having a maximum around 0.6 eV at a density of 3×10^{22} cm^{-3}.

Coker et al (117) employed a Feynman path integral Monte Carlo method to study the electron in helium and xenon. For the electron-atom interaction, a pseudopotential was employed:

$$V(r) = (A/r^4)\{[B/(C+r^6)]-1\}. \qquad 11.$$

In helium, the physical size of the electron chain decreases with increasing density, consistent with localization in a cavity (see Figure 7, *left*). The results for xenon show the dominance of the long-range (attractive) polarization part of the potential, because of which the electron is quasifree (see Figure 7, *right*).

LOCALIZATION NEAR CRITICAL DENSITIES These path integral studies also provide insight into the low mobility observed at densities near the critical density. The phenomenon suggests the existence of quasilocalized states in xenon (23), argon (21), krypton (25), and methane (118). Localization in clusters is suggested as an explanation. For small xenon clusters, the electron is weakly bound (119); induced dipole-induced dipole interactions are very important and tend to destabilize such cluster anions (120). In larger clusters the electron is, however, confined within the cluster, i.e. a bulk state is formed. This is largely a consequence of the electron-xenon pseudopotential, which has a minimum roughly half way between xenon atoms in the cluster. These minima form channels throughout the cluster through which the electron can move. As a cluster becomes larger, the number and depth of these channels increase. Near critical densities, the low density regions surrounding clusters act as barriers, and escape of electrons from clusters is unfavorable. Electron transport most likely occurs when the cluster breaks up. At such densities the mobility is low

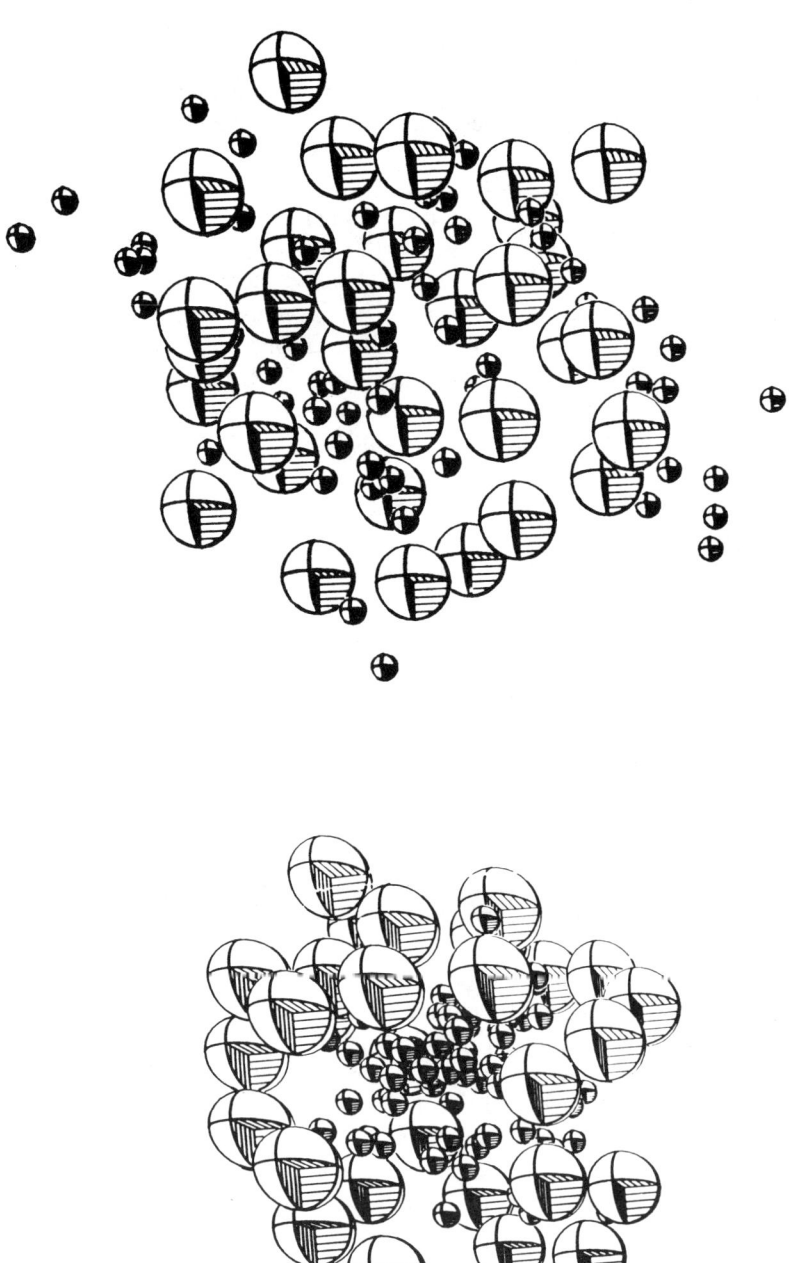

Figure 7 Configuration of electron chain (*small spheres*) in He (*left*) and Xe (*right*) from path integral calculations at 309K for $\rho\sigma^3 = 0.5$. *Large spheres* represent solvent atoms. (Reproduced by permission from Ref. 117.)

due to this unusual localization. The electron, once localized in a cluster, affects its structure; a sharpening of the first peak in the Xe–Xe pair correlation function is observed. This effect, which is electrostriction, has been detected experimentally for localized electrons in some alkanes (45).

Electron Transport at High Fields

In the theoretical treatment of electron transport at high field strengths, where the electrons have a mean electron energy $\langle \varepsilon \rangle$ greater than $k_B T$, the original concepts of Cohen & Lekner (101) have been maintained, and attempts have been made to calculate $\sigma(\varepsilon)$. The electron drift velocity in liquid argon, krypton, and xenon as a function of the electric field strength is characterized by three regions: At low field strength the drift velocity is proportional to the electric field strength; this is followed by a region in which the drift velocity increases less than proportionally with the electric field; finally, at very high field strength a constant drift velocity is observed. Basically, the drift velocity of electrons in these fluids near the triple point exceeds the drift velocity in the rarefied gas at any given value of the reduced electric field strength E/N. The difference is large at lower values of E/N, and it becomes smaller at higher values of E/N for which heating of electrons by the electric field occurs. In the liquids, saturation of the drift velocity with respect to electric field occurs at high values of the field strength. Cohen & Lekner (101) interpreted the higher electron mobility in the liquid phase as compared to the gas phase as resulting from a reduction of the scattering of the slow electrons due to the influence of spatially correlated atoms around the scattering center. Saturation of the electron drift velocity had not been explained within the context of the Cohen-Lekner theory. A few quantitative attempts along different lines were published.

Recently, new approaches have been taken to this problem of hot electrons in liquefied rare gases within the framework of the Cohen-Lekner theory. Atrazhev et al (121–123) assumed that the influence of the liquid structure on the electron-scattering process decreases as the mean electron energy increases, and at energies above a few electron volts the cross section in the liquid approaches that of the single atom in the gas. With this assumption, Atrazhev et al were able to explain the observed drift velocity vs field data.

Sakai et al (124, 125) determined, within the context of the Cohen-Lekner theory, the collision cross sections as a function of electron mean energy for liquid argon. To explain the saturation drift velocity they had to introduce an inelastic collision process at high electric field strengths (hot electron regime), for which they found indications in other experimental observations. Kaneko et al (126) used a gas kinetic approach and found that a constant cross section accounted for the mobility data in liquid argon and methane except at very high fields. The existence of hot electrons in the heavier liquefied rare gases was questioned by Ascarelli (127) but defended by Freeman (128). The experimental evidence for hot electrons was summarized by Schmidt et al (129).

MOLECULAR DYNAMICS CALCULATIONS A completely different model for electron mobility in the heavy rare gas liquids (130, 131) and for electron-positive ion recombination in liquid methane (93) uses the method of molecular dynamics. Here, the electron is thought to be temporarily localized at one argon atom (transient negative ion formation). It is separated from adjacent sites by steep potential barriers. Due to collisions with nearest neighbors, the potential barriers are lowered and the electron hops to a neighboring site. One-dimensional computer simulations on chains of 130 to 150 atoms reproduced the main features of the field dependence of the electron drift velocity in liquid argon, krypton, and xenon. Details of the electron-atom potential energy at the collision are extracted from measured drift velocities as a function of electric field strength. The effect of density on the electron mobility are well reproduced.

In liquid methane, the homogeneous electron-ion recombination coefficient was also calculated as a function of density by molecular dynamics. In the simulations (93), 50 electron-ion pairs were placed in a cubic cell of 1710 nm length, which corresponds to an initial electron density of 10^{13} cm^{-3}. Electron-electron interactions are negligible. The numerical results were in reasonable agreement with the measured (89) density dependence of the recombination coefficient. The rate-determining step in the recombination rate is the energy transfer rate rather than diffusion.

Effective Mass

The mean free path of the electron is quite long in many liquids. Nevertheless, because liquids are quite dense and lack long-range order, the electron is subjected to multiple scattering. To take these properties into account, the effective mass approximation is often assumed in calculating the electron mobility (48) and the conduction band energy (40). However, in the past because of lack of experimental data it has often been assumed that $m^*/m = 1$.

Data on the ratio m^*/m have been derived for a few liquids either from analysis of exciton spectra or from photoionization measurements in liquids. This ratio was evaluated from the energy dependence of ionization cross sections near threshold for argon and some branched molecular liquids and found to be 0.27 (132). From peak positions in exciton spectra, m^*/m was determined in liquid argon and xenon (133, 134) and values between 0.3 and 0.5 obtained.

Electron Localization in Liquid Helium and Neon

Localization of quasifree electrons by strong electron-medium interaction manifests itself as dramatic changes of electron mobility. In liquid helium, low mobility electron bubbles are observed. The localization time was estimated by Iakubov & Khrapak (135) to be less than 0.1 ns. The stability of an electron bubble in liquid neon remained in doubt. New theoretical approaches to the localization criteria were developed: Ebner et al (136, 137) discussed the stability of the electron bubble in liquid neon and found it to be stable in the whole range of the gas/liquid coexistence curve. Hernandez et al (138, 139) treated the transition from quasifree electron states to bubble states in helium.

Electron Transport in Molecular Liquids

Although an immense amount of experimental data on electron mobility in hydrocarbons and related compounds is available, the theoretical description of electron transport in hydrocarbons has not advanced at the same pace as that for the rare gas liquids. This may be due mainly to the fact that the mobility values cover a range from 0.016 to more than 100 cm^2 V^{-1} s^{-1}.

A two-state model for electron transport in liquid hydrocarbons was developed by Berlin et al (140). Localized and extended states are invoked in the transport. The observed mobility is then given as

$$\mu_D = \mu_t P + \mu_f(1-P), \qquad 12.$$

where P denotes the probability of localization and μ_t and μ_f are the mobilities in the localized and extended states, respectively. Estimates for μ_f, obtained by assuming that scattering is due to density fluctuations, were about 30 cm^2 V^{-1} s^{-1} for n-hexane and 400 cm^2 V^{-1} s^{-1} for neopentane. Later, Vertes (141) introduced the method of partial waves, thus circumventing the Born approximation. The mobility was related to the average size of fluctuations in hydrocarbons, which were taken to be in the range 1–3 nm. The method worked for molecules that exhibited spherical symmetry but did not give meaningful results for n-hexane, n-pentane, and cyclohexane. Berlin & Schiller (142) found similar values of these fluctuation radii and concluded that an electron in a hydrocarbon interacts with 60–120 molecules in scattering events.

General theoretical treatments have been attempted (143) by assuming a two-state model in which electrons move in extended states interrupted by residence in traps. The electron drift mobility was calculated with the Hall mobility, the electron effective mass, and the photocurrent threshold as determining quantities. This theory describes the conditions in tetramethylsilane and neopentane quite well, and it has the potential of being tested for other liquids once the pertinent experimental data become available.

The electron mobility in mixtures of two different hydrocarbons was analyzed by Schiller & Nyikos (144) by means of a percolation model. Here, spatial fluctuations are treated only while the physical properties of the subsystems are thought to be independent of time. The model explains the experimental findings well if the electron mobility is high in one component and low in the other. In a later paper (145), temporal fluctuations in conductivity were included and the results were also in reasonable accord with the same experimental data on mixtures.

The optical absorption and photo-induced dc-conductivity of trapped electrons in hydrocarbons were treated by Funabashi (146). The Kubo linear response theory was employed, which is indicated when the band picture is not applicable.

OUTLOOK

Significant advances have been made and interest remains strong in the field of electron transport in nonpolar liquids. A large number of liquids have already been investigated and drift mobilities have been measured over a wide range of conditions. The concept of a conduction band on the one hand seems well established; on the other hand, evidence indicates trapped or localized states also occur. New experimental discoveries as well as improvements to theory are anticipated.

Considerable progress is expected in the future from the utilization of short pulses and very fast time resolution. Ionization studies on the femtosecond scale are already in progress (147, 148). Other nonpolar liquids may be found with fast carriers. More examples may be found of liquids in which the electron moves in a "radical-ion band."

On the theoretical side, we note that methods such as path integral simulations, molecular dynamics, and fractal trajectories are giving additional insight into the electron medium interactions. Some groups (B. J. Berne and D. Chandler, private communications) are currently applying such techniques to molecular liquids, where further development of theory is needed.

Finally, we expect that the use of nonpolar liquids in ionization detectors in medicine, industry, astronomy, and high energy physics will grow. Such detectors can be used either to measure the total energy of particle showers (149) or for medical imaging (150) and can even provide pictures of particle tracks (151, 152). The latter application requires electron drift distances of a few meters and therefore liquids with extremely low concentrations of electron capturing impurities.

ACKNOWLEDGMENT

Preparation of this review was done both at Brookhaven National Laboratory under contract DE-AC02-76CH00016 with the US Department of Energy and supported by its Division of Chemical Sciences, Office of Basic Energy Sciences, and at Hahn Meitner Institut, which is supported by the Federal Ministry of Research and Technology, Federal Republic of Germany. We wish to thank all our colleagues who supported our effort by sending us preprints and reprints of their papers.

Literature Cited

1. Freeman, G. R. 1987. In *Kinetics of Nonhomogeneous Processes*, ed. G. R. Freeman, pp. 19–87. New York: Wiley
2. Schmidt, W. F. 1984. *IEEE Trans. Electr. Insul.* 19: 389–418
3. Braun, C. L., Scott, T. W. 1988. *Radiat. Phys. Chem.* 32: 315–17
4. Gee, N., Freeman, G. R. 1987. *J. Chem. Phys.* 86: 5716–21
5. Holroyd, R. A. 1987. In *Radiation Chemistry: Principles and Applications*, ed. M. A. J. Rodgers, Farhataziz, pp. 201–35. New York: VCH Publ.
6. Allen, A. O. 1976. Natl. Stand. Ref. Data Syst., Natl. Bur. Stand. 58
7. Casanovas, J., Guelfucci, J. P. 1988. *IEEE Trans. Electr. Insul.* 23: 515–28
8. Cohen, M. H. 1977. *Can. J. Chem.* 55: 1906–15
9. Yakovlev, B. S. 1979. *Russ. Chem. Rev.* 48: 615–30
10. Yakovlev, B. S., Lukin, L. V. 1985. In *Advances in Chemical Physics*, ed. K. P. Lawley, 60: 99–160. Chichester: Wiley
11. Warman, J. M. 1982. In *The Study of Fast Processes and Transient Species by Electron Pulse Radiolysis* Baxendale, ed. J. H. Busi, pp. 433–533. Dordrecht, Holland: Reidel
12. Nyikos, L., Schiller, R. 1988. In *The Chemical Physics of Solvation, Part C*, ed. R. R. Dogonadze, E. Kalman, A. A. Kornyshev, J. Ulstrup. Amsterdam: Elsevier
13. Gee, N., Senanayake, P. C., Freeman, G. R. 1988. *J. Chem. Phys.* 89: 3710–17
14. Holroyd, R. A., Ehrenson, S., Preses, J. M. 1985. *J. Phys. Chem.* 89: 4244–49
15. Gonidec, A., Rubbia, C., Schinzel, D., Schmidt, W. F. 1985. CERN/UA1, Internal memo
16. Takeda, S. S., Houser, N. E., Jarnagin, J. 1971. *J. Chem. Phys.* 54: 3195–3206
17. Sakai, Y., Böttcher, H., Schmidt, W. F. 1982. *J. Electrostat.* 12: 89–96
18. Sakai, Y., Böttcher, H., Schmidt, W. F. 1983. *J. Inst. Electr. Engr. Jpn. A* 61: 499–506
19. Gushchin, E. M., Kruglov, A. A., Lebedev, A. N., Obodovski, I. M. 1980. *Sov. Phys. JETP* 51: 775–79
20. Gushchin, E. M., Kruglov, A. A., Obodovskii, I. M. 1982. *Sov. Phys. JETP* 55: 650–55
21. Huang, S. S.-S., Freeman, G. R. 1981. *Phys. Rev. A* 24: 714–24
22. Jacobson, F. M., Gee, N., Freeman, G. R. 1986. *Phys. Rev. A* 34: 2329–35
23. Huang, S. S.-S., Freeman, G. R. 1978. *J. Chem. Phys.* 68: 1355–62
24. Buckley, E., Campanella, M., Carugno, G., Cattadoria, C., Gonidec, A., et al. 1989. *Nucl. Instrum. Methods A* 275: 364–72
25. Floriano, M. A., Freeman, G. R. 1986. *J. Chem. Phys.* 85: 1603–12
26. Floriano, M. A., Gee, N., Freeman, G. R. 1987. *J. Chem. Phys.* 87: 4829–34
27. Gyorgy, I., Freeman, G. R. 1979. *J. Chem. Phys.* 70: 4769–77
28. Holroyd, R. A. Cipollini, N. E. 1979. In *Proc. 6th Int. Congr. Radiat. Res.*, pp. 228–35. Tokyo: Toppan Print.
29. Döldissen, W., Schmidt, W. F., Bakale, G. 1980. *J. Phys. Chem.* 84: 1179–86
30. Itoh, K., Nakagawa, K., Nishikawa, M. 1988. *Radiat. Phys. Chem.* 32: 221–25
31. Itoh, K., Nakagawa, K., Nishikawa, M. 1986. *J. Chem. Phys.* 84: 391–94

32. Nishikawa, M., Holroyd, R. A. 1982. *J. Chem. Phys.* 77: 4678–82
33. Nishikawa, M., Holroyd, R. A., Sowada, U. 1980. *J. Chem. Phys.* 72: 3081–84
34. Lekner, J. 1967. *Phys. Rev.* 158: 130–37
35. Nakagawa, K., Itoh, K., Nishikawa, M. 1987. *Chem. Phys. Lett.* 137: 458–61
36. Nakagawa, K., Itoh, K., Nishikawa, M. 1988. *IEEE Trans. Electr. Insul.* 23: 509–14
37. Asaf, U., Reininger, R., Steinberger, I. T. 1983. *Chem. Phys. Lett.* 100: 363–66
38. Reininger, R., Asaf, U., Steinberger, I. T., Basak, S. 1983. *Phys. Rev. B* 28: 4426–32
39. Reininger, R., Asaf, U., Steinberger, I. T. 1982. *Chem. Phys. Lett.* 90: 287–90
40. Springett, B. E., Jortner, J., Cohen, M. H. 1968. *J. Chem. Phys.* 48: 2720–31
41. Plenkiewicz, B., Jay-Gerin, J.-P., Plenkiewicz, P., Bachelet, G. B. 1986. *Europhys. Lett.* 1: 455–60
42. Plenkiewicz, B., Plenkiewicz, P., Jay-Gerin, J.-P. 1988. *Abstr. Int. Workshop Liquid State Electronics*, p. 123. Berlin
43. Muñoz, R. C., Holroyd, R. A., Nishikawa, M. 1985. *J. Phys. Chem.* 89: 2969–72
44. Muñoz, R. C., Holroyd, R. A. 1986. *J. Chem. Phys.* 84: 5810–15
45. Muñoz, R. C., Holroyd, R. A., Itoh, K., Nakagawa, K., Nishikawa, M., Fueki, K. 1987. *J. Phys. Chem.* 91: 4639–43
46. Holroyd, R. A. 1989. *Mol. Cryst. Liq. Cryst.* In press
47. Muñoz, R. C., Holroyd, R. A. 1987. *Chem. Phys. Lett.* 137: 250–54
48. Basak, S., Cohen, M. H. 1979. *Phys. Rev. B* 20: 3404–20
49. Holroyd, R. A., Gangwer, T. E., Allen, A. O. 1975. *Chem. Phys. Lett.* 31: 520–23
50. Warman, J. M., deHaas, M. P., Zador, E., Hummel, A. 1975. *Chem. Phys. Lett.* 35: 383–86
51. Holroyd, R. A. 1977. *Ber. Bunsenges. Phys. Chem.* 81: 298–304
52. Nishikawa, M., Itoh, K., Holroyd, R. A. 1988. *J. Phys. Chem.* 92: 5262–66
53. Gee, N., Floriano, M. A., Wada, T., Huang, S. S.-S., Freeman, G. R. 1985. *J. Appl. Phys.* 57: 1097–1101
54. Wada, T., Freeman, G. R. 1980. *J. Chem. Phys.* 72: 6726–30
55. Ramanan, G., Freeman, G. R. 1988. *Can. J. Chem.* 66: 1304–7
56. van den Ende, C. A. M., Nyikos, L., Warman, J. M., Hummel, A. 1982. *Radiat. Phys. Chem.* 19: 297–308
57. Warman, J. M. 1987. *Proc. 8th Int. Congr. Radiat. Res.* 2: 272–77. London: Taylor & Francis
58. Gee, N., Freeman, G. R. 1989. *J. Chem. Phys.* 90: 5399–5405
59. Warman, J. M., deHaas, M. P., Hummel, A. 1987. *Proc. 9th Int. Conf. Dielectric Liquids, Salford, England*, pp. 44–49
60. Hummel, A., Luthjens, L. H. 1986. *J. Radioanal. Nucl. Chem.* 101: 293–97
61. Sauer, M. C., Schmidt, K. H. 1988. *Radiat. Phys. Chem.* 32: 281–85
62. Trifunac, A. D., Sauer, M. C., Jonah, C. D. 1985. *Chem. Phys. Lett.* 113: 316–19
63. Balakin, A. A., Yakovlev, B. S. 1979. *Chem. Phys. Lett.* 66: 299–302
64. Lukin, L. V., Tolmachev, A. V., Yakovlev, B. S. 1987. *High Energy Chem. (USSR)* 21: 357–61
65. Braun, C. L., Scott, T. W. 1983. *J. Phys. Chem.* 87: 4776–78
66. Scott, T. W., Braun, C. L. 1986. *Chem. Phys. Lett.* 127: 501–4
67. Lukin, L. V., Tolmachev, A. V., Yakovlev, B. S. 1983. *Chem. Phys. Lett.* 99; 16–20
68. Yamada, S., Yoshida, S., Kawazumi, H., Nagamura, T., Ogawa, T. 1985. *Chem. Phys. Lett.* 122: 391–94
69. Schmidt, K. H. 1983. *Chem. Phys. Lett.* 103: 129–32
70. Novikov, G. F., Yakovlev, B. S. 1985. *High Energy Chem. (USSR)* 19: 226–32
71. Tachiya, M. 1988. *Radiat. Phys. Chem.* 32: 37–41
72. Beck, G. 1979. *Rev. Sci. Instrum.* 50: 1147–50
73. Muñoz, R. C., Ascarelli, G. 1983. *Phys. Rev. Lett.* 51: 215–18
74. Muñoz, R. C., Ascarelli, G. 1983. *Chem. Phys. Lett.* 94: 235–39
75. Muñoz, R. C., Ascarelli, G. 1984. *J. Phys. Chem.* 88: 3712–15
76. Muñoz, R. C. 1988. *Radiat. Phys. Chem.* 32: 169–76
77. Dodelet, J.-P., Freeman, G. R. 1977. *Can. J. Chem.* 55: 2264–77
78. Itoh, K., Muñoz, R. C., Holroyd, R. A. 1989. *J. Chem. Phys.* 90: 1128–32
79. Sowada, U., Warman, J. M., deHaas, M. P. 1982. *Phys. Rev. B* 25: 3434–37
80. Kubota, S., Takahashi, T., Ruan, J. 1982. *J. Phys. Soc., Jpn.* 51: 3274–77
81. Warman, J. M. 1981. *Radiat. Phys. Chem.* 17: 21–24
82. Shibamura, E., Takahashi, T., Kubota, S., Doke, T. 1979. *Phys. Rev. A* 20: 2547–54
83. Bakale, G., Schmidt, W. F. 1982. *J. Electrostatics* 12: 103–6

84. Bakale, G., Schmidt, W. F. 1981. *Z. Naturforsch. Teil A* 36: 802–6
85. Tachiya, M. 1987. *J. Chem. Phys.* 87: 4622–26
86. Bakale, G., Beck, G. 1986. *J. Chem. Phys.* 84: 5344–50
87. Wada, T., Shinsaka, K., Namba, H., Hatano, Y. 1977. *Can. J. Chem.* 55: 2144–55
88. Nakamura, Y., Namba, H., Shinsaka, K., Hatano, Y. 1980. *Chem. Phys. Lett.* 76: 311–14
89. Nakamura, Y., Shinsaka, K., Hatano, Y. 1983. *J. Chem. Phys.* 78: 5820–24
90. Shinsaka, K., Codama, M., Srithanratana, T., Yamamoto, M., Hatano, Y. 1988. *J. Chem. Phys.* 88: 7529–36
91. Tachiya, M. 1987. *J. Chem. Phys.* 87: 4108–13
92. Shinsaka, K., Codama, M., Nakamura, Y., Serizawa, K., Hatano, Y. 1989. *Radiat. Phys. Chem.* In press
93. Morgan, W. L. 1986. *J. Chem. Phys.* 84: 2298–2303
94. Dolgoshein, B. A., Lebedenko, V. N., Rodinov, B. U. 1970. *JETP Lett.* 11: 351–53
95. Schoepe, W., Rayfield, C. W. 1973. *Phys. Rev. A* 7: 2111–21
96. Boriev, I. A., Balakin, A. A., Yakovlev, B. S. 1978. *High Energy Chem.* 12: 16–20
97. Gushchin, E. M., Kruglov, A. A., Obodovskii, I. M. 1982. *Sov. Phys. JETP* 55: 860–62
98. Anderson, D. F., Charpak, G., Holroyd, R. A., Lamb, D. C. 1987. *Nucl. Instrum. Methods A* 261: 445–48
99. Peskov, V., Charpak, G., Miné, P., Sauli, F., Scigocki, D., et al. 1988. *Nucl. Instrum. Methods A* 269: 149–60
100. Schmidt, W. F. 1977. *Can. J. Chem.* 55: 2197–2210
101. Cohen, M. H., Lekner, J. 1967. *Phys. Rev.* 158: 305–9
102. Baird, J. K. 1985. *Phys. Rev. A* 32: 1235–36
103. Polischuk, A. Y. 1984. *J. Phys. B* 17: 4789–95
104. Nishikawa, M. 1985. *Chem. Phys. Lett.* 114: 271–73
105. Ascarelli, G. 1986. *Phys. Rev. B* 33: 5825–33
106. Ascarelli, G. 1986. *Phys. Rev. B* 34: 4278–88
107. Watanabe, S. 1980. *J. Phys. Soc. Jpn.* 49: 38–42; 1981. 50: 1095–1102
108. Vertes, A. 1984. *J. Phys. Chem.* 88: 3722–26
109. Feynman, R. P. 1955. *Phys. Rev.* 97: 660–65
110. Chandler, D., Singh, Y., Richardson, D. M. 1984. *J. Chem. Phys.* 81: 1975–82
111. Nichols, A. L. III, Chandler, D., Singh, Y., Richardson, D. M. 1984. *J. Chem. Phys.* 81: 5109–16
112. Nichols, A. L. III, Chandler, D. 1986. *J. Chem. Phys.* 84: 398–403
113. Nichols, A. L. III, Chandler, D. 1987. *J. Chem. Phys.* 87: 6671–81
114. Sprik, M., Klein, M. L., Chandler, D. 1985. *J. Chem. Phys.* 83: 3042–49
115. Laria, D., Chandler, D. 1987. *J. Chem. Phys.* 87: 4088–92
116. Coker, D. F., Berne, B. J. 1988. *J. Chem. Phys.* 89: 2128–37
117. Coker, D. F., Berne, B. J., Thirumalai, D. J. 1987. *J. Chem. Phys.* 86: 5689–5702
118. Gee, N., Freeman, G. R. 1979. *Phys. Rev. A* 20: 1152–61
119. Martyna, G. J., Berne, B. J. 1988. *J. Chem. Phys.* 88: 4516–25
120. Martyna, G. J., Berne, B. J. 1989. *J. Chem. Phys.* 90: 3744–55
121. Atrazhev, V. M., Iakubov, I. T. 1981. *J. Phys. C* 14: 5139–50
122. Atrazhev, V. M., Dmitriev, E. G. 1986. *J. Phys. C* 19: 4329–38
123. Atrazhev, V. M., Dmitriev, E. G. 1985. *J. Phys. C* 18: 1205–15
124. Sakai, Y., Nakamura, S., Tagashira, H. 1985. *IEEE Trans. Electr. Insul.* 20: 133–37
125. Nakamura, S., Sakai, Y., Tagashira, H. 1986. *Chem. Phys. Lett.* 130: 551–54
126. Kaneko, K., Usami, Y., Kitihara, K. 1988. *J. Chem. Phys.* 89: 6420–26
127. Ascarelli, G. 1980. *J. Phys. Chem.* 84: 1193–96
128. Freeman, G. R. 1981. *J. Chem. Phys.* 74: 3079–81
129. Schmidt, W. F., Sowada, U., Yoshino, K. 1981. *J. Chem. Phys.* 74: 3081–82
130. Leycuras, A., Larour, J. 1982. *J. Phys. B* 15: 6765–77
131. Larour, J. 1986. *Physica* 139/140B: 87–89
132. Baird, J. K. 1983. *J. Chem. Phys.* 79: 316–20
133. Reininger, R., Steinberger, I. T., Bernstoff, S., Saile, V., Laporte, P. 1984. *Chem. Phys.* 86: 189–98
134. Asaf, U., Steinberger, I. T. 1974. *Phys. Rev. B* 10: 4464–68
135. Iakubov, I. T., Khrapak, A. G. 1982. *Rep. Prog. Phys.* 45: 697–751
136. Ebner, C., Punyatiya, C. 1979. *Phys. Rev. A* 19: 856–65
137. Kuan, D.-Y., Ebner, C. 1981. *Phys. Rev. A* 23: 285–93
138. Hernandez, J. P. 1982. *J. Phys. C* 15: 1923–37
139. Smith, D. W., Hernandez, J. P. 1982. *J. Chem. Phys.* 77: 5802–6

140. Berlin, Y. A., Nyikos, L., Schiller, R. 1978. *J. Chem. Phys.* 69: 2401–6
141. Vertes, A. 1983. *J. Chem. Phys.* 79: 5558–62
142. Berlin, Y. A., Schiller, R. 1987. *Radiat. Phys. Chem.* 30: 71–73
143. Baird, J. K., Rehfeld, R. H. 1987. *J. Chem. Phys.* 86: 4090–95
144. Schiller, R., Nyikos, L. 1980. *J. Chem. Phys.* 72: 2245–49
145. Schiller, R., Vertes, A., Nyikos, L. 1982. *J. Chem. Phys.* 76: 678–83
146. Funabashi, K. 1982. *J. Electrostat.* 12: 65–71
147. Bowman, R. M., Lu, H., Eisenthal, K. B. 1988. *J. Chem. Phys.* 89: 606–8
148. Gaudel, Y., Migus, A., Antonetti, A. 1988. In *Chemical Reactivity in Liquids Fundamental Aspects.* New York: Plenum
149. Albrow, M. G., Apsimon, R., Aubert, B., Bacci, C., Bezaquet, A., et al. 1988. *Nucl. Instrum. Methods A* 265: 303–18
150. Meertens, H., van Herk, M., Weeda, J. 1985. *Phys. Med. Biol.* 30: 313–21
151. Aprile, E., Giboni, K. L., Rubbia, C. 1987. *Nucl. Instrum. Methods A* 253: 273–77
152. Bonetti, S., Braggiotti, A., Buckley, E., Campanella, M., Carugno, G., et al. 1989. *Nucl. Instrum. Methods.* In press

THEORETICAL METHODS FOR ROVIBRATIONAL STATES OF FLOPPY MOLECULES

Zlatko Bačić

Department of Chemistry, New York University, 4 Washington Place, New York, New York 10003

John C. Light

The Department of Chemistry and The James Franck Institute, The University of Chicago, Chicago, Illinois 60637

INTRODUCTION

Theoretical calculation of rotation-vibration energy levels of polyatomic molecules is a topic with a long history, characterized by a close, symbiotic relationship with molecular spectroscopy on one side and quantum chemistry on the other. What brings them together is the notion of the potential energy surface, which plays a central role in our understanding of the molecular structure and dynamics. In the case of polyatomic molecules, the experimental spectra cannot be inverted directly to yield potential surfaces [see Ref. (1) for some recent efforts within the semiclassical SCF approach], but they do provide a stringent test for the theoretically obtained potential surfaces and observables derived from them. These surfaces, usually from ab initio calculations, seldom meet the standards of spectroscopic accuracy, especially if more extended, high-energy regions are of interest. The only practical way available to test and improve them is by comparing the calculated and the experimental spectra, and minimizing the difference between the two.

The subject of the theoretical treatment of coupled molecular vibrations has undergone a real renaissance in the past decade. Significant conceptual advances have been made, particularly concerning highly vibrationally and rotationally excited states, and a variety of new computational approaches have been developed. The recent thematic issues of *Computer Physics Communication* (2) and *Faraday Transactions II* (3) testify to the vitality of this research area.

The resurgence of interest in molecular eigenstates owes a great deal to the extraordinary progress in the experimental techniques such as laser spectroscopy and supersonic molecular beams. This has made possible direct probing, in both time and frequency domains, of entirely new classes of molecules and regimes of rovibrational excitation previously beyond reach.

Floppy molecules, i.e. molecules having one or more large amplitude motion (LAM) vibrations, have attracted an exceptional amount of attention from experimentalists and theorists alike. Most of the effort has been focused on the weakly bound, van der Waals and hydrogen-bonded molecular complexes. Reviews by Nesbitt (4) and Miller (5, 6) summarize the experimental methods, notably the high-resolution infrared (IR) spectroscopy, used to study the weakly bound complexes, as well as the wealth of spectroscopic data concerning their structure and vibrational predissociation dynamics. Weakly bound complexes are by no means the only ones having LAM vibrations. In his review on the IR laser spectroscopy of molecular ions (7), Saykally discusses a number of intriguing polyatomic cations that exhibit a complicated interplay of LAM vibrations and tunneling.

LAM vibrations can be found in the more conventional, strongly bound molecules as well, provided that their vibrational modes are sufficiently highly excited, so that significant portions of the potential surface far from the equilibrium are accessible. Preparation and probing of well-defined highly excited vibrational states can now be achieved with many techniques, such as overtone vibration excitation and IR multiphoton excitation (8), proton energy loss spectroscopy (9), IR spectroscopy of molecular species (in particular H_3^+ and its isotopomers) in ion beams (10), stimulated emission pumping (11), and others (12).

The reasons for the exceptional interest in highly vibrationally and rotationally excited molecules and their large amplitude motions are many. The LAM vibrational states, because of the delocalized nature of their wave functions, contain detailed information about large regions of the potential surface beyond the global minimum. Moreover, while executing LAM vibrations, molecules can populate high-energy local minima inaccessible at lower energies, thus permitting detection of new isomers with strange structures and dynamics. In the case of weakly bound complexes, given their shallow minima and low isomerization barriers, all of this can be achieved by exciting the low-frequency van der Waals modes.

On a more fundamental level, studying excited floppy molecules is challenging and rewarding because it forces us to reexamine, modify, and often abandon many of the basic concepts, formalisms, and computational methods applicable to low-energy, small amplitude vibrations (13) but that become inappropriate with increasing excitation energy and amplitudes of vibrational motions. For example, assigning quantum numbers to vibrational eigenstates—a task that is at the heart of the traditional spectroscopic analysis and depends on some approximate mode separability—becomes increasingly difficult and eventually impossible. It is in this regime that experiments come in contact with the question of classical and quantum chaos, which has certainly been one of the most visible and most intensely studied theoretical issues in the past decade [for excellent overviews see Refs. (14, 14a, 15)]. As discussed by Hamilton, Kinsey & Field (11) in their review article on stimulated emission pumping (SEP), highly vibrationally excited molecules, especially ones with LAM vibrations, offer one of the best opportunities for finding experimental manifestations of quantum chaos in molecular spectroscopy or dynamics (11, 16, 17).

Finally, the high-lying LAM vibrational states in many ways fall on the borderline between spectroscopy and chemical dynamics. The energy range in which they lie and the parts of potential surfaces over which they extend are relevant for unimolecular reactions, intramolecular vibrational redistribution (IVR), multiphoton excitation, and laser-molecule interactions in general. Accurate characterization and computation of the LAM rovibrational eigenstates is essential for the quantitative understanding of these and related processes.

The distinct features of the excited LAM vibrational states that make them so interesting and important, like the delocalization over large, anharmonic regions of the potential surfaces and strong coupling to other modes, make the task of calculating their energies and wave functions very difficult. The problem becomes even more complicated if one wants to compute a large number (50–100 or more) of highly excited LAM vibrational states, especially for potential surfaces with several minima separated by isomerization barriers. Approaches based on Watson's normal mode Hamiltonians for bent (18) and linear molecules (19), perhaps best exemplified by the variational method of Whitehead & Handy (20), have been widely used for calculating the low-lying rovibrational states of triatomic molecules. However, the first applications of this method to floppy molecules like CH_2^+ (21, 22) suffered from extremely slow convergence, and even divergence, of the bending energy levels with increasing basis, thus demonstrating clearly that normal coordinates are not appropriate for describing LAM vibrations. Since then, a large amount of effort has been devoted to devising faster and more accurate methods for determining the rovibrational eigenstates of molecules with LAM vibrations.

This review focuses exclusively on excited large amplitude vibrations. We cover only those advances made in the past couple of years which have been aimed at and tested on floppy polyatomic molecules or weakly bound complexes. Consequently, this is not a comprehensive review of all theoretical methods for finding molecular eigenstates. Also, we do not discuss here the (non-, semi-) rigid bender method of Bunker and co-workers, the first approach designed specifically for large amplitude vibrations. This important work has been thoroughly reviewed by Bunker (23) and Jensen (24). Information about the current status of this method can be found in some recent references (25). The present review is largely, though not exclusively, devoted to triatomic molecules, reflecting the fact that most of the methodological advances and the vast majority of applications have dealt with systems of this size.

Preparation of this article has been made much easier by the existence of two excellent earlier reviews, devoted to topics closely related to our own. The first one, by Carter & Handy (26), reviews in broad terms the status of the variational method for calculating the rovibrational levels of molecules, up to 1986. Emphasis is on the techniques that should work best for the potentials with a deep, single minimum, allowing small to medium amplitude motions. The second review, by Tennyson (27), concentrates on the variational method developed by him and Sutcliffe for calculating the rovibrational spectra of floppy triatomics using the scattering coordinates. Many developments discussed in the present review use some aspects of this work. Since both reviews appear in 1986, we confine ourselves to the literature published since then.

COORDINATE SYSTEMS

The first step in formulating any theoretical treatment of the bound state problem is choosing between space-fixed and body-fixed coordinates. Although the space-fixed coordinates were used in some of the early work on floppy molecules (28, 29), all currently used methods employ body-fixed frames. The body-fixed representation, as recognized long ago in the theory of molecular collisions (30, 31), leads to a description that is simpler than in the space-fixed frame, and allows introduction of several decoupling approximations.

Having decided on the body-fixed frame, one then faces the problem that there is no unique way of fixing the coordinate axes to the molecule. Considerable attention has been given to this issue. For lack of space we do not address it here but refer the reader to the review by Tennyson (27) and the papers by Tennyson, Sutcliffe and co-workers, who have considered various coordinate embeddings for triatomic (32, 33) and general polyatomic molecules (34).

After elimination of the center-of-mass coordinates and settling on some definition of the body-fixed axes, 3N-6 internal, purely vibrational degrees of freedom remain for a N atom molecule. For the low-lying, small amplitude vibrations, where the normal mode description (13, 18, 19) is adequate, there is no ambiguity concerning the choice of coordinates in which to perform the energy level calculations. The presence of LAM vibrations changes the situation profoundly. There, one realizes soon that the choice of optimal (or at least good) coordinates depends very much on the particular molecule considered, and that there is no single coordinate system that would be satisfactory for all triatomics, let alone larger molecules. Experience has shown that for any given molecule, the convergence rates and the overall quality of the results, whether obtained by the variational or some approximate method, depend strongly on the coordinates used in the calculation. Therefore, we discuss this problem in some length, for the case of triatomic molecules.

Which criteria should an optimal set of coordinates satisfy? First of all, it must span all of configuration space accessible to the system of interest, regardless of the amplitude of nuclear motion. Second, the coordinates should allow taking advantage, in a simple way, of the highest symmetry present in the system. Third, it is desirable that the coordinates be orthogonal, since that minimizes the number of cross terms in the kinetic energy operator, thus making the Hamiltonian simpler to evaluate. In addition, orthogonal coordinates facilitate partitioning of the full Hamiltonian into Hamiltonians of lower dimensionality, which plays an important role in some of the computational strategies discussed below. Finally, the coordinates should be chosen so as to minimize the mode-mode interaction, i.e. make the Hamiltonian as separable as possible. For more discussion of optimal coordinates we recommend the papers by Wallace (35, 36).

Several coordinate systems have been used in the calculations of the rovibrational energy levels of floppy triatomic molecules:

1. *Jacobi, or scattering coordinates*, R, r *and* θ: For a triatomic molecule A–BC, r is the length of BC diatomic bond, R is the distance of atom A to the center of mass of BC, whereas θ is the angle between R and r. Jacobi coordinates are suitable for describing atom-diatom complexes with two LAM vibrations, and for isomerizing systems. For this reason, they have dominated the calculations of floppy molecules. A mass-scaled version of Jacobi coordinates has existed for a long time (37) and has been extensively

used in molecular scattering (e.g. 38). It has only recently been used in (variational) bound state calculations (39, 40). A disadvantage of Jacobi coordinates is that with them it is difficult to take into account symmetries higher than C_{2v}, e.g. D_{3h}, present in A_3 systems.

2. *Hyperspherical coordinates*: A number of related hyperspherical coordinate systems have been developed. The one we mention here is by Pack (41, 42). It consists of a hyper-radius ρ, which defines the overall size of the system, and two hyper-angles, θ and χ, which describe its shape. This coordinate system and its relationship to Jacobi and other hyperspherical coordinates are discussed in detail by Pack & Parker (42). Papers by Johnson (43–45) are also highly recommended. Slightly different hyperspherical coordinates have been used by Frey & Howard (45a) to calculate the ground state energy of some van der Waals trimers. Hyperspherical coordinates, in contrast to Jacobi, do allow use of the full D_{3h} symmetry. Consequently, they are probably the coordinates of choice for highly symmetric systems, such as H_3^+.

3. *Internal, bond coordinates*, R_1, R_2 *and* θ: Here, R_1 and R_2 are the distances from the central atom to the other two atoms, and θ is the angle between the two bond lengths R_1 and R_2. No bound state calculations of floppy molecules have been reported by using these coordinates. But, since there is nothing that would a priori preclude their applications to LAM vibrations, we include them in this list. The internal coordinates are not orthogonal. They are not suitable for atom-diatom complexes or surfaces with multiple minima. Carter & Handy (26) review their variational procedure based on the R_1, R_2, θ coordinates, and numerous applications to the more rigid, strongly bound triatomics.

4. *Radau coordinates*: The two radial, stretching coordinates R_1 and R_2 measure the distance of the two light atoms from the so-called canonical point, which for a light-heavy-light system (e.g. H_2O) lies very close to the center of the heavy atom. The third coordinate θ represents the angle between R_1 and R_2. For the precise definition, the history, and other aspects of Radau coordinates, the paper by Johnson & Reinhardt (46) should be consulted. Radau coordinates are orthogonal; they may be loosely thought of as the orthogonal version of the bond coordinates, especially when light-heavy-light triatomics are considered.

In addition to the coordinate systems discussed above, we mention that Sutcliffe & Tennyson (47) have formulated a generalized coordinate system for triatomic molecules, which includes both Jacobi and bond coordinates as limiting cases. Also, an alternative set of internal coordinates and the corresponding Hamiltonian, applicable to floppy molecules, have been derived by Secrest and co-workers (48, 49).

HAMILTONIANS

For any chosen set of internal coordinates $\{Q_i\}$, the molecular Hamiltonian H can be written in the form (26, 27)

$$H = T_V + T_{VR} + V(\mathbf{Q}), \qquad 1.$$

where T_V is the purely vibrational kinetic energy operator, T_{VR} is the vibration-rotation kinetic energy operator, and $V(\mathbf{Q})$ represents the potential. For a nonrotating ($J = 0$) molecule, T_{VR} is zero. Derivation of the kinetic energy operators T_V and T_{VR} appearing in Eq. 1, in any coordinate system, is in principle straightforward. It is accomplished by the well-known Podolsky transformation (50). Alternatively, one can use the recently developed and increasingly popular method of Sutcliffe (51). Both approaches involve a lot of tedious algebra. To avoid errors in the derivation, especially for tetratomic and larger molecules, computer codes for symbolic, algebraic manipulation are beginning to be used (52).

Expressions for the full rotation-vibration polyatomic Hamiltonians tend to be quite long, so we prefer not to give them here. The review by Carter & Handy (26) contains detailed derivations of triatomic Hamiltonians in normal and internal (bond) coordinates. Tennyson (27) derives the Hamiltonian in Jacobi coordinates; it is also given by Carter & Handy (26). The triatomic $J = 0$ Hamiltonian in Radau coordinates is presented in the paper by Johnson & Reinhardt (46). Those interested in the Hamiltonian in hyperspherical coordinates are referred to Pack & Parker (42) and Johnson (43–45).

We give here only the $J = 0$ portion of the full triatomic rotation-vibration Hamiltonian in Jacobi coordinates, derived by Tennyson & Sutcliffe (32), since it is essential for our discussion in the later sections. Also, by far the largest number of calculations on floppy molecules, by any of the methods discussed in the present review, have been performed with the Hamiltonian in Jacobi coordinates. For a triatomic molecule A–BC, the $J = 0$ Hamiltonian in Jacobi coordinates can be written as

$$H = -\frac{\hbar^2}{2\mu_1 R^2}\frac{\partial}{\partial R}\left(R^2\frac{\partial}{\partial R}\right) - \frac{\hbar^2}{2\mu_2 r^2}\frac{\partial}{\partial r}\left(r^2\frac{\partial}{\partial r}\right)$$
$$-\frac{\hbar^2}{2}\left(\frac{1}{\mu_1 R^2} + \frac{1}{\mu_2 r^2}\right)\frac{1}{\sin\theta}\frac{\partial}{\partial \theta}\left(\sin\theta\frac{\partial}{\partial \theta}\right) + V(R, r, \theta), \qquad 2.$$

where R is the distance from atom A to the BC center-of-mass, r is the BC bond length, and θ is the angle enclosed by R and r. The reduced masses are

$$\mu_1 = m_A m_{BC}/(m_A + m_{BC}), \qquad \mu_2 = m_B m_C/(m_B + m_C). \qquad 3.$$

The volume element for this Hamiltonian is $d\tau = R^2 r^2 \sin\theta \times dR\,dr\,d\theta$.

For the purpose of later discussion, we find it convenient to partition the Hamiltonian in Eq. 2 as

$$H = h(R,r) + G(R,r)j^2 + V(R,r,\theta), \qquad 4.$$

where $h(R,r)$ stands for the first two partial derivatives in R and r in Eq. 2, $G(R,r) = (\hbar^2/2)(1/\mu_1 R^2 + 1/\mu_2 r^2)$, and j^2 is the angular momentum operator multiplying $G(R,r)$.

Theoretical investigations of floppy molecules with more than three atoms are just beginning. Brocks et al (34) have formulated a Hamiltonian for two interacting polyatomic fragments, as yet not used in any calculation. Treatments of dimers consisting of diatomic molecules (treated as rigid rotors) have been reported (53–55). Hamiltonians appropriate for van der Waals complexes consisting of an atom and a polyatomic molecule (whose vibrational modes are frozen) have been developed (56, 57). Handy (52) has developed Hamiltonians in the internal, bond coordinates for the formaldehyde- and acetylene-like tetratomic molecules.

THE COUPLED-CHANNEL METHOD

Rotation-vibration energy levels of polyatomic molecules can be calculated accurately, without making any simplifying assumptions, by methods that belong to two basic categories. Methods in the first category, which are the subject of the following section, use L^2 basis set representation for all internal degrees of freedom. The desired eigenstates are usually obtained by diagonalizing the Hamiltonian matrix constructed in this basis.

In this section we review the second general approach, which relies on scattering theory. Here, one coordinate is identified as the scattering coordinate R, the remaining degrees of freedom are expanded in a L^2 basis. The Schrödinger equation can then be written as a set of coupled differential equations in R. These are solved with standard techniques of scattering theory, but are subject to the bound state boundary conditions. We illustrate this approach by considering, for simplicity, the $J = 0$ triatomic Hamiltonian in Eq. 2. We rewrite it as

$$H = -\frac{\hbar^2}{2\mu_1 R^2}\frac{\partial}{\partial R}\left(R^2 \frac{\partial}{\partial R}\right) + U(R,\mathbf{X}), \qquad 5.$$

where $U(R,\mathbf{X})$ contains all of the terms of Eq. 2 except for the partial derivative in R, and $\mathbf{X}$ stands for the r and θ coordinates. We want to find the bound state solutions of the Schrödinger equation

$(H - E_i)\Psi_i = 0.$ 6.

The wave function Ψ_i is expanded as

$$\Psi_i(R, \mathbf{X}) = R^{-1} \sum_n g_n^i(R) \Phi_n(\mathbf{X}).$$ 7.

The functions $\{\Phi_n(\mathbf{X})\}$ represent a basis set for the r, θ degrees of freedom, labelled by the collective index n. Substitution of Eq. 7 into Eq. 6, followed by premultiplication by $\Phi_{n'}(\mathbf{X})^*$ and integration over $\mathbf{X}$, results in the standard set of coupled equations. They can be written in matrix form as

$$\left[-\frac{\hbar^2}{2\mu_1} \frac{d^2}{dR^2} \mathbf{I} + \mathbf{W}(R) - E_i \mathbf{I} \right] \mathbf{g}(R) = 0.$$ 8.

Here, $\mathbf{I}$ is the unit matrix and

$$[\mathbf{W}(R)]_{n',n} = \int \Phi_{n'}(\mathbf{X})^* U(R, \mathbf{X}) \Phi_n(\mathbf{X}) \, d\mathbf{X}.$$ 9.

The coupled-channel equations, such as these in Eq. 8, can be integrated numerically by a variety of methods (54, 58–62), mostly developed for scattering calculations, subject to the boundary conditions $\mathbf{g}(R) \sim 0$ for $R \to 0$ and $R \to \infty$. Usually, the trial eigenvalue, E_i, is varied until the two solutions satisfying boundary conditions at $R \to 0$ and $R \to \infty$ match satisfactorily at some intermediate R, defining E_i and Ψ_i. In the more general $J > 0$ case, the coupled equations in Eq. 8 become more complicated (30, 31), but the methods for solving them remain the same.

In spite of its ability to provide accurate energy levels, the coupled-channel method has been used in but a few bound state calculations of polyatomic molecules (61, 62). Its applications to the rovibrational spectra of floppy molecules have so far been limited to the H_2–H_2 dimer (55), with frozen H_2 bond lengths, and atom-diatom van der Waals complexes (54, 63, 64) in which the diatom has been treated as a rigid rotor. We draw attention to Hutson's calculation of the rovibrational spectrum of Ar–HCl (63). It led, by means of least-squares fitting to far-infrared laser (65) and microwave spectra (66) of Ar–HCl, to the determination of a new intermolecular potential surface for this complex.

No coupled-channel calculations have been reported involving highly excited vibrational states of triatomic molecules with LAM vibrations (whether van der Waals complexes or isomerizing, strongly bound molecules), with all modes included. Therefore, the performance of the coupled-channel method relative to some of the variational methods remains to be established.

OUTLINE OF THE VARIATIONAL APPROACH

The variational method (26, 27, 67) in its basic form is readily defined. For a given rotation-vibration Hamiltonian H and suitably chosen expansion functions $\{\Xi_i\}$, a set of algebraic equations

$$\sum_{j=1}^{N} \langle \Xi_i | H - E | \Xi_j \rangle c_j = 0 \quad i = 1, 2, \ldots, N \qquad 10.$$

is solved by diagonalization yielding the desired eigenvalues and eigenvectors. MacDonald's theorem (68) assures us that the eigenvalues obtained in this way are upper bounds to the corresponding exact eigenvalues.

The size of the rovibrational Hamiltonian matrix grows linearly with J. To prevent the calculation from becoming prohibitively time-consuming, variational treatments have generally been confined to low values of J. Recently, Tennyson & Sutcliffe (27, 69) have implemented a promising two-step procedure, which has already extended considerably the range of Js for which variational calculations are practical. In the first step, they generate eigenstates of the Hamiltonian in which only the off-diagonal Coriolis couplings are neglected (69). Due to the presence of the centrifugal distortion term in the first step, the resulting eigenstates provide a well-adapted and compact basis in which, in the second step, the full rovibrational Hamiltonian matrix is constructed and diagonalized (69).

The dimension of the Hamiltonian matrix to be diagonalized ultimately decides whether a certain rotation-vibration problem is amenable to variational treatment. The factor of crucial importance in determining the size of the Hamiltonian matrix is the choice of basis functions for expanding the internal degrees of freedom. When using Jacobi, bond or Radau coordinates mentioned above, the basis set typically consists of product functions of the form

$$H_m(R_1) F_n(R_2) \Theta_{jk}(\theta) D^J_{Mk}(\alpha, \beta, \gamma). \qquad 11.$$

In Equation 11, $H_m(R_1)$ and $F_n(R_2)$ represent some convenient one-dimensional (single-center), orthogonal functions of the radial coordinates, $\Theta_{jk}(\theta)$ is usually the associated Legendre polynomial, and D^J_{Mk} is the rotational (symmetric top) wave function. We denote this representation in terms of orthonormal L^2 basis functions as the finite basis representation, FBR. In many calculations dealing with low-lying, small amplitude vibrational states, the FBR proved satisfactory (26, 27). However, applications to the excited states of floppy molecules revealed serious deficiencies of the FBR in describing LAM vibrations. This prompted the development of a very different, pointwise representation of the internal degrees of freedom, the discrete variable representation, DVR (70). Several demanding calculations have demonstrated that the DVR is well suited for molecules with LAM vibrations. In the following two sections we focus on the methodology and applications of the variational approaches to floppy molecules in FBR and DVR, respectively.

VARIATIONAL CALCULATIONS IN THE FINITE BASIS REPRESENTATION

Before reviewing the applications of the variational approach in the FBR to floppy molecules, and the problems encountered, we discuss briefly various choices of angular and radial basis functions.

Angular Basis Functions

By far the most common choice for angular basis functions $\Theta_{jk}(\theta)$ in Eq. 11 are the associated Legendre polynomials or, for $J = 0$, the Legendre polynomials $P_j(\cos\theta)$ (26, 27). With this basis and the expansion of the potential in Legendre polynomials

$$V(R, r, \theta) = \sum_i V_i(R, r) P_i(\cos\theta), \qquad 12.$$

which is common in both molecular scattering (30, 31) and bound state calculations (26, 27), the angular integration required in evaluating potential matrix elements can be done analytically.

Legendre polynomials have some drawbacks as angular basis sets for general molecular potentials. One problem is that Legendre polynomials provide a "uniform" basis on $(0, \pi)$ and cannot be tailored or optimized easily for any given angular potential. One procedure often used to alleviate this problem is to diagonalize the Hamiltonian in Eq. 2 with the radial coordinates R and r set to their equilibrium values in the Legendre polynomial basis. The resulting eigenstates are used as the new angular (hindered rotor) basis, better adapted to the particular potential (26, 61, 62, 71). An alternative, implemented by Johnson & Reinhardt (46), was to use Jacobi polynomials (of which Legendre polynomials are a special case) as the angular basis. By a judicious choice of their parameters, Jacobi polynomials can be made to emphasize various regions of the angular coordinate (46).

Radial Basis Functions

The one-dimensional basis functions, denoted in Eq. 11 as $H_m(R_1)$ and $F_n(R_2)$, used for representing the radial degrees of freedom fall into two categories. Basis functions in the first category are numerical functions, obtained by numerical integration of a model (pseudo)diatomic radial Schrödinger equation. The model potential used in such calculations has often been defined by the first, isotropic term of the potential expansion in Eq. 12 (with $r = r_e$) (29, 72), or by choosing a particular angular cut through the potential surface $V(R, \theta)$ (setting θ to some fixed value θ_f) (73). In calculations on anisotropic atom-diatom complexes, both approaches have been found to be rather unsatisfactory, since neither could generate a sufficiently accurate and compact representation for the full problem.

In the second, at present much more popular, category are orthogonal polynomial basis functions, which are analytic solutions of model one-dimensional vibrational problems. Most commonly used are the Morse oscillator wave functions (26, 27). They are defined by the well-known Morse potential, whose parameters r_e, D_e, and ω_e can be optimized to produce a reasonably compact representation for a particular calculation. For molecules such as CH_2^+, for which the linear $R = 0$ structure is accessible, spherical oscillator-like functions have been used for the R coordinate (27, 49). Harmonic oscillator wave functions, although less suitable for excited, anharmonic vibrations, have also been employed in calculating the rovibrational spectra of triatomic van der Waals complexes (74).

Applications

Variational FBR calculations on floppy molecules up to 1986 have been reviewed by Tennyson (27) and, to a lesser extent, by Carter & Handy (26). We therefore confine our attention to the relevant work published since 1985.

The rovibrational spectrum of the H_3^+ molecule and its isotopomers has figured prominently in the work of Tennyson, Sutcliffe and co-workers. They have been particularly concerned with the highly rotationally excited states, as evidenced by the calculations on H_2D^+ for $J = 30$ (75) and $J = 11$, 15 (76), and also on H_3^+ and H_2D^+ for $J = 46$ and $J = 54$, respectively (77). These high J calculations, all performed with the two-step procedure (69), are important because they probe, for the first time, the regime of strong rotation-vibration coupling, where the level spacing within the rotational manifold becomes comparable to typical vibrational excitation. In addition, this group has reported rovibrational calculations of the overtone bands of H_3^+ (78) as well as calculations of the rovibrational transition frequencies and line strengths for H_3^+ (79) and H_2D^+ and D_2H^+ (80). Apart from H_3^+, rovibrational states of HeN_2^+ (frozen N_2^+ bond length) (81), and energy levels and transition intensities for Li_3^+ have been calculated (82).

Lee & Secrest have reported two calculations. In the first (82a), they have obtained the rovibrational states of He_2H^+ and have also investigated the possibility for vibrational bonding in this system. Their second paper (49) deals with the rotation-vibration states of CH_2^+.

Reid, Janda & Halberstadt (74), using model potential surfaces, have calculated the rovibrational energy levels and wave functions of the van der Waals complexes $X-Cl_2$ (X = He, Ne, Ar) (with frozen Cl_2), and compared them with the results of some simple models.

We are aware of only a few papers dealing with the rovibrational spectra of tetratomic and larger floppy molecules that have been published during the period covered by this review. In addition to the difficulty of handling a large number of internal degrees of freedom, a major obstacle to more widespread variational treatment of polyatomic complexes is the lack of potentials accurate enough to justify the effort. Brocks & Huygen (56), using an empirical intermolecular potential and employing many of the computational techniques developed for the triatomic systems, have calculated the low-lying van der Waals modes of the atom-molecule complexes Ar-benzene and Ar-tetrazine. Brocks & van Koeven (57) have introduced a method for calculating vibrational and rotational van der Waals states of atom-large molecule complexes, and have tested it on Ar-fluorene.

Leutwyler & Boesiger (57a) have investigated the microsolvent clusters MR_n, consisting of a large aromatic molecule M such as perylene, tetracene, or pentacene, and n rare gas atoms R. The potential used in this study was represented in terms of pair-wise atom-atom interactions between C and H atoms of M and the rare gas atoms. M was treated as a rigid body with no internal degrees of freedom. The energy levels of the intermolecular, van der Waals modes were mostly calculated by solving 1D vibrational Schrödinger equations separately along x, y, and z coordinates.

Problems

Introduction of novel coordinate systems, basis sets, and innovative computational strategies has made new classes of rotationally and vibrationally excited polyatomic molecules amenable to rigorous, variational treatment. At the same time, extensive applications to floppy molecules have shown that the finite basis representation has several shortcomings. One is slow convergence of the excited LAM vibrational states with respect to the FBR consisting of 1D basis functions given in Eq. 11. It has been noted elsewhere (83) that in representative FBR calculations dealing with excited two-mode (fixed diatom bond length) floppy systems such as LiCN/LiNC (84), HCN/HNC (85), KCN (86), Ar–HCl (87), and HeN_2^+ (81) and three-mode systems like HCO/HOC (88) and H_3^+ (89–91), the basis set size was typically a factor 10–40 times the number of converged (to ~ 0.5–1 cm^{-1}) vibrational states.

Such convergence rates make straightforward extension of these methods to higher vibrational states and/or higher dimensional systems very problematic. In fact, in most FBR calculations on floppy triatomics to date, one of the vibrations has been frozen, thus reducing the (vibrational) problem to only two dimensions. In the relatively few cases (H_3^+, Li_3^+, CH_2^+) in which all three modes have been considered, FBR calculations have typically converged (to ~ 1 cm^{-1}) for about the ten lowest vibrational states. In addition, evaluation of potential matrix elements, which involves two- or three-dimensional integration, has required a large number, 10–20, of Gauss quadrature points per dimension, thus making such evaluation computationally very demanding.

The origins of the above problems have been traced (83) to the FBR itself. The variational methods discussed so far have invariably employed products of 1D Morse, harmonic, or numerical oscillator functions to represent the radial coordinates (26, 27). To understand the reasons for their inadequacy when dealing with anisotropic systems having LAM vibrations, it is useful to take a look at Figures 1 and 2. Figure 1 shows a contour plot of the HCN/HNC potential surface (92), while Figure 2 shows three 2D (R, r) cuts made at different angles through the same surface. Strong potential anisotropy is evident in both figures. It is also clear from Figure 2 that radial functions optimized for any one of the 2D (R, r) cuts will provide a poor basis for the cuts at other angles. To compensate for that, a large radial basis is needed. High-order polynomial functions present in a large FBR basis cause the integrands in the potential matrix element integrals to be highly oscillatory, thus making necessary the use of many quadrature points for their evaluation.

The problems just discussed are inherent to the FBR. This suggests that the FBR, although adequate for small amplitude vibrational states, is not the optimal choice for describing delocalized, LAM vibrational states of anisotropic systems. An alternative approach, in which the single-center oscillator basis is abandoned altogether, is described below.

VARIATIONAL AND ADIABATIC CALCULATIONS IN THE DISCRETE VARIABLE REPRESENTATION

The discrete variable representation, DVR (70), is a basis consisting of N discrete points. It is related by a unitary transformation to a set of $N L^2$ basis functions that constitute the familiar finite basis representation, FBR. The DVR appeared first as a strictly numerical technique for evaluating

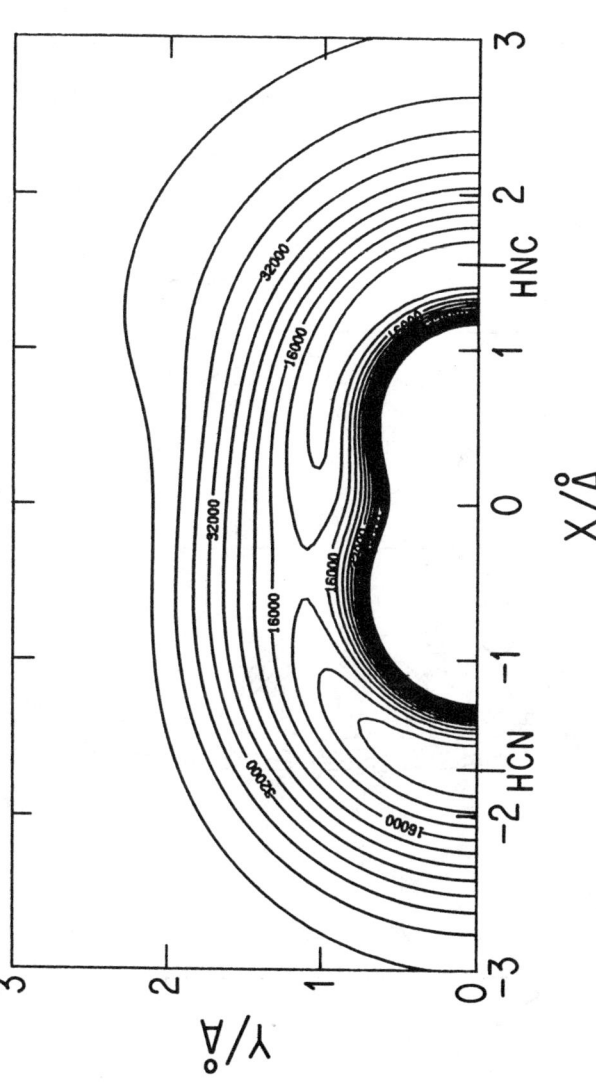

Figure 1 Contour map of the HCN/HNC potential surface by Murrell et al (92) for $r_{CN} = 1.16$ Å (roughly the average of CN bond lengths in the equilibrium HCN and HNC). Contour interval is 4000 cm^{-1}. Reproduced with permission from the *Journal of Chemical Physics*.

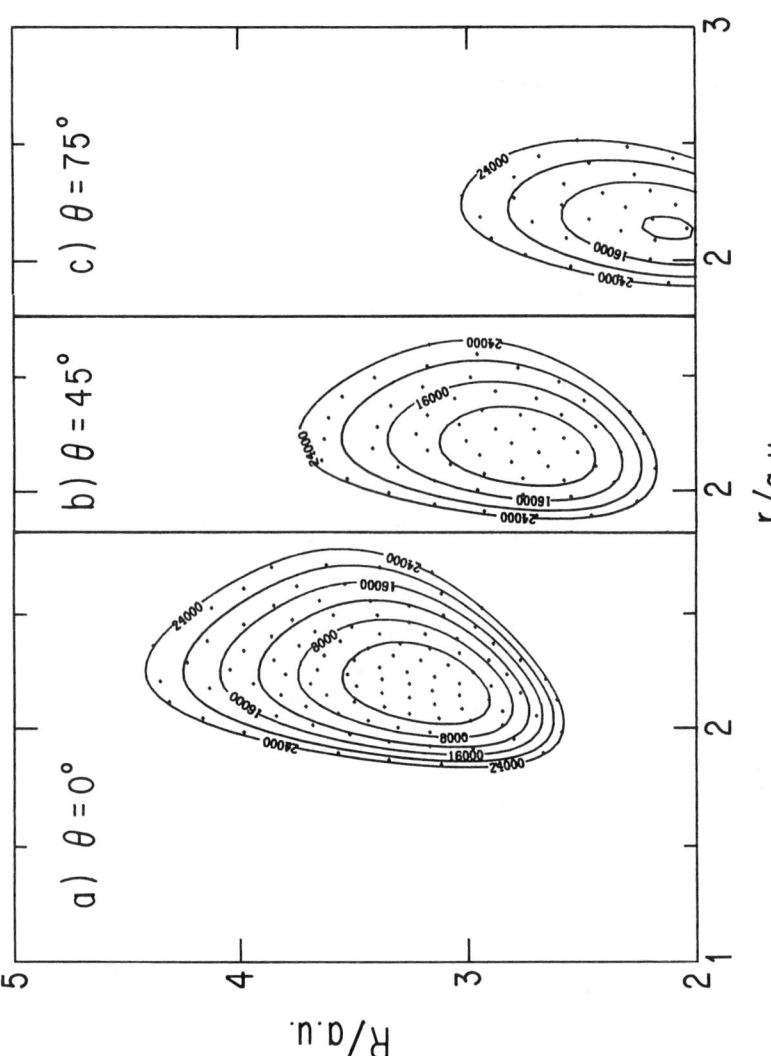

Figure 2 Three typical 2D (R, r) cuts through the HCN/HNC potential surface of Murrell et al (92). Notice how the shapes of the cuts and the positions of the 2D minima vary strongly as a function of angle θ. Also shown are the centers (●) of the 2D distributed Gaussian bases on these cuts [from Ref. (100)]. Reproduced with permission from the *Journal of Chemical Physics*.

potential matrix elements (93). Dickinson & Certain (94) provided the formal basis for the technique, and proved that the discrete points of the DVR were in fact the points of the Gaussian quadrature defined by the corresponding FBR. It has been established only recently that the DVR (70, 83, 95–97), or the equivalent discrete ordinate method (98, 99), can be the representation of choice for formulating and solving more easily a wide range of quantum mechanical problems. Since the basic formalism of the DVR for one dimension has been presented in several papers (70, 98, 99), we state here only two results that are useful for later discussion.

It has been shown (94) that the matrix T of the orthogonal transformation between representations in N Gaussian quadrature points $\{\chi_\alpha\}$ (DVR) and representations in N orthogonal polynomials $\{\phi_i(\chi)\}$ times their weight functions $\{\omega_\alpha\}$ is given by

$$T_{i\alpha} = \omega_\alpha^{1/2} \phi_i(\chi_\alpha). \qquad 13.$$

Also, provided that the elements of the potential matrix in the FBR, V^{FBR}, have been evaluated using the appropriate N point Gaussian quadrature, one can show that

$$(T^T V^{FBR} T)_{\alpha\beta} = V(\chi_\alpha) \delta_{\alpha\beta} \equiv (V^{DVR})_{\alpha\beta}, \qquad 14.$$

i.e. that the potential matrix in the DVR, V^{DVR}, is *diagonal* on the Gaussian quadrature points constituting that DVR. The matrix T appearing in Eq. 14 is defined in Eq. 13. In one dimension, the greater efficiency of the DVR relative to the FBR is largely due to this property of the DVR. In higher dimensions, other advantages appear, as described below.

Within the DVR-based formulation of the bound state problem, a number of complementary computational schemes have been developed (83, 100–104). They can be implemented in any coordinate system in which suitable orthogonal polynomials are defined, so that the DVR transformation in Eq. 13 can be defined. Consequently, it is possible to choose a combination of strategies and coordinate systems that would be most appropriate for a particular molecule. All of the applications to date have involved triatomic molecules, but the same DVR techniques should be applicable to, and perhaps even more important for, larger molecules. Concerning triatomics, there is the possibility of treating one, two, or all three internal coordinates in the DVR. Strategies for doing this are discussed below.

If only one or two internal coordinates are discretized via the DVR, then another basis is needed for the remaining degrees of freedom. We have chosen the distributed (real) Gaussian basis, DGB (105), one- or two-dimensional, depending on the number of coordinates to be represented by the DGB (83, 100–102). The 2D DGB, used in several studies (83, 100–102, 106) to expand the radial r, R coordinates is given by

$$\phi_i = (1/S_{ii})^{1/2}(2A_i/\pi)^{1/4}(2a_i/\pi)^{1/4}\exp[-A_i(R-R_i)^2 - a_i(r-r_i)^2], \qquad 15.$$

where S_{ij} is the overlap between any two Gaussians (83, 100). R_i and r_i define the center of the ith Gaussian, whereas A_i and a_i are the Gaussian exponents. Having defined the necessary ingredients, we now proceed to describe several DVR-based schemes.

DVR–DGB Approach

In this approach, thus far implemented in Jacobi (83, 100, 106) and Radau coordinates (101), the usual set of $j_{max}+1$ coupled-channel equations in the FBR (j_{max} refers to the highest order term in the Legendre polynomial expansion of the angular coordinate) is transformed to the DVR appropriate to Legendre polynomials (83, 100). The LAM angular, bending coordinate χ ($\chi = \cos\theta$) is thereby discretized on a set of points χ_α defined by the ($j_{max}+1$)-point Gauss-Legendre quadrature. On the 3D potential surface $V(R,r,\theta)$, each of the discrete angles χ_α defines a 2D (R,r) cut $V(R,r,\chi_\alpha)$, shown in Figure 2. A two-dimensional DGB, defined in Eq. 15, is distributed on the (R,r) cuts. The Gaussians are placed according to some simple rules regarding the overlap and the exponents A_i and a_i in Eq. 15 (83, 100), resulting in a semiclassical distribution of Gaussian centers, which can be seen in Figure 2. Combining the DVR with the 2D DGB produces a very flexible basis that, in contrast to the FBR, handles easily even strongly anisotropic potentials with multiple minima. It may also be noticed that the DGB is readily tailored to the potential, i.e. it can be distributed *only* over the *relevant* regions of the potential surface, defined by the maximum energy of interest.

At each χ_α, separate matrix representations of the operators $h(R,r)$, $G(R,r)$ and $V(R,r,\chi_\alpha)$, which appear in Eq. 2, are formed in the 2D DGB. The corresponding matrices h and V are block-diagonal in χ_α (V because of Eq. 14); we denote their respective diagonal blocks as h^α and V^α. In the DVR-DGB triatomic Hamiltonian, different χ_αs are coupled only by the transformed angular momentum matrix $T^T j^2 T$ (83, 100).

At this point, the DVR formalism provides a unique way to truncate the DGB, thus decreasing drastically the size of the final Hamiltonian matrix (83, 100, 101). For each of the $j_{max}+1$ χ_αs of the Legendre DVR, a 2D Hamiltonian (which depends on R,r only) is defined

$$H^\alpha = h^\alpha + V^\alpha, \quad (\alpha = 1, 2, \ldots, j_{max}+1), \qquad 16.$$

and diagonalized. A subset of eigenvectors of H^αs, whose eigenvalues lie below a certain energy cutoff (related to the maximum energy of interest), serves as the final, much reduced radial basis in which the full Hamiltonian matrix is formed and diagonalized, thus yielding the desired molecular eigenstates. Experience has shown that including only a small fraction of the total number of eigenvectors of H^αs is sufficient for obtaining converged and accurate final results (83, 100, 101).

An additional benefit of treating the angular coordinate in the DVR is that, as a consequence of Eq. 14, no angular integration is required to evaluate the potential matrix elements. Besides lowering the dimensionality of the integrals, this also eliminates the need for potential surface expansion in Legendre polynomials, given in Eq. 12, which is commonly done in the FBR in order to facilitate integral evaluation and which, in the case of anisotropic potentials, can be a difficult task in itself.

Multidimensional DVR Approach

We have already mentioned the possibility that not only one, but two, or all three internal coordinates (for a triatomic molecule) can be treated in the DVR. At the present time, multidimensional DVRs are really direct product DVRs, based on direct product FBRs consisting of 1D basis functions. Therefore, the multidimensional FBR-to-DVR transformation matrix is a direct product of 1D transformation matrices in Eq. 13 (102, 104, 107).

The multidimensional DVR has several attractive features (102, 104, 107). In the 3D DVR, which is a completely pointwise representation, only those DVR points lying in the relevant, energetically accessible part of the potential surface need to be retained; the rest can be removed from the basis. In addition, there are no numerical integrals to be calculated in the 3D DVR. In order to evaluate potential matrix elements, only the values of the potential at the DVR points are needed. This feature is also shared by the recently proposed collocation method for solving bound state problems (108, 108a, 108b).

In the DVR-DGB procedure described above, the DVR allowed an initial partitioning of the full 3D Hamiltonian into a set of 2D Hamiltonians. If two or three coordinates are treated in the DVR, this process can be carried one step further. By applying the second DVR, each of the 2D Hamiltonians can be decomposed into a set of 1D vibrational problems. A truncated set of 1D eigenvectors is then used as the basis for the 2D Hamiltonians. As in the DVR-DGB approach, these 2D Hamiltonians are diagonalized; a truncated set of their eigenvectors is then recoupled exactly to generate the final, full Hamiltonian matrix. This successive diagonalization-truncation procedure, which has two levels of truncation, is very effective in decreasing the size of the final basis (102, 104, 107). The extra work needed to generate the 1D and 2D eigenvectors is compensated by an impressive reduction of the final Hamiltonian matrix. We note also that the 2D or 3D DVR makes the initial Hamiltonian very sparse, since only "one-dimensional" kinetic energy coupling is present for each of the two or three dimensions.

Adiabatic Approximation in the DVR

In addition to the novel variational techniques described above, the DVR permits a definition of the adiabatic approximation (106), which has several important advantages over the usual formulation in the FBR (109).

The development begins with the realization that the 2D Hamiltonian H^α in Eq. 16 is in fact the adiabatic two-mode stretching Hamiltonian for a particular angle χ_α (100). Consequently, its eigenvalues represent the energies of the stretching vibrations of the triatomic system, at that χ_α, in the adiabatic bend approximation. As a function of χ ($\chi = \cos\theta$), these eigenvalues in turn provide effective angular potentials for the bending motion.

In the DVR approaches discussed above, the entire matrix formed in the 2D eigenvector basis was diagonalized to provide variational estimate to the vibrational eigenstates. We have shown (106) that an adiabatic rearrangement of this full Hamiltonian matrix can be defined, such that the diagonal blocks (labelled by the eigenvalues of the 2D Hamiltonians in Eq. 16) provide the rigorous matrix representation of the adiabatic bend Hamiltonian. Their diagonalization yields progressions of bending levels corresponding to various stretching states.

The off-diagonal blocks of the adiabatically rearranged full Hamiltonian matrix contain all nonadiabatic coupling matrix elements. The nonadiabatic correction to the adiabatic vibrational levels can be readily, and very effectively, taken into account via second-order perturbation theory (106). In this approach, the vibrationally nonadiabatic matrix elements are calculated just as easily as the adiabatic ones. This is not the case with the adiabatic approximation in the FBR (109). There, the presence of derivatives of the radial wave function with respect to the angular coordinate makes the evaluation of nonadiabatic corrections difficult, and is seldom done.

Applications

The DVR-based bound state methods are just a few years old (83), and the number of their applications has not been particularly large. However, almost every one of them has addressed a problem that would be extremely difficult, or even intractable, for the FBR variational or adiabatic methods.

Bačić & Light (100), using the DVR-DGB method (83), have performed the first full three-mode variational calculation of the highly excited, LAM vibrational ($J = 0$) states of HCN/HNC, including states above the isomerization barrier, which show varying degrees of delocalization over both potential minima. Jacobi coordinates and a model, empirical surface by Murrell, Carter & Halonen (92), shown in Figure 1, were employed. Previously, the LAM vibrational spectrum in this energy range important for isomerization could be studied only by approximate methods (23, 24). In addition to producing well over 100 accurate energy levels and wavefunctions, this study has shown that the onset of extensive delocalization over the isomerization barrier is determined by the height of the effective, adiabatic potential, which is more than 2000 cm^{-1} higher than the "bare" potential barrier (100).

The adiabatic approximation has been formulated in the DVR and applied to the two-mode LiCN/LiNC (frozen CN) and three-mode HCN/HNC (106). The accuracy of the adiabatic calculation on LiCN/LiNC relative to the variational DVR-DGB results (83), and the striking improvement caused by including the nonadiabatic correction, can be seen in Figure 3. It is evident that after including perturbatively the nonadiabatic correction, this approach can provide accurate characterization and quantum number assignment for a significant fraction of highly excited LAM states, including the localized states above the isomerization barrier (106).

Bačić, Watt & Light (101) have calculated variationally, in Radau coordinates, the vibrational levels of the (three-mode) water molecule up to $\sim 27{,}000$ cm^{-1} relative to the potential energy minimum. The DVR adiabatic-plus-perturbation treatment (106) has permitted reliable level assignment up to $\sim 18{,}000\text{--}20{,}000$ cm^{-1}.

HCN/HNC (100) and H$_2$O (101) are the first realistic three-mode systems for which the number of accurately calculated energy levels (>100) was large enough to allow their analysis in terms of the level spacings distributions. This statistical measure figures prominently in the theories of classical and quantum ergodicity (11, 15).

Converged energy levels of $J = 0$ H$_3^+$ up to 20,000 cm^{-1} above the potential energy minimum (higher than in any previous H$_3^+$ calculation) have been reported by Whitnell & Light (104). Their calculation was performed in the hyperspherical coordinates of Pack (41, 42) with the 3D DVR and the successive diagonalization-truncation technique (102, 104, 107).

The 3D DVR employed in the above H$_3^+$ calculation was adapted for the full D_{3h} symmetry of this molecule by the procedure for constructing symmetry-adapted DVRs developed by Whitnell & Light (103). This allowed separate calculations for each of the three irreducible representations of H$_3^+$ and automatic, unambiguous symmetry assignment for all the calculated levels.

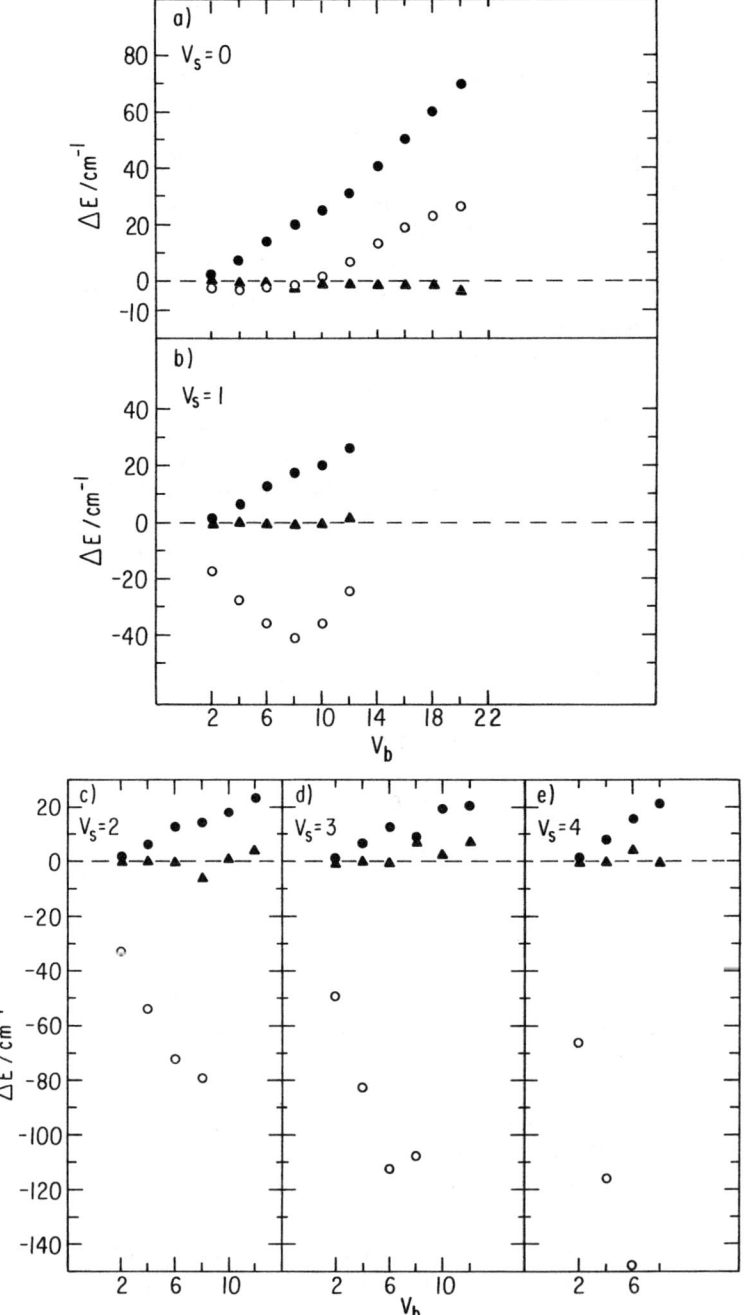

The DVR calculations described here, carried out in several coordinate systems and using different computational schemes, testify to the versatility and power of the DVR-based bound state methods when dealing with highly excited, LAM vibrations in strongly anisotropic, isomerizing, and other floppy molecules.

OTHER METHODS

Recursive Residue Generation Method

The absorption spectrum $I(\omega)$ from an initial state $|0\rangle$ is given by

$$I(\omega) = \sum_{\alpha} |\langle 0|\mu|\alpha\rangle|^2 \delta(\omega - \omega_\alpha), \qquad 17.$$

where μ is the dipole operator, and $\{|\alpha\rangle\}$ and $\{\omega_\alpha\}$ are the eigenvectors and eigenfrequencies, respectively, of the molecular Hamiltonian. The recursive residue generation method (RRGM), is a quantum mechanical method for calculating eigenvalues and residues $|\langle 0|\mu|\alpha\rangle|^2$ of the propagator associated with the time-independent Hamiltonian (111). It is based on the Lanczos algorithm (112) for finding the eigenvalues of sparse matrices. The computational power of the RRGM comes largely from its ability to focus on one row of the eigenvector matrix, consisting of the projections of the initial state onto the eigenvectors, rather than on the whole eigenvector matrix (111). An important feature of the RRGM calculations of spectra is that the most intense transitions are determined quite accurately after relatively few Lanczos recursion steps. Calculation of weaker lines requires many more recursions and is much more time consuming (113).

Very recently, Brunet et al (113), in the first RRGM calculation of the IR absorption spectrum of a realistic three-mode system, calculated vibrational ($J = 0$) eigenvalues and transition intensities of HCN, using the potential surface of Murrell, Carter & Halonen (92) and the ab initio calculated dipole surface of Jørgensen et al (114). Their calculation extended up to $\sim 12,000$ cm^{-1} relative to the zero point energy of HCN. This energy range has also been covered by the variational DVR-DGB calculation (100) on the same potential surface, so that the results of the two calculations can be compared directly. Most energy levels agree to

---

Figure 3 Differences between the adiabatic (●) and adiabatic-plus-perturbation (▲) calculations of the vibrational levels of LiCN (106) and the accurate variational results (83). Results of the adiabatic calculations in the finite basis representation (110) are shown for comparison (○). Symbols V_s and V_b label the stretch and the bend quantum numbers, respectively. Reproduced with permission from the *Journal of Chemical Physics*.

within 2–3 cm^{-1}. Energies of some 15 RRGM-calculated states (out of 93 reported) do differ by 3–10 cm^{-1} from the levels calculated with the DVR-DGB method (100). The reasons for these differences are not clear at this point. They could be caused by some slight inadequacy of the (radial) basis used in the RRGM calculation. Expansion of the highly anisotropic HCN potential surface in terms of Legendre polynomials, done by Brunet et al (113) but not in the DVR-DGB calculation, could conceivably contribute to the differences.

Because the RRGM focused on determining the excitation spectrum from the ground state, the number of vibrational states that the two calculations find below 11,907 cm^{-1}, the energy of the highest state reported by Brunet et al, differs substantially. There are 114 variationally calculated excited states in this energy range (100), whereas Brunet et al (113) find transitions to only 93. The majority of the states not found by the RRGM are in the high-energy part of the calculated spectrum. From the group of 33 variational states $n = 82$–114, only 14 are found in the RRGM calculation. This means that the RRGM, at least for this choice of the initial state, fails to locate a significant fraction of the LAM vibrational states below and above the isomerization barrier (including the $n = 93$ state, the first to be extensively delocalized over both minima). Choice of a different initial state, perhaps one prepared by SEP (11), which would overlap better with highly excited LAM states, may lead to their detection by the RRGM.

Thus, the fact that the effort required from the RRGM to calculate a transition depends primarily on the intensity of the line and not its energy, represents both its strength and weakness. Strong, high-frequency transitions that, because of their energies, may be difficult to calculate variationally, are readily determined by the RRGM. On the other hand, the RRGM may miss entirely the weak lines, even at lower frequencies, that would be easily calculated variationally.

These differences notwithstanding, the RRGM is undoubtedly a promising new method for calculating molecular spectra and should be developed and tested further. It may be especially useful for the high energy portion of the spectrum, which may stretch the limits of the variational approach.

Semiclassical Quantization

In recent years there has been a great deal of interest in developing semiclassical quantization techniques for calculating bound states of multimode systems (115). In spite of such interest, few attempts have been made to calculate vibrational eigenvalues of polyatomic molecules with LAM vibrations by means of semiclassical quantization. One reason may be that the current semiclassical methods of quantization are restricted to regular, quasiperiodic regions of phase space (115). Vibrational states of floppy molecules are very anharmonic, and the mode coupling is strong, characteristics that tend to make the underlying classical mechanics chaotic and hence make floppy molecules rather poor candidates for successful semiclassical quantization.

Smith & Shirts (116), using the semiclassical technique of adiabatic switching, have calculated some vibrational eigenvalues of the three-mode HCN/HNC for the surface by Murrell, Carter & Halonen (92). This has allowed comparison with the variational DVR-DGB calculation (100), which employed the same surface. Judging from the limited number of eigenstates reported by Smith & Shirts (116), the agreement is semiquantitative at best. The differences grow fairly rapidly with increasing excitation of the LAM bending mode. With two quanta in the bend mode, the difference is ~ 10 cm^{-1}, rising to ~ 16 cm^{-1} with four quanta. Since there are many HCN and HNC states below the top of the isomerization barrier with 8, 10, or more quanta in the bending mode (100, 106), semiclassical quantization via adiabatic switching does not appear to provide an accurate description of the most interesting LAM vibrational states near the top or above the isomerization barrier.

Applicability of semiclassical quantization methods to floppy molecules in general cannot be judged from the results of one application. More studies on systems with LAM vibrations that employed a variety of quantization schemes would be of interest.

Optimal-Coordinates Vibrational Self-Consistent Field Method

In applications to realistic multimode systems, additional approximations are often introduced in the rigorous semiclassical quantization approach. The self-consistent field (SCF) approximation (117–119) has been among the more successful and widely used ones. In this approach, each vibrational mode moves in an effective potential that is the average of the full potential over the states of the other modes, all modes being treated self-consistently. The assumption central to the SCF approximation is separability, i.e. that the total vibrational wave function can be written in the Hartree form, as a single product of single-mode functions. This leads to a set of coupled single-mode SCF equations (117–119), which can be solved either quantum mechanically (117) or, in the semiclassical (SC) limit, by applying the Bohr-Sommerfeld quantization condition to the single-mode energies (118, 119). Here, we are concerned with the latter, SC-SCF method.

The quantitative success of the SCF method in any particular application depends on the choice of coordinates (118). Nowhere has this dependence been found to be so strong as in the case of excited LAM vibrations. The SC-SCF study by Bačić, Gerber & Ratner (120) of the coupled bending-stretching energy levels of HCN (frozen CN) represented the first application of the SCF approximation to highly excited LAM vibrations. The focus was on high excitation of the bending mode, because of its association with HCN/HNC isomerization. Physical intuition suggested that in the case of highly excited bending vibrations, a good choice of coordinates might be the one providing good description of the motion along the minimum energy path (MEP) for isomerization. Inspection of the potential surface used in the calculation (92), shown in Figure 1, revealed that the MEP can be roughly described by an ellipse.

This finding suggested the use of ellipsoidal (or spheroidal) coordinates for the HCN calculation, in which the position of the H atom is determined by two distances, r_1 and r_2, from the two foci on the CN axis (120). The interfocal distance provided a parameter a that could be varied to determine ellipsoidal coordinates optimal for a particular stretching-bending state. The importance of coordinate optimization in the SC-SCF approach is evident from Figure 4, in which several energy levels are plotted versus a. Clearly, coordinate optimization always improves the SCF results, but in highly excited bending states, the improvement due to optimization is quite dramatic, and the use of optimal coordinates appears to be essential. Moreover, the optical value of a varies considerably from one state to another. Thus, the choice of optimal coordinates depends not only on the molecule studied, but also on the degree of vibrational excitation (120).

Ellipsoidal coordinates have been found to be the most suitable for SC-SCF calculations on I_2He (121) as well, where the large amplitude motion of the He atom is similar to that of the hydrogen atom in HCN. The ellipsoidal SCF energy levels were within ~ 1 cm^{-1} of the levels from converged CI calculations. In the same paper, a very floppy system $XeHe_2$ was also studied. In this case, hyperspherical coordinates proved to be the optimal choice (121).

It is clear that consideration of factors such as mass ratios and topology of the potential surfaces can provide valuable guidance in selecting optimal coordinates for specific applications. Nevertheless, there is a need for a general algorithm for finding optimal coordinates that would not rely on intuition.

CONCLUDING REMARKS

Significant advances have been made in recent years in the theoretical treatment of rotation-vibration spectra of floppy molecules. Novel

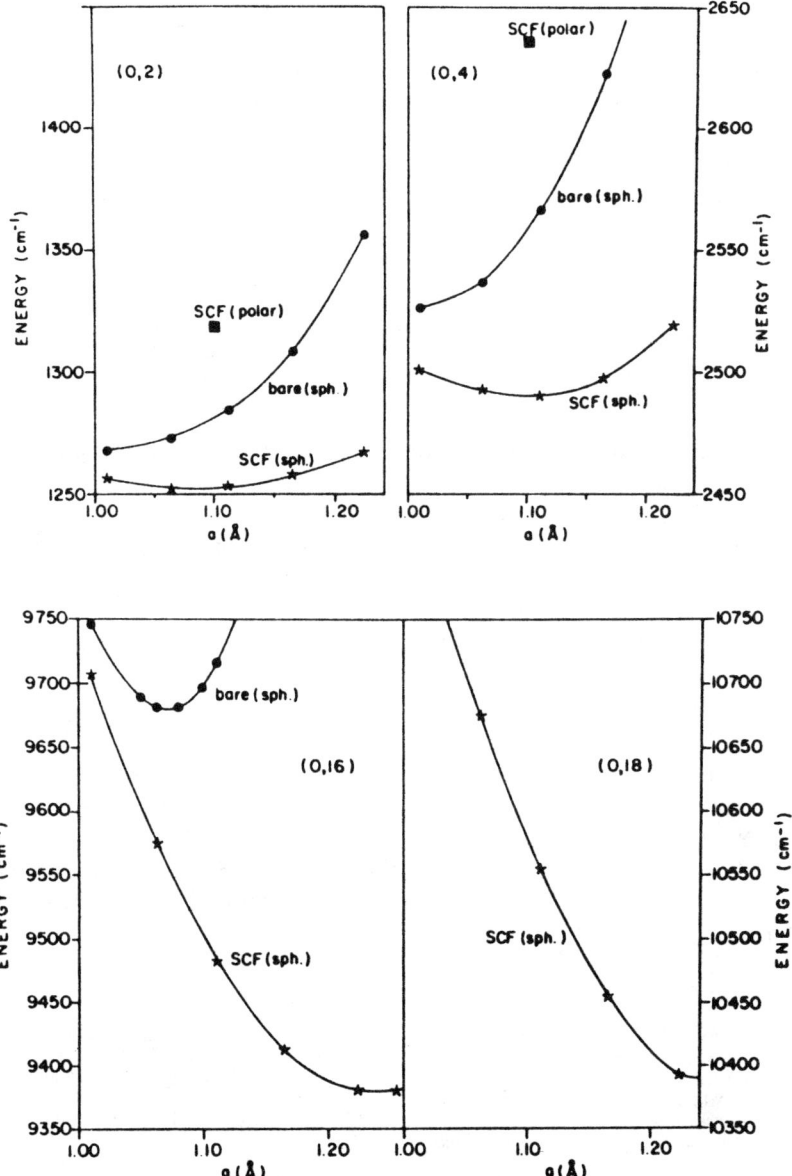

Figure 4 Optimal-coordinate behavior of the SC-SCF energies of the (0, 2) and (0, 4) states (*top*), and the (0, 16) and (0, 18) states (*bottom*). The spheroidal results are plotted versus a, the spheroidal coordinates parameter (half of the interfocal distance). Also shown are the bare-mode results versus the same parameter, and the polar coordinates SCF results (*top*). From Ref. (120).

distributed, pointwise representations have been developed that are more appropriate for delocalized wave functions of large amplitude vibrational states than the usual bases. In the case of floppy triatomic molecules, a large number of excited vibrational states, including those delocalized over multiple minima that may be present on the potential surface, can now be calculated variationally with modest computational effort.

New computational procedures have also allowed variational calculations of rovibrational energy levels of floppy triatomics with very high values of the total angular momentum, where rovibrational interaction is strong and rotational manifolds from many vibrational states overlap. Methodology for dealing with bound states of tetratomic and larger floppy molecules is in the early stages of development but will undoubtedly be an important direction of future efforts. The theoretical treatment of large amplitude vibrations should be generalized to include electronically nonadiabatic bound states, which are receiving increasing attention from experimentalists.

The progress in theoretical methods has coincided with the emergence of several new spectroscopic techniques with unprecedented power to probe directly highly excited, large amplitude vibrational states of molecules and weakly bound clusters. Combined advances in theory and experiment will enable a determination of accurate potential surfaces, including regions far from the equilibrium configuration.

ACKNOWLEDGMENTS

We thank Robert Whitnell for his contributions to some of the original research described in this review. We gratefully acknowledge support from the National Science Foundation through grants CHE-8505001 and CHE-8806514 to J. C. L. In the course of this writing Z. B. has been supported in part by the Camille and Henry Dreyfus Foundation Distinguished New Faculty Grant.

Literature Cited

1. Gerber, R. B., Ratner, M. A. 1988. *J. Phys. Chem.* 92: 3252 and references therein
2. Bowman, J. M., ed. 1988. *Comput. Phys. Commun.* 51: 1
3. 1988. *J. Chem. Soc. Faraday Trans. 2* 84: 1237
4. Nesbitt, D. J. 1988. *Chem. Rev.* 88: 843
5. Miller, R. E. 1988. *Science* 240: 447
6. Miller, R. E. 1986. *J. Phys. Chem.* 90: 3301
7. Saykally, R. J. 1988. *Science* 239: 157
8. Crim, F. F. 1984. *Annu. Rev. Phys. Chem.* 35: 657
9. Chin, Y. N., Friedrich, B., Maring, W., Niedner, G., Noll, M., Toennies, J. P. 1988. *J. Chem. Phys.* 88: 6814 and references therein
10. Carrington, A. 1986. *J. Chem. Soc. Faraday Trans. 2* 82: 1089
11. Hamilton, C. E., Kinsey, J. L., Field, R. W. 1986. *Annu. Rev. Phys. Chem.* 37: 493
12. Ito, M., Ebata, T., Mikami, N. 1988. *Annu. Rev. Phys. Chem.* 39: 123
13. Wilson, E. B., Decius, J. C., Cross, P. C. 1980. *Molecular Vibrations.* New York: Dover

14. Casati, G., ed. 1985. *Chaotic Behavior in Quantum Systems—Theory and Applications*. New York: Plenum
14a. Lefebvre, R., Mukamel, S., eds. 1987. *Stochasticity and Intramolecular Redistribution of Energy*. Dordrecht: Reidel
15. Stechel, E. B., Heller, E. J. 1984. *Annu. Rev. Phys. Chem.* 35: 563
16. Abramson, E., Field, R. W., Imre, D., Innes, K. K., Kinsey, J. L. 1985. *J. Chem. Phys.* 83: 453
17. Sundberg, R. L., Abramson, E., Kinsey, J. L., Field, R. W. 1985. *J. Chem. Phys.* 83: 466
18. Watson, J. K. G. 1968. *Mol. Phys.* 15: 479
19. Watson, J. K. G. 1970. *Mol. Phys.* 19: 465
20. Whitehead, R. J., Handy, N. C. 1975. *J. Mol. Spectrosc.* 55: 356
21. Bartholomae, R., Martin, D., Sutcliffe, B. T. 1981. *J. Mol. Spectrosc.* 87: 367
22. Carter, S., Handy, N. C. 1982. *J. Mol. Spectrosc.* 95: 9
23. Bunker, P. R. 1983. *Annu. Rev. Phys. Chem.* 34: 59
24. Jensen, P. 1983. *Comput. Phys. Rep.* 1: 1
25. Beardsworth, R., Bunker, P. R., Jensen, P., Kraemer, W. P. 1986. *J. Mol. Spectrosc.* 118: 50
26. Carter, S., Handy, N. C. 1986. *Comput. Phys. Rep.* 5: 115
27. Tennyson, J. 1986. *Comput. Phys. Rep.* 4: 1
28. Dunker, A. M., Gordon, R. G. 1976. *J. Chem. Phys.* 64: 354
29. Le Roy, R. J., Carley, J. S. 1980. *Adv. Chem. Phys.* 42: 353
30. Pack, R. T. 1974. *J. Chem. Phys.* 60: 633
31. Lester, W. A. 1976. In *Dynamics of Molecular Collisions, Part A*, ed. W. H. Miller, 1: 1. New York/London: Plenum
32. Tennyson, J., Sutcliffe, B. T. 1982. *J. Chem. Phys.* 77: 4061
33. Tennyson, J., Sutcliffe, B. T. 1983. *J. Mol. Spectrosc.* 101: 71
34. Brocks, G., van der Avoird, A., Sutcliffe, B. T., Tennyson, J. 1983. *Mol. Phys.* 50: 1025
35. Wallace, R. 1984. *Chem. Phys.* 88: 247
36. Leroy, J. P., Wallace, R. 1987. *Chem. Phys.* 111: 11
37. Smith, F. T. 1959. *J. Chem. Phys.* 31: 1352
38. Schatz, G. C., Kuppermann, A. 1976. *J. Chem. Phys.* 65: 4642
39. Bowman, J. M., Wierzbicki, A., Zúñiga, J. 1988. *Chem. Phys. Lett.* 150: 269
40. Bowman, J. M., Zúñiga, J., Wierzbicki, A. 1989. *J. Chem. Phys.* 90: 2708
41. Pack, R. T. 1984. *Chem. Phys. Lett.* 108: 333
42. Pack, R. T., Parker, G. A. 1987. *J. Chem. Phys.* 87: 3888
43. Johnson, B. R. 1980. *J. Chem. Phys.* 73: 5051
44. Johnson, B. R. 1983. *J. Chem. Phys.* 79: 1906
45. Johnson, B. R. 1983. *J. Chem. Phys.* 79: 1916
45a. Frey, J. G., Howard, B. J. 1985. *Chem. Phys.* 99: 415
46. Johnson, B. R., Reinhardt, W. P. 1986. *J. Chem. Phys.* 85: 4538
47. Sutcliffe, B. T., Tennyson, J. 1986. *Mol. Phys.* 58: 1053
48. Estes, D., Secrest, D. 1986. *Mol. Phys.* 59: 569
49. Lee, J. S., Secrest, D. 1988. *J. Phys. Chem.* 92: 1821
50. Podolsky, B. 1928. *Phys. Rev.* 32: 812
51. Sutcliffe, B. T. 1982. *Current Aspects of Quantum Chemistry. Studies in Theoretical Chemistry*, ed. R. Carbo, 21: 99. Amsterdam: Elsevier
52. Handy, N. C. 1987. *Mol. Phys.* 61: 207
53. Tennyson, J., van der Avoird, A. 1983. *J. Chem. Phys.* 77: 5664
54. Danby, G. 1983. *J. Phys. B* 16: 3393
55. Danby, G., Flower, D. R. 1983. *J. Phys. B* 16: 3411
56. Brocks, G., Huygen, T. 1986. *J. Chem. Phys.* 85: 3411
57. Brocks, G., van Koeven, D. 1988. *Mol. Phys.* 63: 999
57a. Leutwyler, S., Boesiger, J. 1987. *Z. Phys. Chem.* 154: 31
58. Dunker, A. M., Gordon, R. G. 1976. *J. Chem. Phys.* 64: 4984
59. Stechel, E. B., Walker, R. B., Light, J. C. 1978. *J. Chem. Phys.* 69: 3518
60. Johnson, B. R. 1978. *J. Chem. Phys.* 69: 4678
61. Shapiro, M., Balint-Kurti, G. G. 1979. *J. Chem. Phys.* 71: 1461
62. Atabek, O., Meret-Artes, S., Jacon, M. 1985. *J. Chem. Phys.* 83: 1769
63. Hutson, J. M. 1988. *J. Chem. Phys.* 89: 4550
64. Hutson, J. M. 1988. *Chem. Phys. Lett.* 151: 565
65. Robinson, R. L., Gwo, D. H., Saykally, R. J. 1988. *Mol. Phys.* 63: 1021
66. Novick, S. E., Janda, K. C., Holmgren, S. L., Waldman, M., Klemperer, W. 1976. *J. Chem. Phys.* 65: 1114
67. Carney, G. D., Sprandel, L. L., Kern, C. W. 1978. *Adv. Chem. Phys.* 37: 305
68. MacDonald, J. K. L. 1933. *Phys. Rev.* 43: 830
69. Tennyson, J., Sutcliffe, B. T. 1986. *Mol. Phys.* 58: 1067

70. Light, J. C., Hamilton, I. P., Lill, J. V. 1985. *J. Chem. Phys.* 82: 1400
71. Carter, S., Handy, N. C. 1986. *Mol. Phys.* 57: 175
72. Le Roy, R. J., van Kranendonk, J. 1974. *J. Chem. Phys.* 61: 4750
73. Brocks, G., Tennyson, J. 1983. *J. Mol. Spectrosc.* 99: 263
74. Reid, B. P., Janda, K. C., Halberstadt, N. 1988. *J. Phys. Chem.* 92: 587
75. Sutcliffe, B. T., Miller, S., Tennyson, J. 1988. *Comp. Phys. Comm.* 51: 73
76. Tennyson, J., Miller, S., Sutcliffe, B. T. 1988. *J. Chem. Soc. Faraday Trans. 2* 84: 1295
77. Miller, S., Tennyson, J. 1988. *Chem. Phys. Lett.* 145: 117
78. Miller, S., Tennyson, J. 1988. *J. Mol. Spectrosc.* 128: 530
79. Miller, S., Tennyson, J. 1988. *Ap. J.* 335: 486
80. Miller, S., Tennyson, J., Sutcliffe, B. T. 1989. *Mol. Phys.* 66: 429
81. Miller, S., Tennyson, J., Follmeg, B., Rosmus, P., Werner, H. J. 1988. *J. Chem. Phys.* 89: 2178
82. Henderson, J. R., Miller, S., Tennyson, J. 1988. *Spectrochim. Acta.* In press
82a. Lee, J. S., Secrest, D. 1986. *J. Chem. Phys.* 85: 6565
83. Bačić, Z., Light, J. C. 1986. *J. Chem. Phys.* 85: 4594
84. Farantos, S. C., Tennyson, J. 1985. *J. Chem. Phys.* 82: 800
85. Founargiotakis, M., Farantos, S. C., Tennyson, J. 1988. *J. Chem. Phys.* 88: 1598
86. Tennyson, J., Farantos, S. C. 1984. *Chem. Phys. Lett.* 109: 160
87. Tennyson, J. 1985. *Mol. Phys.* 55: 463
88. Bowman, J. M., Bittman, J. S., Harding, L. B. 1986. *J. Chem. Phys.* 85: 911
89. Tennyson, J., Sutcliffe, B. T. 1984. *Mol. Phys.* 51: 887
90. Carney, G. D., Adler-Golden, S. M., Lesseski, D. C. 1986. *J. Chem. Phys.* 84: 3921
91. Meyer, W., Botschwina, P., Burton, P. G. 1986. *J. Chem. Phys.* 84: 891
92. Murrell, J. N., Carter, S., Halonen, L. O. 1982. *J. Mol. Spectrosc.* 93: 307
93. Harris, D. O., Engerholm, G. G., Gwinn, W. D. 1965. *J. Chem. Phys.* 43: 1515
94. Dickinson, A. S., Certain, P. R. 1968. *J. Chem. Phys.* 49: 4209
95. Lill, J. V., Parker, G. A., Light, J. C. 1982. *Chem. Phys. Lett.* 89: 463
96. Heather, R. W., Light, J. C. 1983. *J. Chem. Phys.* 79: 147
97. Lill, J. V., Parker, G. A., Light, J. C. 1986. *J. Chem. Phys.* 85: 900
98. Shizgal, B., Blackmore, R. 1984. *J. Comput. Phys.* 55: 313
99. Blackmore, R., Shizgal, B. 1985. *Phys. Rev. A* 31: 1855
100. Bačić, Z., Light, J. C. 1987. *J. Chem. Phys.* 86: 3065
101. Bačić, Z., Watt, D., Light, J. C. 1988. *J. Chem. Phys.* 89: 947
102. Bačić, Z., Whitnell, R. M., Brown, D., Light, J. C. 1988. *Comput. Phys. Comm.* 51: 35
103. Whitnell, R. M., Light, J. C. 1988. *J. Chem. Phys.* 89: 3674
104. Whitnell, R. M., Light, J. C. 1989. *J. Chem. Phys.* 90: 1774
105. Hamilton, I. P., Light, J. C. 1986. *J. Chem. Phys.* 84: 306
106. Light, J. C., Bačić, Z. 1987. *J. Chem. Phys.* 87: 4008
107. Light, J. C., Whitnell, R. M., Park, T. J., Choi, S. E. 1989. In *Supercomputer Algorithms for Reactivity, Dynamics and Kinetics of Small Molecules*, ed. A. Lagana. Holland: Reidel. In press
108. Yang, W., Peet, A. C. 1988. *Chem. Phys. Lett.* 153: 98
108a. Peet, A. C., Yang, W. 1989. *J. Chem. Phys.* 90: 1746
108b. Peet, A. C. 1989. *J. Chem. Phys.* 90: 4363
109. Hough, A. M., Howard, B. J. 1987. *J. Chem. Soc. Faraday Trans. 2* 83: 173 and references therein
110. Farantos, S. C., Tennyson, J. 1986. *J. Chem. Phys.* 85: 6210
111. Friesner, R. A., Brunet, J. Ph., Wyatt, R. E., Leforestier, C. 1987. *J. Supercomputer Appl.* 1: 9
112. Lanczos, C. 1950. *J. Res. Natl. Bur. Stand.* 45: 255
113. Brunet, J. Ph., Friesner, R. A., Wyatt, R. E., Leforestier, C. 1988. *Chem. Phys. Lett.* 153: 425
114. Jørgensen, U. G., Almlöf, J., Gustafsson, B., Larsson, M., Siegbahn, P. 1985. *J. Chem. Phys.* 83: 3034
115. Ezra, G. S., Martens, C. C., Fried, L. E. 1987. *J. Phys. Chem.* 91: 3721 and references therein
116. Smith, B. S., Shirts, R. B. 1988. *J. Chem. Phys.* 89: 2948
117. Bowman, J. M. 1986. *Acc. Chem. Res.* 19: 202
118. Gerber, R. B., Ratner, M. A. 1988. *Adv. Chem. Phys.* 70: 97
119. Gerber, R. B., Ratner, M. A. 1988. *J. Phys. Chem.* 92: 3252
120. Bačić, Z., Gerber, R. B., Ratner, M. A. 1986. *J. Phys. Chem.* 90: 3606
121. Gerber, R. B., Horn, T. R., Ratner, M. A. 1989. In *The Structure of Small Molecules and Ions*, ed. R. Naaman. New York: Plenum. In press

HOLE-BURNING SPECTROSCOPY

Silvia Völker

Center for the Study of Excited States of Molecules,
Gorlaeus and Huygens Laboratories, University of Leiden,
2300 RA Leiden, The Netherlands

INTRODUCTION

The study of optical relaxation processes in solids at low temperature has progressed tremendously in the last 20 years due to the development of coherent optical techniques. The latter would not have been possible without the advent of narrow-band tunable lasers, on the one hand, and lasers with very short pulses (picoseconds, femtoseconds) on the other hand. Such coherent techniques are necessary because spectral line shapes of solids doped with guest molecules are seldom determined by dynamical interactions but principally by strain or structural disorder. This gives rise to inhomogeneous broadening, Γ_{inh}, which in crystalline hosts at low temperature varies between ~ 0.1 and 10 cm^{-1}, whereas in glasses it may amount to ≈ 100–500 cm^{-1}.

One of the techniques to achieve narrowing of spectral lines is site-selection spectroscopy (1–3). A sub-ensemble of molecules within the inhomogeneously broadened absorption band is selectively excited by means of a narrow-band laser, and the fluorescence or phosphorescence signal is detected with a high-resolution monochromator and/or a Fabry-Pérot interferometer. Fluorescence line-narrowing (1) is a special case of this technique (for a review see 3).

The intrinsic or homogeneous spectral linewidth, Γ_{hom}, which yields information on the relaxation processes of the excited state, is given by the effective optical dephasing time, T_2:

$$\Gamma_{hom} = (\pi T_2)^{-1} = (2\pi T_1)^{-1} + (\pi T_2^*)^{-1}, \qquad 1.$$

where T_1 is the population decay time or excited state lifetime (also called longitudinal relaxation time in magnetic resonance), and T_2^* is the pure dephasing time determined by thermally induced fluctuations of the optical transition frequency. The latter term arises from phonon scattering and/or electron and nuclear spin fluctuations, if present. On extrapolation to zero temperature, $T \to 0$, T_1 generally provides a limiting value for T_2 (or Γ_{hom}). Since the homogeneous line is hidden under the inhomogeneously broadened absorption band, Γ_{hom} cannot be obtained by conventional spectroscopy, and coherent laser techniques are needed to solve this problem. Time- and frequency-domain techniques have been developed for this purpose. To the first category belong photon echoes (4–6) and optical free induction decay (7–9); to the second, fluorescence line-narrowing (1–3) and spectral hole-burning (10–17). I concentrate here on the latter.

The essence of spectral hole-burning (HB) is that the transition energy of the guest molecule embedded in a crystalline or glassy solid matrix shifts after absorption of monochromatic light. Irradiation of the inhomogeneously broadened absorption band with a narrow-band laser, of width Γ_l and frequency v_1, can induce molecules that absorb resonantly with the laser to undergo a phototransformation such that the product absorbs at a different frequency. This creates a hole or dip in the original absorption band at v_1 (see Figure 1). The photoproduct may either be stable at low temperature, which leads to a permanent hole, or it can be a metastable state that acts as a population storage level, by which a transient hole is created. The hole is then probed in a second step by scanning a tunable laser over its spectral region with low enough intensity to avoid further burning of the absorption band. The hole represents, under certain conditions (see next section), a negative replica of the homogeneous spectral transition, and the holewidth yields the unknown quantity Γ_{hom}. The optical resolution that can be obtained with HB is 10^3–10^5 times higher than reached with conventional techniques (13–15). This makes HB a powerful tool for spectroscopy in the MHz-regime.

It should be mentioned that because the laser is not selecting molecules in a specific environment, but a set of molecules absorbing at the same frequency v_1, the correlation between transition energy and environmental parameters is, in general, different for the photoproduct and the original molecule. As a consequence, the photoproduct band or "anti-hole" is broader than the hole.

This review is intended to present a critical discussion on specific aspects of spectral hole-burning. First, various HB-mechanisms and techniques are briefly described. Emphasis is then given to optical dephasing processes of electronically excited states of organic molecules diluted as guests in crystalline and glassy hosts at low temperatures (0.3 to 20K). It is shown that the homogeneous linewidth and its temperature dependence differ greatly in the two types of hosts. A critical comparison is subsequently

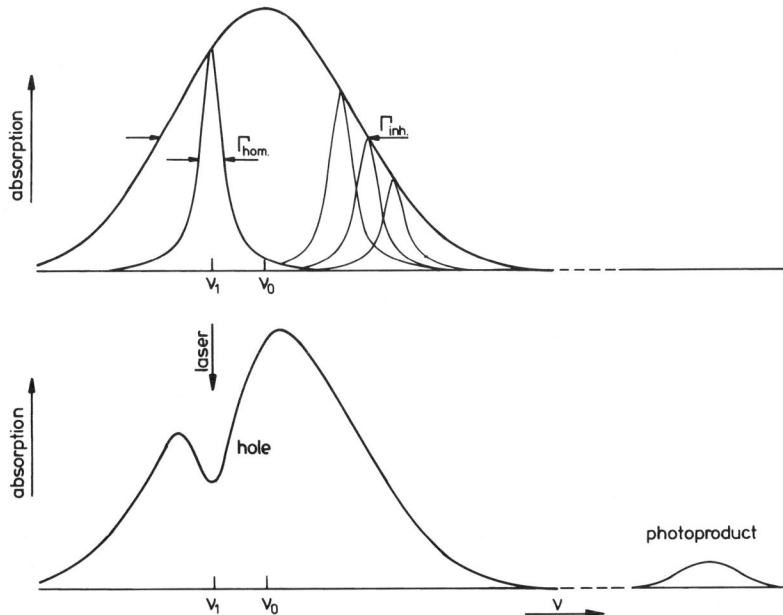

Figure 1 *Top*: Diagram of an inhomogeneously broadened absorption band of width Γ_{inh} consisting of a superposition of individual electronic transitions of homogeneous width Γ_{hom}. *Bottom*: Laser-induced "hole" burnt at frequency v_1 at low temperature. The photoproduct absorbs at a different frequency, which may either be inside or outside the inhomogeneous band.

made between linewidths obtained by hole-burning and photon echoes in amorphous systems. In particular, an actual controversy related to the time scales used in these experiments is addressed, and spectral diffusion processes are discussed. Furthermore, hole-burning experiments on semi-crystalline polymers and mixed organic-inorganic systems are presented. They were carried out with the purpose to test models for optical dephasing in amorphous solids. Recently, hole-burning has also been applied to molecular aggregates and biological complexes. Some of the new findings are briefly considered.

Finally, I comment on a few other applications of permanent hole-burning: the study of small Zeeman, Stark, and pressure effects, and the potential use of HB for optical date storage in the frequency domain.

The reader more deeply interested in these or other aspects of hole-burning may consult the following review articles: for organic systems, (18–26); for inorganic systems, (21, 25–28); for optical spectroscopy of glasses, (18–30); for external field effects, (23, 24, 28, 30, 31); for hole-burning in the IR-region, (25); for technological applications, (25, 26); for optical dephasing in crystals: experiments, (18, 23, 24), and theory, (32a,b); for the theory of optical dephasing in glasses, (29, 30).

HOLE-BURNING MECHANISMS AND TECHNIQUES

Mechanisms

Hole-burning (HB) can be divided in two categories: permanent HB and transient HB. Within the first category, one may distinguish between photochemical hole-burning (PHB) and non-photochemical hole-burning (NPHB). The time scale on which PHB- and NPHB-experiments are performed is usually seconds to minutes, whereas transient HB often occurs in micro- or milliseconds.

PHOTOCHEMICAL HOLE-BURNING (PHB) Two types of photochemical reactions lead to HB: intramolecular reactions within the guest molecule, and intermolecular reactions between the guest and the host. Intramolecular photochemistry may be reversible [e.g. photoautomerism in free-base porphyrins (14, 15, 36–40)], or irreversible [e.g. photodissociation of dimethyl-s-tetrazine (13) and s-tetrazine (35)]. Further, the photochemical process can be the result of the absorption of one photon [e.g. porphyrins (14), quinizarin (17, 41, 42), ionic dyes (43, 44)], or *two photons* [e.g. dimethyl-s-tetrazine (45) and carbazole in boric acid glass (46)]. In all these cases the photoproduct absorption band is usually well separated from the original absorption band (see Figure 1). Photochemical holes are persistent for hours or days, as long as the sample is kept at low temperature.

NON-PHOTOCHEMICAL HOLE-BURNING (NPHB) This process is characteristic for amorphous systems. After excitation with a narrow-band laser, a slight structural rearrangement of the local environment of the guest molecule seems to take place (11, 16, 47–51). Small and co-workers (16, 20, 47) have proposed a mechanism to explain this phenomenon. They assumed that the guest molecule in its ground and excited state is coupled to very low frequency excitations of the glass, the so-called two-level systems (TLS) (see below), and that tunneling occurs in the electronically excited state followed by relaxation to a different ground state configuration. As a consequence, the optical transition is shifted, and a non-photochemical hole appears in the absorption band. The spectral position of the "photoproduct" is expected to be very close (<2 cm^{-1}) to the original molecule (16), usually within the inhomogeneously broadened absorption band. Photostable organic molecules that undergoe NPHB in glassy matrices are, for example, perylene (11), tetracene (16, 20, 47), and pentacene (50,

51). A similar type of photophysical hole-burning has also been observed in inorganic glasses doped with rare earth ions (48).

TRANSIENT HOLE-BURNING (THB) (12, 33, 34, 52) In this process, population is transferred from the ground state through the excited state of interest to a metastable state. The latter may either be the same excited state, a long-living triplet state, or a hyperfine level of the ground state. The lifetime of the hole (usually microseconds to seconds) is determined by the decay time of the metastable state.

Techniques

Persistent holes (PHB and NPHB) are usually burnt with narrow-band dye lasers (width, Γ_l) at a given temperature, and probed either at the same temperature (10–16, 37–40, 50–60), or at a different one (42, 49, 61–63). The most common probing schemes are transmission spectroscopy (11, 42, 47, 50, 59–65) and excitation spectroscopy (10, 13–15, 40, 48, 56, 58). In the latter, the spectrum is generally scanned with the same laser as used for burning, but at 10^{-3}–10^{-1} of its original intensity. In many transmission experiments, a lamp in combination with a monochromator have been used for the hole-probing step, but this introduces significant errors in the determination of the holewidth. The excitation method is more convenient, because samples of low optical quality and density can be measured with high sensitivity at very low laser burning powers (for a critical comparison of the two methods, see 22, 51). The only prerequisite is a luminescent sample. If one wants to get information on the homogeneous linewidth, Γ_{hom}, an additional condition is that $\Gamma_l \ll \Gamma_{hom}$.

The detailed experimental aspects of hole-burning have been discussed in the literature (23, 24) and are not repeated here. Nevertheless, in order to understand what is involved in the experiments, I include a schematic diagram of a PHB and NPBH arrangement in Figure 2 as used in Refs. (39, 51, 66). The sample, immersed in liquid helium, is irradiated with a single-frequency cw dye laser (jitter ≤ 2 MHz) pumped by an Ar^+-ion laser. The apparatus in the lower part of the figure serves for frequency calibration purposes. The holes are probed either by fluorescence excitation spectroscopy or simultaneously by the latter and in transmission through the sample (51). In the absence of emission from the sample, a special type of low temperature photoacoustic spectroscopy has been used, which is based on resonant detection of second sound in superfluid ^4He (67).

Other hole probing techniques that yield signals against zero background are laser frequency modulation (FM) spectroscopy (68–70), ultrasonic modulation of holes (71), polarization spectroscopy (72, 73), and holographic detection (74, 75).

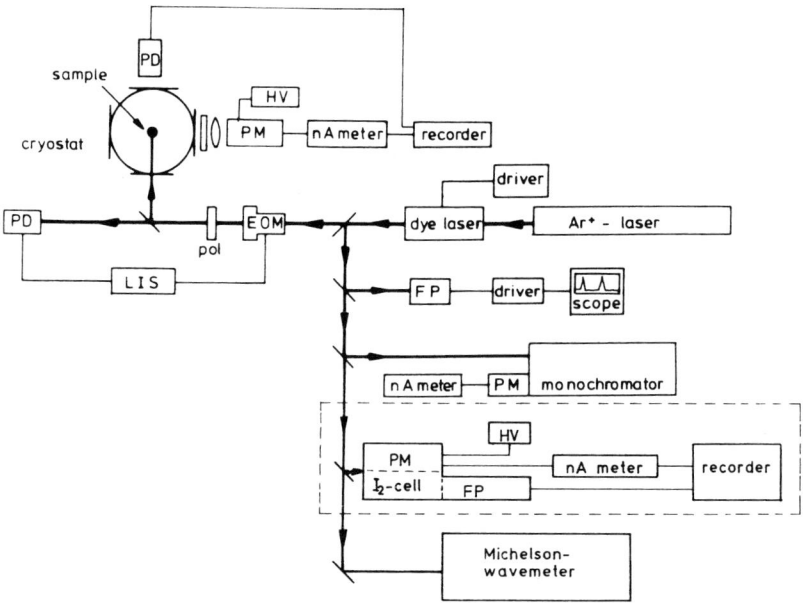

Figure 2 Experimental arrangement for PHB and NPHB [for details see text and Ref. (24)]. FP, Fabry-Pérot interferometer; PM, photomultiplier; EOM, electro-optic modulator; LIS, light intensity stabilizer; PD, photodiode.

Transient hole-burning may either be performed with one or two single-frequency cw dye lasers, depending on the lifetime of the metastable level (12, 33, 34, 52). If two lasers are used, the first one burns the hole at a fixed frequency, whereas the second, at lower intensity, scans the excitation spectrum of the hole either during burning or after a delay time. If one laser is used, the beam is divided in two, and one part is delayed with respect to the other by acousto-optic modulators. For a discussion on cryostats used between 0.3K and 20K, see (76–78).

In order to get an accurate determination of Γ_{hom}, under the assumption that the time scale of the experiment plays no role, a series of holes should be burnt as a function of laser power, P, and burning time, t, and the holewidths extrapolated to $P \to 0$ and $t \to 0$. The value of Γ_{hom} can be estimated from the relation $\Gamma_{hom} \approx 0.5\, \Gamma_{hole} - \Gamma_l$, where Γ_{hole} is the measured total holewidth at half height of an approximately Lorentzian hole, and Γ_l is the laser jitter ($\lesssim 2$ MHz), assuming that $\Gamma_{hom} \ll \Gamma_{inh}$. Since the hole shape is the result of a convolution of a burning step and a probing step, the relation $\Gamma_{hole} = 2\Gamma_{hom}$ is valid for holes burnt and probed at the same temperature (15, 19).

Systematic studies of the influence of laser power and burning time on the holewidths (51, 66, 76) have suggested that many of the holewidths previously reported in the literature (16, 42, 47, 49, 50, 54, 79) may suffer from saturation and/or heating effects. An example is given in Figure 3, where the holewidth, $\frac{1}{2}\Gamma_{hole}$, as a function of burning time, t, for resorufin in glycerol is plotted for two burning powers at 1.2K (76). The values of $\frac{1}{2}\Gamma_{hole}$ on the lowest curve extrapolate smoothly to $\Gamma_{hom} \simeq 105 \pm 10$ MHz for $t \to 0$, whereas the two upper curves only reach the value of Γ_{hom} with a much steeper slope. Thus, shorter burning times are needed to get the right extrapolation value in the latter case. It should be noticed that the holewidths become saturated at longer burning times and/or higher burning powers. It follows that the "true value of Γ_{hom}," at a given temperature, is only obtained at the lowest possible burning fluences Pt. The meaning of the "true value of Γ_{hom}" is discussed below in relation to spectral diffusion processes in glasses.

Local heating of the sample may also be a reason for artifacts in the evaluation of holewidths in amorphous material, since the latter have poor thermal conductivity (80). It is not improbable that in many hole-burning

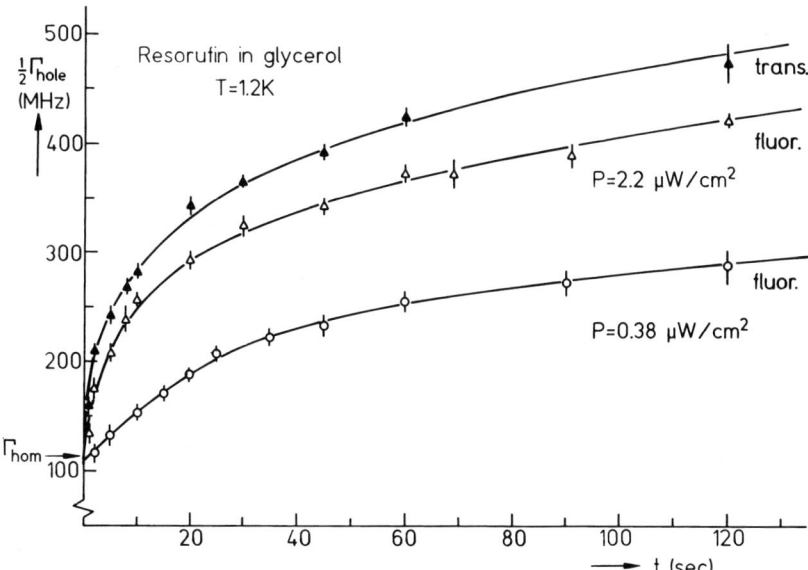

Figure 3 $\frac{1}{2}\Gamma_{hole}$ as a function of burning time for resorufin in glycerol at $T = 1.2$K for two laser burning powers. At $P = 2.2$ μW/cm², the holes were detected simultaneously via fluorescence excitation and transmission through the sample. At $P = 0.38$ μW/cm², holes were observed in fluorescence only. For $t \to 0$, $\frac{1}{2}\Gamma_{hole} = \Gamma_{hom} \simeq 105 \pm 10$ MHz (76).

experiments performed at rather high burning powers in flow cryostats, where the sample was only in contact with helium gas (16, 42, 47, 50, 54, 59, 81) and the holes were probed in transmission, the temperature in the bulk of the sample was higher than assumed.

OPTICAL DEPHASING OF ELECTRONIC TRANSITIONS OF ORGANIC MOLECULES

The high spectral resolution of the order of MHz attainable with hole-burning makes this technique attractive for many applications. In particular, since *permanent* holes at liquid helium temperature represent sharp frequency markers in the absorption spectrum, it is possible to detect frequency shifts, in addition to hole broadening, as a function of temperature, that are many orders of magnitude smaller than the width of the inhomogeneous absorption band. Thus, optical dephasing and relaxation processes of electronically excited states of molecules and ions incorporated in either crystalline or amorphous hosts at low temperature can be studied with high accuracy by HB.

The optical properties and hole-burning characteristics of many of the organic systems to be discussed have been described in detail in (24) and thus are not repeated here.

Crystalline Hosts

EXPERIMENTAL RESULTS The first MHz-resolution HB-experiment performed on an organic system with a holewidth close to the true value of Γ_{hom} was reported in 1976 (13). The system was dimethyl-s-tetrazine (DMST) in a durene crystal. The holewidth at 2K (120 MHz), however, was still significantly broader than the width that corresponds to the decay time of DMST (~ 25 MHz), and was attributed to low temperature electron-phonon coupling (13). In a crystalline system at the lowest temperatures, one would expect, however, that the population decay time T_1 is ultimately the factor limiting the holewidth, because phonon processes are frozen out. This was, indeed, observed for the $S_1 \leftarrow S_0$ 0–0 transition of free-base porphin (H_2P) in an n-octane ($n\text{-}C_8$) crystal, where the holewidth for $T \leq 2K$ was entirely determined by the fluorescence lifetime of H_2P, $T_1 = 17$ ns. Thus, $\Gamma_{\text{hom}} = (2\pi T_1)^{-1}$ when $T \to 0$ (15, 53). The same result was found by HB for other porphin molecules in n-alkane hosts: ZnP in $n\text{-}C_8$ (33), chlorin in $n\text{-}C_6$, $n\text{-}C_8$, and $n\text{-}C_{10}$ (37), H_2P in $n\text{-}C_{10}$ (55), and MgP in $n\text{-}C_8$ (34). By contrast, photon echo decay times, $\frac{1}{2}T_2$, obtained for the $S_1 \leftarrow S_0$ 0–0 transitions of tetracene and pentacene in p-terphenyl (82), naphthalene in durene and perdeutero-naphthalene (83), and pentacene in naphthalene (84) at 1.4K were claimed not to be determined by T_1. It was suggested that not all relaxation processes in these organic crystals were frozen out at the lowest temperature (82). Energy transfer (83) and quadratic electron-phonon coupling (84) processes were held responsible for these discrepancies. Subsequent hole-burning experiments, carried out with a better-stabilized laser (35) and more accurately measured photon echoes (56, 85–87), confirmed that homogeneous linewidths in crystals at 2K are entirely determined by T_1.

As the temperature increases, thermally induced dephasing processes rapidly set in. As mentioned above, *permanent hole-burning* (PHB) has the unique advantage, with respect to coherent transient techniques, to be able to measure not only the homogeneous *linewidth* but also its *frequency shift* as a function of temperature. This is illustrated in Figure 4 for free-base porphin in *n*-octane. Since most of the PHB studies in crystals have been performed on porphin molecules and their derivatives in *n*-alkane hosts, I use these results here as a framework for the understanding of optical dephasing processes in dilute molecular mixed crystals.

The holewidths and frequency shifts of the $S_1 \leftarrow S_0$ 0–0 transition of free-base porphin (H_2P) in *n*-octane were found to depend exponentially on temperature as $\exp(-E/kT)$, with activation energies E between 10 and 35 cm^{-1}, depending on the site (15, 36, 53). Such low activation energies have been interpreted (15, 53) in terms of pseudo-local phonon modes (32a,b, 82), which arise from a distortion of the host crystal due to the presence of the guest molecule (88). In the case of porphin molecules in *n*-alkane crystals, they correspond to librational motions of the guest in the host (15, 53, 89). Localized modes have been observed as phonon sidebands in excitation and fluorescence spectra of mixed molecular crystals, like pentacene in naphthalene (90) and benzoic acid (91), and of various prophin molecules and their derivatives [Mg-porphin (34), Zn-porphin (89), H_2-chlorin (37), H_2-bacteriochlorin (92), H_2-tetra-ter-butylphthalocyanine (93)] in *n*-alkane crystals. The frequencies of these phonon sidebands coincide with the activation energies found from the temperature dependent hole dynamics (94). The frequencies and lifetimes of pseudo-local phonon modes are related to the tightness of the fit of the guest in the host (53, 55, 95). Very similar dephasing behavior as for H_2P was observed for chlorin (H_2Ch) in *n*-alkane hosts (*n*-C_6, *n*-C_8, and *n*-C_{10}). From the very small increase of the holewidth with temperature between 1.6 and 4.2K, and from Zeeman (96–98) and Stark (99) effect measurements, the type of site that H_2Ch occupies in the *n*-alkanes studied was inferred (37).

Optical dephasing of $S_1 \leftarrow S_0$ 0–0 transitions of metal porphins were initially interpreted differently from free-base porphin and its derivatives. Only ZnP (33) and MgP (34) in *n*-C_8 have been studied by *transient hole-*

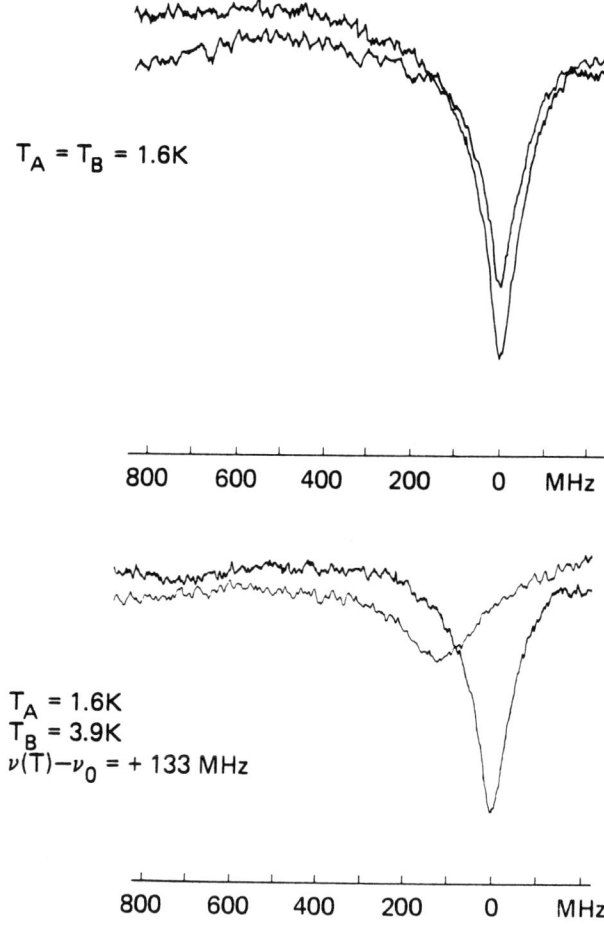

Figure 4 Frequency shift and broadening of a hole burnt in the B_1 site of the 0–0 transition of free-base porphin (H_2P) in n-octane (n-C_8). *Top*: Excitation spectra of two holes burnt at the same frequency at 1.6K in identical samples in two cryostats, A and B. *Bottom*: Excitation spectra of the same holes after raising the temperature of cryostat B to 3.9K. The temperature behavior of the hole is reversible (53).

burning (*THB*) so far. THB occurs here by selective depletion of the ground state population and storage in the metastable triplet state. The temperature dependence of the homogeneous linewidth of ZnP in n-C_8 was attributed to a resonant one-phonon absorption between the S_{1x} and S_{1y} components (33) rather than to coupling to a low-frequency local mode. Subsequent THB experiments on four sites of MgP in n-C_8 proved that optical dephasing is in fact caused by thermal excitation and de-excitation of local phonon modes (34), just as for H_2P in n-alkanes. The one-phonon scattering mechanism (33) could be disregarded, because the crystal-field splittings for the various sites of MgP in n-C_8 differ markedly from the activation energies measured (34).

THEORETICAL MODELS Various theories have been developed to explain optical dephasing of impurities diluted in crystalline hosts. In these models it is assumed that the optical transition is represented by a two-level system coupled to either acoustic, optical, or pseudo-local phonons. For a detailed description of these theories, the reader may consult (32a,b, 100–103). I discuss here the physical implications of one of these models, the exchange model (53, 104–109), because it seems to explain most of the results presented above. It can be summarized as follows (107a,b): Optical dephasing is due to coupling of the electronic oscillator to a specific local mode, which may be thermally excited. In a four-level scheme, the transition between the zero-phonon levels is given by $\langle S_1, 0| \leftarrow \langle S_0, 0|$ with frequency v_0, and between the one-phonon-levels by $\langle S_1, 1| \leftarrow \langle S_0, 1|$ with frequency $v_0 + \Delta/2\pi$ (Δ can be >0 or <0). The electronic oscillation then becomes modulated by absorption and re-emission of a phonon (at $T > 0$) (53). A nonzero value of $\Delta/2\pi = E' - E$ implies that the librational mode frequency in the excited state, E', is different from that in the ground state, E. Another quantity involved in the model is the lifetime τ of the local phonon, which is assumed to be the same in both phonon states, but much shorter (picoseconds) than that of the electronic excitation (nanoseconds) (107a,b). If $kT \ll E$, then the broadening of Γ_{hom} and the frequency shift are given, respectively, by (53, 107b)

$$\Gamma_{\text{hom}}(T) - \Gamma_0 = 2\Delta\tau \frac{\Delta/2\pi}{1+\Delta^2\tau^2} \exp(-E/kT) \qquad 2.$$

and

$$v(T) - v_0 = \frac{\Delta/2\pi}{1+\Delta^2\tau^2} \exp(-E/kT). \qquad 3.$$

The parameters Δ and τ can be independently determined from both frequency shifts and widths in PHB experiments (53). The exchange model was successfully applied to interpret the results of H_2P (53), H_2Ch (37), and MgP (34) in A- and B-sites of n-C_8, and of H_2P in the B-site of n-C_{10} (55). As is shown below, one of the theoretical models for optical dephasing in amorphous systems also assumes that local phonon modes are responsible for the broadening of the homogeneous linewidth in glasses (101, 110, 111).

The first theory that explained optical linewidths and temperature shifts in crystalline systems was reported in 1963 by McCumber & Sturge (112), who predicted different temperature dependences for the linewidth and for the frequency shift. The exchange model (107a,b), in contrast, predicts the same exponential dependence on temperature for both, a prediciton that is consistent with the experimental results obtained for porphin molecules in n-alkane crystals (53, 55, 94).

Another theory for optical dephasing in molecular mixed crystals at low temperatures was presented by De Bree & Wiersma (113), who applied Redfield relaxation theory (114) to a four-level system coupled to an anharmonic phonon bath. The phonon scattering is in this case "uncorrelated," because the lifetime of the local mode is assumed to be different in the two electronic states ($\tau \neq \tau'$) (113). In this model, the homogeneous linewidth Γ_{hom} increases bi-exponentially with temperature (6, 91). Photon echo results of mixed crystals (87, 90, 91) were interpreted in this way. This model was recently also adapted to interpret optical dephasing data in glasses (50).

A third, non-perturbative, theory for optical dephasing in crystals has been proposed by Hsu & Skinner (88). For low temperatures and sufficiently long-lived librational phonon modes, a bi-exponential expression results, which reduces to Eq. 2 of the exchange theory if $\tau \simeq \tau'$ and $\exp(-\Delta/kT) \simeq 1$, and to the "uncorrelated phonon scattering" result of Ref. (113) for $\Delta \tau \gg 1$. Attempts have been made to explain several experimental linewidth results for mixed crystals with this theory (32a, 88, 100, 115).

Amorphous Hosts

The physical properties of amorphous solids at low temperature are very different from those of crystals [for a review on glasses, see (80)]. This is unexpected, because the structure of the material should become unimportant when long wavelength phonons are excited. The most successful theory that has been proposed to explain these anomalies is the tunneling model by Phillips and by Anderson et al (116a,b), based on the assumption that atoms or groups of atoms occupy two energetically inequivalent configurations of the glass, the so-called "two-level-systems" (TLS). A wide distribution of such TLS is found due to the random distribution of atoms or molecules in glasses. Most of the experiments performed at low temperatures since 1972 have been interpreted in the framework of this model (for reviews, see 101, 111, 117).

In the last ten years, much work has been devoted to the study of linewidths of impurity ions and molecules incorporated as guests in amorphous materials in order to understand their dynamics. Since most of the optical dephasing to be discussed here is related to organic glasses, the reader interested in inorganic glasses may consult Refs. (21, 27). The first experiments on the temperature dependence of linewidths on amorphous systems at low temperature were performed in 1976 on a rare earth-doped inorganic glass, Eu^{3+} in silicate between 8 and 90K, by fluorescence line-narrowing (FLN) (118). The linewidths were one to two orders of magnitude larger than those previously found in crystals and, in contrast to the exponential temperature dependence characteristic of the latter, a T^2 dependence was observed. Only one inorganic amorphous system so far has been studied below 1K, Nd^{3+} in fused silica by two-pulse photon echoes, for which a $T^{1.3}$ dependence of Γ_{hom} between 0.1 and 1K was reported (119a–c).

ORGANIC MOLECULES IN ORGANIC GLASSES The first hole-burning experiments on optical linewidths in organic glassy systems were performed by Personov and co-workers in 1974 (11) on perylene in ethanol. Since then most of the optical dephasing results on organic glasses have been obtained with this technique (29). Only a few organic amorphous systems have been investigated by other methods, in particular, accumulated photon echoes (50) and two-pulse picosecond-photon echoes (81, 120, 121).

Holes burnt in $S_1 \leftarrow S_0$ 0–0 transitions of organic molecules in alcoholic glasses were initially reported to be a few cm^{-1} broad at about 2K (11, 16, 41, 42, 47), i.e. orders of magnitude larger than in crystalline hosts (13, 15, 33–37, 56). Furthermore, it was found that the holewidths depended on temperature as T^n, with $n = 1$ (20, 47, 64) or 2 (42, 49, 64), depending on the system studied. In none of these experiments was the fluorescence lifetime-limited value, $\Gamma_0 = (2\pi T_1)^{-1}$ of a few MHz to tens of MHz, reached, neither at low temperature nor by extrapolation to zero temperature, and residual linewidths about three orders of magnitude larger than expected were reported. It was suggested that picosecond dephasing processes would still take place at $T = 0$ (20, 42, 47), which was subsequently proven to be incorrect. The holewidths did not represent the true homogeneous linewidth, because too high laser fluences were used for burning, often combined with too large instrumental bandwidth (66).

Some of the organic molecules that have been studied as guests in organic amorphous hosts by hole-burning are illustrated in Figure 5. Only three so far have also been investigated by other techniques: pentacene by accumulated photon echoes (50), and resorufin and cresyl violet by two pulse picosecond-photon echoes (81, 120, 121).

A systematic hole-burning study of organic amorphous systems was started in the author's group in 1982 (39) with the aim of resolving the contradictions reported in the literature until then. The first molecule

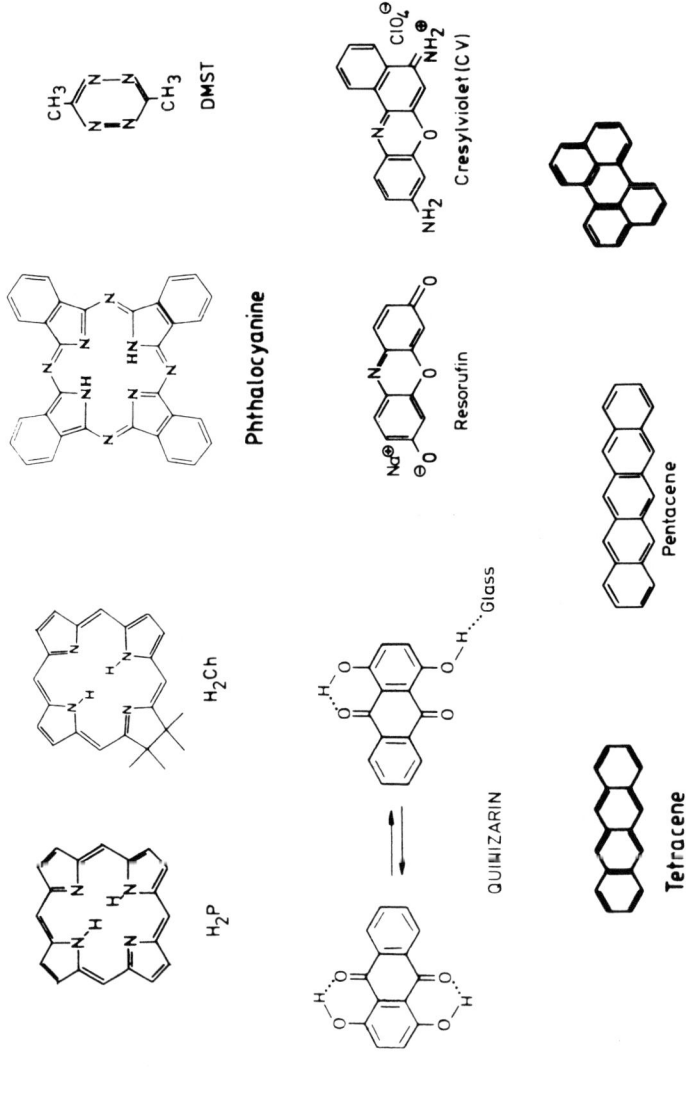

Figure 5 Organic molecules studied as guests in organic amorphous hosts. *Top*: free-base porphin (H$_2$P), free-base chlorin (H$_2$Ch), phthalocyanine, and dimethyl-s-tetrazine (DMST), which all undergo intramolecular-PHB. *Middle*: quinizarin, and the ionic dyes resorufin and cresylviolet (CV), which undergo intermolecular-PHB. *Bottom*: tetracene, pentacene, and perylene, which undergo non-photochemical-HB.

investigated was free-base porphin (H_2P), because of its well-known behavior in crystalline hosts. Hole-burning experiments between 1.2 and 4.2K (39, 40, 66), and later down to 0.3K (43, 76, 77) and up to 20K (76), were performed on the $S_1 \leftarrow S_0$ 0–0 transition in a large variety of glasses and polymers. In strong contrast to the data reported by other groups, the following results were found: (a) narrow holes of about hundreds of MHz to a few GHz at $T \simeq 2K$; (b) the value of $\Gamma_{hom} = \frac{1}{2}\Gamma_{hole}$ (at very low burning fluences) did extrapolate to the fluorescence lifetime-limited value, $\Gamma_0 \simeq 10$ MHz cf H_2P, when $T \to 0$; (c) Γ_{hom}-Γ_0 followed a $T^{1.3}$-temperature law, independent of the organic amorphous host used. This is illustrated in Figure 6, where a log-log plot of Γ_{hom}-Γ_0 versus T for H_2P in various polymers between 0.3 and 20K (66, 76) is shown.

These results, together with those found for the dimer of H_2P, suggested that the $T^{1.3}$ law is independent of the chemical structure of the guest. In fact, many organic amorphous systems subsequently showed that Γ_{hom}-$\Gamma_0 \propto T^{1.3}$, at least between 0.3 and 20K (76), independent of their hole-burning mechanism (40, 43, 52, 76). [See Figure 7 for an example for ionic dyes (43, 76).] In addition, it was verified that Γ_{hom} extrapolates smoothly to $\Gamma_0 = (2\pi T_1)^{-1}$ when $T \to 0$ (43, 44, 76–78), which is summarized in Figure 8.

Hole-burning experiments reported by other groups indicate that a similar power law is obeyed for a variety of organic glasses (60, 120–123) and even for chromophores embedded in proteins (124, 125), at least in a restricted temperature range. Only two organic amorphous systems to our knowledge, free-base octaethylporphyrin (OEP) in polystyrene (PS) and PMMA, have been investigated below 0.3K and down to 0.05K (122, 126a–c). At temperatures between 1.5 and 0.3K, a $T^{1.2}$–$T^{1.3}$ dependence of Γ_{hom}-Γ_0 was observed that apparently crosses over to $T^{1.6}$ (126a) or $T^{2.5}$ (126b), depending on the value of Γ_0, below 0.1K. Hole-burning data above 4K seem to follow a $T^{1.8}$ law (126c), which is discussed below in relation to spectral diffusion processes.

A $T^{1.3}$-dependence of Γ_{hom}-Γ_0 was also observed for pentacene in PMMA between 1.5 and 12K by accumulated photon echoes (50), for cresyl violet in polyvinylalcohol between 6 and 19K by incoherent nanosecond-photon echoes (127a), and for Nd^{3+} glass between 0.1 and 1K by two-pulse nanosecond-photon echoes (119a). It should be mentioned that optical hole-burning experiments with femtosecond lasers have recently been performed by Shank and co-works (127b) on cresyl violet in ethylene glycol at room temperature. Hole-recovery times of about 50 fs were attributed to thermalization of a nonequilibrium distribution of excited states.

The hole-burning vs. photon-echo controversy It has been claimed that

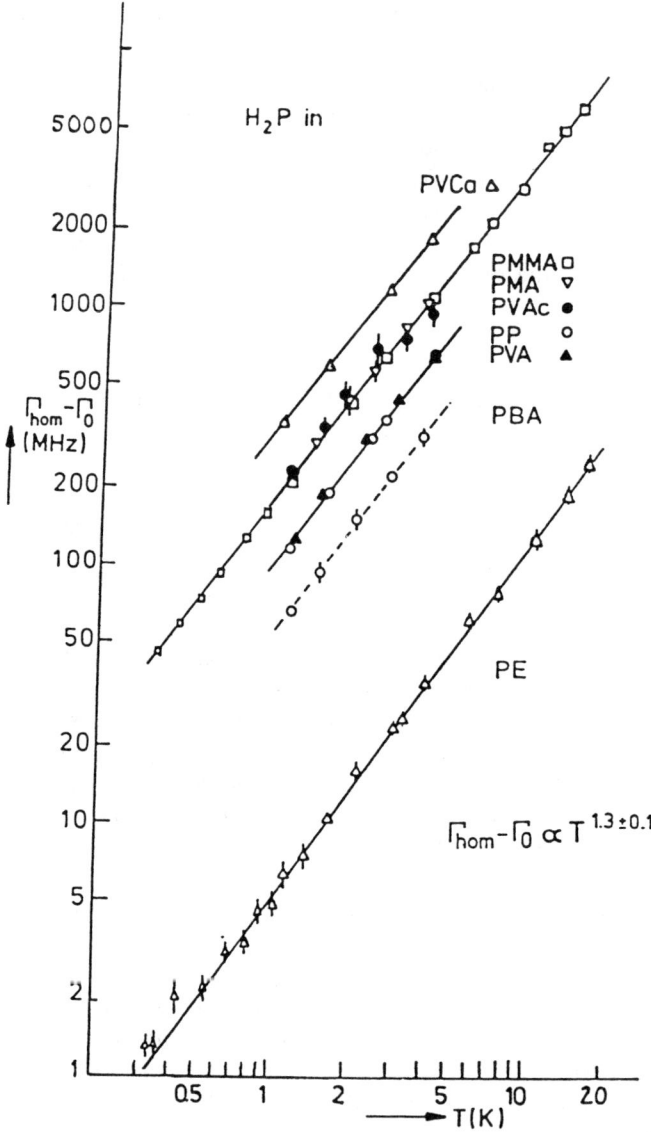

Figure 6 Log-log plot of Γ_{hom}-Γ_0 versus temperature for H_2P in several polymers between 0.3 and 20K (PVCa, polyvinylcarbazole; PMMA, polymethyl-mathacrylate; PMA, polymethylacrylate; PVAc, polyvinylacetate; PP, polypropylene; PVA, polyvinylalcohol; PBA, polybutylacrylate; and PE, polyethylene) (40, 66, 76). Notice that Γ_{hom}-$\Gamma_0 \propto T^{1.3 \pm 0.1}$, independent of the host.

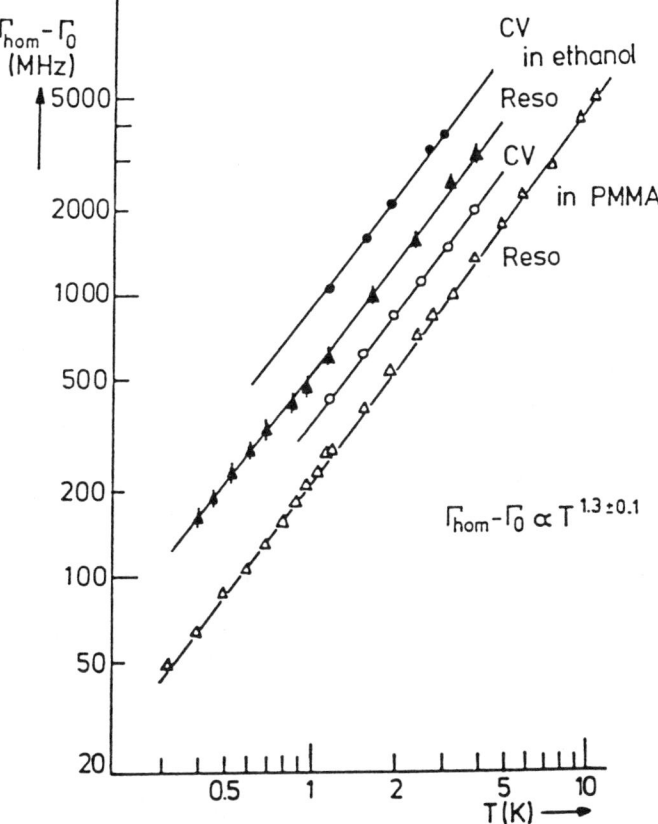

Figure 7 Log-log plot of $\Gamma_{hom}-\Gamma_0$ versus T for the ionic dyes cresyl violet (CV) and resorufin in ethanol and PMMA between 0.3 and 10K. In all cases, $\Gamma_{hom}-\Gamma_0 \propto T^{1.3\pm0.1}$ (43, 76, 128).

linewidths obtained by non-photochemical hole-burning (NPHB) do not reflect the homogeneous linewidth, Γ_{hom}, but are only a measure of slow relaxation processes in the glass (50, 81, 120, 121). This claim was first used to explain contradictory results obtained with accumulated photon-echo and NPHB experiments for pentacene in PMMA (50). Large differences in linewidths were also reported for resorufin in ethanol (120) and glycerol (81), and for cresyl violet in ethanol (121) measured by two-pulse picosecond-photon echoes and hole-burning. In principle, hole-burning and photon echoes may yield different information, because they measure on different time scales (picoseconds to milliseconds for photon echoes, seconds to minutes for permanent HB), and glasses have a very broad distribution of relaxation rates (80, 116a,b).

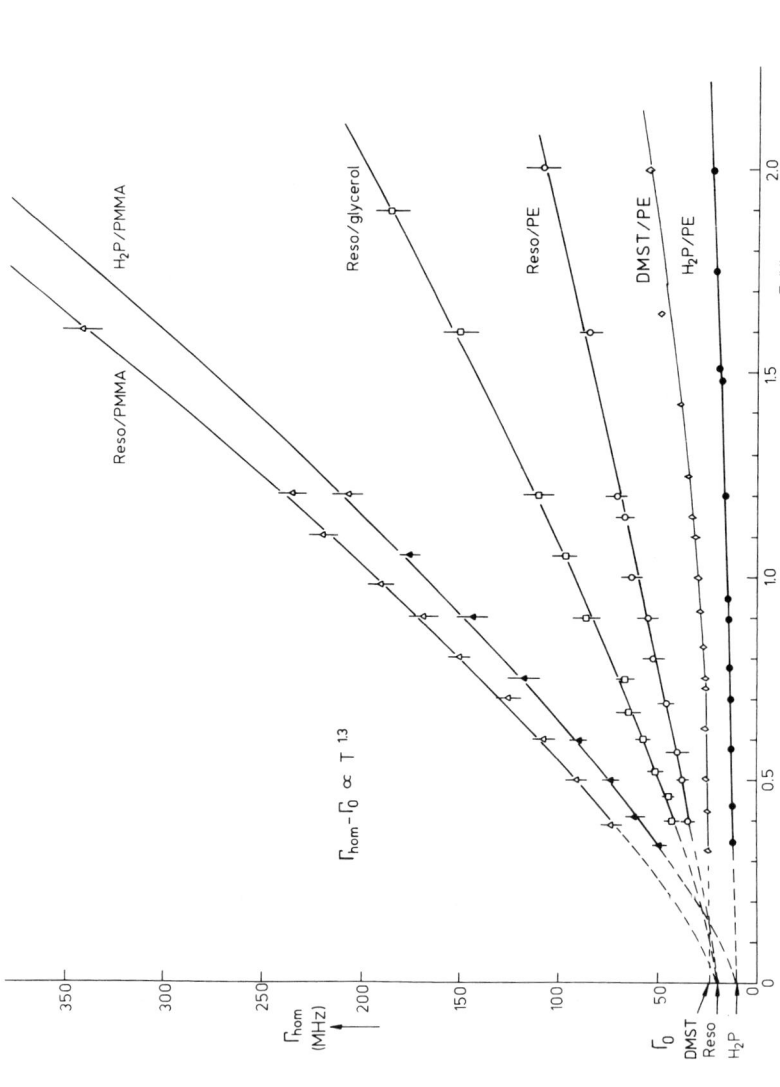

Figure 8 Temperature dependence of the homogeneous linewidth, Γ_{hom}, of resorufin in PMMA, glycerol, and PE, H_2P in PMMA and PE, and dimethyl-s-tetrazine (DMST) in PE at $T \leq 2K$. For $T \to 0$, Γ_{hom} extrapolates to the value of $\Gamma_0 = (2\pi T_1)^{-1}$ for each guest molecule. In PE the value of Γ_0 is almost reached at 0.3K for all three guest molecules. The lines through the data were traced according to $\Gamma_{hom} = \Gamma_0 + aT^{1.3}$ (76, 78).

A reinvestigation by hole-burning of pentacene in PMMA (51), resorufin in ethanol (128), and resorufin in glycerol (76), in particular, of the influences on the holewidth of laser power, burning time, sample preparation, optical density, and sample heating effects, led to results different from those reported in (50, 81, 120). By using 10^3 times lower laser fluence and probing the holes by fluorescence excitation (51), holewidths for pentacene in PMMA at 1.5K were obtained that were more an order of magnitude smaller than in (50). Furthermore, the holewidths showed a $T^{1.3}$ dependence between 1.2 and 4.2K (51), as previously observed for other organic amorphous systems. The HB results were very similar to those obtained by accumulated photon echoes (50), with a small discrepancy attributed to temperature effects (51).

For resorufin in ethanol it was found that the sample preparation, i.e. its cooling rate, is of crucial importance in the determination of the value of Γ_{hom} (128), since ethanol can exist in more than one solid phase. Holewidths that differed by a factor of 8 were found in two phases of ethanol, but in both, Γ_{hom}-$\Gamma_0 \propto T^{1.3}$ between 0.3 and 4.2K (128). The discrepancies between hole-burning and photon echoes (120) were attributed to differences in sample preparation (128).

Hole-burning experiments on resorufin in glycerol performed at very low burning fluences yielded a $T^{1.3}$ dependence of Γ_{hom}-Γ_0 between 0.3 and 22K (76), in contrast to the results obtained in (81) (see Figure 9). The discrepancy between the hole-burning data of the two groups has been explained in terms of differences in burning fluences (76), since more than ten times larger Pt-values have been used in (81). The differences, however, between the hole-burning data of (76) and the photon-echo data of (81) have not been explained in a quantitative manner yet.

Spectral diffusion processes It has been suggested that the extra broadening of the hole-burning with respect to the photon-echo data arises from spectral diffusion, and that it follows a power law dependence T^α with $\alpha \leqq 1$ (81). It is obvious from Figure 9 that the difference between the linewidths determined by hole-burning (76) and by photon echoes (81) for resorufin in glycerol does not obey such a power law, because the data cross each other at $T \sim 12K$ and probably at $T < 1.1K$.

Spectral diffusion processes on a time scale between that of photon echoes (picoseconds to milliseconds) and that of conventional hole-burning experiments (seconds to minutes) have not been unambiguously demonstrated for organic amorphous systems yet. Transient hole-burning experiments on H_2P in PE between 1.2 and 4.2K performed on a time scale of milliseconds (P. J. van der Zaag, unpublished results from the author's laboratory) have yielded results identical to those obtained by conventional

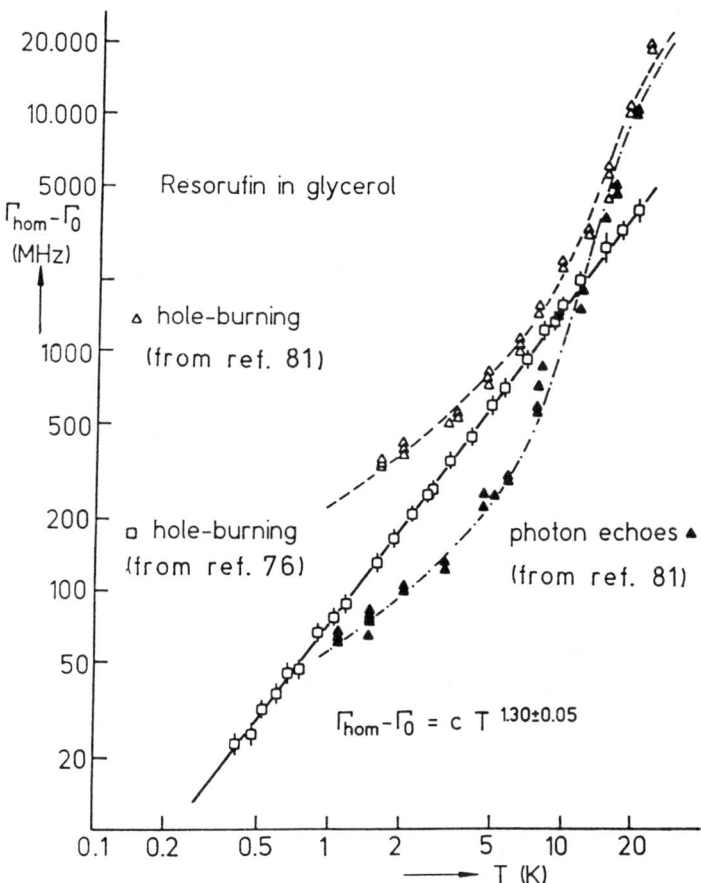

Figure 9 Log (Γ_{hom}-Γ_0) versus log T between ≈ 0.3 and ≈ 25K for resorufin in glycerol. The *dashed curves* have been obtained from hole-burning and photon echo data of Ref. (81). The *full line*, which follows a $T^{1.3}$ power law over two decades in T, is the result of hole-burning experiments (76). All data were plotted with $\Gamma_0 = 20$ MHz.

HB on a time scale of minutes (66, 76). Furthermore, holes burnt in H_2P in PE at 1.2K probed in fluorescence did not broaden as a function of time after burning, not even after many hours, a result that suggests that spectral diffusion does not take place in a time span of seven orders of magnitude (from 10^{-3} to 10^4 s), at least for this system at low temperature. In contrast, hole-burning experiments at 1.5K on a time scale of 10^{-5} s (29, 126b), and resonance fluorescence experiments between 5 and 11K (126c), both on OEP in PS, yielded linewidths that were a factor 7 and 2, respectively, smaller than those obtained by conventional HB in the same group (126a).

It is worth comment that accumulated photon echo and hole-burning experiments on Pr^{3+} in silicate glass between 1.6 and 20K yielded identical linewidths, despite the very different time scales of these two experiments of $\sim 10^{-4}$ s and 10^2 s, respectively (48). Thus, no slow rearrangement of the glass network seems to contribute to the holewidth in this inorganic system. In contrast, three-pulse stimulated nanosecond-photon echo experiments on Nd^{3+} doped-silica at $T \leq 0.1K$ (119b) suggested the presence of short-time spectral diffusion processes.

Slow relaxation processes in organic glasses, occurring on time scales of hours and even days, have been observed by hole-burning (50, 79, 129a,b, 130). This spectral diffusion effect, however, only occurs under special experimental conditions, as has been demonstrated in (51). Holewidths probed simultaneously by fluorescence excitation and in transmission through a sample of pentacene in PMMA at 1.2K were measured as a function of time after burning. The holes detected in fluorescence did not broaden on a time scale of hours, not even at large burning fluences, a finding that indicates that, as long as the molecules are in thermal contact with liquid helium, no spectral diffusion occurs, at least not on a time scale between a few seconds and 10^4 s (51, 76). Similar results have been obtained for resorufin in glycerol (76). However, when holes are probed in transmission through a rather thick amorphous sample [$d = 2$ mm, $OD \approx 0.9$ (51)], they do broaden on a time scale of minutes to hours (50, 51), the effect being more pronounced for higher burning fluences. The conclusion from these experiments was that in the bulk of the glass, which has poor thermal conductivity, the heat of the laser probably induces strain that causes slow structural relaxation of the host. During this process the inhomogeneous linewidth is affected and, as a consequence, broadening of the hole is observed.

Long-time spectral diffusion hole-burning experiments on organic glasses have been interpreted by Friedrich & Haarer in terms of a model based on the concept of spectral diffusion of a dilute spin system. The expression for the holewidth then becomes time dependent (129a,b, 130). From reversible and irreversible hole broadening contributions obtained by thermal cycling experiments, the same group concluded that temperature-induced spectral diffusion is small compared to pure dephasing processes at temperatures between 4 and 25K (63).

Theoretical models Soon after the first experiments on optical dephasing in amorphous materials at low temperature were reported, a number of theories were proposed (47, 118, 131–133). Since then, many more models (50, 110, 119c, 134–140) have been developed to account for subsequent experimental results. All these theories are based on the same physical concept, i.e. the optical transition of the impurity ion or guest molecule interacts with the two-level-systems (TLS) of the glass (80, 116a,b), and the total system interacts with the phonons of the bath. The interaction with the TLS is assumed to be different in the ground and electronically excited states. Furthermore, the density of the TLS-states is assumed to be nearly constant as a function of energy. Since the TLS absorb and reemit phonons of the bath, they modulate the energy levels of the electronic states, and this causes dephasing of the optical transition. The various models differ from each other either in the way the coupling impurity-TLS-phonons are introduced, or in the approximations made to evaluate the linewidth, or in the procedures used to average over the TLS parameters. Some of the models include additional processes in order to be able to explain the experimental results. Small changes in the calculated temperature dependence of Γ_{hom} can easily be achieved by slightly varying either the approximations or the averaging procedure. Each of the theories explains specific aspects of the experimental data, but unfortunately, many of the theoretical assumptions have not been verified experimentally. Reviews of the various theoretical models and the derivation of different temperature dependences of Γ_{hom} can be found in the literature (29, 101, 111, 117, 134, 139).

Since 1983 most of the theories have tried to provide an understanding of the $T^{1.3}$-temperature dependence of the homogeneous linewidth found for many organic amorphous systems between 0.3 and 20K (22, 39, 50, 51, 76) and for one inorganic glassy system (119a). These models are not treated here. The reader interested in a critical comparison between theories and experimental results should consult (21–24).

SEMICRYSTALLINE POLYMERS Hole-burning experiments on organic molecules in semicrystalline polyethylene (PE) were carried out to check whether any of the dephasing models for glasses really applies (78). If librational or localized modes, as predicted in (110), were important for optical dephasing in glasses, one would expect the temperature dependence of $\Gamma_{hom}-\Gamma_0$ in semicrystalline systems to be steeper than in amorphous systems and, conceivably, to approach the exponential function observed for crystalline materials.

In fact, it was found that dimethyl-s-tetrazine (DMST) diffused into previously molten, and subsequently solidified, semicrystalline PE follows a much steeper than $T^{1.3}$ dependence of $\Gamma_{hom}-\Gamma_0$ between 0.3 and 4.2K. The power of T increases with the degree of crystallinity of the host (78). Since DMST diffused into amorphous PMMA follows a $T^{1.3}$ dependence, these results suggested that HB measures the local guest-host interaction, that is, the degree of order in the direct environment of the guest molecule, and not the properties of the host independently of the guest.

Experimental data for amorphous (76, 128) and semicrystalline systems (78) were fitted with two of the theoretical models (107b, 110) in order to estimate the relative influence of localized modes, as compared to TLS modes, on the dephasing process. It appeared that low frequency localized modes may contribute to the dephasing in glasses, but, unfortunately, independent spectroscopic evidence is still lacking, both from site-selection and from hole-burning experiments, for such librations (76, 78, 128).

MIXED ORGANIC-INORGANIC SYSTEMS In order to get a better insight into the excited state dynamics of amorphous materials, two classes of mixed organic-inorganic systems have recently been studied: organic glasses doped with a rare earth ion, and organic molecules as guests in inorganic porous glass.

Organic molecules in porous glass Chlorin and oxazine-4 perchlorate in silicate glass, prepared by the sol-gel method, were studied by hole-burning between 1.7 and 5.7K (141). The homogeneous linewidths of the order of 500 MHz at 1.7K were similar to those observed in pure organic glassy systems, they also broadened as a function of temperature with a $T^{1.3}$ dependence. Preliminary experiments by the same group on dye molecules adsorbed on porous silica at 1.7K indicated that the holewidths were an order of magnitude larger, from which it was concluded that the geometric structure of the environment of the guest seems to be more important than the chemical composition of the amorphous host for the determination of the linewidth. Recent fluorescence line-narrowing and broad-band hole-burning experiments on cresyl violet adsorbed on γ-aluminum and silica gel have suggested that the HB-mechanism is most probably a photo-induced H-transfer from the excited dye to the surface (142).

Narrow-band HB-experiments on resorufin adsorbed on the surface of porous silica (Vycor glass) between 0.3 and 4.2K, and DMST on Vycor glass at 1.2K have shown that Γ_{hom} is in fact one order of magnitude larger than in polymers. Moreover, Γ_{hom} extrapolates to the fluorescence lifetime-limited value of resorufin, $\Gamma_0 = (2\pi T_1)^{-1}$ but is still far from reaching this value at 0.3K. Furthermore, $\Gamma_{hom} - \Gamma_0 \propto T^{1.3}$, as in organic glasses. These results suggest that the substantially broader homogeneous linewidths are probably due to a higher degree of freedom of the guest molecules on the *surface*, as compared to molecules incorporated in the *bulk* of the host (H. van der Laan, unpublished results from the author's group).

Rare earth ions in organic amorphous hosts In contrast to the previous results, HB-experiments performed on Eu^{3+} in organic glasses between 0.3 and 4.2K yielded very different results from those found for either pure organic or inorganic amorphous systems. The inhomogeneous linewidths [from a few GHz to ~ 20 cm^{-1} (143)] are much smaller than in pure glassy systems [a few hundred cm^{-1} (21)]. Furthermore, very narrow holes of a few MHz at liquid helium temperature could be burnt in the $^5D_0 \leftarrow {}^7F_0$ transition of Eu^{3+} in ethanol, water, ether, PMMA, and polyvinylalcohol (PVA), which followed an exponential temperature dependence, like in crystals, with activation energies of a few cm^{-1} (143). Residual linewidths for $T \rightarrow 0$ of the order of $\Gamma_0 \sim 2$–10 MHz (depending on the host) were found, which are about three orders of magnitude larger than the values expected from fluorescence decay. In addition, distinct side-holes and antiholes were observed symmetrically distributed at distances between ~ 20 and 300 MHz from the original hole (see Figure 10) (143). Such multiple hole patterns have been reported for crystalline inorganic systems (28) but not in any type of glassy host.

These results indicate that although the host is a glass in the bulk of the sample, the Eu^{3+} ions are surrounded in their first coordination sphere by *crystalline-like* host molecules that provide an ordered environment in the direct neighborhood of the ion (143). These experiments prove once more that hole-burning is a very sensitive method for detecting the structure of a material in the immediate vicinity of the excited guest molecule.

MOLECULAR AGGREGATES AND BIOLOGICAL COMPLEXES Other types of organic systems that have become attractive subjects for hole-burning in the last few years are molecular aggregates and biological complexes. The former are interesting because they show, in contrast to isolated molecules in a host matrix, exciton-like behavior with coherence lengths delocalized over several tens to hundreds of monomer units (144–147). Such electronically strongly coupled molecules have fast relaxation rates and may induce fast energy transfer processes, as observed in biological aggregates.

The study of biological complexes, like reaction centers of bacteria and plants, and antenna pigments by hole-burning and subpicosecond time-resolved techniques is a challenging task. Its aim: to understand the primary charge separation process in photosynthesis. In contrast to the narrow holes observed for isolated molecules in glasses and even in proteins (124, 125), very broad holes of ~ 100–500 cm^{-1} have been reported for bacterial (148, 149) and green plant (150) reactions centers, with a few exceptions (151a,b). One of the explanations for these surprising results is that the initially excited state decays in ultrashort time (25 fs) (148, 149); another, that strong vibronic mixing takes place with a nearly resonant charge-transfer state (152); and a third one, that strong electron-phonon coupling to a low frequency mode of the protein plays a role (149, 151a,b). The lively debate over these interpretations is still in progress (151a,b, 152).

Hole-burning has also been applied to some photosynthetic antenna

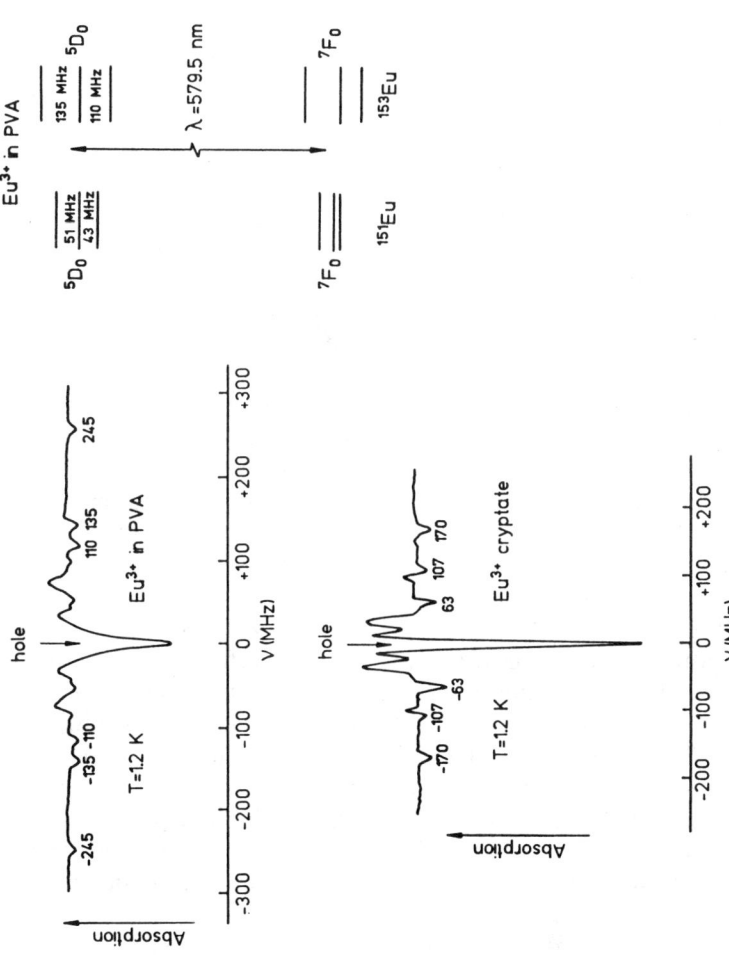

Figure 10 *Left*: Holes burnt in the $^5D_0 \leftarrow {}^7F_0$ transition of Eu^{3+} in PVA at 579.5 nm, and Eu^{3+} cryptate at 580.1 nm, both at 1.2K. The symmetric pattern of side-holes and antiholes corresponds to the nuclear hyperfine splittings in the ground and excited states of Eu^{3+}. The hyperfine splittings of the $^5D_0$-state were determined from the positions of the side-holes. *Right*: Energy level scheme of the $^5D_0 \leftrightarrow {}^7F_0$ transition for the two isotopes of Eu^{3+} in PVA (143).

complexes, from which energy transfer rates between pigments (153) and exciton-phonon coupling constants (154) have been derived. For a general review on hole-burning in biological and chlorophyll-like systems, see (24).

OTHER HOLE-BURNING APPLICATIONS

Vibronic Relaxation

Not only 0–0 electronic transitions, but also excitations involving both electronic and vibrational states, have been studied by hole-burning. The problem of the dissipation of vibrational energy in large organic molecules in solids is not completely clear yet, and hole-burning has helped to further unravel this puzzle. Vibrational relaxation times in the first excited singlet state can be determined directly from the widths of the holes burnt into vibronic bands. Because these times are often very fast (picoseconds or less), they are difficult to measure in the time domain.

The first study of vibronic relaxation processes using PHB was performed on the $S_1 \leftarrow S_0$ transition of free-base porphin in n-octane at 4.2K (155). Subsequent experiments in various n-alkane crystalline hosts have demonstrated that vibronic decay in solids is dominated by guest-host interactions (156). Similar results have been obtained for other organic crystalline systems by accumulated picosecond photon echoes (90, 157). Hole-burning can also be used to determine accurate vibrational frequencies of molecules in hosts with large inhomogeneous widths (158). The method has been applied to various ionic dye systems (44, 159).

External Field Effects

The observation of external field effects on the spectra of complex molecules embedded in crystalline or amorphous hosts is essentially limited by the large inhomogeneously broadened absorption bandwidths ($\Gamma_{inh} \simeq 1-10^2$ cm$^{-1} \simeq 10^4-10^6$ MHz). Frequency shifts induced by the field are generally many orders of magnitude smaller than Γ_{inh} but can be made visible by means of HB. Since the hole represents a sharp frequency marker in the absorption band, Zeeman shifts and Stark splittings of holes can be measured with MHz resolution. Magnetic and electric field effects have been studied by hole-burning in various inorganic crystalline systems (28), and a few organic crystalline and amorphous systems (for reviews, see 23–25, 31). Parameters like the magnetic suseptibility, and the change in magnitude and direction of the electric dipole moment on excitation could be determined in this way.

In the case of organic amorphous systems, the holes burnt in zero field not only shifted by application of a field but also broadened. An analysis of the Zeeman lineshapes of free-base porphin in polyethylene at 4.2K has demonstrated that quantitative information on magneto-optical properties of excited states of complex molecules can be obtained not only from single crystal hosts but also from disordered materials (160).

HB combined with electric field experiments on organic glassy systems have shown a linear Stark effect so far. The hole splitting and broadening was interpreted in Refs. (161, 162) as being due to an intrinsic molecular dipole moment change ($\Delta\mu$) and a matrix-induced dipole moment. Values of $\Delta\mu$ for chlorin in a polymer matrix coincided with those obtained in crystals of n-alkanes (99). Further applications of the Stark effect can be found in Refs. (24, 31).

The influence of external pressure on spectral holes has also been reported. Matrix parameters like compressibility and strain fields could be determined in this way (163a). A microscopic theory of pressure broadening in hole-burning spectra has very recently been developed (163b).

Frequency-Domain Optical Data Storage

A potential technological application of permanent HB is data storage in the frequency domain. The presence of a hole at a particular frequency can be used to store one "bit" of information. The crucial parameter is the number of holes, N, that can be burnt in the inhomogeneously broadened absorption band, Γ_{inh}, at a single spatial spot, i.e. $N \simeq \Gamma_{inh}/\Gamma_{hole}$, where Γ_{hole} represents the holewidth (164a,b). N can be as large as 10^5 in polymer hosts doped with organic molecules (66).

The disadvantage of most of the hole-burning systems studied so far is that they undergo a single-photon process, and the signal-to-noise ratio degrades in each reading scan due to simultaneous burning. "Gated" hole-burning can solve this problem, but it requires two steps, one for optical excitation, and a second to initiate the reaction that leads to hole-burning. The latter may be, for example, a photoionization. The first "gated" hole-burning systems studied were Sm^{2+} in BaClF (165) and carbazole in boric acid (46). A complete review on this subject and other technological applications of persistent hole-burning, like laser pulse shaping combined with electrical field modulation, can be found in chapter 7 of Ref. (25).

CONCLUSIONS

I hope that this review has provided a glimpse of the potential of hole-burning spectroscopy. Hole-burning is a rather general phenomenon occurring in zero-phonon absorption bands of solids at low temperature. I have mentioned various mechanisms responsible for permanent HB and transient HB. Because HB yields, in principle, the homogeneous linewidth of a transition, it is a straightforward method to study optical dephasing and relaxation processes of molecules and ions doped in solid hosts. The pure dephasing term $\Gamma_{hom}-\Gamma_0$ in organic crystals follows an exponential temperature dependence at low temperatures, whereas for organic molecules in glasses it follows a $T^{1.3}$ dependence. Low frequency localized phonon modes are responsible for the dynamics in crystalline materials, but no clear interpretation for glassy systems has yet been found. Many theories have been proposed to explain the $T^{1.3}$ power law, but none of them has predicted it. Rather, the theoretical models have tried to establish the consistency of the calculations with the experimental results. Obviously, more work, experimental as well as theoretical, is needed to understand optical dephasing in amorphous solids.

Because HB allows the determination not only of holewidths but also of frequency shifts with MHz-accuracy it is an excellent technique to study the effects of external fields (magnetic, electric, pressure) on systems with broad inhomogeneous absorption bands. Permanent HB has further potential technological applications such as optical data storage.

It should be emphasized once more that hole-burning is a rather simple spectroscopic technique, with the advantage that it does not need samples of good optical quality and it can be applied to weak electronic transitions. It is ideally suited for studying selective photochemical processes and local guest-host interactions in complex disordered systems with MHz-resolution.

ACKNOWLEDGMENTS

I am grateful to all the co-workers quoted in this review for their invaluable contributions. Further, I would like to thank J. H. van der Waals for his continuous encouragement. L. de Kler is acknowledged for her patience while typing the manuscript. The investigations were supported by the Netherlands Foundation for Physical Research (FOM) and Chemical Research (SON) with financial aid from the Netherlands Organization for Scientific Research (NWO).

Literature Cited

1. Szabo, A. 1970. *Phys. Rev. Lett.* 25: 924; 1971. *Phys. Rev. Lett.* 27: 323
2. Personov, R. I., Al'shits, E. I., Bykovskaya, L. A. 1972. *Opt. Commun.* 6: 169
3. Personov, R. I. 1983. In *Spectroscopy and Excitation Dynamics of Condensed Molecular Systems*, ed. V. M. Agranovich, R. M. Hochstrasser, p. 555 and refs. therein. Amsterdam: North Holland
4. Kurnit, N. A., Abella, I. D., Hartmann, S. R. 1964. *Phys. Rev. Lett.* 13: 567
5. Brewer, R. G. 1977. In *Frontiers in Laser Spectroscopy*, ed. R. Balian, S. Haroche, S. Lieberman, 1: 342. Amsterdam: North Holland
6. Hesselink, W. H., Wiersma, D. A. 1983. See Ref. 3, p. 249 and refs. therein
7. Brewer, R. G., Shoemaker, R. L. 1971. *Phys. Rev. Lett.* 27: 631
8. Genack, A. Z., Macfarlane, R. M.,

1. Brewer, R. G. 1976. *Phys. Rev. Lett.* 37: 1078
9. Burns, M. J., Liu, W. K., Zewail, A. H. 1983. See Ref. 3, p. 301
10. Gorokhovskii, A. A., Kaarli, R. K., Rebane, L. A. 1974. *J. Exp. Theor. Phys. Lett.* 20: 216; 1976. *Opt. Commun.* 16: 282
11. Kharlamov, B. M., Personov, R. I., Bykovskaya, L. A. 1974. *Opt. Commun.* 12: 191
12. Szabo, A. 1975. *Phys. Rev. B* 11: 4512
13. De Vries, H., Wiersma, D. A. 1976. *Phys. Rev. Lett.* 36: 91
14. Völker, S., Van der Waals, J. H. 1976. *Mol. Phys.* 32: 1703
15. Völker, S., Macfarlane, R. M., Genack, A. Z., Trommsdorff, H. P., Van der Waals, J. H. 1977. *J. Chem. Phys.* 67: 1759
16. Hayes, J. M., Small, G. J. 1978. *Chem. Phys.* 27: 151
17. Graf, F., Hong, H. K., Nazzal, A., Haarer, D. 1978. *Chem. Phys. Lett.* 59: 217
18. Rebane, L. A., Gorokhovskii, A. A., Kikas, J. V. 1982. *Appl. Phys. B* 29: 235 and refs. therein
19. Friedrich, J., Haarer, D. 1984. *Angew. Chem.*, Int. Ed. 23: 113
20. Small, G. J. 1983. See Ref. 3, p. 515 and refs. therein; Jankowiak, R., Small, G. J. 1987. *Sciences* 237: 618
21. Macfarlane, R. M., Shelby, R. M. 1987. *J. Lumin.* 36: 179
22. Völker, S. 1987. *J. Lumin.* 36: 251 and refs. therein
23. Völker, S. 1987. In *Excited State Spectroscopy in Solids, XCVI Course of the Enrico Fermi Summer School of Physics*, Varenna, Italy, ed. U. M. Grassano, N. Terzi, p. 363. Amsterdam: North Holland
24. Völker, S. 1989. In *Relaxation Processes in Molecular Excited States*, ed. J. Fünfschilling, p. 113 and refs. therein. Dordrecht: Kluwer
25. Moerner, W. E., ed. 1988. *Topics in Current Physics*, Vol. 44. *Persistent Spectral Hole-Burning: Science and Applications*, and refs. therein. Berlin: Springer-Verlag
26. Sild, O., Haller, K., eds. 1988. *Zero-Phonon Lines*, and refs. therein. Berlin: Springer-Verlag
27. Yen, W. M. 1986. See Ref. 30, p. 23, and refs. therein
28. Macfarlane, R. M., Shelby, R. M. 1987. In *Spectroscopy of Solids Containing Rare Earth Ions*, ed. A. A. Kaplyanskii, R. M. Macfarlane, p. 51. Amsterdam: North Holland
29. Weber, M., ed. 1987. *Optical Linewidths in Glasses*, Special issue of *J. Lumin.* 36: 179–329
30. Zschokke, I. ed. 1986. *Optical Spectroscopy of Glasses*, Ser. C: *Molecular Structures.* Dordrecht: Reidel, and refs. therein
31. Maier, M. 1986. *Appl. Phys. B* 41: 73 and refs. therein
32a. Skinner, J. L., Hsu, D. 1986. *Adv. Chem. Phys.* 65: 1 and refs. therein
32b. Skinner, J. L. 1988. *Annu. Rev. Phys. Chem.* 39: 463 and refs. therein
33. Shelby, R. M., Macfarlane, R. M. 1979. *Chem. Phys. Lett.* 64: 545
34. Dicker, A. I. M., Johnson, L. W., Völker, S., Van der Waals, J. H. 1983. *Chem. Phys. Lett.* 100: 8
35. De Vries, H., Wiersma, D. A. 1977. *Chem. Phys. Lett.* 51: 565
36. Völker, S., Macfarlane, R. M. 1979. *IBM J. Res. Dev.* 23: 547 and refs. therein
37. Völker, S., Macfarlane, R. M. 1980. *J. Chem. Phys.* 73: 4476
38. Macfarlane, R. M., Völker, S. 1980. *Chem. Phys. Lett.* 69: 151
39. Thijssen, H. P. H., Dicker, A. I. M., Völker, S. 1982. *Chem. Phys. Lett.* 92: 7
40. Thijssen, H. P. H., Van den Berg, R. E., Völker, S. 1983. *Chem. Phys. Lett.* 97: 295
41. Drissler, F., Graf, F., Haarer, D. 1980. *J. Chem. Phys.* 72: 4996
42. Friedrich, J., Wolfrum, H., Haarer, D. 1982. *J. Chem. Phys.* 77: 2309
43. Thijssen, H. P. H., Van den Berg, R. E., Völker, S. 1985. *Chem. Phys. Lett.* 120: 503
44. Van den Berg, R. E., Völker, S. 1988. *Chem. Phys.* 128: 257
45. Burland, D. M., Carmona, F., Pacansky, J. 1978. *Chem. Phys. Lett.* 56: 221
46. Lee, H. W., Gehrtz, M., Marinero, E. E., Moerner, W. E. 1985. *Chem. Phys. Lett.* 118: 611
47. Hayes, J. M., Stout, R. P., Small, G. J. 1981. *J. Chem. Phys.* 74: 4266
48. Macfarlane, R. M., Shelby, R. M. 1983. *Opt. Commun.* 45: 46
49. Jankowiak, R., Bässler, H. 1983. *Chem. Phys. Lett.* 95: 310
50. Molenkamp, L. W., Wiersma, D. A. 1985. *J. Chem. Phys.* 83: 1
51. Van den Berg, R., Völker, S. 1986. *Chem. Phys. Lett.* 127: 525
52. Macfarlane, R. M., Shelby, R. M. 1983. In *Laser Spectroscopy VI*, ed. H. P. Weber, W. Lüty, p. 113. Berlin: Springer-Verlag
53. Völker, S., Macfarlane, R. M., Van der Waals, J. H. 1978. *Chem. Phys. Lett.* 58: 8

54. Carter, T. P., Fearey, B. L., Hayes, J. M., Small, G. J. 1983. *Chem. Phys. Lett.* 102: 272
55. Dicker, A. I. M., Dobkowski, J., Völker, S. 1981. *Chem. Phys. Lett.* 84: 415
56. Olson, R. W., Lee, H. W. H., Patterson, F. G., Fayer, M. D., Shelby, R. M., et al. 1982. *J. Chem. Phys.* 77: 2283
57. Burkhalter, F. A., Suter, G. W., Wild, U. P., Samoilenko, V. D., Rasumova, N. V., Personov, R. I. 1983. *Chem. Phys. Lett.* 94: 483
58. Walsh, C. A., Fayer, M. D. 1985. *J. Lumin.* 34: 37
59. Lee, H. W. H., Huston, A. L., Gehrtz, M., Moerner, W. E. 1985. *Chem. Phys. Lett.* 114: 491
60. Breinl, W., Friedrich, J., Haarer, D. 1986. *Phys. Rev. B* 34: 7271
61. Friedrich, J., Haarer, D., Silbey, R. 1983. *Chem. Phys. Lett.* 95: 119
62. Gutiérrez, A., Castro, G., Schulte, G., Haarer, D. 1983. In *Organic Molecular Aggregates*, ed. P. Reineker, H. Haken, H. C. Wolf, p. 206. Berlin: Springer-Verlag
63. Schulte, G., Grond, W., Haarer, D., Silbey, R. 1988. *Chem. Phys.* 88: 679
64. Cuellar, E., Castro, G. 1981. *Chem. Phys.* 54: 217
65. Moerner, W. E., Carter, T. P., Braüchle, C. 1987. *Appl. Phys. Lett.* 50: 430
66. Thijssen, H. P. H., Völker, S. 1985. *Chem. Phys. Lett.* 120: 496
67. Thijssen, H. P. H., Van den Berg, R., Völker, S., Van der Waals, J. H., Husson, L. P. J. 1984. *Chem. Phys. Lett.* 111: 121
68. Bjorklund, G. C. 1980. *Opt. Lett.* 5: 15
69. Bjorklund, G. C., Levenson, M. D., Lenth, W., Ortiz, C. 1983. *Appl. Phys. B* 32: 145
70. Gehrtz, M., Bjorklund, G. C., Whittaker, E. A. 1985. *J. Opt. Soc. Am. B* 2: 1510
71. Huston, A. L., Moerner, W. E. 1984. *J. Opt. Soc. Am. B* 1: 349
72. Levenson, M. D., Macfarlane, R. M., Shelby, R. M. 1980. *Phys. Rev. B* 22: 4915
73. Dick, B. 1988. *Chem. Phys. Lett.* 143: 186
74. Renn, A., Meixner, A. J., Wild, U. P., Burkhalter, F. A. 1985. *Chem. Phys.* 93: 157
75. Saari, P., Kaarli, R., Rebane, A. 1986. *J. Opt. Soc. Am. B* 3: 527 and refs. therein
76. Van den Berg, R., Visser, A., Völker, S. 1988. *Chem. Phys. Lett.* 144: 105
77. Thijjsen, H. P. H., Van den Berg, R., Völker, S. 1983. *Chem. Phys. Lett.* 103: 23
78. Thijssen, H. P. H., Völker, S. 1986. *J. Chem. Phys.* 85: 785
79. Breinl, W., Friedrich, J., Haarer, D. 1984. *J. Chem. Phys.* 80: 3496
80. Anderson, A. C. 1981. In *Amorphous Solids. Low Temperature Properties*, ed. W. A. Phillips, p. 65. Berlin: Springer-Verlag
81. Berg, M., Walsh, C. A., Narasimhan, L. R., Littau, K. A., Fayer, M. D. 1987. *Chem. Phys. Lett.* 139: 66; 1988. *J. Chem. Phys.* 88: 1564
82. Aartsma, T. J., Wiersma, D. A. 1976. *Chem. Phys. Lett.* 42: 520
83. Aartsma, T. J., Wiersma, D. A. 1978. *Chem. Phys. Lett.* 54: 415
84. Cooper, D. E., Olson, R. W., Fayer, M. D. 1980. *J. Chem. Phys.* 72: 2332
85. Morsink, J. B. W., Aartsma, T. J., Wiersma, D. A. 1977. *Chem. Phys. Lett.* 49: 34
86. Duppen, K., Molenkamp, L. W., Morsink, J. B. W., Wiersma, D. A., Trommsdorff, H. P. 1981. *Chem. Phys. Lett.* 84: 421
87. Molenkamp, L. W., Weitekamp, D. P., Wiersma, D. A. 1983. *Chem. Phys. Lett.* 99: 382
88. Hsu, D., Skinner, J. L. 1987. *J. Chem. Phys.* 87: 54 and refs. therein
89. Jansen, G., Noort, M., Canters, G. W., Van der Waals, J. H. 1978. *Mol. Phys.* 35: 283
90. Hesselink, W. H., Wiersma, D. A. 1980. *J. Chem. Phys.* 73: 648
91. Molenkamp, L. W., Wiersma, D. A. 1984. *J. Chem. Phys.* 80: 3054
92. Thijssen, H. P. H., Völker, S. 1981. *Chem. Phys. Lett.* 82: 478
93. Gorokhovskii, A. A., Rebane, L. A. 1977. *Opt. Commun.* 20: 144
94. Dicker, A. I. M., 1982. PhD thesis, Univ. Leiden, chapt. 2
95. Koehler, T. R. 1980. *J. Chem. Phys.* 72: 3389
96. Dicker, A. I. M., Noort, M., Völker, S., Van der Waals, J. H. 1980. *Chem. Phys. Lett.* 73: 1
97. Dicker, A. I. M., Noort, M., Thijssen, H. P. H., Völker, S., Van der Waals, J. H. 1981. *Chem. Phys. Lett.* 78: 212
98. Dicker, A. I. M., Dobkowski, J., Noort, M., Völker, S., Van der Waals, J. H. 1982. *Chem. Phys. Lett.* 88: 135
99. Dicker, A. I. M., Johnson, L. W., Noort, M., Van der Waals, J. H. 1983. *Chem. Phys. Lett.* 94: 14
100. Skinner, J. L., Hsu, D. 1986. *J. Phys. Chem.* 90: 4931 and refs. therein
101. Silbey, R. 1989. See Ref. 24, p. 235 and refs. therein

102. Jones, K. E., Zewail, A. H. 1978. In *Advances in Laser Chemistry*, ed. A. H. Zewail, and refs. therein. Berlin: Springer-Verlag
103. Osad'ko, I. S. See Ref. 3, p. 437 and refs. therein
104. Kubo, R., Tomita, T. 1954. *J. Phys. Soc. Jpn.* 9: 888
105. Anderson, P. W. 1954. *J. Phys. Soc. Jpn.* 9: 316
106. McConnell, H. M. 1958. *J. Chem. Phys.* 28: 430
107a. Harris, C. B. 1977. *J. Chem. Phys.* 67: 5607
107b. Shelby, R. M., Harris, C. B., Cornelius, P. A. 1979. *J. Chem. Phys.* 70: 34
108. Van 't Hof, C. A., Schmidt, J. 1975. *Chem. Phys. Lett.* 36: 457; 1976. *Chem. Phys. Lett.* 42: 73
109. Schmidt, J. 1989. See Ref. 24, pp. 3, 51 and refs. therein
110. Jackson, B., Silbey, R. 1983. *Chem. Phys. Lett.* 99: 331
111. Silbey, R., Kassner, K. 1987. *J. Lumin.* 36: 283 and refs. therein
112. McCumber, D. E., Sturge, M. D. 1963. *J. Appl. Phys.* 34: 1682
113. De Bree, Ph., Wiersma, D. A. 1979. *J. Chem. Phys.* 70: 790
114. Redfield, A. G. 1957. *IBM J. Res. Dev.* 1: 19; 1965. *Adv. Magn. Res.* 1: 1
115. Hsu, D., Skinner, J. L. 1985. *J. Chem. Phys.* 83: 2107
116a. Phillips, W. A. 1987. *Rep. Progr. Phys.* 50: 1657; Phillips, W. A. 1972. *J. Low Temp. Phys.* 7: 351
116b. Anderson, P. W., Halperin, B. I., Varma, C. M. 1972. *Phil. Mag.* 25: 1
117. Reineker, P., Kassner, K. 1986. See Ref. 30, p. 65 and refs. therein
118. Selzer, P. M., Huber, D. L., Hamilton, D. S., Yen, W. M., Weber, M. J. 1976. *Phys. Rev. Lett.* 36: 813
119a. Hegarty, J., Broer, M. M., Golding, B., Simpson, J. R., MacChesney, J. B. 1983. *Phys. Rev. Lett.* 51: 2033
119b. Broer, M. M., Golding, B., Haemmerle, W. H., Simpson, J. R., Huber, D. L. 1986. *Phys. Rev. B* 33: 4160
119c. Huber, D. L., Broer, M. M., Golding, B. 1984. *Phys. Rev. Lett.* 52: 2281; 1986. *Phys. Rev. B* 33: 7297
120. Walsh, C. A., Berg, M., Narasimhan, L. R., Fayer, M. D. 1986. *Chem. Phys. Lett.* 130: 6; 1987. *J. Chem. Phys.* 86: 77
121. Narasimhan, L. R., Pack, D. W., Fayer, M. D. 1988. *Chem. Phys. Lett.* 152: 287
122. Breinl, W., Friedrich, J. 1988. *Chem. Phys. Lett.* 145: 107
123. Carter, T. P., Small, G. J. 1985. *Chem. Phys. Lett.* 120: 178
124. Boxer, S. G., Gottfried, D. S., Lockhart, D. J., Middendorf, T. R. 1987. *J. Chem. Phys.* 86: 2439
125. Köhler, W., Friedrich, J., Fischer, R., Scheer, H. 1988. *Chem. Phys. Lett.* 146: 280
126a. Gorokhovskii, A. A., Korrovits, V., Palm, V., Trummal, M. 1986. *Chem. Phys. Lett.* 125: 355
126b. Gorokhovskii, A. A. 1988. See Ref. 26, p. 102
126c. Gorokhovskii, A. A., Palm, V. V. 1988. *Chem. Phys. Lett.* 153: 328
127a. Nakatsuka, H., Sugiyama, H., Matsumoto, Y. 1987. *J. Lumin.* 38: 31
127b. Brito Cruz, C. H., Fork, R. L., Knox, W. H., Shank, C. V. 1986. *Chem. Phys. Lett.* 132: 341
128. Van den Berg, R., Völker, S. 1987. *Chem. Phys. Lett.* 137: 201
129a. Friedrich, J., Haarer, D. 1986. See Ref. 30, p. 149
129b. Köhler, W., Friedrich, J. 1988. *J. Chem. Phys.* 88: 6655 and refs. therein
130. Breinl, W., Friedrich, J., Haarer, D. 1984. *Chem. Phys. Lett.* 106: 487; 1984. *J. Chem. Phys.* 81: 3915
131. Reinecke, T. L. 1979. *Solid State Commun.* 32: 1103
132. Lyo, S. K., Orbach, R. 1980. *Phys. Rev. B* 22: 4223
133. Reineker, P., Morawitz, H. 1982. *Chem. Phys. Lett.* 86: 359
134. Lyo, S. K. 1982. *Phys. Rev. Lett.* 48: 688; Lyo, S. K. 1986. See Ref. 30, p. 1
135. Huber, D. L. 1982. *J. Non-Cryst. Solids* 51: 241
136. Hunklinger, S., Schmidt, M. 1984. *Z. Phys. B* 54: 93
137. Lyo, S. K., Orbach, R. 1984. *Phys. Rev. B* 29: 2300
138. Reineker, P., Morawitz, H., Kassner, K. 1984. *Phys. Rev. B* 29: 4546
139. Osad'ko, I. S. 1985. *Chem. Phys. Lett.* 115: 411; 1986. *J. Exp. Theor. Phys. Lett.* 90: 1453
140. Kassner, K., Reineker, P. 1987. *Phys. Rev. B* 35: 828
141. Locher, R., Renn, A., Wild, U. P. 1987. *Chem. Phys. Lett.* 138: 405
142. Basché, Th., Bräuchle, C. 1988. *J. Phys. Chem.* 92: 5069
143. Van den Berg, R., Völker, S. 1988. *Chem. Phys. Lett.* 150: 491
144. De Boer, S., Vink, K. J., Wiersma, D. A. 1987. *Chem. Phys. Lett.* 137: 99
145. Hirschmann, R., Köhler, W., Friedrich, J., Daltrozzo, E. 1988. *Chem. Phys. Lett.* 151: 60
146. Trommsdorff, H. P., Zeigler, J. M., Hochstrasser, R. M. 1989. *Chem. Phys. Lett.* 154: 463

147. De Boer, S., Wiersma, D. A. 1989. *Chem. Phys.* 131: 135
148. Meech, S. R., Hoff, A. J., Wiersma, D. A. 1985. *Chem. Phys. Lett.* 121: 287; 1986. *Proc. Natl. Acad. Sci. USA* 83: 9464
149. Boxer, S. G., Lockhart, D. J., Middendorf, T. R. 1986. *Chem. Phys. Lett.* 123: 476; 1986. *FEBS Lett.* 200: 237
150. Vink, K. J., De Boer, S., Plitjer, J. J., Hoff, A. J., Wiersma, D. A. 1987. *Chem. Phys. Lett.* 142: 433
151a. Hayes, J. M., Gillie, J. K., Tang, D., Small, G. J. 1988. *Biochim. Biophys. Acta* 932: 287
151b. Tang, D., Jankowiak, R., Small, G. J., Tiede, D. M. 1989. *Chem. Phys.* 131: 99 and refs therein
152. Won, Y., Friesner, R. A. 1988. *J. Phys. Chem.* 92: 2214
153. Köhler, W., Friedrich, J., Fischer, R., Scheer, H. 1988. *J. Chem. Phys.* 89: 871 and refs. therein
154. Johnson, S. G., Small, G. J. 1989. *Chem. Phys. Lett.* 155: 371
155. Völker, S., Macfarlane, R. M. 1979. *Chem. Phys. Lett.* 61: 421
156. Dicker, A. I. M., Völker, S. 1982. *Chem. Phys. Lett.* 87: 481
157. Hesselink, W. H., Wiersma, D. A. 1981. *J. Chem. Phys.* 74: 886
158. Kharlamov, B. M., Bykovskaya, L. A., Personov, R. I. 1977. *Chem. Phys. Lett.* 50: 407
159. Feary, B. L., Carter, F. P., Small, G. J. 1983. *J. Phys. Chem.* 87: 3590
160. Van den Berg, R., Van der Laan, H., Völker, S. 1987. *Chem. Phys. Lett.* 142
161. Meixner, A. J., Renn, A., Wild, U. P. 1986. *J. Phys. Chem.* 90: 6777
162. Sesselmann, Th., Kador, L., Richter, W., Haarer, D. 1988. *Europhys. Lett.* 5: 351
163a. Sesselmann, Th., Richter, W., Haarer, D., Morawitz, H. 1987. *Phys. Rev. B* 36: 7601 and refs. therein
163b. Laird, B. B., Skinner, J. L. 1989. *J. Chem. Phys.* 90: 3274, 3880
164a. Castro, G., Haarer, D., Macfarlane, R. M., Trommsdorff, H. P. 1978. U.S. patent 4101976
164b. Haarer, D., Pole, R. V., Völker, S. 1978. U.S. Patent 4103346
165. Winnacker, A., Shelby, R. M., Macfarlane, R. M. 1985. *Opt. Lett.* 10: 350

ATOMIC-RESOLUTION SURFACE SPECTROSCOPY WITH THE SCANNING TUNNELING MICROSCOPE

R. J. Hamers

IBM Research Division, T. J. Watson Research Center, Yorktown Heights, New York 10598

INTRODUCTION

The study of surfaces has enjoyed an explosive growth during the last 25 years, due largely to the development of new techniques for probing the symmetry, chemical composition, and the electronic and vibrational states of surfaces and of adsorbed atomic and molecular species. Although a veritable arsenal of surface science tools is available, the study of surfaces is often so complex that even when several tools are applied simultaneously, unambiguous results may not be obtained. The study of surfaces has been greatly advanced during the last five years by the newly developed technique of scanning tunneling microscopy (STM) (1–6). In this technique, a very sharp tip (usually of tungsten) is brought to within a few atomic diameters of the surface under investigation without actual physical contact, so that there is a very small overlap of the wavefunctions of the surface with the nearest atom of the tip. When a small bias voltage (10 mV–4V) is applied between the sample and tip, electrons tunnel across this gap with a probability that increases exponentially as the tip approaches the sample. This exponential dependence of the tunneling current on the sample-tip separation provides an extremely sensitive way of detecting the small changes in the surface height due to the individual atoms, thus providing the basis for the scanning tunneling microscope.

The images obtained in STM are often strongly dependent on the sample-tip bias voltage in a nontrivial manner. Although early STM studies focused on the application of tunneling as a microscope for observing topographic structure, the sensitivity to bias voltage also means that the scanning tunneling microscope is sensitive to the energy states of the sample and tip. This has two main implications: (a) Tunneling microscopy does *not* reveal the positions of atoms themselves; (b) with STM, it is possible to obtain spectroscopic information with atomic spatial resolution. This paper explores the techniques and applications of atomic-resolution spectroscopy with the scanning tunneling microscope.

Compared with other surface spectroscopy techniques such as ultraviolet photoemission spectroscopy (UPS), inverse photoemission spectroscopy (IPS), electron energy loss spectroscopy (EELS), and infrared reflection-absorption spectroscopy (IRRAS), STM has a unique advantage. Whereas these other techniques only provide information averaged over a large region of the surface, the tunneling current in STM flows through a region only $\simeq 5$ Å in diameter (7), so that spectroscopic information can be obtained on an *atom-by-atom* basis. This is an important advance, since many chemical and physical phenomena at surfaces are associated with "active sites" such as dopants, impurities, steps, or defects that occupy only a small fraction of the total surface area. Most experimental techniques average over a large surface area, making study of such local surface properties difficult. With the scanning tunneling microscope, it is now possible to measure directly changes in electronic structure resulting from surface irregularities such as impurities, steps, and defects, as well as to examine the electronic structure of chemically inequivalent atoms in complex ordered structures such as the (7×7) reconstruction of Si(111).

A number of excellent review articles on STM are already available (2, 3, 5, 8). Likewise, the theoretical foundations of tunneling spectroscopy are also well developed from the study of bulk tunnel junctions (8, 9) and have also been applied to tunneling spectroscopy with the scanning tunneling microscope (10–16). This paper presents a selective review of the various ways in which information about the local electronic structure or surfaces may be obtained by using the scanning tunneling microscope, together with the interpretation of these data and some representative examples of the application.

ONE-DIMENSIONAL TUNNELING

Figure 1 shows an energy level diagram for the system consisting of the sample and tip, which are separated by a vacuum. Here, the tip is considered to be a metal with a constant density of states, and the sample also

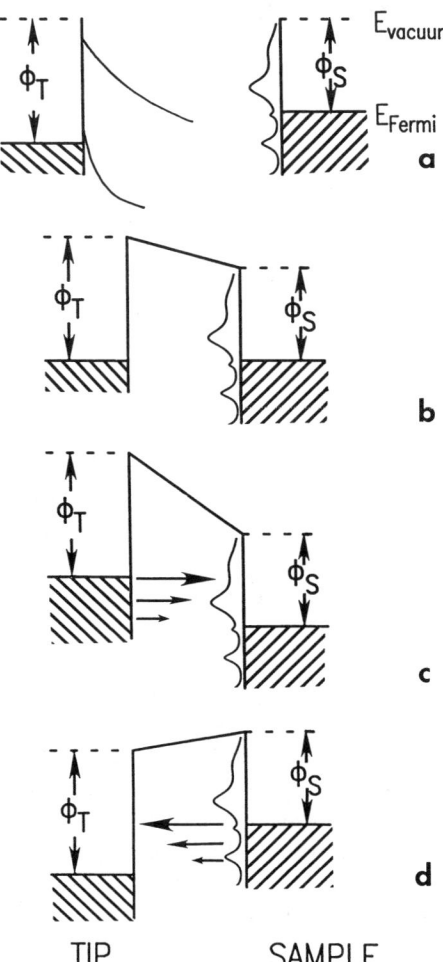

Figure 1 Energy level diagrams for sample and tip. (*a*) Independent sample and tip. (*b*) Sample and tip at equilibrium, separated by small vacuum gap. (*c*) Positive sample bias: electrons tunnel from tip to sample. (*d*) Negative sample bias: electrons tunnel from sample into tip.

contains a distribution of surface states as shown. When the sample and tip are independent, their vacuum levels are considered to be equal, as in Figure 1*a*, and their respective Fermi energies (or levels), E_F, lie below the vacuum level by their respective work functions ϕ_s and ϕ_t. The quantum-mechanical wavefunctions of the electrons are periodic in the solid and decay exponentially into the vacuum region like $\Psi = A \exp[-2\sqrt{2m(\phi-E)}\,(Z/\hbar)]$, where Z is the distance perpendicular to the surface plane and E is the energy measured with respect to the Fermi level. This energy-dependent decay of the wavefunctions is also illustrated in Figure 1a, for different states of the tip. More strongly bound electrons have large negative values for E and so decay quickly into the vacuum, whereas high energy states lying close to the vacuum level decay very slowly. The exponential decay of the wavefunctions into the vacuum is often written in terms of an "inverse decay length" κ as $\Psi = A\exp(-\kappa Z)$, where $\kappa = 2\hbar^{-1}\sqrt{2m(\phi-E)}$. If the sample and tip are in thermodynamic equilibrium, their Fermi levels must be equal, as illustrated in Figure 1b. Electrons attempting to pass from sample to tip (or vice versa) encounter a potential barrier that is approximately trapezoidal in shape, but electrons can tunnel through if the barrier is sufficiently narrow.

When a voltage V is applied to the sample, its energy levels will be rigidly shifted upward or downward in energy by the amount $|eV|$, depending on whether the polarity is negative or positive, respectively. At positive sample bias, the net tunneling current arises from electrons that tunnel from the occupied states of the tip into unoccupied states of the sample, as in Figure 1c. At negative sample bias, the situation is reversed, and electrons tunnel from occupied states of the sample into unoccupied states of the tip, as in Figure 1d. Since states with the highest energy have the longest decay lengths into the vacuum, most of the tunneling current arises from electrons lying near the Fermi level of the negative-biased electrode.

For any given lateral position of the tip above the sample (r), the tunneling current (I) is determined by the sample-tip separation (Z), the applied voltage (V), and the electronic structure of the sample and tip, which is quantitatively described by their respective density of states $[\rho(E)]$.

Spectroscopic information relies on changing the voltage V, but can be obtained in a number of ways, depending on which of the other variables are held constant and which are measured. Voltage dependent STM imaging is the simplest way of obtaining spectroscopic information, by acquiring conventional STM "topographic" information at different applied voltages and comparing the results. More quantitative information regarding the symmetry properties and spatial localization of electronic states is obtained by using modulation techniques to measure dI/dV (usually at constant average tunneling current) as a function of V, usually referred to as "Scanning Tunneling Spectroscopy" or "STS." More complete information can be obtained over a wider energy range simultaneously from complete I-V measurements, at the expense of considerably more complicated data acquisition electronics and data handling.

VOLTAGE-DEPENDENT STM IMAGING

Experimental Technique

When a voltage V is applied to the sample (with the tip at ground), only those states lying between E_F and $E_F + eV$ contribute to the tunneling process. The sign and magnitude of the applied voltage, then, determine which states can contribute to the resulting topographic images. Changes in the symmetry of the STM constant-current topographs (CCTs) as a function of the applied bias can often provide information about the symmetry and energy of the electronic states. These changes can be identified either by acquiring images sequentially at different bias voltages or by changing the applied bias on a line-by-line basis during the raster-scanning of the tip, providing two (or more) interleaved images at different bias voltages. The latter procedure provides a good registration between the images at different voltages, thus eliminating uncertainty due to thermal drift and creep of the piezoelectric scanners.

Interpretation of Voltage-dependent STM Images

The interpretation of voltage-dependent imaging is based almost entirely on the work of Tersoff & Hamann (11, 12), who applied the Bardeen tunneling formalism (17) to the specific geometry encountered in STM, and that of Lang (7, 15, 18–21). A major result of Tersoff's theory is that it predicts that the path followed by the tip in constant-current STM is a contour of constant density of states, measured at the center of curvature of the tip.

Lang used numerical calculations to study the changes in the density of states induced by adsorption of different atoms onto a theoretical "jellium" surface, in the limit of low coverage. Figure 2 summarizes the results of calculations for several specific cases. Each curve represents the *change* in the density of states induces by a single atom of the particular indicated element. This change is determined by first calculating the total state density for the surface including the adsorbed atom, and then subtracting the state density of the original bare surface. Upon adsorption of an alkali metal, for example, charge is transferred from the alkali s-states to the jellium substrate. As a result, the alkali metal has an associated unoccupied state (for Li, a $2s$ state) lying slightly above E_F; this then makes these and other electropositive atoms appear as protrusions at positive sample bias when electrons can easily tunnel into these empty states. Electronegative adsorbates usually increase the density of occupied states and are expected to appear as protrusions at negative sample bias.

A surprising result of Lang's calculations is that for transition metals, with partially filled d-bands lying near E_F, the contribution of these d-

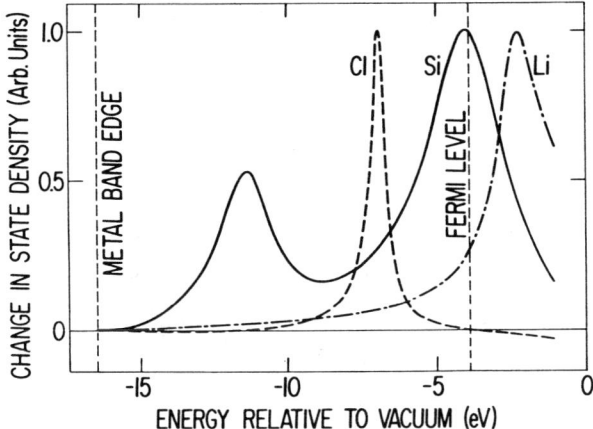

Figure 2 Changes in the density of states induced by adsorption of various atoms as a function of energy (from Ref. 7).

bands to the tunneling current is usually greatly exceeded by the s- and p-band contributions because the occupied d-bands generally have a smaller principal quantum number ($4d$, for Mo) than the nearby s- and p-bands ($5s$ and $5p$, for Mo). The smaller principal quantum number states decay faster in the vacuum, and so their overlap with the wavefunctions of the tip is often negligible. This insensitivity to d-bands indicates that distinguishing between various transition metal atoms at a surface will likely be difficult, although experimental confirmation is still lacking. Lang's calculations indicate that even adsorbed species that have no electronic states near E_F (such as helium) might still be observed in the STM due to electrostatic screening effects.

Applications of Voltage-dependent STM Imaging

ELEMENTAL SEMICONDUCTORS In the case of ordered surfaces, voltage-dependent imaging can provide information on the relative spatial locations of the various electronic states at the surface. Some of the surfaces that have been extensively studied by this method include Si(001) (22–24), Ge(001) (25), Si(111)–(7 × 7) (26–28), the metastable cleaved Si(111)–(2 × 1) surface (29–31), GaAs(110) (32–34), and graphite (35, 36).

Figure 3 shows CCTs of the Si(001) surface obtained at negative and positive bias, reflecting the spatial distribution of filled and empty surface states, respectively. The geometric structure of this surface is well understood (24, 37–39) and consists of pairs of silicon atoms, each of which is bonded to two Si atoms in the next lower atomic layer and its dimer partner; the resulting (2 × 1) unit cell is outlined in Figure 3. At negative

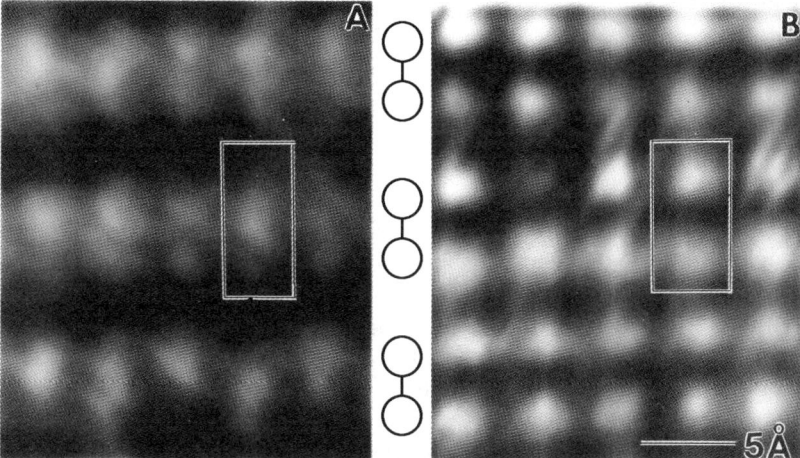

Figure 3 Constant-current topographs of Si(001) at negative (*A*: −1.6 eV) and positive (*B*: +1.6 eV). The (2 × 1) unit cell is outlined, and the relative locations of the dimers are shown in the center.

sample bias (Figure 3*A*), the STM-CCTs show bean-shaped structures, whereas at positive sample bias (Figure 3*B*), the images show a weak minimum between the dimer rows and a deep trough along the center of the dimer row. These differences demonstrate a spatial separation of the filled and empty electronic states on the Si(001) surface.

These differences can be understood based on electronic structure calculations (38, 39). These show that the electronic structure of the dimers can be described in terms of a π bonding state slightly below E_F and a π-antibonding state slightly above E_F, with occupied σ and unoccupied σ^* states lying far from E_F, outside the energy range accessible by tunneling. The occupied π state is predicted to be symmetric with respect to reflection through a mirror plane bisecting the dimer bond (A_1 symmetry), whereas the π^*-antibonding state is antisymmetric with respect to this reflection (B_1 symmetry) and therefore must have a *node* in the wavefunction at the center of the dimer bond.

The STM contours at negative bias reflect the contours of the occupied π state, whereas at positive bias the STM tip follows the contour of the empty π^* state. The deep trough observed in the positive-bias STM images corresponds to the location of the node in the wavefunction of the π^* antibonding state. Thus, the STM images directly reflect the spatial symmetry of these surface-state wavefunctions.

Changes in the spatial distribution of the surface-state wavefunctions can often be induced by species chemisorbed on the surface. Dosing the Si(001) surface with NH_3 significantly changes the negative-bias STM images, and produces a slight depression in the center of the dimer bond, where images on the clean surface showed a maximum (23, 37). These changes were attributed to hydrogen atoms produced by NH_3 dissociation interacting with the dimers. On the clean surface, each surface Si atom is double-bonded via both a σ and a π-bond to its dimer partner, whereas on the hydrogen-covered surface, each surface Si atom is bonded via a single σ-bond to its dimer partner and via another σ-bond to a hydrogen atom. The changes in the negative-bias images induced by the reaction with NH_3 result from the different spatial distribution of the Si–Si π-state and the Si–H σ-state. Thus, the STM images alone can be used to distinguish dimers that have reacted with NH_3 from those that have not (23, 37).

STM images of Si(111) surfaces are also strongly dependent on the bias voltage. The (2×1) reconstruction of the cleaved Si(111) surface (30, 31) shows a shift between negative- and positive-bias STM images due to the different spatial location of occupied and unoccupied electronic states. This shift provides strong confirmation for the π-bonded chain model proposed by Pandey (40).

More complicated voltage-dependence is observed for the Si(111)–(7×7) surface. This surface undergoes an extensive reconstruction extending several layers into the bulk, and the atomic arrangements have only recently been determined (41), based in part on early STM results (4). At most positive bias voltages, STM images reveal 12 adatoms of equal height in each unit cell, as shown in Figure 4A. In a narrow voltage range around +1.4 eV sample bias, the two triangular subunits in each unit cell appear to have slightly different heights. Differences between the two triangular subunits can be attributed to a difference in the stacking sequence of the underlying atomic planes, leading to a stacking "fault" in one half, but not the other. The difference in stacking sequence leads to different interactions between the Si atoms in the first full atomic layer and those lying 4.6 Å below. Around +1.4 eV sample bias, the "unfaulted" half of the unit cell appears higher (indicating a higher density of states) than the faulted half. This situation reverses at negative sample bias voltages, as in Figure 4B, where the adatoms in the faulted half of the unit cell appear higher than those in the unfaulted half. Additionally, the adatoms nearest the deep corner holes appear higher than the central adatoms. These latest differences are attributed to the different environment of the corner adatoms (which are next to one dimer and one rest atom) and the central adatoms (which are next to two dimers and two rest atoms). Such voltage-dependent STM images (26), together with local tunneling spectroscopy measure-

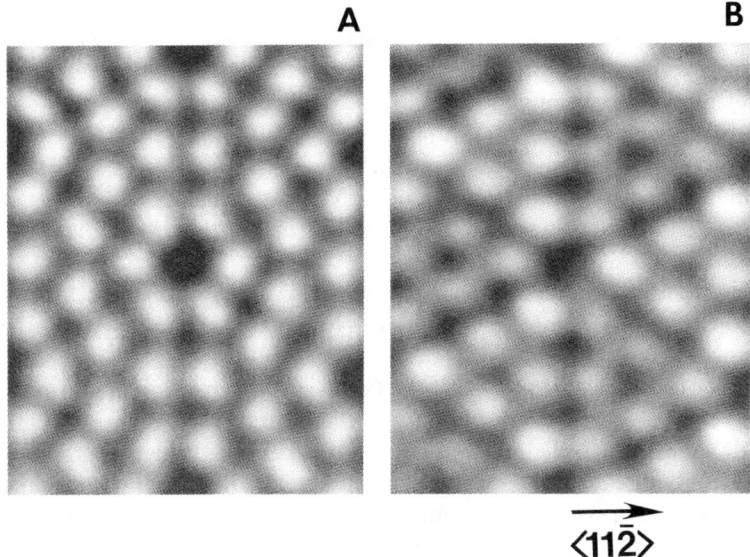

Figure 4 Constant-current topographs of Si(111)–(7 × 7) surface at positive (4A: +2 V) and negative (B: −2 V) sample bias (from Ref. 26). The area is 47 Å × 54 Å.

ments to be discussed below (22, 27), have provided a great deal of insight into the electronic structure of this complicated reconstruction.

COMPOUND SEMICONDUCTORS In compound semiconductors, voltage-dependent contrast changes can be directly related to charge transfer between surface atoms. The (110) surfaces of III–V semiconductors, such as GaAs(110), contain equal numbers of cations (Ga) and anions (As) in the surface layer; in GaAs, they are arranged in a chain-like structure. Charge transfer from Ga to As results in an occupied state centered at the As atoms and an empty state centered at the Ga atoms, and also results in a slight vertical displacement of the Ga atoms with respect to the As atoms. STM images (33) reveal a shift between the peaks observed at positive and negative bias. The peaks at negative bias correspond to the positions of the As atoms, and those at positive bias reveal the Ga atoms. A comparison of this bias-dependent shift with theoretical calculations also allowed the tilt angle between Ga and As atoms at the surface to be quantitatively determined (33, 42).

ADSORBATES AND IMPURITIES Voltage-dependent STM imaging is not restricted to ordered surfaces as discussed above, but is also very useful for identifying and studying the properties of impurities and defects at surfaces. On semiconductor surfaces, local charging due to adsorbates (34, 43) and defects (44) can shift the surface-state bands with respect to the Fermi level, thus resulting in apparent height changes that correspond to the electrostatic screening length. For strongly charged adsorbates and/or lightly doped semiconductors, these effects can dominate the appearance of defects, as in the case of oxygen on n-type GaAs(110) (43). Adsorbates on metals and covalently bonded adsorbates on semiconductors give rise to local changes in the surface electronic structure that allow them to be imaged.

The ability to differentiate between chemically inequivalent atoms is demonstrated in Figure 5, which shows an STM image of Al adatoms organized atop a Si(111) latice in an overlayer with ($\sqrt{3} \times \sqrt{3}$) symmetry, obtained at $+1.2$ V sample bias (44). Each bright protrusion in the ordered structure is an Al adatom. In many of the locations where an Al adatom would be expected, the unit cell appears darker but still has a protrusion in the center. These darker unit cells consist of Si adatoms, rather than the expected Al adatoms, situated atop a bulk-like Si(111) surface. Switching the bias to -2 eV causes the contrast to reverse, so that the Si adatoms appear $\simeq 1$ Å higher than the Al adatoms. By choosing the bias polarity, then, either the Al adatoms or the Si adatoms can be selectively imaged. The origin of these contrast changes are revealed from theoretical cal-

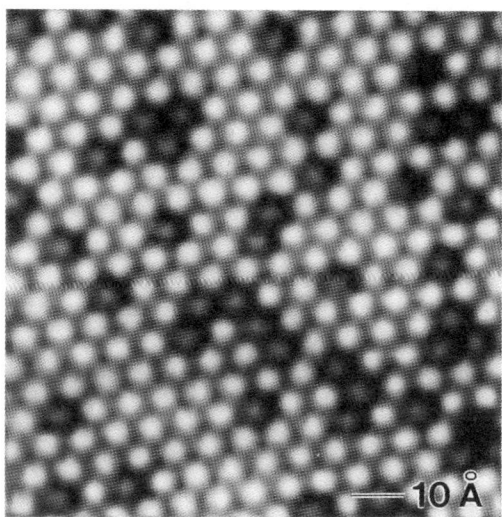

Figure 5 Constant-current topograph of nominally Si(111)–($\sqrt{3} \times \sqrt{3}$)Al overlayer with defects. *Bright protrusions* are individual aluminum adatoms; *darker protrusions* are substitutional defects consisting of silicon adatoms instead of aluminum adatoms.

culations (45) that show that the Al adatoms have an unoccupied p_z orbital, which in Si is shifted below E_F. These changes in the state density directly give rise to contrast in the STM constant-current topographs.

A number of other systems have also been studied by voltage-dependent STM imaging. On graphite, interactions of the surface carbon atoms with those in deeper layers leads to two kinds of inequivalent carbon atoms in the surface layer, and produces interesting voltage-dependent tunneling behavior (35, 36, 46). Metal overlayers on semiconductors have also proven to provide rich voltage-dependent structure; those studied include Ag/Si(111) (47), Al/Si(111) (48), and Cu/Si(111) (49). Most of these metal-on-silicon systems show strongly voltage-dependent changes in constant-current STM topography resulting from complex electronic structure. For such systems, an understanding of the dependence of the STM images on applied bias is crucial for understanding the atomic features that give rise to the STM "topographic" structure.

MODULATION TECHNIQUES—SCANNING TUNNELING SPECTROSCOPY

Constant-current topographs often reveal electronic structure information, but separating the contributions of electronic and geometric structure is not straightforward. One way of obtaining quantitative spectroscopic information in the STM is the use of modulation techniques.

Experimental Techniques

The most common modulation technique is the Scanning Tunneling Spectroscopy (STS) technique first used by Binnig & Rohrer (1, 6). The STS technique involves the application of a constant DC bias V_{DC} with a superimposed high-frequency sinusoidal modulation voltage $V_{mod}\omega_{mod}$ between sample and tip. The component of the tunneling current that is in-phase with the applied voltage modulation is measured with a lock-in amplifier while the feedback loop controller maintains a constant average tunneling current. This provides a measurement of $dI/dV|_{V=V_{DC}}$ simultaneously with the sample topography. Probing the electronic structure at different energies requires separate scans with different values of V_{DC}.

Interpretation of Scanning Tunneling Spectroscopy Results

HIGH VOLTAGES: BARRIER STATE SPECTROSCOPY The STS method was first applied to tunneling spectroscopy investigations of field, or barrier, states (1, 28, 50). Barrier states arise at relatively high voltages (in the field emission regime), where the tunneling barrier is triangular, rather than trapezoidal in shape. Under these conditions, there is a region between the sample and tip in which the potential is low, so that the electrons are nearly free electrons but are bound by the triangular barrier and the positively biased electrode (which can be either sample or tip). The boundary conditions lead to the formation of standing waves, which depend on the detailed shape of the tunneling barrier and the electron energy. As a function of energy, the transmission probability varies depending on the nodal character of these standing wave states, so that measurements of dI/dV vs. V show oscillations. No information about the detailed electronic structure is obtained from these measurements, but Becker et al (28) showed that by numerical integration of Schrödinger's equation, the oscillations in dI/dV could be used to determine an *absolute* sample-tip separation, which is very difficult to determine otherwise.

LOW VOLTAGES: SURFACE STATE SPECTROSCOPY At bias voltages lower than the work function of both tip and sample, structure in dI/dV as a function of V may arise from surface states associated with critical points in the surface-projected bulk band structure or associated with surface states. The interpretation of these low-bias dI/dV measurements is generally based on the WKB approximation for the tunneling current, which can be expressed as

$$I = \int_0^{eV} \rho_s(r, E) \rho_t(r, \pm eV \mp E) T(E, eV, r) \, dE \qquad 1.$$

where the upper signs correspond to positive sample bias ($eV > 0$) and the lower signs to negative sample bias ($eV > 0$), $\rho_s(r, E)$ and $\rho_t(r, E)$ are the density of states of the sample at location r and the energy E, measured with respect to their individual Fermi levels. For negative sample bias, $eV < 0$ and for positive sample bias, $eV > 0$. In the WKB approximation for planar electrodes, the tunneling transmission probability $T(E, eV, r)$ for electrons with energy E and applied bias voltage V is given by

$$T(E, eV) = \exp\left(-\frac{2Z\sqrt{2m}}{\hbar}\sqrt{\frac{\phi_s + \phi_t}{2} + \frac{eV}{2} - E}\right). \qquad 2.$$

Then,

$$\frac{dI}{dV} = \rho_s(r, eV)\rho_t(r, 0)T(eV, eV, r)$$

$$+ \int_0^{eV} \rho_s(r, E)\rho_t(r, \pm eV \mp E)\frac{dT(E, eV, r)}{dV} dE. \qquad 3.$$

The first term in Eq. 3 is the product of the density of states of the sample, the density of states of the tip, and the tunneling transmission probability T. The second term contains the voltage-dependence of the tunneling transmission factor.

At any fixed location, the transmission factor $T(E, eV, r)$ increases monotonically with V and so contributes a smoothly varying "background" on which the spectroscopic information is superimposed. Because the increase is smooth and monotonic, structure in dI/dV as a function of V can usually be assigned to changes in the state density via the first term, thus permitting the density of states to be determined as a function of energy at any particular location on the surface.

When comparing information obtained at different spatial locations or when interpreting images of dI/dV as a function of position, a more careful interpretation must be made. According to the theory of Tersoff & Hamann (11, 12), the tip is expected to follow a contour of constant state density. At vertical separations that are large compared to the distance between the individual atoms on the surface, the atomic corrugations die away and the sample becomes laterally isotropic. In order for this to occur, the effective decay length of the wavefunctions above a local "peak" in the topography (κ_p^{-1}) must be shorter than those above a local "valley" (κ_v^{-1}).

This effect can be seen more quantitatively by using an approximation given by Tersoff & Hamann (11) for the Z-dependent corrugation $\Delta(Z)$,

$$\Delta(Z) \simeq \frac{2}{\kappa} \exp\left(-\frac{\pi^2 Z}{a^2 \kappa}\right) \qquad 4.$$

where a is the lattice constant and κ is the average inverse decay length. A simple analysis based on this formula leads to the approximate expression,

$$\kappa_v \simeq \kappa_p - \frac{2\pi^2}{a^2 \bar{\kappa}} \exp\left(-\frac{\pi^2 \bar{Z}}{\kappa a^2}\right). \qquad 5.$$

All other factors being equal, the effective decay length (and the tunneling transmission factor T) will be greater when measured at a location corresponding to a valley in the topography than when measured over a peak. This spatial variation in the transmission probability shows up in measurements of dI/dV as a background that is essentially an "inverted" topography. Thus, images showing the spatial variation of dI/dV obtained under conditions of constant average tunneling current always contain some topographic information convoluted in with the electronic information.

STS measurements of dI/dV at constant average tunneling current also contain a strongly voltage-dependent background whose origin can be seen by first writing the tunneling current as

$$I = \int_0^{eV} \rho_s(r, E)\rho_t(r, \pm eV \mp E)\exp\left(-\frac{2\sqrt{2mZ}}{\hbar}\sqrt{\phi + \frac{eV}{2} - E}\right)dE. \quad 6.$$

For simplicity, we can assume that the density of states ρ is constant for both sample and tip. Then, by neglecting the influence of the applied voltage on the tunneling barrier, the derivative dI/dV is given by

$$\frac{dI}{dV} = e\rho_s\rho_t \exp\left(-A\sqrt{\phi + \frac{eV}{2}}\right)Z \quad 7.$$

where $A = 2\hbar^{-1}\sqrt{2m}$. In modulation experiments conducted under conditions of constant average tunneling current, the sample-tip separation Z increases as the DC voltage $\bar{V}$ is increased; this variation in Z must be explicitly included. In the low-bias limit, Eq. 6 reduces to

$$\bar{I} = \rho_s\rho_t\bar{V}\exp(-A\sqrt{\phi}Z). \quad 8.$$

Solving this for Z and substituting the result into Eq. 7 gives $dI/dV = \bar{I}/\bar{V}$. Under conditions of constant average tunneling current $\bar{I}$, the quantity dI/dV diverges like $1/V$ as V approaches zero and presents a background term on which the desired spectroscopic information is superimposed.

Experimental Applications of STS

The relative simplicity of the STS technique has lead to its relatively widespread application. Early applications (50, 51) on clean and oxidized Ni(100) surfaces showed evidence for true surface states in addition to the barrier resonances described above. Salvan et al (52) applied this technique to the ($\sqrt{3} \times \sqrt{3}$)R30° overlayer of Au on Si(111) and observed a strong peak near 1 eV above E_F associated with a surface state, as well as field emission resonances at higher voltages. Becker et al (53) applied this technique to Si(111)–(7 × 7) and observed structure near +1.5 and +2.8 V associated with surface states and also observed a difference in energy between the surface states in the faulted and unfaulted halves of the unit cell. In later work (54), they applied STS to the study of surface states on various alloys of Si, Ge, and Sn and observed structure in dI/dV associated with the surface states of these alloys. Other STS studies have investigated the electronic structure of GaSa (55) and stepped Ni(111)/H (56) surfaces.

Many of these studies reveal two important disadvantages to the standard STS technique at constant average tunneling current. The first is the $1/V$ dependence of dI/dV, which makes it difficult to observe structure at low voltages. The second is that at lower voltages, the tip plunges toward the surface in order to maintain constant tunneling current, as shown in Becker's plots of tip height as a function of voltage (53, 54). This is a particularly severe problem on semiconductor surfaces, since it means that the STS technique is not able to probe the electronic structure of states lying near E_F, which are often of primary importance.

The $1/V$ divergence problem can be significantly improved by instead operating under conditions of constant *resistance*. This type of operation was first utilized by Kaiser & Jaklevic (57) to observe surface states on clean Au(111) and Pd(111) surfaces. Figure 6a shows their constant-resistance dI/dV spectra for Pd(111), together with the projected bulk band structure (Figure 6b) and the results of ultraviolet photoemission spectroscopy measurements of the density of states (Figure 6c). The peaks near $+1.0$ and -1.3 V in dI/dV closely correspond to critical points in the 2-D projected band structure, whereas the dI/dV peak near -0.6 eV is in reasonable agreement with a surface state observed in photoemission measurements. At constant average tunneling current, this information would likely have been obscured by the $1/V$ background.

The STS technique works well in cases where states lying near E_F are not of primary interest, but STS cannot probe electronic structure near the Fermi level. As in the case of voltage-dependent STM imaging, a complete mapping of the surface electronic structure as a function of energy and position requires many repeated measurements over the same area. This procedure is tedious at best, and is usually unsuccessful due to instability in the tip as well as thermal drifts in the microscope.

LOCAL I-V MEASUREMENTS

Early tunneling spectroscopy work by Feenstra et al (31) on the cleaved Si(111)–(2 × 1) surface showed that spectroscopic information could be obtained by acquiring the tunneling *I-V* curve and later numerically differentiating this curve. These early studies were performed with blunt tips and probed the area-averaged electronic properties of the surface. To take full advantage of the spatial resolution of the STM, the *I-V* curves need to be measured with "atomic" resolution and at well-defined locations on the surface in order to be able to correlate the surface "topography" with the local electronic structure. Such a measurement would also be best performed at a fixed sample-tip separation to eliminate the Z-dependence of the tunneling probability.

Experimental Method

Several methods of acquiring such local *I-V* information are available. These differ primarily in the details of the data acquisition hardware,

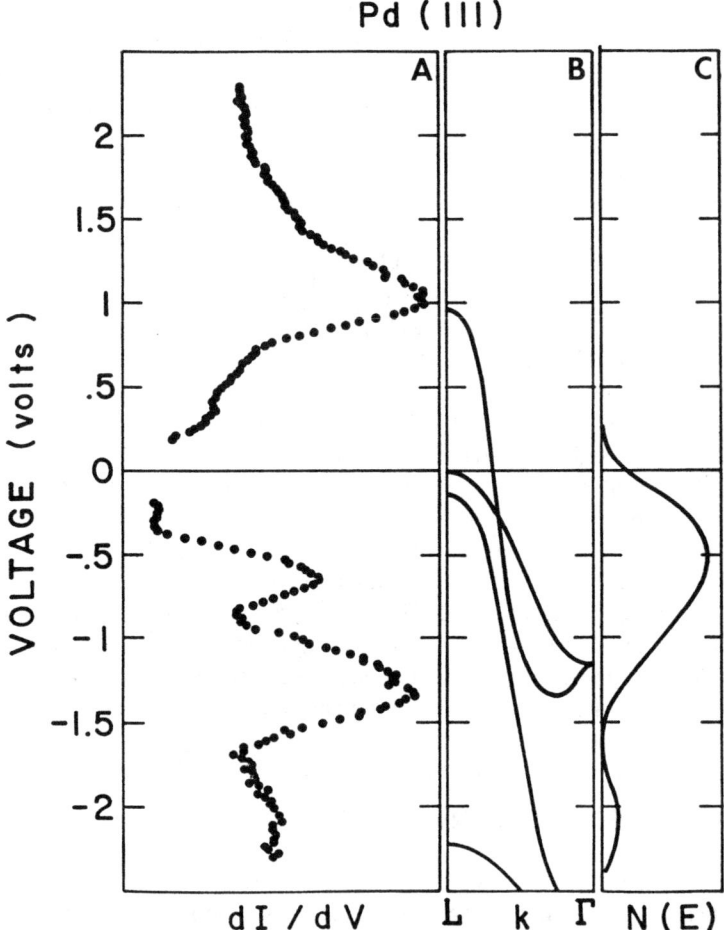

Figure 6 Surface states on Pd(111) obtained by using a modulation technique that maintains constant average conductance (*A*). These data are compared to the band structure for Pd(111) (*B*) and to the density of states determined from ultraviolet photoemission experiments (*C*) (from Ref. 57).

but all provide essentially identical information. Atomically resolved *I-V* measurements were first performed by Hamers et al (27) by rapidly acquiring *I-V* curves (each at a fixed sample-tip separation) while simultaneously slowly scanning the tip position. This technique was used to map out the complete electronic structure of the Si(111)–(7 × 7) unit cell. The method is denoted as Current Imaging Tunneling Spectroscopy (CITS) because in addition to the conventional image of the surface height as a function of position, I-V spectroscopy information is also obtained at each location, thereby permitting the local electronic characteristics to be directly imaged from the current measurements.

In the CITS method, a sample-and-hold circuit is installed in the feedback controller to gate the feedback control system on and off. The gating circuit usually consists of a commercial sample-and-hold amplifier, which can be located before, after, or within the feedback controller. When the feedback system is active, a constant voltage V_{stab} is applied to the sample, and the feedback control adjusts the tip height to maintain a constant tunneling current. When the feedback system is deactivated by means of the sample-and-hold circuit, it no longer corrects for changes in the tunneling current but instead keeps the tip fixed. The applied voltage is then linearly ramped between two voltages, and the tunneling I-V curve is measured. The applied voltage is finally returned to the value V_{stab}, and the feedback controller reactivated. By acquiring the I-V curves rapidly compared to the scan speed of the tip, both the sample "topography" and spatially resolved tunneling I-V characteristics are measured at each location in a two-dimensional raster scan over the surface. The high speed eliminates both vertical and lateral drift of the tip position, providing a one-to-one correspondence between each point in the topography and an I-V curve.

In practice, many modifications of this technique are possible (31, 32, 58, 59). The I-V curves can be measured with waveform recorders, digital oscilloscopes, or analog-to-digital converters in the computer itself. Likewise, the generation of the bias waveform and the timing can be controlled externally or internally to the computer. The I-V curves may also be measured at each surface location or only at a few selected locations. Despite these differences, the various methods generally provide the same essential information: measurement of I as a function of V at constant sample-tip separation at known locations on the surface.

The stabilization voltage V_{stab} plays a special role by determining the contour that the tip follows as it scans across the surface, and usually it must be chosen arbitrarily. The choice of V_{stab} affects the spatially-dependent I-V curves, since the sample-tip separation may be different at each location. Berghaus et al (60) and Feenstra (31) have explicitly demonstrated the changes in local I-V spectra on Si(111)–(7 × 7) and Si(111)–(2 × 1) surfaces at different stabilization voltages. Fortunately, the influence of the choice of stabilization voltage can be almost completely eliminated by proper normalization of the spectroscopic data to eliminate the Z-dependence by using a normalization procedure suggested by Feenstra (31). This normalization procedure is discussed in more detail below.

Analysis and Interpretation of I-V Data

Analysis of the I-V information can be performed in a number of ways. For example, the tunneling current I resulting from some bias voltage V can be directly imaged—a "current image." These images are dependent on the contour that the tip follows, and so cannot be directly interpreted in a quantitative fashion. However, the *symmetry* of the images provides information about the different spatial symmetry of the electronic states of the surface. The degree to which the choice of stabilization voltage influences the CITS images can be easily assessed by comparing data acquired at different stabilization voltages. On the Si(001) surface, Hamers et al (22) showed that the equivalent information was obtained either at positive or negative sample bias. On Si(111)–(7 × 7), changes in the symmetry of the current images coincided with the energies at which steep increases were observed in the local conductance and also coincided with the energies of the surface states known from photoemission studies (27, 61, 62). In some other cases, however, changes in the tunneling probability can dominate the electronic structure information (30, 60), particularly if the energy does not correspond to the energy of a known surface state. In any case, interpretation of the CITS images must be done carefully and must be corroborated by studying the I-V curves at selected locations or by normalizing the data to remove the Z-dependence, as is discussed below.

Identifying the surface-state energies is usually done by analyzing the I-V curves only at selected locations or averaged over a large region. Figure 7 shows plots of the conductance (I/V) vs. V measured at different specific locations within the Si(111)–(7 × 7) unit cell, together with the surface-states observed in ultraviolet photoemission (61) and inverse photoemission (62) spectroscopies. The conductance curves in Figure 7a show steep "onsets" at particular voltages corresponding to the energies of the surface states. The energies of these onsets correspond exactly with the positions of the surface states determined by photoemission shown in Figure 7b, and also correspond to the energies at which symmetry changes are observed in current images. The atomically resolved tunneling spectroscopy measurements directly reveal the atomic origins of the various electronic states. The states near -0.35 eV and $+0.5$ eV arise from the 12 adatoms within each unit cell, whereas the state near -0.8 eV arises from the 6 "rest" atoms. The states near -0.35 and $+1.4$ eV also appear to have some contribution from underlying layers, since they show an asymmetry between the faulted and unfaulted halves of the (7 × 7) unit cell.

Since dI/dV is small except at the "onset" energies, spatial maps of dI/dV at these energies can be loosely interpreted as "images" of the

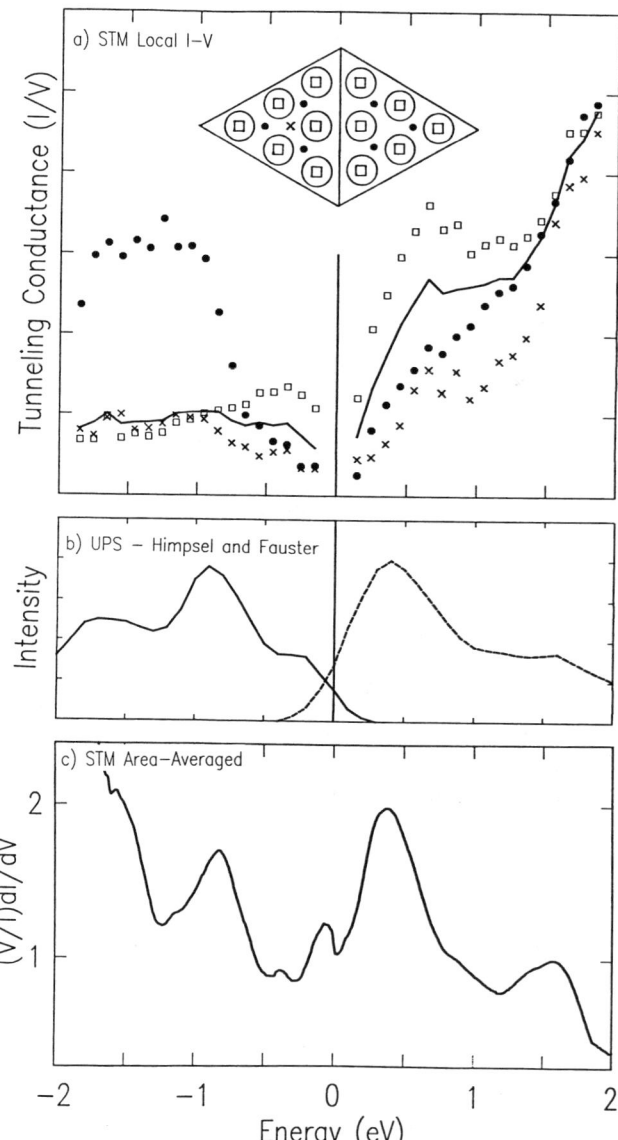

Figure 7 Plots of the conductivity (I/V) measured at various locations within the Si(111)–(7 × 7) unit cell (*a*), compared with the structure observed in photoemission and inverse photoemission spectroscopies (*b*). Also included is the normalized tunneling spectrum averaged over an area encompassing many unit cells (*c*).

surface states themselves. However, just as in modulation experiments, the spatial maps of dI/dV also contain a background contribution in the form of an "inverted topography." Only when dI/dV is large and corresponds to the energy of a surface state can surface states be imaged in this way. This is the case, for example, in the surface-state images of Si(111)–(7 × 7) (27), where images of dI/dV were presented only at energies at which the rapid increase in dI/dV overwhelmed this background effect.

Extracting quantitative information about the sample density of states is difficult because the density of the states of the tip ρ_t and voltage-dependent tunneling transmission probability $T(E, eV)$ are almost always unknown. This can create confusion when spectroscopy results obtained at different lateral positions are compared, since the sample-tip separation (and consequently, the transmission probability) are dependent on the contour that the tip follows, which is determined by the feedback stabilization voltage V_{stab}. The voltage dependence of the transmission probability may also vary as a function of position due to variations in the local work function, band-bending effects, and other phenomena.

In the case of superconductors, the energy range of interest is small enough that the voltage dependence of the tunneling barrier is only a minor effect. In the study of surface states of metals and semiconductors, however, the energy range of interest usually extends several eV on either side of E_F. Nevertheless, since $T(E, eV)$ is a slowly varying function, semi-quantitative electronic structure information can still be obtained and is useful to consider.

Dividing both sides of Eq. 3 by the static conductivity I/V, and then dividing both number and denominator by the factor $T(eV, eV)$ gives

$$\frac{dI/dV}{I/V} = \frac{\rho_s(eV)\rho_t(0) + \int_0^{eV} \frac{\rho_s(E)\rho_t(\pm eV \mp E)}{eT(eV, eV)} \frac{dT(E, eV)}{dV} dE}{\frac{1}{eV}\int_0^{eV} \rho_s(E)\rho_t(\pm eV \mp E)\frac{T(E, eV)}{T(eV, eV)} dE}. \qquad 9.$$

Feenstra (31) has argued that since $T(eV, eV)$ and $T(E, eV)$ appear as ratios in the second term in the numerator and in the denominator, their dependences on separation and applied voltage tend to cancel. Thus, this normalization effectively reduces the data to a form like

$$\frac{dI/dV}{I/V} = \frac{d(\log I)}{d(\log V)} = \frac{\rho_s(eV)\rho_t(0) + A(V)}{B(V)}. \qquad 10.$$

This quantity is equal to unity at $V = 0$. Assuming that $A(V)$ and $B(V)$ vary slowly with voltage, this provides a convenient normalization that works well when the density of states at the Fermi level is moderately high. Feenstra et al (31) acquired I-V curves at different sample-tip separations on Si(111)–(2 × 1) and verified that this normalization minimizes the influence of the sample-tip separation Z.

Figure 7c shows the results of this analysis for area-averaged tunneling I-V data for Si(111)–(7 × 7) (63). A comparison of this curve with the atomic-resolution conductivity measurements in Figure 7a and the photoemission results in Figure 7c shows a close correspondence, with surface-state peaks occurring at -1.5, -0.8, -0.2, $+0.45$, and $+1.55$ eV. Surprisingly, even the relative intensities of the tunneling data and the photoemission data appear to be similar, except for the state near -1.5, which appears quite small in the tunneling data.

Comparison of Normalized I-V Data with True Density of States

In order to show how the tunneling spectra resulting from this normalization compare with the local state density, an artificial density of states (DOS) function was created and then, by using Simmons' formulas (64) (which also includes effects of the image potential), numerically integrated to obtain the I-V curve predicted from tunneling theory. This I-V curve was then analyzed according to standard procedures, and the results compared with the "true" density of states. The shapes and locations of the peaks in this density of states function were chosen so that the calculated $d(\log I)/d(\log V)$ was in reasonable agreement with the experimental curve for Si(111)–(7 × 7) shown in Figure 7c. The other parameters used in the calculation were $\phi_{\text{sample}} = 2.5$ eV, $\phi_{\text{tip}} = 2.5$ eV, and $Z = 8.0$ Å.

Figure 8 shows the results of this simulation. The DOS function shown in Figure 8a was used with Simmons' formulas (assuming a constant DOS for the tip) to generate the I-V curve expected for this distribution, shown in Figure 8b. This was then numerically differentiated to give the plot of dI/dV vs. V shown in Figure 8c, and normalized by the static conductivity to produce the normalized spectrum shown in Figure 8d. Comparing this normalized spectrum in the bottom panel with the starting DOS function shown in the top panel shows that they have peaks in nearly the sample position. However, the intensities of the peaks are quite different. Unoccupied states of the sample are observed much more clearly and with higher intensity in the normalized tunneling spectra than are occupied states, a result of the fact that most of the tunneling electrons arise from states near the Fermi energy of the negatively-biased electrode. This numerical simulation demonstrates that the intensities of the peaks observed in $(dI/dV)/(I/V)$, or equivalently, $d(\log I)/d(\log V)$ are *not* proportional to the surface density of states, particularly at negative sample

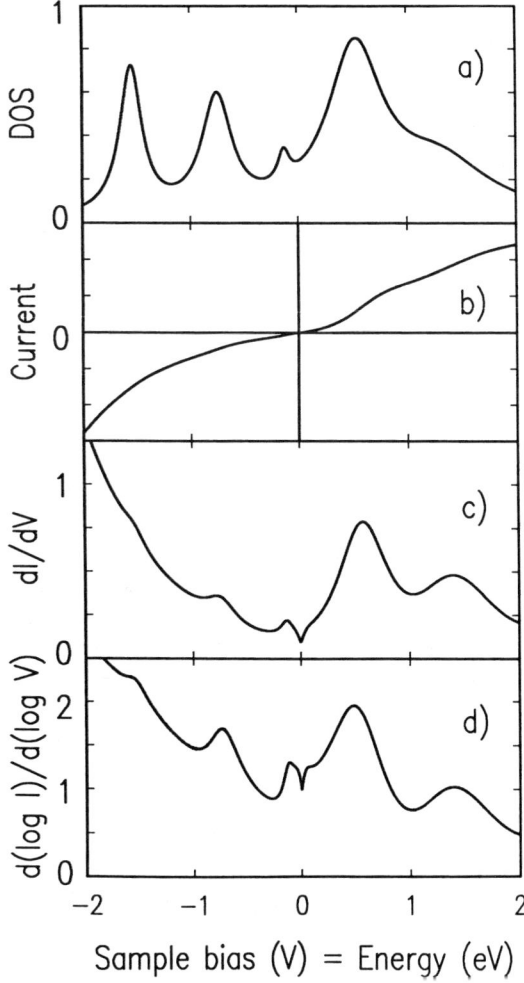

Figure 8 Numerical simulation of tunneling *I-V* spectrum and normalized results: (*A*) Original density of states function. (*B*) Tunneling *I-V* curve calculated by numerical integration of tunneling equations. (*C*) First derivative spectrum. (*D*) Normalized derivative $(dI/dV)/(I/V) = d(\log I)/d(\log V)$ spectrum.

bias. However, this procedure does provide a *convenient* normalization for the spectroscopy data and is relatively independent of sample-tip separation.

The close correspondence between the simulated spectrum in Figure 8*d* and the normalized experimental data for Si(111)–(7 × 7) shown in Figure

7c indicates that the density of states of Si(111)–(7 × 7) is similar to the function shown in Figure 8a.

When normalizing tunneling spectra of surfaces with a surface-state bandgap, artificial discontinuities arise in this normalization due to the strong voltage-dependence of the $B(V)$ term in the denominator of Eq. 9, particularly near the band edges. One alternative procedure is then to divide the dynamic conductivity (dI/dV) by the static conductivity (I/V) extrapolated from measurements at higher voltages (outside the surface-state bandgap). This provides compensation for voltage-dependent changes in the tunneling barrier but prevents the denominator in Eq. 9 from vanishing (32). If the structure at the edges of the surface-state bands are of primary importance, direct plots of I vs. V are satisfactory (65).

Applications of Local I-V Spectroscopy

In addition to Si(111)–(7 × 7) work discussed above, several other surfaces have been extensively studied by using local I-V spectroscopy. Tunneling measurements on Si(111)–(2 × 1) (29–31, 66) show a close correspondence to the theoretical band structure for the π-bonded chain model proposed by Pandey (67). Measurements of the inverse decay length as a function of applied bias also have revealed interesting behavior resulting from dispersion of the surface-state bands, which disperse toward E_F as the parallel momentum $k_\parallel$ increases. At low voltage, tunneling can only occur from states with large $k_\parallel$, which have a short decay length. At higher voltages, tunneling occurs from states with $k_\parallel = 0$, which have the slowest decay. From measurements of the inverse decay length as a function of voltage, the dispersion of the surface-state bands could be inferred from the tunneling measurements.

On Si(001)–(2 × 1) (23, 37), normalized spectra reveal a surface-state bandgap of $\simeq 0.7$ eV, with peaks at -0.85 eV and $+0.35$ eV corresponding to the dimer π-bonding and π^* antibonding states discussed above, in agreement with photoemission results (61, 62). Tunneling measurements show that exposure of this surface to NH_3 eliminates these states and replaces them with an intense state lying 1.1 eV above E_F associated with Si–H antibonding states (37) due to dissociation of the NH_3. Voltage-dependent imaging also provides a contrast between reacted and unreacted dimers, due to changes in the spatial distribution of occupied states upon H adsorption. The Ge(001) surface (25) undergoes a reconstruction similar to that of Si(001), and tunneling results have identified the occupied and unoccupied states of this surface as well.

Figure 9 demonstrates the capability of atomic-resolution I-V measurements to probe directly the electronic properties of defects. This figure shows I-V curves acquired at different distances from a small defect,

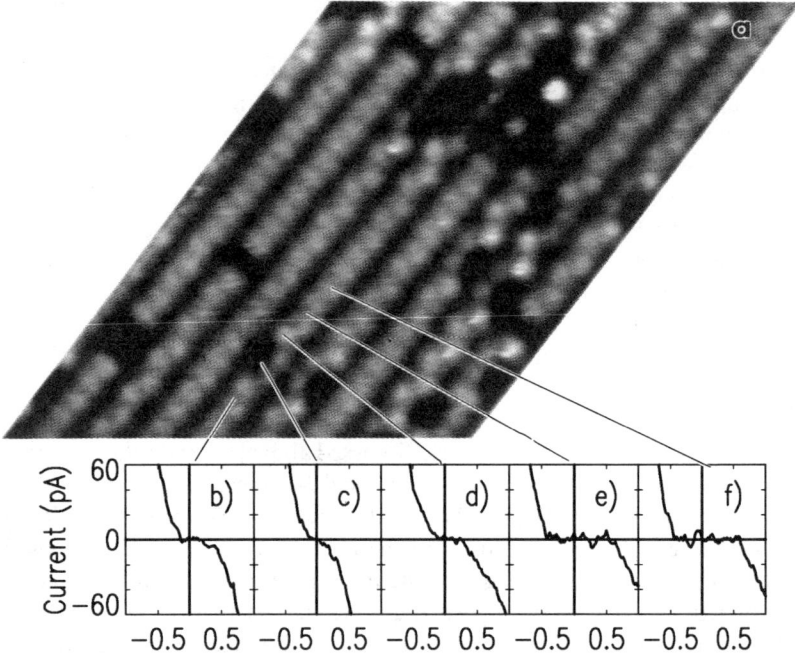

Figure 9 Topographic image of Si(001) surface including defects (*a*) and spatially-resolved *I-V* spectra showing variation in the surface-state bandgap at different lateral separations (*b–f*).

measured simultaneously with the constant-current topograph (at -1.5 eV) (65). Far away from the defect (Figure 9*e*,*f*), the tunneling *I-V* curve shows a clear gap as revealed from the sharp turn-on of the tunneling current near -0.45 eV and $+0.25$ eV [in agreement with the gap edges observed in area-averaged, logarithmically differentiated spectra (37)]. As the defect is approached, the sharp gap edges disappear and, directly over the defect (Figure 9*c*), the tunneling curve exhibits a strong exponential increase both above and below E_F, thus demonstrating a high density of states at E_F at the defect site (65). Such spatially dependent measurements also provide a direct measure of the spatial extent of the wavefunctions associated with defects and impurities.

Single-atom defects occurring in Al overlayers on Si(111) have also been studied (44, 63, 68). Substituting Si for Al shifts the energy of a p_z state from $+1.1$ eV to -0.35 eV, which is readily detectable in the *I-V* spectra and also leads to large contrast changes in constant-current topographs, which can be seen in Figure 5. Unique localization effects are also observed in tunneling spectra at these defects, which are effectively isolated from one another by the large surface atom spacing in the $\sqrt{3}$ structure (44, 68).

The ability to probe the local electronic structure of defects and chemically inequivalent atoms at surfaces has great potential for the study of surface chemical reactivity on an atom-by-atom basis. An example is the decomposition of NH_3 on Si(111)–(7 × 7) studied by Avouris & Wolkow (69, 70). The Si(111)–(7 × 7) structure has several types of chemically inequivalent surface atoms exposed. By combining voltage-dependent STM topographs with local I-V measurements before and after exposing the surface to NH_3, variations in chemical reactivity between the various types of chemically inequivalent atoms were identified. Thus, local I-V spectroscopy provides a way of directly correlating chemical reactivity with local electronic structure.

Other local I-V spectroscopy studies have concentrated primarily on graphite (46) or superconductors. Several studies have reported spatial variations in superconducting energy gaps. Fein et al (71) and Kirtley et al (58) measured I-V curves at each location while scanning over granular superconductors, and later analyzed the data to provide an image of the spatial dependence of the energy gap. Hess et al (59) have used a slightly modified technique to image the electronic state density around a superconducting flux core in $NbSe_2$.

BALLISTIC ELECTRON EMISSION MICROSCOPY (BEEM)

An important variation on conventional tunneling spectroscopy is the Ballistic Electron Emission Microscopy (BEEM) technique invented by Kaiser & Bell (72, 73). This technique probes the electronic properties of buried interfaces, which are essentially invisible to conventional tunneling spectroscopy. This is accomplished by using a three-terminal measurement in which the tunneling tip emits electrons into a thin film of metal on a semiconducting substrate. Electrons tunneling through the vacuum gap pass into the metallic film, where most of them scatter and are collected at the second terminal. However, a small percentage of the electrons ($\simeq 1$–5%) propagate through the metal film and pass into the semiconducting substrate, where they are collected at a third terminal. There is usually a potential (Schottky) barrier at the metal-semiconductor interface that the electrons must overcome in order to pass into the semiconductor. While the tunneling current from the tip to the metal film is kept constant, the bias between the metal film and the semiconducting substrate is varied, and the small current arising at the semiconductor from unscattered electrons is measured. The resulting I-V curve is characteristic of the electronic structure at the metal-semiconductor *interface*. Scanning the tip then provides spatial resolution as well.

The spatial resolution of this technique is remarkably good because of boundary-matching conditions arising at the metal-semiconductor interface (73). When entering the semiconductor, the electrons lose most of their momentum perpendicular to the interface but retain the parallel component, so that the electrons refract. Electrons impinging from steep angles are reflected back into the metal film and do not contribute to the measured current. For Au/GaAs, only those electrons approaching the buried interface from within a few degrees of the normal direction can pass into the GaAs. As a result, lateral resolution of less than 10 Å is possible to achieve even through $\simeq 100$ Å thick metal films!

This technique is particularly useful for directly measuring the potential (Schottky) barrier for electrons at the metal-semiconductor interface but is also capable of probing the bulk band structure of the collector; recent BEEM spectra on GaAs (73) show additional structure that can be related to critical points in the GaAs bulk band spectrum. This very new technique has great potential for the spectroscopic study of buried interfaces.

INELASTIC TUNNELING SPECTROSCOPY

The electronic spectroscopy discussed thus far provides some chemical contrast, but it is not capable in most cases of actually identifying chemical species adsorbed on a surface. For such purposes, measurement of the vibrational spectra of adsorbed species would be more useful, since vibrational spectra are usually very sharp and show features characteristic of particular molecular functional groups. Theoretical calculations (13, 14, 74, 75) have predicted that under certain circumstances, changes in dI/dV as large as 10% might be observed in tunneling measurements. Observing such structure is not an easy task because the vibrational features are so sharp that a low-temperature STM is required in order to avoid thermal broadening of the Fermi levels. Hansma (8) has estimated an effective resolution of 5.4 kT for inelastic tunneling, whereas vibrational features are only a few meV wide.

The only experimental results currently available are for phonons on graphite (76) and for sorbic acid adsorbed on graphite (77), both obtained on a surface that was immersed in liquid helium. In the case of graphite phonons, a good correspondence was observed between peaks in d^2I/dV^2 vs. V and the known phonon energies determined from various other methods. For sorbic acid adsorbed on graphite, peaks were observed in the *first* derivative spectrum instead of the expected second derivative spectrum. Additionally, the peaks were very intense and the energies of the peaks were different from those measured in bulk tunnel junctions. The origins of these discrepancies are not yet resolved and may arise from a strong coupling with graphite bulk states. More work is required before STM will be used to perform molecular identification by means of vibrational spectroscopy.

SUMMARY

In this paper, several of the methods developed for acquiring and interpreting tunneling spectroscopy data have been presented and discussed, together with some representative examples of how each of these techniques can be applied to study the structure, bonding, and reactivity of surfaces. The unique ability of the STM to probe directly the electronic structure of surfaces promises to open yet another dimension in our understanding of surfaces by allowing us to study the energetics of the electronic states at the surface on an atom-by-atom basis and to correlate directly the geometric positions of the atoms with the resulting electronic structure. Advances in the acquisition and interpretation of tunneling spectroscopy data continue to improve our ability to make efficient use of this capability. Because electronic structure is intimately linked with chemical reactivity, tunneling spectroscopy studies in future years should yield new insight into the role of local electronic structure of chemical reactivity as well as a wide variety of other physical and chemical processes at surfaces.

ACKNOWLEDGMENTS

I would like to express my appreciation to J. Demuth for many stimulating discussions of scanning tunneling microscopy/spectroscopy. The partial support of the US Office of Naval Research is also gratefully acknowledged.

Literature Cited

1. Binnig, G., Frank, K. H., Fuchs, H., Garcia, N., Reihl, B., Rohrer, H., Salvan, F., Williams, A. R. 1985. *Phys. Rev. Lett.* 55: 991–94
2. Binnig, G., Rohrer, H. 1986. *IBM J. Res. Dev.* 30: 355–69
3. Binnig, G., Rohrer, H. 1982. *Helv. Phys. Acta* 55: 726–35
4. Binnig, G., Rohrer, H., Gerber, C., Weibel, E. 1983. *Phys. Rev. Lett.* 50: 120–23
5. Binnig, G., Rohrer, H. 1983. *Surf. Sci.* 126: 236–44
6. Binnig, G., Rohrer, H. 1985. *Surf. Sci.* 157: L373–78
7. Lang, N. D. 1986. *IBM J. Res. Dev.* 30: 374–79
8. Hansma, P. K. 1982. *Tunneling Spectroscopy: Capabilities, Applications, and New Techniques.* New York: Plenum
9. Wolf, E. L. 1986. *Electron Tunneling Spectroscopy.* Cambridge: Oxford Univ. Press
10. Garcia, N., Flores, F., Guinea, F. 1988. *J. Vac. Sci. Technol. A* 6: 323–26
11. Tersoff, J., Hamann, D. R. 1985. *Phys. Rev. B* 31: 805
12. Tersoff, J., Hamann, D. R. 1983. *Phys. Rev. Lett.* 50: 1998–2001

13. Baratoff, A., Persson, B. N. J. 1988. *J. Vac. Sci. Technol. A* 6: 331–35
14. Persson, B. N. J., Baratoff, A. 1987. *Phys. Rev. Lett.* 59: 339–42
15. Lang, N. D. 1986. *Phys. Rev. B* 34: 5947–50
16. Chen, C. J. 1988. *J. Vac. Sci. Technol. A* 6: 319–22
17. Bardeen, J. 1961. *Phys. Rev. Lett.* 6: 57–59
18. Lang, N. D. 1987. *Phys. Rev. Lett.* 58: 45–48
19. Lang, N. D. 1988. *Phys. Rev. B* 37: 10395–98
20. Lang, N. D. 1985. *Phys. Rev. Lett.* 55: 230–33
21. Lang, N. D., Williams, A. R. 1978. *Phys. Rev. B* 18: 616–36
22. Hamers, R. J., Tromp, R. M., Demuth, J. E. 1987. *Surf. Sci.* 181: 346–55
23. Hamers, R. J., Avouris, P., Bozso, F. 1988. *J. Vac. Sci. Technol. A* 6: 508–11
24. Hamers, R. J., Tromp, R. M., Demuth, J. E. 1987. *Phys. Rev. B* 34: 5343–57
25. Kubby, J. A., Griffith, J. E., Becker, R. S., Vickers, J. S. 1987. *Phys. Rev. B* 36: 6079–93
26. Tromp, R. M., Hamers, R. J., Demuth, J. E. 1986. *Phys. Rev. B* 34: 1388–91
27. Hamers, R. J., Tromp, R. M., Demuth, J. E. 1986. *Phys. Rev. Lett.* 56: 1972–75
28. Becker, R. S., Golovchenko, J. A., Swartzentruber, B. S. 1985. *Phys. Rev. Lett.* 55: 987–90
29. Stroscio, J. A., Feenstra, R. M., Fein, A. P. 1986. *Phys. Rev. Lett.* 57: 2579–82
30. Stroscio, J. A., Feenstra, R. M., Newns, D. M., Fein, A. P. 1988. *J. Vac. Sci. Technol. A* 6: 499–507
31. Feenstra, R. M., Stroscio, J. A., Fein, A. P. 1987. *Surf. Sci.* 181: 295–312
32. Feenstra, R. M., Martensson, P. 1988. *Phys. Rev. Lett.* 61: 447–50
33. Feenstra, R. M., Stroscio, J. A., Tersoff, J., Fein, A. P. 1987. *Phys. Rev. Lett.* 58: 1192–95
34. Stroscio, J. A., Feenstra, R. M., Fein, A. P. 1987. *Phys. Rev. Lett.* 58: 1668–71
35. Selloni, A., Carnevali, P., Tosatti, E., Chen, C. D. 1985. *Phys. Rev. B* 31: 2602–5
36. Reihl, B., Gimzewski, J. K., Nicholls, J. M., Tosatti, E. 1986. *Phys. Rev. B* 33: 5770–73
37. Hamers, R. J., Avouris, P., Bozso, F. 1987. *Phys. Rev. Lett.* 59: 2071–74
38. Appelbaum, J. A., Baraff, G. A., Hamann, D. R. 1975. *Phys. Rev. Lett.* 35: 729–32
39. Chadi, J. 1979. *Phys. Rev. Lett.* 43: 43–47
40. Pandey, K. C. 1981. *Phys. Rev. Lett.* 47: 1913–17
41. Takayanagi, K., Tanishiro, Y., Takahashi, M., Takahashi, S. 1985. *J. Vac. Sci. Technol. A* 3: 1502–6
42. Tersoff, J., Feenstra, R. M., Stroscio, J. A., Fein, A. P. 1988. *J. Vac. Sci. Technol. A* 6: 497–98
43. Stroscio, J. A., Feenstra, R. M., Fein, A. P. 1987. *Phys. Rev. B* 36: 7718–21
44. Hamers, R. J. 1988. *J. Vac. Sci. Technol. B* 6: 1462–67
45. Northrup, J. E. 1984. *Phys. Rev. Lett.* 53: 683–86
46. Bando, H., Morita, N., Tokumoto, H., Mizutani, W., Watanabe, K., Homma, A., Wakiyama, S., Shigeno, M., Endo, K., Kajimura, K. 1988. *J. Vac. Sci. Technol. A* 6: 344–48
47. van Loenen, E. J., Demuth, J. E., Tromp, R. M., Hamers, R. J. 1987. *Phys. Rev. Lett.* 58: 373–76
48. Hamers, R. J. 1989. *Phys. Rev. B* 39: 5091–5100
49. Demuth, J. E., Koehler, U. K., Hamers, R. J., Kaplan, P. 1989. *Phys. Rev. Lett.* 62: 641–45
50. Garcia, N. 1986. *IBM J. Res. Dev.* 30: 533–42
51. Garcia, R., Saenz, J. J., Garcia, N. 1986. *Phys. Rev. B* 33: 4439–42
52. Salvan, F., Fuchs, H., Baratoff, A., Binnig, G. 1985. *Surf. Sci.* 162: 634–39
53. Becker, R. S., Swartzentruber, B. S., Vickers, J. S. 1988. *J. Vac. Sci. Technol.* 6: 472–77
54. Becker, R. S., Klitsner, T., Vickers, J. S. 1988. *Phys. Rev. B* 38: 3537–40
55. Humbert, A., Salvan, F., Mouttet, C. 1987. *Surf. Sci.* 181: 307–12
56. Van de Walle, G. F. A., Van Kempen, H., Wyder, P., Flipse, C. J. 1987. *Surf. Sci.* 181: 27–36
57. Kaiser, W. J., Jaklevic, R. C. 1985. *IBM J. Res. Dev.* 30: 411–16
58. Kirtley, J. R., Raider, S. I., Feenstra, R. M., Fein, A. P. 1987. *Appl. Phys. Lett.* 50: 1607–9
59. Hess, H. F., Robinson, R. B., Dynes, R. C., Valles, J. M., Waszczak, J. V. 1989. *Phys. Rev. Lett.* 62: 214–18
60. Berghaus, T., Brodde, A., Neddermeyer, H., Tosch, S. 1988. *J. Vac. Sci. Technol. A* 6: 483–87
61. Himpsel, F. J., Fauster, T. 1984. *J. Vac. Sci. Technol. A* 2: 815–21
62. Fauster, T., Himpsel, F. J. 1983. *J. Vac. Sci. Technol. A* 1: 1111–14
63. Hamers, R. J., Demuth, J. E. 1988. *J. Vac. Sci. Technol. A* 6: 512–16
64. Simmons, J. G. 1963. *J. Appl. Phys.* 34: 1793–1803
65. Hamers, R. J. 1989. *J. Vac. Sci. Technol. A*. In press
66. Feenstra, R. M., Thompson, W. A.,

Fein, A. P. 1986. *Phys. Rev. Lett.* 56: 608–14
67. Pandey, K. C. 1985. *Proc. 17th Int. Conf. on Phys. Semiconductors.* New York: Springer-Verlag
68. Hamers, R. J., Demuth, J. E. 1988. *Phys. Rev. Lett.* 60: 2527–30
69. Avouris, P., Wolkow, R. 1989. *Phys. Rev. B* 39: 5091–5100
70. Wolkow, R., Avouris, P. 1988. *Phys. Rev. Lett.* 60: 1049–52
71. Fein, A. P., Kirtley, J. R., Feenstra, R. M. 1987. *Rev. Sci. Inst.* 58: 1806–10
72. Kaiser, W. J., Bell, L. D. 1988. *Phys. Rev. Lett.* 60: 1406–9
73. Bell, L. D., Kaiser, W. J. 1988. *Phys. Rev. Lett.* 61: 2368–71
74. Binnig, G., Garcia, N., Rohrer, H. 1985. *Phys. Rev. B* 32: 1336–38
75. Persson, B. N. J., Demuth, J. E. 1986. *Solid State Commun.* 57: 769–72
76. Smith, D. P. E., Binnig, G., Quate, C. F. 1986. *Appl. Phys. Lett.* 49: 1641–43
77. Smith, D. P. E., Kirk, M. D., Quate, C. F. 1987. *J. Chem. Phys.* 86: 6034–38

ORIENTED MOLECULE BEAMS VIA THE ELECTROSTATIC HEXAPOLE: Preparation, Characterization, and Reactive Scattering

David H. Parker

Department of Chemistry, University of California, Santa Cruz, California 95064

Richard B. Bernstein

Department of Chemistry, University of California, Los Angeles, California 90024

INTRODUCTION

The steric effect is one of the oldest and most intuitive concepts in chemical kinetics, yet our quantitative understanding of it has been quite limited. Chemists recognized very early that the need for "proper" mutual orientation of reactants is second only to energetic requirements for a "successful" collision. "Steric factors" were included long ago in the pre-exponential term of simple rate expressions for elementary reactions; these factors reflected the probability of achieving proper orientation given the random nature of molecular encounters. However, measuring the reactivity for specific collision geometries, i.e. selected impact parameters and reagents' mutual orientation in an elementary reaction, appeared to be impossible.

Control of the impact parameter is still beyond reach for bimolecular reactions of isolated molecules in the gas phase, but control of the reactant orientation before collision became possible with the development of molecular beam techniques during the 1960s. Molecular beam studies of the steric effect, using reagent molecules oriented by the electrostatic hexapole technique, were first reported in 1966. In these first-generation experiments, the orientation of the reagents was only qualitatively defined. A second generation of experiments, with quantitatively known orientation of reactants in single rotational states, began earlier in this decade, owing largely to the availability of supersonic seeded beam sources for producing cold molecules and to the construction of better hexapoles. Application of modern laser detection schemes promises to open up a third generation of state-to-state oriented molecule scattering studies by the 1990s.

In this article, we appraise the current status of research on the steric effect, but have limited our scope to *oriented* molecule scattering studies, using molecular beam M-state selection by the electrostatic hexapole field technique. The term *oriented* implies that "heads" and "tails" of the subject molecule have been distinguished, in contrast to *aligned*, meaning "end-on" vs. "broadside." (Photoselection-based *alignment* experiments compare reactivity for end-on vs. broadside collisions, where "end-on" is an *average* of heads and tails. For asymmetric molecules, with distinguishable heads and tails, alignment experiments miss the main steric effect.) Besides creating beams of oriented molecules for use as reagents, the hexapole technique can also probe the orientation of nascent species.

A general introduction to the field (1) and a progress report on oriented molecule reactions (2) appeared in a recent issue of the *Journal of Physical Chemistry* that was devoted to stereochemical dynamics. Brooks (3), Stolte (4), and Bernstein (5) provided earlier reviews, and Stolte (6) has recently presented a thorough overview of scattering experiments that employ state-selection. Related reviews, including scattering of aligned atoms and molecules, have appeared recently in a book edited by Scoles (7).

The *modus operandi* of oriented molecule research utilizing the hexapole technique is evident by inspection of a generic experimental layout. Figure 1 shows the apparatus used by the authors to study the reaction of oriented methyl iodide with Rb atoms (8a,b). Although many features have improved over time, the basic experiment consists of the same three phases: *before* reaction, reagent focusing and state-selection by the electrostatic hexapole field; *during* reaction, orientation of the reactants in the collision zone; and *after* reaction, detection of product(s), as a function of laboratory scattering angle. Most of the recent improvements have been in reagent preparation, yet the most interesting and difficult questions about the method itself center on the "during" stage. Considerable insight into the reaction dynamics can be gleaned from the post-reaction observations

ORIENTED MOLECULE BEAMS 563

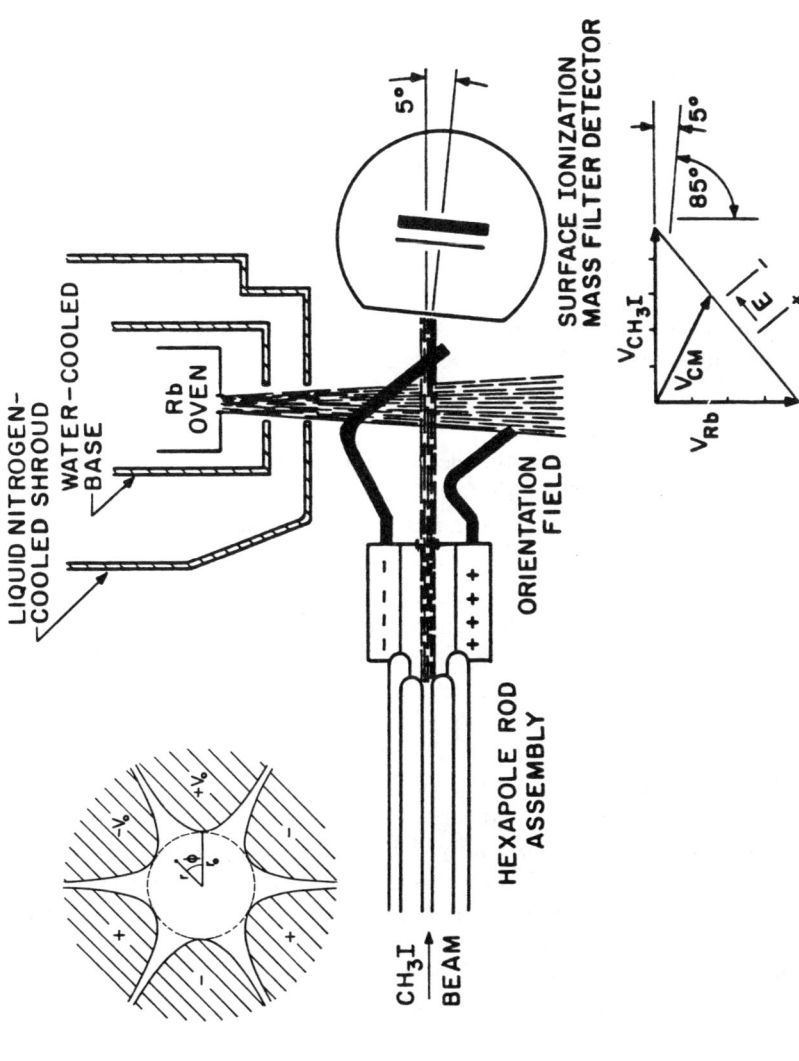

Figure 1 (a) Cross section of an ideal hexapole (12). The voltages applied to the rods are $+V_0, -V_0, +V_0$, etc. (b) Schematic diagram of the beam intersection zone of a hexapole-oriented crossed-beam apparatus. The so-called "tilted" orientation field is shown, as is the velocity vector diagram for the reaction ($CH_3I + Rb \rightarrow RbI + CH_3$). [From Ref. (8a,b).]

by measuring different properties of one or another of the nascent products from reaction of oriented reagents.

PREPARATION OF ORIENTED MOLECULE BEAMS

Polar molecules can be rotationally state-selected and oriented by the electrostatic hexapole technique via their first-order Stark effect. Suitable molecules are symmetric tops, asymmetric tops with a dipole moment along the A or C axis, linear polyatomic molecules excited in bending vibrations, and diatomics with electronic angular momentum.

Theory

The electrostatic hexapole technique works under the principle of rotational selection by focusing (9–11). In an electric field, $\mathbf{E}$, polar symmetric-top molecules in a given $|JKM\rangle$ state precess about $\mathbf{E}$ with an average orientation given by

$$\langle \cos \theta_{\mu,E} \rangle = KM/J(J+1), \qquad 1.$$

where θ is the angle between the electric dipole moment $\boldsymbol{\mu}$ and $\mathbf{E}$, and J, K, M are the usual symmetric-top rotational quantum numbers. An inhomogeneous electric field exerts a radial force on the molecule

$$\mathbf{F}_r = -\partial W/\partial r = \mu_{\text{eff}}(\partial \mathbf{E}/\partial r), \qquad 2.$$

where $\mu_{\text{eff}} = -\partial W/\partial |\mathbf{E}|$, which for symmetric-top molecules is $\mu \langle \cos \theta \rangle$, a constant. Since, for a hexapole, $\partial \mathbf{E}/\partial r$ is proportional to r, the radial force is linear in r, so the molecule's trajectory is governed by an equation for harmonic motion. (In weak fields, Eq. 1 is inapplicable for many molecules, due to the hyperfine coupling effect. This is discussed in the next section.)

The hexapole field acts as a thick lens, focusing molecules with negative values of KM that are initially diverging from the axis to a focal point downstream of the hexapole assembly. All molecules of a given JKM state take trajectories of the same "wavelength," provided that their speeds are the same. (Molecules with positive KM take exponentially diverging trajectories and are ejected radially out of the field.) The "working equation" for standard half-wave hexapole focusing is

$$-\langle \cos \theta \rangle = \frac{\pi^2}{6} \frac{r_0^3}{l} \frac{mv^2}{\mu V_0}, \qquad 3.$$

where r_0 and l are the internal radius and length, respectively, of the hexapole; m and v are the mass and speed, respectively, of the molecule, and V_0 is the hexapole rod voltage. Nonfocusable states, i.e. those with J,

K, or M equal to zero, are usually blocked from reaching the focal point by an on-axis beam stop. As seen from Eq. 3, the best-oriented states focus at the lowest hexapole voltages. The minimum rod voltage at which molecules can be focused, i.e. $V_{\text{threshold}}$, is given by Eq. 3 when $\langle \cos\theta \rangle = -1$.

In 1965, Kramer & Bernstein (12) pointed out that hexapole-focused state-selected molecules have their molecular axes all oriented in the same direction with respect to the internal hexapole **E** field and that laboratory-frame orientation can be achieved by adiabatically redirecting **E** to parallel field lines. Reactive scattering studies by the groups of Brooks (13) and Bernstein (14) employing oriented molecule beams immediately followed. Studies up through the mid-1980s attained only coarsely defined orientation. The ability to select single rotational states came only with the utilization of supersonic beams of rotationally cold molecules formed in nozzle expansions.

Two conditions must be met to produce a beam of molecules populated in a single $|JKM\rangle$ state: (a) the beam velocity spread must be small (e.g. $\Delta v/v \lesssim 0.1$), and (b) the rotational temperature of the beam molecules must be low enough that essentially only a single state with a given $KM/J(J+1)$ value is appreciably populated. Supersonic seeded beams (15) are ideal, with their low ($\lesssim 10\text{K}$) translational and rotational temperatures. Examples of high-resolution rotational JKM-state selection for symmetric-top molecules follow.

Experiment

Figure 2 is a schematic diagram of an oriented-molecule beam machine. Figure 3 illustrates the degree of rotational state selection achievable with modern hexapole-focusing machines. Individual peaks (corresponding to single rotational states) are observed in the focusing curves (i.e. in plots of focused beam intensity versus hexapole rod voltage V_0) for prolate symmetric-top molecules. Low rotational temperatures ($\sim 5\text{K}$) result from the strong expansion conditions, so low that even the $|JKM\rangle = |222\rangle$ state is essentially unpopulated. For simplicity, minus signs on K or M are usually dropped. Figure 3(a), for CH_3F/He, was obtained with the oriented-molecule beam machine at Nijmegen, by using a cw nozzle source and electron impact mass spectrometer detector (16); Figure 3(b), for CH_3Cl/Kr, with the UCLA apparatus (Figure 2) (17a,b). The latter uses a pulsed-valve seeded beam source and pulsed-laser ionization time-of-flight detection. A major advantage to pulsed-laser detection is that the degree of orientation of the ensemble of oriented molecules in the detection volume can be measured directly, as discussed below.

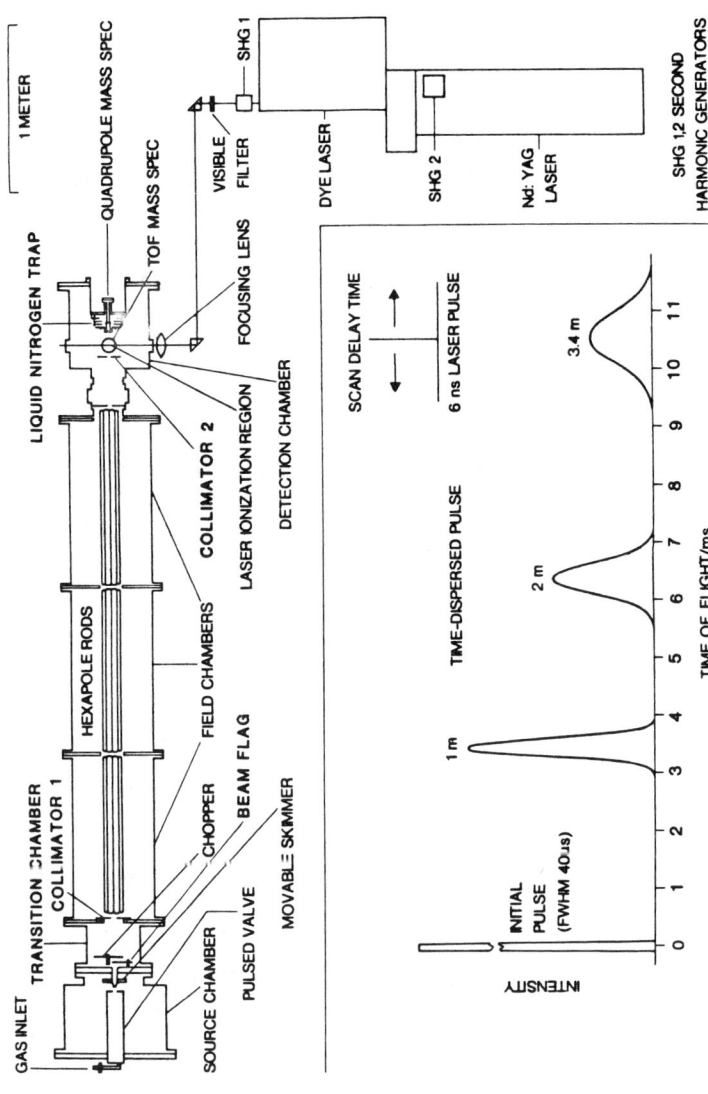

Figure 2 Schematic diagram of a pulsed-beam hexapole-focusing machine for production of oriented molecule beams in single rotational states. Not shown are the guiding and orienting fields, located between the hexapole rods and the laser ionization region. The *insert* below shows the temporal dispersion of the original 40 μs pulse of a CH$_3$I beam (seeded in Kr) as it passes down the hexapole system and arrives in the detection zone. By varying the delay time of the laser pulse with respect to the beam pulse, molecules with different velocities are detected selectively. [From Ref. (17a).]

Ideal reactant selection—essentially monovelocity, highly collimated beams of oriented molecules with all rovibronic quantum numbers specified—can now be achieved by hexapole focusing. A wide range of beam velocity is covered by varying the carrier gas composition and source temperature. The main drawback of the technique is that, in these cold beams, only low quantum number states are populated, and these have rather broad orientational distributions (Figure 4). In addition, it is not yet possible to achieve single-state resolution of oblate-top molecules owing to rotational congestion, even at 5K.

A wider range of reactant states can be accessed by state-specifically exciting the molecular beam by an infrared laser before it enters the hexapole fields. Figure 3(a) includes a focusing curve from the Stolte group for the $|JKM\rangle = |111\rangle$ state of CH_3F with one quantum of the C–F stretch excited by a tunable CO_2 laser. Stark fields are used to tune CH_3F into resonance with the laser; pumping of the unfocusable $M = 0$ states into $M = 1$ levels is detected by an enhancement of the intensity of the focused beam. It has been possible to pump a large fraction of groundstate CH_3F molecules into the excited state (18). Reactive scattering studies using the laser-excited hexapole-focused reactants are described below.

Following the hexapole field, an adiabatic transition is made from the sixpole field to parallel plate "guiding" fields along the beam direction, overlapping the hexapole and scattering center assembly. The guiding and orientational fields preserve the axis of quantization. Voltage differences from one field to the next must be small compared to the spacing between Stark levels to avoid "flips" in the selected quantum numbers (19). In the scattering center, the applied electric field direction **E** (defining M) is set along the relative velocity vector by tilting the electric fields appropriately. Simple bent-rod fields are shown in Figure 1, along with a velocity vector diagram for the $CH_3I + Rb$ reaction, in which the relative velocity $\mathbf{v}_r$ defines the proper **E** direction. In more recent oriented-molecule experiments (in which product chemiluminescence is detected), a more open structure, a so-called "harp" field, is used (2).

CHARACTERIZATION OF ORIENTATION

Theory

Molecules of a given average orientation $\langle \cos\theta \rangle$, governed by the hexapole voltage V_0, Eq. 3, pass into the guiding field region. The average orientation, however, is only the first moment of the full orientational probability distribution function (pdf), $P_{JKM}(\cos\theta)$, which can be calculated from properties of rotational state wavefunctions (20, 21a). The low $|LKM\rangle$ states behave nonclassically: They precess about a large average angle and,

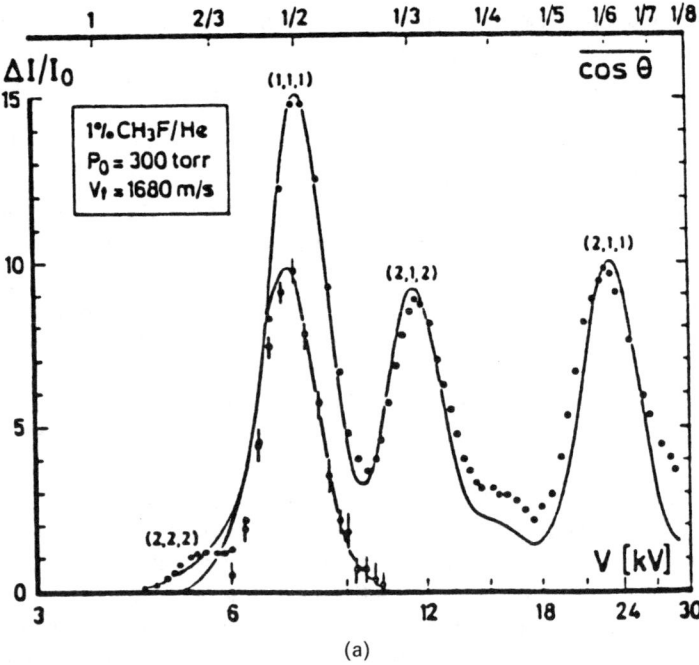

(a)

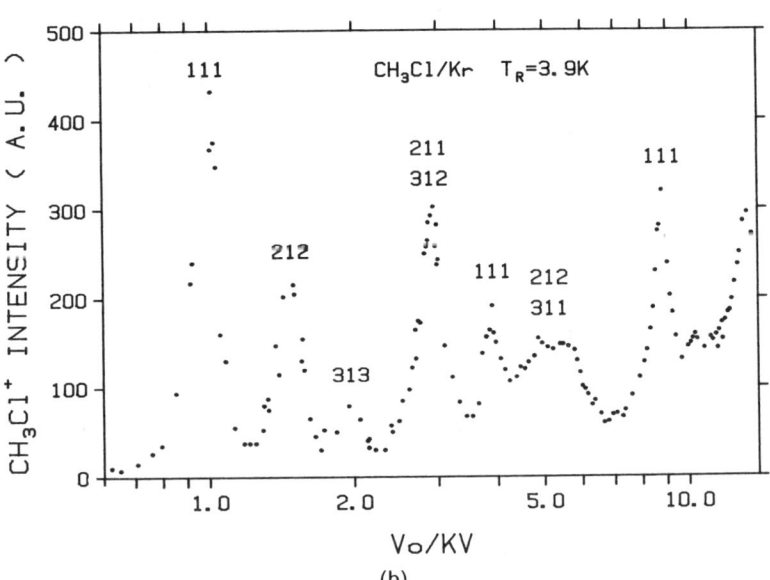

(b)

due to quantum effects, about a large range of angles. Choi & Bernstein (21a) have described in detail the classical and quantal characteristics of state-selected symmetric-top molecules. For the rotational states of CH_3Cl identified in the focusing curve of Figure 3(b), the corresponding $P(\cos\theta)$ distributions are plotted in Figure 4. Thus, for these low $|JKM\rangle$ states, the pre-collision orientation is already spread over a wide range of angles, and is likely to be further smeared by other effects (to be discussed below) during the approach trajectory preceding reaction.

The molecular axis orientational distribution function for a selected $|JKM\rangle$ state is usually represented by a Legendre expansion (4):

$$P_{JKM}(\rho) = \frac{2J+1}{2} \sum_{n=0}^{2J} C_n(JKM) P_n(\rho), \qquad 4.$$

where P_{JKM} is the quantum probability distribution function, P_n are the Legendre polynomials, and $\rho \equiv \cos\theta$. Coefficients C_n have been tabulated (21) for $|JKM\rangle$ values up to $J = 4$. Equation 4 can also be expressed (22) in terms of the Legendre moments of the pdf, i.e. $\bar{P}_n \equiv \langle P_n(\rho)\rangle$, as follows:

$$P_{JKM}(\rho) = \sum_{n=0}^{2J} \left(\frac{2n+1}{2} \bar{P}_n\right) P_n(\rho). \qquad 5.$$

Values of the Legendre moments $\bar{P}_n$ for symmetric tops with $J \leq 3$ have been tabulated (22). The J value determines the number of moments present in the summation (namely, $2J$). Thus, the $|111\rangle$ state can be defined by only a P_1 and a P_2 moment, but many moments are required to represent the ensemble of states in a thermal beam source.

Experiment

Until recently, it has not been possible to demonstrate experimentally the degree (or even the direction) of orientation produced by hexapole focusing. (Indirect evidence has come from computer simulation of the experimental focusing curves, and, of course, from the direction and magnitude

---

Figure 3 (a) Focusing curve for a specified He-seeded beam of CH_3F. Plotted is the incremental beam intensity ΔI relative to the nonfocused beam, I_0, as a function of the hexapole rod voltage, $V = 2V_0$. The *upper horizontal axis* indicates the corresponding average orientation $\langle\cos\theta\rangle$, where θ is the angle between the molecular axis **r** and **E** within the hexapole. A computer-simulated curve is plotted along with the experimental points (*dots*). The lower data set is the beam intensity after populating the $v_3 = 1$, $|JKM\rangle = |111\rangle$ state by pumping the $v_3 = 0$ $|110\rangle$ state (in Stark resonance) with a CO_2 laser. [From Ref. (16).] (b) Focusing curve for a pulsed Kr-seeded beam of CH_3Cl (velocity 0.34 km s^{-1}, with a rotational temperature of 3.9K). The JKM state assignments are given. "Overtone peaks," multiple loops for the $|111\rangle$ and $|212\rangle$ states, are also shown. [From Ref. (17a).]

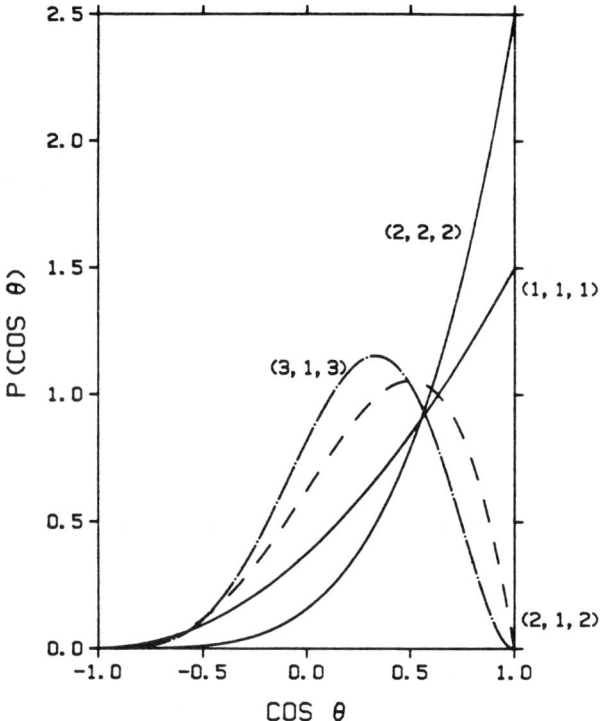

Figure 4 Calculated orientational probability distribution functions for the specified *JKM* states of a symmetric-top molecule. [From Ref. (1), based on Ref. (21a).] Note that the same distributions are obtained for *J–K–M* states (since only the *KM* product occurs in Eq. 1). The average orientations, i.e. $\langle \cos \theta \rangle$ for the states shown are $\frac{2}{3}$, $\frac{1}{2}$, $\frac{1}{3}$, and $\frac{1}{4}$ for the states $|222\rangle$, $|111\rangle$, $|212\rangle$, and $|313\rangle$, respectively.

of the chemical reactive asymmetry effect itself.) This situation has now been remedied by polarized laser photoionization techniques, first (qualitatively) by Heinzmann and co-workers (23), who used VUV direct ionization and, more recently (quantitatively), by Gandhi et al (24a–d) who used laser ionization time-of-flight mass spectrometry of fragments from polarized laser photodissociation of the oriented molecules. Confirmation of orientation is important because (*a*) the sign itself of the dipole moment of some reagent molecules (such as CF_3I and CF_3Br) was uncertain; and (*b*) the $|JKM\rangle$ representation for the symmetric-top states is appropriate only in the strong-field limit (in which the nuclear spins are decoupled from the molecular rotational angular momentum), and is not applicable in regions of smaller field strength, such as in the guiding or orientation fields. (This is discussed below.)

Prompt photodissociation of an oriented molecule ejects fragments preferentially with respect to the initial molecular orientation direction. For example, the probability density function for the iodine atom fragment of oriented CH_3I or t-butyl iodide excited via a parallel transition should be that of the parent molecule, modulated by a $\cos^2\theta$ factor from the absorption of the laser radiation, polarized parallel to the E field (21). Since the I atom of the oriented t-butyl iodide molecule points toward the negative plate of the orientation field, the photodissociation ejects the I in this direction and the t-butyl radical to the positive plate. The neutral fragments are then photoionized (by subsequent photons within the same laser pulse) and the ions detected by time-of-flight.

Figure 5 shows the time-of-flight (TOF) arrival patterns (24b) for the

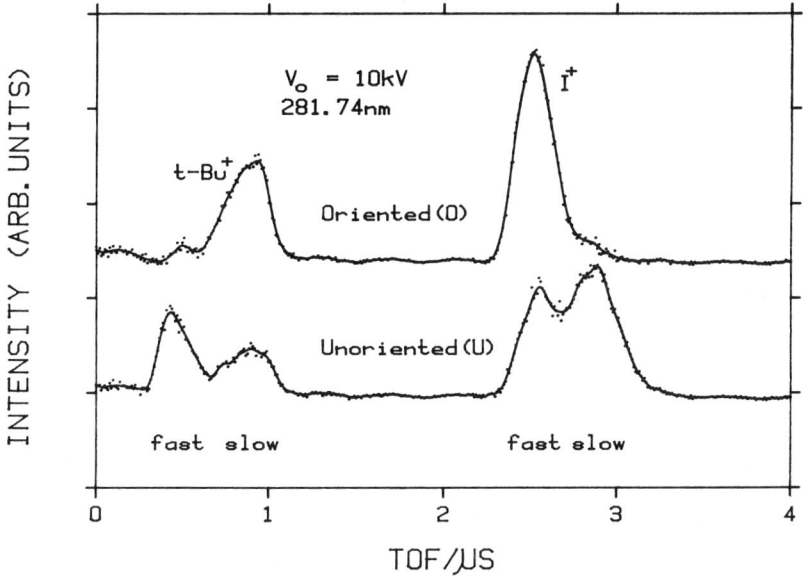

Figure 5 Time-of-flight mass spectra of pulsed beams of oriented (O) and unoriented (U) t-butyl iodide obtained with polarized laser photofragmentation and multiphoton ionization at 281.74 nm [Ref. (24b).] This wavelength ionizes both the alkyl and iodine fragments. The first doublet is for $m/z = 57$ ($C_4H_9^+$), the second for $m/z = 127$ (I^+). Within each doublet, the early component comes from upward-ejected fragments, the later from downward. The orientation field **E** points upward, in the direction of the TOF ion detector. When the orientation field is applied, most of the molecules point with I-end up and t-Bu down, so they photofragment such that the principal I^+ peak is "early," whereas that of the t-Bu$^+$ is "late." [For details, see Ref. (24b); also Ref. (90) on t-butyl bromide.]

t-butyl and I fragments from oriented t-butyl iodide photodissociated by laser radiation, with wavelength chosen to ionize the I atom by a two-photon, resonant, overall three-photon ionization (25); i.e. a (2+1) resonance-enhanced multiphoton ionization (REMPI) process. Dramatic differences are seen in the TOF spectra from the oriented and unoriented molecule beams. Details of the extraction of the degree of orientation from the doublets in the TOF patterns have been published (24a–d). Analysis of Figure 5 indicates that hyperfine recoupling has slightly lowered the average orientation from its expected strong-field value.

For the favorable case of prompt dissociation of the target molecules, the polarized laser-induced detection method is ideal for analysis of orientation. The hexapole orientation technique also prepares molecules in a unique condition for subsequent studies of photodissociation. Future applications using angle- and velocity-distribution-sensitive REMPI methods such as ion-imaging (26) are expected.

INFLUENCE OF ELECTRIC FIELDS UPON ORIENTATION

Moderate electric fields ($\gtrsim 200$ V/cm) suffice to decouple the nuclear hyperfine effect for most symmetric-top molecules except for iodides. Decoupling of hyperfine and rotational or vibrational interactions in other types of molecules with a linear Stark effect often requires much higher field strengths. The existence of a focusing curve *per se* is a good indication that μ_{eff} is essentially constant within the strong field of the hexapoles ($E \gtrsim 10$ kV/cm), cf. Eq. 2. Lower ($\lesssim 2$ kV/cm) field strengths are, however, desirable in the guiding field and orientation plates regions to avoid discharges and to allow rapid switching of field direction. Thus, the possibility arises that μ_{eff}, and thus the average orientation $\langle \rho \rangle$, depends on E. Field strength effects are a concern in ensuring that the optimal orientation (i.e. that of Eq. 1) is attained. However, for some molecules such as NO, the influence of field strength is sufficiently great (and theoretically well known) that varying **E** offers an elegant means of controlling NO orientation and thus the collision geometry.

Theory

Three types of angular momentum (nuclear spin, electronic, and vibrational) can couple with rotation and reduce the average orientation. Although the theory of such coupling is well known (27), detailed molecular spectroscopic information is required in order to predict the E-dependence. Hexapoles focus molecules populated in the upper levels of the high-field-limit Stark spectrum, i.e. those with the largest positive Stark effect. On leaving the strong field of the hexapole, the populations may relax down and cross over into other M states. The time-scale and extent of these level crossings is often difficult to predict. Zare and co-workers (28) have treated the dynamics of nuclear spin coupling effects on diatomic molecules (e.g. HF) aligned via polarized laser excitation. Their analysis is relevant to the present case of the hexapole field-to-guiding-field transition.

A strong electric field decouples the low-field mixed l doublets in N_2O, producing a pure $l = \pm 1$ state. Since the rotational properties of N_2O are known from molecular beam electric resonance studies (29), it has been possible to predict (30) the E-field dependence of $\langle \cos \gamma_0 \rangle$. Similar calculations were reported by Jones & Brooks (31) in treating symmetry splitting of $\pm K$ levels in slightly asymmetric tops.

Nuclear hyperfine effects are dominant in NO. Since the $^2\Pi_{3/2}$ electronic state is hexapole-focused (32), the selected strong-field states $\pm \Omega = J = \pm M = \frac{3}{2}$, $I = 1$, $M_I = -1$, 0 and 1, having the strongest Stark effect ($\langle \cos \theta \rangle = 0.6$), yield, at zero field, all M_F components of the upper λ-doubled level of a single F state: $|\Omega| = J = \frac{3}{2}$, $I = 1$, $F = \frac{5}{2}$ ($M_F = -\frac{5}{2}$, $-\frac{3}{2}, \ldots, \frac{5}{2}$) if the transfer occurs adiabatically. A simple transformation from the high-field (uncoupled) limit to zero field is possible, allowing a prediction of the admixture of M states ($\frac{3}{2}, \frac{1}{2}, -\frac{1}{2}, -\frac{3}{2}$) as a function of E. By measuring the reactivity at several different orientation field voltages, the contribution of each M state can be deduced.

Hyperfine effects can have a deleterious effect on the attainable orientation of symmetric tops in weak fields when the nuclear coupling is strong (as for Br and I compounds). The focused molecules contain an atom with nuclear spin, I, which couples with J to give a total angular momentum $F = J+I, J+I-1, \ldots, |J-I|$. Then M becomes M_F, the projection of angular momentum on a chosen direction, so a given $|JKM\rangle$ parent state becomes an ensemble of states $|JKIFM_F|\rangle$, whose average orientation is reduced. Choi (33) has recently carried out calculations of this disorientation effect for the methyl halides.

Experiment

Jalink et al (30) have measured the E-dependence of the disorientation due to recoupling of l-doubling in the $|jlm\rangle = |111\rangle$ and $|212\rangle$ states of N_2O, using data from the reactive scattering of N_2O with Ba, and found good agreement between experiment and theory. Van den Ende & Stolte (32) have measured the reactivity of oriented NO plus O_3 as a function of the applied electric field strength. (However, confirmation of the variation of the NO M-state admixture with E was not possible.) As discussed below, results of an independent study of this reaction (34) were found to be consistent with the results of the oriented NO experiments.

Gandhi et al (24d, 35) have used the polarized photofragmentation TOF technique to measure directly the effect of the applied electric field upon the degree of orientation of state-selected CH_3I.[1]

Earlier studies of oriented CH_3I reactions (8) employed field strengths less than 200 V/cm, too low to decouple J and I fully. The observed steric effects (discussed below) may be substantially stronger under fully decoupled conditions.

There has also been a concern that the anisotropic long-range forces and fields induced by the collision partner may "reorient" the initially oriented molecule on the collisional time scale (of a few ps). This issue is discussed below.

APPLICATIONS OF THE HEXAPOLE TECHNIQUE

Hexapoles have now been used for reagent preparation and product analysis in reactive, nonreactive, and surface-scattering studies. Orientation fields can direct **E** parallel or antiparallel with **v** (for heads vs. tails), or perpendicular to **v** (for broadside collisions). Scattered intensities for each setting are compared to yield information on the steric dependence. $P(\cos\theta)$ now becomes $P(\cos\gamma_0)$, where γ_0 is the initial (pre-collision) angle of attack on the molecular axis by the collision partner. Reorientation and other processes may change γ_0 to γ, the angle of attack at "impact." The fact that strong steric effects are observed in most of the systems studied suggests that reorientation effects are secondary. Reorientation and alignment models are discussed below.

Beam Focusing and State Selection

Hexapole-focusing curves have yielded information on dipole moments (58b, 91, 92), intramolecular orientational dynamics in processes such as inversion doubling in NH_3 (36–38), and the "degradation" of rotational quantum numbers in large molecules at high temperature (39). In addition, rotational relaxation pathways in supersonic expansions can be deduced from the J, K populations of state-resolved focusing curves (16, 17) (e.g. Figure 3), e.g. different J and K manifold temperatures (40) have been deduced from focusing curves for symmetric-top molecules in jet-expanded beams (93). Hexapole focusing has been used by Kuwata and co-workers

---

[1] Another result of Ref. (24c) is that the extremely high field strength ($\gtrsim 10^6$ V/cm) of the focused, pulsed (5 ns) photofragmentation-detection laser at optical frequencies did not cause detectable sample alignment.

(41) to determine the rotational temperature of chemically produced SH radicals.

Oriented Molecule Beam–Surface Scattering

Steric effects on gas-surface scattering have recently been reported. Novakoski & McClelland (42) used a hexapole field to detect orientation in CF_3H molecules desorbed from Ag. Specular scattering of CF_3H by Ag yielded no measurable polarization. Measurements at off-specular angles (trapping-desorption) showed that as the surface temperature is increased a growing fraction of the trapped molecules leave the surface with the orientation $F_3CH \cdots Ag$.

Specular scattering can also have a strong orientation dependence, as shown by Kuipers et al (43a) in a study of the scattering of oriented NO by a Ag(111) surface. When the O-end of NO attacks first, more NO rotational excitation is expected to occur (resulting in scattering into smaller angles) compared to the N-end approach. Figure 6 (43b) shows the relative difference in O-end and N-end attack for trapping-desorption. As the collision energy increases, trapping becomes increasingly favored for O-end approach. This is opposite to the known preference for N-end trapped equilibrium geometry, which was observed in a recent study of the scattering of oriented NO by a Pt(111) surface (43c).

Heinzmann and co-workers (44) have measured the sticking probability of hexapole-oriented NO on Ni(100): a higher sticking probability was observed for NO approaching with the N-end toward the surface. Curtiss & Bernstein (94) measured the steric effect in the scattering of oriented CH_3F by graphite. The specularly scattered intensity is greater for incident molecules oriented with CH_3 toward the surface. The effect increases with the degree of orientation.

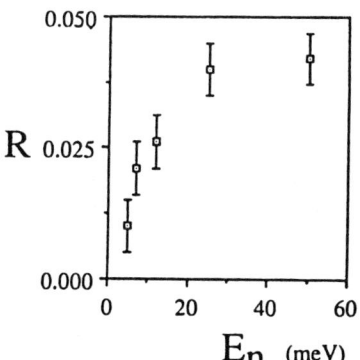

Figure 6 Relative (fractional) difference R in intensity of NO molecules scattered off a Ag(111) surface, for O-end-first orientation versus N-end-first, plotted as a function of the component of translational energy of the NO normal to the surface, E_n. [From Ref. (43b).]

Crossed-Beam Scattering with Oriented Molecules

The first "raw data" from the crossed-beam reaction of Rb with oriented methyl iodide (14) is shown in Figure 7, and clearly demonstrates that collisions of Rb with the I-end of CH_3I are more reactive than collisions with the methyl end. Selective surface ionization was used to detect reactively back-scattered RbI. (The steric effect for nonreactive scattering is in the opposite direction, because of competition between reactive and nonreactive contributions to the total back-scattering cross section.)

Surface ionization is unique in its high detection sensitivity (for alkali and alkali halide systems) but it offers little information on product state distributions. Chemiluminescence detection has replaced surface ionization in recent oriented molecule crossed-beam scattering studies. It has

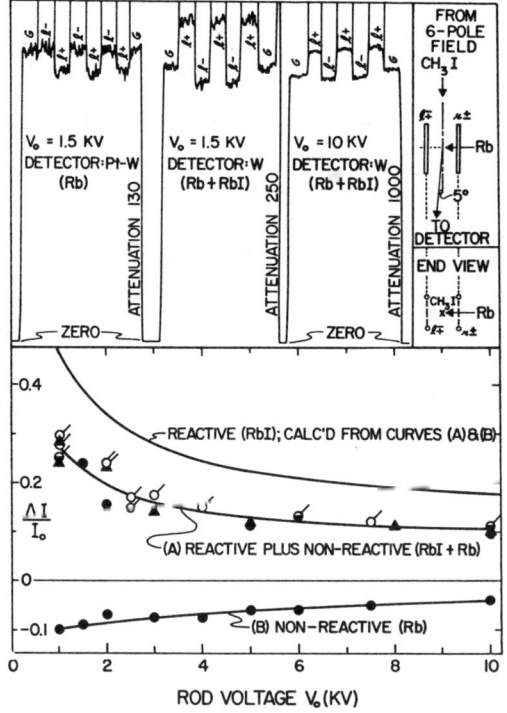

Figure 7 First reactive asymmetry data for the reaction of oriented CH_3I molecules with Rb, showing "raw" chart recording, a sketch of the orientation field configuration (*upper*) and a plot of the fractional asymmetry, $\Delta I/I_0$, vs. hexapole rod voltage (*lower*). [From Ref. (14).]

a high sensitivity and can provide polarization information (and some internal state resolution). Laser-induced fluorescence or ionization techniques offer ideal product-state resolution but have not yet been employed in detecting products of oriented reactants.

Representation of Orientation-Dependent Reactivity

Orientation-dependent reaction cross sections, $\sigma(\cos \gamma_0)$, can be represented as a Legendre moment expansion in $\cos \gamma_0$ (45),

$$\sigma(\cos \gamma_0) = \sum_{n=0}^{\infty} \sigma_n P_n(\cos \gamma_0) \qquad 6.$$

where γ_0 is the initial angle of attack and P_n is the nth Legendre polynomial. Steric opacity functions, defined by Stolte et al (20) as the dimensionless ratio $\sigma(\cos \gamma_0)/\sigma_0$ [analogous to impact parameter opacity functions, $P(b)$], are more commonly measured and are obtained directly from the σ_n moments of Eq. 6.

Usually only σ_1/σ_0 and σ_2/σ_0, the dependence on reactant orientation and alignment, can be extracted from the data. For the $|JKM\rangle = |111\rangle$ state, it is possible to determine both σ_1/σ_0 and σ_2/σ_0 from two scattered product intensity ratios, I_F/I_0 and I_U/I_0 (where F ≡ favorable orientation, i.e. "heads," and U ≡ unfavorable, "tails"; I_0 is the reference; i.e. unoriented reagent beam):

$$\sigma_1/\sigma_0 = (I_F - I_U)/I_0 \qquad 7a.$$

and

$$\sigma_2/\sigma_0 = 5(I_F + I_U - 2I_0)/I_0. \qquad 7b.$$

Related expressions for chemiluminescence polarization measurements are given in Ref. (45).

There is a problem with truncating $\sigma(\cos \gamma_0)/\sigma_0$ after the first two moments, however, especially if there are substantial "cones of nonreactivity"; the procedure can yield an artifact, "negative reactivity," in these zones.

Using surprisal analysis, Engel & Levine (46) suggested an unbiased model $\sigma(\cos \gamma_0)$ function (an expansion of exponentiated Legendre polynomials) that avoids the negative reactivity artifact. The results reproduce the essential features of other model functions but are incapable of switching off reactivity completely. Simple step, linear and trapezoidal model functions (2, 20, 45, 66) with cones of nonreaction have also been employed successfully.

Vector Properties

Orientation is a vector property of reaction. In a far-sighted series of papers, Herschbach and co-workers (47–50) showed how vector property measurements can yield insight into the details of dynamical stereochemistry. In modern orientation experiments, the initial and final relative velocity vectors **v** and **v'** (or wave vectors **k** and **k'**), one reactant's molecular axis direction **r** and one product's rotational angular momentum vector **j'**, can be specified. Such measurements can regain information on the azimuthal asymmetry of the reaction that is lost by averaging over impact parameters (48). For example, Herschbach and co-workers (51a,b) have measured the three-vector correlation (**k**, **k'**, **j'**) in the reaction of Cs with CH_3I and found a strong correlation of the **k'** and **j'** azimuthal orientations (of the CsI product) about **k** (**k'** and **j'** were determined simultaneously): **j'** was found to be preferentially oriented perpendicular to the **k**, **k'** plane.

The three-vector correlation (**r**, **k**, **k'**) has been measured for several reactions using hexapole-oriented reagents. The reactant orientation specifies **r** with respect to **k**, while product angular-velocity distributions (**k**, **k'**) are measured, for the reactions of alkyl halides with atoms, as discussed below. The three-vector correlation (**r**, **k**, **j'**) has been measured by using hexapole orientation for **r** with respect to **k**, and product rotational angular momentum alignment of **j'** with respect to **k** was measured via the polarization of the product chemiluminescence (30, 45), discussed below.

Along with vector correlations, orientation-dependent excitation functions (differential or integral reaction cross section versus collision energy) and partial cross sections in recoil, angular momentum polarization, and internal energy have also been measured, and are discussed below.

REVIEW OF EXPERIMENTAL SCATTERING STUDIES

"First-Generation Experiments"

REACTIONS OF ORIENTED ALKYL HALIDE+ALKALI ATOMS The reactions of alkali metals with oriented methyl iodide were reported 23 years ago by Brooks & Jones (13) and by Beuhler et al (14). More refined measurements of the reaction of oriented CH_3I with Rb (using a characterized mixture of reactant states) were later carried out by Parker et al (8a,b) and have been amply reviewed (5). Classical trajectory studies sensitive to the methyl iodide orientation have also been reported (52), along with models of angle-dependent barriers to reaction (53) and calculations of the effects of anisotropic forces on the "at-barrier" reagent orientation (54–56).

Coarse product angular distributions (8b) for the oriented methyl iodide + Rb reaction are shown in Figure 8. A strong (three-vector) correlation of the RbI angular distribution with the initial orientation is apparent: Direct backscattering (small impact parameters) shows a pronounced steric effect, whereas side- and forward-scattering (large impact parameters) are essentially independent of CH_3I orientation. Perhaps the most significant finding is the existence of a substantial steric "cone of non-reaction" (8a, 20, 57). A semiquantitative comparison of the steric effects in the reactions of K atoms with CH_3I vs. tert-butyl iodide has been reported by Marcelin & Brooks (58a).

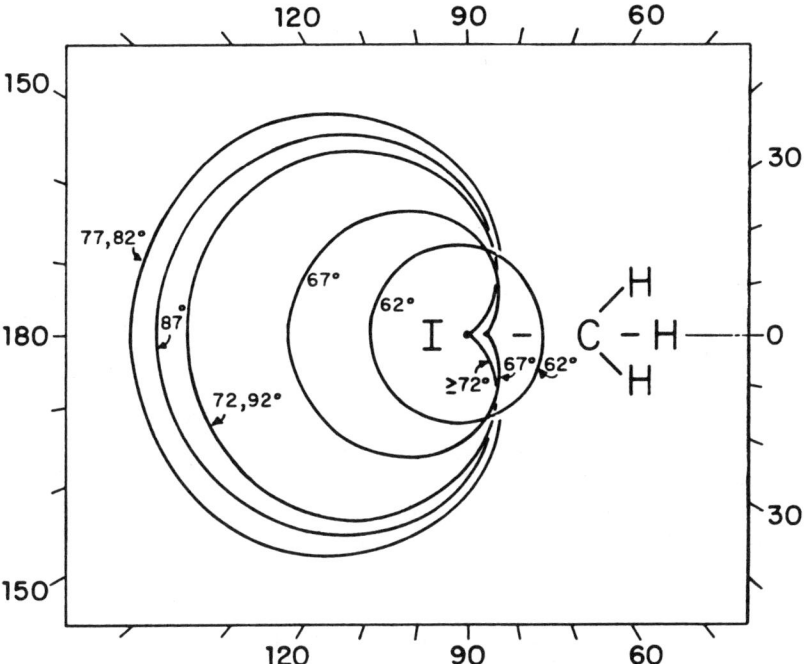

Figure 8 Polar plots of the orientation dependence of the intensity of reactively scattered RbI at different (indicated) laboratory scattering angles. More forward scattering (e.g. at a lab angle of 62°) comes from collisions with large impact parameters, which lead to much less reactive asymmetry. [From Ref. (8b).]

REACTIONS OF ORIENTED TRIFLUOROMETHYL HALIDE+ALKALI ATOMS Angular distributions of products from oriented trifluoromethyl halide reactions with K have been reported by Brooks and co-workers (3, 58b–d). Brooks et al have also investigated the azimuthal steric dependence (4) of the reactions of $K + CF_3I$ (58d) and CF_3Br (58e) by orienting the molecule along the direction of the detector instead of along **k**. Analysis of these data is less straightforward than for the methyl halide reactions because of the possibility of a KF product (not distinguishable by surface ionization from KBr or KI). [From separate experiments, it is believed that KF is *not* formed (58f,g).] There was also some question, now resolved (24c), about the sign of the dipole moment of the reagent and thus of the direction of the pre-collision orientation.

The product recoil distributions are shown in Figure 9. The *upper panel* plots the KBr product recoil distributions for K reacting with heads- and tails-oriented CF_3Br; the *lower panel* is similar for the KI from oriented CF_3I.

An overall null steric factor was found for the $K + CF_3I$ reaction, yet there is a gross difference in the product angular distribution between heads and tails orientations. (Heads yields backscattered KI, whereas tails results in side- and forward-scattered product.) This propensity is different for the CF_3Br reaction; here, the tails orientation yields much less product, so the overall steric effect is larger.

When a "harpoon" mechanism (59) involving "migration" (3) is invoked, this behavior is explained (58a–g) by a long-range (orientation-independent) electron jump to CF_3I. Following this, the CF_3I^- instantly dissociates and K^+I^- is ejected along the initial CF_3I axis direction. An electron jump to CF_3Br is expected at shorter distances, for which the orientation with the Br pointing toward the incoming K atom is assumed to be more favorable for reaction. The diatomic product angular distributions and total reactivity are consistent with this harpoon mechanism.

Further experiments by Harland et al (60a) demonstrated the effect of molecular orientation on electron transfer. They have used a charge-exchange source of fast K atoms (ca. 5–15 eV) and detected collisional ionization. For both CH_3I and CF_3I the ion yield is greatest when the K atoms are incident upon the I end of the molecule, despite the fact that in CH_3I the I end is negative and in CF_3I the I end is positive. The "harpooning" electron thus appears to jump to the chemically reactive end of the molecule, and not merely to the end that is positive (60).

REACTION OF ORIENTED $CF_3H + Ar^* \rightarrow CF_3^* + Ar + H$ Kuwata and co-workers (61) have reported observations of CF_3^* emission produced from the reaction

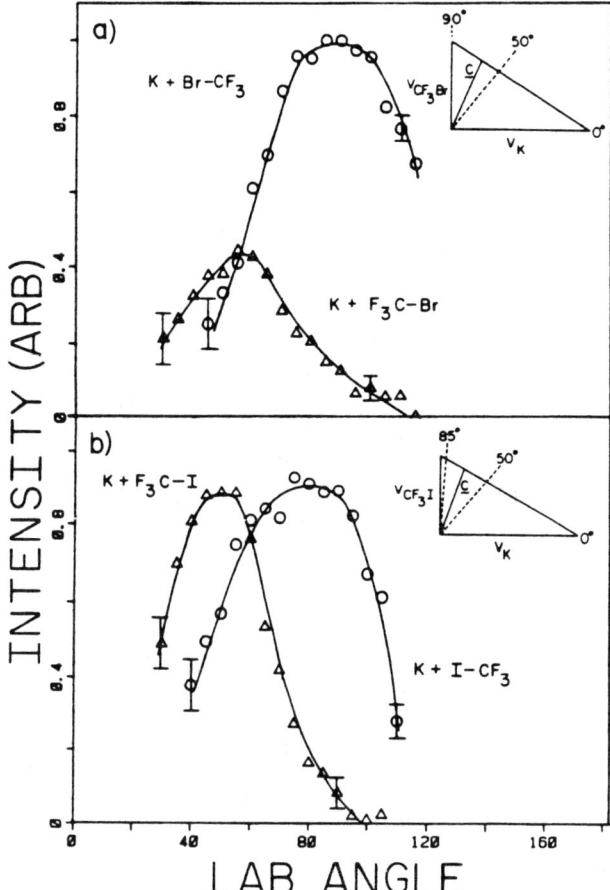

Figure 9 (a) Laboratory angular distributions of reactively scattered KBr from K+CF$_3$Br, oriented favorably (K-Br) vs. unfavorably (K-F$_3$), as indicated. (b) Laboratory angular distributions of reactively scattered KI from K+CF$_3$I, similarly. *Insets*: velocity vector diagrams showing the lab angles for which the KBr (and KI) product intensities maximize. [From Ref. (58e).]

of oriented CF$_3$H with metastable Ar($^3P_{2,0}$), produced by electron impact. A short (40 cm) hexapole field was used to focus a mixture of states from a pulsed supersonic expansion of neat CF$_3$H, yielding an unstructured focusing curve. A large steric effect was observed with the F-end (and sideways) orientations favored over reaction at the H-end of CF$_3$H. However, the steric effect increased with increasing hexapole rod voltage, V_0; this is unexpected because higher V_0 corresponds to focused molecules of lower orientation. Interpretation of the observed steric effects dependence awaits further experiments and analysis.

"Second-Generation" Experiments

REACTION OF ORIENTED $NO + O_3 \rightarrow NO_2^* + O_2$ Van den Ende & Stolte (32) have investigated the reaction of a hexapole-focused beam of NO molecules in the $\Omega = J = M = \frac{3}{2}$ state with O_3 gas at $T = 120K$ by detecting the visible NO_2^* chemiluminescence. The results, at a collision energy of 0.71 eV, show preferential reaction for two approach geometries: "heads," where the O-end of the NO molecule points toward the relative velocity vector; and "broadside-tail," with an average orientation $\langle \cos \theta \rangle = -0.3$.

Valentini and co-workers (34) measured the angle-velocity distribution of NO_2 from the reaction of NO with O_3 by using a crossed-beam machine with mass spectrometric detection. The results showed that the reaction is bimodal, with a narrow backward-scattered NO_2 peak and a broad sideways-scattered NO_2 lobe clearly separated.

This structure in the differential reactive cross section for the total NO_2 product from the unoriented reaction can be associated with the bimodal steric opacity function of the NO_2^* chemiluminiscence. The "heads" orientation is believed (34) to yield backwards scattering, and the broadside-tail orientation, sideways scattering. Further analysis and comparison of the two studies has been presented (2).

REACTION OF ORIENTED $N_2O + Ba \rightarrow BaO^* + N_2$ The most detailed study of the orientation dependence of reactivity has been that of Stolte and co-workers (2, 30, 45, 62–66) for the reaction of Ba with N_2O. Figure 10 is a schematic diagram of the apparatus. Chemiluminescence from the BaO^* product has been studied for two N_2O rotational levels ($J = 1, 2$) of the $v = 1$ state of the N_2O bending vibration, over a wide range of collision energies. The BaO chemiluminescence yield (62a,b, 63), polarization (30, 64), and internal energy (65) were measured for three approach geometries, favorable (Ba approaching the O-end of N_2O), unfavorable (opposite), and "broadside." The raw data have been deconvoluted (66) for the effects of velocity spread, lack of parallelism between **E** and **k**, lack of full l decoupling, plus polarization and wavelength bias in the detection system. Deconvolution for the wide range of precession angles (Figure 4) of the selected states $|jlm\rangle = |111\rangle$ and $|212\rangle$ was also carried out. Figures 11–14 display the key results for the reactions of oriented N_2O $|111\rangle$ molecules.

Figure 11 shows the dependence of the observed reactive asymmetry upon the strength of the orientation field. For N_2O, $E \gtrsim 1000$ V cm^{-1} was thus used for all steric experiments.

Excitation functions for several orientations (30) are shown in Figure

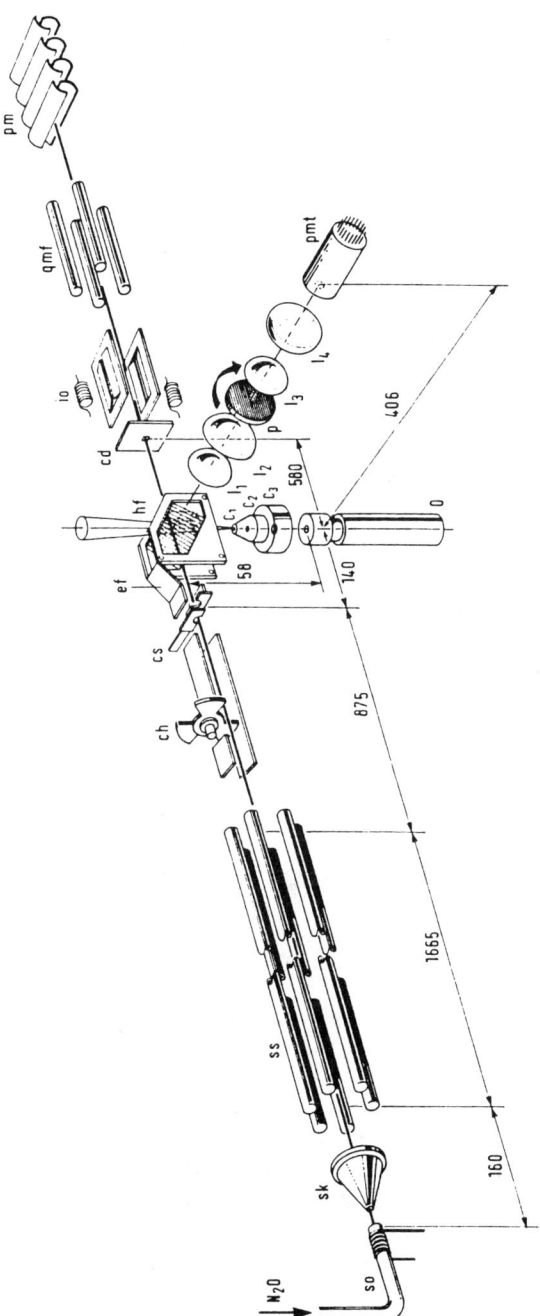

Figure 10 Schematic diagram of the oriented molecule beam machine of Stolte and co-workers (2, 16, 63) for the study of the orientation dependence of the reaction of N_2O with Ba. Dimensions are in mm. Key abbreviations are as follows: so = source, sk = skimmer, ss = hexapole, ch = chopper, cs, c_1, c_2, c_3, cd = collimators, hf = "harp field" for orientation, o = oven (Ba), p = polarizer, pmt = photomultiplier tube (to detect chemiluminescence), qmf = quadrupole mass filter. [See Refs. (2, 16, 63, 66) for details.]

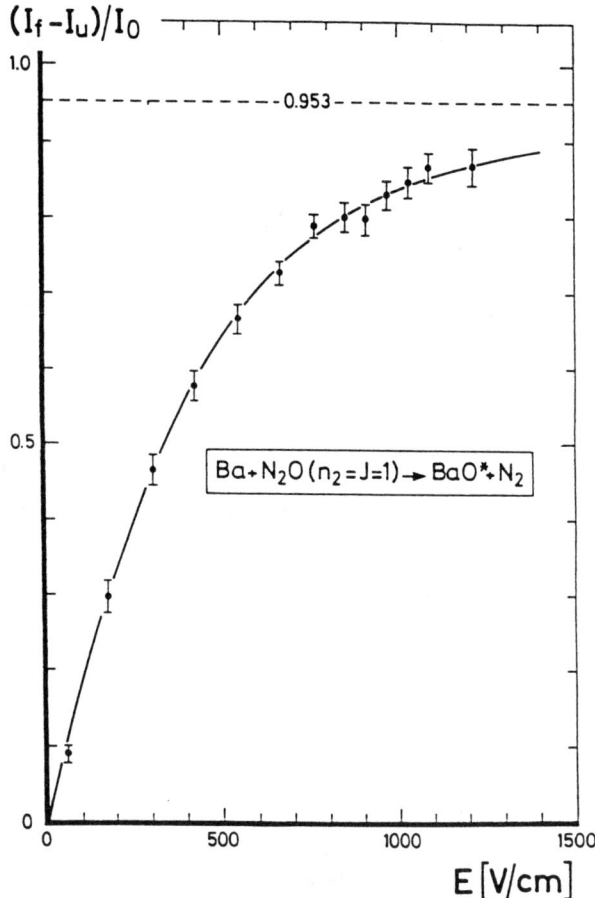

Figure 11 Dependence of the chemiluminescence reactive asymmetry (for the reaction of N_2O $|111\rangle$ with Ba at a collision energy of 0.1 eV) as a function of the orientation field strength E (the Stark energy, which decouples the l-doubled level, $|111\rangle$, of the N_2O). *Solid curve*: theoretical fit (with the asymptotic value shown). [From Ref. (2).]

12. Note that the reactivity reaches a maximum at ca. 0.09 eV and declines rapidly thereafter. Most of the decline is due to a reduction in the product intensity for the "favorable" orientation as the collision energy increases. This behavior has been attributed to "recrossing" [cf. trajectory simulations (67)], i.e. a return to the reactant channel for collisions with excess energy above the angle-dependent barrier to reaction. Analogous excitation functions for the $|212\rangle$ state show similar steric behavior but an overall 10–25% higher reactivity. Figure 13 shows the energy dependence

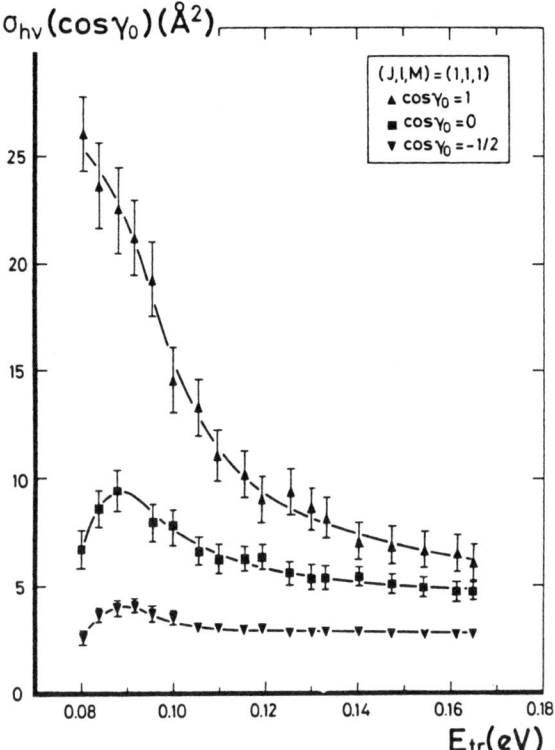

Figure 12 (a) Orientation-specified excitation functions for the reaction of oriented $N_2O + Ba \rightarrow BaO^* + N_2$. All data curves refer to N_2O in the $|111\rangle$ state but deconvoluted for spread in orientation. *Lower*, for the "unfavorable" direction; the *middle* for broadside; the *upper* represents extrapolated values for "perfect favorable" orientation; $\cos \gamma_0 = 1$. [Adapted from Ref. (30).]

of the first and second steric moments. The zeroth moment σ_0 differs significantly for the two reactant states; σ_1/σ_0 is large ($\cong 1.3$) and declines with increasing collision energy for both states. However, σ_2/σ_0 is small and less energy-dependent (for both states).

Figure 14 (65) shows how the dispersed chemiluminescence spectrum varies with orientation, at two collision energies. There is found to be a small steric effect for molecules of high internal energy (low kinetic energy), identified by a red shift, in the spectrum for favorable orientations. The explanation proposed (65) is as follows: Those collisions with favorable orientations experience a lower barrier to reaction than unfavorable orientations. Higher excess energy above the barrier (i.e. for the favorable

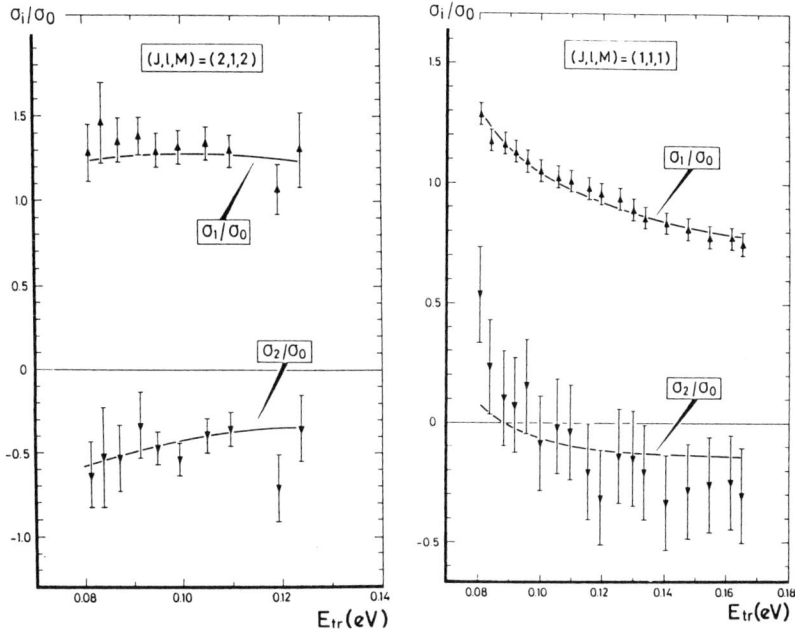

Figure 13 First and second moments of the steric opacity function for the reaction of Ba with N_2O in the $|212\rangle$ state *(left)* and $|111\rangle$ *(right)*, i.e. σ_1/σ_0 and σ_2/σ_0 vs. collision energy. The most definitive results are for the $|111\rangle$ reactant, and especially for the P_1-moment, i.e. the orientation effect σ_1/σ_0, which shows a significant magnitude, decreasing with increasing translational energy. [From Ref. (66).]

orientations) may translate to higher recoil energy, thus lower internal energy (red-shift). At higher collision energies, the recrossing effect (above) reduces the net reaction more for the higher recoil, lower internal energy products, thus leading to more pronounced red shifts.

A good example of vector correlations is the dependence of the product rotational angular momentum alignment $(\hat{\mathbf{k}} \cdot \hat{\mathbf{J}}')$ on reactant orientation. Figure 15 (45) shows the steric opacity function and the related product alignment vs. the (deconvoluted) orientation at several collision energies. Both the reactivity and the degree of product alignment drop rapidly as the collision energy is increased (for the unfavorable geometry). At the highest collision energy, the product alignment approaches the maximum allowed value ($A = -0.5$) for perfect "heads" orientation ($\cos \gamma_0 = 1$), while the alignment (as well as the reactivity) approaches zero for "tails" ($\cos \gamma_0 = -1$). An orientation-dependent modified DIPR-DIP model has been proposed (68) to account for these trends.

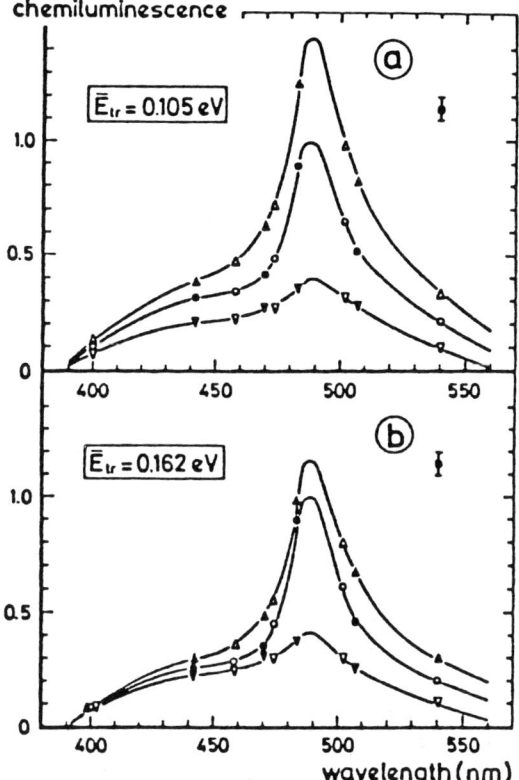

Figure 14 BaO* chemiluminescence spectra from the reaction of Ba with oriented N_2O $|111\rangle$ reactants at the two specified collision energies. The three curves in each graph (a and b) refer to "favorable" (*upper*), unoriented (*middle*), and "unfavorable" (*lower*) orientations. [From Ref. (65).]

REACTION OF ORIENTED $CH_3F + Ca^* \rightarrow CH_3 + CaF^*$ Experiments on the reaction of oriented methyl halides with metastable Ca atoms have been reported by Janssen et al (69). Chemiluminescence from CaF* $A(^2\Pi)$ was detected (16) from the crossed-beam reaction of CH_3F in the $|111\rangle$ state (cf. focusing curve in Figure 4) with a metastable Ca beam. Eliminating the strong Ca* lines with narrow bandpass filters isolated the weak CaF* chemiluminescence signals sufficiently to make measurement of the orientation effect possible. For the reaction producing CaF(A-X) emission, the steric effect increases with increasing collision energy. Figure 16 shows the orientation dependence and the overall excitation function. (As with the $Rb + CH_3I$

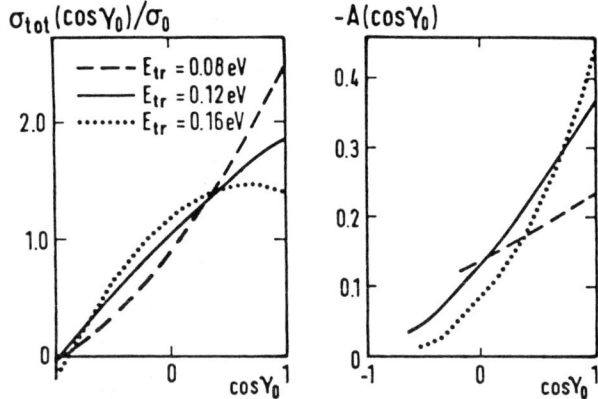

Figure 15 Left: Steric opacity function for the chemiluminescent reaction of oriented N_2O with Ba at the three specified collision energies. Right: The calculated product alignment A based on the polarization data, similar plot. Cos γ_0 is the deconvoluted orientation. The limit $A = -0.5$ corresponds to $\mathbf{J}' \perp \mathbf{k}'$ while $A = 0$ means random alignment. [From Ref. (45).]

and $Ba + N_2O$ reactions, the excitation function shows a maximum and then a rapid decline with increasing collision energy.) The orientation results suggest some "reorientation" at low collision energies (low steric effect, higher reaction cross section), and more direct collisions at higher energies (larger steric effect, lower cross section).

Experiments with CH_3F $|111\rangle$, but vibrationally excited in the C–F stretch ($v = 1$) show qualitatively the same reactivity as the ground-state reactant, although there is a small difference in the orientation dependence of the $v = 0$ and $v = 1$ excitation functions. Selective excitation of the bond that is to break during reaction appears to be ineffective in promoting this reaction. Analogous experiments with oriented CH_3Cl and CH_3Br have been carried out (70).

THEORETICAL ASPECTS: ORIENTATION DEPENDENCE OF CHEMICAL REACTIVITY

Theoretical efforts in this field have accompanied experimental work from the very beginning (71), ranging from one-dimensional models to full-scale classical trajectory simulations on accurate potential energy surfaces (52, 72). Recent papers have addressed several pressing experimental issues, such as the interpretation of experimental opacity functions (53), the "reorientation" question (54–56), and the "recrossing" phenomenon (67).

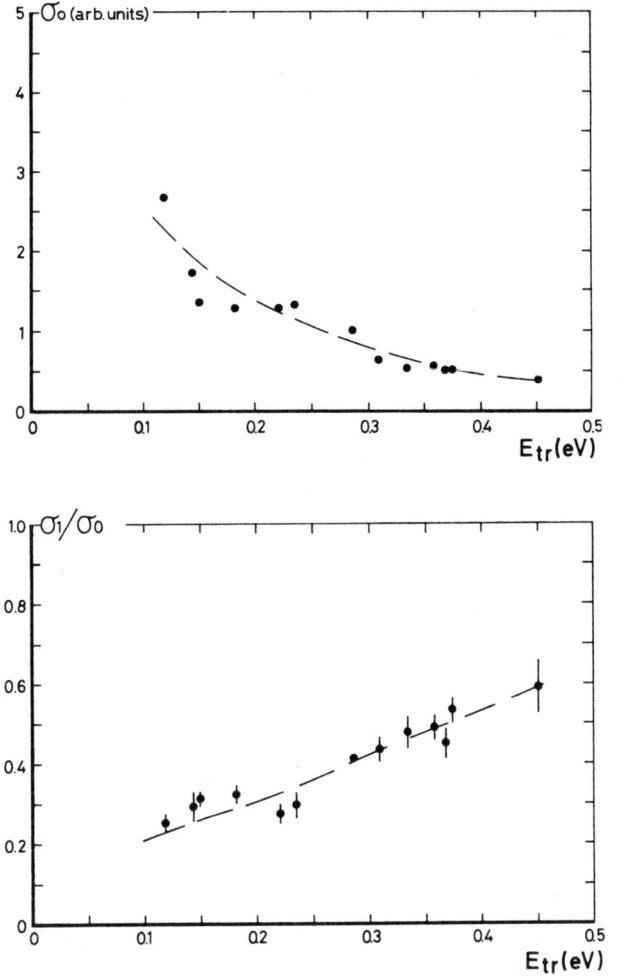

Figure 16 Experimental results for the reaction of metastable Ca* atoms with oriented CH_3F ($|111\rangle$). *Upper*: CaF* chemiluminescence cross section vs. collision energy (from ca. 0.1 to 0.5 eV). *Lower*: first moment (σ_1/σ_0) of the steric opacity function, plotted similarly. In contrast to the expected negative energy dependence of this orientation moment (cf. Figure 13) for this reaction, σ_1/σ_0 increases with collision energy. [From Ref. (69).]

Of course, steric concepts are built into almost all theoretical studies, but here we review only work directly related to dynamical stereochemistry.

From an experimental viewpoint, there are at least three levels at which to compare steric theory (or models) with observations (73). The crudest is the overall steric factor for reaction and how it agrees with expectation based on the bulkiness of blocking groups. More quantitative is the shape of the cross-section function just above the threshold (averaged over all orientations). Most refined are comparisons of predicted and measured steric effects, i.e. the orientational opacity function and its dependence upon collision energy.

Classical Rigid-Body Models

"Painted spheres" (74), the first steric models, have a single parameter (a solid angle of "acceptance"); when integrated over impact parameters, they can often recover a given experimental steric opacity function at one collision energy. The model serves to scale the reactive vs. nonreactive "areas" of the molecule, and provides an approximate relation between the scattering angle and the impact parameter (74, 75). A more realistic description comes from the angle-dependent line-of-centers (ADLOC) model for reactive scattering (76–78). Incorporated into the venerable line-of-centers model (59) is a barrier to reaction that depends on the angle of attack, γ, evaluated at the barrier, often taken to be the surface of a sphere or convex rigid body. Both impact parameter and collision energy determine the line-of-centers (LOC) energy; only collisions with LOC energy in excess of the barrier react. The relation between γ_0 and the at-barrier angle, γ, is b-dependent (even when intermolecular forces are neglected). At large impact parameters, the two angles differ substantially (53). These models predict a decline in the steric effect as the collision energy is raised. Several studies (73, 79) have dealt with the shape of the post-threshold orientation-averaged cross section function.

Evans et al (73) have incorporated orientation-dependent barriers and angular momentum considerations into the classical theory of convex rigid-body (i.e. ellipsoids) collisions. Simple analytical expressions resulted for fitting the steric opacity function by a "colliding-pair anisotropy" parameter (related to the reactive asymmetry). This approach also recovers the orientation-averaged excitation function behavior. However, it is not really a "predictive" theory, since the parameters are not known *ab initio*. Improvements have included allowing for nonrigidity and correcting for the effects of "loading" (shift from the geometrical center-of-mass during reaction). Predictions of product vibrational and rotational excitation based on a modified ADLOC model have been reported by Schechter et al (80). In this approach, the barrier height and shape are adjusted to recover the experimental steric opacity function at a given collisional energy.

An angle-dependent barrier has been deduced (66) for the $Ba + N_2O$ reaction. The barrier is lowest for head-on collision with the O-end of N_2O

and very large for the opposite orientation near the N-end. Excitation functions (Figure 12) for this reaction show a rapid dropoff in reactivity for collision energies above 0.09 eV, arising mainly from a lower "favorable" reactivity. By modifying the ADLOC model (30, 66) to include an empirical "recrossing" correction that attenuates reactivity for collisions with excess energy above the line-of-centers barrier, both the steric ratios and the excitation functions could be simulated.

Angle-Dependent Barrier Models: Evaluations by Classical Trajectory Calculations

Since *ab initio* potential energy surfaces (PES) are not available (so far) for the reactions studied experimentally, confirmation of these empirically derived angle-dependent potential barriers is not possible. A semi-empirical PES for the $Rb+CH_3I$ reaction shows an even more drastic angle-dependent barrier to reaction than that derived from experimental data via an ADLOC model (5).

Classical trajectory calculations for the $H+D_2 \rightarrow HD+D$ reaction (78) that use the *ab initio* potential surface, the so-called LSTH PES, *have* confirmed the post-threshold orientation-averaged excitation function predicted by the ADLOC model. Of course, there are no orientation experiments on such systems, involving homonuclear diatomics, but alignment experiments are conceivable. It is clear that, to the extent that quasiclassical trajectory simulations mimic the quantal (and experimental) reactive scattering attributes, they will be useful in explicating orientation dependence observations of the type discussed above.

Anisotropic Force Models, Alignment, and Reorientation

All the above steric models assume an angle-dependent barrier to reaction in the effective potential. The effects of long-range anisotropic intermolecular forces were not included. Yet they have been long recognized as the source of the elusive "reorientation" effect at low collision energies. Luo & Benson (54) and Levine & Bernstein (55, 56) have investigated this problem by evaluating the angle- and distance-dependence of dipole-induced dipole and dispersion forces. For typical atom-dipolar molecule systems, the anisotropic component of the potentials is usually $\lesssim 20\%$ of the spherical part. Since the angular term is proportional to $P_2(\cos\gamma)$, it can lead only to an alignment effect on reactivity. An orientational effect requires a $P_1(\cos\gamma)$ anisotropy, which can arise from a dipole-quadrupole term or from explicit inclusion of an odd-order repulsive interaction.

Even-order dipole forces can align reagents as the collision takes place, equivalent to "orbital-following" in excited atom reactions and "locking" in ion-molecule reactions. This can reduce the influence of pre-orientation of reagents, the so-called "reorientation effect," but under certain circumstances can *enhance* reactive asymmetry (56).

As noted above, for the Ca*+CH$_3$F reaction, the steric effect decreases with decreasing collision energy. This may be due to the strong angular terms in the long-range potential for this system. Further theoretical work is necessary.

Many other aspects of chemical reaction dynamics, e.g. post-threshold excitation functions (81) and the entire energy dependence, including the effects of rotational and vibrational energy (82), need to be better determined and understood before such a detailed attribute as the steric dependence can be explained. Oriented molecule experiments will stimulate continued efforts in all these directions.

RELATED FIELDS AND NEW DIRECTIONS

A reinvigoration of molecular reaction dynamics, and especially stereochemical dynamics, has occurred in the past few years, partly due to the introduction of new experimental techniques (and the refinement of old ones) (83, 84). Alternative approaches to studying the steric effect have been introduced with great success. Noteworthy is the technique for the study of reactions within van der Waals adducts (limited by precursor geometry) developed by the groups of Wittig (85) and Soep (86). A new approach, also taking advantage of constrained geometry, has recently been advanced by Polanyi (87) in studies of reactions of species oriented and trapped on a surface.

Alignment effects on reaction dynamics are also of great current interest (84, 88, 89). A significant alignment effect has been reported by Loesch and co-workers (88) for the reaction of laser-aligned excited HF with K.

Omission of the above elegant, and very relevant, steric studies is based only on space limitations (and, of course, the authors' enthusiasm for hexapole techniques!).

"Third-generation" oriented molecule reactions, using state-selected reagents and state-specific detection of products, are in the offing. These experiments will not only reveal the steric effect in great detail, but will come close to the "ultimate" experiment in which every reactant and product variable is specified. In the meantime, refinements and new applications of first- and second-generation experiments will continue. Orientation studies of gas-surface interactions have just begun, and the results suggest that the effort has considerable merit. Further photofragmentation studies of oriented molecules will provide key diagnostics (on degree of orientation and alignment) and important insights into photodissociation dynamics. Orientation dependence of chemical reactivity (a.k.a. the steric effect) is a subject that will probably occupy the attention of chemists for at least another few decades!

ACKNOWLEDGMENTS

The authors wish to thank Dr. S. Stolte for many significant discussions. D. H. P. gratefully acknowledges the NSF and the NATO Collaborative Research Grants Program for support. R. B. B. acknowledges research support by the NSF under Grant CHE86-15286.

Literature Cited

1. Bernstein, R. B., Herschbach, D. R., Levine, R. D. 1987. *J. Phys. Chem.* 91: 5365–77
2. Parker, D. H., Jalink, H., Stolte, S. 1987. *J. Phys. Chem.* 91: 5427–37
3. Brooks, P. R. 1976. *Science* 193: 11–16
4. Stolte, S. 1982. *Ber. Bunsenges. Phys. Chem.* 86: 413–21
5. Bernstein, R. B. 1987. In *Recent Advances in Molecular Reaction Dynamics*, ed. R. Vetter, J. Vigué, pp. 51–58. Paris: Editions du CNRS
6. Stolte, S. 1988. See Ref. 7, pp. 631–52
7. Scoles, G., ed. 1988. *Atomic and Molecular Beam Methods*, Vol. I. London/New York: Oxford Univ. Press
8a. Parker, D. H., Chakravorty, K. K., Bernstein, R. B. 1981. *J. Phys. Chem.* 85: 466–68
8b. Parker, D. H., Chakravorty, K. K., Bernstein, R. B. 1982. *Chem. Phys. Lett.* 86: 113–17
9. Bennewitz, H. G., Paul, W., Schlier, C. 1955. *Z. Phys.* 141: 6–15
10. Reuss, J. 1988. See Ref. 7, Chap. 11; 1975. *Adv. Chem. Phys.* 30: 389–415
11a. Bernstein, R. B. 1982. *Chemical Dynamics via Molecular Beam and Laser Techniques*, Sect. 3.7. London/New York: Oxford Univ. Press
11b. Thuis, H., Stolte, S., Reuss, J. 1979. *Chem. Phys.* 43: 351–64
12. Kramer, K. H., Bernstein, R. B. 1965. *J. Chem. Phys.* 42: 767–70
13. Brooks, P. R., Jones, E. M. 1966. *J. Chem. Phys.* 45: 3449–50
14. Beuhler, R. J., Bernstein, R. B., Kramer, K. H. 1966. *J. Am. Chem. Soc.* 88: 5331–31
15. Anderson, J. B., Andres, R. P., Fenn, J. B. 1966. *Adv. Chem. Phys.* 10: 275–317
16. Jalink, H., Janssen, M., Harren, F., Van Den Ende, D., Meiwes-Broer, K. H., Parker, D. H., Stolte, S. 1987. See Ref. 5, pp. 41–50
17a. Gandhi, S. R., Curtiss, T. J., Xu, Q.-X., Choi, S. E., Bernstein, R. B. 1986. *Chem. Phys. Lett.* 132: 6–10
17b. Gandhi, S. R., Xu, Q.-X., Curtiss, T. J., Bernstein, R. B. 1987. *J. Phys. Chem.* 91: 5437–41
18. Adam, A. G., Gough, T. E., Isenor, M. R., Scoles, G. 1985. *Phys. Rev. A* 32: 1451–57
19. Maltz, C., Weinstein, N. D., Herschbach, D. R. 1972. *Mol. Phys.* 24: 133–50
20. Stolte, S., Chakravorty, K. K., Bernstein, R. B., Parker, D. H. 1982. *Chem. Phys.* 71: 353–61
21a. Choi, S. E., Bernstein, R. B. 1986. *J. Chem. Phys.* 85: 150–61
21b. Zare, R. N. 1989. *Chem. Phys. Lett.* 156: 1–6
22. Bernstein, R. B., Choi, S. E., Stolte, S. 1989. *Trans. Faraday Symp.* 24: In press
23. Kaesdorf, S., Schönhense, G., Heinzmann, U. 1985. *Phys. Rev. Lett.* 54: 885–87
24a. Gandhi, S. R., Curtiss, T. J., Bernstein, R. B. 1987. *Phys. Rev. Lett.* 59: 2951–54
24b. Xu, Q.-X., Jung, K.-H., Bernstein, R. B. 1988. *J. Chem. Phys.* 89: 2099–2106
24c. Gandhi, S. R., Bernstein, R. B. 1988. *J. Chem. Phys.* 88: 1472–73; 1988. *Z. Phys. D* 10: 179–85
24d. Gandhi, S. R. 1988. PhD dissertation. Univ. Calif., Los Angeles
25. Parker, D. H. 1983. In *Ultrasensitive Laser Techniques*, ed. D. S. Kliger, pp. 233–309. New York: Academic
26. Thoman, J., Chandler, D. W., Janssen, M. H. M., Parker, D. H. 1988. *Laser Chem.* 9: 27–46
27. Townes, C. H., Schawlow, A. L. 1955.

Microwave Spectroscopy. New York: McGraw-Hill
28. Altkorn, R., Zare, R. N., Greene, C. H. 1985. *Mol. Phys.* 55: 1–9
29. Reinartz, J. M. L., Meerts, W. L., Dymanus, A. 1978. *Chem. Phys.* 31: 19–29
30. Jalink, H., Janssen, M. H., Geijsberts, M., Stolte, S., Parker, D. H., Wang, Z. W. 1988. In *Selectivity in Chemical Reactions*, ed. J. C. Whitehead, pp. 195–220. Dordrecht: Kluwer Acad.
31. Jones, E. M., Brooks, P. R. 1970. *J. Chem. Phys.* 53: 55–58
32. Van den Ende, D., Stolte, S. 1980. *Chem. Phys.* 45: 55–64; 1984. *Chem. Phys.* 89: 121–39; 1980. *Chem. Phys. Lett.* 76: 13–15
33. Choi, S. E. 1987. PhD dissertation. Univ. Calif., Los Angeles
34. Van den Ende, D., Stolte, S., Cross, J. B., Kwei, G. H., Valentini, J. J. 1982. *J. Chem. Phys.* 77: 2206–8
35. Gandhi, S. R., Bernstein, R. B. 1989. In preparation
36. Butkovskaya, N. J., Larichev, M. N., Leipunskii, I. O., Morozov, I. I., Tal'rose, V. L. 1976. *Chem. Phys.* 12: 267–71
37. Gandhi, S. R., Bernstein, R. B. 1987. *J. Chem. Phys.* 87: 6457–67
38. Ohashi, K., Kasai, T., Kuwata, K. 1988. *J. Phys. Chem.* 92: 5954–58
39. Farley, F. W., Novakoski, L. V., Dubey, M. K., Nathanson, G. M., McClelland, G. M. 1988. *J. Chem. Phys.* 88: 1460–61
40. Douketis, C., Gough, T. E., Scoles, G., Wang, H. 1984. *J. Phys. Chem.* 88: 4484–87
41. Kasai, T., Ohashi, K., Ohoyama, H., Kuwata, K. 1986. *Chem. Phys. Lett.* 127: 581–84
42. Novakoski, L. V., McClelland, G. M. 1987. *Phys. Rev. Lett.* 59: 1259–62
43a. Kuipers, E. W., Tenner, M. G., Kleyn, A. W., Stolte, S. 1988. *Nature* 334: 420–21
43b. Tenner, M. G., Kuipers, E. W., Kleyn, A. W., Stolte, S. 1988. *J. Chem. Phys.* 89: 6552–53
43c. Kuipers, E. W., Tenner, M. G., Kleyn, A. W., Stolte, S. 1989. *Phys. Rev. Lett.* 62: 2152–55
44. Fecher, G., Volkmer, M., Böwering, N., Pawlitzky, B., Heinzmann, U. 1989. *Trans. Faraday Symp.* 24: In press
45. Jalink, H., Nicolasen, G., Stolte, S., Parker, D. H. 1989. *Trans. Faraday Symp.* 24: In press
46. Engel, Y. M., Levine, R. D. 1984. *Chem. Phys.* 91: 167–71
47. Herschbach, D. R. 1962. *Disc. Faraday Soc.* 33: 281–83
48. Case, D. A., McClelland, G. M., Herschbach, D. R. 1978. *Mol. Phys.* 35: 541–73
49. McClelland, G. M., Herschbach, D. R. 1979. *J. Phys. Chem.* 83: 1445–54
50. Barnwell, J. D., Loeser, J. G., Herschbach, D. R. 1983. *J. Phys. Chem.* 87: 2781–86
51a. Hsu, D. S. Y., McClelland, G. M., Herschbach, D. R. 1974. *J. Chem. Phys.* 61: 4927–28
51b. Case, D. A., Herschbach, D. R. 1975. *Mol. Phys.* 30: 1537–64
52. Blais, N. C., Bernstein, R. B. 1986. *J. Chem. Phys.* 85: 7030–37
53. Janssen, M. H. M., Stolte, S. 1987. *J. Phys. Chem.* 91: 5480–86
54. Luo, Y., Benson, S. W. 1988. *J. Phys. Chem.* 92: 1107–10
55. Levine, R. D., Bernstein, R. B. 1988. *J. Phys. Chem.* 92: 6954–58
56. Bernstein, R. B., Levine, R. D. 1989. *J. Phys. Chem.* 93: 1687–88
57. Choi, S. E., Bernstein, R. B. 1985. *J. Chem. Phys.* 83: 4463–69
58a. Marcelin, G., Brooks, P. R. 1975. *J. Am. Chem. Soc.* 97: 1710–15
58b. Brooks, P. R., Jones, E. M., Smith, K. 1969. *J. Chem. Phys.* 51: 3073–81
58c. Brooks, P. R. 1973. *Faraday Discuss. Chem. Soc.* 55: 299–306
58d. Brooks, P. R., McKillop, J. S., Pippin, H. G. 1979. *Chem. Phys. Lett.* 66: 144–48
58e. Carman, H. S., Harland, P. W., Brooks, P. R. 1986. *J. Phys. Chem.* 90: 944–48
58f. Marcelin, G., Brooks, P. R. 1973. *J. Am. Chem. Soc.* 95: 7885–86
58g. Hardee, J., Brooks, P. R. 1977. *J. Phys. Chem.* 81: 1031–33
59. Levine, R. D., Bernstein, R. B. 1987. *Molecular Reaction Dynamics and Chemical Reactivity.* London/New York: Oxford Univ. Press
60a. Harland, P. W., Carman, H. S., Phillips, L. F., Brooks, P. R. 1989. *J. Chem. Phys.* 90: 5201–3
60b. Carman, H. S., Phillips, L. F., Brooks, P. R. 1988. See Ref. 30, p. 307
61. Ohoyama, H., Kasai, T., Ohashi, K., Kuwata, K. 1987. *Chem. Phys. Lett.* 136: 236–40
62a. Jalink, H., Parker, D. H., Meiwes-Broer, K. H., Stolte, S. 1986. *J. Phys. Chem.* 90: 552–54
62b. Parker, D. H., Jalink, H., Stolte, S. 1987. *Discuss. Faraday Soc.* 84: 184–88
63. Jalink, H., Parker, D. H., Stolte, S. 1986. *J. Mol. Spectrosc.* 121: 236–37
64. Jalink, H., Parker, D. H., Stolte, S. 1987. *J. Chem. Phys.* 85: 5372–73
65. Jalink, H., Stolte, S., Parker, D. H. 1987. *Chem. Phys. Lett.* 140: 215–20

66. Jalink, H., Nicolasen, G., Stolte, S., Parker, D. H. 1989. In preparation
67. Schechter, I., Levine, R. D. 1986. *Int. J. Chem. Kin.* 18: 1023–45
68. Janssen, M. H., Nicolasen, G., Stolte, S. 1989. Work in progress
69. Janssen, M. H., Parker, D. H., Stolte, S. 1989. *Trans. Faraday Symp.* 24: In press
70. Janssen, M. H., Parker, D. H., Stolte, S. 1989. Work in progress
71. Karplus, M., Godfrey, M. 1966. *J. Am. Chem. Soc.* 88: 5332–33
72. La Budde, R. A., Kuntz, P. J., Bernstein, R. B., Levine, R. D. 1973. *Chem. Phys. Lett.* 19: 7–10; 1973. *J. Chem. Phys.* 59: 6286–98
73. Evans, G. T., She, R. S. C., Bernstein, R. B. 1985. *J. Chem. Phys.* 82: 2258–66; She, R. S. C., Evans, G. T., Bernstein, R. B. 1986. *J. Chem. Phys.* 84: 2204–11; Evans, G. T. 1987. *J. Chem. Phys.* 86: 3852–58; 1987. *J. Chem. Phys.* 87: 3865–66; 1988. *J. Chem. Phys.* 88: 3401
74. Beuhler, R. J., Bernstein, R. B. 1968. *Chem. Phys. Lett.* 2: 166–69; 1969. *Chem. Phys. Lett.* 3: 118–18; 1969. *J. Chem. Phys.* 51: 5305–15
75. Bernstein, R. B. 1985. *J. Chem. Phys.* 82: 3656–58
76. Smith, I. W. M. 1982. *J. Chem. Ed.* 59: 9–14
77. Levine, R. D., Bernstein, R. B. 1984. *Chem. Phys. Lett.* 105: 467–71; 1986. *Chem. Phys. Lett.* 132: 11–15
78. Blais, N. C., Bernstein, R. B., Levine, R. D. 1985. *J. Phys. Chem.* 89: 10–13
79. Pollak, E., Wyatt, R. E. 1983. *J. Chem. Phys.* 78: 4464–76
80. Schechter, I., Prisant, M. G., Levine, R. D. 1987. *J. Phys. Chem.* 91: 5472–80
81. González-Ureña, A. 1987. *Adv. Chem. Phys.* 66: 213–335
82. Sathyamurthy, N. 1983. *Chem. Rev.* 83: 601–18
83. Bernstein, R. B., Zewail, A. H. 1988. *Chem. Engr. News* 66(45): 24–43
84. Zhang, R., Rakestraw, D., McKendrick, K. G., Zare, R. N. 1988. *J. Chem. Phys.* 89: 6283–94
85. Beulow, S., Radhakrishnan, G., Wittig, C. 1987. *J. Phys. Chem.* 91: 5409–12
86. Jouvet, C., Boivineau, M., Duval, M. C., Soep, P. B. 1987. *J. Phys. Chem.* 91: 5416–22
87. Polanyi, J. C., Williams, R. J. 1988. *J. Chem. Phys.* 88: 3363–71
88. Hoffmeister, M., Schleysing, R., Loesch, H. 1987. *J. Phys. Chem.* 91: 5441–45
89. Bernstein, R. B. 1988. In Ref. 30, pp. 1–21
90. Xu, Q.-X., Quesada, M. A., Jung, K.-H., Mackay, R. S., Bernstein, R. B. 1989. *J. Chem. Phys.* In press
91. Butkovskaya, N. J., Larichev, M. N., Leipunskii, I. O., Morozov, I. I., Tal'rose, V. L. 1979. *Chem. Phys. Lett.* 63: 375–77
92. Gandhi, S. R., Bernstein, R. B. 1988. *Chem. Phys. Lett.* 143: 332–36
93. Choi, S. E., Bernstein, R. B. 1989. *Israel J. Chem.* In press
94. Curtiss, T. J., Bernstein, R. B. 1989. *Chem. Phys. Lett.* In press

MEASUREMENT OF FORCES BETWEEN SURFACES IN POLYMER FLUIDS

Sanjay S. Patel

AT&T Bell Laboratories, 600 Mountain Avenue, Murray Hill, New Jersey 07974

Matthew Tirrell

Department of Chemical Engineering and Materials Science, University of Minnesota, Minneapolis, Minnesota 55455

INTRODUCTION

The physical origins of the forces exerted among polymer molecules include those arising from sources familiar to small molecules (van der Waals, electrostatic, hydrogen bonding, and others) as well as those that can be traced directly to the chain-like character of the polymer. Macromolecules multiply the effects of interactions among their constituent monomers so that a small intersegmental interaction can produce a large intermolecular effect. This is a key factor in polymer adsorption on solids, interaction of polymer molecules with solvents, and interface formation in polymers via thermodynamic phase separation. Chain connectivity and rotational isomerism give a random (or self-avoiding) walk character to macromolecular configurations (1). Distortion of this "natural" population of conformations by the application of perturbations, such as surface potentials, mechanical strain, or electromagnetic fields, produce "elastic" or confinement forces arising from the tendency to maximize the randomness of the conformational distribution. Macromolecular fluids, as they are viscous and frequently highly entangled, can also exhibit nonequilibrium forces, which arise from incomplete adaptation to changes in their environment. Such forces can be very slow to relax (2).

This review summarizes the current status of efforts to measure directly the forces exerted between polymer-coated *surfaces*, with the emphasis on the polymer-related forces discussed above. *Direct measurement* here refers to techniques by which two semi-macroscopic surfaces are brought to within relative separations of molecular dimensions, so that forces between surfaces, not between individual molecules, are measured (3). These direct force measurement methods enable the collection of data on intermolecular forces under well-controlled experimental conditions in a physical configuration that leads directly to conclusions about the role of these intermolecular forces in applications such as the stabilization of colloidal dispersions, adhesion, lubrication, chromatography, mechanical integrity of polymer composites, and biocompatibility phenomena. In certain situations, force measurement provides the means to probe, via direct mechanical action, the configurations of polymer molecules. The polymer surfaces of interest include polymer-bearing surfaces such as adsorbed layers interacting across a liquid medium and thin bulk polymer layers interacting in a neat condition. Forces of interest include symmetrical interaction between identical layers and asymmetrical interaction between different polymers or between polymers and other solids.

We exclude from consideration here a significant body of interesting, related work on techniques for the measurement of intermolecular forces. In this category we place: (*a*) imaginative, but indirect, methods to measure forces as a function of distance between surfaces, such as by means of hydrodynamics (4), compression cell techniques (5), or X-ray measurements of spacings between osmotically compressed layers, which have been applied to study forces between DNA molecules (6), or clay lamellae (7); (*b*) techniques such as film drainage (8) that have been developed to measure forces across thin fluid films; (*c*) scanning tunnelling microscopy (9) and atomic forces microscopy (10), which give tremendous lateral resolution of structure on a surface but are inherently incapable of investigating the forces between two distinct, semi-macroscopic surfaces of controllable and variable composition. In developing a comprehensive picture of surface forces and structure, these techniques are strongly complementary to direct force measurement. As we discuss below, force measurements on polymers are frequently interpretable in terms of the configuration and structure of the macromolecules at the interface. In this vein, force measurement is also complementary to other techniques that give such data, including ellipsometry, microscopy, scattering, and spectroscopy (11, 12).

Lodge (13) has given an excellent historical review of the development of direct force measurement, to which the reader may turn for more complete information, that describes past considerations of alternative designs of direct force measuring instruments. The present review emphasizes the work of the last five years. Direct force measurement between solids has its roots in early (1920s and 1930s) studies of adhesive contact forces between crossed filaments (14) and between glass spheres (15). It was not until the 1950s that the measurement of forces between solids at discrete separations was performed. These measurements were made by groups in the Netherlands (16) and the USSR (17), where the two principal experimental configurations in current use were developed. One method, based on spring deflection to measure force and optical interferometry to measure separation, was developed in the Netherlands, then refined considerably by several groups in the UK, most notably by Tabor's group at the Cavendish Laboratory (18). The second technique, developed and still used primarily in the USSR (19), uses torsional deflection of a filament to measure forces. Both optical and electrical capacitance techniques are used to determine separation (20).

A major advance came with the development of the version of the apparatus introduced by Israelachvili & Tabor (21), which produced enormous improvements in the resolution in the measurement of intersurface separation and, perhaps even more significantly, in the flexibility of the kinds of surfaces that can be studied. The apparatus uses cleaved mica crystal surfaces to provide an atomically smooth substrate, on which other surfaces of interest, such as polymers, can be constructed. It is now possible to measure forces with sensitivity of 10 nN and separations of better than 1 Å, smaller than any intervening fluid molecules (22). The Israelachvili technique (23, 24), which has been used for all the polymer work we review here, has been described in detail several times, so we do not repeat discussion of technique here. We point out, however, that variations on the basic method have been developed for particular purposes (25–28). Other designs of direct force measurement apparatus have been applied to polymers (29, 30), but they have not, in our view, produced reliable results.

One aspect of the Israelachvili technique that is crucial to further discussion is that the forces are measured between surfaces that are curved along cylindrical contours and brought together in an orthogonal crosscylinder configuration to avoid difficulty of achieving parallel alignment of two flat plates separated by molecular distances (22). Therefore, force (F) vs. separation (D) data are presented as $F(D)/R$ vs. D, where R is the (geometric mean of the two principal radii of) curvature of the cylinders and D is the distance of closest approach. The measured force at a certain D is the net result of interactions occurring over a range of separations. Upon integrating these force contributions over the separations between the cylindrical surfaces, Derjaguin (31) showed that $F(D)/R$ was equal to

2π times the energy per unit area of interaction between the surfaces, a geometry-independent quantity. Plotting the data in this format gives the basic interaction energy and, from a pragmatic viewpoint, provides the means to normalize data from different experiments. It also means that the force sensitivity quoted above is not a particularly crucial element of this technique, since the absolute force is related to the curvature or effective area of contact. Spring deflection as a means of measuring forces is convenient (relative to electrobalance methods, for example), since it is not necessary to know a priori whether forces are attractive or repulsive or change sign. However, the heart of this technique is the level of precision of separation measurement by optical interferometry (32).

Though mica surfaces are the primary surfaces used in these measurements, it is possible to deposit on or coat these surfaces with polymer layers, metal films (33–37), surfactant layers (23), and Langmuir-Blodgett films (38) to alter the physical and chemical nature and the chemistry of the interacting surfaces while keeping them smooth by virtue of the atomically smooth mica underneath. The object of this review is polymer surfaces. The largest part of the work published to date concerning direct force measurement with polymer surfaces has been done with adsorbed, neutral polymer layers on mica, brought together in a medium of pure solvent or very dilute polymer concentration. Consequently, this work constitutes the main discussion in the first part of this article. Within this area of interactions between adsorbed polymer layers, we devote considerable attention to categories of measurements, about which there is now a sufficient body of data from different laboratories to make a comparative analysis of the data worthwhile. These categories are (a) data on forces between adsorbed homopolymer (that is, polymers made from a single type of monomer unit) layers near or below the Flory theta temperature (1), and (b) data on forces between adsorbed layers of amphiphilic macromolecules—multicomponent macromolecules in which one portion of the molecule is strongly adsorbing and the other adsorbs negligibly or not at all. We show what general conclusions can be drawn from this body of data and then focus attention on certain broad remaining questions. The second part of this review covers more wide-ranging, and as yet less well-established areas of research with direct force measurement applied to polymers. That section includes work on polyelectrolytes, polypeptides, proteins, and other biological polymers, and measurements in polymer melts.

All forces described in this review are believed to be equilibrium forces unless we explicitly state otherwise. That is to say that for most of the data reported here, the force reported at a certain separation is that corresponding to effective equilibrium, with effects of history of deformation relaxed on the timescales relevant to the measurements reported. Examples of hysteresis in the forces, where it has been established that this relaxation has not occurred, are mentioned. We discuss experimental and theoretical definition of effective equilibrium for adsorbed polymer layers in the next section. In the last section, we discuss measurements whereby dynamic effects are studied in their own right. In the concluding discussion we cover some of the future possibilities for work on direct force measurement between surfaces in polymers. Because of page length constraints we do not separately and systematically review all the relevant theoretical developments, but confine ourselves to discussion of those theoretical results with most immediate application to the experiments. A theoretical review has been completed recently (39). This article should be viewed as a review of direct force measurement rather than a comprehensive review of polymer adsorption, though this has been the main application of direct force measurement on polymers to date. Several good reviews of polymer adsorption are available (40–44).

FORCES BETWEEN ADSORBED LAYERS OF HOMOPOLYMERS

General Considerations

Typical protocol for direct force measurement between adsorbed polymer layers is to immerse the mica substrate surfaces in a dilute (10 to 100 μg/mL), filtered (<0.2 μm) polymer solution for a time sufficient (e.g. hours) to achieve the equilibrium adsorbed amount (typically 1 to 10 mg/m^2 of surface area). The mica surfaces have previously been cleaved, placed in the apparatus, and tested for contamination by assuring that they exhibit clean, adhesive contact. Pure solvent is then introduced and $F(D)$ measured to determine that no contamination was introduced with the solvent. With clean organic solvents, the contamination-free condition is usually indicated again by monotonic attraction into adhesive contact between the mica sheets. Contamination by unwanted dust, surfactants, or polymers frequently gives long-range repulsion. Very pure, dry solvents, especially those with high symmetry of the molecules, can give repulsive structure forces (45–48), due to molecular packing effects at short range (<10 solvent diameters); electrolyte solutions as solvents can give long-range electrostatic repulsion (49–51).

Essentially all work to date has been done on layers adsorbed from dilute solution. Immersion of the surfaces in concentrated solutions risks the complications of a viscous environment in which forces are slow to relax to equilibrium. In most cases, the adsorbed layers have been saturated nonetheless, since adsorption isotherms typically rise very steeply at low concentration. This means that it is very difficult to study experimentally the theoretically convenient situation of an isolated adsorbed chain. This is in contrast to polymers in free solution, where isolation can be achieved by dilution.

Most work on homopolymers has been done in the so-called "weak adsorption" limit (52). The resolution of the two seemingly conflicting statements about high affinity adsorption isotherms and "weak adsorption" lies in the second sentence of this review. Individual polymer segments may be weakly attracted to a surface with a binding energy, $E_b \ll kT$; the large number of segments per molecule still favors binding of the polymer molecule to the surface, in effect producing an amplified binding energy, $nE_b \gg kT$, where n ($\gg 1$) is the number of bound segments. Under these circumstances, the molecules in the adsorbed layers adopt a conformation different from, but with a characteristic size comparable to, the coil conformation in bulk solution. This is accomplished by having "trains" of adsorbed segments as anchor points between "loops" and "tails" of segments extending into solution. (Stronger segmental binding, $E_b \gg kT$, produces a flattened conformation.)

This weak adsorption condition produces effectively irreversible adsorption. Adsorption of one segment is sufficient for a polymer to be adsorbed. With one segment attached, others are tied near the surface in proximity for binding. With several segments bound, the Boltzmann factor for desorption, $\exp(-nE_b/kT)$, can be very small. These are the conditions that produce the effective equilibrium discussed above. It is reasonable under many practical circumstances to treat adsorption as irreversible. An adsorbed layer exposed to an infinite bath of pure solvent may still profitably be considered as an effective equilibrium system, but with the total adsorbed amount constrained to be constant, invariant to the reduced polymer concentration in the bulk over the time scales of experimental interest. This situation, which we will henceforth refer to as *constrained equilibrium*, is more the rule than the exception in the situations we describe, and so we take pains only to point out exceptions. By this we mean that the adsorbed amount is stable against washing with pure solvent but we do not rule out rearrangement of the molecules in situ.

The effective irreversibility of adsorption should not be taken to mean that the molecules in these layers are static objects, however. The weak segmental binding energy implies a small barrier to detachment of segments. This property makes it possible for polymers on the surface to exchange with those in solution (53), via progressive displacement of the former by the latter, especially if the latter are higher molecular weight, and raises interesting, unexplored questions about the mobility such molecules may have along the surface without complete detachment (41).

In what follows, the quality of a solvent for a polymer is measured by the excluded volume parameter:

$$v = \int [1 - \exp(-V(\mathbf{r})/kT)] \, d\mathbf{r} \qquad 1.$$

where $V(\mathbf{r})$ is the effective mean interaction potential between a pair of segments separated by $\mathbf{r}$ (54). This excluded volume can be thought of as the volume of an effective repulsive (toward other segments) shell around a segment created by solvation. The excluded volume becomes zero at a characteristic temperature, $T = \theta$, known as the theta temperature. The self-avoiding walk performed by the chain due to excluded volume of the segments becomes (up to logarithmic corrections) a random walk for $v = 0$. In this "ideal state," the mean square end-to-end distance of the random walk performed by the chain, $\langle r^2 \rangle$, and the mean square radius of gyration, $\langle r_g^2 \rangle$, are given by:

$$\langle r^2 \rangle = 6 \langle r_g^2 \rangle = Na^2 \qquad 2.$$

where N is the number of segments in the polymer and a is the size of a segment. Around $T = \theta$, Eq. 1 may be expanded to give:

$$v = Ka^3(1 - \theta/T) \qquad 3.$$

with K a positive constant for the usual case of increasing solvent quality (V or v) with T. Therefore, $v > 0$ for $T > \theta$, and $v < 0$ for $T < \theta$; v is related directly to the Flory-Huggins χ parameter. The two parameters v and a, along with the chain length N, provide an adequate representation of the dilute solution equilibrium properties of polymer solutions for $v > 0$ in the form of relationships between dimensionless groups such as: $\langle r^2 \rangle / Na^2 = f(vN^{1/2}/a^3)$ (39, 54). The regimes of behavior for the coil characteristic size can be broken down roughly like

$$\langle r^2 \rangle \approx N^{6/5} a^{4/5} v^{2/5}, \quad \text{for } vN^{1/2}/a^3 > 1, \text{ swollen coil} \qquad 4.$$

$$\langle r^2 \rangle \approx Na^2, \quad \text{for } -1 < vN^{1/2}/a^3 < 1, \text{ ideal coil} \qquad 5.$$

$$\langle r^2 \rangle \approx N^{2/3} a^2, \quad \text{for } -1 > vN^{1/2}/a^3, \text{ collapsed coil.} \qquad 6.$$

For solution concentrations beyond the dilute limit, higher order interactions beyond the binary interaction embodied in v must be included. For the purposes of this review, the third virial coefficient parameter w can be added to account for three-body interactions and an adequate level of description obtained in terms of the set of parameters, N, a, v, and w. This three-body interaction is essential also to describe accurately single chain behavior at or below $v = 0$, since the coil segment density gets high in the interior as the coils collapse. Thus, Eq. 6 above should be modified for a more thorough treatment of poor solvent conditions, though the basic functional dependence given there is correct.

FORCE MEASUREMENT IN POOR SOLVENTS

Adsorption from solution is a competition between adsorption and dissolution. Poor solvents give an edge to adsorption and thus produce significant adsorbed amounts. The first well-defined direct force measurements on adsorbed polymer layers were made by Klein (55, 56) on polystyrene (PS) adsorbed on mica from cyclohexane (CH) at 23°C. Figure 1 includes these data, along with several other comparable sets (57, 58). On reducing the separation between the surfaces, the first interaction detected is attractive, setting in at a separation around 2 to 3 times r_g ($=\langle r_g^2 \rangle^{1/2}$). The design of the Israelachvili apparatus does not permit measuring on portions of the force-separation curve where the gradient exceeds the spring constant of the force measuring spring, hence the gap in these data. The spring can readily be stiffened to eliminate these gaps, if desired.

It has been established experimentally, by ellipsometric and hydrodynamic measurements made on adsorbed layers under these conditions (42), that the average layer thickness varies approximately like r_g, or like $N^{1/2}$. The radius of gyration is therefore a natural length scale to use in attempting to unify all the data in a single plot. In Figure 1, the ability of this scaling to reduce five different data sets for several molecular weights and two chemically different polymers is illustrated. Table 1 shows the relevant information on experimental conditions, molecular size, and adsorbed amount. To the precision permitted by the data, this scaling of the range of the forces works well. It is interesting to note that the self-consistent field theory of Scheutjens & Fleer (59, 60), based on a lattice model, predicts that the range of the forces should vary like Γ, the total adsorbed amount, measured say in mg/m^2. The exact variation of Γ with molecular weight (M) is not established definitively experimentally but it is not widely at odds with the relation, $\Gamma \approx M^{1/2}$ (42), that is necessary for Γ scaling to be consistent with r_g scaling. The data on adsorbed amount reported with the force measurements are not precise enough to decide this question. Klein & Luckham (61) have offered more detailed consideration of how the range of the forces extracted from direct force measurements should vary with the molecular weight of the adsorbed polymer.

The quantitative explanation of these data is incomplete but the general features are understood (62). Attraction sets in as the segments begin to intermingle at $D = 2$ to $3\, r_g$. One source of attraction is the fact that $v < 0$

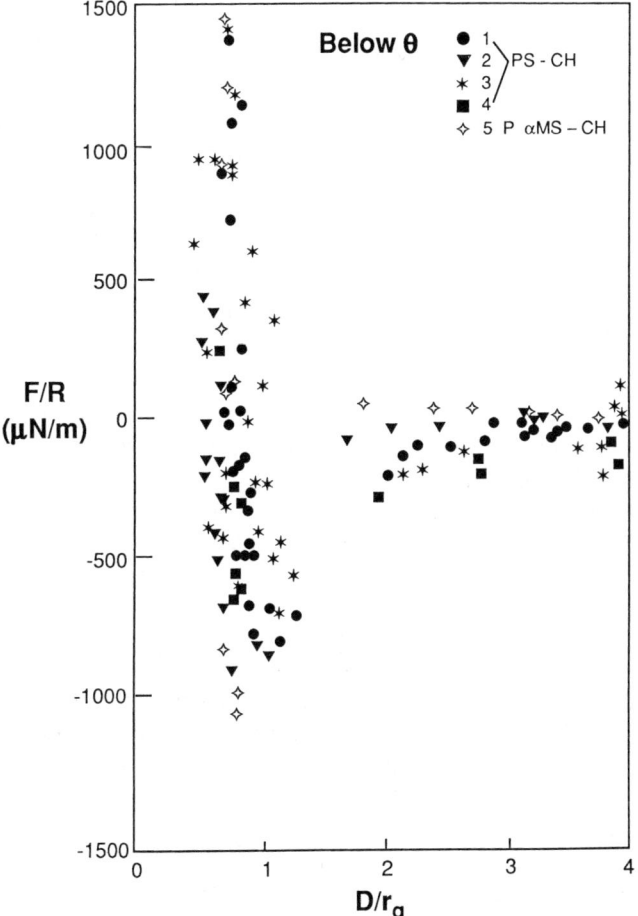

Figure 1 Force vs. distance data for polystyrene and poly α-methylstyrene in cyclohexane below the θ point. Other information on the samples is given in Table 1.

for PS in CH at 23°C. That is to say that PS segments prefer contact with one another to contact with solvent. We refer to this interaction from now on as the *osmotic interaction*. In general, this may be attractive or repulsive, depending on the sign of v, which in turn depends on solvent quality and temperature. A second source of attraction is that, as D approaches a small multiple of r_g, individual polymer molecules can become simultaneously adsorbed on both surfaces, resulting in an interaction known as *bridging*. Bridging is always attractive, since the most probable end-to-end separ-

Table 1 Force measurements on adsorbed layers, $T < \theta$. Data on samples of Figure 1

Sample no.	Ref.	System[a]	Molecular weight	Temperature (°C)	r_g^b (nm)	Γ (mg/m^2)[c]
1	(56)	PS-CH	6×10^5	24	22.5	5.5 ± 1.0
2	(57)	PS-CH	6×10^5	23	22.5	5.5 ± 1.0
3	(56)	PS-CH	1×10^5	21	9.2	—
4	(57)	PS-CH	9×10^5	26	27.5	3.6 ± 0.6
5	(58)	PαMS-CH[d]	9×10^4	25	9.0	3.0 ± 0.3

[a] Symbols for sample designations given in text.
[b] All r_g values are θ condition values from data of Schmidt & Burchard (66).
[c] These are the values reported with the force measurements.
[d] Theta temperature for PαMS-CH is approximately the same as for PS-CH (34.5°C) (58).

ation for any portion of the chain, such as the spanning portion, is zero; attachment to each surface tethers them a distance D apart and produces a tension of entropic origin in the bridge.

The attractive minimum is situated at about $D = r_g$. At smaller separations, $D \approx (2/3)r_g$, the force turns repulsive. One evident source of this repulsion is that as D becomes less than r_g, under constrained equilibrium conditions, chain conformations become compressed, resulting in an entropic elastic repulsion. Full equilibrium would drive polymer molecules from the gap at small D and would result in an attractive $F(D)$ curve at all D. The repulsion at short distances, in spite of the increasing confinement of the chains, is one of the most important manifestations of the constrained equilibrium condition. Klein (55, 56) has suggested in addition that, at high compression, the repulsive force has a contribution from three-body and higher order interactions than those embodied in v (e.g. w and coefficients of higher order terms). Finally, incompressibility of the bulk polymer would produce a very steep barrier at a small separation determined by the adsorbed amount and the density of the material.

The most important question raised by the ensemble of data in Figure 1 is that of the relative magnitudes of the contributions to the attraction of osmotic effects and of bridging. If osmotic effects are dominant, direct force measurement may become an interesting means to explore polymer solvent interactions. From a practical standpoint, in the stabilization of colloidal dispersions in which the attraction is to be made small, it is important to know whether to focus on the solvent quality effects that dictate the osmotic interactions or on the surface accessibility effects that relate to bridging, since, to bridge, a macromolecule must gain access to each surface (63). The two effects are not entirely separable, because bridging requires a certain adsorption affinity, which is diminished by a good solvent tending to pull segments into solution.

FORCE MEASUREMENT NEAR $T = \theta$

The first experimental approach to this problem was to measure the interactions between adsorbed layers under conditions where $v > 0$ ($T > \theta$). This would have the effect of making the osmotic interactions repulsive, whereas bridging, if it occurs, would always produce an attraction. As a pragmatic matter, it was desired to work with the same or closely related polymer-solvent mixtures so that the adsorbed amounts and other circumstances of the experiment would be comparable to the $T < \theta$ case. This led to work at or just above $T = \theta$, in the same solvents in which work had been done below θ (57, 64). The current design of the equipment does not readily permit measurements at temperatures above about 50°C, so pushing the PS-CH system well above θ has not been feasible. On the other hand, PS does not adsorb readily on mica from any of its common room temperature good solvents [e.g. toluene (65), tetrahydrofuran]. This is not to say that adsorption from good solvents cannot occur (42). Mica exhibits weaker affinity for PS than, say, chromium, on which much ellipsometric work has been done. There is a small disagreement in the literature concerning whether PS adsorbs a small amount or not at all on mica from toluene; there is complete accord that any adsorption is slight and weak. This point is mentioned again below in discussing good solvents and amphiphiles.

Figure 2 shows a compilation of published data on forces between PS layers adsorbed on mica from CH and CP (cyclopentane) just above θ (34.5°C for CH, 19.6°C for CP). The collapse of the data with the separation scaled by r_g does not seem to work quite as well as in Figure 1, though we use the unperturbed radius of gyration for scaling in both of the Figures, with no adjustment for temperature sensitivity of chain dimensions over this rather small temperature range (66). (See Table 2.) Few reliable data for Γ of PS on mica under these conditions are available. However, the clearest point made by Figure 2 is that, at $T > \theta$, some data show attraction and some do not. Both molecular weights of PS in CH (57) show attraction at $T > \theta$, whereas two sets of PS-CP (64) data show no attraction. The presence of attraction for the PS-CH data at 37°C indicates that bridging is a major contributor. One caveat that should be inserted at this point is that 37°C is only 3.5°C above θ for the bulk solution. Since the theta point is a rather delicately balanced condition, consideration of whether the bulk θ point is the proper reference point in a narrow gap may be in order (57). On the other hand, leaving aside for the moment the data of curve 3 in Figure 2, the data on PS-CP show essentially no attraction. Recent data of H.-W. Hu and S. Granick (unpublished) show that the data at $T > \theta$ in PS-CP do scale well with r_g. Klein

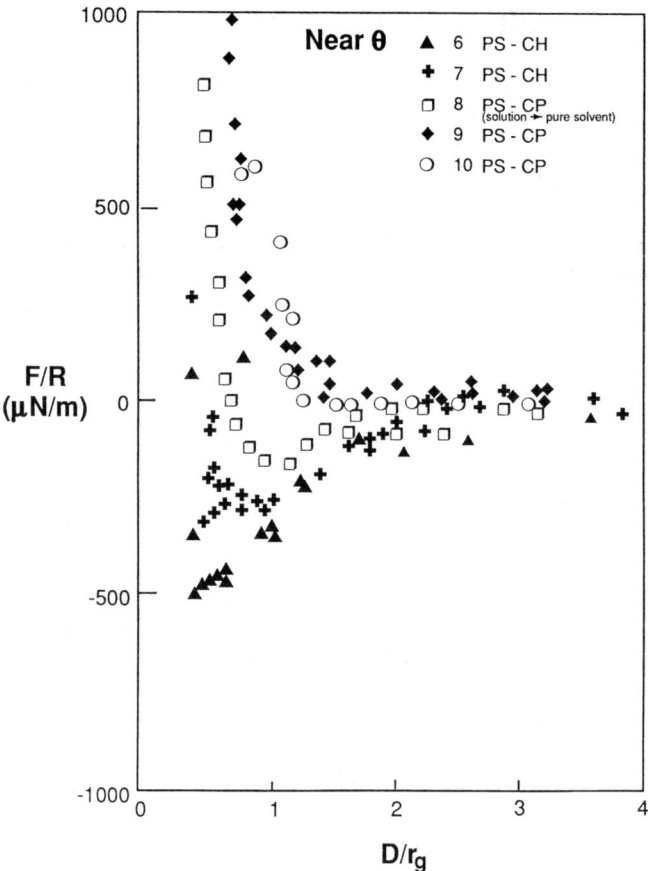

Figure 2 Force vs. distance data for polystyrene and cyclohexane and cyclopentane near the θ point. Other information on the samples is given in Table 2.

& Pincus (67) and Ingersent et al (68, 69) have suggested that the distinction between PS-CH and PS-CP may be at the third virial coefficient level. As we pointed out in the introduction, these effects must be considered when $v \approx 0$. Data on w exist for the system PS-CH, and calculations incorporating its effects succeed in reproducing the qualitative spatial features of the PS-CH data in Figure 2. In particular, the attractive portion of the curve is seen in the calculations, in agreement with the idea that bridging attraction is the dominant effect. In the case of PS-CP, w may produce sufficient osmotic repulsion to outweigh the bridging. Data on w for this system are not available.

Table 2 Force measurements on adsorbed layers, $T \cong \theta$. Data on samples of Figure 2

Sample no.	Ref.	System[a]	Molecular weight	Temperature (°C)	r_g[b] (nm)	Γ (mg/m^2)[c]
6	(57)	PS-CH	9×10^5	37	27.5	e
7	(57)	PS-CH	6×10^5	37	22.5	e
8	(64)	PS-CP[d]	2×10^6	23	41.0	—
9	(64)	PS-CP	2×10^6	23	41.0	—
10	(64)	PS-CP	6×10^5	23	22.5	2.5 ± 1.5

[a] Symbols for sample designations given in text.
[b] All r_g values are θ condition values from data of Schmidt & Burchard (66).
[c] These are the values reported with the force measurements.
[d] Sample 8 conditions are exactly those of sample 9 except that 8 was measured after the incubation solution of sample 9 was washed out and replaced with pure solvent.
[e] No adsorbed amounts reported but clear evidence given that Γ at 37°C was less than Γ at 23–26°C.

Granick and co-workers (70) have recently made direct force measurements on the PS-CP system in some detail. They examined effects of molecular weight and also of the presence of isomers such as n-pentane in the CP used. For a PS sample of molecular weight 5×10^5, long-range forces were monotonically repulsive at 28°C and mildly attractive at 18°C when a CP containing 5% linear pentanes was used. The range of forces was 2–3 r_g at both temperatures. This contrasts with experiments in 99% pure CP, in which the long-range forces were strongly attractive at 18°C and weakly attractive at 28°C. The range of the forces was 2–3 r_g at 18°C and 5–6 r_g at 28°C. These results underscore the importance of what might have been thought to be rather small variations in solvent quality. The fact that the range of the forces is smaller near θ (18°C) than farther away (28°C) is qualitatively consistent with the idea that the range of the forces should be smaller in the temperature range where v $\approx$ 0. Klein (62) pointed to this idea to explain the difference in range between layers of PS measured in CH and CP at 23°C. (Compare the relevant curves in Figures 1 and 2.) The range is longer in the poor solvent than in the theta solvent, since the binary interaction occurring on first overlap when v = 0 produces no force, whereas attractions are generated from binary interactions at $T < \theta$. Clearly variation in adsorbed amount will affect the range of the forces as well.

The approach of using mixed solvents purposefully as media for force measurements in which v can be varied smoothly from positive to negative has been explored by Marra & Hair (71). They show that adsorbed amount increases as the proportion of poorer solvent increases. Other than that, it is difficult to draw quantitative conclusions from their data, in part because it may happen that the adsorbed layer is preferentially swollen with the better solvent of the mixture, thus making the local solvation conditions in the adsorbed layers different from the global mixture. On the other hand, many practical applications in which adsorbed layers are used for steric protection against colloidal flocculation (paints, for example) contain several solvents to get the right balance of solvating power, volatility, and viscosity.

Curve 8 in Figure 2 requires special discussion (64). These data were obtained under conditions identical to those of Curve 9, except that the incubation solution, which contained 15 μg/mL of PS, was present in the case of 9 and was replaced by pure solvent for Curve 8. No data are given for the adsorbed amount either before or after washing, but the shorter range of the close repulsive barrier, and the appearance of an attractive portion of the curve, which may be attributed to bridging, suggest that the explanation of these data is that some desorption has been induced by washing, which has, in turn, resulted in additional bridging due to increased accessibility of the surface. This system is an exception to the general rule that, due to the constrained equilibrium principle, the adsorbed amount in saturated surfaces is invariant to changes in the polymer concentration surrounding the surfaces. PS-CH conforms to the generality, but desorption can be induced by raising the temperature at constant concentration. (Compare Curve 4 in Figure 1 with Curve 6 in Figure 2.)

These results show that, although to a certain extent they can be discussed separately, solvent quality and bridging produce an integrated effect. Some efforts have been made to delineate their separate effects. Israelachvili et al (57) demonstrated that, in spite of the high affinity isotherms typically exhibited in polymer adsorption, less than saturated coverage of the mica surface in the force-measuring apparatus can be produced by incubating in the polymer solution with the two surfaces close to one another (<100 μm), in effect limiting the diffusional access to the surfaces. In this way, they produced a 9×10^5 molecular weight PS layer in CH with $\Gamma = 1.1 \pm 0.2$ mg/m^2, which is about 30% of the saturation coverage reported for this polymer in Table 1. The range of the forces here was shorter ranged than for the saturated surfaces, but interestingly, the depth of the attractive minimum was approximately twice that for saturated surfaces. The precision of the data in this experiment enabled a demonstration that the long-range part of the attraction conforms quite closely to an exponential decay, with a decay length numerically equal to the position of the attractive minimum. Such an exponential variation in the long range is anticipated from any theoretical considerations that view the adsorbed coil as a random walk perturbed by weak segmental interactions with the surface. The diffusion-like equation for the conformational distribution function tails off exponentially far from the surface (72). Exponential variation with a decay length depending on surface coverage has been predicted recently in a detailed calculation by Ingersent et al (69).

The deeper attractive minimum arises from the enhancement in bridging possible at lower surface coverage. Almog & Klein (64) explored this phenomenon, also by using the diffusion starvation technique, for the PS-CP system. In this case, the interaction that is monotonically repulsive for saturated surfaces is markedly attractive when the surface coverage is limited, an effect qualitatively similar to that of reducing the coverage by washing the surfaces. Granick et al (58) studied the effects of bridging directly by comparing the forces between two saturated layers of poly-α-methylstyrene (PαMS) with those between one layer of PαMS and bare mica. In the latter case there is no impediment to bridge formation. The results of this comparison are shown in Figure 3. Several features are of interest in comparing the two parts of this Figure. Foremost is the increase by a factor of about 25 in the depth of the attractive minimum when one surface is left uncovered. The positions of the long-range attraction, the minimum, and the short-range repulsion are important as well. Although the minimum and the repulsion are seen at separations of about half in the one-sided coverage case compared to the two-sided, the long-range attraction is scarcely diminished in range. This result may be due to the fact that although two polymer layers just beginning to overlap near $T = \theta$ produce very small osmotic forces, substantially more attraction can be generated if the tails and loops extending from one surface can uninhibitedly form bridges with the other surface. Granick et al also observed that the adhesive minimum continued to deepen during several minutes in contact, thus indicating that some dynamic phenomena were involved in the development of the bridging forces. This work also led to a semiquantitative estimate of the segmental binding energy, E_b, for the PαMS segment on mica as $(1/3)kT$, consistent with the weak adsorption condition.

We close this section with some discussion of the status of quantitative theoretical modeling of these results (39). Two categories of models have been developed to the point of being able to make detailed comparisons of predictions with data such as those of Figures 1–3. One is the self-consistent, mean field (SCF) formalism originating with Edwards (73) and Helfand (74). The fullest development of this path has been made by Scheutjens, Fleer and co-workers (59, 60, 75–79), who have used a discrete, lattice model, by Ploehn, Russel and co-workers (39, 80, 81), and by Muthukumar & Ho (81a), who have pursued the mathematical analogy between the continuous SCF equations and the Schrödinger equation. The former facilitates the calculation of quantities such as distributions of loops, tails, and bridges, at the price of an artificial discretization of space

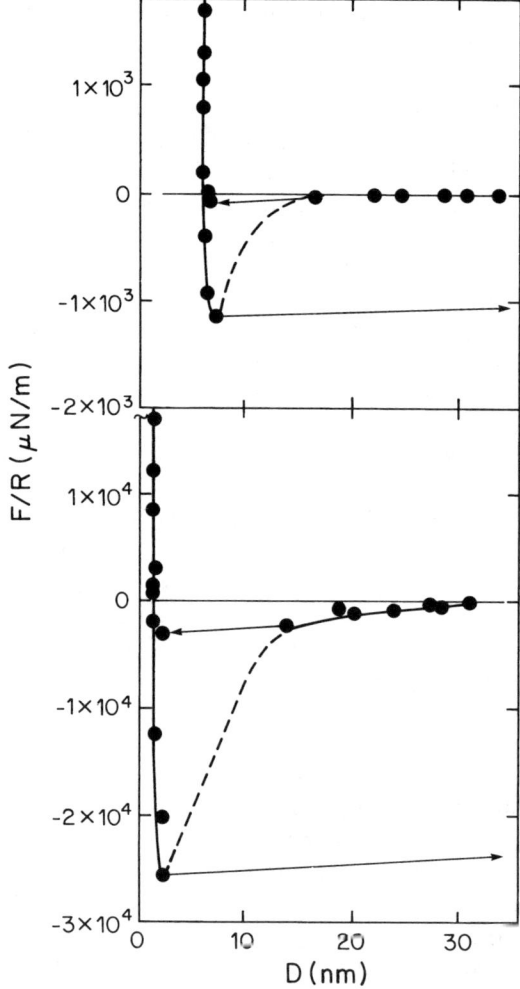

Figure 3 Force vs. distance data between mica surfaces immersed in cyclohexane at 25°C. *Upper curve*: Each mica sheet bears 3 mg/m^2 of poly α-methylstyrene. *Lower curve*: One mica sheet is bare, the other bears 3 mg/m^2 of poly α-methylstyrene. *Arrows* indicate jump instabilities described in the text.

and the difficulties inherent in a statistical mechanical lattice model in simply extracting physical insights. The continuous SCF approach has its main advantage in the power and versatility of mathematical analysis that can be brought to bear to illuminate the physics of the results. Both contain fundamentally the same physical view, particularly in their reliance on the usual mean field assumption that a segment placed at a certain location is acted upon by an average potential created by the distribution of solvent and segments. Correlations and fluctuations in the local distribution of segment density are left out of the usual, mean-field SCF models.

In the second category are the density functional theory (DFT) models, proposed for polymers at interfaces by de Gennes (52, 82), based on earlier work by Cahn (83), and developed most fully by Pincus and co-workers (67–69). The idea here is to write the free energy of the interfacial system as a spatial integral of a function consisting of the sum of a free energy of a hypothetical system homogeneous at the local density of polymer segments, plus a term in gradients or some other measure of the variation of the local density, arising directly from the inhomogeneity of the interfacial profile. The calculation seeks the profile of segment density that minimizes the integral free energy of the system. The Euler-Lagrange equations for this functional minimization are differential equations for the profile, akin to those of the SCF theory discussed above. It is straightforward within this formalism to incorporate correlations (non-mean field effects) into the local free energy term; the down side is that because the theory is formulated in terms of the density profile, it can give only indirect information on the arrangement of individual chain conformations.

Both types of models can make detailed predictions of the $F(D)$ between two surfaces bearing adsorbed homopolymers below and near $T = \theta$, and both have to date achieved about the same degree of success (67–69, 75, 77, 84). All the qualitative features of Figures 1–3 can be reproduced by both classes of theories, at least to within the level of uncertainty in current knowledge of certain key parameters, such as Γ and w. Most important among the successes of the theories is their ability to reproduce the pattern of interactions and trends such as the shortening of the range and deepening of the attractive minimum with decreasing surface coverage. As mentioned above, with accurate thermodynamic inputs, they can also reproduce the observed switch from attraction to monotonic repulsion with increasing solvent quality. Their most important shortcoming with respect to the body of data represented in Figures 1–3 is that, with the fairest estimates of the necessary parameters, including some direct fitting to force data to get the ranges correct, the theories underpredict the magnitudes of the attractive interactions by factors of at least 2 to 3. The discrepancies may still be attributable in part to imperfect knowledge of Γ and of the thermodynamic parameters. Efforts on the part of experimentalists to establish these quantities more precisely are needed. In addition, one of the weakest points of all current theory is in the lack of detailed treatment of the polymer-surface interaction. Both SCF and DFT theories treat this via a single surface energy parameter. This parameter is typically taken to be independent of the polymer-solvent interaction, when in reality they must be connected. It is difficult at this stage to suggest how to do better; few experimental numbers are available for the single parameter, let alone data to support more detailed treatments. Marra & Hair (85) have attempted to manipulate the surface interaction parameter experimentally by examining forces between adsorbed layers in mixtures of polar and nonpolar poor solvents, thus playing on the competition between adsorption and dissolution in determining the effective surface interaction energy. The results are interesting but do not clarify the situation quantitatively.

FORCE MEASUREMENTS IN GOOD SOLVENTS

Although polymers tend to dissolve rather than adsorb in good solvents, they can still adsorb if the affinity for the surface is high enough. This is the condition of most interest for colloidal stabilization by adsorbed homopolymers, since good solvent can be expected to produce the strongest osmotic repulsion on overlap of two adsorbed layers. In this connection, it should also be borne in mind that if polymers can adsorb on one surface, they can bridge between two in close proximity. Attraction can still exist in this regime, especially if the coverage is low enough to permit access to the second surface.

As discussed above, systems in which substantial adsorption occurs above θ are relatively rare. Measurements on poly(ethylene oxide) (PEO) in both aqueous electrolyte and in toluene attempt to study these conditions. Israelachvili and co-workers (86) published the first data on forces in the PEO system in water containing 0.04 M $MgSO_4$. They observed very long-range forces ($\gg 10\ r_g$, though it is important to note that, in contrast to all studies discussed above, this work was done with a sample of broad molecular weight distribution), but more importantly, they found substantial nonequilibrium relaxation and hysteresis effects. The forces were repulsive on first approach, rising steeply at onset, then less sharply at smaller D. On separating after slight compression, the forces dropped much more sharply than they had risen, and on recompression, rose along the decompression curve, not along that of the original compression. Stronger compression followed by decompression gave rise to adhesive forces on separation, in contrast to the pure repulsion seen on approach. Very strong compression forced the polymer from the gap and the mica into adhesive van der Waals contact. Finally, they found that over a period of 24 hours, the adsorbed amount increased such that the longest range forces were detectable at greater than 30 r_g. Their conclusion was unavoidably vague, namely that their adsorbed layers exhibited "gel-like properties," possibly arising from microaggregates.

The precedent for unusual aggregation phenomena in PEO solutions is well-established (87–89). In particular, it has been proposed that PEO aqueous solutions contain both microcrystallites (below the bulk melting temperature of about 55°C) and noncrystalline microgel particles. As particles such as these are readily deformable, they are difficult to remove by filtration (90). Whether they are equilibrium structures is unclear; that they are long-lived and dependent on the history of sample preparation, especially heating during attempted dissolution, has been established (87–90).

Klein & Luckham (91) studied two samples of PEO of narrow molecular weight distribution (molecular weights = 1.6×10^5 and 4.0×10^4) produced by anionic polymerization, in contrast to the broad distribution sample examined by Israelachvili and co-workers (86). Force measurements were made in a medium of 0.1 M KNO_3. They, too, found hysteresis in the form of differences between approaching and separating force curves. More rapid compression-decompression cycles gave lower $F(D)$ values than slower cycles nearer to equilibrium. Rapid compression, followed by maintenance of constant force, produced relaxation manifested by increase of D to the slow compression curve. The longest range of the forces was at about $6 \pm 1\ r_g$, and Klein & Luckham found evidence for neither attraction nor adhesion in their data. The general shape of the $F(D)$ curves—rising steeply (in a log-log plot) at small compression, then less steeply closer in—is similar to the data of Israelachvili et al (86), except for the adhesion and more complex relaxation effects. Klein & Luckham (91) showed that the constrained equilibrium condition applies to the aqueous PEO system.

Luckham & Klein (65) also studied forces in the PEO-toluene system, another good solvent situation. The pattern of $F(D)$ behavior they report for this system strongly resembles that for PEO-electrolyte. The range of the forces they report is longer in toluene ($\approx 8.5 \pm 1\ r_g$) and scales approximately with $M^{0.4}$ (61). Toluene has been reported to have a more favorable Flory-Huggins χ parameter (0.39) (92) with PEO than aqueous electrolyte. Klein and co-workers (93) were led to conclude that the repulsion seen in these PEO experiments is probably osmotic in origin. This interpretation relies on the constrained equilibrium nature of polymer adsorption, even under these good solvent conditions, so that when the adsorbed layers are brought into overlap the positive v values generate repulsion. De Gennes has developed a model of this situation (52, 83), based in part on the DFT method mentioned above, since in good solvent it is more important to use non-mean field scaling then it is for T near θ. The calculation predicts that F/R varies like D^{-2} on weak overlap and, closer in, like $D^{-5/4}$ for stronger compression, consistent with the qualitative observation of steep repulsion followed by more gradual repulsion. Klein & Luckham have shown that their data are tangent to these power law predictions in reasonable ranges, although the very weakest overlap data varies even more strongly than the D^{-2} above.

Most recently, Marra & Hair (94) have reexamined the PEO-toluene system by using a polymer of the same molecular weight (40,000) and origins (Toyo-Soda Co.) as one used by Klein & Luckham. They obtain data that are much more reminiscent of those of Israelachvili et al (86) than those of Klein & Luckham (91). In particular, they see stronger hysteresis effects, the development of adhesion on strong compression, and instabilities in the approaching repulsive force curve, all of which suggest that they are squeezing material from the gap, possibly aggregate particles. They described their solution preparation methods somewhat more fully than any of the other reports on forces in PEO systems. They heated the initial solutions above the melting temperature of PEO (60°C) in an attempt to get homogeneity; they nonetheless observed some precipitation on cooling to the 22°C measurement temperature. They did not use filters to clarify their solutions, since they worried that toluene may leach other impurities into the solution. This may have permitted the introduction of particles and microgels. [Israelachvili et al (86), and essentially all other groups working with organic solvents, do use filters with rare ill effects (95).]

The results on this system by Klein and co-workers admit a reasonable theoretical interpretation, though they do not give in their papers any information on solution preparation from which one could assess the likelihood of aggregation in light of ample independent evidence of this possibility; they saw no striking evidence of it in their force measurements. For the hysteresis they saw, and the difference between rapid and slow compression-decompression cycles, they have offered the following reasonable explanation. In order for substantial adsorption to occur in a good solvent, there must be an appreciable surface affinity per segment. Klein and co-workers estimate 4.5 ± 1 mg/m^2 for PEO on mica in the aqueous electrolyte. On the other hand, the long range of the forces, owing to swelling by the surrounding good solvent, stretches many segments out away from the surface into solution. On compression, more segments are forced on to the attractive mica surface and do not have time to move off after a rapid decompression (93). Barford & Ball (96) have elaborated these ideas into a more complete theory of nonequilibrium effects in polymer adsorption. Klein (62) has further suggested, based on their measurements and this interpretation, that this sort of sensitivity to timescale of compression-decompression is a "signature" of forces between layers of polymer adsorbed on surfaces from good solvents. However, given that only one polymer has been examined to any significant extent under good solvent conditions, that there is much evidence of its aggregation in solution, and that there is disagreement concerning its hysteretic behavior, it may also be that this range of behavior is peculiar to PEO.

In any case, Klein & Luckham (97) have used the PEO-aqueous electrolyte system to examine forces between adsorbed layers with less than saturation coverage. With the starvation technique introduced by Israelachvili et al (57), they have measured forces during the build-up of Γ on the mica. The demonstrated inter alia that, for a level of coverage intermediate between zero and saturated, attractive forces can exist between polymer-bearing surfaces immersed in a solvent of good quality for the adsorbed polymer. The form of the $F(D)$ profiles under these conditions is like that of Figure 1, but with the depth of the attractive minimum of order -50 μN/m, considerably less than the attractions near θ. The attraction under these circumstances is almost certainly due to bridging but is reduced relative to the poor solvent saturated layers, since chains must here find their way through an osmotically and sterically repulsive barrier to form bridges. Klein & Luckham (97) have pointed out that, though this attractive minimum is small by comparison to the poor solvent data, it corresponds to an attractive energy of about 100 kT between two 1 μm diameter colloidal particles situated 50 nm apart. Therefore, this level of attraction is clearly of practical significance in dispersion stability. Even well above θ, attraction and flocculation could occur if the conditions of surface energy and Γ lead to bridging.

One other system for which adsorption on mica and forces have been measured under good solvent conditions has also produced conflicting data among different groups. This is the PS-toluene system mentioned above. Three groups (65, 71, 98) have specifically looked for and reported no adsorption of PS on mica from toluene. In contrast, Watanabe et al (99) found repulsive forces for two different molecular weights of PS in toluene solutions exposed to mica. The repulsion set in at 1 to 2 r_g, scaled roughly with r_g, and no attraction was seen. This work also found that the PS could be rapidly displaced by a more strongly adsorbing polymer. These facts add up to suggest that under the conditions of these experiments, PS adsorbed weakly both in quantity and energetics. Bridges were not formed due to the sparse but nonetheless osmotically repulsive barrier. Why one group sees weak adsorption while others none at all has not been established with certainty. It could be that the system is right on the edge between adsorption and dissolution; subtle changes in impurity levels or dryness of the solvent may tip the balance one way or the other, similar to the sensitivity of the PS-CP system near θ to isomeric content of the solvent.

FORCES BETWEEN ADSORBED LAYERS OF POLYMERIC AMPHIPHILES

General Considerations of Polymeric Amphiphiles

Homopolymers have one characteristic segmental interaction with the surface and one with the solvent. Multicomponent macromolecules introduce the possibility of more than one characteristic surface and solvent interaction per chain. Amphiphilic polymers are those copolymers in which the segments with different characteristics are segregated into blocks along the chain. Gast has recently reviewed some of the interesting physical chemistry of block copolymers (100). Polymeric amphiphiles share with small molecule surfactants the tendency to straddle phase boundaries owing to the selectivities of interactions built into the macromolecules. They create phase boundaries by self-assembly into micelles in selective solvents and microstructured morphologies in the bulk melt state.

From the point of view of adsorption, the most interesting situation is selective adsorption from selective solvents, where one block of segments adsorbs strongly and the other is strongly solvated and adsorbs negligibly. In the extreme case, the adsorbing part is more strongly attracted to the surface than envisioned by the weak adsorption limit. This may produce in some situations a flattened conformation of the adsorbing block on the surface, while the nonadsorbing segments extend into solution. We refer to this as ideal amphiphile behavior. Most of the discussion here will be directed to two-part amphiphiles, called *diblock copolymers*. Chains in which one block is made very small, providing an isolated "sticky" point on the chain, for example at the end, are considered as part of this class. Part of the long-term interest in block copolymers consists in the many possibilities beyond simple, two-part amphiphiles, for example, in multi- and nonlinear block copolymers.

Homopolymers adsorb onto surfaces in the form of trains, loops, and tails, since any segment can potentially adsorb. In ideal amphiphile behavior, all the nonadsorbing segments extend from the surface in cilia. The adsorbing and nonadsorbing segments work in concert to set the adsorbed amount. The former compete for space on the surface; the latter crowd one another in the swelling solvent near the surface. Under different circumstances one part or another may dominate the competition. De Gennes (41) has coined the terms anchor-dominated and buoy-dominated for the two regimes.

The first experiments on force measurement between adsorbed polymer amphiphiles were done with the aim of understanding better the results on adsorbed homopolymers. As described above, many polymers do not adsorb significantly, at least on relatively passive surfaces like mica, from good solvents. In poorer solvents, significant adsorption occurs but the resulting attraction contains contributions from bridging and osmotic effects that are difficult to assess independently. Block copolymers afford the opportunity to build the stronger adsorption characteristic of poor solvents into one part of the molecule and use that to fasten another part of the molecule, with good solvent interactions, to the surface. Furthermore, if the adsorbing blocks cover each surface densely, it will be difficult for the strongly solvated blocks to reach the other bare surface. That, plus their negligible relative affinity for the surface over the solvent, should effectively eliminate the possibility for bridging in this situation. One could then attempt to interpret the resulting forces in terms of the osmotic interactions between the solvated chains in the region between the surfaces. Hadziioannou et al (101) showed that block copolymers of poly(2-vinyl pyridine) and polystyrene (PVP-PS) adsorbed in a way resembling this ideal amphiphile behavior. In toluene, the PS is strongly solvated and the PVP very poorly solvated and strongly adsorbed. Watanabe et al (99) showed that the weak adsorption of PS from toluene, discussed above, played no role in the block copolymer adsorption. PVP-PS adsorption was completely insensitive to the presence of PS in the solution or to PS preadsorbed on the surface. The copolymer rapidly displaced PS and formed layers that were indistinguishable from those adsorbed from solutions containing only copolymer. The data on two different molecular weights of PVP-PS are included in the dimensionless plot, along with other more recent data on polymeric amphiphiles, in Figure 4.

Without yet explaining the particular scaling used in Figure 4, it is clear that bringing the solvated cilia of these polymeric chains together in a good solvent produces a monotonic repulsive force, as is qualitatively to be expected for forces arising from osmotic pressure alone. Hadziioannou et al (101) went on to show that if these layers were bathed in cyclohexane at 38°C (after original adsorption from toluene), the forces were shorter ranged but still monotonically repulsive, in marked contrast to the behavior represented for PS in CH in Figure 2. One interpretation of this contrast is that bridging has been eliminated in the amphiphile system; shorter ranged repulsion relative to toluene is generated by the weaker osmotic pressure near θ. Attractive forces were seen between the amphiphile layers in CH at 21°C, with an attractive minimum of $<100\ \mu N/m$, substantially less than is seen in Figure 1, again pointing to a dominant contribution of bridging in the case of an adsorbed homopolymer near the θ point.

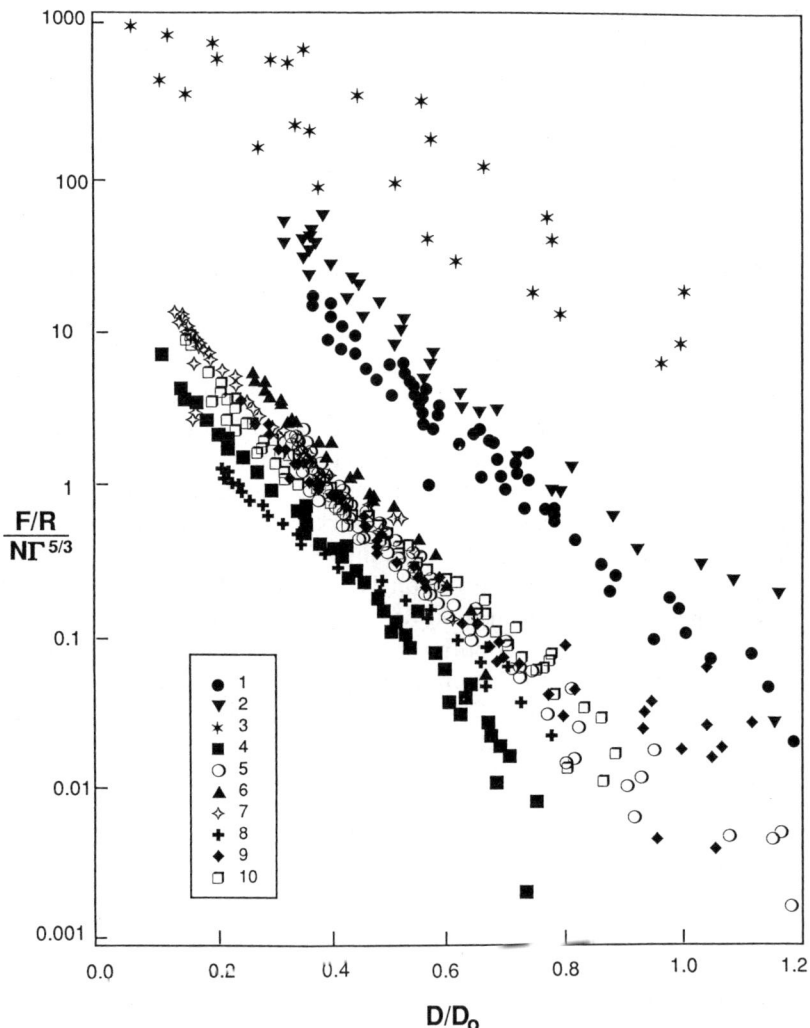

Figure 4 Normalized force vs. distance data for polystyrene-containing polymer amphiphiles immersed in toluene. Data are normalized as described in the text. Sample designations and other information are given in Table 3.

Comparison of Data

Table 3 and Figure 4 give the data on all the polymer systems resembling ideal amphiphiles on which forces between adsorbed layers have been made (98, 101–105). All these systems exhibit the long-range, monotonic

Table 3 Force measurements on adsorbed polymer amphiphiles. Data on samples in Figure 4 (Solvent in all cases is toluene)

Sample no.	Ref.	System[a]	Mw/Mn	D_0 (nm) (see text)	Γ (mg/m^2)[b]
1	(102)	PVP-PtbS 2–31	1.5	49	2.87
2	(102)	PVP-PtbS 6–15	2.2	32	2.45
3	(102)	PVP-PtbS 2–5	2.3	12	1.17
4	(103)	PEO-PS 3–181	1.1	200	2.87
5	(104)	PEO-PS 18–90	1.4	90	0.65
6	(104)	PEO-PS 8–92	1.4	110	1.02
7	(104	PEO-PS 20–25	1.4	243	1.50
8	(98)	X-PS[c] x–141	1.02	144	2.33
9	(101)[d]	PVP-PS 60–60	1.1	65	1.63
10	(101)	PVP-PS 60–95	1.1	98	1.80

[a] Symbols for sample designations are given in text. First is adsorbing block; second is solvated block. Numbers under designation are molecular weights of respective blocks in thousands.

[b] These are values estimated by the procedure given in text. No Γ values have been reported for most of the tabulated force measurements; in such cases, where Γ has been reported or estimated, these values fall within the range of experimental error.

[c] X refers to a monomeric, zwitterionic sticking group attached to the end of the PS chain.

[d] Data reported in this reference for 60–150 polymer showed some hysteresis at high compression, but never showed hysteresis for maximum compression < 5000 μN/m. To assure no nonequilibrium effects in the data of Figure 4, data on this polymer have been omitted. They do, however, fit with the other data on PVP-PS in Figure 4.

repulsion described above. An important feature of the dimensionless representation introduced in Figure 4 is that the separation between the surfaces is scaled by a parameter D_0, which is proportional to $L_c = Na$, the total contour length of the chain. This parameter is discussed more fully below, but it is clear that this scaling is very different from the r_g scaling of Figures 1 and 2.

This scaling, plus a quantitative comparison of the ranges of the forces with the characteristic length scales, r_g and L_c, suggests that the non-adsorbing blocks of these polymers have adopted a conformation that is considerably extended from the random coil dimensions characteristic in solution or in adsorbed homopolymer layers. All of the data [with the exception of some of the data on PVP-poly(*t*-butylstyrene) (PtbS) (102), to be discussed in more detail] conform to this scaling. The stretching implied by this scaling introduces another element into the interpretation of these $F(D)$ curves that is not present in the homopolymer data. The energetics of stretching the distribution of polymer conformation must be factored into the energetics of the $F(D)/R$ interaction.

The scaling in Figure 4 comes out of recent theoretical considerations of the roles of osmotic pressure and stretching in the assembly of adsorbed layers of polymer amphiphiles. The general idea of the factors at play in the interpretation of these data can be gained from the model proposed by Alexander (106) for the structure of layers of terminally grafted chains. De Gennes later (107) extended the analysis to study the interaction of a grafted layer with polymer in solution and with another grafted layer. Application of this model to data on block copolymers implies that the anchoring block can be viewed as the entity that grafts the solvated block to the surface, but it exerts no other direct influence on the range profile of segments extending away from the surface. This keeps the analysis simple for the time being; we elaborate subsequently. In particular, we present the results using simplified Flory arguments (72) for the interactions. Alexander's original work included scaling treatments of the solution interaction (106).

The controlling parameter that produces the interesting behavior characteristic of these grafted chain systems, which have come to be called *polymer brushes*, is the density of grafting of chains to the surface. In dimensionless terms, this density will be expressed as $\sigma = a^2/\delta^2$, where δ is the average spacing between graft points on the surface. For the time being we take this graft density to be given; in fact σ is proportional to Γ, and Γ is related to the adsorption affinity of the adsorbing block and the interactions among the chains on the surface. If $\delta < r_g$, or equivalently, if $\sigma > N^{-6/5}$, each cilium on the surface will begin to develop significant overlap with its neighbors. In a good solvent environment, this will, in turn, develop osmotic pressure in the brush. One means for the brush to reduce its osmotic pressure is to swell into the ocean of pure solvent with which it is in contact. This swelling involves the stretching of chain conformations. Alexander (106) and de Gennes (107) balanced the energetics of osmotic pressure and elastic resistance to stretching by using the simplest conceivable representations of each effect. The free energy per chain in a high density brush was written:

$$E_{\text{chain}}/kT \approx vN(Na\sigma/L_0) + L_0^2/Na^2 \qquad 7.$$

where L_0 is the equilibrium layer thickness to be determined. The first term is the excluded volume parameter (measuring effects of binary segment contacts) times the number of segments per chain times the mean volume fraction of segments averaged over the entire layer; the second term is the energetics of a Hookean spring distorted from its resting mean extension (Eq. 2). Averaging the segment volume fraction over the entire layer is equivalent to assuming that the concentration is constant in the layer, so this approach has sometimes been called the *step-function* model. This is clearly the coarsest mean field approximation. Minimization of the free energy of Eq. 7 gives:

$$L_0 \approx Na(\sigma v)^{1/3} \approx L_c(\sigma v)^{1/3}. \qquad 8.$$

For the data on PVP-PS it has been established that for constant σ, the experimentally determined range of the $F(D)$ profiles, which is the D_0 parameter used to scale separation in Figure 4, varies as L_c and therefore as L_0. From Eqs. 7 and 8, we see that E_{chain} varies as $N\sigma^{2/3}$. The energy of a layer of density σ, E_{layer} will be $\approx \sigma E_{chain} \sim N\sigma^{5/3}$.

These ideas have been extended by Patel et al (108) to calculation of the $F(D)$ profile with the use of one additional simplifying idea, the noninterpenetration hypothesis. If, on initial overlap, the two brushes compress but do not interpenetrate, then Eq. 7 can be converted directly to an equation for the energy of two brushes interacting at a separation D by substituting $D/2$ for L_0. The result for the energy of two interacting layers can be written:

$$E_{\text{two layers}}/kT \approx N\sigma^{5/3} f(D/L_0) \qquad 9.$$

where $f(D/L_0)$ is a dimensionless function of scaled distance. Since F/R is proportional to the energy per unit area of interaction via the Derjaguin approximation, we expect F/R to be proportional to $E_{\text{two layers}}$, so that Eq. 9 rationalizes the scaling of the force axis in Figure 4.

An argument for the noninterpenetration hypothesis is that the stretching occurred in the single layers because of the ability of the layers to reduce their osmotic pressure by swelling into the pure solvent bath. Juxtaposition of another layer removes this incentive, so the layers compress themselves to reduce their stretching energy. Whether the noninterpenetration condition is strictly obeyed is arguable. It is clear, however, that this line of argument says that the force measured at some D will be less than that from osmotic pressure alone, due to the shrinking of extended chains, until a high enough compression is reached, so that the chains are no longer stretched.

These ideas have been pursued by using more rigorous self-consistent mean field theory. Hirz (109) and Cosgrove and co-workers (110) have used the Scheutjens-Fleer lattice model to calculate segment profiles for one and two layers. In this calculation, at high enough σ, one finds the scaling of the mean layer thickness suggested by Eq. 8; however, these calculations also showed that the profile of nonadsorbing segments is not step-like but is clearly more gradual at the periphery of the layer. Milner and co-workers (111) explained this puzzling observation with a clever, elegant, analytical SCF theory that lent itself to more physical interpretation. They asked, "What potential (to be determined self-consistently from the segment density profile) will permit an ensemble of walks (the chain contours) to begin (the free ends of the chains) anywhere within the layer but satisfy the condition that every walk makes it to a certain point (the grafting surface) in an equal number of steps (monodisperse chains)?" They recognized that a harmonic potential has this property; the period of an oscillator in a harmonic field is not dependent on the amplitude of oscillation. The harmonic potential for the distribution of segments near the surface corresponds to a parabolic segment density profile, in contrast to the step profile of the Alexander-de Gennes approach.

Milner and co-workers (MWC) carried through much more detailed analysis based on their recognition that the polymer brush was an example of the "strongly stretched" limit of the SCF theory, first discussed by Semenov (112). This strong stretching limit in SCF theory corresponds mathematically, and physically, to the classical limit of the Schrödinger equation. An important conclusion of the MWC calculation is that the simpler Alexander-de Gennes analysis based on the step profile gives the right scaling relationships. In particular, Eqs. 8 and 9 are retained with different numerical prefactors, and with different functional forms for the dimensionless force distance relationship, $f(D/L_0)$.

This analysis of polymer brushes has enabled investigation of a number of factors that are beyond the reach of other SCF calculations. It has been shown that the free ends of the chains are distributed throughout the brush, with a maximum in the probability of finding an end at about $0.7 L_0$, in contrast to the step profile, where all the ends are located at the outside edge (L_0) of the layer. The calculated profiles correspond well to those computed by Hirz (109) under similar conditions. Their results produce an argument in favor of the noninterpenetration hypothesis, advanced above, for the symmetric interaction between identical layers. The self-consistent profile in these calculations has a symmetric maximum between the plates, which in turn produces a density minimum midway between the two plates. These results have been generalized to layers that are polydisperse in chain length (113); effects of nonuniform chain length are most prominent at the periphery of the layer and in the weak compression regime of the $F(D)$ data, causing the force to rise more gradually than in monodisperse layers. The polydisperse brush is predicted to be taller than the monodisperse brush by an amount of order $(M_w/M_n - 1)^{1/2}$. The MWC model has also been applied to calculation of the bending moduli of polymer brushes grafted to liquid interfaces (114), as in polymeric microemulsions.

Figure 4 shows that all the data published so far on polymeric amphiphiles, except for the data on PVP-PtbS, can be unified into a narrow band by using the scaling of the force axis suggested by Patel et al (108) and by the MWC model. This is especially true of the data on PEO-PS and PVP-PS, samples 5–7, 9, 10. The data on PVP-PtbS (samples 1–3) were obtained on samples of comparatively high polydispersity and so may not conform to this scaling for that reason. Furthermore, we have made no allowance for the fact that the interaction parameter v in toluene may be different between PS and PtbS. In order to assess the importance of conclusions to be drawn from Figure 4, it must be explained how the values of Γ were deduced, since precise values of Γ have not generally been published with the $F(D)$ data.

Equations 8 and 9 illustrate that with molecular weight known, the surface density σ, which is directly proportional to Γ, is directly related to both L_0 and to the $F(D)$ profile. We have taken the MWC model as the basis in our construction of Figure 4. When Eq. 9 is transformed into an equation for $F(D)/R$, the result from the MWC model is:

$$F/RkT = CN(L_0/L_c)^2\sigma[(5/9)(1/u+u^2-u^5/5)-1] \qquad 10.$$

where the function in square brackets is the specific form of the dimensionless function $f(D/L_0)$ resulting from the MWC model ($u = D/L_0$). Notice that since $L_0 \sim \sigma^{1/3}$, this equation corresponds to the scaling of Figure 4. The constant C in Eq. 10 depends on the type of polymer (through v); however, it should be the same for all the amphiphiles with nonadsorbing PS blocks (samples 4–10 in Table 3 and Figure 4). The differences in these scaled profiles for PtbS and PS amphiphiles (Figure 4) may be, at least in part, due to a difference between C_{PtbS} and C_{PS}. The procedure followed to ascertain consistent values of Γ with which to scale the various data was to begin by picking out from the data an experimental estimate of the range of the forces, D_0. Our first estimate of this was the point at which the measured forces rose to about 25 μN/m, a working estimate of the minimum force distinguishable from zero in the apparatus. The ratio L_0/L_c was then calculated from $L_0/L_c = D_0/2Na$. Γ was then calculated from the MWC version of Eq. 8. The data were then plotted as $F/RNkT$ vs. $[f(u)]$, as suggested by Eq. 10. The slope of this plot gives a second, independent estimate of σ and therefore of Γ. If this value of Γ was different from that coming from the estimate of the experimental range of the forces, D_0 was adjusted to achieve consistency. Corrections for polydispersity were made according to the MWC model; these were very substantial for the PVP-PtbS data, and only for these data were the final D_0 values used for scaling very different from reasonable experimental estimates from the ranges of the forces. In defense of our decision to adjust the latter, we feel that the value of Γ obtained from the profile itself should be more reliable than that from the range.

It seems to us that a fair conclusion on examining the ensemble of data plotted in Figure 4 is that they conform well to the relationships among the parameters predicted by the MWC version (111) of SCF theory, and furthermore, at this level of discussion, the scaling behavior is quite well represented by the cruder arguments advanced by Alexander (106), by de Gennes (107), and later by Patel et al (108). This is useful information to have in order to assess the reliability of similar arguments applied to different systems involving grafted chains, such as star polymers, polymer micelles and microemulsions, and polymers made from macromonomers. Physically, the reason that mean field theory works so well, even in its crudest form, is that the stretching of the chains in a polymer brush damps the fluctuations that require corrections to mean field.

The degree to which the functional form, $f(u)$, predicted by the MWC theory fits the data quantitatively is a separate question. Milner (115) has explored this in a detailed comparison of his model with the data of Taunton et al (98) on X-PS (PS end-functionalized with a zwitterionic group: $-N^+(CH_3)_2(CH_2)_3SO_3^-$) adsorbed on mica from toluene (sample 8 in Figure 4). Taunton et al (98) reported Γ for this polymer in the range 3 ± 0.5 mg/m^2. Using $\Gamma = 3.5$ mg/m^2 (at the high end of this range, and somewhat higher than the estimated adsorbance we deduced by the means above, reported in Table 3), a value of v deduced from independent data on osmotic pressure in PS-toluene, and independent data on r_g to get a, Milner (115) found very good agreement between his model and the $F(D)$ data. A small, but nonnegligible, correction for the very small polydispersity was used to fit the longest range part of the data. The agreement of the model with data reinforces the qualitative success of the model in predicting the right scaling among the variables. This X-PS system is arguably the best system for testing these theories since Taunton et al (98) proved via control experiments that this polymer adsorbs under their conditions only by its functionalized end. It remains to be seen, via examination of a series of molecular weights, whether this polymer conforms to the linear scaling of the range of the forces with molecular weight, Eq. 8, embodied in this model. This linearity in N of the PS block has been reported for the PVP-PS copolymers (108). It may be that in practice diblock polymers are better embodiments of the idea of polymer brushes, since they can have more tenacious anchors.

In Figure 4, it is possible to bring all the PS data reported there into very close accord in the scaled coordinates by adjusting the Γ values without going far outside the reported range of errors. This means that to a fair degree of accuracy, the MWC model gives a reasonable quantitative representation of all the available data on forces between layers of adsorbed polymeric amphiphiles (though it does seem to require surface densities at the high end of what has been reported). Milner (115) also compared the data with the $F(D)$ predictions of the step function model. The MWC model is clearly quantitatively superior in the weak compression, long-range part of the data. The two models become indistinguishable in the high compression regime, since as the chains become squeezed, the stretching term is reduced and the situation is dominated by the osmotic pressure of the segments in the gap, which is handled identically in the two models. An important open question is, "To what extent are the parabolic profile and proper scaling excluded volume exponents necessary for a quantitative fit to the data?"

Another major question remaining in this area does not relate directly to force measurement as such, but does relate to what determines Γ, and therefore the effective graft density σ, in these polymeric amphiphiles. Clearly, when layers such as these self-assemble from solution, the resultant σ is controlled by interactions among the polymers, the surface, and the solvent. The data are few so we cannot review the situation definitively; we can sketch possibilities. There are several competing theories of adsorption of polymeric amphiphiles (116–118).

For strong adsorption and poor solvation of one block, combined with high molecular weight of this block to multiply the surface interaction, we expect an anchor-dominated regime. That means that σ will be dictated by the anchoring block and will be insensitive to the molecular weight of the solvated block. This appears to be the situation obtaining in the series of PVP-PS polymers reported by Tirrell et al (105). Constant molecular weight of the PVP produced constant σ as the molecular weight of the PS was increased. Within the anchor-dominated regime there is the question of how the adsorbing blocks interact on the surface to dictate σ. Hadziioannou et al (101), suggested a "pancake" model in which the adsorbing block covers an amount of surface corresponding to the projected area of the poorly solvated coil of the adsorbing block in solution, estimated from Eq. 6. This leads directly to $\sigma = N_{\text{anc}}^{-2/3}$, the graft density diminishing as the amount of surface per adsorbing block increases. Marques et al (117) have suggested that the anchoring block will cover the surface in a thin (possibly slightly swollen) melt layer whose thickness is determined by the balance between van der Waals attraction to and spreading on the surface by the anchoring block. This leads to $\sigma \sim N_{\text{anc}}^{-1}$. They also envisioned an anchor-dominated regime in which the anchor blocks were adsorbed strongly enough to dominate, but still in the weak adsorption limit, so that there might be loops, trains, and tails in the anchoring block (118). In fact, the PEO-PS samples of Figure 4 may be in this limit. All the PEO blocks studied are quite small compared to the PS buoys, so any effects of these loops, such as the hysteresis seen in homo-PEO, may be negligible in these amphiphiles.

In the buoy-dominated regime, the cumulative adsorption energy in the anchors is not enough to overwhelm completely the osmotic interaction among the buoys as they attempt to crowd into the brush. In this regime, σ may depend still on N_{anc} but will also depend somehow inversely on N_{buoy}. Direct surface force measurements, in conjunction with the more traditional tools of polymer adsorption, such as ellipsometry, radiolabeling, and spectroscopy, may be useful in sorting out these dependencies. Other interesting issues remaining in the field of polymer amphiphiles include interaction between dissimilar layers and the study of more elaborate self-assembled layers, such as from molecular weight or chemical mixtures, or multicomponent block copolymers. The detailed knowledge of the arrangement of segments in these layers, from direct force measurement combined with theory, should be put to work in clever means of using these layers to deliver chemistry between a surface and its environment.

RECENT EXPLORATORY WORK IN DIRECT FORCE MEASUREMENT APPLIED TO POLYMERS

Whereas the research on adsorbed polymers described in the first part of this review has reached a stage at which it is amenable to quantitative theory and detailed interpretation, there is a growing body of work in less well-known territory. We discuss such work in this concluding section as a means of bringing the farthest reaching frontiers some attention. A unified synthesis is more difficult to achieve in the presentation of these results.

Polyelectrolytes

All work on adsorbed polyelectrolytes has been done in aqueous media, so it is necessary to understand the physical chemistry of the mica surface in aqueous electrolyte (49–51) to interpret the data on polyelectrolytes. The principal feature for this discussion is that in water, potassium ion is dissolved from the mica surface and, while other ions may adsorb, the surface takes on a negative charge typically in the range of one charge per 10 to 100 nm^2. This in turn means, inter alia, that cationic polyelectrolytes adsorb more readily than anionic ones. Luckham & Klein (119), and later

Dix et al (120), have studied poly-L-lysine adsorbed on mica from varying ionic strengths of KNO_3. They found monotonically increasing, long-range repulsion of first approach (compression) of the layers. On decompression, and subsequent compression cycles, they found no such long-range repulsion, but rather a steeper, shorter range repulsion that appeared to be reversible on the timescales examined. Their interpretation was that the strong electrostatic attraction between segments and surface fixes a loose, "loopy" structure (extending to some $15\, r_g$) at first. On compression, more segments are brought into contact with the surface and stick strongly, fixing the crushed conformation. This is a more irreversible version of the ideas invoked to explain the hysteresis of the PEO data (65, 91). A theory of nonequilibrium polymer adsorption has been motivated by these results (96, 121).

Marra & Hair (122) looked at forces between PVP layers adsorbed on mica in acid solutions (where PVP is charged) and alkaline solutions where PVP is neutral. They found that the charged polymer changed the electrolytic environment near the surface, and therefore the $F(D)$ profiles, but they saw none of the prominent hysteresis effects that occur with polylysine. Marra & Hair (123) have also studied forces between mica surfaces in the presence of poly(styrene sulfonate) (PSS), an anionic polyelectrolyte. In this case they found no adsorption of the polymer but a strong modification of the electrolytic environment in the gap, which they attribute to counterion condensation on the PSS backbone. Charged polyelectrolytes can adsorb on similarly charged surfaces if the nonelectrostatic attraction is strong enough or if the electrostatic repulsion can be diminished, such as by raising the ionic strength (43). Both the lattice version (124) and the continuous version (125) of SCF calculations have been applied to model adsorbed polyelectrolytes. The latter group has developed the analog of the MWC model for grafted polyelectrolytes. These mean field theories may encounter difficulties at low ionic strength where the electrostatic interactions may introduce correlations.

Proteins and Biological Polymers

Interactions between proteins attached to surfaces and membranes play fundamental roles in many processes of biological and medical significance. Following early work by Klein (56), Luckham and co-workers (126, 127) have taken the lead in applications of direct force measurements applied to biological systems. They have examined proteins and model polypeptides, including polylysine, cytochrome C, concavalin A, and myelin basic protein. Each of these systems produces forces that depend on its detailed structures and on the importance, in each, of electrostatic and hydrophobic forces. Lee & Belfort (128) have recently completed a study of the enzyme ribonuclease adsorbed on mica, where surface force measurements have been used to determine that the elliptically shaped protein adsorbs with its long axis parallel to the surface at low Γ, then rotates progressively to an end-on posture as Γ increases. Although these results are difficult to generalize at present, direct force measurements have great potential for exploring specific biological interactions via these techniques if a suitable surface on the mica substrate required for the optics of the Israelachvili type device can be recreated.

Polymer Melts: Equilibrium and Dynamic Measurements

Three studies of interactions between mica surfaces immersed in polymer melts have been published recently (129–131). These studies examined polybutadienes (PB) of low molecular weight (~ 1000 and ~ 3500) (129), a perfluorinated polyether (FPE) of several low molecular weights (130), and a commercial polydimethylsiloxane (PDMS) (131). Only the first study used monodisperse polymer. The typical protocol for these experiments was to trap a droplet between the surfaces rather than to immerse the two surfaces in a sea of viscous polymer.

In each of the three studies of polymer melts, a monotonic repulsive force extending to about ten times the nominal r_g of the polymer was observed. Horn & Israelachvili (131) also observed structural forces at $D \ll r_g$. Such ranges of forces are comparable to those described above for adsorbed layers in solvents. Monte Carlo simulations (132, 133) and theory show that no long-range equilibrium forces should exist unless the polymers are pinned to the wall (134) on the timescale of the measurements, and even then they should produce forces extending no more than about $2\, r_g$.

Horn and co-workers (135) have recently resolved this puzzle via a more careful examination of viscous relaxation effects. They have shown that the previous results on polymer melts could be explained based on the time it takes for a viscous liquid to drain from the gap on squeezing down to a few molecular diameters. Their realization was based on the analysis of draining from small gaps by Chan & Horn (136) and on new measurements of a monodisperse sample of PDMS. At very small separations, they still find evidence of structural forces and an apparently higher viscosity of the fluid in the layers directly adjacent to the wall, which suggests that some molecules may be temporarily "pinned" at the surface, perhaps by dynamic effects such as entanglement with other molecules.

The analysis of Chan & Horn (136), on which this investigation was based, was also one of the starting points for a growing effort to use the direct force measurement equipment to measure dynamic properties of fluids between solid surfaces spaced at molecular distances. Chan & Horn proposed doing this via a drainage experiment in which the two surfaces were continuously driven toward one another and the dynamic properties of the fluid determined from the lag required for the spring-mounted driven surface to catch up with the rate of driving. Nearly simultaneously, Israelachvili (137) showed that the two surfaces could be driven in an oscillatory fashion, normal to one another. Each of these techniques was applied to the measurement of the effective viscosities of small molecule fluids in gaps of molecular dimensions. They each found that the effective viscosity was nearly equal to that in bulk, except for a thin layer near the wall of apparently higher viscosity. Israelachvili (138) used the oscillatory method to measure the effective viscosity with an adsorbed PS layer in CH. From these measurements he calculated a hydrodynamic layer thickness that agreed well with the equilibrium layer thickness from the equilibrium force measurements. Christenson & Israelachvili (139) applied the same method to measure the viscosity of very thin films of crude oils in an application to petroleum production. Interpretation of these viscosity results is complicated somewhat by contributions to the effective viscosity from fluid in regions of separation beyond the point of closest approach (140).

Most recently, methods have been developed to shear (move the mica surfaces parallel to one another) with the mica surfaces at molecular separations (25, 27, 141). Here the results are very preliminary but seem to indicate discontinuous variation of the effective viscosity with D as the mica surfaces approach. Bitsanis et al (142), have studied this situation via nonequilibrium molecular dynamics and molecular theory to show that at molecular scale values of D, the effective viscosities from squeezing and shearing experiments should be very different. The two measurements average the total resistance to movement, which is what is measured in these experiments, very differently over the strongly inhomogeneous density profile in the narrow gap.

CONCLUSIONS

We have tried to make this review as timely as possible given the acceleration in rate of progress in this area. The field of measurements on adsorbed polymer layers is maturing to the point that detailed molecular theories can be used to model and interpret the results. Many other areas of exploration are plainly in very preliminary stages. Exciting developments with this technique are in the offing as surfaces other than mica come into play (33–37) [see also recent work with sapphire single crystal platelets (143)] and as new developments in the instrument are made (144). Direct force measurement constitutes an important complement to near-field microscopy techniques in the exploration of surfaces and provides a unique avenue to explore molecular level interactions at surfaces.

ACKNOWLEDGMENT

The authors acknowledge with gratitude the assistance, collaboration, and discussions we have had with the pioneers of modern methods of direct force measurement applied to fluids: Jacob Israelachvili, Roger Horn, and Jacob Klein. We have had many useful discussions concerning the theoretical interpretations of these results with Pierre-Gilles de Gennes, Barry Ninham, Phil Pincus, Avi Halperin, Scott Milner, and Tom Witten. Our co-workers, especially Georges Hadziioannou, Hiroshi Watanabe, Steve Granick, Ed Parsonage, Suzan Hirz, and Yen-Lane Chen have been very important in building our understanding of these phenomena through their experimental work and insights. Financial support for this work has been generously supplied to the University of Minnesota by the National Science Foundation (Polymers Program DMR-8115733 and PYI Program CBT-8352364), the DuPont Marshall Laboratories, the 3M Company, and the Shell Companies Foundation.

Literature Cited

1. Flory, P. J. 1953. *Principles of Polymer Chemistry*. Ithaca: Cornell Univ. Press; 1969. *Statistical Mechanics of Chain Molecules*. New York: Interscience
2. Ferry, J. D. 1980. *Viscoelastic Properties of Polymers*. New York: Wiley. 3rd ed.
3. Israelachvili, J. N. 1985. *Intermolecular and Surface Forces*. London: Wiley
4. Alexander, B. M., Prieve, D. C. 1987. *Langmuir* 3: 788–95
5. Hamola, A., Robertson, A. A. 1976. *J. Colloid Interface Sci.* 34: 286–97
6. Rau, D. R., Lee, B. K., Parsegian, V. A. 1984. *Proc. Natl. Acad. Sci. USA* 81: 2621–25
7. Viani, B. E., Low, P. F., Roth, C. B. 1983. *J. Colloid Interface Sci.* 96: 229–44
8. Lyklema, J., van Vliet, T. 1978. *Faraday Discuss. Chem. Soc.* 65: 25–32
9. Binnig, G., Rohrer, H. 1982. *Helv. Phys. Acta* 55: 726–35
10. Binnig, G., Quate, C. F., Gerber, C. 1986. *Phys. Rev. Lett.* 56: 930–33
11. Wu, S. 1982. *Polymer Interface and Adhesion*. New York: Marcel Dekker
12. Feast, W. J., Munro, H. S., eds. 1987. *Polymer Surfaces and Interfaces*. New York: Wiley
13. Lodge, K. B. 1983. *Adv. Colloid Interface Sci.* 19: 27–73
14. Tomlinson, G. A. 1928. *Philos. Mag.* 6: 695–712
15. Bradley, R. S. 1932. *Philos. Mag.* 13: 853–62
16. Overbeek, J. T. G., Sparnaay, M. J. 1954. *Discuss. Faraday Soc.* 18: 12–24
17. Derjaguin, B. V., Titijevskaia, A. S., Abrikossova, I. I., Malkina, A. D. 1954. *Discuss. Faraday Soc.* 18: 24–41
18. Tabor, D., Winterton, R. H. S. 1969. *Proc. R. Soc. London Ser. A* 312: 435–50
19. Derjaguin, B. V. 1987. *Langmuir* 3: 601–6
20. Tirrell, M. 1988. *Opportunities and Research Needs in Adhesion Science and Technology*, ed. G. G. Fuller, K. L. Mittal, 1: 23–42. St. Paul: Hitex
21. Israelachvili, J. N., Tabor, D. 1972. *Proc. R. Soc. London Ser. A* 331: 19–38
22. Israelachvili, J. N., Adams, G. E. 1978. *J. Chem. Soc. Faraday Trans. I* 74: 975–99
23. Israelachvili, J. N. 1987. *Acc. Chem. Res.* 20: 415–21
24. Israelachvili, J. N. 1987. *Proc. Natl. Acad. Sci. USA* 84: 4722–24

25. Israelachvili, J. N., McGuiggan, P. M., Homola, A. M. 1988. *Science* 240: 189–91
26. Peterson, I. 1988. *Sci. News* 133: 283
27. van Alsten, J., Granick, S. 1988. *Phys. Rev. Lett.* 61: 2570–73
28. Tonck, A., Georges, J. M., Loubet, J. L. 1988. *J. Colloid Interface Sci.* 126: 150–63
29. Götze, T., Sonntag, H. 1988. *Colloids Surf.* 31: 181–201
30. Belouschek, P., Maier, S. 1986. *Progr. Colloid Polymer Sci.* 72: 43–50
31. Derjaguin, B. V. 1934. *Kolloid. Zh.* 69: 155–64
32. Israelachvili, J. N. 1973. *J. Colloid Interface Sci.* 44: 259–72
33. Maeda, M., White, H. S., McClure, D. J. 1986. *J. Electroanal. Chem. Interfacial Electrochem.* 200: 383–87
34. Smith, C. P., Maeda, M., Atanasoska, Lj., White, H. S., McClure, D. J. 1988. *J. Phys. Chem.* 92: 199–205
35. Fan, F. F., Bard, A. J. 1987. *J. Am. Chem. Soc.* 109: 6262–68
36. Parker, J. L., Christenson, H. K. 1988. *J. Chem. Phys.* 88: 8013–14
37. Coakley, C. J., Tabor, D. 1978. *J. Phys. D* 11: L77–82
38. Israelachvili, J. N., Marra, J. 1986. *J. Methods Enzymol.* 127: 353–60
39. Ploehn, H. J., Russel, W. B. 1989. *Adv. Chem. Eng.* In press
40. Silberberg, A. 1985. *Encyclopedia of Polymer Science and Engineering*, 1: 577–94. New York: Wiley. 2nd ed.
41. de Gennes, P.-G. 1987. *Adv. Colloid Interface Sci.* 27: 189–209
42. Takahashi, A., Kawaguchi, M. 1982. *Adv. Polym. Sci.* 46: 1–66
43. Fleer, G. J. 1988. *Surfactants Sci. Ser.* 27: 105–58
44. Cohen-Stuart, M. A., Cosgrove, T., Vincent, B. 1986. *Adv. Colloid Interface Sci.* 24: 143–239
45. Horn, R. G., Israelachvili, J. N. 1981. *J. Chem. Phys.* 75: 1400–11
46. Christenson, H. K., Horn, R. G. 1985. *Chem. Scr.* 25: 37–41
47. Christenson, H. K., Israelachvili, J. N. 1984. *J. Chem. Phys.* 80: 4566–67
48. Christenson, H. K., Blom, C. E. 1987. *J. Chem. Phys.* 86: 419–24
49. Israelachvili, J. N. 1985. *Chem. Scr.* 25: 7–14
50. Pashley, R. M., Israelachvili, J. N. 1984. *J. Colloid Interface Sci.* 101: 511–23
51. Pashley, R. M. 1981. *J. Colloid Interface Sci.* 80: 153–62; 83: 531–46
52. de Gennes, P.-G. 1981. *Macromolecules* 14: 1637–44
53. Pefferkorn, E., Carroy, A., Varoqui, R. 1985. *Macromolecules* 18: 2252–58; 1986. *Macromolecules* 19: 944
54. Yamakawa, H. 1971. *Modern Theory of Polymer Solutions*. New York: Harper & Row
55. Klein, J. 1980. *Nature* 288: 248–50
56. Klein, J. 1983. *J. Chem. Soc. Faraday Trans. 1* 79: 99–118
57. Israelachvili, J. N., Tirrell, M., Klein, J., Almog, Y. 1984. *Macromolecules* 17: 204–9
58. Granick, S., Patel, S., Tirrell, M. 1986. *J. Chem. Phys.* 85: 5370–71
59. Scheutjens, J. M. H. M., Fleer, G. J. 1979. *J. Phys. Chem.* 83: 1619–35; 1980. *J. Phys. Chem.* 84: 178–90
60. Fleer, G. J., Scheutjens, J. M. H. M. 1982. *Adv. Colloid Interface Sci.* 16: 341–57
61. Klein, J., Luckham, P. F. 1986. *Macromolecules* 19: 2007–10
62. Klein, J. 1988. *Studies in Polymer Science*, ed. M. Nagasawa, 2: 333–52. Amsterdam: Elsevier
63. Napper, D. H. 1983. *Polymeric Stabilization of Colloidal Dispersions*. London: Academic
64. Almog, Y., Klein, J. 1985. *J. Colloid Interface Sci.* 1985: 33–44
65. Luckham, P. F., Klein, J. 1985. *Macromolecules* 18: 721–28
66. Schmidt, M., Burchard, W. 1981. *Macromolecules* 14: 210–11
67. Klein, J., Pincus, P. 1982. *Macromolecules* 15: 1129–35
68. Ingersent, K., Klein, J., Pincus, P. 1986. *Macromolecules* 19: 1375–81
68a. Rossi, G., Pincus, P. 1988. *Europhys. Lett.* 5: 641–47
69. Ingersent, K., Klein, J., Pincus, P. 1989. *Macromolecules* 22: In press
70. Hu, H.-W., van Alsten, J., Granick, S. 1989. *Langmuir* 5: 270–72
71. Marra, J., Hair, M. L. 1988. *Macromolecules* 21: 2349–55
72. de Gennes, P.-G. 1979. *Scaling Concepts in Polymer Physics*. Ithaca: Cornell Univ. Press
73. Edwards, S. F. 1965. *Proc. Phys. Soc. London* 85: 613–24
74. Helfand, E. 1975. *J. Chem. Phys.* 62: 999–1005
75. Scheutjens, J. M. H. M., Fleer, G. J. 1985. *Macromolecules* 18: 1882–1900
76. Scheutjens, J. M. H. M., Fleer, G. J. 1982. *The Effects of Polymers on Dispersion Properties*, ed. Th. F. Tadros, pp. 145–68. London: Academic
77. Scheutjens, J. M. H. M., Fleer, G. J. 1986. *J. Colloid Interface Sci.* 111: 504–15
78. Cosgrove, T., Vincent, B., Crowley, T.

L., Cohen Stuart, M. A. 1984. *ACS Symp. Ser.* 240: 147–59
79. Cohen Stuart, M. A., Waajen, F. H. W., Cosgrove, T., Vincent, B., Crowley, T. L. 1984. *Macromolecules* 17: 1825–30
80. Ploehn, H. J., Russel, W. B. 1989. *Macromolecules* 22: 266–76
81. Ploehn, H. J., Russel, W. B., Hall, C. K. 1988. *Macromolecules* 21: 1075–85
81a. Muthukumar, M., Ho, J.-S. 1989. *Macromolecules* 22: 965–73
82. de Gennes, P.-G. 1982. *Macromolecules* 15: 492–500
83. Cahn, J. W. 1977. *J. Chem. Phys.* 66: 3667–72
84. Rossi, G., Pincus, P. A. 1988. *Europhys. Lett.* 5: 641–46; 1989. *Macromolecules* 22: 276–83
85. Marra, J., Hair, M. L. 1988. *Macromolecules* 21: 2356–62
86. Israelachvili, J. N., Tandon, R. K., White, L. R. 1979. *Nature* 277: 120–21; 1980. *J. Colloid Interface Sci.* 78: 430–43
87. Brown, W. 1985. *Polymer* 26: 1647–50
88. Polik, W. F., Burchard, W. 1983. *Macromolecules* 16: 978–82
89. Daoust, H., St.-Cyr, D. 1984. *Macromolecules* 17: 596–601
90. Chauveteau, G. 1982. *J. Rheol.* 26: 111–25
91. Klein, J., Luckham, P. F. 1982. *Nature* 300: 429–30; 1984. *Macromolecules* 17: 1041–48
92. Ansorena, F. L., Fernandez-Berridi, M. J., Barandiaran, M. J., Guzman, G. M., Iruin, J. J. 1981. *Polymer Bull.* 4: 25–30
93. Klein, J., Almog, Y., Luckham, P. F. 1984. *ACS Symp. Ser.* 240: 227–44
94. Marra, J., Hair, M. L. 1988. *J. Colloid Interface Sci.* 125: 552–60
95. Toprakcioglu, C., Klein, J., Luckham, P. F. 1987. *J. Chem. Soc. Faraday Trans. 1* 83: 1703–9
96. Barford, W., Ball, R. C. 1987. *J. Chem. Soc. Faraday Trans. 1* 83: 2515–23
97. Klein, J., Luckham, P. F. 1984. *Nature* 308: 836–37
98. Taunton, H. J., Toprakcioglu, C., Fetters, L. J., Klein, J. 1988. *Nature* 332: 712–14
99. Watanabe, H., Patel, S., Tirrell, M. 1988. *Polym. Prepr. Am. Chem. Soc. Div. Polym. Chem.* 29: 370; 1989. *ACS Symp. Ser.* In press
100. Gast, A. P., Aug. 1988. *Proc. NATO Adv. Study Inst.*, Strasbourg
101. Hadziioannou, G., Patel, S., Granick, S., Tirrell, M. 1986. *J. Am. Chem. Soc.* 108: 2869–76
102. Ansarifar, M. A., Luckham, P. F. 1988. *Polymer* 29: 329–35
103. Taunton, H. J., Toprakcioglu, C., Klein, J. 1988. *Macromolecules* 21: 3333–36
104. Marra, J., Hair, M. L. 1989. *Colloids Surf.* 34: 215–26
105. Tirrell, M., Patel, S., Hadziioannou, G. 1987. *Proc. Natl. Acad. Sci. USA* 84: 4725–28
106. Alexander, S. 1977. *J. Phys. Paris* 38: 983–87
107. de Gennes, P.-G. 1980. *Macromolecules* 13: 1069–75
108. Patel, S., Tirrell, M., Hadziioannou, G. 1988. *Colloids Surf.* 31: 157–79
109. Hirz, S. J. 1986. *Modeling of interactions between adsorbed block copolymers.* Masters thesis. Univ. Minn., Minneapolis
110. Cosgrove, T., Heath, T., van Lent, B., Leermakers, F., Scheutjens, J. 1987. *Macromolecules* 20: 1692–96
111. Milner, S. T., Witten, T. A., Cates, M. E. 1988. *Europhys. Lett.* 5: 413–18; 1988. *Macromolecules* 21: 2610–19
112. Semenov, A. N. 1985. *Sov. Phys. JETP* 61: 733–42 (1985. *Zh. Eksp. Teor. Fiz.* 88: 1242–56)
113. Milner, S. T., Witten, T. A., Cates, M. E. 1989. *Macromolecules* 22: 853–61
114. Milner, S. T., Witten, T. A., Cates, M. E. 1988. *J. Phys.* 49: 1951–62
115. Milner, S. T. 1988. *Europhys. Lett.* 7: 695–99
116. Gast, A. P., Munch, M. 1988. *Macromolecules* 21: 1360–66
117. Marques, C. M., Leibler, L., Joanny, J.-F. 1988. *Macromolecules* 21: 1051–59
118. Marques, C. M., Joanny, J.-F. 1989. *Macromolecules* 22: 1454–58
119. Luckham, P. F., Klein, J. 1984. *J. Chem. Soc. Faraday Trans. 1* 80: 865–78
120. Dix, L. R., Toprakcioglu, C., Davies, R. J. 1988. *Colloids Surf.* 31: 147–49
121. Barford, W., Ball, R. C., Nex, C. M. M. 1986. *J. Chem. Soc. Faraday Trans. 1* 82: 3233–44
122. Marra, J., Hair, M. L. 1988. *J. Phys. Chem.* 21: 6044–51
123. Marra, J., Hair, M. L. 1989. *J. Colloid Interface Sci.* 128: 511–22
124. Papenhuizen, J., van der Schee, H. A., Fleer, G. J. 1985. *J. Colloid Interface Sci.* 104: 540–52
125. Miklavich, S., Marcelja, S. 1988. *J. Phys. Chem.* 92: 6718–22
126. Klein, J., Luckham, P. F. 1984. *Colloids Surf.* 10: 65–76
127. Afshar-Rad, T., Bailey, A. I., Luckham, P. F., McNaughton, W., Chap-

man, D. 1987. *Colloids Surf.* 25: 263–77; 1988. *Colloids Surf.* 31: 125–46; 1987. *Biochim. Biophys. Acta* 915: 101–11
128. Lee, C. S., Belfort, G. 1989. *Proc. Natl. Acad. Sci. USA.* In press
129. Israelachvili, J. N., Kott, S. J. 1988. *J. Chem. Phys.* 88: 7162–66
130. Montfort, J. P., Hadziioannou, G. H. 1988. *J. Chem. Phys.* 88: 7187–96
131. Horn, R. G., Israelachvili, J. N. 1988. *Macromolecules* 21: 2836–41
132. ten Brinke, G., Ausserré, D., Hadziioannou, G. H. 1988. *J. Chem. Phys.* 89: 4374–80
133. Kumar, S. K., Vacatello, M., Yoon, D. Y. 1988. *J. Chem. Phys.* 89: 5206–15
134. de Gennes, P.-G. 1987. *CR Acad. Sci. Paris* 305: 1181–83
135. Horn, R. G., Hirz, S. J., Hadziioannou, G. H., Frank, C. W., Catala, J. M. 1989. *J. Chem. Phys.* 90: 6767–74
136. Chan, D. Y. C., Horn, R. G. 1985. *J. Chem. Phys.* 83: 5311–24
137. Israelachvili, J. N. 1986. *J. Colloid Interface Sci.* 110: 263–71
138. Israelachvili, J. N. 1986. *Colloid Polym. Sci.* 264: 1060–65
139. Christenson, H. K., Israelachvili, J. N. 1988. *J. Colloid Interface Sci.* 119: 194–202
140. Granick, S., van Altsen, J., Israelachvili, J. N. 1988. *J. Colloid Interface Sci.* 125: 739–40
141. van Alsten, J., Granick, S. 1989. *Tribology Trans.* In press
142. Bitsanis, I., Vanderlick, T. K., Davis, H. T., Tirrell, M. 1988. *J. Chem. Phys.* 89: 3152–62
143. Horn, R. G., Clarke, D. R., Clarkson, M. T. 1988. *J. Mater. Res. Sci.* 3: 413–16; 1988. *J. Mater. Sci.* 46: 413–16
144. Luesse, C., van Alsten, J., Granick, S. 1988. *Rev. Sci. Instrum.* 59: 811–12

VACUUM UV PHOTOPHYSICS AND PHOTOIONIZATION SPECTROSCOPY

Tomas Baer

Department of Chemistry, University of North Carolina, Chapel Hill, North Carolina 27599-3290

INTRODUCTION

Vacuum ultraviolet frequencies of the electromagnetic spectrum lie in the region in which air absorbs radiation, so experiments must be carried out in a vacuum. In practice, this region extends from about 180 nm to about 30 nm. Below 110 nm is the windowless region in which a LiF window no longer transmits light.

The field of vacuum UV (VUV) photophysics is difficult to review because it is so broad. It encompasses experiments in the gas, liquid, and solid phases. It covers, dynamics, structure, and spectroscopy. To review the field so broadly would clearly be inappropriate. Instead, this review is limited to the major recent advances made in the areas of gas-phase spectroscopy and dynamics. The experiments of interest are those carried out with conventional UV light sources, synchrotron radiation, VUV lasers, and multiphoton ionization. The latter is not technically a VUV source, since the photon energy of interest is achieved by exciting the molecule or atom with two or more photons of lower energy. However, as long as the final states investigated lie in the VUV region, it seems appropriate to include a discussion of these studies. On the other hand, the experiments on ion spectroscopy carried out with IR lasers, are not discussed even though the information obtained by this method is very similar to that obtained by high resolution VUV spectroscopy (e.g. threshold photoelectron spectroscopy).

THE TYPES OF VUV LIGHT SOURCES

The choice of light sources available in the VUV region has increased dramatically in the past ten years. The important characteristics of intensity, tunability, resolution, and time structure are outlined in Table 1.

Conventional Light Sources

Conventional light sources consist of gas discharge lamps, whose characteristic spectra are determined by the type of gas used and the experimental conditions (1, 2). Thus, an electrical discharge in a low-pressure He lamp gives rise to the well-known He(I) emission line at 58.4 nm, which corresponds to the $2p \rightarrow 1s$ emission of neutral excited He. At still lower pressures, the He^+ ions can become the dominant emitters, thus giving rise to the $2s \rightarrow 1s$ He(II) line at 30.4 nm. Both of these sources are continuous in time. On the other hand, at the considerably higher pressures of about 50 to 200 Torr and when excited with a pulsed electrical discharge (10 to 90 kHz), the He lamp provides a broad continuum, known as the Hopfield continuum, which extends from 105 to 60 nm (2, 3). A similar behavior is shown by the low- and high-pressure Ar sources. However, the latter two sources are seldom used. Both the Hopfield and the high pressure Ar source require substantial differential pumping of the light source and are thus rather inconvenient. They also require a monochromator to specify the desired wavelength. The flux is thus a function of the monochromator resolution.

A strong source that does not require differential pumping is the H_2 many-line source (2). Its major drawback is that it is not continuous in wavelength, so it is not suitable for high resolution scanning. However, its set of closely spaced lines is very convenient when the precise energy is not of great interest. As with the Hopfield continuum, the many-line source requires a monochromator.

Synchrotron Sources

We have witnessed remarkable growth in work carried out at synchrotron radiation facilities over the past ten years (4-7). Light is generated when an electron bunch, circulating in a storage ring, is deflected by one of several bending magnets located around the ring. The newer generation rings also contain wigglers and undulators, which consist of closely spaced magnets that cause the electron bunches to undulate as they pass through the straight sections of the ring (8-10). Interference effects resulting from these undulations concentrate the emission energy in a wavelength region $\Delta\lambda = \lambda/n$ where $n =$ the number of oscillations. The rings at which the bulk of the work in chemical physics is currently pursued are Berlin

Table 1 Comparison of various VUV light source characteristics

Source	Pressure (Torr)	Repetition rate	Pulse length	Photon Flux/sec	Resolution (meV)	Wavelength range (nm)
He(I)	<1	cw	na	10^{12}	0.1	58.4
He(II)	<0.1	cw	na	10^{11}	0.1	30.8
He (Hopfield)	40–200	10–80 kHz	2–0.1 μs	10^9	5	60–105
H$_2$ many line	1	cw	na	10^9	15	86–150
Synchrotron	na	$2 \times 10^6 – 2 \times 10^8$	0.1–1 ns	$10^{12}–10^{13}$	1	all
VUV laser	na	10–100 Hz	5 ns	10^{11}	0.01	80–180
Free electron	na	30 or 10^7 Hz	20 ps	$10^{16}–10^{20}$	0.1	IR–?
MPI	na	10–100 Hz	5 ns	10^{17}	0.01	all

Elektronen Speicherring für Synchrotron Radiation (BESSY) in Berlin, ACO and super-ACO in Orsay, the National Synchrotron Light Source (NSLS) in Brookhaven, Aladan in Madison, Wisconsin, Stanford Synchrotron Radiation Laboratory (SSRL) at Stanford University, California, and the Photon Factory in Osaka, Japan. The NSLS and Aladan and super-ACO are already using some undulators. The new rings, which are specifically designed to incorporate undulators, are under construction at Berlin and at Lawrence Berkeley Laboratory (LBL) in Berkeley. These synchrotron radiation sources have several properties that make them unique. Among them is the broad tunability from the visible to the X-ray regions, their pulsed structure, their brightness (typical divergence of a bending magnet source is about 20 mrad and for an undulator much less), and their clean polarization characteristics.

VUV Lasers

A number of methods exist for generating VUV laser light. Raman shifting via the anti-Stokes lines in H_2 results in a series of lines shifted toward higher energy by the vibrational spacing of the H_2 molecule (11). However, the intensity drops off dramatically with the number of shifts. More intense VUV laser light can be generated by doubling and tripling visible light in a nonlinear medium. One approach is through third harmonic generation in such gases as Xe or Ar (12–14). Visible light at a wavelength λ is focused into a cell of Xe or at the output of a pulsed expansion of Xe and converted to a wavelength $\lambda/3$ with an efficiency of about 10^{-6}. A typical scanning range for Ar is 97.3 to 102.3 nm (14).

Another approach to the generation of VUV laser light is by four-wave mixing in such gases as Hg or Mg (15, 16). Laser number one is in near resonance with an allowed two-photon transition, while laser number two is scanned. The system responds by the emission of a photon with an energy that is the sum of the three laser photons. A typical scanning range with Hg is 112 to 120 nm.

Multiphoton Processes

The use of multiphoton transitions (17) to investigate highly excited states is not strictly a use of VUV radiation. However, the only difference between such studies and those that use VUV lasers is that in the former, the molecule being investigated is acting as a nonlinear medium by absorbing two or more photons, whereas in the later, the VUV photons are first generated with some other medium. Nevertheless, there are some important differences. The laser must be tightly focused in a multiphoton process, so that there is a very high probability that further photon absorption takes place. This often results in a complicated series of nonlinear absorption events that makes analysis of the data difficult. The other important difference in the two approaches lies in the selection rules for photon absorption. The spectroscopic information obtained is thus enhanced by the use of several n-photon processes.

Free Electron Lasers (FEL)

When the undulator length in a synchrotron is increased and the oscillation kept small, it is possible to generate stimulated emission from the oscillating electron current (18, 19). Such free electron lasers (FEL) have been demonstrated in the infrared (20) and in the visible and near UV region (21, 22) of the spectrum. It is predicted, but not yet demonstrated, that the FEL can be made to operate in the vacuum UV with substantial power and excellent resolution (see Table 1).

THE IONIZATION PROCESS

Threshold Photoionization and Dissociation

The photoionization and photodissociation processes that take place within a few electron volts of a molecule's ionization energy are remarkably rich (23–25). This is in contrast to the events following the absorption of high energy photons (20–30 eV), which is best described as a direct ionization into the continuum state, modified perhaps by generally broad resonances. On the other hand, at lower energies, many long-lived neutral states located near the molecule's ionization energy take part in the photon-absorption step, so that this process is often dominated by bound-to-bound transitions between two neutral states. Once excited, these states then decay by either ejection of the electron (autoionization or preionization) or by dissociation (predissociation), as shown in Eqs. 1 and 2.

$$AB + h\nu \rightarrow AB^* \begin{array}{l} \rightarrow AB^+ + e^- \\ \rightarrow A + B^* \end{array} \quad \begin{array}{l} 1. \\ 2. \end{array}$$

If it has sufficient energy, the AB^+ ion may subsequently dissociate. One of the neutral products in Reaction 2 is often electronically excited and can thus decay by fluorescence.

Reaction path 1 differs from direct ionization in two ways. The first lies in the distribution of ion vibrational states that are produced as a result of the different Franck-Condon factors (FCF) involved. In the direct ionization process, the appropriate FCF is given by $\langle \phi(n) | \phi(i) \rangle$, where the two vibrational wavefunctions refer to the ground state neutral and the final ion state, respectively. On the other hand, in the autoionization process with no interference from dissociation channels, the FCF is expressed as $\langle\phi(r)|\phi(i)\rangle$, where the $\phi(r)$ refers to the vibrational wavefunction of the autoionizing Rydberg state (26, 27). Attempts have been made to test this by measuring photoelectron spectra (PES) resulting from the excitation of various autoionizing Rydberg states (28–32). The first systematic efforts were carried out by Kinsinger & Taylor (28) at the synchrotron facility in Madison, Wisconsin on the oxygen molecule. Although these distributions of vibrational peaks in the PES follow quite well the distributions calculated with the FCF between Rydberg and final ion states, the conclusions remain somewhat clouded by the uncertain assignment of the Rydberg states, and by the insufficient energy resolution in exciting them. Many of the peaks in O_2 (33) consist of several overlapping states, so the identity of the Rydberg state is less than certain.

Whether the simple theory of Bardsley & Smith (26, 27) is valid is difficult to determine, because another ionization path invariably interferes with the simple autoionization process. This second process arises from a coupling between the dissociation and ionization steps (23, 24, 32, 34–39). As pointed out by Guyon et al (35) for the case of N_2O and Baer & Guyon (32) for the case of CO_2, ionization can take place during the course of dissociation. The distribution of final ion states is then affected in two ways. First, ionization of a dissociating species will lead to a broad distribution of final ion states. However, as has been experimentally observed, there is a considerable propensity for producing the ion in the highest possible internal energy state (32, 35). The proposed mechanism, shown in Figure 1, involves the following scheme:

$$AB(\tilde{X}) + h\nu \rightarrow AB^*(R_A) \rightarrow AB^{**}(R_X) \rightarrow AB^+ + e^-. \qquad 3.$$

The initially formed Rydberg state converging to an upper excited state (e.g. the ion $\tilde{A}$ state) is converted into a Rydberg state converging to the ground electronic state (the ion $\tilde{X}$ state) with an accompanying increase of the vibrational energy. Autoionization from this highly excited Rydberg state can then lead to the formation of highly vibrationally excited ions by vibrational autoionization (40). This process, which involves the conversion of a minimum amount of vibrational energy into electronic energy, was first observed in the autoionization of H_2 and has been found in the autoionization of NH_3 as well (41). The inclusion of rotational autoionization in the Guyon/Baer model has recently been proposed by Chupka et al (42). This was introduced in order to explain the experimental observation (32, 35) of very sharply peaked probabilities for the production of ions in the highest possible internal energy states. Vibrational autoionization alone could not explain the sharp features, because the density of states in an ion, such as CO_2 or N_2O, is not great enough.

Rotational autoionization involves the conversion of rotational energy

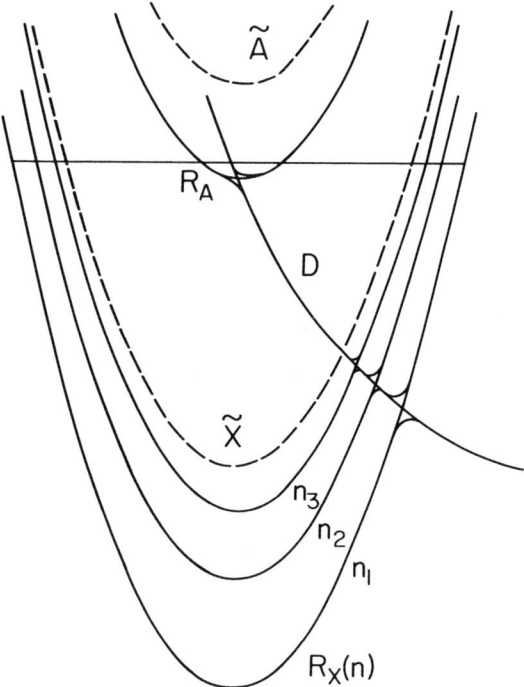

Figure 1 The potential energy diagram for resonant autoionization. The *dashed lines* are ion states, the *solid lines* are neutral Rydberg states. Excitation of the molecule is initially to the R_A state from which it dissociates/ionizes. Taken from Guyon et al (35) with permission.

to the Rydberg electron in a very high n level, thereby ejecting it into the continuum. How fast is rotational autoionization? Can it compete with vibrational autoionization when the two are both energetically possible? A beautiful experiment has addressed this problem for the case of the H_2 molecule. O'Halloran et al (43) excited the H_2 molecule with two photons to the third vibrational level of the inner $\tilde{E}$ well of the $\tilde{E}, \tilde{F}$ state. A second laser color was then used to reach autoionizing Rydberg states with the H_2^+ ($v = 2, N = 3$) core at energies between the $v = 1$ and $v = 2$ ion states. For principal quantum number $n > 21$, the Rydberg state could decay either by vibrational autoionization to produce a $H_2^+(v = 1)$ ion, or by rotational autoionization to produce a $H_2^+(v = 2)$ ion. The resulting electron kinetic energies were measured with a magnetic bottle time-of-flight (TOF) electron energy analyzer (44). The partial cross sections for formation of the $v = 1$ and $v = 2$ ions are shown in Figure 2. They exhibit an interesting structure that appears to be consistent with the multiquantum defect

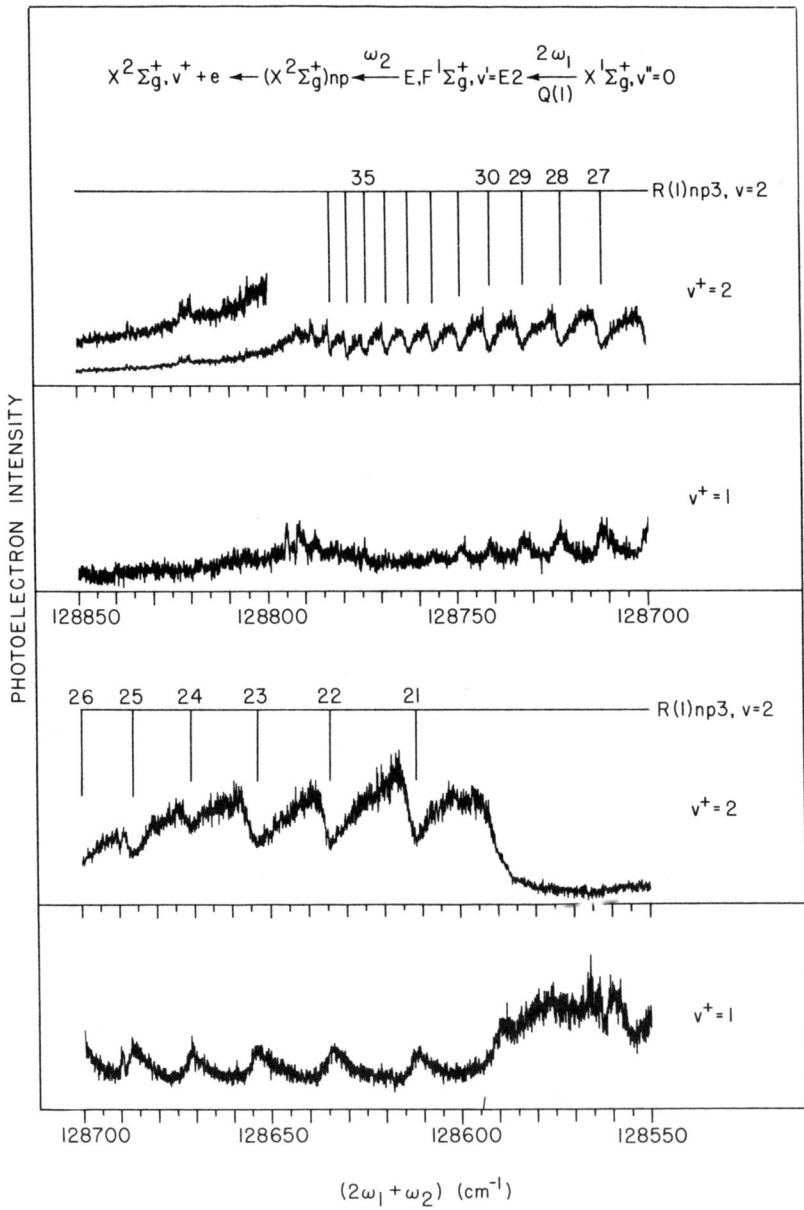

Figure 2 Excitation scans over autoionizing Rydberg states of H_2 using 2+1 REMPI while scanning ω_2. $v^+ = 1$ and $v^+ = 2$ refer to the product ion vibrational states. They are formed by vibrational and rotational autoionization, respectively. Taken from O'Halloran (43) with permission.

theory (MQDT) (45). As evident in Figure 2, vibrational autoionization contributes only 10 to 18%; this result indicates that the coupling between the Rydberg electron and the angular momentum of the H_2 molecule is stronger than coupling between the electron and the vibrational degree of freedom.

The autoionization of a molecule via the dissociative states (Figure 1) has two important consequences. The first is the experimental observation of dissociation onsets in Franck-Condon gaps of ions. The threshold for the O^+ ion production from N_2O^+ is experimentally observed at its thermochemical dissociation limit, even though this energy lies in the middle of a very large Franck-Condon gap, which is inaccessible by direct, or by the Bardsley-Smith type of autoionization (46). This type of autoionization takes place in essentially all molecules within about 3 eV of the ionization energy (47–49). The coupling of dissociative and ionizing channels has also been noted in the photoelectron spectra following resonance-enhanced multiphoton ionization (REMPI) (50–52).

Partial Photoionization Cross Sections

The photoionization event can lead to a multitude of final states. Among them are the parent ion in the various internal energy states, E and v, product ions, and product neutrals.

$$ABC + hv \rightarrow ABC^* \rightarrow ABC^+(E,v) + e^- \qquad 4.$$

$$\rightarrow AB^+(v) + C + e^- \qquad 5.$$

$$\rightarrow A + BC(E,v). \qquad 6.$$

Because the branching ratios to the final states are dependent upon the nature of the excited ABC^* state, it is possible to investigate the coupling between the excited and the various continuum states by measuring partial photoionization cross sections. These are often called *excitation spectra*.

Variable wavelength photoelectron spectroscopy (PES) with constant final state detection has been one of the major approaches employed (53–56). In this method, the exciting photon energy is varied along with the pass energy of the electron energy analyzer, so that a constant final ion state is monitored. Morin et al (55) investigated the production of $O_2^+(b\,{}^4\Sigma_g^-, v = 0$–$4)$ as a function of the excitation energy from 18 to 22 eV. This is a region of the oxygen spectrum that is rich in autoionizing states that converge to the $\tilde{B}\,{}^2\Sigma_g^-$ state of O_2^+. The very complex structure in the total photoionization spectrum is considerably simplified in the partial spectra, and the results indicate that in the Rydberg to final ion state transition, a $\Delta v = 0$ propensity rule is followed. Such a rule is generally associated with autoionization in which the bond distance in the Rydberg and final ion states are similar. This is nearly the case for these two states, whose ion cores differ only in their electron spin. These partial cross sections have been fit reasonably well with just three adjustable parameters with the multiquantum defect theory (57).

Another example of final ion state excitation spectroscopy is in the case of CO in the 19 to 27 eV region. In a classic paper, Stockbauer et al (58) described the effect of a shape resonance on the production of $v = 0, 1, 2$, and 3 in the ground ionic state as the photon energy is varied over the broad resonance located some 10 eV above the final ion state. Photoionization spectra are often dominated by two types of resonances referred to as *Feshbach* and *shape resonances*. The former is associated with the autoionization of a bound state whose width, determined by the lifetime of the pre-ionizing state, is generally narrow and whose shape is characterized by the Fano profile as calculated by the MQDT (59, 60). A shape resonance (23, 53, 61) is generally much broader, extending over several electron volts. It is associated with a state whose electron is bound by a centrifugal well resulting from its angular momentum. The branching ratio to the various final ion vibrational states varies significantly when the molecule is excited in the vicinity of this shape resonance. These effects are a direct consequence of the enhanced coupling of electronic and nuclear motion caused by the temporary trapping of the photoelectron in the molecular core. A comprehensive review (53) discusses the theory as well as its application to the N_2 molecule (62). A major advance in recent years has been the experimental set-up that measures triple differential partial cross sections in which the electron is dispersed with respect to ejection angle as well as kinetic energy (63). Molecules such as H_2 (64), CO (65), NO (66), O_2 (67), N_2O (68), and CO_2 (69) have been measured.

The threshold photoelectron spectrum (TPES) can also be considered a partial cross section or excitation spectrum. A comparison of the CO_2 photoionization efficiency (PIE) and TPES (32) in the large Franck-Condon gap between the ionic $\tilde{X}$ and $\tilde{A}$ states is shown in Figure 3. Unlike the normal PES, the TPES exhibits substantial structure in this region. The origin of the signal has already been discussed in terms of the interaction with neutral dissociative states. The structure that is weakly evident in the PIE is greatly enhanced in the TPES. A similar relationship between the PIE and the TPES has been reported by Delwiche et al (70) for COS and by Hubin-Franskin (69) for CS_2. Because the TPES structure is also evident in the ion spectrum, it is associated with autoionizing states that give preferentially zero energy electrons. The analysis of this structure led the authors to propose that these states belong to a Rydberg series converging to the $\tilde{A}$ states of the CO_2^+, COS^+, and CS_2^+, respectively (3, 69, 70). The vibrational progression in these Rydberg states is associated with the

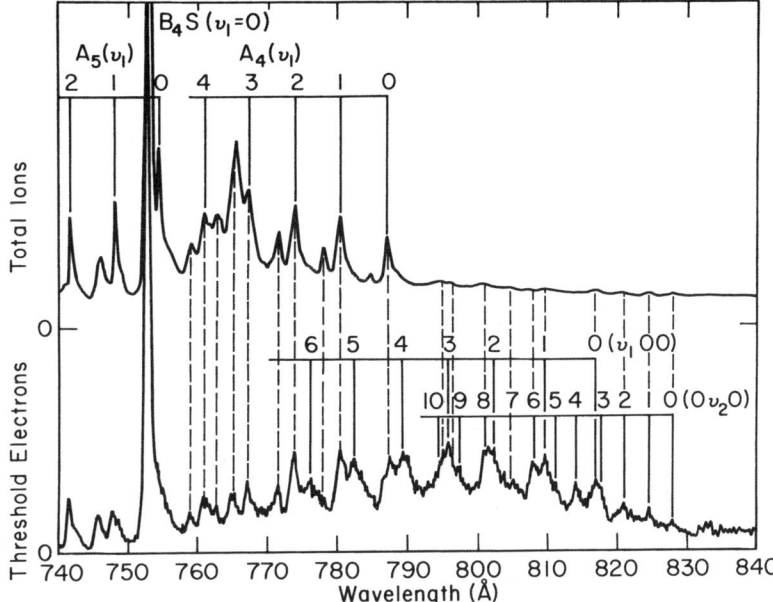

Figure 3 The total ion scan and the TPES of CO_2 in the Franck-Condon gap region between the X and the A states. The large peaks labeled A_4 and A_5 are Rydberg states converging to the ion's A state. Taken from Baer & Guyon (32) with permission.

bending mode. Within the framework of the model proposed above, these states interact strongly with dissociative neutral states, an interaction that enhances the production of threshold electrons. In fact, as much as 8% of all the ionization signal from these excited states goes to the production of threshold electrons. This represents a very large fraction in the small 10 meV band pass of the threshold analyzer.

The final partial cross section to be mentioned here is the fluorescence excitation spectrum. These spectra are of two sorts. Either an excited state of the ion fluoresces to a lower ionic state, or the neutral fragments following dissociation fluoresce. Thus, by tuning the monochromator or filter to one or the other, the production of specific excited ion states or neutral dissociation channels can be investigated as a function of the excitation energy. Among the molecules so investigated are H_2O (71), H_2 (72–74), COS (75), Cl_2 (76), N_2O (77), the rare gas dimers (78), and rare gas hydrides (79, 80). The fluorescence investigated for the rare gas dimers and hydrides were those involving a bound-free transition.

Fluorescence techniques are becoming more refined [see Nenner & Beswick (23)]. For instance, Poliakoff et al (81) measured the polarization of the fluorescence in the autoionization decay of CS_2. Such measurements provide information about the symmetry and alignment of the fluorescing states.

High Resolution Rydberg State Spectroscopy

High resolution Rydberg spectroscopy of states lying above and below the ionization limit has undergone a revolution with the application of lasers by means of both multiphoton and one-photon ionization. The use of highly monochromatic lasers, often with two colors, in exciting cold samples produced in pulsed molecular beams has produced highly resolved spectra of numerous high-lying electronic states. Although relatively few systems have been investigated so far, it is clear that this approach can be applied rather generally and in particular to large molecules that could not be studied previously.

Not unexpectedly, the H_2 molecule has received considerable attention, and it is in this molecule that we understand best the interaction between different electronic potential curves, and the competition between dissociation and ionization. The early high resolution (7 cm^{-1}) photoionization spectra of Chupka & Berkowitz (82) obtained with the Hopfield continuum lamp, and the series of papers by Glass-Maujean and coworkers (70–72) on the fluorescence of H atoms from the dissociation of excited H_2 molecules, have greatly expanded our knowledge of the states below and above the ionization energy. However, all of these studies were carried out via one-photon absorption, so the only states accessed were of ungerade symmetry.

Certainly the most interesting of the gerade states is the $\tilde{E}, \tilde{F}\ ^1\Sigma_g^+$ double-well state, whose ground state vibrational level lies at about 98,000 cm^{-1}. It can be reached by two-photon absorption from the H_2 ground state by using light with a wavelength between 190 and 200 nm, a wavelength region that can be generated by anti-Stokes Raman shifting. Marinero et al (83) reported one of the first studies of this double-well state by absorption from the ground state by using a 2+1 REMPI detection scheme with a resolution of about 0.9 cm^{-1}. Several of the low levels of both the inner E and outer F well are detected. This state is of considerable interest because of the perturbation of the vibrational levels in each of the two wells. The higher vibrational levels of the outer state are difficult to reach from the ground state H_2 molecule because the average internuclear distance is about 2.4 Å. However, recently Steadman & Baer (84) have found that the photodissociation of H_2S at about 70,000 cm^{-1} produces large quantities of vibrationally excited $H_2(v)$ ($v = 10$–14) molecules. From these high levels, with an average internuclear bond distance of nearly 2 Å, it is possible to excite the $\tilde{F}(v) = 1, 2, 4, 5, 7, 8, 10$ levels. The observed energies agree very well with the latest calculations of Senn & Dressler (85).

The vibrational wavefunctions of the $\tilde{E}$, $\tilde{F}$ state can be explored by measuring the vibrational energy distribution of the final H_2^+ produced by ionizing via the $\tilde{E}$, $\tilde{F}$ intermediate state. The average internuclear distance of the $\tilde{E}$ well (1.012 Å) is very similar to that of H_2^+ (1.052 Å) (86), so the $\Delta v = 0$ propensity for ionization should follow. This was in fact observed, although many other vibrational levels were populated in a manner that depended on both the intermediate state vibrational and rotational levels (87). These aberations were ascribed to electronic transition moment variations with the internuclear distance.

The non-Franck Condon transition in the ionization via the $\tilde{C}\,^1\Pi_u$ state has also been the subject of a number of recent studies (88–92). The dramatic departure from the expected Franck Condon distribution of H_2^+ ion states has been explained (92) in terms of the interaction between the H_2^+ ion state and the doubly excited dissociative states that lie in the same energy region by the mechanism discussed above. These dissociative states in H_2 can only be accessed by multiphoton transitions because they involve the excitation of two electrons.

Other studies of higher lying H_2 states have been concerned with ion pair production (93) by the use of two VUV photons, and the production of H atoms from various dissociative states (94–97).

Triplet states of H_2 cannot be accessed from the singlet ground state. However, they can be produced by electron impact excitation and the fluorescence analyzed. Alternatively, they can be investigated by laser-induced dissociation into neutral products. With this approach, van der Koot et al (98) passed a 3 keV beam of H_2^+ ions through a charge exchange chamber to produce fast-moving neutral H_2 molecules. Those that were in the $\tilde{c}\,^3\Pi_u^-$ state were then excited with a visible dye laser to one of several higher-lying triplet states, the $\tilde{g}\,^3\Sigma_g^+$, $\tilde{i}\,^3\Pi_g$, $\tilde{j}\,^3\Delta_g$, states whose origin lies just below the H $(1s)$ + H $(n=2)$ dissociation limit. Because these states are predissociated by the repulsive $\tilde{b}\,^3\Sigma_u^+$ state leading to ground state H atoms, the absorption can be monitored by the arrival of the H atoms. The products are formed with considerable kinetic energy, which can be determined from the spacial distribution of the signal on a position sensitive detector. Undulations observed in the kinetic energy spectrum were attributed to the shape of the excited state vibrational wavefunction. It was determined that some of the levels dissociate by tunneling through the π barrier (99).

Another well-studied molecule is NO. Its one-photon autoionization structure was recorded and analyzed most completely by Ono et al (100), with a resolution of 10 cm^{-1} (0.14 Å). Over 150 states were identified and assigned. More recently, Pratt and co-workers (101) investigated some of the same energy region by using two-photon absorption at a laser resolution of about 0.15 cm^{-1}. However, because of the two-photon selection rules, the observed autoionizing states were not those previously investigated and assigned, but were rather the *ndsp* states. They were found to autoionize via vibrational autoionization.

Two other molecules have recently been investigated by VUV lasers: H_2O (102) and HI (103). In the water study, photons between 99 and 93 nm were generated by frequency tripling of 297 to 279 nm laser light in either Ar or Xe. Large variations in the VUV laser output were normalized by dividing the H_2O^+ signal by the signal from acetylene ions, whose photoionization signal has little structure in this energy region. The resolution of 1.5 cm^{-1} [versus 50 cm^{-1} for the previous study by Katayama et al (104)] and the colder sample temperature allowed for rotational resolution of the peaks. A number of Rydberg states were identified or reassigned more definitively. The Rydberg state lifetimes were determined from the peaks widths to be about 1 ps (102).

Threshold Photoelectron Spectroscopy

The pulsed nature of synchrotron radiation and VUV lasers makes these ideal light sources for threshold photoelectron spectroscopy (TPES). Time-of-flight analysis of initially zero energy electrons, coupled with the rejection of energetic electrons with off-axis velocity components, can yield resolutions that far exceed those obtainable from dispersive analyzers used to analyze energetic electrons. The resolution measured by TOF is best for low-energy electrons, and it improves as the extraction field is reduced. The practical limit of the resolution when synchrotron radiation is used (between 1 and 10 cm^{-1}) is the time between photon pulses (typically less than 100 ns) and the band pass of the photon monochromator. In order to reach this energy resolution, it will be necessary to measure the electron TOF with a time resolution of about 100 ps; this should be possible [see Moller & Zimmerer (105)] with the new generation of synchrotrons that have light pulse widths of less than 100 ps. The major limiting factor will be the photon flux from the monochromator as the slits are narrowed. This is not a factor in the laser-based experiment in which flux and resolution are not inversely related. In the 10 Hz, 5 ns laser experiment, which can be either REMPI or single-photon VUV, the resolution is about 0.1 cm^{-1}, determined principally by the laser resolution (106).

SYNCHROTRON RADIATION STUDIES A number of molecules have been investigated by means of TPES using synchrotron radiation. Among them are N_2 (107), COS (70), and the previously mentioned CO_2 (32). This latter study of the $^{12}CO_2$ and $^{13}CO_2$ TPES covered the range from the ionization onset at 13 eV to 23 eV. The combined threshold analyzer and light source resolution of 10 meV was sufficient to observe the isotope shifts in the asymmetric stretch and the bending modes. These states are not accessible via the usual selection rules, but because of the interaction between the dissociative ionizing states, the bend in particular can be excited. On the other hand, the asymmetric stretch is hardly excited in any of the bands and remains completely unknown for the $\tilde{A}$ state.

The TPES in most of the $\tilde{A}$ state is totally ordinary, consisting mostly of the $(v, 0, 0)$ progression. The exceptions are two sharp and strong peaks in the $^{12}CO_2^+$ between the $(4, 0, 0)$ and $(5, 0, 0)$ peaks. These interloper peaks were replaced in the $^{13}CO_2^+$ TPES by a single peak displaced in energy from the two other peaks. The fluorescence lifetime of one of the $^{12}CO_2^+$ peaks was measured by Schlag et al (108) and found to be different from the normal $\tilde{A}$ state lifetimes. On the basis of these results, Baer & Guyon (32) concluded that these peaks are part of a triplet state that is excited by accidental resonances between a particular rotational level of the initially excited Rydberg state and a hidden autoionizing state.

LASER STUDIES A major improvement in threshold photoelectron spectroscopy has come about as a result of the use of pulsed lasers. The high photon resolution and flux, and the longer time between photon pulses, which allows for the use of very long electron drift tubes, serve to increase the electron energy resolution dramatically. Electron TOF analysis allows collection of either the complete electron energy spectrum or the threshold electron spectrum. The information obtained is complementary. The total PES gives information about the final ion states produced, whereas the TPES yields exceptionally good resolution but gives unclear information about the peak intensities because of the interference from autoionization states.

The most thoroughly studied molecule is NO. Riley et al (109, 110), using normal REMPI-PES, achieved a resolution of 24 cm^{-1}, which was sufficient to resolve the rotational levels of NO$^+$ for the higher rotational levels (which are more widely spaced). On the other hand, Muller-Dethlefs and co-workers (106, 111) have achieved a resolution of nearly 1 cm^{-1} with the threshold electron technique. They used a two-color $1+1$ REMPI process via the A state and were able to resolve completely the rotational levels of NO$^+$ for $J = 0$–4. The first laser made possible the selection of a single intermediate state rotational level. Yet they found that the rotational state distribution of the NO$^+$ product ion was rather broad. For $N_A = 0$, they observed the NO$^+$ rotational levels, 0, 1, 2, and 3. However, the distribution narrowed considerably as the $\tilde{A}$ state rotational level was increased. Thus for $N_A = 3$, nearly 90% of the ions were produced in the

$N = 3$ level. However, as the authors point out, the intensities in such a threshold electron spectrum must be interpreted with caution because of the role of autoionizing states in determining the peak intensities.

THE DISSOCIATION POLYATOMIC IONS

Photoionization (PI) remains one of the major tools for studying the dissociation of polyatomic ions. Much of this work involves collecting mass-analyzed photoion signal as a function of the photon energy. To discuss and review all of this work would require more space than is available. However, some of the new developments can be listed. These include the PI of free radicals (112–114), which provides information about the ionization energy and structure of the free radical. The spectroscopy of the methyl radical has also been investigated by laser MPI (115). Ions of stable molecules recently investigated by photoionization mass spectrometry include acetylene and ethylene up to 20 eV (116), methyl fluoride (117), sulfur dibromide (118), the dibromobenzenes (119), several molecules that dissociate to methylallyl ions (120), silane (114), tropone (121), isopropyl methyl ether (122), anisole (123), nitromethane (124), and borane and diborane (125).

Cluster ions continue to be the subject of numerous studies. However, definitive information is often difficult to extract because the ions are generally present in a distribution of cluster sizes. Exceptions are dimers, especially when studied with the aid of electron-ion coincidence techniques (126–128), or higher order clusters when studied by the use of REMPI in which a particular cluster is selected by its spectroscopy at the intermediate level (129). The field of cluster ion research as studied with synchrotron radiation has recently been reviewed (130). An interesting class of clusters is the rare gas-molecule dimers, which appear to decay via a Penning ionization following the ionization of the rare gas atom (131). The clusters of water (132), allene (133), methylfluoride (134), SO_2-benzene and -butene (135), ethylene-HCl (136), benzene (137), benzene-HCl (138), and others (139) have also been investigated by photoionization.

Studies of Energy-Selected, Singly Charged Ions

METHODS OF STATE SELECTION   A general technique for energy selecting ions is that of photoelectron photoion coincidence (PEPICO) (140–142). In this approach, molecules are photoionized by a single photon from a dispersed cw light source. The ions are produced in a distribution of internal energy states from the ion ground state up to a maximum energy of $hv - IE$, where IE is the ionization energy of the molecule. By conservation of energy and momentum, an ion of a particular internal energy is associated with an electron of energy $hv - IE - E_{ion}$. Thus, ions can be state selected by collecting them in coincidence with electrons of a given energy. The advantages of the PEPICO approach is that it is extremely general, and ions can be prepared with virtually any internal energy. The major disadvantage is that the ion energy resolution is limited to about 10 meV. This limitation is imposed by the applied electric field necessary to extract the ions.

A far more precise method of ion energy selection is by the use of lasers. Laser state selection of ions is best carried out with pulsed lasers and a pulsed molecular beam source. The following steps are required for good state selection:

$$AB \xrightarrow{nhv} AB^+ + e^- \qquad \qquad 7.$$

$$AB^+ \xrightarrow{hv'} AB^{+*} \rightarrow A^+ + B. \qquad \qquad 8.$$

One method of preparing the ion in the ground state in the first reaction is with a single VUV laser photon tuned to the molecule's ionization energy. In that case, the laser does not need to be tightly focused, and photon absorption stops at one photon. The second laser color is used to promote the ion to an excited state above the dissociation limit. If a VUV laser of the appropriate energy is not available, then the ionization step will require two or more photons, which are absorbed in two or more one-photon absorption steps. In general, the last photon absorbed, which causes ionization, will bring the total energy above the ionization energy, so that some of the energy will be lost to the photoelectron. This electron energy distribution must be monitored in order to determine the distribution of ion energies created in the MPI process. With the appropriate choice of the intermediate state, a narrow energy distribution of the ion can often be produced.

The major advantages in using lasers for state selection are the high resolution for energy selection and the ability to extend the rate measurements to very high rates. The latter benefit comes about because the state selection process is not hurt by the presence of strong electric fields, which are necessary to investigate very rapid reactions. Aside from the difficulty in preparing some ions in the ground state, the major disadvantage of the laser approach is that many excited states cannot be reached, among them the high vibrational levels of the ground electronic state and certain excited states, transitions to which are forbidden by selection rules. The two methods, PEPICO and laser preparation of ions, are thus complementary.

Both the PEPICO and the pulsed laser techniques use the ion time of flight for determining the ion lifetime. In order to have a measurable dissociation rate, the ion must dissociate during the course of its flight to the ion detector. When a parent ion of kinetic energy, E_p, and mass M_p, dissociates, the nascent daughter ion of mass M_d will have a kinetic energy of $E_p(M_d/M_p)$. Since the parent ion kinetic energy varies in time because of the applied acceleration fields, the daughter ion kinetic energies will also vary with time. The two methods commonly employed for determining ion lifetimes depend on whether the ion dissociates in a field free drift region or in an acceleration region. If an ion dissociates in an acceleration region, the daughter ion TOF distribution will be asymmetrically broadened toward longer times of flights. This asymmetry can be modeled by assuming a decay curve, such as an exponential or bi-exponential. In this method the decay rate is monitored in a more or less continuous fashion over a range of dissociation times. It is possible to determine whether the decay is best described by a single or by several exponentials.

The rate can also be determined from a single measurement if the dissociation takes place in a field free region followed by an acceleration region. Because the daughter and parent ions have a different kinetic energy (but equal velocities) as they emerge from the drift region, an acceleration of the two groups of ions will disperse them in time of flight. Since the parent ion time of arrival at the drift region, t_a, and the time spent in the drift region, t_d, are well known, the rate can be determined from the following relationship:

$$R = \frac{D^+}{P^+} = \frac{\int_{t_a}^{t_a+t_d} \exp(-kt)}{\int_{t_a+t_d}^{\infty} \exp(-kt)}. \qquad 9.$$

Determination of whether the decay is of a single or a multiple exponential form is not possible from a single measurement such as this.

REMPI STUDIES OF DISSOCIATION RATES The laser technique was pioneered by Zare et al (142, 143) in the measurement of the aniline and chlorobenzene ion dissociations. However, the use of only one laser color greatly limited these first experiments. The use of two laser colors has recently been exploited by Neusser and co-workers (144–146) for the investigation of the benzene ion dissociation. The dissociation rate was measured by the use of a reflectron TOF mass spectrometer. The reflectron, a special case of an acceleration region following a drift region, consists of a long (ca. 1 m) drift region and an electrostatic mirror that reflects the ions. The dynamical information in combination with the high mass resolution of the reflectron make this a particularly good method for an ion such as benzene in which the neutral fragments of interest are H, H_2, C_2H_2, and C_3H_3. In fact, this was the first measurement of the H and H_2 loss rates from the benzene ion since the early work of Andlauer & Ottinger (147).

Within 2.5 eV of the first dissociation limit, the benzene ion decays via four dissociation channels to $C_6H_5^+$, $C_6H_4^+$, $C_4H_4^+$, and $C_3H_3^+$. Kuhlewind et al (146) measured the total dissociation rate over a range extending nearly three orders of magnitude. The individual rate for $C_3H_3^+$ formation, which is determined from the total rate and the branching ratios of the four dissociation channels, now has a measured range that extends to nearly four orders of magnitude, from 10^2 to 5×10^5 sec^{-1}.

Of particular note in this series of benzene investigations is that the rotational levels of the ions could be partially selected because the ionization took place via the rotationally selected intermediate S_1 state (148). Because large changes in J are not expected in the final ionizing step, Kiermeier et al (148) were able to produce the benzene ions in rather narrow ranges of rotational energy ($\Delta J = \pm 5$). A plot of the rate for a constant total energy (vibrational plus rotational) showed that the dissociation rate was reduced by about 30% as J varied from 1 to 60. This confirms previous findings that ionic unimolecular reactions are not rotationally adiabatic (149). That is, the rotations are coupled to the total heat bath. The unexpected result from these experiments is that the rates decrease even faster than expected based simply on the increase in the activation energy due to the centrifugal barrier in the transition state. This extra decrease was explained by the stronger effect of K-mixing on the density of states $\rho(E, J)$ in the precursor ion than on $W^{\ddagger}(E, J)$ in the activated complex (148).

Although the benzene ion dissociation is interesting and the rate measurements are of very high quality, this system, unfortunately, is currently of limited usefulness in advancing our understanding of unimolecular reactions. The reason is that we know so little about the structure and energetics of the dissociation products, and even less about the transition states. How is H_2 lost from the benzene ion? Do C_2H_2 and C_3H_3 losses proceed via the same transition state? What is the structure of the $C_4H_4^+$ ion? Why is it that the "simple" H loss channel has a lower activation entropy than the "complicated" rearrangement channels leading to $C_4H_4^+$ and $C_3H_3^+$? The many unanswered questions limit the analysis of the data to the statistical theory in its simplest form, i.e. RRKM (150) or the quasiequilibrium theory (QET) (151), in which more or less arbitrary vibrational frequencies and energies are used. In order to apply the more sophisticated versions of the theory, such as the adiabatic channel theory (152), or the transition state switching model (153) to the excellent data of Neusser and colleagues, it is necessary to have more information about the mechanism of the benzene ion dissociation. This information will surely be forthcoming from high level ab initio calculations in the next few years.

A variation of the MPI/reflectron technique has been reported by Tai & El-Sayed (154) in which ions are extracted from the ion source at times delayed from the laser pulse. In this manner, the decay rates toward seven dissociation channels of dichlorobenzene ions were measured. Lemaire et al (155) have carried out the only three-color MPI rate measurement. Two laser colors were used to prepare phenetole ions in the ground electronic and vibrational state. The third laser excited the ions to energies above the dissociation limit.

PEPICO STUDIES OF DISSOCIATION RATES The bulk of the ion dissociation dynamics is still being produced by the PEPICO technique. Among the small ions recently investigated is the dissociation of O_2^+ in the $\tilde{c}^4\Sigma_u^-$ state (156). This state, residing at 25 eV, has a very small well that holds only two vibrational levels because its potential energy curve is homogeneously perturbed by a dissociative state leading to the $O(^4S)+O(^1D)$ dissociation products. However, the $v = 0$ level of the $\tilde{c}$ state dissociates to the lower energy $O^+(^4S)+O(^3P)$ state as well. Richard-Viard et al (156) found that the branching ratio between these two dissociation limits varies significantly when $^{16}O_2^+$ is replaced by $^{18}O_2^+$. In addition they estimated the lifetimes with respect to predissociation to be about 7×10^{-10} and 1.8×10^{-10} sec for the light and heavy isotopes, respectively. The results agreed with a tunneling model that used several different calculated and estimated potential functions.

Another unusual reaction of a small ion is the dissociation of CF_3I^+ investigated by Low et al (157), who used a He(I) light source and analyzed the electrons with a 180° hemispherical analyzer. At an energy of about 13.5 eV, this ion dissociates from a repulsive $\tilde{A}$ state via two channels to produce $CF_3^+ + I$ and $CF_3 + I^+$ with the release of considerable kinetic energy. The large kinetic energy release is evidenced in the TOF distribution by a split in the forward and backward scattered product ions. The novel result for this dissociation is that there exists a forward/backward asymmetry in the TOF distribution, and that it is in opposite directions for the two product channels. That is, the CF_3^+ product is preferentially forward scattered with respect to the ion detector, whereas the I^+ peak is preferentially backward scattered. Low et al (157) interpreted this in terms of a CF_3I^+ ion that is aligned by the ejection of the electron so that the iodine end is preferentially directed toward the electron detector. This is the only such case known to date.

The dissociation rates of numerous and more complex ions have been studied by PEPICO during the past few years. Among them are methyl formate (158), dimethyl sulfoxide (159), the di-halobenzenes (160), nitrobenzene (161), anisole (162), Chromiumhexacarbonyl (163), o-nitrotoluene (164), m- and p-nitrotoluene (165), n-butylbenzene (166), the butyl-alcohol isomers (167), n-propanol (168), and tetrazene (169).

Many of these ions dissociate by mechanisms that are at least qualitatively understood and that can be modeled by the use of just two parameters in the statistical theory. These parameters are the activation energy and the activation entropy. The latter is evaluated from the assumed, or known, vibrational frequencies of the molecular ion and the transition state. Figure 4 shows a classic competition between a simple bond rupture and a rearrangement reaction for the case of the n-butylbenzene ion. The simple bond break to produce the m/z 91 fragment ion has a higher activation energy, but a positive activation entropy (loose transition state), hence it dominates at high energy (166).

Many dissociation reactions are more complex. It is often difficult to model a reaction in a meaningful manner unless the mechanism is known. Our understanding of more complex ionic unimolecular dissociation reactions is currently expanding as a result of advances made in ab initio

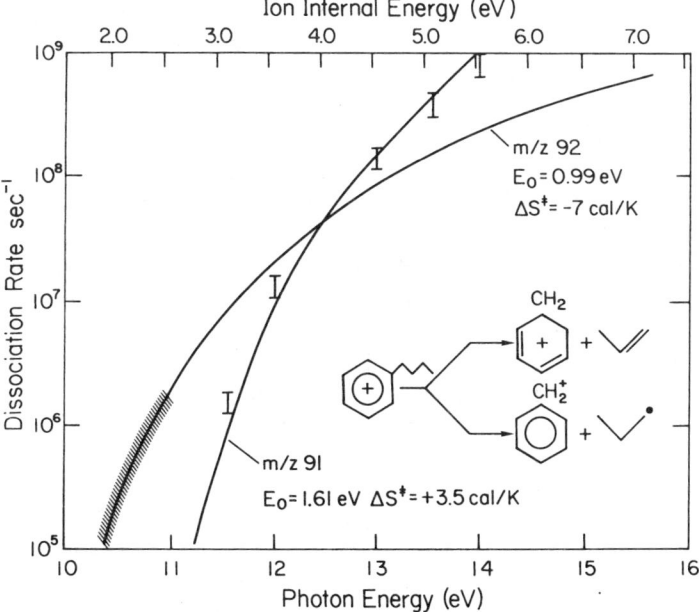

Figure 4 The measured and calculated (RRKM/QET) dissociation rates of n-butylbenzene as a function of the ion's internal energy. Taken from Baer et al (166) with permission.

calculations (170–174). The ready availability of supercomputers has made it possible to determine ionic and neutral structures and energies with high precision. In addition, vibrational frequencies are also being calculated and used to correct the calculated energies for the zero-point energy. Finally, it is now readily possible to calculate the transition state structure and vibrational frequencies as well. This makes testing of the RRKM theory possible with no adjustable parameters (170, 171).

Heinrich et al (170) have recently carried out a theoretical study of the acetone ion reaction in which the following two reactions take place with an activation energy of about 0.3 eV.

$$CH_3COCH_3^+ \rightarrow \begin{matrix} CH_3CO^+ + CH_3 \\ CH_2O^+ + CH_4 \end{matrix} \qquad 10.$$

The aim of the study was to calculate the potential energy surface of acetone ions and its isomers in order to understand the relationship between the two competing low energy reactions as well as the unusual slowness of the CH_4 loss channel. These calculations showed how the two dissociation channels are related by a long-lived complex involving the acetyl ion (CH_3CO^+) and CH_3. At ion energies slightly below the onset energy for CH_3 loss, the methyl radical moves rather freely along the backbone of the acetyl ion. During this time, a hydrogen transfer can take place via tunneling through a small barrier, thereby producing the CH_4 product. The tunneling step is rate-determining and thus explains the slow dissociation rate via CH_4 loss.

What is particularly interesting about this study is the derived shape of the potential energy surface. The potential energy rises in a normal manner as the CH_3–$COCH_3^+$ bond is broken. However, it quickly becomes very flat over a very large range of CH_3 positions along the acetyl ion backbone. This type of surface, which appears to be rather common in ionic dissociation reactions, raises serious questions about the applicability of the simple RRKM/QET statistical model. In this theory, we treat the molecule only in terms of the molecular ion and the transition state. The various contortions of the ion in between these two states is not generally considered. Yet it hardly seems possible that a strange potential energy surface, such as exists in the case of the acetone ion dissociation, should have no effect on the dissociation rate.

The absolute rate measurements carried out with the PEPICO technique have focused attention recently on another unique aspect of ionic dissociations. This is the role of the long-lived complexes, in particular the ion-dipole complex (167, 168). These long-range forces appear to have a profound effect on the dissociation rate. The interaction potential between the products from a neutral reaction are generally described by an attractive $1/r^6$ potential, the functional form appropriate for an ionic reaction with an ion-induced dipole attraction is $1/r^4$, and interaction between an ion and a nonrotating dipole is given by $1/r^2$. This extremely long-range interaction has two interesting features. First, a $1/r^2$ potential does not support a centrifugal barrier, so the location of the transition state is difficult to determine. More importantly, the number of vibrational energy levels supported by such a nonrotating ion dipole potential is infinite (168, 175). Even with a rotating dipole the number of states is extremely large, and this has the effect of decreasing the dissociation rate by increasing the density of states in the denominator of the RRKM/QET expression:

$$k(E) = \frac{W^{\ddagger}(E-E_0)}{h\rho(E)}.$$ 11.

An example of a reaction involving an ion-dipole complex is the dissociation of the *n*-propanol ion that loses H_2O. This ion passes over (or tunnels through) an isomerization barrier and rearranges into the ion dipole complex between the cyclopropane ion and the dipolar water molecule (Figure 5) (168, 176). This complex, which has been calculated at the Hartree-Fock level with a Gaussian 86 program at the 6-31G* level, has all of the characteristics expected of an ion-dipole complex: Each unit has the structure and vibrational frequencies associated with the free ion or dipole, and the water dipole points precisely in the direction of the cyclopropane ion center of charge. The dissociation mechanism via this complex is actually expected on the basis of microscopic reversibility because the two products, as they approach each other, will certainly fall into the 0.5 eV ion-dipole well.

Shao et al (168) measured the dissociation rate of the *n*-propanol ion and found it to be nearly four orders of magnitude slower than predicted by RRKM/QET calculations. The RRKM/QET statistical theory calculations of the cyclopropane ion-water complex were carried out with the calculated vibrational frequencies of the complex and assumed frequencies of the transition state. (In the absence of a real or centrifugal dissociation barrier, no unique structure exists for the transition state, so these vibrational frequencies could not be calculated by ab initio methods.) The sizable discrepancy between experimental and calculated rates was attributed to the large density of vibrational states associated with the extremely anharmonic $1/r^2$ potential.

The final interesting aspect of the ion-dipole complex dissociation is the question of free energy flow. Energy flow in complex molecules is aided by coupling among the various vibrational modes. This coupling is generally

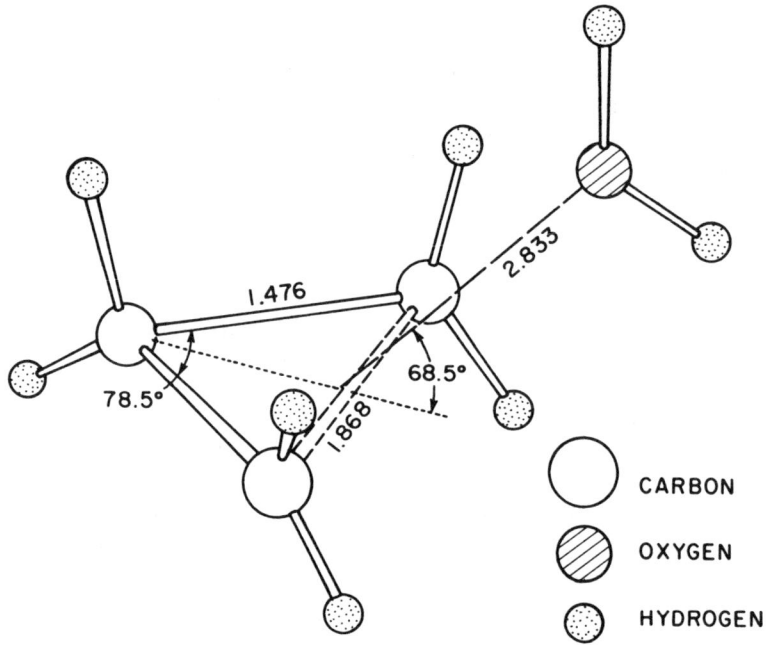

Figure 5 Calculated (Gaussian 86 6–31G*) structure of the ion-dipole complex formed between the cyclopropane ion and H_2O. Taken from Shao et al (168) with permission.

strongest when the vibrational frequencies are of about equal magnitude. In the case of the energized ion-dipole complex in which the two units are some distance apart, the anharmonic ion-dipole stretch mode has vibrational energy intervals that become vanishingly small as the energy increases. Thus, we would not expect this motion to couple very well with the remaining frequencies of the separated fragments. This lack of energy flow may further inhibit the dissociation reaction.

The Dissociation Dynamics of Doubly Charged Ions

The study of doubly charged ion dissociations, a topic reviewed by Eland et al (177), is more complicated than the study of singly charged ions because three particles are produced in the ionization process. Thus, energy selection of the doubly charged ion can be achieved only by a simultaneous measurement of two electron energies. In addition, two ions are produced in the dissociation reaction, and thus provide the opportunity of carrying out a coincidence study between the two ions. The overall reactions are as follows:

$$ABC + h\nu \to ABC^{+2} + 2e^- \qquad 12.$$

$$ABC^{+2} \to AB^+ + C^+ \quad \text{or} \quad A^+ + B + C^+. \qquad 13.$$

Although Reactions 13 are always very exothermic because of the Coulomb repulsion, the doubly charged ion is often stable over a small energy range. In addition, metastable doubly charged ions are rather common in mass spectrometry. These are ions that have lifetimes with respect to dissociation between 0.1 and 100 μs. The reason for both the stability and the metastability of these ions lies in the shape and electronic structure of the potential energy curves. The dissociative state is strongly repulsive and its potential energy rises rapidly as the two ions approach each other. This rise is so rapid that it intersects the stable state at large ion separation and thus at a high potential energy. The intersection of curves provides a well for the bound doubly charged ion state. High level ab initio calculations with configuration interaction have shown that the doubly charged ground states of the CX_2^{2+} (X = O, S) ions are triplets that can be reasonably described by a simple two-hole configuration (178). However, higher excited states require multiconfiguration methods. Some work on the spectroscopy of doubly charged ions has been reported (179, 180). Whether the slow dissociation rate of some ions is a result of curve crossing (i.e. electronic pre-dissociation) or of tunneling through the barrier has not been established. On the basis of unrestricted Hartree-Fock calculations, Gill & Radom (181) have found that the loss of H^+ from AlH^{2+} or $N_2H_2^{2+}$ proceeds via H atom loss followed by a charge transfer step when the two fragments are some distance apart.

Doubly charged ions and their formation are interesting for several reasons. They may be important in interstellar space (182). These ions are produced by a formally non-allowed two-electron excitation step. Very little is known concerning the correlation between the two ejected electrons. How is the excess energy partitioned? The coincidence experiment between the two electrons, which might provide some answers, has not been done. This is a difficult experiment because both particles are dispersed not only in velocity but in direction as well. On the other hand, threshold behavior of the double ionization signal has been measured in photoionization (183). It rises linearly with the excess photon energy, which is consistent with the Wannier threshold law (184). (In single photoionization events the threshold is a step function.) The linear threshold law is based on a statistical partitioning of the electron energy in the allowable phase space, which implies that the two electrons are more likely to be ejected with more or less equal energies. However, one would expect that shape or Feshbach resonances could alter such energy partitioning.

Most of the experiments on the dissociation of doubly ionized ions have been carried out with synchrotron and He(II) radiation because of the high energies involved (30–50 eV). Photoion photoion coincidence (PIPICO) events between the two singly charged fragment ions have been recorded by TOF mass spectrometry for such ions as CO_2^{2+} (185), CS_2^{2+} (185), $N_2O_2^{2+}$ (186), Cl_2^{2+} (187), H_2O^{2+} (188, 189), CH_4^{2+} (190), N_2^{2+} (191, 192), and NH_3^{2+} (193). In addition, the double ionization of H_2 has been investigated in order to learn about electron correlation (194). Very recently, the dissociation of the triply charged CO^{3+} ion has been investigated at energies in the vicinity of 125 eV (195).

In the PIPICO experiment, the circular ion collector is divided into two semicircles to collect the two charged product ions that are ejected into opposing semicircles by the Coulomb repulsion (23). The data then consist of the time difference between the arrival of the two positively charged ion fragments. This TOF difference is a function of the fragment ion masses, their kinetic energy, and the angle at which they are ejected.

A major advance in the study of doubly charged ions has been the triple coincidence experiment, photoelectron, photoion, photoion, (PEPIPICO), in which one of the electrons and both of the ions are measured in coincidence. In this experiment, pioneered by Eland and co-workers (196), the electron provides the start signal for measuring the ions' times of flight. The electron is not energy selected, and in fact could advantageously be replaced by a photon pulse if the light source were pulsed. Among the advantages of PEPIPICO is the ability to distinguish the products $O_2^+ + S^+$ from $O^+ + O^+$, both of which are produced from the dissociation of SO_2^{2+} (194).

The PEPIPICO experiment is providing new and interesting information about the mechanism for three-body reactions. A convenient and powerful method to display the data is by a contour diagram in which the times of flight of the two ion pairs are plotted on the two Cartesian coordinates, as shown in Figure 6 for CS_2^{2+} and COS^{2+}, which were investigated with a He(II) lamp. The large kinetic energy release manifests itself in the oblong contours. Since the kinetic energy release acts in opposite directions for the two departing fragments, the contours have a negative slope. For single dissociation steps, such as $CS^+ + S^+$, the slope is -1. However, the contour for the $C^+ + S^+$ fragments has a slope of -3.15, a value that is consistent with a sequential reaction in which the dication first breaks apart to $CS^+ + S^+$, after which the CS^+ dissociates to $C^+ + S$. On the other hand, in the COS^{2+} dissociation, $S^+ + C^+$ slope is 1.92, which is significantly less than the expected 2.33. Eland (197) analyzed these results in detail and was able to conclude that the initially produced CO^+ dissociated within the Coulomb field of the S^+ at a distance of about 8 Å. This indicates that the second dissociation occurred within 10^{-13} s of the first dissociation.

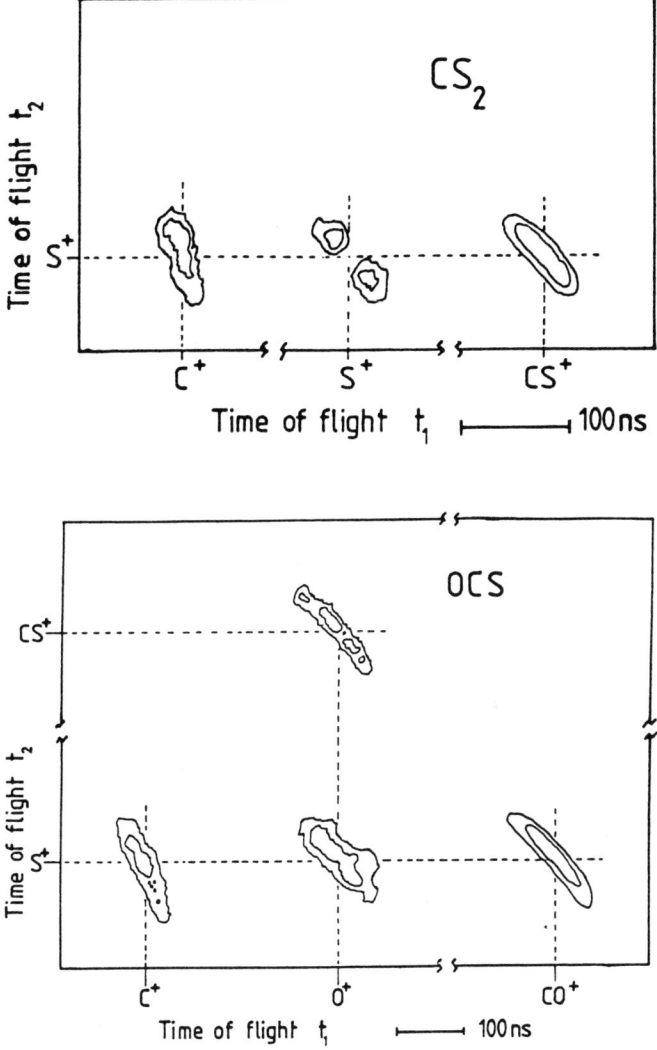

Figure 6 PEPIPICO contours for charged fragments from COS^{2+} and CS_2^{2+}. The contours are drawn at 66 and 33% of maximum intensity for each process. Taken from Eland (197) with permission.

These data are thus able to answer the interesting questions about the nature of multiple dissociations. Are they sequential or simultaneous? If the latter, are they statistical, isotropic, or colinear?

Another molecule investigated by PEPIPICO is NH_3^{2+} (198). The double ionization onsets for the production of $NH_2^+ + H^+$, $NH^+ + H^+$, and $N^+ + H^+$ were determined, using synchrotron radiation as a variable energy light source, to be 35.7, 43.3, and 49.2 eV, respectively. The advantage of the variable energy light source is that it allowed Stankiewicz et al (198) to perform a wavelength scan of the various ion production rates and thereby determine onsets for new dissociation channels. The first channel proceeds with a release of kinetic energy of 5 eV, which is consistent with the formation of the product ions, NH_2^+ ($^1A_1$). The data suggest very little vibrational excitation of this triatomic ion.

The two higher dissociation channels involve three products. The kinetic energy released (7.7 eV) between the NH^+ and H^+ in the second dissociation is most revealing. It is consistent with the formation of NH^+ in the ground state, and with zero kinetic energy of the H atom. This rules out a sequential reaction in which the H atom is released after the two ions, say $NH_2^+ + H^+$, have separated, because in that case the H atom would carry off kinetic energy. Rather, the H atom must have been released either simultaneously with the two charged fragments, or more likely prior to their formation.

At much higher energies, in the region in which Auger electrons are ejected, it is possible to collect coincidences between the energy-selected Auger electrons and the doubly charged ions or products, i.e. PEPICO. The removal of an inner core electron by radiation of about 1200 eV results in an Auger transition in which a higher energy electron falls into the empty core hole and the Auger electron is ejected at a specific kinetic energy. Unlike the PEPIPICO or PIPICO experiments discussed above, this PEPICO approach selects the internal energy of the doubly charged ion. In fact, the particular electronic configuration can be selected.

Such PEPICO experiments have been pioneered in recent years by Eberhardt and co-workers (199–201). A number of decay channels are possible at these high energies even for simple diatomic ions such as the N_2^{2+} ion, which is observed first at an energy of 43.2 eV. These are as follows:

$$N_2^{2+} \rightarrow \begin{cases} N^+ + N^+ & 38.8\,eV \\ N^{2+} + N & 56.3\,eV \\ N^{2+} + N^+ + e^- & 69\,eV \end{cases}$$

14.
15.
16.

The indicated energies are the dissociation limits and refer to the products in their lowest internal energy state. By measuring the kinetic energy of the fragment ions, the final ion states can be determined.

Because the initial electron is ejected from a core hole localized at a particular atom, it would seem that the subsequent dissociation might be site specific. This has been found to be true to some extent in the case of N_2O (200). When the 2π electron is removed in the Auger decay, the N–N bond is preferentially broken. On the other hand, "complete dissociation into three atomic fragments occurs when the innermost valence σ electrons participate in the Auger decay" (200). However, competing dissociation paths cannot be eliminated.

ACKNOWLEDGMENTS AND APOLOGIES

It is a pleasure to acknowledge my co-workers in the field of VUV photochemistry, among them P. M. Guyon, O. Dutuit, I. Nenner, S. Leach, T. Govers, R. Botter, J. Steadman, J. D. Shao, J. C. Morrow, and S. Olesik. I also thank the National Science Foundation and the Department of Energy for financial support. Finally, I apologize to all of those who have contributed to the field of VUV photophysics, whose works I have intentionally or unintentionally ignored in this review.

Literature Cited

1. Rabalais, J. W. 1977. *Principles of Ultraviolet Photoelectron Spectroscopy*, pp. 18–48. New York: Wiley
2. Samson, J. A. R. 1967. *Techniques of Vacuum Ultraviolet Spectroscopy*, pp. 94–175. New York: Wiley
3. Berkowitz, J. 1979. *Photoabsorption, Photoionization, and Photoelectron Spectroscopy*, pp. 408–61. New York: Academic
4. Kunz, C. 1979. *Synchrotron Radiation, Techniques and Applications*. New York: Springer
5. Koch, E. E. 1983. *Handbook on Synchrotron Radiation*, Vol. 1. Amsterdam: North Holland
6. Marr, G. V. 1987. *Handbook on Synchrotron Radiation*, Vol. 2. Amsterdam: North Holland
7. Baumgartel, H., Jochims, H. W., Brutschy, B. 1987. *Z. Phys. Chem.* 154: 1–30
8. Attwood, D., Halbach, K., Kim, K. J. 1985. *Science* 228: 1265–72
9. Attwood, D. T., Kim, K. J. 1986. *Nucl. Instrum. Methods Phys. Res. A* 246: 86–90
10. Ortega, J. M., Billardon, M., Jezesquel, G., Thiry, P., Petroff, Y. 1984. *J. Phys.* 45: 1883–88
11. Wilke, V., Schmidt, W. 1979. *Appl. Phys.* 18: 177–81
12. Page, R. H., Larkin, R. J., Kung, A. H., Shen, Y. R., Lee, Y. T. 1987. *Rev. Sci. Instrum.* 58: 1616–20
13. Miller, J. C., Compton, R. N., Cooper, C. D. 1982. *Laser Techniques for Extreme Ultraviolet Spectroscopy*, ed. T. R. McIlrath, R. R. Freeman. New York: Am. Inst. Phys.
14. Marinero, E. E., Rettner, C. T., Zare, R. N., Kung, A. H. 1983. *Chem. Phys. Lett.* 95: 486–91
15. Hilbig, R., Wallenstein, R. 1983. *IEEE J. Quant. Electron.* QE-19: 1759–70
16. Payne, M. G., Garrett, W. R., Ferrell, W. R. 1984. *Inst. Phys. Conf. Ser.* 71: 195–204
17. Lin, S. H., Fujimura, Y., Neusser, H. J., Schlag, E. W. 1984. *Multiphoton Spectroscopy of Molecules*. Orlando, Fla: Academic. 206 pp.
18. Madey, J. M. J. 1971. *J. Appl. Phys.* 42: 1906–13
19. Madey, J. M. J. 1984. *Free Electron Generation of Extreme Ultraviolet Coherent Radiation*, ed. J. M. J. Madey, C. Pelligrini. New York: Am. Inst. Phys.
20. Rohatgi, R., Schwettman, H. A., Smith, T. I., Swent, R. L. 1988. *Nucl. Instrum. Methods Phys. Res. A* 272: 32–36
21. Prazeres, R., Ortega, J. M., Bazin, C., Bergher, M., Billardon, M., et al. 1987. *Europhys. Lett.* 4: 817–22
22. Leach, S. 1988. *Frontiers of Laser Spectroscopy of Gases, NATO ASI Ser. C* 234: 89–152
23. Nenner, I., Beswick, J. A. 1987. In

Handbook on Synchrotron Radiation, ed. G. V. Marr, 2: 355–465. Amsterdam: Elsevier
24. Guyon, P. M., Nenner, I. 1980. *Appl. Opt.* 19: 4068–79
25. Dehmer, J. L., Dill, D., Parr, A. C. 1985. *Photophysics and Photochemistry in the Vacuum Ultraviolet. NATO ASI Ser. C,* Vol. 142
26. Bardsley, J. N. 1967. *Chem. Phys. Lett.* 1: 229–32
27. Smith, A. L. 1970. *Philos. Trans. R. Soc. London Ser. A* 268: 169–75
28. Kinsinger, J. A., Taylor, J. W. 1973. *Int. J. Mass Spectrom. Ion Phys.* 11: 461–74
29. Ferreira, L. P. 1984. PhD thesis. Univ. Paris, Orsay
30. Eland, J. H. D. 1980. *J. Chem. Phys.* 72: 6015–19
31. White, M. G., Grover, J. R. 1985. *Proc. Conf. Autoionization of Atoms and Small Molecules,* ed. J. Berkowitz. Argonne, Ill: Argonne Nat. Lab.
32. Baer, T., Guyon, P. M. 1986. *J. Chem. Phys.* 85: 4765–78
33. Dehmer, P. M., Chupka, W. A. 1975. *J. Chem. Phys.* 62: 4525–34
34. Baer, T., Guyon, P. M., Nenner, I., Tabche-Fouhaile, A., Botter, R., Ferreira, L. F. A., Govers, T. R. 1979. *J. Chem. Phys.* 70: 1585–92
35. Guyon, P. M., Baer, T., Nenner, I. 1983. *J. Chem. Phys.* 78: 3665–72
36. Hubin-Franskin, M. J., Delwiche, J., Guyon, P. M. 1987. *Z. Phys. D* 5: 203–16
37. Morin, P., Nenner, I. 1986. *Phys. Rev. Lett.* 56: 1913–16
38. Giusti-Suzor, A., Jungen, C. 1984. *J. Chem. Phys.* 80: 986–1000
39. Chupka, W. A., Miller, P. J., Eyler, E. E. 1988. *J. Chem. Phys.* 88: 3032–36
40. Berry, R. S., Nielson, S. E. 1970. *Phys. Rev. A* 1: 383–402
41. Miller, P. J., Chupka, W. A., Eland, J. H. D. 1988. *Chem. Phys.* 122: 395–401
42. Chupka, W. A., Miller, P. J., Eyler, E. E. 1988. *J. Chem. Phys.* 88: 3032–36
43. O'Halloran, M. A., Pratt, S. T., Tomkins, F. S., Dehmer, J. L., Dehmer, P. M. 1988. *Chem. Phys. Lett.* 146: 291–96
44. Kruit, P., Read, F. H. 1983. *J. Phys. E* 16: 313–24
45. Cornaggia, C., Guisti-Suzor, A., Jungen, C. 1987. *J. Chem. Phys.* 87: 3934–41
46. Nenner, I., Guyon, P. M., Baer, T., Govers, T. R. 1980. *J. Chem. Phys.* 72: 6587–92
47. Murray, P. T., Baer, T. 1979. *Int. J. Mass Spectrom. Ion Phys.* 30: 165–74
48. Dujardin, G., Leach, S., Dutuit, O., Govers, T., Guyon, P. M. 1983. *J. Chem. Phys.* 79: 644–57
49. White, M. G., Leroi, G. E., Ho, M. H., Poliakoff, E. D. 1987. *J. Chem. Phys.* 87: 6553–58
50. Sappey, A. D., Harrington, J. E., Weisshaar, J. C. 1988. *J. Chem. Phys.* 88: 5243–45
51. Miller, P. J., Li, L., Chupka, W. A., Colson, S. D. 1988. *J. Chem. Phys.* 88: 2972–75
52. Lefebvre-Brion, H., Dehmer, P., Chupka, W. A. 1986. *J. Chem. Phys.* 85: 45–50
53. Dehmer, J. L., Parr, A. C., Southworth, S. H. 1987. *Handbook on Synchrotron Radiation,* ed. G. V. Marr, 2: 241–353. Amsterdam: Elsevier
54. Southworth, S. H., Parr, A. C., Hardis, J. E., Dehmer, J. L. 1987. *J. Chem. Phys.* 87: 5125–30
55. Morin, P., Nenner, I., Adam, M. Y., Hubin-Franskin, M. J., Delwiche, J., Lefebvre-Brion, H., Giusti-Suzor, A. 1982. *Chem. Phys. Lett.* 92: 609–14
56. Miller, P. J., Li, L., Chupka, W. A., Colson, S. D. 1988. *J. Chem. Phys.* 89: 3921–22
57. Rault, M., LeRouzo, H., Raseev, G., Lefebvre-Brion, H. 1983. *J. Phys. B* 16: 4601–17
58. Stockbauer, R., Cole, B. E., Ederer, D. L., West, J. B., Parr, A. C., Dehmer, J. L. 1979. *Phys. Rev. Lett.* 43: 757–61
59. Fano, V. 1970. *Phys. Rev. A* 2: 353–65
60. Lefebvre-Brion, H., Giusti-Suzor, A., Raseev, G. 1985. *J. Chem. Phys.* 83: 1557–66
61. Dehmer, J. L., Dill, D. 1979. *Electron-Molecule and Photon-Molecule Collisions,* ed. T. Rescigno, V. McKoy, B. Schneider, pp. 225–66. New York: Plenum
62. Southworth, S. H., Parr, A. C., Hardis, J. E., Dehmer, J. L. 1986. *Phys. Rev. A* 33: 1020–23
63. Southworth, S. H., Parr, A. C., Hardis, J. E., Dehmer, J. L., Holland, D. M. P. 1986. *Nucl. Instrum. Methods Phys. Res. A* 246: 782–86
64. Parr, A. C., Hardis, J. E., Southworth, S. H., Feigerle, C. S., Ferrett, T. H., et al. 1988. *Phys. Rev. A* 37: 437–43
65. Hardis, J. E., Ferrett, T. A., Southworth, S. H., Parr, A. C., Roy, P., Dehmer, J. L., Dehmer, P. M., Chupka, W. A. 1988. *J. Chem. Phys.* 89: 812–13
66. Southworth, S. H., Parr, A. C., Hardis, J. E., Dehmer, J. L. 1987. *J. Chem. Phys.* 87: 5125–30
67. Miller, P. J., Chupka, W. A., Winnic-

zek, J., White, M. G. 1988. *J. Chem. Phys.* 89: 4058–61
68. Poliakoff, E. D., Ho, M. H., White, M. G., Leroi, G. E. 1986. *Chem. Phys. Lett.* 130: 91–97
69. Hubin-Franskin, M. J., Delwiche, J., Guyon, P. M. 1987. *Z. Phys. D* 5: 203–16
70. Delwiche, J., Hubin-Franskin, M. J., Guyon, P. M., Nenner, I. 1981. *J. Chem. Phys.* 74: 4219–27
71. Dutuit, O., Tabche-Fouhaile, A., Nenner, I., Frohlich, H., Guyon, P. M. 1985. *J. Chem. Phys.* 83: 584–96
72. Glass-Maujean, M., Breton, J., Guyon, P. M. 1987. *Z. Phys. D* 5: 189–201
73. Glass-Maujean, M., Breton, J., Guyon, P. M. 1985. *J. Chem. Phys.* 83: 1468–70
74. Glass-Maujean, M., Guyon, P. M., Breton, J. 1986. *Phys. Rev. A* 33: 346–50
75. Tabche-Fouhaile, A., Hubin-Franskin, M. J., Delwiche, J. P., Frohlich, H., Ito, K., Guyon, P. M., Nenner, I. 1983. *J. Chem. Phys.* 79: 5894–99
76. Wormer, J., Moller, T., Stapelfeldt, J., Zimmerer, G., Haaks, D., et al. 1988. *Z. Phys. D* 7: 383–95
77. Poliakoff, E. D., Ho, M. H., Leroi, G. E., White, M. G. 1986. *J. Chem. Phys.* 85: 5529–34
78. Moller, T., Stapelfeldt, J., Beland, M., Zimmerer, G. 1985. *Chem. Phys. Lett.* 117: 301–6
79. Moller, T., Beland, M., Zimmerer, G. 1985. *Phys. Rev. Lett.* 55: 2145–48
80. Moller, T., Beland, M., Zimmerer, G. 1987. *Chem. Phys. Lett.* 136: 551–56
81. Poliakoff, E. D., Dehmer, J. L., Parr, A. C., Leroi, G. E. 1987. *J. Chem. Phys.* 86: 2557–62
82. Chupka, W. A., Berkowitz, J. 1969. *J. Chem. Phys.* 51: 4244–68
83. Marinero, E. E., Vasudev, R., Zare, R. N. 1983. *J. Chem. Phys.* 78: 692–99
84. Steadman, J., Baer, T. 1989. *J. Chem. Phys.* In press
85. Senn, P., Dressler, K. 1987. *J. Chem. Phys.* 87: 6908–14
86. Huber, K. P., Herzberg, G. 1979. *Molecular Spectra and Molecular Structure*, Vol. 4, *Constants of Diatomic Molecules*. New York: Van Nostrand Reinhold. 715 pp.
87. Anderson, S. L., Kubiak, G. D., Zare, R. N. 1984. *Chem. Phys. Lett.* 105: 22–27
88. Pratt, S. T., Dehmer, P. M., Dehmer, J. L. 1986. *J. Chem. Phys.* 85: 3379–85
89. O'Halloran, M. A., Pratt, S. T., Dehmer, P. M., Dehmer, J. L. 1987. *J. Chem. Phys.* 87: 3288–98
90. Xu, E. Y., Tsuboi, T., Kachru, R., Helm, H. 1987. *Phys. Rev. A* 36: 5645–53
91. Hickman, A. P. 1987. *Phys. Rev. Lett.* 59: 1553–56
92. Chupka, W. A. 1987. *J. Chem. Phys.* 87: 1488–98
93. Kung, A. H., Page, R. H., Larkin, R. J., Shen, Y. R., Lee, Y. T. 1986. *Phys. Rev. Lett.* 56: 328–31
94. Bonnie, J. H. M., Eenshuistra, P. J., Los, J., Hopman, H. J. 1986. *Chem. Phys. Lett.* 125: 27–32
95. Parker, D. H., Buck, J. D., Chandler, D. W. 1987. *J. Phys. Chem.* 91: 2035–37
96. Verschuur, J. W. J., Noordam, L. D., Bonnie, J. H. M., van Linden van den Heuvell, H. B. 1988. *Chem. Phys. Lett.* 146: 283–90
97. Buck, J. D., Parker, D. H., Chandler, D. W. 1988. *J. Phys. Chem.* 92: 3701–5
98. Koot, W., van der Zande, W. J., Los, J. 1987. *Phys. Rev. Lett.* 58: 2746–49
99. Koot, W., van der Zande, W. J., Los, J., Keiding, S. R., Bjerre, N. 1989. *Phys. Rev. A* 39: 590–604
100. Ono, Y., Linn, S. H., Prest, H. F., Ng, C. Y., Miescher, E. 1980. *J. Chem. Phys.* 73: 4855–61
101. Pratt, S. T., Dehmer, P. M., Dehmer, J. L. 1986. *J. Chem. Phys.* 85: 5535–40
102. Page, R. H., Larkin, R. J., Shen, Y. R., Lee, Y. T. 1988. *J. Chem. Phys.* 88: 2249–63
103. Hart, D. J., Hepburn, J. W. 1989. *Chem. Phys.* 129: 51–64
104. Katayama, D. H., Huffman, R. E., O'Bryan, C. L. 1973. *J. Chem. Phys.* 59: 4309–19
105. Moller, T., Zimmerer, G. 1987. *Phys. Scr.* T17: 177–85
106. Muller-Dethlefs, K., Sander, M., Schlag, E. W. 1984. *Z. Naturforsch. Teil A* 39: 1089–91
107. Zubek, M., King, G. C., Rutter, P. M. 1988. *J. Phys. B* 21: 3585–94
108. Schlag, E. W., Frey, R., Gotchev, B., Peatman, W. B., Pottak, H. 1977. *Chem. Phys. Lett.* 51: 406–8
109. Wilson, W. G., Viswanathan, K. S., Sekreta, E., Reilly, J. P. 1984. *J. Phys. Chem.* 88: 672–43
110. Viswanathan, K. S., Sekreta, E., Davidson, E. R., Reilly, J. P. 1986. *J. Phys. Chem.* 90: 5078–84
111. Sander, M., Chewter, L. A., Muller-Dethlefs, K., Schlag, E. W. 1987. *Phys. Rev. A* 36: 4543–46
112. Berkowitz, J. 1988. *Radiat. Phys. Chem.* 32: 23–30

113. Berkowitz, J., Mayhew, C. A., Ruscic, B. 1988. *J. Chem. Phys.* 88: 7396–7404
114. Berkowitz, J., Greene, J. P., Cho, H., Ruscic, B. 1987. *J. Chem. Phys.* 86: 1235–48
115. Chen, P., Colson, S. D., Chupka, W. A. 1988. *Chem. Phys. Lett.* 147: 466–70
116. Shiromaru, H., Achiba, Y., Kimura, K., Lee, Y. T. 1987. *J. Phys. Chem.* 91: 17–19
117. Locht, R., Momigny, J., Ruehl, E., Baumgartel, H. 1987. *Chem. Phys.* 117: 305–13
118. Minkwitz, R., Lekies, R., Jochims, H. W., Ruehl, E., Baumgartel, H. 1986. *Z. Naturforsch. Teil B* 41: 784–86
119. Moini, M., Leroi, G. E. 1986. *J. Phys. Chem.* 90: 4002–6
120. Traeger, J. C. 1986. *J. Phys. Chem.* 90: 4114–18
121. Ziesel, J. P., Malinovich, Y., Ohmichi, N., Lifshitz, C. 1987. *Chem. Phys. Lett.* 136: 81–86
122. McAdoo, D. J., Traeger, J. C., Hudson, C. E., Griffin, L. L. 1988. *J. Phys. Chem.* 92: 1524–30
123. Ziesel, J. P., Lifshitz, C. 1987. *Chem. Phys.* 117: 227–35
124. Lifshitz, C., Rejwan, M., Levin, I., Peres, T. 1988. *Int. J. Mass Spectrom. Ion Proc.* 84: 271–82
125. Ruscic, B., Mayhew, C. A., Berkowitz, J. 1988. *J. Chem. Phys.* 88: 5580–93
126. Cordis, L., Gantefor, G., Hesslich, J., Ding, A. 1986. *Z. Phys. D* 3: 323–27
127. Mitsuke, K., Ohno, K. 1989. *J. Phys. Chem.* 93: 501–3
128a. Norwood, K., Guo, J. H., Luo, G., Ng, C. Y. 1988. *J. Chem. Phys.* 88: 4098–99
128b. Norwood, K., Guo, J. H., Luo, G., Ng, C. Y. 1989. *Chem. Phys.* 129: 109–23
129. Dehmer, P. M., Pratt, S. T., Dehmer, J. L. 1987. *J. Phys. Chem.* 91: 2593–98
130. Brutschy, B., Bisling, P., Ruehl, E., Baumgartel, H. 1987. *Z. Phys. D* 5: 217–31
131a. Ruehl, E., Bisling, P., Brutschy, B., Beckmann, K., Leisin, O., Morgner, H. 1986. *Chem. Phys. Lett.* 128: 512–16
131b. Kamke, B., Kamke, W., Wang, Z., Ruehl, E., Brutschy, B. 1987. *J. Chem. Phys.* 86: 2525–29
132. Shiromaru, H., Shinohara, H., Washida, N., Yoo, Y.-S., Kimura, K. 1987. *Chem. Phys. Lett.* 141: 7–11
133. Ruehl, E., Brutschy, B., Bisling, P., Baumgartel, H. 1988. *Ber. Bunsen Phys. Chem.* 92: 194–200
134. Ruehl, E., Bisling, P., Brutschy, B., Baumgartel, H. 1986. *J. Electron Spectrosc. Relat. Phenom.* 41: 411–18
135. Grover, J. R., Walters, E. A., Newman, J. K., White, M. G. 1985. *J. Am. Chem. Soc.* 107: 7329–39
136. Walters, E. A., Grover, J. R., White, M. G. 1986. *Z. Phys. D* 4: 103–10
137. Grover, J. R., Walters, E. A., Hui, E. T. 1987. *J. Phys. Chem.* 91: 3233–37
138. Walters, E. A., Grover, J. R., White, M. G., Hui, E. T. 1987. *J. Phys. Chem.* 91: 2758–62
139. Grover, J. R., Walters, E. A., Arneberg, D. L., Santandrea, C. 1988. *Chem. Phys. Lett.* 146: 305–9
140. Baer, T. 1979. In *Gas Phase Ion Chemistry*, ed. M. T. Bowers, 1: 153–97. New York: Academic
141. Baer, T. 1986. *Adv. Chem. Phys.* 64: 111–202
142. Proch, D., Rider, D. M., Zare, R. N. 1981. *Chem. Phys. Lett.* 81: 430–34
143. Durant, J. L., Rider, D. M., Anderson, S. L., Proch, F. D., Zare, R. N. 1984. *J. Chem. Phys.* 80: 1817–25
144. Neusser, H. J., Kuhlewind, H., Boesl, U., Schlag, E. W. 1985. *Ber. Bunsenges. Phys. Chem.* 89: 276–81
145. Neusser, H. J. 1987. *Int. J. Mass Spectrom. Ion Proc.* 79: 141–81
146. Kuhlewind, H., Kiermeier, A., Neusser, H. J., Schlag, E. W. 1987. *J. Chem. Phys.* 87: 6488–98
147. Andlauer, B., Ottinger, C. 1971. *J. Chem. Phys.* 55: 1471–72
148. Kiermeier, A., Kuhlewind, H., Neusser, H. J., Schlag, E. W., Lin, S. H. 1988. *J. Chem. Phys.* 88: 6182–90
149. McCulloh, K. E., Dibeler, V. H. 1976. *J. Chem. Phys.* 64: 4445–50
150. Marcus, R. A., Rice, O. K. 1951. *J. Phys. Coll. Chem.* 55: 894–907
151. Rosenstock, H. M., Wallenstein, M. B., Wahrhaftig, A. L., Eyring, H. 1952. *Proc. Natl. Acad. Sci. USA* 38: 667
152. Troe, J. 1988. *Ber. Bunsenges. Phys. Chem.* 92: 242–52
153. Chesnavich, W. J., Bowers, M. T. 1984. *Prog. React. Kinet.* 11: 139–267
154. Tai, T. L., El-Sayed, M. A. 1986. *Chem. Phys. Lett.* 130: 224–30
155. Lemaire, J., Dimicoli, I., Botter, R. 1987. *Chem. Phys.* 115: 129–42
156. Richard-Viard, M., Dutuit, O., Ait-Kaci, M., Guyon, P. M. 1987. *J. Phys. B* 20: 2247–54
157. Low, K. G., Hampton, P. D., Powis, I. 1985. *Chem. Phys.* 100: 401–13
158. Nishimura, T., Zha, Q., Meisels, G. G. 1987. *J. Chem. Phys.* 87: 4589–97
159. Zha, Q., Nishimura, T., Meisels, G. G. 1988. *Int. J. Mass Spectrom. Ion Proc.* 83: 1–12

160. Olesik, S., Baer, T., Morrow, J. C. 1986. *J. Phys. Chem.* 90: 3563–68
161a. Panczel, M., Baer, T. 1984. *Int. J. Mass Spectrom. Ion Proc.* 58: 43–61
161b. Nishimura, T., Das, P. R., Meisels, G. G. 1986. *J. Chem. Phys.* 84: 6190–99
162. Das, P. R., Gilman, J. P., Meisels, G. G. 1986. *Int. J. Mass Spectrom. Ion Proc.* 68: 155–65
163. Das, P. R., Nishimura, T., Meisels, G. G. 1985. *J. Phys. Chem.* 89: 2808–12
164. Shao, J. D., Baer, J. 1988. *Int. J. Mass Spectrom. Ion Proc.* 86: 357–67
165. Baer, T., Morrow, J. C., Shao, J. D., Olesik, S. 1988. *J. Am. Chem. Soc.* 110: 5633–38
166. Baer, T., Dutuit, O., Mestdagh, H., Rolando, C. 1988. *J. Phys. Chem.* 92: 5674–79
167. Shao, J. D., Baer, T., Lewis, D. K. 1988. *J. Phys. Chem.* 92: 5123–28
168. Shao, J. D., Baer, T., Morrow, J. C., Fraser-Monteiro, M. L. 1987. *J. Chem. Phys.* 87: 5242–50
169. Nenner, I., Dutuit, O., Richard-Viard, M., Morin, P., Zewail, A. H. 1988. *J. Am. Chem. Soc.* 110: 1093–98
170. Heinrich, N., Louage, F., Lifshitz, C., Schwartz, H. 1988. *J. Am. Chem. Soc.* 110: 8183–92
171. Caballor, R., Poblet, J. M., Sarasa, J. P., Lovella, S., Sole, A. 1988. *J. Phys. Chem.* 92: 3336–41
172. Morrow, J. C., Baer, T. 1988. *J. Phys. Chem.* 92: 6567–71
173. Hehre, W. J., Radom, L., Schleyer, P. v. R., Pople, J. A. 1986. *Ab Initio Molecular Orbital Theory*. New York: Wiley
174. Yates, B. F., Bouma, W. J., Radom, L. 1984. *J. Am. Chem. Soc.* 106: 5805–8
175. Case, K. M. 1950. *Phys. Rev.* 80: 797–806
176. Holmes, J. L., Mommers, A. A., Szulejko, J. E., Terlouw, J. K. 1984. *J. Chem. Soc. Chem. Commun.*, pp. 165–67
177. Eland, J. H. D., Wort, F. S., Lablanquie, P., Nenner, I. 1986. *Z. Phys. D* 4: 31–42
178. Millie, P., Nenner, I., Archirel, P., Lablanquie, P., Fournier, P., Eland, J. H. D. 1986. *J. Chem. Phys.* 84: 1259–69
179. Leach, S. 1988. *Radiat. Phys. Chem.* 32: 563–72
180. Eland, J. H. D.,. Price, S. D., Cheney, J. C., Lablanquie, P., Nenner, I., Fournier, P. G. 1988. *Philos. Trans. R. Soc. London Ser. A* 324: 247–55
181. Gill, P. M. W., Radom, L. 1988. *J. Am. Chem. Soc.* 110: 5311
182. Leach, S. 1986. *J. Electron Spectrosc. Relat. Phenom.* 41: 427–38
183. Lablanquie, P., Nenner, I., Eland, J. H. D., Delwiche, J., Hubin-Franskin, M. J., Morin, P. 1985. *Photophysics and Photochemistry Above 6 eV*, ed. F. Lahmani, p. 53. Amsterdam: Elsevier
184. Wannier, G. 1953. *Phys. Rev.* 90: 817–25
185. Lablanquie, P., Nenner, I., Millie, P., Morin, P., Eland, J. H. D., Hubin-Franskin, M. J., Delwiche, J. 1985. *J. Chem. Phys.* 82: 2951–60
186. Price, S. D., Eland, J. H. D., Fournier, P. G., Fournier, J., Millie, P. 1988. *J. Chem. Phys.* 88: 1511–15
187. Fournier, P. G., Fournier, J., Salama, F., Stark, D., Peyerimhoff, S. D., Eland, J. H.D. 1986. *Phys. Rev. A* 34: 1657–66
188. Richardson, P. J., Eland, J. H. D., Fournier, P. G., Cooper, D. L. 1986. *J. Chem. Phys.* 84: 3189–94
189. Winkoun, D., Dujardin, G., Hellner, L., Besnard, M. J. 1988. *J. Phys. B* 21: 1385–94
190. Leach, S. 1987. *J. Mol. Struct.* 157: 197–214
191. Hellner, L., Besnard, M. J., Dujardin, G., Malinovich, Y. 1988. *Chem. Phys.* 119: 391–97
192. Besnard, M. J., Hellner, L., Dujardin, G., Winkoun, D. 1988. *J. Chem. Phys.* 88: 1732–36
193. Winkoun, D., Dujardin, G. 1986. *Z. Phys. D* 4: 57–64
194. Dujardin, G., Besnard, M. J., Hellner, L., Malinovitch, Y. 1987. *Phys. Rev. A* 35: 5012–19
195. Lablanquie, P., Delwiche, J., Franskin-Hubin, M. J., Nenner, I., Eland, J. H. D., Ito, K. 1988. *J. Mol. Struct.* 174: 141–46
196. Eland, J. H. D., Wort, F. S., Royds, R. N. 1986. *J. Electron Spectrosc. Relat. Phenom.* 41: 297–309
197. Eland, J. H. D. 1987. *Mol. Phys.* 61: 725–45
198. Stankiewicz, M., Hatherly, P. A., Frasinski, L. J., Codling, K., Holland, D. M. P. 1989. *J. Phys. B* 22: 21–31
199. Eberhardt, W., Plummer, E. W., Lyo, I. W., Carr, R., Ford, W. K. 1987. *Phys. Rev. Lett.* 58: 207–10
200. Murphy, R., Eberhardt, W. 1988. *J. Chem. Phys.* 89: 4054–57
201. Larkins, F. P., Eberhardt, W., Lyo, I. W., Murphy, R., Plummer, E. W. 1988. *J. Chem. Phys.* 88: 2948–55.

PROTON TRANSLOCATION IN PROTEINS

Robert A. Copeland[1] *and Sunney I. Chan*

Arthur Amos Noyes Laboratory of Chemical Physics,
California Institute of Technology, Pasadena, California 91125

INTRODUCTION

The active transport of protons across the low dielectric barrier imposed by biological membranes is accomplished by a plethora of proteins that span the ca. 40 Å of the phospholipid bilayer. The free energy derived from the proton electrochemical potential established by the translocation of these protons can subsequently be used to drive vital chemical reactions of the cell, such as ATP synthesis and cell locomotion. Membrane-bound proton translocating proteins have now been found for a variety of organisms and tissues (1). The driving force for proton pumping in these proteins is supplied by numerous mechanisms, including light absorption (e.g. bacteriorhodopsin) (2a,b), ligand binding (e.g. ATPase) (3), and electrochemistry (e.g. electron transfer through cytochrome c oxidase) (4). Thus nature has devised a variety of methods for supplying the energy required for proton pumping by these proteins. Such diversity notwithstanding, the proteins most likely share some common elements of structure and mechanism that allow them to function as proton pumps. A number of theoretical mechanisms have been put forth for both general proton translocation (5–7) and for energy coupling in specific proton pumps. However, despite almost three decades of intensive research, the details of the mechanism(s) and structural requirements for proton pumping remain largely unresolved. To some extent this is the result of the paucity of structural information available for integral membrane proteins. This situation may soon improve as a result of advances in protein methodologies

---

[1] Present Address: Department of Molecular Pharmacology and Biochemistry, Merck Sharp and Dohme Laboratories, P.O. Box 2000, Rahway, NJ.

that have allowed several integral membrane proteins to be successfully crystalized (8), and the increased use of genetic engineering to obtain recombinant proton translocating proteins that will offer an opportunity to assess the importance of specific amino acids for the proton translocation process (9).

A number of reviews of protein-mediated proton translocation have appeared, mostly dealing with various of theoretical aspects of proton pumping (5–7, 10–12). However, the subject of the structural requirements for integral membrane proton translocators has not been reviewed in the recent literature. In this chapter we thus concentrate on the structural aspects of proton translocation by proteins. Rather than attempting a comprehensive review of all proton translocating proteins, we first focus on some of the more important theoretical mechanisms for protein-facilitated movement of protons across membranes and, where possible, show how well these theoretical mechanisms fit with experimental data for particular proteins: bacteriorhodopsin and F_o/F_1 ATP synthase. In the remainder of the chapter we provide a detailed account of the current state of knowledge for a particular proton-translocating protein, cytochrome c oxidase, which has been the focus of research in our laboratory for a number of years.

MECHANISMS FOR PROTON MOTION THROUGH PROTEINS

One can envisage a variety of mechanisms by which a protein could mediate transport of protons across a biological membrane. Several such mechanisms are considered in Figure 1 (10).

Figure 1A illustrates what might be referred to as the water wheel mechanism for proton pumping. Here a particular group within the protein picks up a proton while in contact with one side of the membrane. A conformational transition of the protein then ensues that moves the protonated group into contact with the other side of the membrane where the proton is released. Reversal of the protein conformational change completes the cycle and primes the protein for the next proton-pumping event. A mechanism of this type would require major rearrangements of a large portion of the protein matrix, and would thus require a significant free energy dissipation to drive the pumping cycle. Although conformational transitions clearly occur in proton-translocating proteins, we know of no experimental evidence at the moment for alternating access of portions of the protein matrix to the two sides of the membrane, as would be required for the water wheel mechanism.

Figure 1B illustrates the "gated channel" mechanism for proton pump-

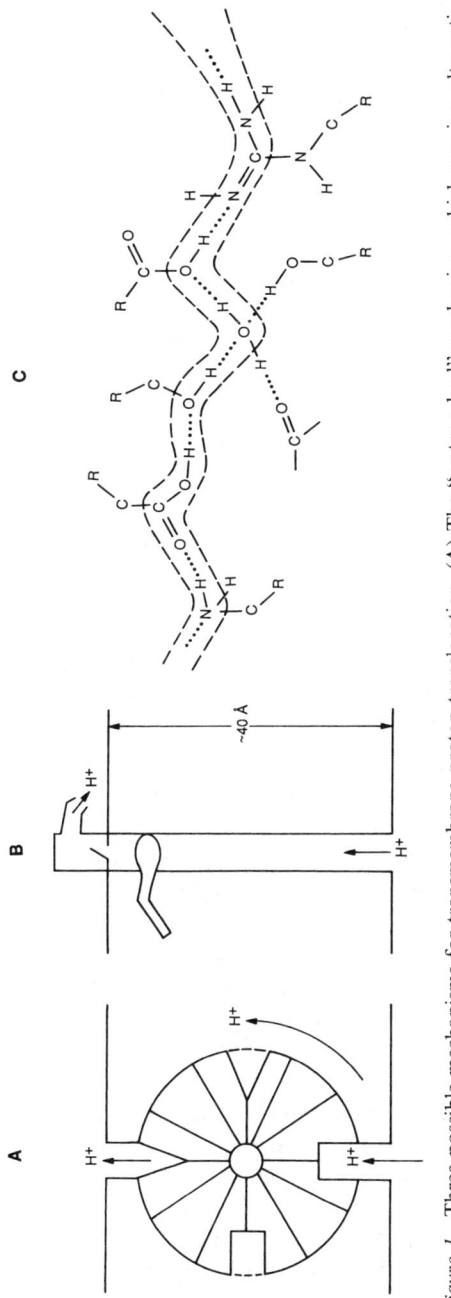

Figure 1 Three possible mechanisms for transmembrane proton translocation. (A) The "water wheel" mechanism, which requires alternating access of a particular portion of the protein matrix to the two sides of the membrane. (B) The "gated channel" mechanism in which some group or groups within the protein provide a physical barrier to proton movement across the bilayer. In this mechanism the barrier is displaced in response to some conformational transition of the protein. (C) The "hydrogen-bonded chain" mechanism in which a chain of hydrogen bonds is set up among amino acid side chains that spans the membrane bilayer. This figure was adapted from Refs. (10, 11).

ing in which protons passively travel through a pore in the protein until they encounter a group or groups that block their further movement to the other side of the membrane. This proton gate acts as a turnstile, allowing the proton alternate access to the two sides of the membrane in response to some signal from the protein matrix (i.e. a conformational change). As first pointed out by Nagle & Tristam-Nagle (10), pores through integral membrane proteins cannot be simple water channels because the electric field in such a case would exceed the dielectric breakdown field of most materials (ca. 10^6 V/cm). Thus, in order to avoid dielectric breakdown, one must restrict attention to *narrow* channels.

Figure 1C illustrates the hydrogen-bonded chain (HBC) mechanism for proton translocation that has been championed by Nagle and co-workers (10, 11). Here the side chains of certain amino acids are used to construct a hydrogen-bonding network that traverses the bilayer. The HBC hypothesis has been well received by the biophysical community and we thus describe it in some detail. Figure 2 illustrates the mechanism of proton pumping for a HBC constructed of hydroxyl side chains (10). In the resting state of the protein, the hydrogen bonds along the HBC occur in a particular configuration that minimizes their collective potential energy; this is represented in Figure 2 by having all of the protons on the left side of the hydrogen bonds (a). A conformational change then occurs that alters the relative energies of this resting state and some alternative configuration (c; represented by having all of the protons now on the right side of the bonds) so that the alternative configuration is now favored, and the protons along the chain move to the other side of the bonds in response to the configurational energy differential. The protein, which has up to now assumed an intermediate excited state, must ultimately relax back to its original conformation, as the input impulse is dissipated. This decay can be coupled energetically to endergonic translocation of protons across the membrane: An ion can enter the chain on the right side and initiate the tandem propagation of a charge, from left to right, along the chain. In this manner the protein reverts back to the original "resting" configuration, completing the cycle. An in-depth discussion of the HBC mechanism can be found in the recent review by Nagle & Tristam-Nagle (10).

In Figure 2 the HBC is constructed by using only hydroxyl side chains as the proton conductors. In principle, any groups capable of simultaneously serving as a proton donor and acceptor could form part of a HCB, including amide backbone protons. Nagle and co-workers however, argue that backbone protons are unlikely to participate in the HCB because of the requirement for bond "turning" during pumping (10). "Turning" of amide groups, particularly in regions of defined secondary structure, would

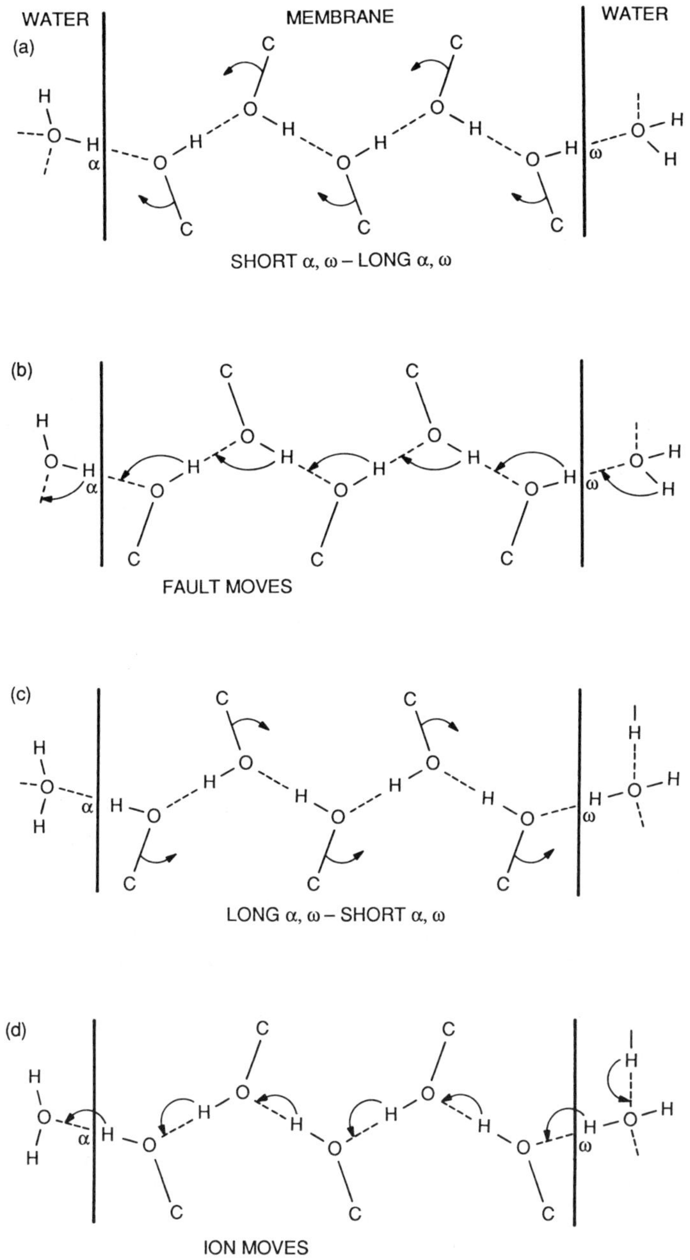

Figure 2 Mechanism for ion movement across the membrane bilayer facilitated by a hydrogen-bonded chain of amino acids. Adapted from Ref. (10).

involve a prohibitively large activation energy. Thus HCB are most likely restricted to the side chains of certain amino acids. The amino acid side chains capable of HBC involvement are illustrated in Figure 3. Assuming an average length of 2.5–3.5 Å for each hydrogen bond, and making allowance that not all hydrogen bonds will be parallel to the membrane normal, Nagle & Morowitz conclude that a HBC would require 20 amino acid residues to traverse the 40 Å distance of a typical biological membrane (11). Such a HBC should be quite stable: Nagle & Morowitz estimate an enthalpy of formation of ca. 120 kcal/mol.

One attractive aspect of the HBC hypothesis is that it is amenable to testing. The hypothesis provides a rationale for locating the pumping machinery of a proton-translocating protein from a knowledge of the protein's primary and secondary structure. We now briefly review how well the HBC hypothesis is standing up against the structural data available for two proton-translocating proteins, namely bacteriorhodopsin (BR) and F_o/F_1 ATP synthase.

BACTERIORHODOPSIN

Bacteriorhodopsin is a proton-translocating protein for which a great deal of structural information is available, and thus it provides a good test of the HBC hypothesis (2a). BR is the major protein component of the purple membranes of *Halobacterium halobium*. The protein consists of a single polypeptide chain of 248 amino acids whose complete primary structure has been determined (13). The protein contains a single cofactor, all-*trans* retinal covalently linked to the protein via a Schiff base to lys 216, which serves as a photon sink to provide the driving force for proton pumping (2a). An electron diffraction map has been obtained to 7 Å resolution for two-dimensional ordered arrays of BR within purple membranes. This map suggests seven transmembrane helical segments (14), whose location along the protein's primary sequence has been inferred from hydropathy analysis and chemical modification experiments (2a). Figure 4 summarizes the most widely accepted structural model for BR. HBC-forming amino acid residues located within the seven transmembrane helices of BR are circled in Figure 4. There is clearly an ample number of these residues for the formation of a membrane-spanning HBC, particularly when one recalls that a single HBC may be composed of residues from more than one transmembrane helix in the folded protein. Spectroscopic data have suggested that during the proton pumping photocycle of BR, protonation/deprotonation events occur at several potential HBC residues, tyrosine and aspartate (15); environmental change also occurs about tryptophan residues (16). Most recently, Khorana and co-workers have expressed the

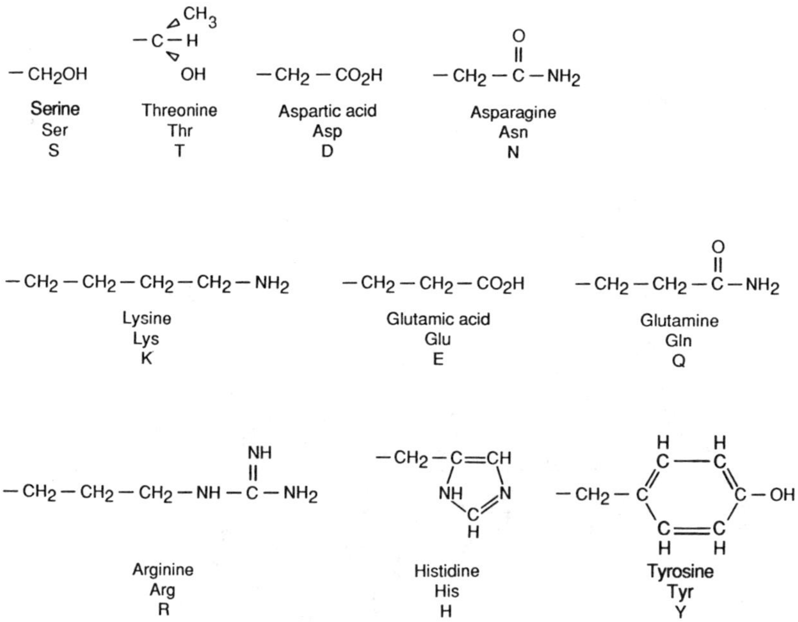

	CH₃		
—CH₂OH	—C—H	—CH₂—CO₂H	—CH₂—C(=O)—NH₂
	OH		
Serine	Threonine	Aspartic acid	Asparagine
Ser	Thr	Asp	Asn
S	T	D	N

—CH₂—CH₂—CH₂—CH₂—NH₂   —CH₂—CH₂—CO₂H   —CH₂—CH₂—C(=O)—NH₂

Lysine   Glutamic acid   Glutamine
Lys   Glu   Gln
K   E   Q

—CH₂—CH₂—CH₂—NH—C(=NH)—NH₂   —CH₂—C=CH—NH—CH=N   —CH₂—C₆H₄—OH

Arginine   Histidine   Tyrosine
Arg   His   Tyr
R   H   Y

Figure 3 The side-chain structures for hydrogen-bond-chain-forming amino acids. Adapted from Ref. (76).

gene for BR in *E. coli* and have begun systematic site-directed mutagenesis studies to identify the key amino acid residues required for proton pumping. Among the mutants studied so far, several involve residues at potential HBC sites, within the transmembrane helices: Tyr 185 → Phe; Asp 85 → Asn; Asp 96 → Asn; and Asp 212 → Glu, Asn, or Ala. These residues are highlighted in Figure 4 by *darker circles* (2a,b). As expected, the substitutions made here lead to dramatic effects on the proton pumping efficiency of the mutant BRs. In particular, mutants affecting Asp 85 or 96 completely abolish proton pumping activity!

ATP SYNTHASE

The ATP synthases of mitochondria, chloroplasts, and bacteria represent a second family of proton-translocating proteins for which significant structural information is available (17). These proteins are multi-subunit complexes that share the common structural motif illustrated in Figure 5. The proteins can be divided into two functional domains: a transmembrane proton-conducting domain, F_o, and an extramembrane nucleotide-binding domain, F_1. In the intact complex proton translocation by the F_o domain

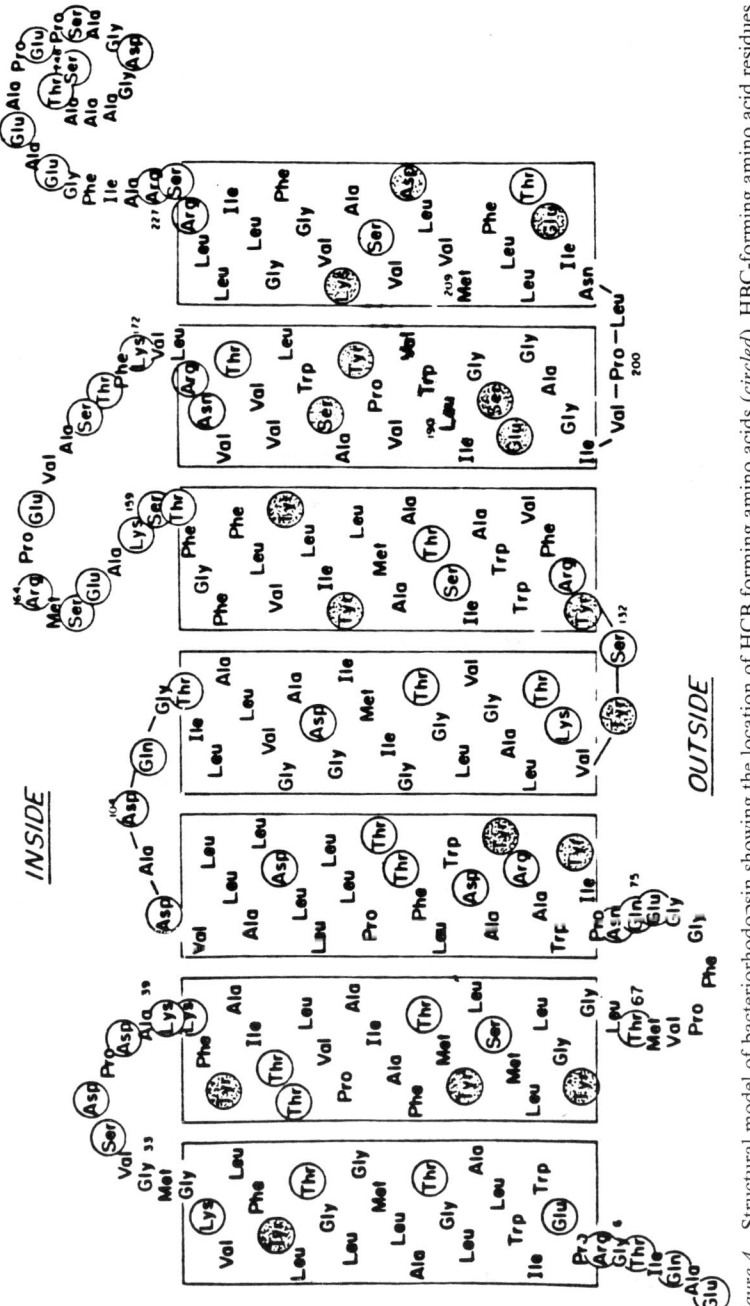

Figure 4 Structural model of bacteriorhodopsin showing the location of HCB forming amino acids (*circled*). HBC-forming amino acid residues that have been altered via site directed mutagenesis are highlighted by *shaded circles*. Adapted from Ref. (2a).

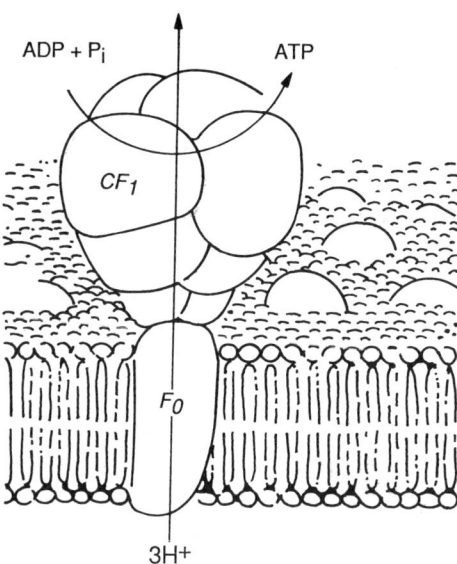

Figure 5 Structural model for the F_o/F_1 ATP synthase within a membrane bilayer.

is ATP-dependent, suggesting some type of allosteric communication between the F_o and F_1 domains (17). The F_o domain can be isolated free of the F_1 domain and reconstituted into artificial phospholipid vesicles. Under these conditions, the isolated F_o domain can also translocate protons in response to an ionophore-induced potassium diffusion potential.

The proton-translocating F_o domain consists of multiple copies of three subunits designated *a*, *b*, and *c*. Complete amino acid sequences have been reported for the *a*, *b*, and *c*, subunits of the F_o domain of the *E. coli* protein (18). Recently, isolated subunit *c* from the F_o domain of lettuce chloroplasts and yeast mitochondria has been reconstituted into phospholipid vesicles and shown to conduct protons. Thus there is good reason to suspect that subunit *c* constitutes at least part of the proton-translocating apparatus of the F_o domain (18). Unfortunately, there are only six conserved HBC-forming amino acid residues in subunit *c*. Sebald & Wachter accordingly have argued against the HBC hypothesis in the case of the ATP synthase (19). However, as first noted by Nagle & Tristam-Nagle, one must bear in mind that multiple copies of subunit *c* occur in the F_o domain and together these would provide more than the requisite 20 HBC-forming residues (10). Recently, Cox et al reported that subunit *a* from a variety of sources also contains a membrane-spanning amphipathic helix with a number of conserved polar residues (20). These workers suggest that these polar residues on subunit *a* combine with Asp 61 of subunit c to form a transmembrane proton-conducting network, as illustrated in Figure 6. This model is consistent with predictions of the secondary and tertiary structure for the F_o domain of ATP synthase.

It should be clear from the above discussion of the structural data for bacteriorhodopsin and ATP synthase that no direct experimental support for the HBC-hypothesis has yet been reported. The best one can say at this point is that the available structural information is not inconsistent with this hypothesis.

Protons are not unique in their role as the coupling ions in energy transducing systems. Recently, Skulachev reviewed the evidence for sodium

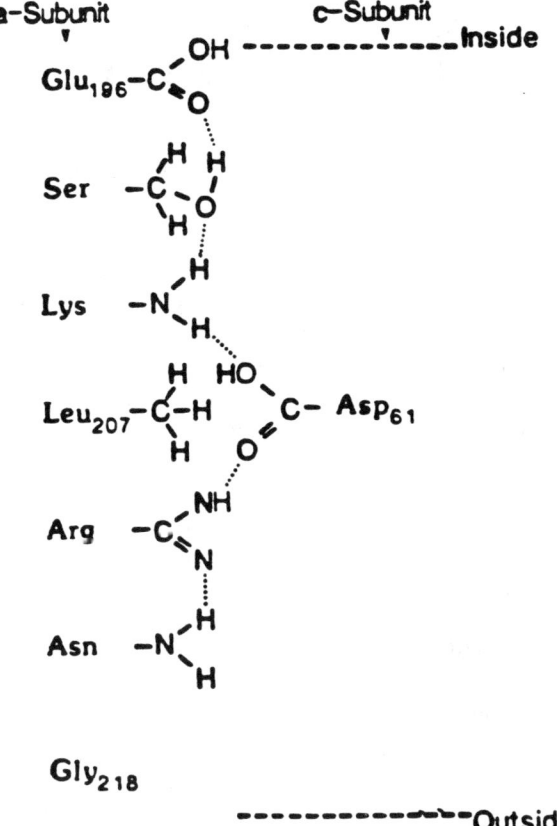

Figure 6 Model of intersubunit HBC formed in the F_o domain of the F_o/F_1 ATP synthase. Based on the model of Cox et al (20).

ion-based bioenergetics (21). Onsager originally proposed a form of the HBC-hypothesis to account for sodium and potassium ion transport in nerve axons, but quickly realized that such ions could not be efficiently transported in this fashion (22). Thus if one assumes that similar structural elements are needed for proton and sodium ion translocation, an alternative to the HBC hypothesis is required. That we consider mechanisms suitable for either proton or sodium ion translocation is imperative in light of the fact that several systems are now known to use either protons or Na^+ translocation for energy coupling. These observations have led Boyer to propose recently that the hydronium ion (H_3O^+) is the actual translocated species in what have traditionally been considered as *proton*-translocating proteins (23). No specific molecular mechanisms have been offered for how hydronium ion transport might occur; however, the concept of protein-mediated hydronium ion transport certainly deserves greater attention.

CYTOCHROME *c* OXIDASE

The remainder of this chapter is devoted to an in-depth review of a third proton-translocating protein, cytochrome *c* oxidase. The choice to highlight this particular protein is in part due to our personal biases, since a significant amount of our research efforts have been focused on this enzyme. Recently we have begun to investigate the structural elements required for proton pumping in cytochrome *c* oxidase. An excellent review of the relationship between structure and function in cytochrome *c* oxidase has recently been provided by Wikström et al (24).

Cytochrome *c* oxidase is the terminal enzyme in the respiratory electron transport chain of mitochondria and many aerobic prokaryotes (4). In mammals the enzyme is composed of 12–15 subunits that assemble to form a Y-shaped complex spanning the inner mitochondrial membrane, as illustrated in Figure 7. The enzyme accepts four electrons from ferrocytochrome *c* and transfers them, via its four redox active metal centers, to molecular oxygen. During the course of dioxygen reduction, four protons are consumed from the mitochondrial matrix to form two molecules of water. This scalar proton consumption results in an electrochemical gradient across the inner mitochondrial membrane. Additionally, the enzyme can actively pump up to one proton for each electron that traverses the membrane from ferrocytochrome *c* to molecular oxygen. This proton pumping is intimately coupled to the electron transfer activity of the oxidase, and the enzyme is thus referred to as a *redox-linked proton pump* (24).

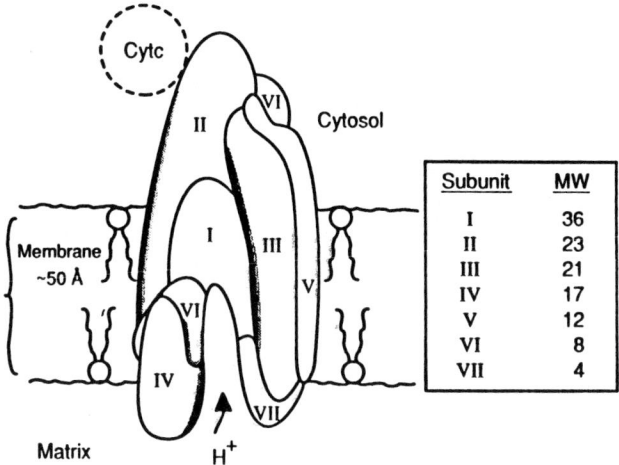

Figure 7 Structural model of mammalian cytochrome c oxidase. Adapted from Ref. (4).

Structure of the Redox-Active Metal Centers

Cytochrome c oxidase is a metallo-enzyme that contains two iron, three copper, one zinc, and one magnesium ion. The zinc, magnesium, and one of the copper ions (Cu_X) do not appear to participate in the electron transfer mechanism of the enzyme, and it is doubtful that they play any direct role in the redox-linked proton translocation. The four remaining metal centers consist of two iron-containing heme A chromophores (referred to as heme a and heme a_3) and two copper ions (referred to as Cu_A and Cu_B). Two of these metal centers, heme a and Cu_A, serve as the initial electron acceptors from ferrocytochrome c. The other two metals, heme a_3 and Cu_B, form a binuclear site for dioxygen binding and reduction. The chemistry of dioxygen reduction by cytochrome c oxidase has been worked out in some detail, and offers a fascinating view of the complexity of enzyme catalysis. This aspect of cytochrome c oxidase's enzymatic activity has been reviewed several times in recent years, and the reader is referred to these excellent accounts for further details (25–27). It is now widely accepted that one or both of the low-potential metal centers provide the link between redox activity and proton pumping. The most compelling evidence for this has come from the recent work of Wikström & Casey on proton pumping in inhibited submitochondrial particles (28).

Heme a is a six-coordinate, low-spin heme in both its ferric and ferrous states. Comparative ENDOR measurements on the yeast oxidase and the enzyme in which all the imidazole side chains of the histidine residues have been replaced by ^{15}N-substituted imidazole have provided unequivocal evidence for bis-imidazole coordination in this heme center (29), at least in the ferric state. Comparison of optical and resonance Raman spectra of heme a with those of heme A model compounds has also led Babcock and co-workers to suggest that heme a is bis-imidazole coordinated in both the ferric and ferrous oxidation states of the iron (30). Studies of fluorescence energy transfer from bound Zn-substituted cytochrome c suggest that heme a is within 25 Å of the heme of cytochrome c (31). The standard entropy of reduction for heme a is quite large, -50.8 eu, and suggests significant structural alterations about this metal or in the protein upon reduction (32). Since no ligation changes occur upon reduction, one possibility is that these structural changes involve relative motions of the entire heme group, or its peripheral substituents, with respect to the surrounding polypeptide. Evidence for structural perturbations of heme a is provided by optical spectroscopy. Both hemes contribute to the enzyme's characteristic absorption spectrum in the 350–700 nm region. The allowed (Soret) $\pi-\pi^*$ transition of heme a is red shifted relative to that of heme a_3 and heme A model compounds. Babcock's group has suggested that this red shift is the result of hydrogen bonding between the formyl oxygen of heme a and a protonated amino acid side chain, most probably tyrosine, from the protein matrix (33a,b). Using resonance Raman spectroscopy, this group has shown that formyl-stretching frequency for heme a is significantly downshifted relative to that of heme a_3, as expected if the former were involved in a strong hydrogen bond as a hydrogen acceptor. Babcock & Callahan (33a,b) have estimated the strength of this hydrogen-bonding for ferric and ferrous heme a as 3.0 and 5.3 kcal/mol, respectively. This change in the formyl group's hydrogen-bond strength upon reduction of heme a has formed the basis of a mechanism for redox-linked proton translocation by cytochrome c oxidase, based on heme a, as discussed below. The suggestion that the heme a formyl moiety is involved in hydrogen-bonding to an amino acid side chain is supported by the observation of Copeland & Spiro (34) that the frequency of the Raman band assigned to the heme a formyl stretch is shifted when the enzyme is incubated in ^2H_2O buffer. Copeland & Spiro reported that the rate of hydrogen/deuterium exchange at the heme a formyl group's proton donor was unaffected by enzyme turnover (34); however, Babcock's group has recently shown that this result is an artifact of the reductant used by Copeland & Spiro, and that in fact the rate of exchange is accelerated by turnover (G. T. Babcock et al, personal communication).

Thus the heme a binding pocket appears to be at least transiently in contact with the aqueous phase. Artzatbanov et al have shown that the midpoint reduction potential of heme a displays a ca. 30 mv/pH unit dependence, which is specifically associated with the matrix pH (35). Heme

a can also be perturbed by addition of Ca^{2+} or protons (36). The heme a group also appears to undergo some structural perturbation upon membrane energization. Wikström and co-workers have shown that in reduced, well-coupled submitochondrial particles, addition of ATP results in a red shift of the enzyme's heme absorption bands (37). Based on studies of isolated cytochrome c oxidase in detergent solution and on heme A model compounds, Wikström and co-workers concluded that the energy-linked absorption change of heme a is due to ATP-induced protonation of the heme a propionate group (38).

Cytochrome c oxidase forms a tight 1:1 complex with its physiological electron transfer partner, cytochrome c, in low ionic strength solution. Recently Weber et al studied this 1:1 complex by circular dichroism (CD) and magnetic circular dichroism (MCD) in both the fully oxidized and fully reduced forms of the complex (39). Significant spectroscopic changes in the c heme were observed upon complexation of either the oxidized or reduced proteins, and an additional change was seen in the oxidized complex for heme a. The origin of the spectroscopic change in heme a could not be elucidated from the data of Weber et al; it could be due to an electronic rearrangement at the heme or a structural change in relative heme orientation with respect to the surrounding polypeptide. Whatever the nature of this heme a perturbation, it seems to be unique to cytochrome c binding as it could not be mimicked by molecules known to inhibit binding of cytochrome c to the oxidase: apocytochrome c, porphyrin cytochrome c, or spermine (39). Thus although heme a does not undergo any redox-induced ligation changes, this chromophore does seem to show some structural flexibility, with respect to the surrounding polypeptide, under a variety of conditions relevant to the enzyme's catalytic activities.

Cu_A is structurally and spectroscopically unique among biological copper centers (40). The weak ($\varepsilon = 2$ mM^{-1} cm^{-1}) near IR band seen at ca. 830 nm in the oxidized enzyme's spectrum has been suggested to arise exclusively from Cu_A, and has been assigned to a ligand-to-metal charge transfer transition involving Cu_A and a sulfur ligand (41). This band disappears in the fully reduced and CN-bound, mixed-valence enzyme, consistent with the above assignment. MCD and optically detected magnetic resonance spectroscopy reveal additional absorption bands of Cu_A at 455, 470, 520, 560, 580, and 790 nm. All of these bands are x, y polarized, suggesting a high degree of axial symmetry for the metal center, and are assigned as sulfur (cysteine) to copper charge transfer transitions (42). ENDOR studies of isotopically enriched yeast cytochrome c oxidase have provided definitive evidence that Cu_A is ligated by at least one cysteine sulfur and one histidine nitrogen in its oxidized state (43a,b).

The EPR spectrum of Cu_A is also atypical for biological copper ions in that it shows virtually no hyperfine coupling, and its g-values (1.99, 2.03, and 2.18) are unusually small. These unusual EPR characteristics have been interpreted as arising from a tetragonally distorted ligand geometry about the Cu_A center (44). Chan and co-workers have further suggested that the Cu_A EPR spectrum indicates significant delocalization of a sulfur σ-electron onto the copper ion so that this center might more correctly be viewed as a Cu(I)-sulfur radical complex rather than a simple Cu(II) center (44).

As with heme a, reduction of Cu_A is associated with a large negative entropy change ($\Delta S^{o'} = -49.7$ eu), suggesting a conformational change associated with the redox activity at this metal center (45). Recently, preliminary evidence for such a conformational transition has been provided by this laboratory from Cu EXAFS studies of native and Cu_A-depleted cytochrome c oxidase in both the fully oxidized and fully reduced states. These data have suggested a bis-dithiolate coordination for the Cu_A site, and that one of the Cu–S bonds becomes elongated when the site is reduced (P. M. Li, personal communication).

Structural Aspects of the Protein

As mentioned above, mammalian cytochrome c oxidase is composed of 12–15 subunits, whereas the enzyme from prokaryotic sources usually contains far fewer subunits. It is now generally agreed that the catalytic core of the enzyme, in terms of both its electron transfer and proton pumping activity, is made up of two or three subunits (subunits I, II, and III); the other subunits in the enzyme from higher organisms most likely play some type of regulatory function. All of the redox-active metal ions are located in subunits I and II. Based on the spectroscopic evidence (vide supra), the Cu_A binding site requires two cysteine residues and at least one histidine. Comparison of the amino acid sequences for subunits I–III of various organisms suggests that the Cu_A binding site is located within subunit II, since this is the only subunit with two highly conserved cysteine residues (46a,b). Supporting evidence has come from recent chemical modification labeling experiments (47). Photoaffinity cross-linking studies have shown that subunit II also contains the high-affinity binding site for cytochrome c (48). The binding of cytochrome c to the enzyme is electrostatic and involves a cluster of lysine residues on one surface of cytochrome c and carboxylate residues on the oxidase. Two highly conserved carboxylate residues within subunit II have been implicated in the electrostatic interaction of the enzyme with cytochrome c, and these are located close to the suggested Cu_A binding domain (49). Since the cytochrome c binding site must be in contact with the cytosolic aqueous phase, these results indicate that Cu_A must also be close to the cytosol. Only two histidine residues are highly conserved within subunit II, and both of these are within the proposed Cu_A binding site. Since both heme A chromophores require histidine residues as axial ligands, it is unlikely that either heme is located within subunit II. Thus both hemes are suggested to be located within subunit I, which does contain enough conserved histidines to accommodate both heme chromophores. Since Cu_B is known to be within 5 Å of heme a_3, it is believed that this metal center is also contained within subunit I (46a).

Subunits I and II are clearly necessary for the proper functioning of cytochrome c oxidase. Previously, it was believed that subunit III was also required for the proton-pumping activity of the enzyme. This inference was based on the observation that the proton pumping activity could be impaired by binding DCCD to a specific glutamate residue within subunit III (50). However, it has more recently been demonstrated that subunit III-depleted bovine cytochrome c oxidase retains its proton pumping activity when reconstituted into phospholipid vesicles, albeit with a reduced proton/electron stoichiometry (51). Likewise the two-subunit enzyme isolated from *Paracoccus denitrificans* pumps protons in reconstituted vesicles (52). Thus it now seems that subunit III plays an ancillary role, if any, in proton pumping; the necessary molecular machinery for proton translocation must therefore reside within subunits I and II.

The method of Kyte & Doolittle has been used to locate transmembrane segments within the primary sequences of subunits I and II; Figure 8 depicts the proposed structures for these two subunits that result from such analysis (46a). Only strictly conserved residues are shown in this figure. The proposed metal-ligating residues are highlighted, and those residues capable of participating in HBC formation are circled. In Figure 8 a large portion of the subunit I polypeptide is buried within the lipid bilayer, in agreement with the experimental observation that this subunit does not react with water-soluble chemical or immunological labels presented to either the matrix (inner) or cytosolic (outer) surfaces (4). There are 12 transmembrane helices within subunit I, ten of which contain HBC-forming amino acids. Segments VI and X are particularly rich in HBC-competent residues and also invariant histidines that could serve as the axial ligands to heme a. Wikström and co-workers have proposed a model for the heme a binding site that uses one histidine from segment VI and one from segment X as the axial ligands to the iron, and an invariant tyrosine residue within segment X as the hydrogen-bond partner of the heme a formyl oxygen (46a). This model provides a structural basis for a heme a based mechanism of redox-linked proton translocation that utilizes intersubunit hydrogen-bonds to make up a HBC network that includes the tyrosine-heme a formyl hydrogen bond. The idea that the heme a

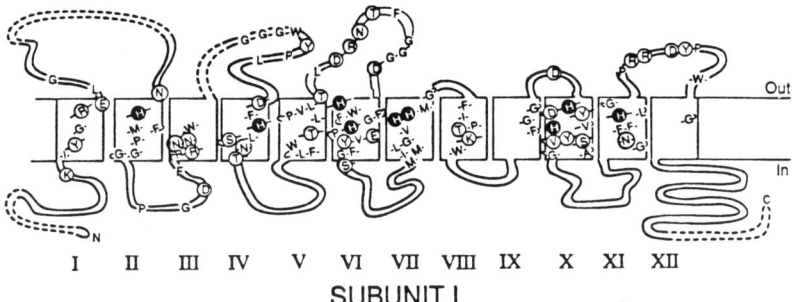

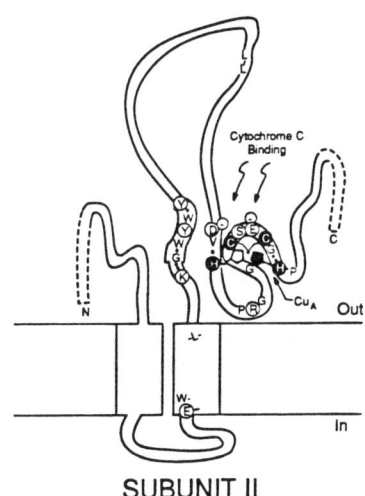

Figure 8 Structural models for subunits I and II of cytochrome *c* oxidase highlighting the locations of conserved HBC-forming amino acid residues (*open circles*) and the proposed metal-ligating amino acid residues (*closed circles*). Adapted from Ref. (46a).

formyl hydrogen bond might be involved in proton translocation was originally proposed by Babcock & Callahan (33a,b); the details of such a mechanism are discussed in the next section.

Subunit II is thought to contain two transmembrane helices. The majority of the polypeptide of this subunit is exposed to the cytosol, again in agreement with both chemical and immunological labeling studies (4). Only one HBC-forming residue, a glutamate at the matrix terminus of the second transmembrane helix, is strictly conserved in this subunit. However, the oxidase from every species thus far sequenced shows between four and six HBC-forming residues within the second transmembrane helix of subunit II. Even accounting for the nonconserved HBC-forming residues, a membrane-spanning HBC network cannot be constructed from subunit II alone. This does not, by any means, exclude a role for subunit II in proton translocation. Intersubunit hydrogen bonding between subunits I and II could be involved in proton pumping. Since the functional unit of cytochrome c oxidase appears to be a dimer (4), one also cannot exclude a HBC involving subunit II–subunit II intersubunit hydrogen bonds.

Mechanisms of Redox-Linked Proton Translocation

A redox-linked proton pump has specific requirements beyond those so far discussed for a generalized proton pump. To illustrate this, we consider simple models in which oxido-reduction of a single metal center (either heme a or Cu_A) provides the energetic link between electron transfer and proton translocation. In these models, a proton is taken up or ejected as the metal ion alternates between two valence states, oxidized and reduced. Clearly one also needs to specify the protonation state of an acidic group linked to the proton pumping redox center (4, 53). In addition, two protein conformations must exist for each of the valence states; one providing access for the proton pumped to the cytosol (proton output), and one providing access of the proton from the matrix side of the membrane (proton input). Thus one requires a total of eight states to describe fully the proton pumping cycle. These eight states are represented in the now familiar cubic scheme of Wikström et al (4) in Figure 9.

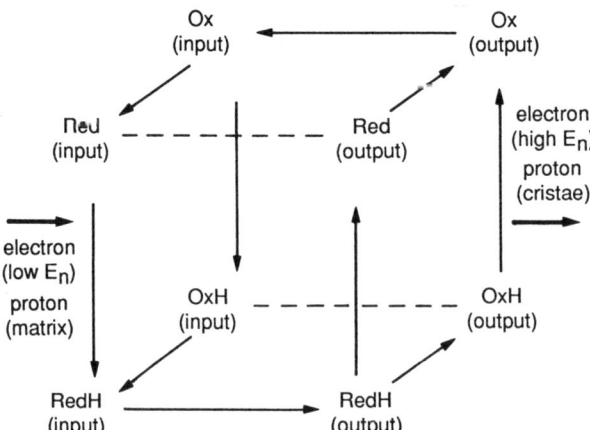

Figure 9 The eight-state "cubic" model for redox-linked proton translocation by cytochrome c oxidase as first proposed by Wikström et al (4).

Any of the vectorial transitions from the top face of the cube in Figure 9 to the bottom face must proceed via protonation of the linked acidic group. It should be emphasized that this acidic group need not be in close proximity to the redox center. If the acidic group were in direct spatial contact with the metal center, this would correspond to the most direct form of coupling that one can imagine. On the other hand, the acidic group could be far removed from the metal center. The coupling between redox activity and protonation in this extreme would be indirect and mediated by conformational transition of the protein. The simplicity of the direct coupling mechanism is quite appealing, but we must point out that increasing evidence favors the indirect mechanism for two other proton pumps, ATP synthase (3) and bacteriorhodopsin (54). As far as cytochrome c oxidase is concerned, no compelling evidence favours either coupling mechanism at this juncture (but see the next section on redox-linked conformational transitions). At first glance one might expect the model in Figure 9 to impose a strong pH dependence on the redox potential of the pump site. In fact, it has been argued that such a requirement favors heme a as the site of redox linkage over Cu_A, since the former metal center shows a pH-dependent midpoint potential. Blair et al have specifically addressed this issue and have shown on theoretical grounds that a pH-mediated midpoint potential is not obligatory for a redox center linked to proton translocation (55). The recent spectroelectrochemical studies of Ellis et al (32) and Blair et al (56) have shown that neither heme a nor Cu_A exhibits a strong pH dependence of their redox potentials. Thus, at present, no compelling theoretical or experimental data exist to distinguish between heme a and Cu_A as the more likely site of linkage on these grounds. Despite the paucity of data, however, two structurally detailed models for redox-linked proton translocation in cytochrome c oxidase have emerged, one involving heme a (33a,b), and the other involving Cu_A (57).

The heme a-based model was originally proposed by Babcock & Callahan, and is based on the observation that the strength of hydrogen bonding between the formyl oxygen of heme a and some proton donor(s) in the protein varies between the oxidized and reduced states of the heme iron (33a,b). Measurement of the heme a formyl's C=O stretching frequency from resonance Raman spectroscopy suggests that the hydrogen-bond strength differs by ca. 110 mV (2.5 kcal/mol) between the ferric and ferrous states of heme a. In the model of Babcock & Callahan this energy difference is used to provide part of the driving force for proton translocation against the ca. 200 mV electrochemical gradient of the inner mitochondrial membrane. Figure 10 outlines the proton pumping mechanism proposed by Babcock & Callahan based on these ideas (33a,b). In the stable oxidized state, the formyl oxygen is hydrogen-bonded to a proton donor group

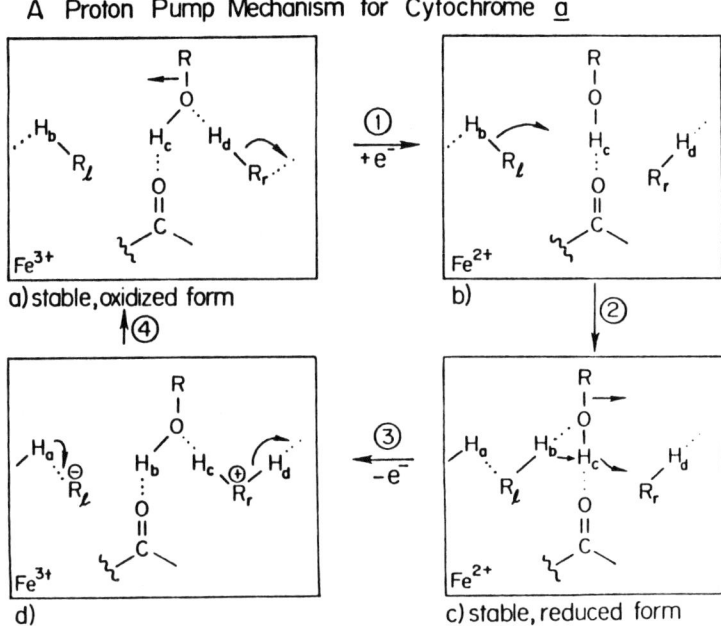

Figure 10 A heme *a*-based mechanism for redox-linked proton translocation by cytochrome *c* oxidase, as first proposed by Babcock & Callahan (33a,b).

(ROH) that is intermediate between two HBCs, one connected to the matrix side of the membrane and the other to the cytosol. Upon reduction of the heme iron, the hydrogen-bond strength increases between the now electron-rich formyl oxygen and the proton of the donor group (H_c). Babcock & Callahan proposed that this change in hydrogen-bond strength causes a geometry change in the donor group, allowing it to interact with the hydrogen of another acidic residue in close juxtaposition at the end of the matrix side HBC (H_b). As the cycle continues, the hydrogen-bond strength between the conjugate base RO^- and H_b increases at the expense of the formyl-H_c bond. Eventually H_c is replaced by H_b as the proton hydrogen-bonded to the heme *a* formyl, as H_c is transferred to the cytosolic hydrogen-bonded chain. To complete the cycle, there must be tandem proton migration within the HBC on the matrix side to replenish the proton hole originally occupied by H_b, followed by subsequent uptake of a proton at the opening of the channel from the matrix aqueous phase. This scheme offers a simple mechanism for providing alternating access of the pump site to the two sides of the membrane. That is, the heme *a* formyl hydrogen bond acts as a redox-linked proton gate. Unfortunately, the scheme makes no provisions for the gating of electron flow to obviate futile cycles. A number of treatments of the enzyme have also been recently reported that disrupt the proton-pumping activity of the enzyme without perturbation of the heme a environment. In this connection, Babcock & Callahan have offered a modified version of their model in which the proton pumping machinery is not in close proximity to heme a. Here the change in hydrogen-bond strength between the formyl and the proton donor upon reduction of heme a is proposed to result in a global conformational change in the protein that is transmitted to the proton-translocating element of the enzyme.

An alternative model for proton pumping in cytochrome c oxidase has been put forth by Gelles et al (57) based on Cu_A as the site of redox coupling. As in the model of Babcock & Callahan (33a,b), the site is hypothesized to be an electron-driven proton gate. In contrast to Babcock & Callahan (33a,b), however, Gelles et al (57) allowed for the gating of electron flow. Because of electron leaks, Gelles et al (57) argued that electron gating should be an essential element of any model of a redox-linked proton pump. The protein must be able to tune the rate constants of the relevant electron transfer processes to enhance the coupled process and suppress the uncoupled pathway. Gelles et al (57) proposed conformational switching as a means of achieving this. In their proposal, the electron enters the Cu_A site in one conformation of the enzyme, and is transferred out of the site in a different conformation during the coupled reaction. When the electron transfer is not coupled to proton pumping, the electrons can leak from the Cu_A center to the dioxygen reduction site in the same protein conformation. Clearly, to obviate the futile cycle, the conformational switching must be kinetically more facile than the electron leak. Gelles et al (57) pointed out that the electron gating itself need not involve a global conformational change. A local structural change at the site of redox-linkage might suffice so long as the necessary gating ratios are achieved. On the other hand, more global structural changes probably accompany the proton translocation steps of the proton-pumping cycle (the gating of proton flow).

The details of the Gelles et al model are outlined in Figure 11 (57). Two proton conducting channels are implicated, one leading to the cytosol from the Cu_A site and the other in communication with the matrix. In the oxidized state of Cu_A, the copper ion is ligated by two histidine nitrogens and two cysteine sulfurs arranged in a distorted tetrahedral geometry. This is the electron input state. Upon electron reduction, the bis-dithiolate cysteine coordination is expected to become asymmetric; i.e. one of the Cu–S bonds becomes more elongated relative to the other (43b). Gelles et al (57) proposed that a tyrosine (or a residue with a similar pK_a) is in close

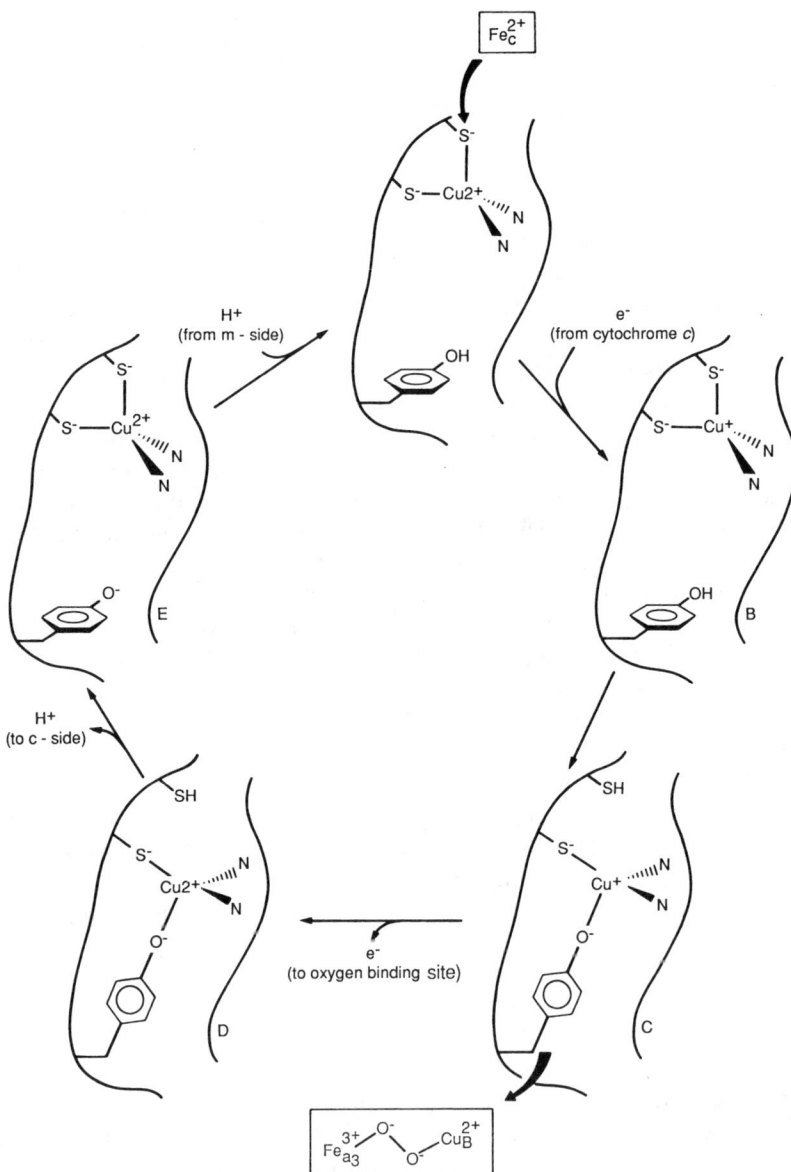

Figure 11 A Cu_A-based mechanism for redox-linked proton translocation by cytochrome c oxidase, as first proposed by Gelles et al (57).

proximity to the copper ion in the matrix HBC, and it can interact with the copper ion when it becomes reduced. If this interaction becomes sufficiently strong, the tyrosine oxygen can displace the cysteine sulfur in the already elongated Cu–S bond away from the copper ion toward the cytosolic HBC. The change in pK_as of the incoming tyrosine oxygen and outgoing cysteine sulfur accompanying this ligand rearrangement can lead to ionization of the tyrosine and protonation of the displaced cysteine. In this manner, part of the redox energy of the Cu_A site is expended in transferring the proton from the matrix HBC to the cystosolic side of the membrane, i.e. in gating the proton flow. The Cu_A site is now in the electron output state and poised for facile electron transfer to the dioxygen reduction site. To restore the original ligand arrangement following reoxidation of the copper, the displaced cysteine must become coordinated to the copper ion and give up its proton to the cystosol, and the ionized tyrosine must be re-protonated from the matrix HBC. Protein conformational changes must play a role here, to ensure tandem proton migrations within the matrix HBC to neutralize the tyrosinate anion, as does the positive charge at the Cu_A site, which provides a barrier for proton slippage from the cytosol to the matrix. The model proposed by Gelles et al is consistent with the data available on the ligand structure about Cu_A. As discussed above, the coordinated ligands for cupric Cu_A are almost certain to include two cysteine sulfurs and at least one histidine nitrogen. That a conformational transition occurs upon electron input to Cu_A is suggested by the large entropic change associated with reduction of this metal center (45), as well as some recent preliminary Cu EXAFS results on the reduced enzyme (P. M. Li, personal communication).

Although no direct experimental evidence supports the model of Gelles et al, some circumstantial data suggest a role for the Cu_A center in proton pumping by the enzyme. Several methods for perturbing the local environment of Cu_A, and their effects on the enzyme's proton pumping activity, have recently been reported. Gelles & Chan have shown that treatment of the oxidase with the sulfhydl reagent *p*-hydroxymercuri benzoate (pHMB) disrupts the coordination sphere of Cu_A and converts this metal center into a form resembling a type 2 copper center in terms of its EPR spectrum and redox potential (58). Nilsson et al found that a similar $Cu_A \rightarrow$ type 2 Cu conversion could be induced by heating the enzyme to ca. 43°C in the presence of a zwitterionic detergent (59). In both of these modified Cu_A enzymes, low temperature EPR and resonance Raman spectroscopies indicate little, if any, perturbation of heme *a* (59, 60). Heating the enzyme in a similar fashion in the presence of nonionic detergents also leads to a disruption of the Cu_A coordination sphere, but in this case one obtains a mixture of Cu_A forms (61). When these modified forms of cytochrome *c* oxidase were reconstituted into phospholipid vesicles, they were found to be no longer competent in terms of proton pumping (61) or capable of sustaining a transmembrane proton gradient when compared to vesicles containing the native enzyme (59, 62, 63). In the latter, the modified enzyme vesicles showed a significantly increased permeability toward protons (i.e. they were leaky). This proton permeability could arise from the fusion of the matrix and cystosol HBCs into one continuous transmembrane proton-conducting channel due to disruption of the proton gating machinery at the Cu_A center (63). Although a variety of control experiments have been conducted in the studies of the Cu_A-modified enzymes, one cannot completely rule out the possibility that disruption of the Cu_A site has led to creation of a transmembrane aqueous or proton channel in a region of the protein that is only remotely connected to the perturbed metal center. Such a scenario is likely in indirect coupling models in which the coupling between the electron and proton flow must be transmitted over some distance between the redox center and proton-translocating machinery. Nevertheless, taken together these results do provide some indication that the Cu_A center of cytochrome c oxidase plays a role in proton translocation.

Redox-Linked Conformational Transitions of Cytochrome c Oxidase

One common feature of all the hypothetical mechanisms for redox-linked proton translocation in cytochrome c oxidase is an obligatory conformational transition that provides the switch between proton accessibility from one side of the membrane to the other; i.e. a proton-input state to proton-output state conversion. In the direct coupling models this conformational change may be restricted to the immediate vicinity of the involved redox center, and may involve only slight motions of the amino acid side chains associated with the metal center. In the indirect coupling models the conformational change must be more global in nature, since here the driving force for proton translocation must be communicated over some distance between the redox center and the proton-translocating machinery. One might therefore expect that cytochrome c oxidase would display some type of redox-linked conformational change associated with the low potential metal centers of the enzyme. Indeed a number of results now suggest that such a conformational transition does occur in this enzyme.

In the early 1970s Yamamoto & Okunuki showed that reduction of the metal centers of cytochrome c oxidase results in a significant stabilization of the enzyme toward proteolytic digestion (64). Cabrel & Love also showed via sedimentation studies that reduction of the enzyme resulted in a ca. 3% increase in the protein's volume (65). The majority of this volume change was attributed to reduction of the low potential metal centers by comparison of the relative volumes of the fully reduced and CO mixed-valence forms of the enzyme. Since reduction of both heme a and Cu_A are associated with large, negative entropy changes, such redox-linked changes of the enzyme are reasonable. These conformational changes are apparently not limited to the immediate vicinity of the metal centers. Several groups have shown that reduction of the low-potential metal centers greatly accelerates the rate of inhibitory cyanide binding at the oxygen binding site (66-68). This suggests that cytochrome c oxidase is an allosteric enzyme capable of transmitting redox-induced structural changes at the low potential metal centers between subunits and over a large distance to the oxygen-binding site. There is also evidence that intramolecular electron transfer from the low-potential metal centers to the oxygen-binding site is mediated by a conformational transition of the enzyme. Malmström and co-workers have reported data that suggest that such intramolecular electron transfer does not occur until two electrons have entered the enzyme and a conformational transition occurs (69).

It should be pointed out, however, that this conformational transition was introduced here for the purpose of electron gating. Electron gating is a necessary but not sufficient condition for a redox-linked proton pump. Additional structural and/or conformational rearrangements need to be invoked to provide for energetic linkage between the electrons and the protons and to achieve the gating of proton flow. When all these requirements are incorporated into the model of Malmström et al (70, 71), their scheme becomes formally equivalent to the mechanism advocated by Gelles et al (57).

Unfortunately, at this juncture, we do not have more direct handles on the conformational transitions that occur during turnover of the enzyme. Copeland et al have recently reported evidence for a redox-linked conformational transition in cytochrome c oxidase based on steady-state and stopped-flow tryptophan fluorescence spectroscopy (71, 72). Inasmuch as we, as well as others (B. Hill, personal communication; D. Rousseau, personal communication) have obtained evidence that the steady-state fluorescence results are artifactual, these findings must now be discounted.

Future Prospects for Cytochrome c Oxidase

The prospects for further elucidation of the mechanism of proton translocation in cytochrome c oxidase are quite promising, in view of several recent advances that should provide more detailed structural information on the enzyme as a whole, and the proton pumping machinery in particular. New spectroscopic handles, such as picosecond laser spectroscopy, may provide new insights into the nature and extent of structural perturbations of the protein matrix associated with proton pumping. The recent cloning of the three subunits of the *P. denitrificans* enzyme holds the promise of identifying key amino acid residues via site directed mutagenesis (73). There has also been a recent report that the bacterium *Thermus thermophilus* can express a single subunit terminal oxidase in which heme *a* is replaced with a *b*-type heme (74). Whether this enzyme pumps protons when reconstituted into phospholipid vesicles is not yet known, but experiments to address this issue can potentially provide a means of determining the low-potential metal center involved in proton pumping. Finally, we note with great interest the report by Caughey and co-workers on the preparation of crystals of bovine cytochrome *c* oxidase that diffract to 8 Å resolution (75). If the quality and size of these crystals can be improved in the near future, the possibility for a three-dimensional structure for at least the oxidized enzyme will soon be within reach.

ACKNOWLEDGMENTS

We wish to thank the many collaborators who have contributed to the work reported here. In particular we wish to thank P. M. Li, P. A. Smith, J. Gelles, D. F. Blair, S. N. Witt, J. Morgan, N. E. Gabriel, T. Nilsson, W. R. Ellis, Jr., H. B. Gray, H. Wang, M. Ma, R. Larsen, M. Ondrias, T. G. Spiro, and C. Martin. We also gratefully acknowledge helpful discussions with W. Woodruff, G. Babcock, B. Malmström, M. Wikström, M. Brunori, and P. Sarti. The work reported here from our laboratory was supported by grant GM22432 from the National Institute of General Medical Sciences, US Public Health Service to S. I. C. R.A.C. acknowledges support from a Chaim Weizmann Research Fellowship.

Literature Cited

1. Nicholls, D. G. 1982. *Bioenergetics*. New York: Academic
2a. Khorana, H. G., Braiman, M. S., Chao, B. H., Doi, T., Flitsch, S. L., et al. 1987. *Chem. Scr. B* 27: 137
2b. Khorana, H. G. 1988. *J. Biol. Chem.* 263: 7439
3. Boyer, P. D. 1987. *Biochemistry* 26: 8503
4. Wikström, M., Krab, K., Saraste, M. 1981. *Cytochrome Oxidase, A Synthesis*. New York: Academic
5. Freund, F. 1981. *Trends Biochem. Sci.* 6: 142
6. Scheiner, S. 1985. *Acct. Chem. Res.* 18: 174
7. Glasser, L. 1975. *Chem. Rev.* 75: 21
8. Michel, H. 1983. *Trends Biochem. Sci.* 8: 56
9. Kabach, H. R. 1987. *Biochemistry* 26: 2071
10. Nagle, J. F., Tristam-Nagle, S. 1983. *J. Membr. Biol.* 74: 1
11. Nagle, J. F., Morowitz, H. J. 1978. *Proc. Natl. Acad. Sci. USA* 75: 298
12. Kayalan, C. 1979. *J. Membr. Biol.* 45: 37
13. Dunn, R. J., Hackett, N. R., Huang, K. S., Jones, S. S., Khorana, H. G., et al. 1983. *Cold Spring Harbor Symp. Quant. Biol.* 48: 853
14. Henderson, R., Unwin, P. N. T. 1975. *Nature* 257: 28
15. Roepe, P., Ahl, P. L., Das Gupta, S. K., Herzfeld, J., Rothschild, K. J. 1987. *Biochemistry* 26: 6696
16. Rothschild, K. J., Roepe, P., Ahl, P. L.,

Earnest, T. N., Bogomolni, R. A., Das Gupta, S. K., Mulliken, C. M., Herzfeld, J. 1986. *Proc. Natl. Acad. Sci. USA* 83: 347
17. Senior, A. E., Wise, J. G. 1983. *J. Membr. Biol.* 73: 105
18. Hoppe, J., Sebald, W. 1984. *Biochim. Biophys. Acta* 768: 1
19. Sebald, W., Wachter, E. 1979. In *Energy Conservation in Biological Membranes. Colloq. Mosbach Ser.*, ed. G. Schafer, M. Klingenberg, 29: 228-36. New York: Springer-Verlag
20. Cox, G. B., Fimmel, A. L., Gibson, F., Hatch, L. 1986. *Biochim. Biophys. Acta* 849: 62
21. Skulachev, V. P. 1985. *Eur. J. Biochem.* 151: 199
22. Onsager, I. 1969. *Science* 166: 1359
23. Boyer, P. D. 1988. *Trends Biochem. Sci.* 13: 5
24. Wikström, M., Saraste, M., Penttilä, T. 1985. *The Enzymes of Biological Membranes*, ed. A. N. Martonosi, 4: 111. New York: Plenum
25. Malmström, B. G. 1982. *Annu. Rev. Biochem.* 52: 21
26. Naqui, A., Chance, B. 1986. *Annu. Rev. Biochem.* 55: 137
27. Chan, S. I., Witt, S. N., Blair, D. F. 1988. *Chem. Scr. A* 28: 51
28. Wikström, M., Casey, R. P. 1985. *J. Inorg. Biochem.* 23: 327
29. Martin, C. T., Scholes, C. P., Chan, S. I. 1985. *J. Biol. Chem.* 260: 2857
30. Babcock, G. T., Callahan, P. M., Ondrias, M. R., Salmeen, I. 1981. *Biochemistry* 20: 959
31. Dockter, M. E., Steinemann, A., Schatz, G. 1978. *J. Biol. Chem.* 253: 311
32. Ellis, W. R., Wang, H., Blair, D. F., Gray, H. B., Chan, S. I. 1986. *Biochemistry* 25: 161
33a. Callahan, P. M., Babcock, G. T. 1983. *Biochemistry* 22: 457
33b. Babcock, G. T., Callahan, P. M. 1983. *Biochemistry* 22: 2314
34. Copeland, R. A., Spiro, T. G. 1986. *FEBS Lett.* 197: 239
35. Artzatbanov, V. Y., Konstantinov, A. A., Skulachev, V. P. 1978. *FEBS Lett.* 87: 180
36. Wikström, M., Saari, H. 1975. *Biochim. Biophys. Acta* 408: 170
37. Wikström, M. K. F. 1972. *Biochim. Biophys. Acta* 283: 385
38. Saari, H., Penttilä, T., Wikström, M. 1980. *J. Bioenerg. Biomembr.* 12: 325
39. Weber, C., Michel, B., Bosshard, H. R. 1987. *Proc. Natl. Acad. Sci. USA* 84: 6687
40. Chan, S. I., Bocian, D. F., Brudvig, G. W., Morse, R. H., Stevens, T. H. 1979. In *Cytochrome Oxidase*, ed. T. E. King, Y. Orii, B. Chance, K. Okunuki, p. 177. Amsterdam: Elsevier
41. Beinert, H., Shaw, R. W., Hansen, R. E., Hartzell, C. R. 1980. *Biochim. Biophys. Acta* 591: 458
42. Thomson, A. J., Greenwood, C., Peterson, J., Barrett, C. P. 1986. *J. Inorg. Biochem.* 28: 195
43a. Stevens, T. H., Martin, C. T., Wang, H., Brudvig, G. W., Scholes, C. P., Chan, S. I. 1982. *J. Biol. Chem.* 257: 12106
43b. Martin, C. T., Scholes, C. P., Chan, S. I. 1988. *J. Biol. Chem.* 263: 8420
44. Chan, S. I., Bocian, D. F., Brudvig, G. W., Morse, R. H., Stevens, T. H. 1978. In *Frontiers of Biological Energetics*, ed. P. L. Dutton, J. S. Leigh, A. S. Scarpa, Vol. 2. New York: Academic
45. Wang, H., Blair, D. F., Ellis, W. R., Gray, H. B., Chan, S. I. 1986. *Biochemistry* 25: 167
46a. Holm, L., Saraste, M., Wikström, M. 1987. *EMBO J.* 6: 2819
46b. Steffens, G. J., Buse, G. 1979. See Ref. 40, pp. 153-59
47. Hall, J., Moubarak, A., O'Brien, P., Pan, L. P., Cho, I., Millett, F. 1988. *J. Biol. Chem.* 263: 8142
48. Bisson, R., Jacobs, B., Capaldi, R. 1980. *Biochemistry* 19: 4173
49. Millett, F., deJong, C., Poulson, L., Capaldi, R. 1983. *Biochemistry* 22: 546
50. Casey, R. P., Thelen, M., Azzi, A. 1980. *J. Biol. Chem.* 255: 3994
51. Saraste, M., Penttilä, T., Wikström, M. 1981. *Eur. J. Biochem.* 115: 261
52. Solioz, M., Carafoli, E., Ludwig, B. 1982. *J. Biol. Chem.* 257: 1579
53. Krab, K., Wikström, M. 1987. *Biochim. Biophys. Acta* 895: 25
54. Fodor, S. P. A., Ames, J. B., Gebhard, R., van den Berg, E. M. M., Stoeckenius, W., Lugtenburg, J., Mathies, R. A. 1988. *Biochemistry* 27: 7097
55. Blair, D. F., Gelles, J., Chan, S. I. 1986. *Biophys. J.* 50: 713
56. Blair, D. F., Ellis, W. R., Wang, H., Gray, H. B., Chan, S. I. 1986. *J. Biol. Chem.* 261: 11524
57. Gelles, J., Blair, D. F., Chan, S. I. 1987. *Biochim. Biophys. Acta* 853: 205
58. Gelles, J., Chan, S. I. 1985. *Biochemistry* 24: 3963
59. Nilsson, T., Copeland, R. A., Smith, P. A., Chan, S. I. 1988. *Biochemistry* 27: 8254
60. Larsen, R. W., Ondrias, M. R., Copeland, R. A., Li, P. M., Chan, S. I. 1989. *Biochemistry*. In press
61. Sone, N., Nicholls, P. 1984. *Biochemistry* 23: 6550

62. Li, P. M., Morgan, J. E., Nilsson, T., Ma, M., Chan, S. I. 1988. *Biochemistry* 27: 7538
63. Nilsson, T., Gelles, J., Li, P. M., Chan, S. I. 1988. *Biochemistry* 27: 296
64. Yamamoto, T., Okunuki, K. 1970. *J. Biochem.* 67: 505
65. Cabrel, F., Love, B. 1972. *Biochim. Biophys. Acta* 283: 181
66. Jones, M. G., Bickar, D., Wilson, M. T., Brunori, M., Colosimo, A., Sarti, P. 1984. *Biochem. J.* 220: 57
67. Jensen, P., Wilson, M. T., Aasa, R., Malmström, B. G. 1984. *Biochem. J.* 224: 829
68. Scholes, C. P., Malmström, B. G. 1984. *FEBS Lett.* 198: 125
69. Brzezinski, P., Thörnström, P. E., Malmström, B. G. 1986. *FEBS Lett.* 194: 1
70. Brzezinski, P., Malmström, B. G. 1987. *Biochim. Biophys. Acta* 894: 29
71. Copeland, R. A., Smith, P. A., Chan, S. I. 1987. *Biochemistry* 26: 7311
72. Copeland, R. A., Smith, P. A., Chan, S. I. 1988. *Biochemistry* 27: 3552
73. Raitio, M., Jalli, T., Saraste, M. 1987. *EMBO J.* 6: 2825
74. Zimmermann, B. H., Nitsche, C. I., Fee, J. A., Rusmak, F., Münch, E. 1988. *Proc. Natl. Acad. Sci. USA* 85: 5779
75. Yoshikawa, S., Tera, T., Takahashi, Y., Tsukihara, T., Caughey, W. S. 1988. *Proc. Natl. Acad. Sci. USA* 85: 1354
76. Creighton, T. E. 1984. *Proteins, Structures and Molecular Properties.* New York: Freeman

AUTHOR INDEX

A

Aarons, J., 94, 98
Aartsma, T. J., 506, 507
Aasa, R., 695
Abarenkov, I., 67
Abdel-Meguid, S. S., 228
Abe, H., 218, 219, 221, 223, 225
Abe, Y., 364, 366
Abella, I. D., 500
Abrahams, E., 18, 46
Abrahams, E. A., 309
Abramson, E., 471
Abrikossova, I. I., 599
Achiba, Y., 652
Adam, A. G., 418, 432, 567
Adam, G., 109
Adam, M. Y., 645
Adamic, K. J., 91, 101, 109
Adams, A., 414, 420
Adams, G. E., 599
Adkins, C. J., 310
Adler, D., 250
Adler, S. E., 376
Adler-Golden, S. M., 482
Adolf, D., 364
Aeppli, G., 297, 303
Afshar-Rad, T., 629
Agmon, N., 134
Ahl, P. L., 676
Ahrens, T. J., 238, 239, 242, 244, 245, 252, 256
Aidun, J., 247
Ait-Kaci, M., 656
Akhter, S., 262, 265, 266
Akimoto, S., 250
Alario-Franco, M. A., 303, 318
Alben, R., 371
Albers, J., 55
Albery, W. J., 134
Albridge, R. G., 201
Albrow, M. G., 464
Alcock, C. B., 238, 239
Alder, B. J., 7, 237, 238
Alekseev, V. A., 63
Alexander, B. M., 598
Alexander, M. H., 380
Alexander, S., 622, 626
Alexandrov, I. V., 131, 132
Algra, H. A., 42
Alivisatos, A. P., 190

Allcock, H. R., 105
Allegra, G., 361
Allen, A. O., 440, 446, 455
Allen, G., 367
Allen, W. N., 85
Almlöf, J., 491
Almog, Y., 604, 607, 609, 610, 615-17
Al'shits, E. I., 499, 500
Altfield, J. P., 319
Altkorn, R., 379, 573
Al'tshuler, L. V., 239
Alvey, M. D., 174, 180
Amano, T., 399, 400
Ambroseo, J. R., 148
Ames, J. B., 689
Amis, E. J., 230
Andeen, C. G., 101, 109
Andersen, T., 417, 418, 430
Anderson, A. B., 301
Anderson, A. C., 36, 505, 510, 515, 520
Anderson, D. F., 455
Anderson, J. B., 565
Anderson, J. S., 298, 301, 302
Anderson, P. W., 18, 33, 35, 42, 46, 69, 262, 267, 279, 309, 317, 509, 510, 515, 520
Anderson, S. L., 649, 654
Andersson, S., 274, 275, 302
Andlauer, B., 655
Andreani, R., 92
Andres, R. P., 565
Andresen, P., 183, 385, 386, 401
Anfinsen, C. B., 207, 208
Angell, C. A., 24, 109, 252, 254
Ansarifar, M. A., 620-22
Ansorena, F. L., 615
Antal, J., 427
Antić-Jovanović, A., 419
Antonetti, A., 129, 464
Antonides, E., 189
Antoniewicz, P. R., 176, 180
Aoki, K., 248
Appel, J., 310
Appelbaum, J. A., 536, 537
Appelblad, O., 419, 433, 434
Appling, J. R., 379, 395

Aprile, E., 464
Apsimon, R., 464
Aragon, R., 304
Aravamudan, G., 299, 303
Archirel, P., 661
Arias, J., 202
Arikawa, T., 386, 394
Ariyasu, J. C., 144, 159
Armand, M., 85, 86, 88, 107, 108
Armand, M. B., 92, 94, 104-6
Armentrout, P. B., 416-18
Armstrong, R. D., 89
Arneberg, D. L., 652
Artzatbanov, V. Y., 683
Asaf, U., 444, 445, 457, 462
Asano, S., 74
Asano, T., 311
Asaumi, K., 244-47
Ascarelli, G., 450, 457, 458, 462
Ashburn, J. R., 311
Ashcroft, N. W., 63, 67, 80, 81, 168, 237, 238, 243
Ashfold, M. N. R., 375, 383, 385, 391
Askew, T. R., 314
Aspect, A., 380
Atabek, O., 477, 479
Atanasoska, Lj., 600, 631
Athénour, C., 414, 420
Atkins, C. G., 385
Atrazhev, V. M., 461
Attwood, D., 638
Aubert, B., 464
Aubry, A., 301
August, J., 389, 399
Auker, B. H., 250
Aumiller, G. D., 342, 346, 347
Aurivillius, B., 298
Ausserré, D., 630
Austin, P., 105
Auty, R. P., 10, 15, 16
Averbach, R. A., 129
AVOURIS, P., 173-206; 176, 180, 183, 184, 186, 188, 190, 191, 194, 196-98, 201, 536, 538, 553-55
Azria, R., 197
Azuma, Y., 418, 419, 430, 432
Azzi, A., 686

AUTHOR INDEX

B

Baba, M., 380
Babcock, G. T., 683, 687, 689
BAČIĆ, Z., 469-98; 481, 482, 484, 485, 487, 489, 491, 494, 495
Bacci, C., 464
Bachelet, G. B., 445
Bacis, R., 419
BAER, T., 637-69; 642, 643, 645-48, 650-52, 657, 658
BAGCHI, B., 115-41; 117, 118, 120, 124, 127, 130, 131, 133, 134, 136, 137
Baggott, J. E., 375
Baghi, B., 134
Bagus, P. S., 176, 180, 275, 421, 427, 430
Bahadur, D., 297, 298, 310
Bahar, I., 364, 367
Bai, Y. Y., 387
Bailey, A. I., 629
Bain, A. J., 379
Baird, J. K., 456, 462, 463
Bakale, G., 444, 453
Baker, R. T. K., 162, 163
Bakshiev, N. G., 117
Balakin, A. A., 449, 454
Balchan, A. S., 248
Baldini, G., 245
Balducci, G., 416
Baldwin, K. R., 90
Baldwin, R. L., 208, 209
Balint-Kurti, G. G., 378, 385, 386, 401, 477, 479
Ball, R. C., 363, 616, 629
Baller, T., 400
Balzarin, D. A., 80
Bamford, D. J., 396
Band, E., 168
Bando, H., 541, 555
Bando, Y., 303
Danct, M. J., 126
Banks, E., 319
Bannister, D. J., 105
Bar, I., 394
Baraff, G. A., 281, 282, 536, 537
Barandiaran, M. J., 615
Baratoff, A., 532, 544, 556
Barbara, P. F., 116, 127-29, 132
Barbara, T. M., 299
Barbargie, Z., 134
Barbee, T. W. III, 237, 238
Bard, A. J., 600, 631
Bardeen, J., 535
Bardsley, J. N., 642
Barford, W., 616, 629
Barker, J. A., 275
Barnett, J. D., 242, 243
Barnwell, J. D., 578

Baroni, S., 247
Barrett, C. P., 684
Barrow, R. F., 415, 416, 423, 425, 426, 429
Barry, J. A., 418, 432
Barry, M. D., 400
Barsch, G. R., 238, 239, 245
Bartholomae, R., 471
Bartko, J., 85
Bartosch, C. E., 199
Bartsch, R. G., 228
Basak, S., 444-46, 450, 457, 462
Basché, Th., 521
Bashford, D., 230
Baskaran, G., 317
Baskes, M. I., 192, 263, 278-80
Bass, A. M., 417, 419, 430, 433
Bass, J., 238, 242, 244, 245, 256
Basset, J. M., 162
Bassett, D. W., 261, 264, 285
Bassett, W. A., 238, 240, 242
Bässler, H., 502, 503, 505, 511
Bastide, J., 368
Bath, A., 385
Batlogg, B., 311, 314, 315, 317
Batra, I. P., 275
Battle, P. D., 303
Bauer, E., 193, 301
Bauerle, R. J., 148
Baughcum, S. L., 400
Baumann, C. A., 417, 428
Baumgartel, H., 638, 652
Baumgartner, A., 212, 223
Bauschlicher, C. W. Jr., 275, 408, 412, 416, 420, 421, 427, 433
Bayers, R., 311
Bayley, J. M., 383
Bazin, C, 641
Beardsworth, R., 472
Beauchamp, J. L., 416-18
Beck, G., 449, 453
Becker, K. H., 389
Becker, R. S., 536, 541, 542, 544, 545, 553
Beckerle, J. D., 143, 166
Beckman, K., 652
Bednorz, J. G., 291, 310, 315
Beech, F., 312
Behm, R. J., 276
Beigang, R., 188
Beinert, H., 684
Beland, M., 647
Belfort, G., 629
Bell, A. T., 143, 144, 159, 162
Bell, L. D., 555, 556
Bell, P. M., 237, 238, 240, 242, 243, 254

Bell, S. E., 105
Bello, I., 85
Belouschek, P., 599
Beltzung, M., 368
Bendler, J. T., 18
Benedict, W. S., 417, 430
Benjamin, G. S., 353, 364, 365
Benna, B., 148
Bennewitz, H. G., 564
Beno, M. A., 312, 315
Benoist d'Azy, O., 394
Benoit, H., 25
Ben-Reuven, A., 380
Benson, S. W., 579, 588, 591
Berberian, J. G., 10, 11, 17, 20
Berendsen, H. J. C., 209, 210, 212, 224
Berg, M., 506, 511, 513, 515, 517, 518
Berg, R. A., 414, 420
Berggren, K. F., 74
Berghaus, T., 547, 548
Bergher, M., 641
Bergsman, J. P., 134
Berkovic, G., 343, 346
Berkowitz, J., 638, 646, 648, 652
Berlin, Y. A., 463
Bernage, P., 419
Berne, B., 136
Berne, B. J., 459, 460
Bernholc, J., 281
BERNSTEIN, R. B., 561-95; 58, 400, 562-67, 569-72, 574-80, 588, 591, 592
Bernstoff, S., 462
Berret, J. F., 36, 42, 43, 45
Berry, R. S., 642
Bersohn, R., 389, 390, 398, 400
Bertel, E., 196
Berthier, C., 92, 94, 105
Bertoni, C. M., 185
Besnard, M. J., 662
Besson, J. M., 244-46
Beswick, J. A., 380, 641, 642, 646, 647, 662
Betz, O., 201
Beuhler, R. J., 565, 576, 578, 590
Beulow, S., 592
Beushausen, V., 385
Bevan, D. J. M., 301
Beyerlein, R. A., 299
Bezaquet, A., 464
Bhat, S. V., 315
Bhat, V., 314, 317
Bhatt, R. N., 72
Bhattacharja, S., 108
Bhattacharya, S., 36, 45, 49
Bhattacharyya, K., 343
Bhide, V. G., 305, 310
Bianchi, E., 102

AUTHOR INDEX 701

Bianconi, A., 316, 320
Bickar, D., 695
Bierstedt, F. E., 297, 310
Billardon, M., 638, 641
Binder, H., 63
Binder, K., 33, 40, 46, 54, 212, 219
Binnig, G., 264, 269, 320, 531, 532, 538, 541, 544, 556, 598
Birge, N. O., 45, 49
Birgneau, R. J., 297
Birshtein, T. M., 366, 367
Bisling, P., 652
Bisson, C. L., 274, 275
Bisson, R., 685
Bitsanis, I., 631
Bittman, J. S., 482
Bixon, M., 131, 132
Bjerre, N., 649
Bjorklund, G. C., 503
Black, J., 275
Black, J. F., 379, 381, 391, 400, 401
Blackmore, R., 485
Blair, D. F., 682, 689, 691, 692, 695
Blais, N. C., 579, 588, 590, 591
Blanchin, M. G., 302
Blazey, K. W., 315
Blinc, R., 31
Block, H., 118
Block, S., 238, 240, 242-44
Bloembergen, N., 328, 329, 343
Blom, C. E., 601
Blonsky, P. M., 105
Blumberg, A. A., 102
Bocian, D. F., 684, 685
Bodea, S., 243
Bodensteiner, T., 63, 64, 66, 78, 79
Bodo, G., 207
Boesch, L. P., 24
Boesiger, J., 481
Boesl, U., 654
Boeva, R. S., 107
Bogomolni, R. A., 676
Böhmer, R., 32, 37, 42-45, 48, 49, 51, 52
Bohnen, K. P., 262, 280
Boivineau, M., 592
Bonacić-Koutecky, V., 132
Bonetti, S., 464
Bonino, F., 95
Bonnie, J. H. M., 649
Bonzel, H. P., 265
Booth, C., 105
Bor, G., 160
Boriev, I. A., 454
Borik, M. A., 319
Boring, J. W., 193, 201

Börjesson, L., 102, 104
Born, M., 117
Born, R., 32, 48
Bornstein, R., 155
Bose, T. K., 13
Bosshard, H. R., 684
Bossis, G., 7
Bostoen, C., 54
Botschwina, P., 482
Böttcher, H., 441, 442, 448
Botter, R., 642, 656
Bottyan, L., 69
Bouchal, K., 368
Boue, F., 368
Bouma, W. J., 658
Bourdon, E. B. D., 184, 202
Bouridah, A., 107
Bourson, P., 37
Bowen, H. K., 250
Bower, J. H., 328
Böwering, N., 575
Bowers, M. T., 655
Bowman, D. R., 181
Bowman, J. M., 159, 470, 474, 482, 493
Bowman, R. M., 464
Boxer, S. G., 513, 522
Boyd, G. T., 337, 346
Boyer, P. D., 671, 681, 689
Bozso, F., 159, 188, 202, 536, 538, 553, 554
Bradley, R. S., 599
Bradshaw, A. M., 143, 144, 159, 162, 168
Braggiotti, A., 464
Braiman, M. S., 671, 676
Brako, R., 187
Brand, J. L., 143, 262, 266
Brandes, G. R., 414, 415
Brandow, B. H., 250
Brandts, J. F., 208, 223
Brant, P., 85
Bräuchle, C., 503, 521
Brault, J. W., 415
Braun, C. L., 440, 449
Braunstein, G., 85
Brearley, A. M., 127, 128
Bredig, M. A., 250
Breinl, W., 503, 505, 513, 519
Brenig, W., 174, 191
Brennert, G. F., 63, 69
Brereton, M. G., 369
Breton, J., 647, 648
Brewer, J. H., 320
Brewer, R. G., 500
Bridges, C., 107
Briggs, A. G., 415, 425
Briggs, R. G., 385
Brinkman, W. F., 69, 308
Brinkmann, D., 92
Brister, K. E., 244-47
Brito-Cruz, C. H., 342, 346, 347, 513

Brittain, R., 414, 415, 422
Brocks, G., 473, 476, 480, 481
Brodde, A., 547, 548
Brodwin, M., 96
Broer, M. M., 511, 513, 519, 520
Brohan, L., 319
Broida, H. P., 415, 417, 421, 429, 432
Brooks, P. R., 562, 565, 573, 574, 578-81
Brot, C., 7
Brotzman, R. W., 364, 369
Brouard, M., 385, 389, 393, 399-401
Brout, R., 34
Brown, C. R., 418, 430, 432, 433
Brown, D. S., 102, 485, 489
Brown, F., 328
Brown, J. M., 63, 249, 415, 419, 427, 434
Brown, N. L., 15
Brown, W. L., 173, 201, 615
Browne, M., 33
Broyden, C. G., 284
Broyer, M., 380
Bruccoleri, R. E., 211
Bruce, P. G., 102, 107, 108
Brucker, C. F., 168
Brudle, C. R., 180
Brudvig, G. W., 684, 685
Brühlmann, U., 377, 379, 387, 394, 395
Brundle, C. R., 143, 144, 159, 162, 275
Brunet, J. Ph., 491, 492
Bruno, A. E., 387
Brunori, M., 695
Brusius, U., 70, 72, 74
Brutschy, B., 638, 652
Brzezinski, P., 695
Bucher, B., 314, 319
Büchler, A., 414, 420
Buck, J. D., 649
Buckingham, A. D., 12-14
Buckley, E., 441, 464
Budde, F., 183
Bugyi, L., 265
Buhse, L. F., 127
Bukowinski, M. S. T., 247
Bunker, P. R., 472, 489
Buntin, S. A., 183, 202
Bunting, E. N., 240
Burak, I., 377, 378, 387, 388, 391, 396, 397, 400, 401
Burch, K. D., 168
Burchard, W., 606, 607, 609, 615
Burdett, J. K., 303, 317
Burke, N., 267
Burkhalter, F. A., 503
Burland, D. M., 502

Burns, A., 11
Burns, A. R., 176, 182, 183, 193
Burns, M. J., 500
Bursill, L. A., 302
Burton, P. G., 482
Busch, G. E., 376
Buse, G., 685
Butenhoff, T. J., 398
Butkovskaya, N. J., 574
Butler, L. J., 393
Buttrey, D. J., 297, 303, 304
Buzier, M., 368
Bykovskaya, L. A., 499, 500, 502, 503, 511, 524
Byrne, C. A., 352, 355

C

Caballor, R., 658
Cabaud, B., 75
Cabrel, F., 694
Cahn, J. W., 613, 615
Caignaert, V., 303
Cailletet, L., 80
Calabrese, J. C., 311, 314
Calame, J. P., 101, 107, 109
Calef, D. F., 120, 121, 125, 127, 131, 132
Callahan, P. M., 683, 687, 689
Callear, A. B., 416, 417, 428, 432
Cameron, G. C., 102
Campana, J. E., 85
Campanella, M., 441, 464
Campion, A., 148, 190, 320
Cannella, V., 33
Canters, G. W., 507
Capaldi, R., 685
Carafoli, E., 686
Cardy, H., 390
Carel, C., 301
Carleton, K. L., 398
Carley, J. S., 472, 480
Carlson, K. D., 414, 420
Carman, H. S., 580, 581
Carmeli, B., 130
Carmesin, H. O., 30, 35, 54
Carmona, F., 502
Carnevali, P., 536, 541
Carney, G. D., 478, 482
Carosella, C. A., 85
Carpy, A., 298
Carr, R., 340, 664
Carrington, A., 470
Carroy, A., 602
Carter, E. A., 426
Carter, F. P., 524
Carter, S., 471, 472, 474, 475, 478, 482, 489, 491, 493, 494
Carter, T. P., 503, 505, 506, 513

Carugno, G., 441, 464
Casanova, R., 265, 285
Casanovas, J., 440
CASASSA, M. P., 143-71; 145, 146, 151, 162, 163, 165, 166
Casati, G., 471
Case, D. A., 377, 379, 578
Case, K. M., 659
Casey, R. P., 682, 686
Castleman, A. W., 70, 77
Castles, J. R., 301
Castner, E. W. Jr., 118, 126-28
Castro, G., 503, 511, 525
Catala, J. M., 630
Cates, M. E., 624-26
Catlow, C. R. A., 99, 301, 302
Cattadoria, C., 441
Caucheteux, C., 32
Caughey, W. S., 696
Cava, R. J., 92, 311, 314, 315, 317, 318
CAVANAGH, R. R., 143-71; 145, 146, 151, 160, 162, 163, 165, 166, 183, 202
Cecchi, J. L., 174
Ceperley, D. M., 237, 238
Certain, P. R., 485
Ceyer, S. T., 143
Chabagno, J. M., 85, 86, 88, 108
Chabal, Y., 143
Chabal, Y. J., 159, 265
Chabanel, M., 100
Chadi, J., 536, 537
Chadwick, A. V., 99, 107, 302
Chagagno, J. M., 92, 94, 105
Chakraverty, B. K., 316, 320
Chakraverty, D., 319
Chakravorty, K. K., 562, 563, 567, 574, 577-79
Chalek, C. L., 414, 420
Chamberlain, G. A., 377, 380, 383
Chan, D. Y. C., 630
CHAN, S. I., 671-98; 682-85, 689, 691-95
Chance, B., 682
Chance, R. R., 144, 159, 163, 167, 184, 190
Chandler, D., 130, 458, 459
Chandler, D. W., 400, 572, 649
Chandra, A., 120, 124, 127, 136, 137
Chandrasekhar, G. V., 316
Chang, M. H. W., 74
Chang, R. K., 328
Chang, Z. P., 238, 239, 245
Chao, B. H., 671, 676
Chapman, D., 629
Charpak, G., 455
Chase, W. J., 129

Chatani, Y., 98, 99
Chauveteau, G., 615
Chawla, D., 378, 396
Chawla, G., 379, 396, 397
Cheeseman, P. A., 252, 254
Cheetham, A. K., 305, 319
Chen, C. D., 536, 541
Chen, C. J., 532
Chen, C. K., 328, 343
Chen, J. M., 328
Chen, K., 105
Chen, M. H. W., 80
Chen, P., 652
Chen, S. P., 192, 278
Chen, W., 329, 333, 334, 338
Chenavas, J., 319
Cheney, J. C., 661
Cheradame, H., 102, 103, 105, 107, 109
Cherry, D. W., 201
Chesnavich, W. J., 655
Chesser, N. J., 37
Cheung, A. S.-C., 411, 415-17, 421-27, 430, 431
Cheval, G., 424
Chevaldonnet, C., 390
Chewter, L. A., 651
Chiang, C. K., 90, 94, 95, 97, 98
Chidsey, C. E. D., 344
Childs, W. J., 412, 414, 420, 424
Chin, Y. N., 470
Ching, W. Y., 264
Chiodelli, G., 92, 93, 107
Cho, H., 652
Cho, I., 685
Choi, S. E., 400, 487, 489, 565-67, 569-71, 573, 574, 579
Chojnacki, T. P., 162
Chothia, C., 216
Chow, K. S., 44, 45
Chrenko, R. M., 240, 243
Christensen, N. E., 246, 247, 317
Christenson, H. K., 600, 601, 631
Christmann, K., 276
Chryssikos, G. D., 11
Chu, C. W., 311
Chuang, M.-C., 398
Chuang, T. J., 143, 174, 183
Chupka, W. A., 642, 645, 646, 648, 649, 652
Churchwell, E., 4'5, 418, 422
Cichanowski, S. W., 15
Ciferri, A., 102
Cini, M., 189
Cipollini, N. E., 443, 444
Clancy, S., 90, 101
Clark, J. H., 127
Clark, S. P. Jr., 253

AUTHOR INDEX 703

Clarke, D. R., 631
Clarke, M. D., 89
Clarkson, M. T., 631
Clarson, S. J., 355
Clearfield, A., 319
Cline, R. E. Jr., 131
Coakley, C. J., 600, 631
Codama, M., 454
Codling, K., 663, 664
Coffin, R. L., 208, 223
Cohen, C., 368
Cohen, J. B., 301
Cohen, M. H., 440, 445, 446, 450, 456, 457, 461, 462
Cohen, M. L., 237, 238, 243, 248
Cohen, S. A., 174
Cohen-Addad, J. P., 103, 105, 109
Cohen Stuart, M. A., 601, 611
Coker, D. F., 459, 460
Cole, B. E., 646
Cole, K. S., 15, 16, 49
COLE, R. H., 1-28; 6, 10, 11, 13-17, 19, 20, 22, 25, 49, 118
Coletti, F., 201
Collete, C., 368
Collins, B. T., 299
Collins, W. E., 201
Colonomos, P., 22
Colosimo, A., 695
Colson, S. D., 645, 652
Coltrim, M. E., 193, 194
Comes, F. J., 377, 386, 389, 390, 392, 393, 398
Compton, R. N., 640
Condon, E. U., 409
Conrad, H., 143, 144, 159, 162
Cooper, C. D., 640
Cooper, D. E., 507
Cooper, D. L., 662
COPELAND, R. A., 671-98; 683, 693-95
Cordis, L., 652
Corenzwit, E., 249
Corish, J., 302
Cornaggia, C., 645
Cornelius, P. A., 509, 521
Cosgrove, T., 601, 611, 623
Costello, S. A., 184
Cotter, D., 146
Courdurier, G., 162, 163
Courtens, E., 31, 32, 35, 36
Cowan, B. P., 30, 34, 35
Cowan, P., 261, 264
COWIE, J. M. G., 85-113; 102, 103, 105-7, 109
Cowin, J. P., 184, 201, 202
Cox, D. E., 319, 320
Cox, G. B., 679, 680
Crabtree, G. W., 315
Craig, B. B., 200

Craig, J. H., 198
Craven, J. R., 105
Crawford, M. K., 129
CREE, S. H., 85-113
Creighton, T. E., 207, 220, 222, 677
Cremaschi, P., 276
Creutz, C., 131, 132
Crim, F. F., 375, 393, 400, 470
Crippen, G. M., 218, 221
Crist, B., 95
Cross, J. B., 574, 582
Cross, L. E., 300
Cross, P. C., 471, 473
Crosswhite, H. M., 424
Crowley, T. L., 611
Cuddihy, E. F., 86
Cuellar, E., 503, 511
Cui, S., 193, 201
Curro, J. G., 364, 371
Curtiss, T. J., 565, 566, 569, 570, 572, 574, 575
Cusack, N. E., 63, 70
Cusanovich, M. A., 228
Czuba, P., 201
Czurylo, E. A., 364, 366
Czyzewski, J. J., 173

D

Dabrowski, B., 320
Dai, H. L., 341
Dai, Y., 320
Dalal, N. S., 31
Dalard, F., 107
Daltrozzo, E., 522
Damien, J. C., 46
Danby, G., 476, 477
Dandrea, R. G., 243
Daoust, H., 615
Dargelos, A., 390
Darken, L. S., 250
Darling, C., 319
Darshane, V. S., 299
Das, P., 184
Das, P. R., 657
Das Gupta, S. K., 676
Dasgupta, S. R., 162
Daubert, J., 46, 47
David, F., 419
David, W. I. F., 311, 319
Davidson, D. W., 10, 16, 118
Davidson, E. R., 651
Davies, G. R., 105, 106
Davies, R. J., 629
Davis, D. N., 427
Davis, E. A., 63, 69, 309
Davis, G. T., 90, 94, 95, 97, 98
Davis, H. T., 631
Davis, R. F., 275

Davis, S. P., 414, 415, 420, 421
Daw, M. S., 192, 263, 278-80
Day, P., 304, 311
Dayringer, H. E., 228
Deam, R. T., 363
Debarre, D., 396
Debever, J. M., 201
De Boer, S., 522
DeBolt, L. C., 364, 366
De Bree, Ph., 510
deBruyn, J. R., 80
Debye, P., 3
de Castro, A. R. B., 328
Decius, J. C., 471, 473
Deckert, A. A., 262, 266
Declemy, A., 127, 128
DeCorpo, J. J., 85
de Gennes, P.-G., 601, 602, 611, 613, 615, 618, 622, 626, 630
DeGrado, W. F., 228
De Groot, S. R., 78
deHaas, M. P., 446, 448, 452
Dehmer, J. L., 641, 643-47, 649, 652
Dehmer, P. M., 642-46, 649, 652
deJong, C., 685
De Jongh, L. H., 42
Delaval, J. M., 419, 433, 434
Deloche, B., 371
Delsemme, A., 417
Delwiche, J., 642, 645, 646, 648, 650, 661, 662
Delwiche, J. P., 647
Demuth, J. E., 183, 186, 191, 275, 536, 538, 539, 541, 546, 548, 550, 551, 554-56
Denney, D. J., 17
De Pristo, A. E., 278
Derjaguin, B. V., 599
Deroo, P., 107
Descamps, M., 32
De Sourtis, M., 316, 320
Deutch, J. M., 136
Deutsch, A. J., 413, 422
Devillers, M. A. C., 74
DeVore, T. C., 415, 416, 427
Devoret, M., 30, 34, 35
de Vries, A. E., 400
De Vries, H., 500, 502, 503, 506, 507, 511
DeVries, R. C., 240, 243
DeYoreo, J. J., 36, 42, 44, 45
Dhingra, M. M., 213, 214
Dibeler, V. H., 655
Dick, B., 503
Dicker, A. I. M., 502-4, 506, 507, 509-11, 513, 520, 524, 525
Dickie, R. A., 352, 355
Dickinson, A. S., 485

AUTHOR INDEX

Diebold, G. J., 380
Diesen, R. W., 376
DiFoggio, R., 261, 266
Dill, D., 641, 646
DiMarzio, E. A., 109
Dimicoli, I., 379, 386, 656
Dimov, D. K., 107
DiNardo, N. J., 183, 276
Ding, A., 652
Dintzis, H. M., 207
Disch, R. L., 14
Distin, P. A., 250
Dix, L. R., 629
Dixon, R. N., 377, 378, 383, 389-91, 393, 394, 396, 398
Dmitriev, E. G., 461
Dobbs, J. N., 36
Dobkowski, J., 503, 506, 507, 509, 510
Dobrowski, S. A., 106
Docker, M. P., 377, 379, 384, 385, 389, 395, 399
Dockter, M. E., 683
Dodelet, J.-P., 451
Dogonadze, R. R., 116, 131
Dohlus, R., 146, 148
Doi, M., 363
Doi, T., 671, 676
Doke, T., 452
Döldissen, W., 444
Dolg, M., 430, 432, 433
Dolgoshein, B. A., 454
Domen, K., 183
Domenges, B., 299, 303
Doniach, S., 180
Dose, V., 186
Douay, M., 424
Douketis, C., 574
Doumerc, J. P., 299
Doussineau, P., 36
Dowrey, A. E., 356, 367
Dransfeld, K., 42
Dresselhaus, M. S., 85
Dresser, M. J., 180
Dressler, K., 648
Drickamer, H. G., 248
Drissler, F., 502, 511
Drowart, J., 416, 417
Druger, S. D., 109
Dübal, H.-R., 393
Dubault, A., 371
Dubey, M. K., 574
Dubois, L. H., 422
Dubs, M., 377, 379, 394
Dubs, R. L., 379, 395
Duckers, L. J., 70
Duclos, S. J., 247
Duclot, M. J., 85, 86, 88, 108
Dufour, C., 424
Dugan, H., 377, 378, 391
Duiser, J. A., 359
Dujardin, G., 645, 662
Dulick, M., 412, 419, 434

Dumesic, J. A., 162, 163
Duncan, T. M., 163
Dunker, A. M., 472, 477
Dunlap, B. D., 315
Dunn, K. J., 240, 243
Dunn, R. J., 676
Dunn, T. M., 413, 414, 420, 425
Duplessix, R., 368
Dupon, R., 96, 100
Duppen, K., 507
Dupree, R., 69
Dupuis, M., 396
Durand, D., 37
Durant, J. L., 654
Dusek, K., 352, 368
Dutton, D. B., 245
Dutuit, O., 645, 647, 648, 656, 657
Duval, M. C., 592
Dyke, J. M., 414-16, 422, 426, 427
Dylla, H. F., 174
Dymanus, A., 573
Dynes, R. C., 42, 547, 555
Dyson, H. J., 208, 224
Dyson, R., 88
Dzvonik, M., 400

E

Earnest, T. N., 676
Easley, W. C., 416, 419, 428
Eastman, D. E., 174, 185, 189
Ebata, T., 399, 400, 470
Ebeling, W., 62
Eberhardt, W., 664, 665
Ebner, C., 463
Eby, J. E., 245
Eckstrom, T., 299
Economu, E. N., 308
Eddy, M. M., 319
Ederer, D. L., 646
Edmundson, A., 228
Edwards, D. F., 241
Edwards, P. P., 299, 305, 308-10
Edwards, S. F., 35, 352, 363, 364, 369, 611
Eenshuistra, P. J., 649
Eesley, G. L., 143
Efrima, S., 131, 132
Egerton, R. F., 319
Eggert, J. H., 246
Eguiluz, A. G., 144, 159, 163
Ehlert, M. K., 303
Ehrenson, S., 118, 441
Ehrlich, D. J., 174
Ehrlich, G., 261, 264, 265, 285
Eichinger, B. E., 352, 359, 360, 370, 371
Einstein, T. L., 262, 263, 267
Eisele, B., 44, 51, 52

Eisenberg, A., 85, 86
Eisenthal, K. B., 129, 134, 343, 464
Ekardt, W., 168
Ekwelundu, E., 199
Elaesser, T., 148
Eland, J. H. D., 642, 660-63
Elber, R., 208, 214
El-Hanany, U., 63, 69, 70
Ellis, W. R., 683, 685, 689, 693
El-Sayed, M. A., 656
Elschner, S., 37, 55
Emery, V. J., 316, 317
Enderle, H. F., 352
Endo, H., 63, 70
Endo, K., 541, 555
Endo, Y., 418, 430
Endoh, Y., 297
Engel, V., 385, 386, 394, 401
Engel, Y. M., 577
Engerholm, G. G., 485
Engleking, P. C., 417, 430
Englman, R., 154, 298
ERMAN, B., 351-74; 351, 359-62, 364, 366-69, 371
Ernst, R. M., 45, 49
Ernst, W. E., 412
Ertl, G., 183, 265, 276
Eskes, H., 317
Estes, D., 474
Esteve, D., 30, 35
Evans, D. F., 22
Evans, G. T., 589, 590
Evans, J., 107, 108
Even, U., 68, 70, 75
Evers, J., 246
Ewen, B., 368
Ewing, C. T., 85
Eyers, A., 186
Eyler, E. E., 642
Eyring, E. M., 100
Eyring, H., 655
Ezra, G. S., 492, 493

F

Faber, T. E., 66
Fan, F. F., 600, 631
Fano, U., 377, 379
Fano, V., 646
Farago, B., 368
Farantos, S. C., 481, 491
Farley, F. W., 574
Farrell, J. N., 248
Farrington, G. C., 107
Farrow, R., 311, 317
Fasman, G., 209
Fatuzzo, E., 136
Faust, W. L., 200
Fauster, T., 548, 553
Fauteux, D., 92-95, 97, 107
Fayard, M., 298

Fayer, M. D., 503, 506, 507, 511, 513, 515, 517, 518
Féménias, J.-L., 414, 420, 424
Fearey, B. L., 503, 505, 506
Feary, B. L., 524
Feast, W. J., 598
Fecher, G., 575
Fee, J. A., 696
Feenstra, R. M., 536, 538-40, 545, 547, 548, 550, 551, 553, 555
FEIBELMAN, P. J., 261-90; 193, 262, 263, 265, 267, 268, 272, 281, 282, 284
Feigerle, C. S., 646
Feile, R., 36, 46-48, 54
Fein, A. P., 536, 538-40, 545, 547, 548, 550, 551, 553, 555
Feinberg, J., 414
Felder, P., 394, 395
Felderhof, B. U., 6
Feldman, L. C., 201
Fellner-Feldegg, H., 11
Fendt, A., 158
Fenn, J. B., 565
Fenton, B. E., 85
Ferguson, R., 103
Ferloni, P., 92, 93, 107
Ferm, P., 183
Fernandez-Berridi, M. J., 615
Ferrante, R. F., 428
Ferreira, L. P., 642
Ferreira, L. R. A., 642
Ferrell, W. R., 640
Ferrett, T. A., 646
Ferrett, T. H., 646
Ferry, J. D., 597
Fetters, L. J., 617, 620, 621, 626
Feulner, P., 180-82, 189, 191, 201
Feynman, R. P., 458
Field, R. W., 408, 412, 414, 419, 433, 470, 471, 489, 492
Filseth, S. V., 396
Fimmel, A. L., 679, 680
Fine, R. M., 211
Fink, H. W., 265, 285
Fink, J., 316, 320
Finkelstein, V. A., 207
Firth, A. M., 105, 106, 109
Fischer, B., 34, 35, 54
Fischer, C. R., 276
Fischer, K., 246
Fischer, R., 513, 522, 524
Fischer, S. F., 148, 158
Fish, D., 105
Fixman, M., 371
Flank, A. M., 316, 320
Fleer, G. J., 601, 604, 611, 613, 629

Fleischmann, M., 328
Fleishman, L., 36, 49
Fleming, G. R., 116-18, 126-29, 133, 134
Fletcher, D. A., 408, 419
Flippen, R. B., 314
Flipse, C. J., 544
Flitsch, S. L., 671, 676
Flodström, S. A., 189, 196
Flores, F., 532
Floriano, M. A., 442, 443, 448, 459
Flory, P. J., 215, 222, 225, 351, 356, 359, 360, 364, 366, 368, 369, 371, 597, 600
Flower, D. R., 476, 477
Förster, T., 166
Fodor, S. P. A., 689
Foiles, S. M., 278, 280
Foley, H. M., 413
Follmeg, B., 480, 481
Foltz, M. F., 396
Fonden, T., 263
Fontaine, A., 316, 320
Fontanella, J. J., 91, 101, 107, 109
Fontcuberta, J., 303
Ford, G. C., 228
Ford, W. K., 664
Fork, R. L., 513
Forman, R. A., 242, 243
Fossum, J. O., 55
Founargiotakis, M., 481
Fournier, J., 662
Fournier, P., 661
Fournier, P. G., 661, 662
Franck, E. U., 70
Frank, C. W., 630
Frank, K. H., 186, 531, 541
Frank, V., 37, 38, 48, 49, 55, 56
Franskin-Hubin, M. J., 662
Franz, G., 64, 78
Franz, J. R., 63, 69, 74
Fraser-Monteiro, M. L., 657-60
Frasinski, L. J., 663, 664
Fratello, V. J., 254
Fraunenfelder, H., 131
Frech, R., 100
Freed, K. F., 363, 390
Freeman, A. J., 237, 238, 262
Freeman, G. R., 439-43, 448, 451, 455, 458, 459, 462
Freeman, H. C., 225
Freire, J., 371
Freund, F., 671, 672
Freund, H.-J., 183
Frey, J. G., 474
Frey, R., 651
Freyland, W., 63, 64, 68-70, 78, 79
Fried, L. E., 492, 493

Friedli, C., 237, 238
Friedman, H. L., 131, 132
Friedrich, B., 470
Friedrich, H., 502, 503, 505, 506, 511
Friedrich, J., 501-5, 513, 519, 522, 524
Friedrich, V., 126
Fries, P. H., 125
Friesner, R. A., 491, 492, 522
Fritz, J. N., 243, 244
Fritzson, P., 74
Frohlich, H., 647, 648
Frosch, R. A., 413
Fu, C. L., 262
Fuchs, H., 531, 541, 544
Fucke, W., 63
Fueki, K., 446, 447, 461
Fuggle, J. C., 180, 316, 320
Fujii, A., 399, 400
Fujii, H., 250
Fujii, Y., 248
Fujimori, A., 317
Fujimura, Y., 640
Fujishiro, I., 242
Fujiwaka, S., 70
Fukunaga, O., 244
Fukuroda, A., 386
Fukutomi, M., 311
Fulcher, G. S., 46
Fulkui, K., 268
Fulton, R. L., 8
Funabashi, K., 464
Furtak, T. E., 328
Furukawa, J., 352

G

Gadzuk, J. W., 144, 159, 163, 167, 180, 184, 186, 191
Gai, P. L., 314
Galehouse, D. C., 414, 415
Gallaher, T. N., 415
Gallezot, P., 162, 163
Gallo, A. R., 190
Galy, J., 298
Gammie, G., 299
Ganapathi, L., 298, 299, 303, 311, 313, 318
Ganapathiappan, S., 105
Gandel, Y., 129
Gandhi, S. R., 565, 566, 569, 570, 572, 574, 580
Gandini, A., 102, 103, 105, 107, 109
Ganguli, A. K., 297, 299, 311, 314, 320
Ganguly, P., 297-99, 303-5, 309-11, 315-17, 320
Gangwer, T. E., 446
Gans, P. J., 221
Gantelfor, G., 652
Gao, L., 311

AUTHOR INDEX

Garcia, A., 237, 238
Garcia, N., 531, 532, 541, 544, 556
Garcia, R., 544
Garel, J. R., 208, 209
Garland, C. W., 46, 55
Garno, J. P., 42
Garrett, M. B., 428
Garrett, W. R., 640
Garrido, L., 355
Garstein, E., 301
Gast, A. P., 618, 627
Gathers, G. R., 63
Gathers, R. G., 63
Gatilov, L. A., 249
Gatterer, A., 416, 417, 419, 427, 433
Gaudel, Y., 464
Gauyacq, J. P., 187
Gavarri, J., 301
Gavoille, G., 305
Gaydon, A. G., 417, 419, 429
Gaylord, R. J., 369, 371
Gebhard, R., 689
Gedanken, A., 400
Gee, N., 440-43, 448, 459
Geerlings, J. J. C., 185-87
Gehring, G. A., 298
Gehring, K. A., 298
Gehrtz, M., 502, 503, 506, 525
Geijsberts, M., 573, 578, 582, 585, 591
Gelles, J., 689, 691-95
Genack, A. Z., 500, 502, 506, 507, 511
Gennett, T., 132
Gent, A. N., 366
George, S. M., 143, 262, 266
George, T. F., 144, 159, 163, 310
Georges, J. M., 599
Gerber, C., 264, 269, 531, 538, 598
Gerber, G., 380
Gerber, R. B., 469, 493-95
Gerdanian, P., 301, 302
Gericke, K.-H., 377, 386, 389, 390, 392, 393, 398
Gerlach-Meyer, U., 390
Gerry, M. C. L., 415, 419, 421, 422, 433
Gerschel, A., 11
Gertner, B. J., 134
Geus, J. W., 162, 163
Ghelis, C., 207
Ghosh, A. P., 194, 197, 201
Gianozzi, P., 247
Gibb, T. C., 303, 305
Gibbs, G. V., 254
Gibbs, J. H., 109
Giboni, K. L., 464
Gibson, F., 679, 680
Giesen, K., 344

Giess, P., 185
Gigli, G., 416
Giles, J. R. M., 105, 106
Gilgen, H. H., 174
Gill, P. M. W., 661
Gillbro, T., 134
Gillespie, R. J., 20
Gillie, J. K., 522
Gillson, J. L., 297, 310
Gilman, J. P., 657
Gilton, T. L., 184, 201
Gimzewski, J. K., 536, 541
Giser, D., 127, 132
Giusti-Suzor, A., 642, 645, 646
Glarum, S. H., 17, 18, 20
Gläser, W., 63, 64, 66, 78, 79
Glasser, L., 671, 672
Glass-Maujean, M., 380, 647, 648
Gleitzer, C., 305
Glötzel, D., 248
Glownia, J. H., 149
Gluck, N. S., 199
Gō, M., 209, 214, 215, 224
Gō, N., 207, 214, 215, 218, 219, 221, 223, 225
Goddard, W. A. III, 273-75, 277, 316, 426
Godfrey, M., 588
Goettel, K. A., 238, 240, 244-46
Goldbart, P., 54
Goldberg, L. S., 146
Goldblatt, M., 160
Goldenberg, D. P., 220
Golder, A. J., 90
Golding, B., 511, 513, 519, 520
Goldstein, R. E., 63, 80, 81
Gole, J. L., 414, 416, 418-20, 422, 427, 428, 432
Golovchenko, J. A., 536, 541, 542
Golovin, M. N., 132
Gölzenleuchter, H., 389, 390
Gomer, R., 174, 181, 261, 266
Goncharov, A. F., 244-46
Gonidec, A., 441
Gonzalez, J., 318
Gonzalez, J. M. G., 303
González-Ureña, A., 592
Goodenough, J. B., 291, 293, 295, 297, 298, 303, 305, 310, 320
Goodman, D. W., 268, 274
Goodman, L. S., 424
Gopalakrishnan, J., 291, 298, 299, 301, 303, 304, 311, 314, 318, 319
Gorczyca, G., 37
Gordon, R. G., 472, 477
Gordon, R. M., 416, 417, 424, 428, 430

Goreand, M., 299
Gorecki, W., 92, 94, 105
Gorokhovskii, A. A., 500-3, 507, 513, 518
Gorry, P. O., 400
Gortel, Z. W., 181, 182, 191
Goshorn, D. P., 311
Gotchev, B., 651
Gottfried, D. S., 513, 522
Gottlieb, M., 353, 364, 365, 369, 371
Gottlieb, Yu. Ya., 366
Götze, T., 599
Götzlaff, W., 63, 70, 77, 78, 81
Gouedard, G., 377, 379
Gough, T. E., 567, 574
Govers, T. R., 642, 645
Graener, H., 146, 148
Graessley, W. W., 359
Graf, F., 500, 502, 511
Graham, W. H., 134
Graham, W. R. M., 428
Grampp, G., 132
Grangier, P., 380
Granick, S., 599, 604, 606, 609, 611, 619, 627, 631
Grannan, E. R., 45, 52
Granneman, E. H. A., 185, 186
Grant, E. R., 386
Gravenor, B. W. J., 414-16, 422, 426, 427
Gray, C. G., 120, 121
Gray, F., 92
Gray, F. M., 106
Gray, H. B., 689
Greatrex, R., 305
Greaves, G. N., 99
Greedan, J. E., 303
Green, D. W., 418, 429, 432, 433
Green, T. A., 193, 194, 201
Greenbaum, S. G., 91, 101, 109
Greenblatt, M., 299, 319
Greene, B. I., 149
Greene, C. H., 377, 379, 573
Greene, J. P., 652
Greenler, R. G., 168
Greenwood, C., 684
Greenwood, N. N., 305
Greiner, N. R., 389
Grenier, J. C., 303
Greppi, G., 102
Griffin, L. L., 652
Griffith, J. E., 536, 553
Grimeland, B., 415
Grimes, R. W., 301
Grimley, T. B., 262, 263, 267
Grimm, H., 30, 35
Grochulski, T., 11
Groenewegen, P. P. M., 15
Grond, W., 503, 519

AUTHOR INDEX 707

Grondey, S., 30, 35
Gross, D. J., 54
Grote, R. F., 130
Grothert, W., 389
Groult, D., 311, 313
Grover, J. R., 642, 652
Grozdanov, T. P., 185
Grubb, S. G., 343, 344
Gruehn, R., 302
Grün, F., 366, 367
Grunewald, A. U., 386, 389, 393
Grzybowski, T. A., 246
Gubbins, K. E., 120, 121
Guelfucci, J. P., 440
Guest, J. A., 380, 381, 399
Guido, M., 416
Guinea, F., 532
Guisti-Suzor, A., 645
Gumhalter, B., 180
Gunnarson, O., 180
Guo, J. H., 652
Guo, Y., 316
Gupta, G., 213, 214
Gurney, R. W., 184
Gurry, R. W., 250
Gurvich, L. V., 415, 421
Gushchin, E. M., 441, 443, 454
Guss, J. M., 225
Gustafson, B., 491
Guth, E., 357, 367
Gutiérrez, A., 503
Gutzwiller, M. C., 308
Guyon, P. M., 641-43, 645-48, 650, 651, 656
Guyot-Sionnest, P., 329, 332-34, 338, 340, 344, 348
Guzman, G. M., 615
Gwinn, W. D., 485
Gwo, D. H., 477
Gyorgy, I., 443

H

Haaks, D., 647
Haarer, D., 500-6, 511, 513, 519, 525
Haber, E., 211
Hackett, N. R., 676
Hadziioannou, G., 619, 620, 623, 625, 627
Hadziioannou, G. H., 630
Haemmerle, W. H., 511, 519
Haerri, H.-P., 379, 388, 391, 400, 401
Hafner, J., 78
Hage, F., 344
Hagenmuller, P., 299, 306, 318, 319
Hagler, A. T., 218
Hagstrum, H. D., 178, 184, 189, 190

Haight, R., 344
Hair, M. L., 143, 144, 159, 162, 609, 614, 616, 617, 620, 621, 629
Halbach, K., 638
Halberstadt, N., 480, 481
Hall, C. K., 611
Hall, D. G., 22, 211
HALL, G. E., 375-405; 129, 130, 377-79, 381, 387, 391, 396, 397, 400, 401
Hall, J., 685
Halle, L. F., 416-18
Haller, K., 501, 502
Halliday, L. A., 127, 129
Halonen, L. O., 482, 489, 491, 493, 494
Halperin, B. I., 33, 42, 510, 515, 520
Halpern, J. B., 385
Halquist, J. O., 238, 240
Hamann, D. R., 262, 263, 268, 269, 272, 320, 532, 535-37, 543
Hamaya, N., 248
HAMERS, R. J., 531-59; 536, 538-41, 546, 548, 550, 551, 553-55
Hamilton, C. E., 470, 471, 489, 492
Hamilton, D. S., 511, 519
Hamilton, I. P., 479, 482, 485
Hamlin, R. C., 228
Hammer, P. D., 421
Hammersley, J. M., 109
Hamnett, A., 310
Hamola, A., 598
Hampton, P. D., 656
Hamza, A. V., 183
Hancock, G., 385
Handy, N. C., 471, 472, 474-76, 478, 482
Hannon, M. J., 90, 101
Hanrahan, C. P., 202
Hansch, T. W., 337, 346
Hansen, P. A., 148
Hansen, R. C., 415, 416, 423, 424, 426, 427
Hansen, R. E., 684
Hansma, P. K., 532, 556
Hardee, J., 580
Harding, C. A., 90, 94, 95, 97, 98
Harding, L. B., 482
Hardis, J. E., 645, 646
Hardy, L. C., 95
Harland, P. W., 580, 581
Harren, F., 565, 569, 574, 583, 587
Harrer, W., 132
Harrington, J. E., 645
Harris, A. B., 32
Harris, A. L., 344

Harris, C. B., 190, 509, 521
Harris, C. S., 90, 95, 97
Harris, D. O., 419, 433, 485
Harris, J., 274
Harris, S. M., 429
Harrison, I., 184, 202
Harrison, J. F., 428
Harrison, M. R., 299, 310
Harrison, P. M., 228
Hart, D. J., 650
Hart, H. R., 30, 34
Hartmann, S. R., 500
Hartzell, C. R., 684
Harvey, S. C., 210, 212
Harvie, J. L., 102
Hase, K., 248
Hasegawa, A., 247
Hasitsume, N., 367
Hasselbrink, E., 381, 401
Hastings, J. B., 319
Hastings, M. P., 415, 426
Hatano, Y., 454, 462
Hatch, L., 679, 680
Hatherly, P. A., 663, 664
Haugen, H. K., 400
Häusler, D., 385, 386, 401
Haussühl, S., 36, 48, 49, 52
Havriliak, S. Jr., 15, 19
Haworth, L., 85
Hayden, B. E., 165
Hayden, C. C., 400
Hayes, J. M., 500, 502, 503, 505, 506, 511, 519, 522
Haymet, A. D. J., 126
Hays, J. F., 254
Hazen, R. M., 250
Heath, T., 623
Heather, R. W., 485
Heatley, F., 105
Hedge, M. S., 320
Hefner, W., 70, 71, 77
Hegarty, J., 511, 513, 520
Hegde, M. S., 304
Hehre, W. J., 658
Heidemann, A., 30, 35
Heiland, W., 201
HEILWEIL, E. J., 143-71; 145, 146, 149, 151, 160, 162, 163, 165, 166
Heine, V., 67
Heinrich, G., 352
Heinrich, N., 658
Heinz, D. L., 238, 242, 253, 254, 256
Heinz, T. F., 328, 329, 341-43
Heinzmann, U., 186, 570, 575
Heisel, F., 132
Helfand, E., 611
Heller, E. J., 471, 489
Hellner, L., 662
Helm, H., 649
Helmis, G., 352

AUTHOR INDEX

Hemley, R. J., 237, 238, 254, 255
Hemminger, J. C., 144, 159
Henderson, J. R., 480
Henderson, R., 676
Hendra, P. J., 328
Hendrickson, W. A., 228
Hennig, S., 394
HENSEL, F., 61-83; 63, 64, 66, 68, 70-72, 74-79, 81
Hepburn, J. W., 387, 388, 391, 396, 397, 400, 401, 650
Herman, F., 420
Hermann, A. M., 311
Hermans, J. J., 367
Hermans, J. Jr., 221
Hernandez, J. P., 63, 70, 77, 463
Herschbach, D. R., 376, 377, 379, 562, 567, 570, 578
Hervieu, M., 299, 303, 311, 313
Herz, J., 368, 371
Herzberg, G., 412, 419, 649
Herzfeld, J., 676
Herzfeld, K. F., 309
Hess, H. F., 547, 555
Hess, W. P., 400
Hesselink, W. H., 500, 507, 510, 524
Hesslich, J., 652
Heuer, A. H., 301
Hibma, T., 92, 94, 97, 98
Hickman, A. P., 649
Hicks, J., 134
Hicks, J. M., 341, 343
Hidaka, Y., 250, 297
Higashi, G. S., 174
Hilbig, R., 640
Hildenbrand, D. L., 414
Hill, R. J., 254
Himpsel, F. J., 185, 189, 262, 316, 548, 553
Hines, M. A., 143
Hinks, D. G., 30, 35, 36, 46, 54, 312, 315, 320
Hioki, T., 85
Hiraoka, Y., 208, 209
Hirlimann, C., 342
Hirota, E., 418, 430
Hirsch, J. E., 316
Hirschmann, R., 522
Hirz, S. J., 623, 624, 630
Hixson, R. S., 63
Ho, J.-S., 611
Ho, K. M., 262, 280
Ho, M. H., 645-47
Ho, W., 174, 199, 200, 265, 276
Hoareau, A., 75
Höchli, U. T., 30, 35, 48, 49, 52

Hochstrasser, R. M., 148, 149, 522
Hock, J. L., 198
Hocking, W. H., 415, 416, 421, 422, 424, 426, 427
Hodgson, A., 377, 383-85, 389, 399
Hoeve, C. A. J., 102
Hoff, A. J., 522
Hoffmeister, M., 592
Hogue, J. V., 180
Hoheisel, W., 202
Höhne, F. E., 77
Holland, D. M. P., 646, 663, 664
Holliday, L., 85
Holloway, S., 143, 144, 159, 162, 186, 263, 269, 272, 280
Holm, L., 685-87
Holmberg, C., 274
Holmes, J. L., 659
Holmgren, S. L., 477
HOLROYD, R. A., 439-68; 440, 441, 443, 444, 446-48, 450, 451, 455, 461
Holzapfel, W., 145
Homma, A., 541, 555
Homola, A. M., 599, 631
Hong, H. K., 500, 502
Hönig, B., 218
Honig, J. M., 297, 303-5, 307, 308, 310, 318, 320
Hopfield, J. J., 134
Hopman, H. J., 649
Hoppe, J., 679
Hor, P. H., 311
Horn, K., 186
Horn, R. G., 601, 630, 631
Horn, T. R., 494
Horowitz, H. S., 299, 319
Hoshino, S., 15
Hosoya, S., 313
Hougen, J. T., 427
Hough, A. M., 488
Houghten, R. A., 208, 224
Houle, F. A., 200
Houser, N. E., 441
HOUSTON, P. L., 375-405; 375, 377-79, 381, 387, 388, 391, 396, 397, 400, 401
Howard, A. E., 209, 210, 212, 228
Howard, B. J., 474, 488
Hoye, J. S., 125
Hsu, D., 502, 507, 509, 510
Hsu, D. S. Y., 578
Hsu, T. C., 317
Hu, H.-W., 609
Huang, G., 411, 415, 416, 418, 421, 422, 425, 426, 432
Huang, K. S., 676

Huang, S. S.-S., 443, 448, 458, 459
Huang, T. C., 316
Huang, T. L., 245, 247
Huang, Z. J., 311
Hubbard, J. B., 22, 136, 137, 308
Hubbard, S. R., 70
Hubbard, W. B., 237
Huber, D. L., 264, 511, 519
Huber, J. R., 377, 379, 387, 394, 395
Huber, K.-P., 412, 419, 649
Hubin-Franskin, M. J., 642, 645-48, 650, 661, 662
Hubner, H.-J., 148
Hudson, B., 129
Hudson, C. E., 652
Huffman, R. E., 650
Hui, E. T., 652
Humbert, A., 544
Hummel, A., 446, 448
Hunklinger, S., 42, 519
Hunnicutt, S. S., 399
Hunt, J. H., 344, 348
Hunt, J. W., 129
Hunt, L. A., 129, 130
Hunter, T. F., 400
Huntington, J. B., 237, 238
Huppert, D., 116, 129, 132
Huq, R., 107
Husinsky, W., 201
Hussain, Z., 194
Husson, L. P. J., 503
Huston, A. L., 503, 506
Hutchinson, M., 144, 159
Hutchinson, P., 79
Hutchison, J. L., 298
Hutson, J. M., 477
Huygen, T., 476, 481
Hybertsen, M. S., 317
Hynes, J. T., 116, 117, 130, 134, 137

I

Iakubov, I. T., 461, 463
Ibach, H., 265, 275
Idiodi, J., 263
Ignatiev, A., 199
Ikeguchi, M., 208, 209
Ikezi, H., 70, 72
Ilavsky, M., 368
Imelick, B., 162, 163
Imoto, H., 299
Imre, D., 471
Ingersent, K., 607, 611, 613
Ingram, M. D., 102
Ingwall, R. T., 364, 366
Innes, K. K., 471
Inoue, G., 394
Inutake, M., 70
Irgens-Defregger, A., 148

AUTHOR INDEX 709

Irish, D. E., 100
Iruin, J. J., 615
Isenor, M. R., 567
Ishida, Y., 74
Ishihara, A., 367
Ishihara, H., 102
Isogai, Y., 211
Israelachvili, J. N., 598-601, 604, 607, 609, 610, 614, 617, 628, 630, 631
Itie, J. P., 244-46
Ito, K., 647, 662
Ito, M., 399, 400, 470
Itoh, K., 444-48, 450, 451, 461
Itoh, N., 200
Ittah, V., 132
Iwamoto, R., 102
Iwata, L., 387
Iwatate, K., 162
Izawa, H., 319

J

Jackel, J. L., 318
Jäckle, J., 39
Jackson, B., 509, 519
Jackson, W. H., 375
Jacob, W., 186
Jacobs, A., 389-91
Jacobs, B., 685
Jacobs, D. C., 379
Jacobs, I. S., 33
Jacobsen, K. W., 278, 279
Jacobsen, T., 93, 108
Jacobson, A. J., 311, 318
Jacobson, F. M., 441, 443
Jacobson, H., 222, 225
Jacon, M., 477, 479
Jadhao, V. G., 305
Jaeger, R., 189, 196, 275
Jaenicke, R., 207
Jaenicke, W., 132
Jakeman, R. J. B., 319
Jaklevic, R. C., 545, 546
Jakubicki, G., 319
Jakubov, I. T., 63
Jalink, H., 562, 565, 567, 569, 573, 574, 577, 578, 582, 583, 585-88, 590, 591
Jalli, T., 696
James, D. B., 93, 102
James, H. M., 12, 357, 359, 367
James, M. N. G., 211
James, R., 302
Jamieson, J. C., 244
Janas, V. F., 368
Janda, K. C., 477, 480, 481
Janev, R. K., 185
Janik, B., 30, 35
Jankowiak, R., 501-3, 505, 511, 522
Janse, E. C., 189

Jansen, G., 507
Jansen, H. J. F., 237, 238
Janssen, M. H. M., 400, 565, 569, 572-74, 578, 579, 582, 583, 585-87, 590, 591
Jarnagin, J., 441
Jarry, J. P., 367
Jarzeba, W., 116, 127-29, 132
Jasien, P. G., 428
Jayaprakash, C., 42
Jayaraman, A., 237, 238, 240, 249, 315
Jay-Gerin, J.-P., 445
JEANLOZ, R., 237-59; 238, 239, 241-47, 250, 253-56
Jeannot, F., 301
Jefferson, D. A., 299, 303, 319
Jennison, D. R., 176, 182, 183, 193, 194
Jensen, P., 472, 489, 695
Jeon, S. J., 246
Jeong, Y. H., 45, 49
Jephcoat, A. P., 240, 242, 255
Jepsen, O., 247
Jernigan, R. L., 216, 221, 225
Jeung, G. H., 420
Jezequel, G., 638
Jha, S. S., 328
Jilek, E., 314, 319
Joanny, J.-F., 627, 628
Jochims, H. W., 638, 652
Johari, G. P., 32, 53
Johnson, A. D., 143
Johnson, A. E., 128, 129
Johnson, A. L., 198
Johnson, B. R., 474, 475, 477, 479
Johnson, D. C., 299
Johnson, H. L., 420
Johnson, I. D., 129
Johnson, L. W., 503, 504, 506, 507, 509, 511, 525
Johnson, M. D., 79
Johnson, M. W., 319
Johnson, R. E., 173, 193, 201
Johnson, S. G., 524
Johnsson, P., 263
Johnston, D. C., 299, 310, 311
Johnston, D. R., 13
Jonah, C. D., 129, 448
Jonas, J., 299
Jones, E. M., 565, 573, 574, 578, 580
Jones, K. E., 509
Jones, L. H., 160
Jones, M. G., 695
Jones, R. O., 262
Jones, R. P., 129
Jones, S. S., 676
Jones, T. B., 101
Jones, V. O., 196
Jones, W. E., 416, 427
Jorgensen, B., 201

Jorgensen, J. D., 312
Jørgensen, U. G., 491
Jortner, J., 70, 125, 129, 134, 154, 445, 462
Josland, G. D., 414, 422
Jostons, A., 301
Joswig, H., 381
Joubert, J. C., 319
Jouvet, C., 592
Joyce, S. A., 198
Julian, M., 49, 51
Jung, K.-H., 570, 571
Jungen, C., 642, 645
Jungmann, K., 202
Jüngst, S., 63, 64, 68, 81
Junkes, J., 416, 417, 419, 427, 433

K

Kaarli, R., 503
Kaarli, R. K., 500, 503
Kabach, H. R., 672
Kachru, R., 649
Kadono, R., 320
Kador, L., 525
Kaesdorf, S., 570
Kahl, G., 78
Kahlow, M. A., 127-29, 132
Kaiser, B., 75
Kaiser, W., 144, 145, 148, 158
Kaiser, W. J., 545, 546, 555, 556
Kajimura, K., 541, 555
Kakeno, M., 85
Kakol, Z., 297, 320
Kaldis, E., 314, 319
Kalman, E., 116
Kamke, B., 652
Kamke, W., 652
Kane, E. O., 282
Kaneko, K., 461
Kang, T.-J., 127, 128, 132
Kano, F., 218, 223, 225
Kanter, I., 54
Känzig, W., 30, 34
Kao, C. N. R., 314
Kao, C. T., 344
Kao, F. J., 199
Kaplan, M. L., 92-96
Kaplan, P., 541
Karim, O. A., 126
Karpinski, J., 314, 319
Karplus, M., 207, 208, 211, 214, 224, 230, 588
Kasai, P. H., 414, 416, 420, 423
Kasai, T., 574, 575, 580
Kasatani, K., 100
Kassner, K., 509, 510, 519, 520
Kasuya, T., 33
Katayama, D. H., 650

AUTHOR INDEX

Kato, H., 88, 380
Katsner, M. A., 297
Katz, A. I., 243
Kauzmann, W., 208, 224
Kawaguchi, M., 601, 604, 607
Kawahara, H., 100
Kawai, R., 176, 183, 198
Kawajima, K., 208, 209
Kawamoto, J., 85
Kawamura, K., 252, 254
Kawasaki, M., 386, 394
Kawazumi, H., 449
Kay, J. G., 418, 428, 429, 432, 433
Kayalan, C., 672
Kayser, R. F., 137
Keenan, P. C., 413, 415, 420, 422, 425
Keery, K. M., 134
Keesee, R. G., 70, 77
Keiding, S. R., 649
Keller, B. A., 394, 395
Kelley, R. D., 268, 274
Kellogg, G. L., 261, 264
Kemnitz, K., 343
Kemp, R. J., 415, 425
Kendelewicz, T., 196, 275
Kendrew, J. C., 207
Kenney-Wallace, G., 116
Kenney-Wallace, G. A., 129, 130
Kern, C. W., 478
Kern, D. P., 174
Kerr, E. A., 389, 390
Kestin, J., 70
Kevan, S. D., 275
Kevin, K., 317
Khan, I. M., 105
Kharlamov, B. M., 500, 502, 503, 511, 524
Khorana, H. G., 671, 676
Khrapak, A. G., 463
Kiermeier, A, 654, 655
Kihlborg, L., 299
Kikas, J. V., 501, 502
Kikkawa, S., 318, 319
Kikoin, I. K., 70
Kikuchi, T., 386, 394
Kilian, H. G., 352
Killis, A., 102, 103, 105, 109
Kim, K. J., 638
Kim, M. W., 343, 344
Kimelman, D., 218, 221
Kimura, K., 652
Kindt, S., 412
King, D. A., 143, 144, 159, 162
King, D. S., 146, 183, 202
King, G. C., 650
King, M., 85, 86
Kinomura, N., 318
Kinsey, J. L., 470, 471, 489, 492

Kinsinger, J. A., 642
Kirk, M. D., 556
Kirkpatrick, T. R., 54
Kirkwood, J. G., 3, 5, 13
Kirtley, J. R., 547, 555
Kishimoto, I., 39
Kisiel, Z., 11
Kiskinova, M., 265, 268, 274
Kiskinova, M. P., 268, 274
Kitazawa, K., 311
Kitihara, K., 461
Kitson, D. H., 211
Kittel, C., 33, 302
Kivelson, D., 126, 136, 137
Kjems, J. K., 32, 46, 48, 54, 55
Klafter, J., 125
Klauser, R., 168
Klee, H., 30, 35
Klee, S., 377, 389, 390, 392, 393, 398
Klein, J., 604, 606, 607, 609-11, 613, 615-17, 620, 621, 626, 628, 629
Klein, M. L., 37, 458
Klein, M. W., 34, 35, 54
Klein, P. G., 369
Kleinermanns, K., 389, 390
Klekamp, A., 201
Klemperer, W., 414, 420, 477
Klenin, M. A., 35
Kley, D., 389
Kleyn, A. W., 575
Klitsner, T., 544, 545
Kloczkowski, A., 359, 360
Knaak, W., 36, 42-45
Knight, A. E. W., 129
Knight, J., 90
Knight, L. B., 416, 419, 428
Knight, W. D., 33
Knittle, E., 239, 242-47, 250, 256
Knockel, H., 417, 418, 430
Knorr, K., 30, 32, 35-37, 39, 40, 46-49, 52-54
Knotek, M. L., 174, 193, 194, 196
Knott, K. F., 85
Knowles, P. J., 432
Knox, W. H., 513
Knuth, B., 63, 81
Kobayashi, S., 386, 394
Kobayashi, T., 88, 89, 104
Kobylyanskii, A. I., 415, 421
Koch, E. E., 638
Koehler, B. G., 143, 262, 266
Koehler, T. R., 507
Koehler, U. K., 541
Koenig, R., 146
Kohler, S. J., 400
Köhler, W., 513, 519, 522, 524
Kohlrausch, R., 128
Kohn, W., 181, 281, 282

Koizumi, M., 318, 319
KOLINSKI, A., 207-35; 212, 222, 228
Kollman, P. A., 209, 210, 212, 218, 221
Kolmeder, C., 148
Kondo, Y., 49, 244, 246
Kondow, T., 380, 398
Konstantinov, A. A., 683
Koons, N., 85
Koos, D. A., 342
Koot, W., 649
Kornbilt, L., 199
Kornyshev, A. A., 116, 118
Korppi-Tommola, J., 155
Korrovits, V., 513, 518
Kosower, E. M., 116, 132
Koster, G. F., 281
Kotelchuck, D., 209, 224
Kott, S. J., 630
Kottis, Ph., 127
Kourouklis, G. A., 315
Koutecký, J., 267, 420
Kovács, I., 427
Krab, K., 671, 681, 682, 686
Kraeft, W. D., 62
Kraemer, W. P., 472
Krajewski, J. J., 311, 314, 317
Kramer, K. H., 563, 565, 576, 578
Kramer, O., 352, 355
Kramers, H. A., 130
Kraus, J. S., 201
Krauss, M., 408, 430, 432
Krautwald, H. J., 385, 391
Kreevoy, M. M., 134
Kremp, D., 62
Kress, J. D., 278
Kretzschmar, K., 168
Kreuzer, H. J., 181, 182, 191
Krieger, T. J., 12
Krigbaum, W. R., 220, 367
Krishnamurthy, G., 416, 427
Kristjansson, K. S., 400
Kröckertskothen, T., 417, 418, 430
Kroger, P., 131, 132
Krohn, C. E., 71
Krok, F., 102, 107
Kruglov, A. A., 441, 443, 454
Kruit, P., 643
Krumhansl, J. A., 62
Kuan, D.-Y., 463
Kubaschewski, O., 238, 239
Kubby, J. A., 536, 553
Kubiak, G. D., 649
Kubo, R., 509
Kubota, S., 452
Kuchitsu, K., 380, 398
Kuech, T. F., 174
Kuge, H., 389
Kuhlewind, H., 654, 655
Kuhn, W., 360, 366, 367

AUTHOR INDEX 711

Kuipers, E. W., 575
Kujino, K., 100
Kuleshova, L. V., 249
Kulikov, A. N., 415, 421
Kulkarni, G. V., 303, 317
Kulkarni, J. A., 299
Kumada, N., 318
Kumar, S. K., 630
Kummel, A. C., 379
Kung, A. H., 640, 649
Kuntz, I. D., 218, 221
Kuntz, P. J., 588
Kunz, C., 638
Kuppermann, A., 474
Kurata, M., 230
Kurizki, G., 380
Kurnit, N. A., 500
Kurtz, R. L., 193, 195-97
Kurz, J. L., 134
Kurz, L. C., 134
Kushick, J. N., 214
Kuwamoto, H., 310
Kuwata, K., 574, 575, 580
Kwakman, L. F. T., 185, 186
Kwei, G. H., 574, 582
Kwiecien, J. Z., 46

L

Labana, S. S., 352, 355
Labbe, Ph., 299
Laberge, N. L., 24
Lablanquie, P., 660-62
La Budde, R. A., 588
Lagally, M. G., 264
Lagarde, P., 316, 320
Lagerqvist, A., 415, 416, 419, 422, 433, 434
Lahmani, F., 394
Laird, B. B., 525
Lal, J., 352, 355
Lamb, D. C., 455
Lambert, P. M., 299, 310
Lanczos, C., 491
Landau, L., 62
Landolt, H. H., 155
Landolt-Börnstein, 407
Lang, N. D., 176, 180, 181, 183, 185, 187, 190, 198, 263, 265, 269, 272, 278, 279, 281, 282, 532, 535, 536
Langhoff, S. R., 408, 412, 416, 420, 433
Langley, N. R., 365
Langlois, J., 316
Langreth, D. C., 144, 159, 167
Lanzerotti, L. J., 173
La Placa, S. J., 311
Laporte, P., 462
Lapp, A., 368
Lardeux, C., 394
Laria, D., 458, 459

Larichev, M. N., 574
Larkin, R. J., 640, 649, 650
Larkins, F. P., 664
Larour, J., 462
Larsen, R. W., 693
Larsson, M., 491
Laubereau, A., 144-46, 148
Lavi, R., 394
Lawley, K. P., 375
Lax, M., 6
Le, P. N., 428
Leach, S., 641, 645, 661, 662
Lebedenko, V. N., 454
Lebedev, A. N., 441
Lebowitz, J. L., 121
Leconte, M., 162
Lee, B. K., 598
Lee, C. C., 92, 96
Lee, C. H., 328
Lee, C. S., 629
Lee, H. W., 502, 525
Lee, H. W. H., 503, 506, 507, 511
Lee, J. S., 474, 480, 481
Lee, L. C., 389
Lee, M. B., 143
Lee, N., 417, 430
Lee, P. A., 309
Lee, P. S., 428
Lee, S., 230
Lee, S. B., 265
Lee, V. Y., 311
Lee, Y. L., 95
Lee, Y. T., 640, 649, 650, 652
Leermakers, F., 623
Lefebvre, M., 396
Lefebvre, R., 471
Lefebvre, Y., 419, 433, 434
Lefebvre-Brion, H., 645, 646
Leffer, J. E., 134
Leforestier, C., 491, 492
Legally, M. G., 261, 264
Lehmann, J. C., 377, 379
Lehto, J., 319
Lehwald, S., 275
Leibler, K., 11
Leibler, L., 627
Leipunskii, I. O., 574
Leisin, O., 652
Lekies, R., 652
Lekner, J., 444, 456, 461
Lemaire, J., 656
LeNest, J. F., 102, 103, 105, 107, 109
Lenth, W., 503
Leone, S. R., 375, 400
Leonhardt, R., 145
Leonowicz, M. E., 303
Lerner, R. A., 208, 224
Leroi, G. E., 645, 647, 652
LeRouzo, H., 646
Leroy, J. P., 473
Le Roy, R. J., 472, 480

Lesseski, D. C., 482
Lessing, H. E., 127
Lester, W. A., 396, 472, 477, 479
Leung, R. C., 46
Leutwyler, S., 481
Levelut, A., 36
Levenson, M. D., 503
Levêque, M., 103, 105, 107
Levich, V. G., 131
Levin, I., 652
Levin, M., 70
Levine, R. D., 562, 570, 577, 579, 580, 584, 588, 590, 591
Levinos, N. J., 344
Levinthal, C., 207, 208, 211
Levitt, M., 216, 217, 221
Levy, R. M., 214
Lewandowski, J. T., 311, 319
Lewis, D. K., 657, 658
Lewis, L. J., 37
Lewis, P. N., 209, 214, 224
Lewis, R. A., 414, 416, 422, 427
Leycuras, A., 462
Li, L., 645
Li, P. M., 693, 694
Li, Z., 213
Liberman, M. H., 364, 366
Licciardello, D. C., 309
Lichtman, D., 198, 199
Liebsch, A., 349
Lifshitz, C., 652, 658
LIGHT, J. C., 469-98; 477, 479, 481, 482, 484, 485, 487, 489, 491
Lightfoot, P., 303
Likar, M. D., 393
Lill, J. V., 479, 482, 485
Lin, S. F., 220
Lin, S. H., 640
Linde, P. F., 389
Lindefelt, U., 282
Lindgren, B., 415
Lineberger, W. C., 417, 418, 430
Linford, R. G., 104
Linn, S. H., 649
Linnebach, E., 390
Linton, C., 414, 415, 421
Lipari, N. O., 281
Lippert, E., 132
Lippincott, E. R., 240
Lipton, M., 209
Littau, K. A., 506, 511, 515, 517, 518
Liu, J. Z., 315
Liu, K., 414
Liu, W. K., 500
Liu, Z.-M., 184
Liu, Z. X., 198
Llorente, M. A., 371

AUTHOR INDEX

Lloyd, K. G., 144, 159
Locher, R., 521
Locht, R., 652
Lockhart, D. J., 513, 522
Lodge, K. B., 598
Lodge, T. P., 230
Loesch, H., 592
Loeser, J. G., 578
Loge, G. W., 380, 398
Loiacono, D. N., 344
LOIDL, A., 29-60; 30, 32, 36-40, 42-49, 51-56
Lomonova, E. E., 319
Longo, J. M., 297, 299, 303, 319
Longworth, G., 303
Loong, C. K., 319
López-Sancho, J. M., 180
López-Sancho, M. P., 180
Loponen, M. T., 42
Lorentz, H. A., 2
Loring, R. F., 125, 126, 137
Los, J., 185-87, 649
Louage, F., 658
Loubet, J. L., 599
Loubriel, G. M., 194, 201
Louie, S. G., 248
Love, B., 694
Lovell, S. E., 22
Lovella, S., 658
Low, K. G., 656
Low, P. F., 598
Loy, M. M. T., 341, 342
Lu, H., 464
Lu, T. M., 261, 264
Lucchese, R. R., 159
Luckham, P. F., 604, 607, 615, 616, 620, 628, 629
Ludeña, E., 414, 420
Ludwig, B., 686
Luesse, C., 631
Lüft, H. W., 383
Lugtenburg, J., 689
Lukin, L. V., 441, 449
Lumry, R., 208, 223
Lundqvist, B. I., 263
Lundqvist, S., 262, 277, 282
Luntz, A. C., 144, 159, 163, 167
Luo, G., 652
Luo, Y., 579, 588, 591
Lupien, M. D., 92, 94, 107
Lüscher, E., 63
Luthjens, L. H., 448
Lüty, F., 32, 36, 37, 39, 45, 49, 51
Lyding, J. W., 299
Lykke, K. R., 417, 418, 430
Lyklema, J., 598
Lyne, M. P. J., 418, 432
Lyo, I. W., 664
Lyo, S. K., 519, 520

Lyyra, A. M., 417, 427, 430, 431
Lyyra, M., 419, 434

M

Ma, M., 694
Ma, S.-K., 120
MacCallum, J. R., 92, 102, 104, 106
MacChesney, J. B., 511, 513, 520
MacDonald, J. K. L., 478
Macek, J. H., 377, 379
Macfarlane, R. M., 500-3, 506, 507, 509, 511, 513, 519, 520, 522, 524, 525
Mackay, R. S., 571
MacKnight, W. J., 93
Mackrodt, W. C., 301
Macosko, C. W., 352, 353, 364, 365
Macpherson, M. T., 375, 383
Madden, P., 136, 137
Madey, J. M. J., 641
Madey, T. E., 143, 144, 159, 162, 173, 174, 189, 191, 193, 195-98, 268, 274
Madhavan, P., 276
Madhavan, P. V., 433, 434
Madison, V., 211
Madix, R. J., 379
Maeda, H., 311
Maeda, M., 600, 631
Magistris, A., 92, 93, 107
Mahganti, S. D., 45
Mahgerefteh, D., 381, 382, 391
Mahoney, R. T., 376
Maier, M., 502, 524, 525
Maier, S., 599
Maignan, A., 311, 313
Maigret, B., 221
Mak, C. H., 143, 262, 266
Makarenko, I. N., 244-46
Malet, L., 417-19, 432
Mali, M., 92
Malik, S. K., 208, 223
Malinovich, Y., 652, 662
Malkina, A. D., 599
Mallard, W. G., 414, 420
Mällo, A., 263
Malmström, B. G., 682, 695
Mal'tsev, A. A., 416
Maltz, C., 567
Mandel, M., 22
Manghnani, M. H., 243, 244, 255
Manninen, M., 278
Manson, M. G., 275
Manthiram, A., 320
Many, A., 199
Mao, H. K., 237, 238, 240, 242, 243, 254, 255

Marathe, V. R., 299
Marcati, F., 160
Marcelin, G., 579, 580
Marcelja, S., 629
March, N. H., 79, 262, 277, 282
Marchese, F., 102
Marcott, C., 356, 367
Marcus, R. A., 131, 133, 134, 655
Mariani, C., 186
Marinero, E. E., 502, 525, 640, 648
Maring, W., 470
MARK, J. E., 351-74; 351, 352, 355, 359-61, 364, 368, 369, 371
Maroncelli, M., 118, 126-28
Marques, C., 627
Marques, C. M., 627, 628
Marr, G. V., 638
Marra, J., 600, 609, 614, 616, 617, 620, 621, 629
Marsh, E. P., 184, 201
Marshall, E. M., 414, 418, 419
Martens, C. C., 492, 493
Martens, J., 76
Martensson, P., 536, 547, 553
Martin, A. C. S., 102, 103, 105-7, 109
Martin, C., 311, 313
Martin, C. T., 683, 684, 691
Martin, D., 471
Martin, G. A., 162, 163
Martin, H., 412
Martin, J. L., 129
Martin, R. M., 202
Martin, S., 238, 240, 244-46
Martinez, M. T., 393, 401
Martino, F., 74
Martins, J. L., 237, 238
Martyna, G. J., 459
Martynuk, M. M., 63
Marucco, J. F., 302
Marx, D. T., 320
Masad, A., 132
Mashimo, S., 11
Mason, E. A., 14
Mason, P. R., 136
Mason, T. O., 301
Masson, C., 250
Mathias, E. C., 80
Mathies, R. A., 689
Matsubara, T., 62, 131, 132
Matsuda, H., 62
Matsui, Y., 252, 254
Matsumoto, Y., 513
Matsuo, T., 39
Matsuoka, M., 328
Matsushita, Y., 230
Matsuyama, H., 315
Mattheis, L. F., 63, 66, 69, 74
Mayhew, C. A., 652

AUTHOR INDEX 713

Maynard, K. J., 100
Mazenko, G., 120
Mazurenko, Yu. T., 117
McAdoo, D. J., 652
McAuliffe, M. J., 129
McBane, G. C., 388
McCaffery, A. J., 379
McCall, S. L., 70, 72
McCammon, J. A., 209, 210, 212, 224
McCarroll, W. H., 299, 319
McCarron, E. W. III, 318
McClelland, G. M., 377, 379, 574, 575, 578
McClure, D. J., 600, 631
McConnell, H. M., 509
McCoustra, M. R. S., 386
McCulloh, K. E., 655
McCumber, D. E., 510
McDonald, R. S., 152
McDowell, R. S., 160
McFeely, F. R., 196
McGhie, A. R., 107
McGill, T. C., 275
McGuiggan, P. M., 599, 631
McGuire, M., 132
McIntyre, G. J., 36, 39, 40
McIntyre, J. E., 105, 106
McIntyre, K. S., 415, 422
McKendrick, C. B., 389, 390
McKendrick, K. G., 592
McKillop, J. S., 580
McKoy, V., 379, 395
McLaughlin, I. L., 78
McLean, A. B., 316
McLendon, G., 132
McLennaghan, A. W., 104
McMahan, A. K., 246, 248
McManis, G. E., 131, 132
McNally, P., 299
McNaughton, W., 629
McQuaid, M. J., 418, 419, 432
McQueen, R. G., 239
McQuillan, A. J., 328
McQuistan, R. B., 199
Meade, C., 243, 255
Medved, D. B., 199
Meech, S. R., 522
Meertens, H., 464
Meerts, W. L., 573
Mehrotra, P. N., 304
Meier, W., 377, 378, 391
Meijer, G., 385
Meir, W., 184, 201
Meirovitch, H., 215
Meisels, G. G., 657
Meissner, B., 42, 43, 45
Meissner, M., 36, 42, 44, 45
Meiwes-Broer, K. H., 565, 569, 574, 582, 583, 587
Meixner, A. J., 503, 525
Melinon, J., 75

Melius, C. F., 273-75, 277, 278, 280
Memmel, N., 186
Men, A. N., 301
Menduiña, C., 371
Meng, R. L., 311
Menzel, D., 174, 176, 180-82, 189, 191, 201
MERER, A. J., 407-38; 408, 411, 415-18, 421-28, 430
Meret-Artes, S., 477, 479
Meriaudeau, P., 162, 163
Mermin, N. D., 81
Merrill, E. W., 353, 364, 365
Merrill, P. W., 413, 422
Mertin, M., 302
Mertz, B., 37, 39, 42-45, 51, 52
Mestdagh, H., 657
Metiu, H., 202
Metropolis, N. A., 213
Meyer, H., 32
Meyer, W., 482
Meyers, K. O., 353, 364, 365
Michel, B., 684
Michel, C., 311, 313
Michel, H., 672
Michel, K. H., 32, 35-37, 39, 45, 54, 55
Middendorf, T. R., 513, 522
Miehé, J. A., 132
Miescher, E., 649
Migus, A., 129, 464
Mikami, N., 470
Miklavich, S., 629
Miller, D. R., 352
Miller, J. C., 640
Miller, P. J., 642, 645, 646
Miller, R. E., 470
Miller, S., 480, 481
Millett, F., 685
Millie, P., 661, 662
Milligan, R. F., 305
Millot, F., 302
Mills, D. L., 144, 159, 275
Milne, C. J., 389, 399
Milner, D. F., 132
Milner, S. T., 624, 626, 627
Milton, D. J., 415, 416, 424, 426, 427
Min, B. I., 237, 238
Miné, P., 455
Ming, L. C., 242, 243, 255
Minichino, C., 63
Minier, M., 92, 94, 105
Minkwitz, R., 652
Minomura, S., 247, 248
Miraglia, S., 312
Misewich, J., 149
Mishra, A. K., 131
Miskovic, Z., 191
Mitchell, A. C., 243, 249, 250
Mitchell, A. W., 320

Mitsuke, K., 652
Miyazawa, S., 216, 221, 225
Mizuno, H., 218, 219, 221, 223, 225
Mizutani, W., 541, 555
Moacanin, J., 86
Mobbs, R. H., 105
Moerner, W. E., 501-3, 506, 524, 525
Mohan Ram, R. A., 297-99, 310, 311, 313
Moini, M., 652
Molenkamp, L. W., 502, 503, 505-7, 510, 511, 513, 515, 517, 519, 520
Möller, R., 380
Moller, T., 647, 650
Momigny, J., 652
Mommers, A. A., 659
Monnerie, L., 362, 367, 371
Mons, M., 379, 386
Montford, J. P., 630
Montroll, E. W., 18
Moog, E. R., 181
Moore, C. A., 146
Moore, C. B., 396, 398
Moore, J. N., 148
Moos, H. W., 154
Mopsik, F. I., 10
Morawetz, H., 351
Morawitz, H., 143, 144, 159, 162, 519, 525
Morgan, J. E., 694
Morgan, W. L., 454, 462
Morgan, W. W., 420
Morgner, H., 652
Mori, T., 245, 247
Moriarty, J. A., 243
Morin, P., 642, 645, 657, 661, 662
Morinaga, M., 301
Morita, A., 26
Morita, N., 541, 555
Moriya, K., 39
Morokuma, K., 400
Moroney, L. M., 99
Morowitz, H. J., 672, 676
Morozov, I. I., 574
Morris, A., 414-16, 422, 426, 427
Morris, G. A., 105
Morris, J. M., 129
Morris, K., 418, 419, 432
Morris, R. J., 201
Morris, S., 256
Morrison, R. J. S., 129, 386
Morrissey, J. J., 311
Morrow, J. C., 657
Morse, M. D., 390
Morse, R. H., 684, 685
Morse, R. I., 376
Morsink, J. B. W., 507
Mory, S., 146

Moser, C., 414, 420
Moss, W. C., 238, 240, 246
Mott, N. F., 62-64, 68, 69, 77, 305, 308-10
Moubarak, A., 685
Moult, J., 211
Mouttet, C., 544
Moy, D., 36
Moynihan, C. T., 24
Moze, O., 311
Muetterties, E. L., 168
Mukamel, S., 125, 126, 131, 137, 154, 471
Müller, J., 277, 310
Müller, K. A., 291, 310, 315
Müller, T., 201
Muller-Dethlefs, K., 650, 651
Mulliken, C. M., 676
Mullin, C., 340
Münch, E., 696
Munch, M., 627
Muñoz, R. C., 446, 447, 450, 451, 461
Munro, B., 102
Munro, H. S., 598
Munro, R. G., 242
Murakami, T., 297
Muraleedharan, K., 299
Muramutu, S., 318
Murchand, R., 319
Murday, J. C., 189
Murphy, D. W., 318
Murphy, R., 664, 665
Murray, D. W., 92-96
Murray, P. T., 645
Murrell, J. N., 482, 489, 491, 493, 494
Murthy, K., 311, 313
Muscat, J., 262, 267
Muthukumar, M., 611
Muto, F., 318
Muyskens, K. J. C., 400
Mydosh, J. A., 33
Mysen, B. O., 254

N

Naccache, C., 162, 163
Nachman, D. F., 408, 417-19
Nadler, I., 381, 382, 391
Nadler, W., 133
Nagabhushana, S., 297, 298, 310
Nagai, K., 364, 366
Nagamura, T., 449
Nagano, M., 319
Nagano, S., 108
Nagaoka, K., 105
Nagarajan, R., 297, 314, 317, 320
Nagarajan, V., 127, 128, 132
Nagasawa, M., 230
Nagata, T., 380, 398

Nagel, S. R., 36, 45, 49, 51, 53
Nagle, J. F., 672, 676, 679
Nakagawa, K., 444-47, 461
Nakamura, S., 461
Nakamura, Y., 454, 462
Nakatsuka, H., 513
Nakayama, N., 303
Nakazawa, H., 244
Namba, H., 454
Nanjundaswamy, K. S., 311, 314
Napper, D. H., 606
Naqui, A., 682
Nara, S., 62
Narasimhan, L. R., 506, 511, 513, 515, 517, 518
Narayanamurti, V., 42
Naruse, H., 105
Nathanson, G. M., 574
Naudts, J., 36, 37, 45
Nazzal, A. I., 311, 316, 500, 502
Neale, F. E., 70
Neddermeyer, H., 547, 548
Nee, T. W., 136
Negami, S., 19
Nelin, C. J., 421, 427
Nellis, W. J., 243, 249, 250
Nelson, D. L., 310
Nelson, J. S., 280
Nemethy, G., 211, 218, 221
Nenner, I., 641-43, 645-48, 650, 657, 660-62
Nesbitt, D. J., 470
Netzel, T. L., 131, 132
Neuburger, N. A., 371
Neumann, M., 276
Neumark, D. M., 417, 418, 430
Neusser, H. J., 640, 654, 655
Newman, J. K., 632
Newnham, R. E., 300
Newns, D. M., 176, 183, 187, 198, 262, 267, 536, 538, 548, 553
Newsam, J. M., 311
Newsham, I. G., 180
Newton, M. D., 131, 132, 433, 434
Nex, C. M. M., 629
Ng, C. Y., 649, 652
Nguyen, N., 303
Niay, P., 419
Nicholas, C. V., 105
Nicholls, D. G., 671
Nicholls, J. M., 536, 541
Nicholls, P., 693, 694
Nichols, A. L. III, 120, 121, 125, 127, 458, 459
Nicolasen, G., 577, 578, 582, 583, 586, 588, 590, 591
Nicoll, J. F., 81

Niedner, G., 470
Nielson, S. E., 642
Nieminen, R. M., 278
Nightingale, J., 391
Nilsson, T., 693, 694
Nimura, N., 15
Ninomiya, M., 416, 427
Nishi, N., 201
Nishikawa, A., 247
Nishikawa, M., 444-48, 457, 461
Nishimura, T., 657
Nitsche, C. I., 696
Nitzan, A., 109, 130, 154
Noack, K., 160
Noble, M., 393
Noda, I., 230, 356, 367
Noda, S., 85
Noll, F., 63, 64, 68
Noll, M., 470
Nomura, S., 297
Nonella, M., 394
Noordam, L. D., 649
Noort, M., 507, 525
Nordlander, P., 185, 274, 280
Norman, D., 275
Norman, G. E., 62
Norrish, R. G. W., 375, 416, 417, 428, 432
Nørskov, J. K., 263, 269, 272, 278-80
Northrup, J. E., 541
Northrup, S. H., 130
Norwood, K., 652
Novakoski, L. V., 574, 575
Novick, S. E., 477
Novikov, G. F., 449
Novotny, A., 211
Nücker, N., 316, 320
Nyholm, R., 196
Nyikos, L., 441, 448, 463, 464

O

Obodovskii, I. M., 441, 443, 454
O'Brien, P., 685
O'Bryan, C. L., 650
Ochrymowycz, L. A., 90, 101
Oda, M., 297
Odagaki, T., 62
Oeser, R., 368
Ogai, A., 387
Ogata, N., 88, 89, 104, 108
Ogawa, T., 62, 66, 449
Ogden, R. W., 356
Ogita, N., 62
Ogorzalek, R., 379
Ogorzalek Loo, R., 379, 388, 391, 400, 401
O'Halloran, M. A., 380, 381, 643, 644, 649
Ohashi, K., 574, 575, 580

AUTHOR INDEX 715

Ohishi, Y., 248
Ohmichi, N., 652
Ohno, K., 652
Ohoyama, H., 575, 580
Ohtaki, Z., 88, 89, 104
Okabe, H., 375
Okabe, Y., 315
Okamoto, T., 250
Okamura, S., 98, 99
Okamura, T., 127
O'Keefee, M., 299
Okita, T., 303
O'Konski, C. T., 25
Okunuki, K., 694
Okuyama, T., 201
Olafson, B. D., 274, 275
Olesik, S., 657
Olsen, J. L., 310
Olson, B. L., 320
Olson, R. W., 503, 507, 511
O'Mahony, J., 393, 399-401
Ondrey, G. S., 385, 389, 390
Ondrias, M. R., 683, 693
Ono, Y., 649
Onodera, A., 248
Onodera, Y., 245, 247
Onsager, L., 3, 22, 116, 117, 121, 681
Oppallo, M., 132
Oppermann, W., 365
Oprysko, M. M., 174
Orbach, R., 519
Orcutt, R. H., 13
Ortega, J. M., 638, 641
Ortiz, C., 503
Ortiz-Lopez, J., 32, 37, 45, 49
Osad'ko, I. S., 509, 519, 520
Osamura, T., 396
Osgood, R. M., 174
Osguthorpe, D. J., 218, 224
O'Shaughnessy, D. J., 193, 201
Osiko, V. V., 319
Ossicini, S., 185
Ottinger, C., 655
Otto, H., 132
Oudemanns, G. J., 13
Overbeek, J. T. G., 599
Overhof, H., 70, 72, 74, 77
Owen, J., 33
Oxtoby, D. W., 117, 127, 130, 133, 134

P

Pacansky, J., 502
Pack, D. W., 511, 513, 515
Pack, R. T., 472, 474, 475, 477, 479, 489
Page, R. H., 640, 649, 650
Paine, G. H., 211-14
Pak, H., 369
Pakkanen, T. A., 276
Palm, V., 513, 518

Palma, A., 385
Palmer, H. B., 414, 422
Palmer, R. G., 18, 46
Pan, L. P., 685
Panczel, M., 657
Pandey, K. C., 538, 553
Pannetier, J., 299, 303
Pantaloni, S., 95
Panteleichuk, O. G., 63
Pantelides, S. T., 281
Papadia, S., 263
Papenhuizen, J., 629
Papke, B. L., 88, 96, 100, 109
Parenteau, L., 197, 198
Park, T. J., 487, 489
PARKER, D. H., 561-95; 400, 562, 563, 565, 567, 569, 572-74, 577-79, 582, 583, 585-91
Parker, G. A., 474, 475, 485, 489
Parker, J. L., 600, 631
Parker, J. M., 85, 92, 96
Parkin, S. S. P., 311
Parks, C. C., 194
Parola, A., 80
Parr, A. C., 641, 645-47
Parrish, R. G., 207
Parsegian, V. A., 598
Parson, J. M., 414
Pashley, R. M., 601, 628
Passerini, S., 95
Passner, A. L., 70, 72
Pasternak, M., 248
Patel, S., 604, 606, 611, 617, 619, 620, 623, 625, 627
PATEL, S. S., 597-635
Patey, G. N., 125
Pathak, C. M., 414, 422
Pathamanathan, K., 32
Patterson, F. G., 503, 507, 511
Paul, W., 564
Pauling, L., 280
Pavitt, N., 211
Pawlitzky, B., 575
Payne, D. R., 88, 96
Payne, M. G., 640
Péalat, M., 396
Pearse, R. W. B., 417, 419, 429
Pearson, D. S., 359, 360, 368
Peatman, W. B., 651
Pedley, J. B., 414, 418, 419
Peet, A. C., 487
Pefferkorn, E., 602
Pehrsson, P. E., 85
Penn, S. M., 400
Penney, T., 320
Penttilä, T., 681, 684, 686
Penzar, Z., 168
Penzkofer, A., 144
Pepper, M., 310
Perahia, D., 214

Percus, J. K., 121
Peres, T., 652
Perry, R. A., 400
Pershan, P. S., 328, 329, 333
Personov, R. I., 499, 500, 502, 503, 511, 524
Persson, B. N. J., 144, 159, 163, 167, 184, 190, 532, 556
Perutz, M. F., 216
Pešić, D. S., 419
Peskov, V., 455
Pestak, M. W., 74, 80
Peterson, I., 599
Peterson, J., 684
Pethrick, R. A., 104
Petroff, Y., 638
Petrucci, S., 100
Pettersson, A. V., 415
Peyerimhoff, S. D., 662
Pfab, J., 386
Philipp, H. R., 241
Philips, L. A., 127
Phillips, J. G., 414, 415, 420, 421
Phillips, L. F., 580
Phillips, W. A., 33, 42, 510, 515, 520
Pian, T. R., 201
Pianelli, A., 301
Picard, C., 302
Picot, C., 368
Piela, L., 216
Piermarini, G., 238, 240
Piermarini, G. J., 240, 242-44
Pinchemel, B., 416, 417, 419, 424, 428, 429, 433, 434
Pincus, P., 607, 611, 613
Pincus, P. A., 613
Pings, C. J., 66
Pinto, G. R., 343
Pippin, H. G., 580
Pirug, G., 265
Plenkiewicz, B., 445
Plenkiewicz, P., 445
Plitjer, J. J., 522
Ploehn, H. J., 601, 603, 611
Plummer, E. W., 276, 341, 664
Poblet, J. M., 658
Podolsky, B., 475
Poeppelmeier, K. R., 303
Pohl, R. O., 33, 36, 42, 44, 45
Poland, D., 222
Polanyi, J. C., 184, 202, 592
Pole, R. V., 525
Poliakoff, E. D., 645, 647
Polian, A., 244-46
Polik, W. F., 615
Polischuk, A. Y., 456
Pollack, S. S., 102
Pollak, E., 590
Pollitt, S., 310
Pollock, E. L., 7

AUTHOR INDEX

Poornimadevi, C. S., 134
Popielawski, J., 72
Pople, J. A., 12, 13, 658
Poradzisz, A., 201
Porter, G.,375
Porter, R. F., 246
Portier, R., 298
Portis, A. M., 315
Postawa, Z., 201
Potenza, J. A., 299
Pottak, H., 651
Pouchard, M., 299
Poulsen, O., 424
Poulson, L., 685
Poumellec, B., 302
Powell, D., 414, 415, 422
Powis, I., 379, 391, 400, 656
Prager, M., 30, 35
Prakash, H., 299, 310
Prakash, O., 297, 298, 310
Praliaud, H., 162, 163
Prassides, K., 311
Pratt, S. T., 643, 644, 649, 652
Prazeres, R., 641
Premilat, S., 221
Preses, J. M., 441
Press, W., 30, 35
Prest, H. F., 649
Preston, H. J. T., 430
Pretzer, W. R., 168
Preuss, H., 430, 432, 433
Prewitt, C., 319
Price, D. L., 36
Price, S. D., 661, 662
Prieve, D. C., 598
Prins, W., 368
Prisant, M. G., 590
Privalov, P., 208, 223, 224
Proch, D., 654
Proch, F. D., 654
Prock, A., 144, 159, 163, 167, 184, 190
Proctor, M. J., 379
Provost, J., 311, 313
Prud'homme, J., 93
Psaro, R., 162
Pszczolkowski, L., 11
Ptitsyn, O. B., 207, 208, 366, 367
Punyatiya, C., 463
Purnell, M. R., 419, 434
Puschmann, A., 201
Puska, M. J., 278, 279
Puxley, D. C., 305

Q

Qadri, S. B., 243
Qian, C. X. W., 387
Quate, C. F., 556, 598
Quesada, M. A., 571
Queslel, J. P., 352, 355, 361, 362, 364, 367, 368, 371

R

Rabalais, J. W., 638
Rademann, K., 75, 76
Radhakrishnan, G., 592
Radom, L., 658, 661
Radousky, H. B., 249
Raftery, J., 421, 434
Raghavachari, K., 159
Rahman, T. S., 275
Raider, S. I., 547, 555
Raitio, M., 696
Rajoria, D. S., 305
Rajumon, M. K., 311, 317
Rakestraw, D., 592
Ramaker, D. E., 174, 189, 193, 196-98
Ramakrishnan, T. V., 51, 53, 308, 309, 315, 317
Ramanan, A., 299, 318
Ramanan, G., 448
Ramanujachary, K. V., 299
Ramarao, G., 305
Ramdani, A., 305
Ramdas, S., 304
Rance, M., 208, 224
Randeria, M., 45, 52
RAO, C. N. R., 291-326; 291, 297-99, 301, 303-5, 308-11, 314, 318-20
Rao, G. R., 311, 313, 314, 316, 317, 320
Rao, K. J., 298, 301, 319
Raseev, G., 433, 434, 646
Rasing, Th., 341, 343, 344, 346
Rasumova, N. V., 503
Ratner, M. A., 86, 88, 90, 95-97, 100, 104, 109, 469, 493-95
Rau, D. R., 598
Rault, M., 646
Raveau, B., 299, 303, 310, 311, 313
Rayfield, C. W., 454
Read, B. F., 367
Read, F. H., 643
Ready, J. F., 143
Reatto, L., 79
Rebane, A., 503
Rebane, L. A., 500-3
Rech, T., 75
Redfield, A. G., 510
Redhead, P. A., 174
Redondo, A., 275
Reedijk, J., 42
Reedy, G. T., 418, 429, 432, 433
Regan, L., 228
Rehfeld, R. H., 463
Rehm, V., 196
Rehner, J., 360
Reichert, M., 127
Reichlin, R., 238, 240, 244-46
Reid, B. P., 480, 481
Reidl, W., 180
Reihl, B., 531, 536, 541
Reil, B., 185, 186
Reilly, J. P., 651
Reimann, C. T., 201
Reinartz, J. M. L., 573
Reinecke, T. L., 519
Reineker, P., 510, 519, 520
Reinhardt, W. P., 474, 475, 479
Reininger, R., 444, 445, 457, 462
Reisler, H., 381, 382, 387, 391, 393
Rejwan, M., 652
Reller, A., 303
Remy, M., 184
Render, R., 77
Renhorn, I., 419, 433, 434
Renker, B., 46, 47
Renn, A., 503, 521, 525
Rennar, N., 365
Rentzepis, P. M., 129
Resseinsky, M., 311
Rettig, W., 132
Rettner, C. T., 640
Reuss, J., 564
Revane, L. A., 507
Riviere, J. C., 320
Rhodin, T. N., 168, 275
Ricard, D., 328, 343
Rice, C. E., 318
Rice, D. W., 228
Rice, M. H., 239
Rice, O. K., 655
Rice, S. F., 408, 412, 414
Rice, T. M., 69, 72, 308, 317
Richards, D. R., 320, 415, 416, 423, 425, 426
Richards, P. M., 194, 201
Richards, W. G., 421, 434
Richardson, D. M., 458
Richardson, J. S., 225, 228, 231
Richardson, N. V., 143, 144, 159, 162
Richardson, P. J., 662
Richard-Viard, M., 656, 657
Richmond, G. L., 328, 342
Richter, D., 368
Richter, L. J., 183, 202
Richter, W., 525
Rider, D. M., 654
Riedel, E. K., 42
Rieder, K. H., 275
Riedl, W., 189
Riess, H. J., 344
Rietman, E. A., 92-96
Rigaud, D., 92, 94, 105
Rigden, S. M., 252, 256
Riggleman, B. M., 248

AUTHOR INDEX 717

Rikukawa, M., 88, 108
Riley, M. E., 193, 194
Riley, S. J., 400
Ripoll, D. R., 216
Rips, I., 125, 134
Risenberg, L. A., 154
Robbins, J. L., 163
Robbins, R. J., 129
Roberts, S., 30, 34
Robertson, A. A., 598
Robinson, G. W., 129
Robinson, J. M., 328, 342
Robinson, R. B., 547, 555
Robinson, R. L., 477
Robitaille, C. D., 92-95, 97, 107
Robota, H. J., 190
Robson, B., 218, 224
Rodinov, B. U., 454
Rodnik, R., 187
Rodriguez, F., 368
Rodriguez, J., 303
Roe, R.-J., 367
Roepe, P., 676
Roger, G., 380
Rogers, F. J., 252
Rohatgi, R., 641
Rohrer, H., 264, 269, 320, 531, 532, 538, 541, 556, 598
Rojhantalab, H. M., 342
Rolando, C., 657
Romberg, H., 316
Ronca, G., 361
Roop, B., 184
Roos, J., 92
Röpke, G., 77
Rose, G. D., 224
Rose, J. H., 69
Rose, S. L., 85
Rosen, B., 416-19, 427, 432, 433
Rosenberg, R. A., 194
Rosenblatt, D. H., 275
Rosenbluth, A. W., 213
Rosenbluth, M. N., 213
Rosenfeld, A., 146
Rosenstock, H. M., 655
Rosenwaks, S., 394
Rosique, C., 318
Rosmus, P., 480, 481
Ross, E., 102
Ross, M., 243-47, 249, 252
Ross, R. G., 70, 74
Rossi, A. R., 176, 180
Rossi, G., 613
Rossky, P. J., 124, 125
Roth, C. B., 598
Roth, R. S., 312
Roth, S., 318
Rothe, E. W., 385
Rothschild, K. J., 676

Rowe, J. M., 30, 35-37, 39, 40, 42, 46, 54, 55
Rowe, M. D., 379, 400
Rowlinson, J. S., 81
Roy, P., 646
Royds, R. N., 662
Ruan, J., 452
Rubbia, C., 441, 464
Rubio, J., 180
Ruckenstein, E., 162, 163
Ruderman, M. A., 33
Rudnick, J., 349
Rudy, A., 245, 247
Ruehl, E., 652
Ruggiero, A. J., 127
Rulliere, C., 127, 128
Ruoff, A. L., 244-47
Rupp, L. W., 311, 314, 317
Ruscic, B., 652
Rush, J. J., 30, 35-37, 42, 46, 54
Rusiecki, S., 314, 319
Rusmak, F., 696
Russek, A., 418
Russel, W. B., 601, 603, 611
Rutkowski, J., 201
Rutter, P. M., 650
Ryberg, R., 159

S

Saalfeld, F. E., 85
Saari, H., 684
Saari, P., 503
Sadagianizadeh, K., 106
Saenz, J. J., 544
Sahu, D., 45
Saile, V., 462
Saito, S., 418, 430
Saito, Y., 102
Sakai, Y., 441, 442, 448, 461
Salama, F., 662
Salmeen, I., 683
Salpeter, E.-W., 416, 417, 419, 427, 433
Salvan, F., 531, 541, 544
Sambe, H., 197, 198
Samoilenko, V. D., 503
Samson, J. A. R., 638
Sanche, L., 197, 198
Sandahl, J., 102, 104
Sander, M., 650, 651
Sandstrom, D. R., 180
Sanesi, M., 92, 93
Sankar, G., 314, 317
Santandrea, C., 652
Santoro, A., 318
Sanui, K., 88, 89, 104, 108
Sappey, A. D., 645
Sarantidis, K., 129, 130
Sarasa, J. P., 658
Saraste, M., 671, 681, 682, 685, 686, 696

Sarma, D. D., 311, 316, 317, 320
Sarma, M. H., 213, 214
Sarma, R. H., 213, 214
Sarti, P., 695
Sassenberg, U., 419, 424, 425, 427, 433
Sathyamurthy, N., 592
Satija, S. K., 35, 48
Sato, H., 100, 386, 394
Sato-Sorensen, Y., 244
Satpathy, S., 247
Sauer, M. C., 448
Sauli, F., 455
Savoy, R., 311
Sawatzky, G. A., 189, 316, 317
Saykally, R. J., 470, 477
Sbrignadello, G., 160
Scanlon, J. C., 303
Schaefer, H. F., 396
Schaefer, M., 243
Schall, H., 412
Schamps, J., 416, 417, 419, 421, 428, 429, 433, 434
Schantz, S., 102
Schartman, R. R., 303
Schatz, G. C., 474, 683, 685, 689, 693
Schawlow, A. L., 572
Schechter, I., 584, 588, 590
Scheer, H., 513, 522, 524
Scheiner, S., 671, 672
Schellman, J. A., 208, 222, 223, 225
Schenck, P. K., 414, 420
Scheraga, H. A., 209, 211, 213-16, 218, 221, 222, 224
Scheutjens, J., 623
Scheutjens, J. M. H. M., 604, 611, 613
Schiferl, D., 243
Schiffer, M. R., 228
Schiller, R., 441, 463, 464
Schinke, R., 375, 385, 386, 391, 393, 394, 396, 401
Schinzel, D., 441
Schlag, E. W., 640, 650, 651, 654, 655
Schleyer, P. v. R., 658
Schleysing, R., 592
Schlier, C., 564
Schlüter, M., 281, 282
Schluter, M., 317
Schmeisser, D., 186, 191
Schmidt, J., 509
Schmidt, K. H., 448, 449
Schmidt, L. D., 143
Schmidt, M., 519, 606, 607, 609
SCHMIDT, W. F., 439-68; 439, 441, 442, 444, 448, 453, 455, 462, 640

Schmiedl, R., 377, 378, 391
Schmitt, R. W., 33
Schmutzler, R. W., 70, 71
Schneemeyer, L. F., 311, 314, 317
Schneider, E., 64, 78
Schneider, M. P., 184, 201
Schnieder, L., 385, 391
Schön, W., 36
Schönhammer, K., 180
Schönhense, G., 186, 570
Schönherr, G., 63, 70, 71, 77, 78, 81
Schober, O., 276
Schoch, K. F., 85
Schoemaker, D., 49
Schoepe, W., 454
Scholes, C. P., 683, 684, 691, 695
Schollhorn, R., 318
Schorl, R., 316
Schräder, T., 32, 37, 38, 48, 49, 55, 56
Schrieffer, J. R., 262, 263, 267, 317
Schröder, J. O., 418, 432
Schroeder, L. W., 415, 425
Schuck, P., 191
Schulte, G., 503, 519
Schulten, K., 133
Schulten, Z., 133
Schwartz, H., 658
Schwartz-Lavi, D., 394
Schwarzenegger, K., 70, 72
Schweizer, K. S., 364
Schwettman, H. A., 641
Scigocki, D., 455
Scoles, G., 562, 567, 574
Scott, P. R., 421, 434
Scott, T. W., 440, 449
Scrosati, B., 95
Sebald, W., 679
Sechenkov, A. P., 70
Secrest, D., 474, 480, 481
Seebauer, E. G., 143
Segal, G. A., 387
Segev, E., 385
Segner, J., 184, 202
Seilmeier, A., 144, 145, 148
Seki, S., 32, 39
Sekreta, E., 651
Selin, L.-E., 415, 416, 422
Selloni, A., 536, 541
Selzer, P. M., 511, 519
Semenov, A. N., 624
Semlyen, J. A., 355
Sen, A., 319
Senanayake, P. C., 441
Senf, F., 196
Senior, A. E., 677, 679
Senior, M., 429
Senn, P., 648
Sennesal, J. M., 421
Serebrennikov, L. V., 416
Serizawa, K., 454
Sesselmann, Th., 525
Sethna, J. P., 29, 44, 45, 51-53
Seydel, U., 63
Shafer, M. W., 316, 320
Sham, L. J., 281, 282
Shaner, J. W., 63, 242, 249
Shank, C. V., 342, 513
Shannon, V. L., 328, 342
Shao, J. D., 657, 658
Shapira, Y., 199
Shapiro, M., 378, 385, 391, 477, 479
Shaw, R. W., 684
She, R. S. C., 589, 590
Shelby, R. M., 501, 503, 506, 507, 509, 511, 513, 519-22, 524, 525
SHEN, Y. R., 327-50; 150, 327-29, 332-34, 337, 338, 340, 341, 343, 344, 346, 348, 640, 649, 650
Sheng, Z. Z., 311
Shenkin, P. S., 211
Shepherd, J. P., 304
Sheppard, N., 143, 144, 159, 162, 168
Sherrington, D., 54
Sherwood, R. C., 249
Shetty, A. N., 101
Shibamura, E., 452
Shieh, H.-S., 228
Shigeno, M., 541, 555
Shimaoka, K., 15
Shimomura, O., 244, 247, 248
Shindo, K., 247
Shinohara, H., 201, 652
Shinohara, I., 105
Shinsaka, K., 454, 462
Shirane, G., 297
Shirk, J. S., 419, 433
Shirley, D. A., 194
Shirley, J. E., 408, 417, 418
Shiromaru, H., 652
Shirts, R. B., 493
Shizgal, B., 485
Shlesinger, M. F., 18
Shoemaker, R. L., 500
Shore, J. E., 19
Shortley, G. H., 409
Shriver, D. F., 86, 88, 90, 95-97, 100, 101, 104, 105, 109
Siegbahn, P., 491
Sienko, M. J., 309
Sikorski, A., 222, 227, 228, 230
Silberberg, A., 601
Silbey, R., 144, 159, 163, 167, 184, 190, 503, 509, 510, 519, 520
Sild, O., 501, 502
Silvera, I. F., 246
Simmons, G., 245
Simmons, J. G., 551, 553, 554
Simon, A., 299
Simon, J. D., 116, 126, 127, 132
Simons, A. L., 70, 72
Simons, J. P., 375, 377, 380, 383-85, 389, 393, 399, 401
Simpson, J. R., 511, 513, 519, 520
Singer, R. E., 352, 355
Singh, K. K., 298, 299
Singh, Y., 458
Singhal, S. R., 414, 415
Sitz, G. O., 379, 400
Sivakumar, N., 377-79, 381, 387, 391, 396, 397
Skarnulis, A. J., 319
Skelton, E. F., 243
Skillman, S., 420
Skinner, J. L., 19, 130, 502, 507, 509, 510, 525
SKOLNICK, J., 207-35; 212, 222, 227, 228, 230
Skulachev, V. P., 681, 683
Slaski, M., 315
Slater, J. C., 281
Sleight, A. W., 297, 299, 310, 314, 319
Small, G. J., 500, 501-3, 505, 506, 511, 513, 519, 522, 524
Smid, J., 105
Smith, A. J., 383
Smith, A. K., 162
Smith, A. L., 642
Smith, B. S., 493
Smith, C. P., 600, 631
Smith, D. J., 302
Smith, D. P. E., 556
Smith, D. W., 463
Smith, F. T., 473
Smith, G. C., 162
Smith, I. W. M., 590
Smith, J. M., 160
Smith, J. M. A., 228
Smith, K., 574, 580
Smith, M. K., 101, 109
Smith, P. A., 693-95
Smith, T. I., 641
Smith, W. W., 228
Smoes, S., 416, 417
Smoot, S. W., 108
Smyth, C. P., 17
Smyth, K. C., 414, 420
Snowdon, K. G., 305
Snowdon, K. J., 201
So, S. K., 199
Soderholm, L., 312, 315
Soep, P. B., 592
Sole, A., 658
Solgadi, D., 394

Solioz, M., 686
Solomon, J., 376
Soltz, D., 105
Solymosi, F., 265
Somorjai, G., 327
Somorjai, G. A., 340
Sompolinsky, H., 54
Sone, N., 693, 694
Sonntag, H., 599
Sontoro, A., 312
Soper, A., 311
Sorensen, P. R., 93, 108
Sorokin, P. P., 149
Sorrie, G. A., 102
Soulethie, J., 46
Southworth, S. H., 645, 646
Sowa, E. C., 280
Sowada, U., 444, 452, 462
Spackman, M. A., 254
Spaepen, F., 254
Spalek, J., 297, 305, 307, 308, 310, 320
Sparaglione, M., 126
Sparnaay, M. J., 599
Sparpaglione, M., 131
Sparrow, T. G., 319
Spiglanin, T. A., 400
Spinrad, H., 422, 425
Spiro, T. G., 683
Spitzmann, K., 42
Sprandel, L. L., 478
Sprik, M., 458
Springett, B. E., 445, 462
Spurling, T. H., 14
Srdanov, V. I., 419, 433
Sreedhar, K., 311, 316, 317, 320
Srithanratana, T., 454
Srivastava, G. P., 284
Srivastava, J. K., 299
St.-Cyr, D., 615
Staemmler, V., 385, 386, 391
Stainer, M., 95, 96
Stamenova, R. T., 107
Stankiewicz, M., 663, 664
Stankowski, J., 31
Stanners, C. D., 184, 202
Stapelfeldt, J., 647
Stark, D., 662
Stauffer, J. L., 414, 420
Staverman, A. J., 359
Steadman, J., 648
Stechel, E. B., 176, 182, 183, 193, 471, 477, 489
Steele, B. C. H., 93
Steele, R. E., 414
Steffens, G. J., 685
Steimle, T. C., 408, 412, 414, 417-20, 433
Stein, D. L., 18, 46
Stein, R. S., 93, 367
Steinberg, D. J., 242-44

Steinberger, I. T., 444, 445, 457, 462
Steinemann, A., 683, 685, 689, 693
Steinmann, W., 344
Stell, G., 124, 125
STEPHENSON, J. C., 143-71; 145, 146, 151, 160, 162, 163, 165, 166
Stepto, R. F. T., 352
Stern, E. A., 349
Sternlieb, B. J., 320
Stevens, J. R., 102, 104
Stevens, T. H., 684, 685
Stevens, W. J., 408, 428, 430, 432
Stevenson, D. J., 237
Stevenson, S. A., 162, 163
Stiles, P. J., 137
Still, W. C., 209
Stishov, S. M., 244, 252
Stockbauer, R. L., 174, 189, 193, 195, 196, 646
Stockmayer, W. H., 222, 225
Stoeckenius, W., 689
Stöhr, J., 189, 196, 275
Stoll, H., 430, 432, 433
Stoller, L., 208, 223
Stolpher, E. M., 252, 256
Stolte, S., 58, 400, 562, 564, 565, 567, 569, 573-75, 577-80, 582, 583, 585-88, 590, 591
Stook, P. W., 240
Stott, M., 263, 278
Stout, R. P., 502, 503, 505, 506, 511, 519
Strange, J. H., 107
Straube, E., 352
Strauss, C. E., 388, 400, 401
Stringat, R., 414, 420
Stroscio, J. A., 536, 538-40, 545, 547, 548, 550, 551, 553
Stulen, R. H., 174
Sturge, M. D., 510
Su, S.-G., 127, 132
Subbanna, G. N., 303, 311
Subbarao, G. V., 291, 299, 303, 304
Subramanian, M. A., 299, 303, 311, 314
Suga, H., 32, 39
Sugai, S., 208, 209
Sugiura, M., 85
Sugiyama, H., 513
Suhr, H., 344
Sullivan, N. S., 30, 34, 35
Sumi, H., 133, 134
Sumita, I., 96
Sumitani, M., 127
Sundberg, R. L., 471
Sundström, V., 134

Superfine, R., 344
Surratt, G. T., 275
Susman, S., 30, 35, 36, 42, 45, 46, 49, 54
Sussner, H., 42
Sutcliffe, B. T., 471, 473, 476, 478, 480, 482
Suter, G. W., 503
Sutin, N., 131, 132
Suto, M., 389
Sutter, H., 14
Suzuki, K., 70
Suzuki, M., 297
Suzuki, T., 245, 247, 250
Svendsen, B., 238, 242, 244, 245, 256
Swallen, S., 127, 132
Swanson, B. I., 160
Swarts, C. A., 275
Swartzentruber, B. S., 536, 541, 542, 544, 545
Swenson, C. A., 249
Swent, R. L., 641
Syassen, K., 246-48
Synowiec, J. A., 129
Syono, Y., 252, 254
Szabo, A., 133, 499, 500, 503, 504
Szczepanska, L., 31
Szeftel, J. M., 275
Szu, S. C., 216
Szulejko, J. E., 659
Szymonski, M., 201

T

Tabche-Fouhaile, A., 642, 647, 648
Tabor, D., 599, 600, 631
Tachiya, M., 449, 453, 454
Tadokoro, H., 96, 102
Tagashira, H., 461
Tai, T. L., 656
Takagi, H., 311, 317
Takahashi, A., 601, 604, 607
Takahashi, M., 538
Takahashi, S., 538
Takahashi, T., 90, 95, 97, 240, 315, 452
Takahashi, Y., 96, 696
Takano, M., 303
Takayanagi, K., 538
Takeda, S. S., 441
Takeda, Y., 303
Takemura, K., 247, 248
Taketomi, H., 218, 219, 221, 223, 225
Taleb-Ibrahimi, A., 196
Tal'rose, V. L., 574
Tam, N.-H., 146
Tammadon, S., 252, 254
Tanaka, S., 311
Tanaka, T., 367, 368

AUTHOR INDEX

Tanaka, Y., 311
Tandon, R. K., 614-16
Tanford, C., 207, 208, 223, 224
Tang, D., 522
Tang, S. L., 143
Tanishiro, Y., 538
Taran, J.-P. E., 396
Tatibana, M., 367
Tatsumoto, E., 250
Taunton, H. J., 617, 620, 621, 626
Taylor, A. W., 411, 415-17, 421, 422, 425, 426, 430, 431
Taylor, J. W., 642
Taylor, R. D., 248
Tealdi, A., 102
Tearesaki, O., 301
Teegarden, K. J., 245
Teeters, D., 100
Teller, A. H., 213
Teller, D., 310
Teller, E., 213
Tembe, B. L., 126, 131, 132
Temmen, M., 428
ten Brinke, G., 630
Tenner, M. G., 575
Tennyson, J., 472, 473, 475, 476, 478, 480-82, 491
Tera, T., 696
Teramoto, E., 230
Terauchi, H., 301
Terlouw, J. K., 659
ter Meulen, J. J., 385
Tersoff, J., 269, 320, 532, 535, 536, 539, 543
Tetot, R., 301, 302
Thelen, M., 686
Theolier, A., 162
Thijssen, H. P. H., 502, 503, 505, 507, 511, 513-16, 518, 520, 521, 523
Thirion, P., 364
Thirumalai, D., 54, 78
Thirumalai, D. J., 459, 460
Thiry, P., 638
Tholence, J. L., 46
Thoman, J. W., 400, 572
Thomas, G. A., 305
Thomas, J. M., 303, 314, 319
Thomlinson, W., 319
Thompson, A. B., 253
Thompson, J. C., 71
Thompson, K. R., 415, 416, 419, 422, 428
Thompson, W. A., 341, 342, 553
Thomson, A. J., 684
Thornber, M. R., 301
Thörnström, P. E., 695
Thuis, H., 564
Tice, D. R., 261, 264, 285

Ticich, T. M., 393
Ticktin, A., 387, 395
Tiede, D. M., 522
Tiemann, E., 417, 418, 430
Tilley, R. J. D., 299, 301, 302
Tippelskirch, H. V., 70
TIRRELL, M., 597-635; 599, 604, 606, 607, 609-11, 617, 619, 620, 623, 625, 627, 631
Titijevskaia, A. S., 599
Titular, U. M., 136
Titze, B., 385
Tobin, J. G., 275
Tobin, R. G., 143, 144
Toda, H., 100
Toennies, J. P., 265, 470
Togo, M., 89
Tokumoto, H., 541, 555
Tokura, Y., 316, 317
Tolk, N. H., 174, 201
Tolmachev, A. V., 449
Tom, H. W. K., 340-43, 346, 347
Tombari, E., 11
Tominaga, T., 22
Tomita, T., 509
Tomkins, F. S., 643, 644
Tomlin, A. S., 102
Tomlinson, G. A., 599
Tonck, A., 599
Topp, M. R., 127, 129
Toprakcioglu, C., 616, 617, 620, 621, 626, 629
Torardi, C. C., 299, 311, 314, 319
Torell, L. M., 102, 104
Torng, C. J., 311
Torrance, J. B., 316
Törring, T., 412
Tosatti, E., 536, 541
Tosch, S., 547, 548
Toupin, R. A., 6
Tourmoux, M., 319
Touzelin, B., 302
Townes, C. H., 572
Traeger, J. C., 652
Träger, F., 202
Traum, M. M., 174, 201
Travis, J. C., 414, 420
Treffry, A., 228
Treichler, R., 180, 189
Treloar, L. R. G., 351, 352, 354, 356, 361, 365, 369, 371
Trenary, M., 159
Trifunac, A. D., 448
Tristam-Nagle, S., 672, 679
Troe, J., 655
Trommsdorff, H. P., 500, 502, 506, 507, 511, 522, 525
Tromp, R. M., 536, 538, 539, 541, 546, 548, 550

Trummal, M., 513, 518
Trybula, Z., 31
Tsai, P. P., 299
Tsong, T. T., 261, 264, 265, 285
Tsuboi, T., 649
Tsukihara, T., 696
Tsvetanov, Ch. B., 107
Tuck, A. F., 390, 394
Tully, J. C., 159, 174, 185, 201, 262
Tups, H., 248
Turnbull, D., 254

U

Uchida, S., 311, 317
UCHTMANN, H., 61-83; 63, 70, 72, 74, 76, 77, 81
Ueda, Y., 218, 219, 221, 223, 225
Uemura, Y. J., 320
Ugo, R., 162
Uhler, U., 419, 434
Ullman, R., 368
Ulstrup, J., 116
Umbach, E., 180
Umezawa, A., 315
Unguris, J., 181
Unseld, K., 352
Untch, A., 394
Unwin, P. N. T., 676
Uppal, M. K., 299, 303
Upton, T. H., 273-75, 277
Uram, K. J., 159
Urbach, L. E., 341
Urbina, C., 30, 34, 35
Usami, Y., 461

V

Vacatello, M., 630
Vala, M., 414, 415, 422
Valenti, B., 102
Valentini, J. J., 574, 582
Valles, J. M., 547, 555
Valletregi, M., 303, 318
van Alsten, J., 599, 609, 631
van Amersfoot, P. W., 185, 186
Van Beek, W. M., 22
Van Bentum, P. J. M., 315
Van Brunt, R. J., 377
van den Berg, E. M. M., 689
Van den Berg, R. E., 502-5, 513-15, 517, 519-22, 525
van den Ende, C. A. M., 448
Van den Ende, D., 565, 569, 573, 574, 582, 583, 587
van den Heuvel, H. B. V., 649
van der Avoird, A., 473, 476
Van der Laan, H., 525
van der Schee, H. A., 629

AUTHOR INDEX 721

van der Veen, J. F., 189
Van der Waals, J. H., 500, 502-4, 506, 507, 509, 511, 525
van der Zande, W. J., 649
van der Zwan, G., 117, 134, 137
Van de Walle, G. F. A., 544
van Dover, R. B., 315
van Gunsteren, W. F., 209, 210, 212, 224
van Herk, M., 464
Van Hove, M. A., 276
Van Kempen, H., 544
van Koeven, D., 476, 481
van Kranendonk, J., 480
van Lent, B., 623
van Loenen, E. J., 541
van Smaalen, S., 144, 159, 163
Van Valkenburg, A., 240
van Veen, G. N. A., 400
van Veen, N., 389, 390
van Veen, N. J. A., 400
Van Vleck, J. H., 3, 5
van Vliet, T., 598
Van Zandt, L. L., 305, 308
Van Zee, R. J., 417, 418, 428, 430, 432, 433
Vanderlick, T. K., 631
Vandersall, M., 134
Van't Hof, C. A., 509
Varma, C. M., 33, 42, 304, 510, 515, 520
Varoqui, R., 602
Vasanthacharya, N. Y., 297
Vasquez, M., 211, 215
Vassilev, K. G., 107
Vasudev, R., 393, 394, 648
Vaughan, R. W., 163
Vaughan, W. E., 136
Venkatachlam, C. M., 224
Verdine, J. C., 162, 163
Vershuur, J. W. J., 649
Vertes, A., 457, 463, 464
Veseth, L., 415, 425
Viani, B. E., 598
Vickers, J. S., 536, 544, 545, 553
Vidyasagar, K., 303
Vigué, J., 380
Vijayakumar, P., 126
Vijayaraghavan, R., 299, 310, 311, 313, 314, 317
Vilgis, T. A., 352, 364, 369, 371
Villeneuve, G., 306
Vincent, B., 601, 611
Vincent, C. A., 92, 93, 102, 104, 106-8
Vink, K. J., 522
Violand, N. V., 228
Visser, A., 504, 505, 513
Viswanathan, K. S., 651

Viswanathan, R., 299, 310
VÖLKER, S., 499-530; 51, 500, 502-7, 509, 511, 513-15, 517, 520, 524, 525
Vogel, H., 46
Vogel, M., 132
Vogt, T., 37, 38, 48, 49, 55, 56
Vohra, Y. K., 244, 246, 247
Volin, K. J., 312
Volkmann, U. G., 46, 48, 49, 52, 54
Volkmer, M., 575
Vollmer, M., 202
Voter, A. F., 192, 278, 280
Vukanic, J., 191

W

Waajen, F. H. W., 611
Wachter, E., 679
Wada, A., 216
Wada, T., 448, 454
Wadsley, A. D., 302
Wagner, A., 174
Wahl, M., 390, 391
Wahr, J. C., 376
Wahrhaftig, A. L., 655
Waits, L. D., 399
Wakiyama, S., 541, 555
Waldeck, D. H., 190
Waldeck, J. R., 379, 381, 401
Waldman, M., 477
Walker, G. C., 128, 129
Walker, R. B., 477
Walker, S. M., 118, 262, 263, 267
WALKUP, R. E., 173-206; 183, 188, 191, 193-97, 201
Wall, F. T., 221, 360
Wallace, R., 473
Wallenstein, M. B., 655
Wallenstein, R., 640
Wallis, R. H., 310
Walmsley, C. M., 418
Walsh, C. A., 503, 506, 511, 513, 515, 517, 518
Walsh, J. M., 239
Walsh, S. P., 275
Walters, E. A., 652
Wandelt, K., 180, 186
Wang, C. C., 329, 334
Wang, C. H., 35, 48
Wang, C. S., 328
Wang, E., 299
Wang, G. C., 261, 264
Wang, H., 211, 245, 574, 684, 689
Wang, Y., 129
Wang, Y. Q., 311
Wang, Z., 100, 652
Wang, Z. W., 573, 578, 582, 585, 591

Wannier, G., 661
Warchol, M. P., 136
Ward, I. M., 105, 106
Ward, K. B., 228
Warman, J. M., 441, 446, 448, 452
Warren, W. W., 63, 66, 69-71, 74
Warshel, A., 217, 221
Waseda, Y., 241, 252
Washida, N., 652
Wasserman, B., 85
Waszczak, J. V., 299, 547, 555
Watanabe, D., 301
Watanabe, H., 26, 617, 619
Watanabe, K., 541, 555
Watanabe, M., 104, 108
Watanabe, N., 105
Watanabe, S., 457
Watanabe, W., 88, 89
Watson, J. K. G., 471, 473
Watt, D., 485, 489
Watters, K. L., 162
Watts, D. C., 128
Weaver, D. L., 207, 208, 230
Weaver, M. J., 131, 132
Webb, M. B., 181
Webb, S. P., 126, 127
Weber, C., 684
Weber, D. C., 85
Weber, M., 502, 511, 518, 520
Weber, M. J., 511, 519
Weber, P. C., 228
Webster, F., 381
Webster, R., 105
Wedig, U., 430, 432, 433
Weeda, J., 464
Wegner, H., 77
Weide, D., 183
Weidenauer, R., 202
Weijs, P. J. W., 316
Weil, T., 364
Weinberg, W. H., 276
Weinert, M., 262
Weinstein, N. D., 567
Weir, C. E., 240
Weir, S. T., 247
Weisman, R. B., 149
Weiss, A., 31
Weiss, M., 265
Weiss, S., 14
Weisshaar, J. C., 396, 645
Weitekamp, D. P., 507, 510
Welge, K. H., 377, 378, 385, 391
Weller, R., 390, 391
Wells, A., 55
Wells, P. B., 162, 163
Weltner, W., 414, 415, 417, 418, 420, 422, 425, 428, 430, 432, 433

AUTHOR INDEX

Wen, X. G., 317
Wentzcovitch, R. M., 243
Werner, H. J., 432, 480, 481
Wertheim, M. S., 7
West, J. B., 417, 429, 432, 646
Western, C. M., 391
Weston, J. E., 93
Wetlaufer, D. B., 207, 208, 220, 223, 224
Wetton, R. E., 102
Whang, G. C., 264
Wharton, L., 414, 420
Wheatley, J. M., 317
Whitaker, B. J., 379
White, C. T., 189
White, D. N. J., 211
White, H. S., 600, 631
White, J. M., 184, 262, 265, 266
White, K. I., 146
White, L. R., 614-16
White, M. G., 379, 395, 642, 645, 646, 652
White, N. M., 407, 420
Whitehead, R. J., 471
Whiteway, S., 250
Whiting, W., 102
Whitman, L. J., 265
Whitmore, D. H., 95, 100, 108
Whitmore, P. M., 190
Whitnell, R. M., 485, 487, 489
Whittaker, E. A., 503
Whitten, J. L., 276
Whittingham, M. S., 310, 318
Whyman, R., 168
Wiebel, E., 264, 269, 531, 538
Wiersma, D. A., 500, 502, 503, 505-7, 510, 511, 513, 515, 517, 519, 520, 522, 524
Wierzbicki, A., 474
Wignor, C. P., 237, 238
Wikström, M., 671, 681, 682, 684-86
Wikström, M. K. F., 684
Wild, U. P., 503, 521, 525
Wilder, J. A., 186
Wilke, V., 640
Wilkerson, J. L., 428
Wilkinson, J. P. T., 389, 390
Williams, A. R., 180, 185, 263, 265, 281, 282, 531, 535, 541
Williams, B. G., 319
Williams, E. M., 174
Williams, G., 17, 25, 128
Williams, H. J., 249
Williams, Q., 238, 242, 244, 245, 254, 256
Williams, R. J., 184, 592
Williams, R. T., 194, 200
Wilson, D. J., 105

Wilson, E. B., 471, 473
Wilson, J. A., 310
Wilson, K. R., 134, 376, 400
Wilson, M. T., 695
Wilson, W. D., 274, 275
Wilson, W. G., 651
Wimmer, E., 262
Windwer, S., 221
Wing, R. F., 407, 420, 422, 425
Winkler, M. A., 63
Winkoun, D., 662
Winnacker, A., 525
Winniczek, J. W., 379, 395, 646
Winsor, P. IV, 22
Winter, R., 63, 64, 66, 68, 78, 79
Winters, R. H., 224
Wintersgill, M. C., 91, 101, 107, 109
Winterton, R. H. S., 599
Winzen, H., 246
Wise, J. G., 677, 679
Wissbrun, K. F., 90, 101
Witt, S. N., 682
Witten, T. A., 624-26
Wittig, C., 381, 382, 391, 393, 592
Wnek, G. E., 85
Wold, A., 319
Wolf, E. L., 532
Wolf, G., 252
Wolfrum, H., 502, 503, 505, 506, 511
Wolfrum, J., 389-91
Wol'kenstein, T., 199
Wolkow, R., 555
Wolynes, P. G., 22, 23, 26, 120, 124, 125, 130-32, 136
Won, Y., 522
Wong, J., 24
Wong, T., 96
Wood, D. M., 168
Woodward, R., 428
Worboys, M. R., 99, 107
Worley, S. D., 163
Wormer, J. L., 199
Wort, F. S., 660, 662
Wortis, M., 42
Wright, P. E., 208, 224
Wright, P. V., 85, 88, 91, 92, 96
Wu, E., 105
Wu, L., 45, 49
Wu, M. K., 311
Wu, S., 598
Wurth, W., 189
Wyatt, J., 102
Wyatt, R. E., 491, 492, 590
Wyckoff, H., 207
Wyder, P., 544

Wyman, J., 3
Wynne, J. J., 343

X

Xia, D. W., 105
Xu, E. Y., 649
Xu, J., 238, 240, 242, 243
Xu, Q.-X., 565, 566, 569-71, 574

Y

Yabushita, S., 400
Yagi, T., 250
Yakovlev, B. S., 441, 449, 454
Yamada, K., 85
Yamada, S., 449
Yamakawa, H. J., 230, 603
Yamamoto, M., 454
Yamamoto, O., 303
Yamamoto, T., 694
Yamaoka, S., 244
Yan, Y. J., 126
Yang, H., 276
Yang, L. L., 107
Yang, Q. Y., 143
Yang, S., 398, 400
Yang, W., 487
Yang, X., 391
Yano, Y., 100
Yao, M., 63, 70, 72, 74, 81
Yardley, J. T., 166
Yaris, R., 212, 222, 228
Yarmoff, J. A., 196
Yarmush, D. L., 211
Yates, B. F., 658
Yates, J. T. Jr., 143, 144, 159, 162, 163, 173, 174, 180, 202, 268, 274
Ye, T.-Q., 148
Yeh, S. W., 127
Yelon, W. B., 303
Yen, R., 342
Yen, W. M., 501, 502, 511, 519
Ying, Z., 199, 200
Yokoyama, M., 102
Yon, J., 207
Yonezawa, F., 62, 66, 74
Yoo, Y.-S., 652
Yoon, D. Y., 630
Yoshida, H. K., 315
Yoshida, K., 33
Yoshida, S., 449
Yoshihara, K., 127
Yoshikawa, S., 696
Yoshino, K., 462
Young, A. P., 33, 40, 46
Young, P. A., 184, 202
Young, W. H., 78
Yu, M. L., 187, 198

Z

Zachariasen, W. H., 299, 310
Zador, E., 446
Zahurak, S. M., 92, 318
Zanderighi, G. M., 162
Zare, R. N., 376, 377, 379-81, 393, 394, 398, 401, 573, 592, 640, 648, 649, 654
Zaremba, E., 263, 278
Zeigler, J. M., 522
Zeldovitch, G., 62
Zeller, R. C., 33
Zeringue, K. J., 418, 430, 432, 433
Zewail, A. H., 401, 500, 509, 592, 657
Zha, Q., 657
Zhang, F. C., 317
Zhang, Q. J., 181
Zhang, R., 592
Zhang, S. C., 317
Zhdanov, V. P., 144, 159, 163, 167
Zhu, X. D., 340, 341, 344
Zichi, D. A., 134
Ziesel, J. P., 652
Ziman, J. M., 73
Zimmerer, G., 201, 647, 650
Zimmermann, B. H., 696
Zinth, W., 145, 148
Zou, Z., 317
Zschokke, I., 502
Zubek, M., 650
Zunger, A., 282
Zúñiga, J., 474
Zusman, L. D., 131, 132
Zwanzig, R. W., 12, 19, 120, 125, 136, 137
Zyrnicki, W., 416, 427

SUBJECT INDEX

A

Acetylene
   photoionization mass spectrometry and, 652
Adsorbates
   dielectric surfaces and, 150-59
   interactions of, 261-88
      Anderson-model-based calculations and, 266-67
      energies required for, 264-66
      first-principles methods and, 272-78
      immersion-energy calculations and, 278-80
      response theory and, 267-72
      scattering theory method and, 281-88
      theoretical approaches to, 266-88
   scanning tunneling microscopy and, 539-41
   vibrational energy transfer in, 143-69
   vibrational spectroscopy and, 344
Adsorption
   at interfaces
      surface second harmonic generation and, 338-41
Affine network theory
   elasticity and, 360-61
Alcohols
   relaxation in, 19
n-Alcohols
   time-dependent fluorescence Stokes shift and, 127
Alkali halides
   experimental scattering studies of, 578-79
   melting curves of
      pressure and, 252
   neutral atom desorption from, 200
   pressure-induced dissociation of, 249
   strain defects in
      orientational freezing and, 34
Alkali metals
   metal-nonmetal transition in, 68-70
   nearly free electron approach for, 66-68
Alkaline earth monohalides
   dipole moments of, 412

Alkanes
   electron transport in, 439-40
   substituted
      complex plane loci for, 20-21
n-Alkanes
   electron mobility in, 441
   density and, 444
   pressure and, 446
Alkyl nitrites
   photodissociation of, 394
Allene
   clusters of
      photoionization and, 652
Aluminum
   fluorine desorption from, 192
American Chemical Society, 1-2
Amides
   time-dependent fluorescence Stokes shift and, 127
Amino acid substitution
   protein conformation and, 207
1-Aminonaphthalene
   time-dependent fluorescence Stokes shift and, 127
8-Amino-1-naphthalenesulphonic acid
   Stokes shift dynamics in, 129
2-Amino-7-nitrofluorene
   Stokes shift relaxation in, 127
4-Aminophthalamide
   time-dependent fluorescence Stokes shift and, 127
Amorphous solids
   impurities diluted in
      optical dephasing of, 510-11
Amphiphiles
   polymeric
      forces between, 618-28
Anderson model
   adsorbate-adsorbate interactions and, 266-67
Anderson transition, 305
Angular basis functions
   vibrations of floppy molecules and, 479
Angular momentum
   molecular orientation and, 572-73
Aniline ion dissociation, 654
Anisole
   photoelectron photoion coincidence and, 657
   photoionization mass spectrometry and, 652
Anisotropy
   recoil, 376-77, 398
Antoniewicz mechanism, 180-82

Apoferritin
   α-helical motif in, 228
Arc melting
   transition metal oxide synthesis and, 319
Argon
   electron attachment reactions in, 453
   electron drift velocity in, 461
   electron emission in, 454-55
   electron mobility in, 444
      conduction band energy and, 444
   electron transport in, 440, 456-57
   homogeneous electron-ion recombination in, 454
   second virial coefficient for, 13
Ashcroft empty core potential, 67
ATP synthase
   hydrogen-bonded chain hypothesis and, 677-81
Auger decay
   ion desorption from ionic solids and, 193-97
Autoionization, 641-45
   Bardsley-Smith, 645
   rotational, 642-43
   vibrational, 643-45

B

Bacteria
   ATP synthase of, 677
Bacteriorhodopsin
   hydrogen-bonded chain hypothesis and, 676-77
Ballistic electron emission microscopy, 555-56
Bardeen tunneling formalism, 535
Bardsley-Smith autoionization, 645
Barium ferrites
   intergrowth structures in, 302-3
Barrier state spectroscopy, 541-42
Basak-Cohen model, 457
Becker-Döring-Zeldovitch theory, 76
Benzene
   clusters of
      photoionization and, 652
Benzene ion dissociation, 655
Benzene-propanol mixtures
   Stokes shift dynamics in, 129

725

Betainphosphate
  glassy state of, 31
Betainphosphite
  glassy state of, 31
Bevan clusters, 301
Bianthryl
  excited
    intramolecular electron transfer reaction in, 116, 132
  time-dependent fluorescence Stokes shift and, 127
Biological polymers
  direct force measurements and, 629-30
Birefringence
  deformed polymer networks and, 356
  Kerr effect, 25
  strain, 366
  rotational isomeric state models and, 364
Bis(4-aminophenyl)sulphonate sulphone
  time-dependent fluorescence Stokes shift and, 127
1,2-Bis(trimethylsilyl)ethane
  electron mobility in, 441
Borane
  photoionization mass spectrometry and, 652
Bovine pancreatic trypsin inhibitor
  folding and unfolding pathways of, 221
  three-dimensional cubic lattice model of, 219
1-Bromopentane
  complex plane loci for, 20-21
3-Bromopentane
  complex plane loci for, 20-21
Buckingham-Disch experiment, 13-14
n-Butane
  electron mobility in
    density and, 444
Butyl-alcohol isomers
  photoelectron photoion coincidence and, 657
n-Butylbenzene
  photoelectron photoion coincidence and, 657

C

Cadmium
  solubility behavior in polymers, 90-91
Calcium nitrates
  solubility behavior in polymers, 90-91
Calcium oxide
  bonding in, 411-12

Calcium thiocyanate
  solubility behavior in polymers, 91
Calorimetry
  differential scanning
    polymer-electrolyte mixtures and, 92
Carbon dioxide
  quadrupole moment for, 13
Carbon monoxide
  adsorbed on transition metals, 159-67
  methanation on nickel, 274
  second harmonic generation and, 340
  orientational disorder in, 29
Cesium
  Knight shift in, 69
  metal-nonmetal transition in, 64
Cesium bromide
  pressure-induced dissociation of, 249
Cesium dioxide
  defect formation in, 302
Cesium iodide
  metallization of, 244-47
  pressure-induced decomposition of, 247-49
  thermodynamic state of ultrahigh pressures and, 238
Charge transfer reactions
  solvation dynamics and, 130-35
Chemical reactivity
  oriented molecule beams and, 588-92
Chemical vapor deposition technique
  transition metal oxide synthesis and, 319
Chemiluminescence
  chromium oxide and, 427
  crossed-beam scattering studies and, 576-77
  titanium oxide and, 422
Chemisorption theory, 261-88
Chlorin
  hole-burning experiments and, 521
Chlorobenzene
  dielectric measurements in, 11
Chlorobenzene ion dissociation, 654
Chloroplasts
  ATP synthase of, 677
Chromiumhexacarbonyl
  photoelectron photoion coincidence and, 657
Chromium oxide
  bonding in, 411
  $3d$ spectra of, 427-28

Clausius-Clapeyron equation, 252, 254
Clausius-Mossotti function, 3, 73
  virial expansion of, 12
Cluster calculations
  adsorbate-adsorbate interactions and, 273-77
  cuprate superconductors and, 316
Cobalt oxide
  bonding in, 412
  $3d$ spectra of, 432
Cohen-Lekner theory, 461
Cole-Davidson function, 16, 19
Computer simulation
  orientational correlations of simple molecules and, 7
  protein folding and, 207-32
  small, constrained peptides and, 210-11
  small, unconstrained polypeptides and, 211-16
  solvation dynamics and, 126-27
  tertiary structure in globular proteins and, 216-32
Concavalin A
  direct force measurements and, 629
Constrained junction theory
  elasticity and, 361-64
Copper
  cuprate superconductors and, 315-17
  hydrogen dissociation on, 276
  solubility behavior in polymers, 90-91
  surface phonon spectrum of, 280
Copper oxide
  bonding in, 410-12
  $3d$ spectra of, 433-34
Coumarins
  time-dependent fluorescence Stokes shift and, 127
Coupled-channel method
  vibrations of floppy molecules and, 476-77
Crown ethers
  metal cation solvation and, 88
Crucible free method
  transition metal oxide synthesis and, 319
Cryogenic liquids
  electron mobility in, 441-42
Cryptands
  metal cation solvation and, 88
Crystalline melts
  structural transformations in ultrahigh pressure and temperature and, 252-56

SUBJECT INDEX 727

Crystallographic shear
  point defects in oxides and, 302
Crystals
  dipolar
    glassy states of, 31
    relaxation of, 14-20
  glassy states of, 30-31
  impurities diluted in
    optical dephasing of, 506-10
  KBr:KCN
    dipolar freezing in, 48-52
    elastic behavior of, 35
    glass state in, 39
    phase and glass transition
      anomalies in, 39-42
    phase transitions in, 37-39
    quadrupolar freezing in, 45-48
    specific heat in, 42-45
  NaCN:KCN
    random strain theory and, 55-56
    orientational disorder in, 29-30
  surfaces of
    electronegativity and, 267-68
Cuprates
  superconducting
    properties of, 315
    synthesis of, 319
Cuprate superconductors
  properties of, 311-14
Current imaging tunneling spectroscopy, 546-47
Cyanide glasses, 36-56
  glass transition in, 36-39
  nature of, 53-56
  relaxation dynamics at, 45-53
  phase transitions in, 36-39
  specific heat in, 29, 39-45
Cyanoadamantane
  supercooled plastic phase in, 32
Cyanogen halides
  photodissociation dynamics and, 380-81
Cyclohexane
  electron transport in, 448
  force measurement in polymers and, 604
Cyclohexanol
  orientational disorder in, 32
Cyclohexene
  electron transport in, 448
Cytochrome c
  direct force measurements and, 629
Cytochrome c oxidase, 681-96

  redox-active metal centers in, 682-85
  redox-linked conformational transitions of, 694-95
  redox-linked proton translocation in
    mechanisms of, 688-94
    structure of, 685-88
Cytochrome c'
  α-helical motif in, 228

D

Debye-Falkenhagen effect, 22, 26
Debye relaxation time
  time dependence of solvation energy and, 116
Decaglycine
  entropy in
    anharmonicity and, 214
cis-Decalin
  electron transport in, 448
trans-Decalin
  electron transport in, 448
Defect oxides, 300-3
Dehydroxylation
  topotactic
    transition metal oxide synthesis and, 319
Density functional theory, 613
Desorption
  at interfaces
    surface second harmonic generation and, 338-41
Deuterated methanes
  electron mobility in, 442
Deuterium
  liquid
    electron mobility in, 441
Diamond-anvil cell, 237
Diamond cell
  high-pressure, 240-44
Diblock copolymers, 618
Diborane
  photoionization mass spectrometry and, 652
Dibromobenzenes
  photoionization mass spectrometry and, 652
Dicholorobenzene ion dissociation, 656
Dielectric constant
  permanent dipole moment and, 3
  polarizability and, 3
  time dependence of solvation energy and, 116
Dielectrics, 1-26
  dipolar liquids and crystals and, 14-20
  electrolytes and conducting glasses and, 20-24

  field problem and, 2-8
  gases and, 11-14
  nonlinear dynamics and, 24-26
Dielectric spectroscopy
  range and capabilities of, 8-11
Dielectric surfaces
  adsorbates on, 150-59
Diethylaminobenzonitrile
  electron transfer reactions in, 132
3,3'-Diethyloxadicarbocyanine iodide
  photoisomerization of, 348
Differential scanning calorimetry
  polymer-electrolyte mixtures and, 92
Diffusion starvation technique, 611
Dimethylaminobenzonitrile
  electron transfer reactions in, 132
7-Dimethylaminocoumarin-4-acetate
  solvation dynamics of, 128
4-Dimethylamino 4'-nitrostilbene
  kinetics of solvation in polar solvents, 127
Dimethylaminophenyl sulfone
  excited
    intramolecular electron transfer reaction in, 116
4,4'-Dimethylaminophenyl sulphone
  electron transfer reactions in, 132
4-(9-anthryl)-N,N-Dimethylaniline
  time-dependent fluorescence Stokes shift and, 127
2,2-Dimethylbutane
  Hall effect and, 450
Dimethyl formamide
  time-dependent fluorescence Stokes shift and, 128
Dimethyl nitrosamine
  photodissociation of, 394
Dimethyl sulfoxide
  photoelectron photoion coincidence and, 657
Dimethyl-s-tetrazine
  hole-burning spectroscopy and, 506
Dipolar crystals
  glassy states of, 31
  relaxation of, 14-20
Dipolar liquids
  collective orientational relaxation in, 135-36
  relaxation in, 14-20

728 SUBJECT INDEX

dielectric, 136-37
  polarization, 121-22
  solvation dynamics in, 115-38
Dipole coupling, 5
  on a rigid lattice in an applied electric field, 6-7
Dipole moments
  dielectric constant and, 3
Dipoles
  solvation dynamics of, 124, 127-29
  inhomogeneous dielectric response and, 118-19
Discrete variable representation
  vibrations of floppy molecules and, 479, 482-91
Dissociative electron attachment process, 197-99
DMSO
  time-dependent fluorescence Stokes shift and, 128

E

Effective medium theory, 278-80
Elasticity
  rubber-like, 351-73
    elastomeric networks and, 352-55
    experiments in, 355-56
    stress-strain relationships and, 369-72
Elasticity theory, 356-68
  molecular, 357-61
  non-Gaussian, 361-64
  phenomenological, 356-57
Elastomers
  rubber-like elasticity and, 352-55
Electric fields
  dipole coupling on a rigid lattice in, 6-7
  electric moment density P and, 2
  Lorentz, 2-3
  oriented molecule beams and, 572-74
Electric moment density P
  electric field and, 2
Electrolytes
  dielectric properties of, 20-24
  dissolved in polymers, 85-109
    criteria for, 86-92
    ion conductivity and, 104-9
    phase diagrams for, 92-95
    spectroscopy and, 100-4
    structure of, 96-100
Electromagnetic spectrum
  dielectric region of, 5, 8
Electron attachment reactions
  field-dependent, 452-53

Electron diffusion coefficient, 452
Electron drift mobility, 441-49
  conduction band energy and, 444-45
  cryogenic liquids and, 441-42
  density and, 442-44
  high pressure and, 446
  photoassisted, 448-49
  temporary negative ions and, 446-48
Electronegativity
  work function of crystal surfaces and, 267-68
Electron energy loss spectroscopy
  adsorbate-adsorbate interactions and, 275
  transition metal oxides and, 319
Electronic transitions
  desorption by, 173-203
    Auger decay and, 193-97
    Menzel-Gomer-Redhead model of, 174-76
    negative ion, 197-99
    nuclear motion and, 191-93
    substrate excitation and, 202-3
    valence excitation and, 176
    valence ionization and, 176-83
  fragmentation at surfaces and, 183-84
  of organic molecules
    optical dephasing of, 506-24
    quenching processes and, 184-91
Electron microscopy
  transition metal oxides and, 319
Electrons
  superthermal
    energy relaxation of, 450-52
Electron solvation
  dynamics of, 129-30
Electron-stimulated-desorption-ion-angular-distribution, 173-74
Electron transfer reactions
  adiabatic/nonadiabatic parameter and, 131-32
  solvent reorganization dynamics and, 132
  zero barrier intramolecular, 132-33
Electron transport
  drift mobility and, 441-49
  Hall mobility and, 450
  high field strengths and, 461-62

hot electrons and, 450-53
  nonpolar fluids and, 439-65
  theory of, 455-64
Electrostatic hexapole technique
  applications of, 574-78
  polar molecules and, 564-65
Electrostatic interactions
  protein folding and, 216-17
Embedded atom method, 278-80
Equilibrium thermodynamics
  protein conformation and, 208
Escherichia coli
  bacteriorhodopsin of, 677
Ethane
  electron attachment reactions in, 453
  electron mobility in
    density and, 444
Ethers
  crown
    metal cation solvation and, 88
Ethylene
  clusters of
    photoionization and, 652
    photoionization mass spectrometry and, 652
Euhler-Lagrange equations, 613
Excited states
  lifetime of
    quenching processes and, 184-91
Expanded metals
  liquid-vapor transition in, 78-81
  metal-insulator transition in, 61-81
Extended X-ray absorption fine structure
  cytochrome c oxidase and, 685
  oxygen adsorption on nickel and, 275
  polymer-electrolyte mixtures and, 99-100
  transition metal oxides and, 319
External field effects
  hole-burning and, 524-25

F

Ferrites
  barium
    intergrowth structures in, 302-3
Ferrobielectricity, 300
Ferrobimagnetism, 300
Ferrocytochrome c, 682
Ferroics, 300
Ferromagnetoelectricity, 300

SUBJECT INDEX 729

Field emission microscopy
  adsorbate-adsorbate interaction energies and, 265-66
Field ion microscopy
  adsorbate-adsorbate interaction energies and, 264-65
Finite basis representation
  vibrations of floppy molecules and, 479-82
First-principles scattering theory
  adsorbate-adsorbate interactions and, 281-88
Flash photolysis
  vanadium monoxide and, 425-26
Floppy molecules
  vibrations of, 469-96
    coordinate systems for, 472-74
    coupled-channel method and, 476-77
    Hamiltonians for, 475-76
    optimal-coordinates vibrational self-consistent field method and, 493
    recursive residue generation method and, 491-92
    semiclassical quantization and, 492-93
    variational method and, 478-91
Fluids
  nonpolar
    electron transport in, 439-65
    homogeneous electron-ion recombination in, 453-55
  See also Liquids
Fluorescence polarization
  polymer networks and, 356
Fluorine
  desorption from aluminum, 192
Formaldehyde
  photofragmentation of, 396-98
Formic acid
  photodissociation of, 399-400
Franck-Condon factors, 641-42
Free electron lasers, 641
Frequency-domain optical data storage
  hole-burning and, 525

G

Gallium arsenide
  ballistic electron emission microscopy and, 556
  scanning tunneling microscopy and, 536-39

Gas discharge lamps
  vacuum ultraviolet light and, 638
Gases
  dielectric virial coefficients of, 11-14
  rare
    electron emission in, 454-55
Generalized Smoluchowski equation
  solvation dynamics and, 119-24
Genetic engineering, 207
Germanium
  local I-V spectroscopy and, 553
  scanning tunneling microscopy and, 536
Glasses
  conducting
    dielectric properties of, 20-24
  cyanide
    glass transition in, 53-56
    phase transitions in, 36-39
    relaxation dynamics in, 45-53
    specific heat in, 29, 39-45
  lithium fluoroborate
    dielectric properties of, 24
  low-temperature specific heat in, 33
  organic molecules in
    optical dephasing of, 511-21
  orientational, 29-57
    dielectric behavior of, 34-36
  silicate
    hole-burning experiments and, 521
  sodium trisilicate
    dielectric properties of, 24
  spin
    freezing process in, 33
Globular proteins
  fluctuating secondary structure of, 208
  tertiary structure prediction in, 216-32
Glycerol
  relaxation in, 16
Glycerol triacetate
  time-dependent fluorescence Stokes shift and, 128
Glyoxal
  photodissociation of, 395
Gold
  surface reconstruction behavior in, 280

Graphite
  local I-V spectroscopy and, 555
  scanning tunneling microscopy and, 536
  sorbic acid adsorbed on
    inelastic tunneling spectroscopy and, 556

H

Halides
  alkali
    experimental scattering studies of, 578-79
    melting curves of, 252
    neutral atom desorption from, 200
    pressure-induced dissociation of, 249
    strain defects in, 34
  cyanogen
    photodissociation dynamics and, 380-81
  hydrogen
    phase transitions in, 14
  nitrosyl
    photodissociation of, 387
Hall effect, 68, 70
Hall mobility, 440, 450
Halobacterium halobium
  bacteriorhodopsin of, 676
di-Halobenzenes
  photoelectron photoion coincidence and, 657
Hamiltonians
  vibrations of floppy molecules and, 475-76
Hartree-Fock approximation, 267, 308
Havriliak-Negami function, 19
Helium
  desorption from tungsten, 191
  liquid
    electron emission in, 454
    electron localization in, 463
    electron mobility in, 441
    second virial coefficient for, 13
Hellmann-Feynman formula, 284
Heme *a*, 682-85
Hexamethyldisilane
  electron mobility in, 441
Hexane
  dilution studies and, 20
  electron transport in, 449
n-Hexane
  electron mobility in pressure and, 446
Hexapole-focusing technique
  applications of, 574-78
  polar molecules and, 564-65

SUBJECT INDEX

High resolution Rydberg spectroscopy, 648-50
Hole-burning
  applications of, 524-25
  mechanisms of, 502-3
  nonphotochemical, 502-3
  photochemical, 502
  techniques of, 503-6
  transient, 503, 507-8
Hole-burning spectroscopy, 499-526
  organic molecules and, 506-24
Homopolymers
  adsorbed layers of
    forces between, 601-4
  See also Polymers
Hot electron relaxation, 450-52
Hydrocarbons
  electron attachment reactions in, 452-53
  electron emission in, 454-55
  electron mobility in, 440
  conduction band energy and, 444
  density and, 444
  pressure and, 446
Hydrogen
  adsorption on nickel, 276
  liquid
    electron mobility in, 441
  metallization of
    pressures required for, 237
  molecular
    triplet states of, 649
Hydrogen-bonded chain hypothesis, 674-76
  ATP synthase and, 677-81
  barteriorhodopsin and, 676-77
Hydrogen bonding, 3, 6, 16
Hydrogen bromide
  quadrupole moment for, 14
  relaxation in, 15
Hydrogen chloride
  relaxation in, 15
Hydrogen cyanide
  photodissociation dynamics and, 380-81
Hydrogen halides
  phase transitions in, 14
Hydrogen iodide
  relaxation in, 15
Hydrogen mixtures
  ortho-para
    glassy states of, 30
Hydrogen peroxide
  ultraviolet photodissociation of, 388-93
Hyperfine structure
  manganese oxide and, 428
  scandium oxide and, 413-20

vanadium monoxide and, 423-27
Hyperspherical coordinates, 474

I

Ice
  hexagonal
    orientational disorder in, 29
    relaxation in, 15-16
Immersion-energy calculations
  adsorbate-adsorbate interactions and, 278-80
Immitance spectroscopy, 8
Immunoglobulins
  random "tweak" algorithm and, 211
Impedance spectroscopy, 8
Inelastic tunneling spectroscopy, 556-57
Infinitely adaptive structures, 302
Infrared reflection-absorption spectroscopy, 532
Infrared spectroscopy
  broadband probe and, 148-49
  narrowband probe and, 148
  polymer-electrolyte mixtures and, 101-4
  polymer networks and, 356
  single-frequency transient, 145-46
  sum-frequency generation probe and, 149-50
Insulators
  band-gap excitation of, 199-201
  molecular adsorption on
    second harmonic generation and, 340
Interfaces
  harmonic generation at, 327-49
  molecular adsorption and desorption at, 338-41
  molecular orientations at
    measurements of, 342-44
  structure symmetry of, 341-42
Internal bond coordinates, 474
Inverse photoemission spectroscopy, 532
1-Iodoctane
  complex plane loci for, 20-21
Ionic bonding
  ultrahigh pressures and, 238
Ionic solids
  ion desorption from
    Auger decay and, 193-97
Ions
  solvation dynamics of, 122-24
  inhomogeneous dielectric response and, 118-19

lattice model calculation of, 125-26
Iron oxide
  bonding in, 412
  $3d$ spectra of, 429-32
Isomerization
  solvation dynamics and, 134
Isopentane
  electron mobility in
    density and, 442-44
Isopropanol
  2-amino-7-nitrofluorene in
    Stokes shift relaxation in, 127
Isopropyl methyl ether
  photoionization mass spectrometry and, 652
Israelachvili technique, 599, 604

J

Jacobi coordinates, 473-74

K

Kerr effect birefringence, 25
Kinetic trapping
  protein conformation and, 208
Kirkwood-Frölich equation, 8
Knight shift, 69-71, 74, 77
Knotek-Feibelman mechanism, 193-94
Koch-Cohen clusters, 301
Kohlrausch-Williams-Watts function, 17-19, 128
Kohn-Sham equation, 281-82
Kramers-Kronig relation, 73
Krypton
  electron drift velocity in, 461
  electron mobility in, 444
  conduction band energy and, 444
  electron transport in, 440, 456-57
Kubo-Green linear response theory, 24

L

Lanthanide oxides
  $4f$
    electronic states of, 412
Laser frequency modulation spectroscopy, 503
Laser-heated diamond cell, 240-44
Laser-induced desorption
  adsorbate-adsorbate interaction energies and, 266
Laser-induced fluorescence
  crossed-beam scattering studies and, 577
  iron oxide and, 431

SUBJECT INDEX 731

polyatomic molecules and, 389-90
triatomic molecules and, 387-88
Lasers
  free electron, 641
  threshold photoelectron spectroscopy and, 651-52
  vacuum ultraviolet, 640
LDS-750
  time-dependent fluorescence Stokes shift and, 127
Lekner theory, 454
Linear response theory, 136
Liouville equation, 126, 134
Liquid metals
  metal-insulator transition in, 61-81
Liquids
  cryogenic
    electron mobility in, 441-42
    dielectric measurements in, 11
  dipolar
    collective orientational relaxation in, 135-36
    dielectric relaxation in, 136-37
    polarization relaxation in, 121-22
    relaxation in, 14-20
    solvation dynamics in, 115-38
  nonpolar
    electron transport in, 439-65
    homogeneous electron-ion recombination in, 453-55
  polar
    electron solvation in, 120
    thermal expansion of short-range order and, 66
Lithium fluoroborate glasses
  dielectric properties of, 24
Lithium perchlorate
  dissolved in polypropylene glycol, 86
Local density functional theory, 262
Lorentz local field, 2-3
Low energy electron diffraction
  adsorbate-adsorbate interaction energies and, 264-65
  adsorbate superlattice structures and, 266-67

M

Magnetic circular dichroism
  titanium oxide and, 422
Magnetoferroelectricity, 300
Manganese oxide
  bonding in, 412

$3d$ spectra of, 428-29
Maxwell-Boltzmann statistics, 456
Maxwell equations
  macroscopic
    electric field E of, 7-8
Melts
  polymer
    direct force measurements and, 630-31
  silicate
    structural transformations in, 252-56
Menzel-Gomer-Redhead model, 174-76
Mercury
  metal-semiconductor transition in, 70-78
Metals
  alkali
    metal-nonmetal transition in, 68-70
    nearly free electron approach for, 66-68
  expanded
    liquid-vapor transition in, 78-81
    metal-insulator transition in, 61-81
  hot electron effects in, 201-2
  monovalent, 63-70
    metal-nonmetal transition in, 68-70
  noble
    adsorbed noble metal clusters on, 280
  transition
    carbon monoxide adsorbed on, 159-67
    scanning tunneling microscopy and, 535-36
Met-enkephalin
  backbone conformation of, 211-14
Methanes
  deuterated
    electron mobility in, 442
    electron transport in, 440
  halogenated
    dielectric measurements in, 11
    homogeneous electron-ion recombination in, 454
Methylcyclohexane
  electron transport in, 449
Methyl fluoride
  photoionization mass spectrometry and, 652
Methyl formate
  photoelectron photoion coincidence and, 656-57
3-Methyl indole
  solvation of, 129

Methyl iodide
  photodissociation of, 400-1
Methyl iodide + rubidium reaction
  oriented
    coarse product angular distributions for, 579
Methyl nitrites
  photodissociation of, 394-95
3-Methylpentane
  dilution studies and, 20
  electron mobility in
    pressure and, 446
Metropolis criterion, 213, 217
Micas
  crystalline
    hydroxyl ion vibrational relaxation in, 157-58
Microscopy
  adsorbate-adsorbate interaction energies and, 264-66
  ballistic electron emission, 555-56
  electron
    transition metal oxides and, 319
  optical
    polymer-electrolyte mixtures and, 92
  scanning tunneling, 531-57
    one-dimensional, 532-34
    voltage-dependent, 535-41
  surface monolayer, 346
  See also Spectroscopy
Mitochondria
  ATP synthase of, 677
  respiratory electron transport chain of
    cytochrome c oxidase in, 681
Mixed valence
  transition metal oxides and, 303-5
Molecule beams
  oriented, 561-93
    characterization of, 567-72
    chemical reactivity and, 588-92
    electric fields and, 572-74
    experimental scattering studies with, 578-88
    hexapole-focusing technique and, 574-78
    preparation of, 564-67
Monohalides
  alkaline earth
    dipole moments of, 412
Monovalent metals, 63-70
  metal-nonmetal transition in, 68-70
Monoxides
  $3d$ transition
    bonding in, 408-12

732 SUBJECT INDEX

Monte Carlo simulation
  crystalline melts and, 252
  electron-ion recombination process and, 454
  protein folding and, 212-14
  relaxation in orientational glasses and, 54
  thermoelasticity and, 364
Mooney-Rivlin equation, 356-57
Morin transition, 294
Mott transition, 308-9
Multiphoton transitions
  vacuum ultraviolet radiation and, 641-42
Multiquantum defect theory, 643-46
Myelin basic protein
  direct force measurements and, 629
Myohemrythrin
  α-helical motif in, 228

N

1-Naphthylamine
  Stokes shift dynamics in, 127
National Defence Research Council, 6
Nearly free electron approach, 66-68
Neon
  liquid
    electron localization in, 463
    electron mobility in, 441
Neopentane
  electron mobility in, 440
  density and, 444
  Hall effect and, 450
Neutron diffraction
  transition metal oxides and, 319
Neutron scattering
  elastomers and, 356
New York Academy of Sciences, 5
Nickel
  carbon monoxide on, 274
    second harmonic generation and, 340
  chemisorption on
    cluster calculations of, 274-75
  hydrogen adsorption on, 276
  hydrogen dissociation on, 280
  oxygen adsorption on, 275
Nickel oxide
  bonding in, 410-12
  $3d$ spectra of, 432-33
Nile Red
  time-dependent fluorescence Stokes shift and, 127
Nitric acid
  photodissociation of, 399

n-Nitriles
  time-dependent fluorescence Stokes shift and, 127
Nitrites
  alkyl
    photodissociation of, 394
  methyl
    photodissociation of, 394-95
Nitrobenzene
  photoelectron photoion coincidence and, 657
p-Nitrobenzoic acid
  adsorbed
    second harmonic generation and, 343
Nitrogen
  liquid
    electron mobility in, 448
Nitromethane
  photoionization mass spectrometry and, 652
Nitrosyl halides
  photodissociation of, 387
Nitrotoluene
  photoelectron photoion coincidence and, 657
Noble metals
  adsorbed noble metal clusters on
    migration of, 280
Nonpolar fluids
  electron transport in, 439-65
  drift mobility and, 441-49
  Hall mobility and, 450
  hot electrons and, 450-53
  properties of, 453-55
  theory of, 455-64
  homogeneous electron-ion recombination in, 453-55
Nuclear magnetic resonance
  globular proteins and, 208
  Knight shift in cesium and, 69
  polymer-electrolyte mixtures and, 92
  polymer networks and, 356
  transition metal oxides and, 319

O

Optical microscopy
  polymer-electrolyte mixtures and, 92
Optical multichannel analyzer detector, 149
Optical second harmonic generation, 327-49
Optimal-coordinates vibrational self-consistent field method
  vibrations of floppy molecules and, 493-94

Organic glasses
  organic molecules in
    optical dephasing of, 511-21
Organic molecules
  electronic transitions of
    optical dephasing of, 506-24
Orientational glasses, 29-57
  dielectric behavior of, 34-36
Oriented molecule beams, 561-93
  characterization of, 567-72
  chemical reactivity and, 588-92
  electric fields and, 572-74
  experimental scattering studies with, 578-88
  hexapole-focusing technique and, 574-78
  preparation of, 564-67
Orthorhombicity
  cuprate superconductors and, 313
Osmotic interaction, 605
Oxazine-4-perchlorate
  hole-burning experiments and, 521
Oxide pyrochlores
  A-site vacancies in, 303
  electronic properties of, 299
Oxides
  lanthanide
    electronic states of, 412
  rare earth
    mixed valent, 304
  transition metal, 291-320
    characterization of, 319-20
    metal-nonmetal transitions in, 305-10
    mixed valence in, 303-5
    point defects in, 300-3
    properties of, 292-300
    $3d$ spectra of, 407-35
    superconductivity in, 291, 310-17
    synthesis of, 318-19
Oxide spinels
  properties of, 298-99
Oxygen
  adsorption on nickel, 275
  cuprate superconductors and, 315-17

P

Pancreatic trypsin inhibitor
  backbone residues in, 217-18
  bcc lattice model of, 220-21
  native conformation of, 218
Paracoccus denitrificans
  cytochrome c oxidase of, 686

SUBJECT INDEX 733

Paramagnetism
  perovskites and, 296-97
Pauli paramagnetism
  perovksites and, 296-97
Pentacene
  nonphotochemical hole-
    burning and, 502
n-Pentane
  electron mobility in, 441
  density and, 442-44
  pressure and, 446
Peptides
  small, constrained
    structure prediction in, 210-11
Perfluorobenzene
  electron transport in, 448
Perovskites
  A-site vacancies in, 303
  properties of, 294-98
Perylene
  nonphotochemical hole-
    burning and, 502
Phantom network theory
  elasticity and, 357-60
Phonons
  on graphite
    inelastic tunneling spectros-
      copy and, 556
Photochemical hole-burning, 502
Photodissociation
  threshold, 641-45
Photodissociation dynamics
  polyatomic molecules and,
    388-401
  triatomic molecules and, 380-88
  vector correlations in, 375-401
  E-μ-(v-J), 377-78
  E-μ-J, 377
  E-μ-v, 376-77
  quantifying, 379-80
  v-J, 377-78
Photoelectron photoion coincidence, 652-59
  dissociation rates and, 656-60
Photoelectron photoion photoion
  coincidence, 662-64
Photofragment spectroscopy,
  375
Photoionization
  dissociation of polyatomic
    ions and, 652-65
  partial, 645-48
  threshold, 641-45
Photoionization efficiency, 646
Photoionization mass spectrometry, 652
Photoionization spectroscopy,
  637-65
Photoion photoion coincidence,
  662-64

Photons
  desorption by electronic tran-
    sitions and, 173
Physical chemistry
  dielectrics in, 1-26
  ultrahigh pressures and
    temperatures and, 237-56
Picosecond laser spectroscopy,
  695
Polarizability
  dielectric constant and, 3
Polarization spectroscopy, 503
Polarized infrared spectroscopy
  polymer networks and, 356
Polar liquids
  electron solvation in, 120
Polar molecules
  4π/3 catastrophe and, 3
  electric fields at
    induction of, 3
Polar solvents
  dynamics of, 135-37
Poly(L-alanine)
  α-helix conformation of, 217
Poly(alkylene sulphide)
  electrolyte solvation and, 90
Polyatomic ions
  dissociation of
    photoionization and, 652-65
Polyatomic molecules
  photodissociation of, 388-401
  rotation-vibration energy
    levels of, 469
  calculation of, 476-79
Poly(dimethylsiloxane)
  bimodal networks of, 370
Polyelectrolytes
  adsorbed
    direct force measurement
      and, 628-29
Poly(epichlorohydrin)
  electrolyte solvation and, 88
Polyethylene
  semicrystalline
    hole-burning experiments
      and, 520-21
Poly(ethylene adipate)
  electrolyte solvation and, 89
Poly(ethylene imine)
  electrolyte solvation and, 90
Poly(ethylene oxide)
  force measurements in, 614-17
  interactions of, 86-91
Poly(ethylene succinate)
  electrolyte solvation and, 88
Polylysine
  direct force measurements
    and, 629
Poly-L-lysine
  adsorbed
    direct force measurement
      and, 629

Polymer brushes, 622
Polymeric amphiphiles
  adsorbed layers of
    forces between, 618-28
Polymer melts
  direct force measurements
    and, 630-31
Polymers
  biological
    direct force measurement
      and, 629-30
  direct force measurement and,
    628-31
  electrolytes dissolved in, 85-109
  criteria for, 86-92
  ion conductivity and, 104-9
  phase diagrams for, 92-95
  spectroscopy and, 100-4
  structure of, 96-100
  measurement of forces in,
    597-32
  good solvents and, 614-18
  near $T = \theta$, 607-14
  poor solvents and, 604-6
  random coil
    configurational entropy of,
      215
  semicrystalline
    hole-burning experiments
      and, 520-21
  See also Homopolymers
Polypeptides
  direct force measurements
    and, 629
  small, unconstrained
    structure prediction in, 211-16
Poly(β-propiolactone)
  electrolyte solvation and, 89
Poly(propylene glycol)
  electrolyte solvation and, 88
  lithium perchlorate dissolved
    in, 86
Polystyrene
  block copolymers of
    adsorption of, 619
  force measurement in, 604-6
Polythioethers
  interactions of, 86-87
Poly(vinyl acetate)
  complexes with inorganic
    electrolytes, 101
Poly(2-vinyl pyridine)
  block copolymers of
    adsorption of, 619
Porcine growth hormone
  α-helical motif in, 228
Porphin
  free-base
    electronic transitions in,
      506-7

SUBJECT INDEX

Potassium iodide
 pressure-induced dissociation of, 249
Pressure
 ultrahigh
  metallization of cesium iodide and, 244-47
  physical chemistry and, 237-56
Prokaryotes
 respiratory electron transport chain of
  cytochrome c oxidase in, 681
Propane
 electron attachment reactions in, 453
 electron mobility in density and, 444
1-Propanol
 relaxation in, 16
n-Propanol
 photoelectron photoion coincidence and, 657
n-Propanol ion
 dissociation of
  ion-dipole complex in, 659
Propylene carbonate
 time-dependent fluorescence Stokes shift and, 128
Propylene glycol
 relaxation in, 16
Protein conformation
 amino acid substitution and, 207
 kinetic trapping and, 208
Protein folding
 computer simulation of, 207-32
 conformational tree search algorithm for, 209
 electrostatic interactions and, 216-17
 simplified models for, 217-32
Proteins
 direct force measurements and, 629-30
 globular
  fluctuating secondary structure of, 208
  tertiary structure prediction in, 216-32
 proton translocation in, 671-96
  mechanisms for, 672-76
 reversible renaturation of, 208
Proton translocation, 671-96
 mechanisms for, 672-76
 redox-linked
  mechanisms of, 688-94
Pyrochlores
 A-site vacancies in, 303
 electronic properties of, 299

Q

Quantum probability distribution function, 569
Quartz
 amorphous
  hydroxyl impurities embedded in, 157
Quasiatom theory, 278
Quasiequilibrium theory, 655

R

Radau coordinates, 474, 486
Radial basis functions
 vibrations of floppy molecules and, 479-80
Raman spectroscopy
 polymer-electrolyte mixtures and, 100-1
Rare earth ions
 hole-burning experiments and, 521-22
Rare earth oxides
 mixed valent, 304
Rare earths
 ferromagnetic pyrochlores of, 299
Rare gases
 electron emission in, 454-55
Rare gas-molecule dimers
 photoionization and, 652
Recoil anisotropy, 376-77, 398
Recursive residue generation method
 vibrations of floppy molecules and, 491-92
Redox-linked proton pump, 681
Reflectometry
 time domain, 11, 20, 22
Resonance-enhanced multiphoton ionization, 645
Resonance Raman spectroscopy, 683
Resonance spectroscopy, 8
Resorufin
 hole-burning experiments on, 517
Respiratory electron transport chain
 cytochrome c oxidase in, 681
Response theory
 adsorbate-adsorbate interactions and, 267-72
Ribonuclease
 direct force measurements and, 630
Ronca-Allegra theory
 elastic free energy of, 361-62
Rotational autoionization, 642-43
Rotational isomeric state models, 364

Rotational isomerization, 367-68
 rotational isomeric state models and, 364
Rubber-like elasticity, 351-73
 elastomeric networks and, 352-55
 experiments in, 355-56
 stress-strain relationships and, 369-72
Rubidium iodide
 pressure-induced dissociation of, 249
Rubredoxin
 native conformation of, 218
Ruby-fluorescence technique, 243-44
Ruderman-Kittel-Kasuya-Yosida interaction, 30, 33

S

Scandium oxide
 bonding in, 410-12
 $3d$ spectra of, 413-20
Scanning tunneling microscopy, 531-57
 adsorbate-adsorbate interaction energies and, 264
 one-dimensional, 532-34
 voltage-dependent, 535-41
Scanning tunneling spectroscopy, 541-45
 experimental applications of, 544-45
 interpretation of results, 541-44
 I-V measurements and, 545-55
 techniques of, 541
Scattering theory method
 adsorbate-adsorbate interactions and, 281-88
Scheutjens-Fleer lattice model, 623
Schrödinger equation, 409, 476, 611
Second harmonic generation, 327-49
 surface
  applications of, 338-48
  experimental considerations in, 337-38
  surface nonlinear susceptibilities and, 332-37
  theory of, 329-32
Segmental orientation, 366-67
 rotational isomeric state models and, 364
Semiclassical quantization
 vibrations of floppy molecules and, 492-93

SUBJECT INDEX 735

Semiconductors
  ballistic electron emission microscopy and, 555-56
  band-gap excitation of, 199-201
  molecular adsorption on second harmonic generation and, 340
  scanning tunneling microscopy and, 536-39
Semicrystalline polymers
  organic molecules in hole-burning experiments on, 520-21
Silane
  photoionization mass spectrometry and, 652
Silica
  adsorbates on model systems for, 155-58
  hydroxyl relaxation on, 150-55
Silicate glasses
  organic molecules in hole-burning experiments on, 521
Silicate melts
  structural transformations in ultrahigh pressure and temperature and, 252-56
Silicon
  local I-V spectroscopy and, 553-55
  scanning tunneling microscopy and, 536-38
Siloxanes
  electrolyte solvation and, 89-90
Silver
  surface reconstruction behavior in, 280
Site-selection spectroscopy, 499
Slip-link theory
  elasticity and, 364
Smoluchowski equation, 119-24, 133, 136, 449
Sodium trisilicate glasses
  dielectric properties of, 24
Solids
  impurities diluted in
    optical dephasing of, 510-11
  ionic
    ion desorption from, 193-97
    optical relaxation processes in, 499
Solitons, 19
Solvation
  dynamics of, 115-38
    charge transfer reactions and, 130-35

  computer simulation of, 126-27
  continuum model theories of, 117-19
  generalized Smoluchowski equation approach to, 119-24
  homogeneous dielectric models of, 117-18
  inhomogeneous dielectric models of, 118-19
  molecular theories of, 119-27
  nonequilibrium mean spherical approximation model of, 124-25
  theoretical studies of, 125-26
  time-resolved experimental studies of, 127-30
electron
  dynamics of, 129-30
energy of
  time dependence of, 116
Solvents
  good
    force measurements in polymers and, 614-18
  polar
    dynamics of, 135-37
  polarity of
    reaction potential surface and, 115
  poor
    force measurement in polymers and, 604-6
Sorbic acid
  adsorbed on graphite
    inelastic tunneling spectroscopy and, 556
Spectrometry
  photoionization mass, 652
Spectroscopy
  barrier state, 541-42
  current imaging tunneling, 546-47
  diatomic 3d transition metal oxides and, 407-35
  dielectric
    range and capabilities of, 8-11
  electron energy loss, 532
  high resolution Rydberg state, 648-50
  hole-burning, 499-526
    organic molecules and, 506-24
  immittance, 8
  impedance, 8
  inelastic tunneling, 556-57
  infrared
    broadband probe and, 148-49

  narrowband probe and, 148
  polymer-electrolyte mixtures and, 101-4
  polymer networks and, 356
  single-frequency transient, 145-46
  sum-frequency generation probe and, 149-50
  infrared reflection-absorption, 532
  inverse photoemission, 532
  laser frequency modulation, 503
  photofragment, 375
  photoionization, 637-65
  picosecond laser, 695
  polarization, 503
  polymer-electrolyte mixtures and, 100-4
  Raman
    polymer-electrolyte mixtures and, 100-1
    resonance, 683
  resonance, 8
  scanning tunneling, 541-45
    experimental applications of, 544-45
    interpretation of results, 541-44
    I-V measurements and, 545-55
    techniques of, 541
  site-selection, 499
  surface, 344
  surface state, 542-44
  threshold photoelectron, 650-52
  transition metal oxides and, 319-20
  tryptophan fluorescence, 695
  ultraviolet photoemission, 532
  vibrational, 144
    adsorbates and, 344
Spinels
  properties of, 298-99
Spin glasses
  freezing process in, 33
Spin paramagnetism
  metal-nonmetal transition in alkali metals and, 69
Stark effect, 564, 572
Statistical Mechanical Algorithm for Predicting Protein Structure, 212-14
Step-function model, 623
Steric effect, 561-62
Stokes shift relaxation, 127
Strain birefringence, 366
  rotational isomeric state models and, 364
Strain-dichroism, 367-68
  rotational isomeric state models and, 364

736 SUBJECT INDEX

Sulfur dibromide
  photoionization mass spectrometry and, 652
Sulfuric acid
  dielectric properties of, 22
Superconductivity
  oxide perovskites and, 297
  theoretical approaches to, 317
  transition metal oxides and, 291, 310-17
Superconductors
  cuprate
    properties of, 311-14
    local I-V spectroscopy and, 555
Superlattice calculations
  adsorbate-adsorbate interactions and, 277-78
Surface dynamics
  second harmonic generation and, 346-48
Surface monolayer microscopy, 346
Surface spectroscopy, 344
Surface state spectroscopy, 542-44
Synchrotron radiation
  threshold photoelectron spectroscopy and, 650-51
  vacuum ultraviolet light and, 638-40

T

Temperature
  ultrahigh
    physical chemistry and, 237-56
Tetracene
  nonphotochemical hole-burning and, 502
Tetradecane
  electron mobility in, 441
n-Tetradecane
  electron mobility in, 440
2,2,4,4-Tetramethylpentane
  Hall effect and, 450
Tetramethylsilane
  electron mobility in
    pressure and, 446
  Hall effect and, 450
  homogeneous electron-ion recombination in, 454
Tetramethyltin
  electron emission in, 454
Tetrazene
  photoelectron photoion coincidence and, 657
Thermodynamics
  equilibrium
    protein conformation and, 208

Thermoelasticity
  rotational isomeric state models and, 364
Threshold photoelectron spectroscopy, 650-52
Threshold photoelectron spectrum, 646
Time domain reflectometry, 11, 20, 22
Titanium
  hydrogen dissociation on, 276
Titanium oxide
  $3d$ spectra of, 420-22
Transient hole-burning, 503, 507-8
Transition metal oxides, 291-320
  characterization of, 319-20
  $3d$ spectra of, 407-35
  metal-nonmetal transitions in, 305-10
  mixed valence in, 303-5
  point defects in, 300-3
  properties of, 292-300
  superconductivity in, 291, 310-17
  synthesis of, 318-19
Transition metals
  carbon monoxide adsorbed on, 159-67
  scanning tunneling microscopy and, 535-36
Transition monoxides
  $3d$
    bonding in, 408-12
Transition state theory, 130
Triatomic molecules
  photodissociation of, 380-88
  rovibrational energy levels of
    coordinate systems for, 473-74
Trifluoromethyl halide + alkali atoms
  oriented
    reactions of, 580
2,2,4-Trimethylpentane
  electron transport in, 449
  Hall effect and, 450
Tropone
  photoionization mass spectrometry and, 652
Tryptophan fluorescence spectroscopy, 695
Tungsten
  desorption of helium from, 191

U

Ultraviolet light
  See Vacuum ultraviolet light
Ultraviolet photoemission spectroscopy, 532

Uranium oxide
  defect formation in, 302

V

Vacuum ultraviolet lasers, 640
Vacuum ultraviolet light
  sources of, 638-41
Vacuum ultraviolet photophysics, 637-65
  ionization process in, 64152
Vanadium
  stable isotopes of, 422-23
Vanadium monoxide
  bonding in, 410-11
  $3d$ spectra of, 422-27
Van der Waals complexes
  triatomic
    rovibrational spectra of, 480
Vector correlations, 375-401
  E-μ-(v-J), 377-78
  E-μ-J, 377
  E-μ-v, 376-77
  quantifying
    developments in, 379-80
  v-J, 377-78
Verwey transition, 304
Vibrational autoionization, 643-45
Vibrational spectroscopy, 144
  adsorbates and, 344
Vibronic relaxation
  hole-burning and, 524
Virial coefficients
  of gases, 11-14
Vogel-Fulcher law, 36, 46
Vogel-Tammann-Fulcher equation, 16, 19, 108

W

Wadseley defect, 302
Water
  clusters of
    photoionization and, 652
  liquid
    Kirkwood g-factor of, 6
  photodissociation of, 382-86
Wigner-Seitz model, 445
Williams-Watts function, 17-18
Willis clusters, 302
Wüstite
  antiferromagnetic, 250
  Koch-Cohen clusters and, 301
  phase relations in
    ultrahigh pressure-temperature conditions and, 249-51

SUBJECT INDEX

X

Xenon
  electron attachment reactions in, 453
  electron drift velocity in, 461
  electron emission in, 454-55
  electron mobility in, 444
    conduction band energy and, 444
  electron transport in, 440, 456-57

X-ray crystallography
  transition metal oxides and, 319
X-ray diffraction
  polymer-electrolyte mixtures and, 92, 96-100
  transition metal oxides and, 319
X-ray scattering
  elastomers and, 356

Z

Zener double exchange mechanism, 297
Ziman formula, 66
Zinc
  solubility behavior in polymers, 90-91
Zinc oxide
  bonding in, 412

CUMULATIVE INDEXES

CONTRIBUTING AUTHORS, VOLUMES 36–40

A

Anderson, J. G., 38:489–520
Asher, S. A., 39:537–88
Ausserré, D., 38:317–47
Avouris, P., 40:173–206

B

Bačić, Z., 40:469–98
Baer, T., 40:637–69
Bagchi, B., 40:115–41
Baldridge, K. K., 38:211–52
Bauschlicher, C. W. Jr., 39:181–212
Ben-Shaul, A., 36:179–211
Benson, S. W., 39:1–37
Berne, B. J., 37:401–24
Bernstein, R. B., 40:561–95
Boatz, J. A., 38:211–52
Borden, W. T., 39:212–36
Brown, J. K., 39:341–66
Budil, D. E., 38:561–83

C

Cahill, D. G., 39:93–121
Campion, A., 36:549–72
Casassa, M. P., 40:143–71
Castleman, A. W. Jr., 37:525–50
Cavanagh, R. R., 40:143–71
Ceyer, S. T., 39:479–510
Chan, S. I., 40:671–98
Chang, C.-H., 38:561–83
Chen, S. H., 37:351–99
Christiansen, P. A., 36:407–32
Clouter, M. J., 39:69–91
Cole, R. G., 37:105–25
Cole, R. H., 40:1–28
Copeland, R. A., 40:671–98
Cowie, J. M. G., 40:85–113
Cowley, J. M., 38:57–88
Cree, S. H., 40:85–113

D

Dalton, L. R., 37:459–91
DeBacker, M. G., 38:271–301
de Leeuw, S. W., 37:245–70
Dill, K. A., 39:425–61

Dlott, D. D., 37:157–87
Doll, J. D., 38:413–31
Drobny, G. P., 36:451–89
Dye, J. L., 38:271–301

E

Ebata, T., 39:123–47
Elson, E. T., 36:379–406
Erman, B., 40:351–74
Ermler, W. C., 36:407–32
Ertl, G., 37:587–615
Evans, G. T., 37:105–25
Ezra, G. S., 36:277–320

F

Feibelman, P. J., 40:261–90
Field, R. W., 37:493–524
Fleming, G. R., 37:81–104
Flynn, G. W., 37:551–85
Fredrickson, G. H., 39:149–80
Frei, H., 36:491–524

G

Gadzuk, J. W., 39:395–424
Gast, P., 38:561–83
Gelbart, W. M., 36:179–211
Glaeser, R. M., 36:243–75
Goodman, D. W., 37:425–57
Gordon, M. S., 38:211–52
Grant, E. R., 36:277–320
Griffin, R. G., 39:511–35
Griffiths, J. F., 36:77–104

H

Hall, G. E., 40:375–405
Hamers, R. J., 40:531–59
Hamilton, C. E., 37:493–524
Harding, J. H., 37:53–80
Harris, A. L., 39:341–66
Harris, C. B., 39:341–66
Harris, S. J., 36:31–52
Haymet, A. D. J., 38:89–108
Hearst, J. E., 39:291–315
Heilweil, E. J., 40:143–71
Hensel, F., 40:61–83
Hervet, H., 38:317–47
Herzberg, G., 36:1–30; 38:27–56

Hirota, E., 36:53–76
Holroyd, R. A., 40:439–68
Houk, K. N., 39:213–36
Houston, P. L., 40:375–405
Huppert, D., 37:127–56
Hynes, J. T., 36:573–97

I

Ito, M., 39:123–47

J

Jasinski, J. M., 38:109–40
Jeanloz, R., 40:237–59
Jovin, T. M., 38:521–60

K

Kawaguchi, K., 36:53–76
Keesee, R. G., 37:525–50
Kinsey, J. S., 37:493–524
Klein, M. L., 36:525–48
Kolinski, A., 40:207–35
Kollman, P., 38:303–16
Kommandeur, J., 38:433–62
Koseki, S., 38:211–52
Kosower, E. M., 37:127–56

L

Langhoff, S. R., 39:181–212
Levelt Sengers, J. M. H., 37:189–222
Light, J. C., 40:469–98
Lin, M. C., 37:587–615
Loidl, A., 40:29–60
Loncharich, R. J., 39:212–36
Lowe, M. A., 36:213–41

M

Majewsky, W. A., 38:433–62
Mark, J. E., 40:351–74
Marqusee, J. A., 39:425–61
McCaffery, A. J., 37:223–44
McIntosh, L. P., 38:521–60
Meakin, P., 39:237–67
Meerts, W. L., 38:433–62
Merer, A. J., 40:407–38
Meyerson, B. S., 38:109–40

Mikami, N., 39:123–47
Milligan, R. F., 36:139–58
Munoz-Rojas, A., 38:191–210

N

Naghizadeh, J., 39:425–61
Nibler, J. W., 38:349–81
Norris, J. W., 38:561–83

O

O'Brien, M. P. 38:383–411

P

Parker, D. H., 40:561–95
Patel, S. S., 40:597–635
Patterson, G. D., 38:191–210
Perram, J. W., 37:245–70
Peters, K., 38:253–70
Pimentel, G. C., 36:491–524
Pitzer, K. S., 36:407–32; 38:1–25
Pohl, R. O., 39:93–121
Porschke, D., 36:159–78
Pratt, D. W., 38:433–62
Pratt, L. R., 36:433–49
Prestegard, J. H., 38:383–411
Proctor, M. J., 37:223–44

R

Rao, C. N. R., 40:291–326
Ravishankara, A. R., 39:367–94

Reid, B. R., 36:105–37
Reisler, H., 37:307–49
Rondelez, F., 38:317–47
Rossky, P. J., 36:321–46

S

Schatz, G. C., 39:317–40
Schiffer, M., 38:561–83
Schinke, R., 39:39–68
Schmidt, J., 38:141–61
Schmidt, W. F., 40:439–68
Schmitz, K. S., 37:271–305
Schneider, F. W., 36:347–78
Schroeder, J., 38:163–90
Schurr, J. M., 37:271–305
Scott, B. A., 38:109–40
Sengers, J. V., 37:189–222
Shen, Y. R., 40:327–50
Sheppard, N., 39:589–644
Shlesinger, M. F., 39:269–90
Singel, D. J., 38:141–61
Skinner, J. L., 39:463–78
Skolnick, J., 40:207–35
Slichter, C. P., 37:25–51
Smith, E. R., 37:245–70
Smith, S. O., 39:511–35
Soumpasis, D. M., 38:521–60
Stephens, P. J., 36:213–41
Stephenson, J. C., 40:143–71
Stoneham, E. M., 37:53–80
Stout, J. W., 37:1–23
Sturtevant, J. M., 38:463–88

T

Thirumalai, D., 37:401–24
Thomas, G. A., 36:139–58
Tirrell, M., 40:597–635
Troe, J., 38:163–90

U

Uchtmann, H., 40:61–83

V

Vistnes, A. I., 37:459–91
Völker, S., 40:499–530
Voter, A. F., 38:413–31

W

Walkup, R. E., 40:173–206
Weiner, A. M., 36:31–52
Wemmer, D. E., 36:105–37
Weston, R. E. Jr., 37:551–85
Whetten, R. L., 36:277–320
Whitaker, B. J., 37:223–44
Whitney, D., 37:459–91
Wittig, C., 37:307–49

Y

Yang, J. J., 38:349–81
Young, C. L., 37:459–91

CHAPTER TITLES, VOLUMES 36–40

BIOPHYSICAL CHEMISTRY

High Resolution NMR Studies of Nucleic Acids and Proteins	D. E. Wemmer, B. R. Reid	36:105–37
Effects of Electric Fields on Biopolymers	D. Porschke	36:159–78
Electron Crystallography of Biological Macromolecules	R. M. Glaeser	36:243–75
Fluorescence Correlation and Photobleaching Recovery	E. L. Elson	36:379–406
Dynamic Light Scattering Studies of Biopolymers: Effects of Charge, Shape, and Flexibility	J. M. Schurr, K. S. Schmitz	37:189–222
Molecular Modeling	P. Kollman	38:303–16
Membrane and Vesicle Fusion	J. H. Prestegard, M. P. O'Brien	38:383–411
Biochemical Applications of Differential Scanning Calorimetry	J. M. Sturtevant	38:463–88
The Transition Between B-DNA and Z-DNA	T. M. Jovin, D. M. Soumpasis, L. P. McIntosh	38:521–60
Three-Dimensional X-Ray Crystallography of Membrane Proteins: Insights into Electron Transfer	D. E. Budil, P. Gast, C.-H. Chang, M. Schiffer, J. R. Norris	38:561–83
A Photochemical Investigation of the Dynamics of Oligonucleotide Hybridization	J. E. Hearst	39:291–315
High-Resolution Solid-State NMR of Proteins	S. O. Smith, R. G. Griffin	39:511–35
UV Resonance Raman Studies of Molecular Structure and Dynamics: Applications in Physical and Biophysical Chemistry	S. A. Asher	39:537–88
Computer Simulations of Globular Protein Folding and Tertiary Structure	J. Skolnick, A. Kolinski	40:207–35
Proton Translocation in Proteins	R. A. Copeland, S. I. Chan	40:671–98

CHEMICAL KINETICS—CONDENSED PHASE

Periodic Perturbations of Chemical Oscillators: Experiments	F. W. Schneider	36:347–78
Chemical Reaction Dynamics in Solution	J. T. Hynes	36:573–97
Excited State Electron and Proton Transfers	E. M. Kosower, D. Huppert	37:127–56
Elementary Reactions in the Gas-Liquid Transition Range	J. Schroeder, J. Troe	38:163–90
Three-Dimensional X-Ray Crystallography of Membrane Proteins: Insights into Electron Transfer	D. E. Budil, P. Gast, C.-H. Chang, M. Schiffer, J. R. Norris	38:561–83
Synchronicity in Multibond Reactions	W. T. Borden, R. J. Loncharich, K. N. Houk	39:213–36
Fractal Time in Condensed Matter	M. F. Shlesinger	39:269–90
The Nature of Simple Photodissociation Reactions in Liquids on Ultrafast Time Scales	A. L. Harris, J. K. Brown, C. B. Harris	39:341–66
Dynamics of Solvation and Charge Transfer Reactions in Dipolar Liquids	B. Bagchi	40:115–41
Transport of Electrons in Nonpolar Fluids	R. A. Holroyd, W. F. Schmidt	40:439–68

CHAPTER TITLES

CHEMICAL KINETICS—GAS PHASE

Chemical Kinetics of Soot Particle Growth	S. J. Harris, A. M. Weiner	36:31–52
Thermokinetic Interactions in Simple Gaseous Reactions	J. F. Griffiths	36:77–104
Elementary Reactions in the Gas-Liquid Transition Range	J. Schroeder, J. Troe	38:163–90
Free Radicals in the Earth's Atmosphere: Their Measurement and Interpretation	J. G. Anderson	38:489–520
Kinetics of Radical Reactions in the Atmospheric Oxidation of CH_4	A. R. Ravishankara	39:367–94

CHEMICAL KINETICS—PHOTOCHEMISTRY AND RADIATION CHEMISTRY

Infrared Induced Photochemical Processes in Matrices	H. Frei, G. C. Pimentel	36:491–524
Hot Atoms Revisited: Laser Photolysis and Product Detection	G. W. Flynn, R. E. Weston, Jr.	37:551–85
Picosecond Organic Photochemistry	K. Peters	38:253–70
Transport of Electrons in Nonpolar Fluids	R. A. Holroyd, W. F. Schmidt	40:439–68

CHEMICAL KINETICS—REACTION DYNAMICS

Mechanistic Studies of Chemical Vapor Deposition	J. M. Jasinski, B. S. Meyerson, B. A. Scott	38:109–40
Pyrazin: An "Exact" Solution to the Problem of Radiationless Transitions	J. Kommandeur, W. A. Majewski, W. L. Meerts, D. W. Pratt	38:433–62
Rotational Distributions in Direct Molecular Photodissociation	R. Schinke	39:39–68
Quantum Effects in Gas Bimolecular Phase Chemical Reactions	G. C. Schatz	39:317–40
Vector Correlations in Photodissociation Dynamics	G. E. Hall, P. L. Houston	40:375–405
Oriented Molecule Beams Via the Electrostatic Hexapole: Preparation, Characterization, and Reactive Scattering	D. H. Parker, R. B. Bernstein	40:561–95

COLLOIDS

Models for Colloidal Aggregation	P. Meakin	39:237–67

ELECTROCHEMISTRY

Physical and Chemical Properties of Alkalides and Electrides	J. L. Dye, M. G. DeBacker	38:271–301

GEOCHEMISTRY AND COSMOCHEMISTRY

Free Radicals in the Earth's Atomosphere: Their Measurement and Interpretation	J. G. Anderson	38:489–520
Physical Chemistry at Ultrahigh Pressures and Temperatures	R. Jeanloz	40:237–59
Rubber-like Elasticity	B. Erman, J. E. Mark	40:351–74

LASER CHEMISTRY, ENERGY TRANSFER AND RELAXATION

Stimulated Emission Pumping: New Methods in Spectroscopy and Molecular Dynamics	C. E. Hamilton, J. S. Kinsey, R. W. Field	37:493–524
Rotational Energy Transfer: Polarization and Scaling	A. J. McAffery, M. J. Proctor, B. J. Whitaker	37:223–44
Photo-initiated Unimolecular Reactions	H. Reisler, C. Wittig	37:307–49
The Nature of Simple Photodissociation Reactions in Liquids on Ultrafast Time Scales	A. L. Harris, J. K. Brown, C. B. Harris	39:341–66

LIQUID STATE—SIMPLE FLUIDS

Dynamics of Polyatomic Fluids: A Kinetic Theory Approach	R. G. Cole, G. T. Evans	37:105–25
Vibrational Raman Spectra of Simple Fluids	M. J. Clouter	39:69–91

LIQUID STATE—SOLUTIONS OF ELECTROLYTES; FUSED SALTS

Electrolytes Dissolved in Polymers	J. M. G. Cowie, S. H. Cree	40:85–113

LIQUID STATE—STRUCTURE

The Structure of Polar Molecular Liquids	P. J. Rossky	36:321–46
Theory of Hydrophobic Effects	L. R. Pratt	36:433–49
Computer Simulation of the Static Dielectric Constant of Systems with Permanent Electric Dipoles	S. W. de Leeuw, J. W. Perram, E. R. Smith	37:245–70
The Metal-Insulator Transition in Expanded Fluid Metals	F. Hensel, H. Uchtmann	40:61–83
Dynamics of Solvation and Charge Transfer Reactions in Dipolar Fluids	B. Bagchi	40:115–41

MAGNETIC RESONANCE (ELECTRON SPIN, NUCLEAR, QUADRUPOLE)

Multiple Quantum NMR: Studies of Molecules in Ordered Phases	G. P. Drobny	36:451–89
Fashioning Electron Spin Echoes into Spectroscopic Tools: A Study of Aza-aromatic Molecules in Metastable Triplet States	J. Schmidt, D. J. Singel	38:141–61
High-Resolution Solid-State NMR of Proteins	S. O. Smith, R. G. Griffin	39:511–35

MISCELLANEOUS

Vibrational Circular Dichroism	P. J. Stephens, M. A. Lowe	36:213–41
High Resolution Electron Microscopy	J. M. Cowley	38:57–88
Dielectrics in Physical Chemistry	R. H. Cole	40:1–28
Measurement of Forces Between Surfaces in Polymer Fluids	S. S. Patel, M. Tirrell	40:597–635

MOLECULAR STRUCTURE

Clusters: Properties and Formation	A. W. Castleman, Jr., R. G. Keesee	37:525–50
Molecular Modeling	P. Kollman	38:303–16

PHYSICAL ORGANIC

Picosecond Organic Photochemistry	K. Peters	38:253–70
Synchronicity in Multibond Reactions	W. T. Borden, R. J. Loncharich, K. N. Houk	39:213–36

PHYSICAL PHENOMENA—MISCELLANEOUS

Vibrational Circular Dichroism	P. J. Stephens, M. A. Lowe	36:213–41

POLYMERS AND MACROMOLECULES

Theory of Chain Packing in Amphiphilic Aggregates	A. Ben-Shaul, W. M. Gelbart	36:179–211
ESR and ENDOR of Conducting Polymers	C. L. Young, D. Whitney, A. I. Vistnes, L. R. Dalton	37:459–91
Dynamic Light Scattering Near the Glass Transition	G. D. Patterson, A. Munoz-Rojas	38:191–210
Experimental Studies of Polymer Concentration Profiles at Solid-Liquid and Liquid-Gas Interfaces by Optical and X-Ray Evanescent Wave Techniques	F. Rondelez, D. Ausserré, H. Hervet	38:317–47

Chain Molecules at High Densities at Interfaces	K. A. Dill, J. Naghizadeh, J. A. Marqusee	39:425–61
Electrolytes Dissolved in Polymers	J. M. G. Cowie, S. H. Cree	40:85–113
Computer Simulations of Globular Protein Folding and Tertiary Structure	J. Skolnick, A. Kolinski	40:207–35
Measurement of Forces Between Surfaces in Polymer Fluids	S. S. Patel, M. Tirrell	40:597–635

PREFATORY CHAPTERS

Molecular Spectroscopy: A Personal History	G. Herzberg	36:1–30
The Journal of Chemical Physics: The First 50 Years	J. W. Stout	37:1–23
Of Physical Chemistry and Other Activities	K. S. Pitzer	38:1–25
50 Years of Physical Chemistry, a Personal Account	S. W. Benson	39:1–37
Dielectrics in Physical Chemistry	R. H. Cole	40:1–28

QUANTUM CHEMISTRY

Molecular Dynamics Beyond the Adiabatic Approximation: New Experiments and Theory	R. L. Whetten, G. S. Ezra, E. R. Grant	36:277–320
Relativistic Effects in Chemical Systems	P. A. Christiansen, W. C. Ermler, K. S. Pitzer	36:407–32
On the Simulation of Quantum Systems: Path Integral Methods	B. J. Berne, D. Thirumalai	37:401–24
Theoretical Studies of Silicon Chemistry	K. K. Baldridge, J. A. Boatz, S. Koseki, M. S. Gordon	38:211–52
Ab Initio Studies of Transition Metal Systems	S. R. Langhoff, C. W. Bauschlicher, Jr.	39:181–212
Theoretical Methods for Rovibrational States of Floppy Molecules	Z. Bačić, J. C. Light	40:469–98

QUANTUM MECHANICS

Quantum Ergodicity and Spectral Chaos	E. B. Stechel, E. J. Heller	35:563–89

SCATTERING PHENOMENA—DYNAMICAL

Dynamic Light Scattering Near the Glass Transition	G. D. Patterson, A. Munoz-Rojas	38:191–210

SCATTERING PHENOMENA—STRUCTURAL

Small Angle Neutron Scattering Studies of the Structure and Interaction in Micellar and Microemulsion Systems	S. H. Chen	37:351–99
Experimental Studies of Polymer Concentration Profiles at Solid-Liquid and Liquid-Gas Interfaces by Optical and X-Ray Evanescent Wave Techniques	F. Rondelez, D. Ausserré, H. Hervet	38:317–47
Three-Dimensional X-Ray Crystallography of Membrane Proteins: Insights into Electron Transfer	D. E. Budil, P. Gast, C.-H. Chang, M. Schiffer, J. R. Norris	38:561–83

SOLIDS AND ORDERED ARRAYS—STRUCTURE AND DYNAMICS

The Metal-Insulator Transition	R. F. Milligan, G. A. Thomas	36:139–58
Computer Simulation Studies of Solids	M. L. Klein	36:525–48
Interatomic Potentials in Solid State Chemistry	A. M. Stoneham, J. H. Harding	37:53–80
Optical Phonon Dynamics in Molecular Crystals	D. D. Dlott	37:159–87

Theory of the Equilibrium Liquid-Solid Transition	A. D. J. Haymet	38:89–108
Physical and Chemical Properties of Alkalides and Electrides	J. L. Dye, M. G. DeBacker	38:271–301
Three-Dimensional X-Ray Crystallography of Membrane Proteins: Insights into Electron Transfer	D. E. Budil, P. Gast, C.-H. Chang, M. Schiffer, J. R. Norris	38:561–83
Lattice Vibrations and Heat Transport in Crystals and Glasses	D. G. Cahill, R. O. Pohl	39:93–121
Recent Developments in Dynamical Theories of the Liquid-Glass Transition	G. H. Fredrickson	39:149–80
Orientational Glasses	A. Loidl	40:29–60
Physical Chemistry at Ultrahigh Pressures and Temperatures	R. Jeanloz	40:237–59
Transition Metal Oxides	C. N. R. Rao	40:291–326

SPECTROSCOPY—ELECTRONIC AND PHOTOELECTRONIC

Subpicosecond Spectroscopy	G. R. Fleming	37:81–104
Rydberg Molecules	G. Herzberg	38:27–56
Pyrazine: An "Exact" Solution to the Problem of Radiationless Transitions	J. Kommandeur, W. A. Majewski, W. L. Meerts, D. W. Pratt	38:433–62
Laser Spectroscopy of Large Polyatomic Molecules in Supersonic Jets	M. Ito, T. Ebata, N. Mikami	39:123–47
Theory of Pure Dephasing in Crystals	J. L. Skinner	39:463–78
Spectroscopy of the Diatomic 3d Transition Metal Oxides	A. J. Merer	40:407–38
Hole-Burning Spectroscopy	S. Völker	40:499–530
Vacuum UV Photophysics and Photoionization Spectroscopy	T. Baer	40:637–69

SPECTROSCOPY—VIBRATIONAL

High Resolution Infrared Studies of Molecular Dynamics	E. Hirota, K. Kawaguchi	36:53–76
Raman Spectroscopy of Molecules Adsorbed on Solid Surfaces	A. Campion	36:549–72
Nonlinear Raman Spectroscopy of Gases	J. W. Nibler, J. J. Yang	38:349–81
Vibrational Raman Spectra of Simple Fluids	M. J. Clouter	39:69–91
UV Resonance Raman Studies of Molecular Structure and Dynamics: Applications in Physical and Biophysical Chemistry	S. A. Asher	39:537–88
Vibrational Spectroscopic Studies of the Structure of Species Derived from the Chemisorption of Hydrocarbons on Metal Single-Crystal Surfaces	N. Sheppard	39:589–644
Picosecond Vibrational Energy Transfer Studies of Surface Adsorbates	E. J. Heilweil, M. P. Casassa, R. R. Cavanagh, J. C. Stephenson	40:143–71

STATISTICAL MECHANICS

Thermodynamic Behavior of Fluids Near the Critical Point	J. V. Sengers, J. M. H. Levelt Sengers	37:189–222
Theory of the Equilibrium Liquid-Solid Transition	A. D. J. Haymet	38:89–108
Fractal Time in Condensed Matter	M. F. Shlesinger	39:269–90
Chain Molecules at High Densities at Interfaces	K. A. Dill, J. Naghizadeh, J. A. Marqusee	39:425–61
Theory of Pure Dephasing in Crystals	J. L. Skinner	39:463–78
Orientational Glasses	A. Loidl	40:29–60

The Metal-Insulator Transition in Expanded Fluid Metals	F. Hensel, H. Uchtmann	40:61–83
SURFACES—ADSORPTION AND CATALYSIS		
Catalytic Studies with Metal Single Crystals	D. W. Goodman	37:425–57
Mechanistic Studies of Chemical Vapor Deposition	J. M. Jasinski, B. S. Meyerson, B. A. Scott	38:109–40
Dissociative Chemisorption: Dynamics and Mechanisms	S. T. Ceyer	39:479–510
Picosecond Vibrational Energy Transfer Studies of Surface Adsorbates	E. J. Heilweil, M. P. Casassa, R. R. Cavanagh, J. C. Stephenson	40:143–71
Fundamental Mechanisms of Desorption and Fragmentation Induced by Electronic Transitions at Surfaces	P. Avouris, R. E. Walkup	40:173–206
Theory of Adsorbate Interactions	P. J. Feibelman	40:261–90
SURFACES—STRUCTURE AND DYNAMICS		
Laser Probing of Molecules Desorbing and Scattering from Solid Surfaces	M. C. Lin, G. Ertl	37:587–615
Probing Phenomena at Metal Surfaces by NMR	C. P. Slichter	37:25–51
Recent Developments in the Theory of Surface Diffusion	J. D. Doll, A. F. Voter	38:413–31
The Semiclassical Way to Molecular Dynamics at Surfaces	J. W. Gadzuk	39:395–424
Vibrational Spectroscopic Studies of the Structure of Species Derived from the Chemisorption of Hydrocarbons on Metal Single-Crystal Surfaces	N. Sheppard	39:589–644
Optical Second Harmonic Generation at Interfaces	Y. R. Shen	40:327–50
Atomic-Resolution Surface Spectroscopy with the Scanning Tunneling Microscope	R. J. Hamers	40:531–59
THERMOCHEMISTRY AND THERMODYNAMICS		
Biochemical Applications of Differential Scanning Calorimetry	J. M. Sturtevant	38:463–88

Annual Reviews Inc.
A NONPROFIT SCIENTIFIC PUBLISHER

4139 El Camino Way
P.O. Box 10139
Palo Alto, CA 94303-0897 • USA

ORDER FORM

ORDER TOLL FREE
1-800-523-8635
(except California)

Telex: 910-290-0275

Annual Reviews Inc. publications may be ordered directly from our office by mail, Telex, or use our Toll Free Telephone line (for orders paid by credit card or purchase order*, and customer service calls only); through booksellers and subscription agents, worldwide; and through participating professional societies. Prices subject to change without notice. ARI Federal I.D. #94-1156476

- **Individuals:** Prepayment required on new accounts by check or money order (in U.S. dollars, check drawn on U.S. bank) or charge to credit card—American Express, VISA, MasterCard.
- **Institutional buyers:** Please include purchase order number.
- **Students:** $10.00 discount from retail price, per volume. Prepayment required. Proof of student status must be provided (photocopy of student I.D. or signature of department secretary is acceptable). Students must send orders direct to Annual Reviews. Orders received through bookstores and institutions requesting student rates will be returned. You may order at the Student Rate for a maximum of 3 years.
- **Professional Society Members:** Members of professional societies that have a contractual arrangement with Annual Reviews may order books through their society at a reduced rate. Check with your society for information.
- **Toll Free Telephone orders:** Call 1-800-523-8635 (except from California) for orders paid by credit card or purchase order and customer service calls only. California customers and all other business calls use 415-493-4400 (not toll free). Hours: 8:00 AM to 4:00 PM, Monday-Friday, Pacific Time. *Written confirmation is required on purchase orders from universities before shipment.
- **Telex: 910-290-0275**

Regular orders: Please list the volumes you wish to order by volume number.
Standing orders: New volume in the series will be sent to you automatically each year upon publication. Cancellation may be made at any time. Please indicate volume number to begin standing order.
Prepublication orders: Volumes not yet published will be shipped in month and year indicated.
California orders: Add applicable sales tax.
Postage paid (4th class bookrate/surface mail) **by Annual Reviews Inc.** Airmail postage or UPS, extra.

ANNUAL REVIEWS SERIES		Prices Postpaid per volume USA & Canada/elsewhere	Regular Order Please send:	Standing Order Begin with:
			Vol. number	Vol. number
Annual Review of **ANTHROPOLOGY**				
Vols. 1-14	(1972-1985)	$27.00/$30.00		
Vols. 15-16	(1986-1987)	$31.00/$34.00		
Vol. 17	(1988)	$35.00/$39.00		
Vol. 18	(avail. Oct. 1989)	$35.00/$39.00	Vol(s). _____	Vol. _____
Annual Review of **ASTRONOMY AND ASTROPHYSICS**				
Vols. 1, 4-14, 16-20	(1963, 1966-1976, 1978-1982)	$27.00/$30.00		
Vols. 21-25	(1983-1987)	$44.00/$47.00		
Vol. 26	(1988)	$47.00/$51.00		
Vol. 27	(avail. Sept. 1989)	$47.00/$51.00	Vol(s). _____	Vol. _____
Annual Review of **BIOCHEMISTRY**				
Vols. 30-34, 36-54	(1961-1965, 1967-1985)	$29.00/$32.00		
Vols. 55-56	(1986-1987)	$33.00/$36.00		
Vol. 57	(1988)	$35.00/$39.00		
Vol. 58	(avail. July 1989)	$35.00/$39.00	Vol(s). _____	Vol. _____
Annual Review of **BIOPHYSICS AND BIOPHYSICAL CHEMISTRY**				
Vols. 1-11	(1972-1982)	$27.00/$30.00		
Vols. 12-16	(1983-1987)	$47.00/$50.00		
Vol. 17	(1988)	$49.00/$53.00		
Vol. 18	(avail. June 1989)	$49.00/$53.00	Vol(s). _____	Vol. _____
Annual Review of **CELL BIOLOGY**				
Vol. 1	(1985)	$27.00/$30.00		
Vols. 2-3	(1986-1987)	$31.00/$34.00		
Vol. 4	(1988)	$35.00/$39.00		
Vol. 5	(avail. Nov. 1989)	$35.00/$39.00	Vol(s). _____	Vol. _____

ANNUAL REVIEWS SERIES	Prices Postpaid per volume USA & Canada/elsewhere	Regular Order Please send:	Standing Order Begin with:
		Vol. number	Vol. number

Annual Review of **COMPUTER SCIENCE**
 Vols. 1-2   (1986-1987)................$39.00/$42.00
 Vol. 3   (1988) ....$45.00/$49.00
 Vol. 4   (avail. Nov. 1989)..........$45.00/$49.00   Vol(s). _____   Vol. _____

Annual Review of **EARTH AND PLANETARY SCIENCES**
 Vols. 1-10   (1973-1982)................$27.00/$30.00
 Vols. 11-15 (1983-1987)................$44.00/$47.00
 Vol. 16   (1988) ....$49.00/$53.00
 Vol. 17   (avail. May 1989)..........$49.00/$53.00   Vol(s). _____   Vol. _____

Annual Review of **ECOLOGY AND SYSTEMATICS**
 Vols. 2-16   (1971-1985)................$27.00/$30.00
 Vols. 17-18 (1986-1987)................$31.00/$34.00
 Vol. 19   (1988) ....$34.00/$38.00
 Vol. 20   (avail. Nov. 1989)..........$34.00/$38.00   Vol(s). _____   Vol. _____

Annual Review of **ENERGY**
 Vols. 1-7   (1976-1982)................$27.00/$30.00
 Vols. 8-12   (1983-1987)................$56.00/$59.00
 Vol. 13   (1988) ....$58.00/$62.00
 Vol. 14   (avail. Oct. 1989)..........$58.00/$62.00   Vol(s). _____   Vol. _____

Annual Review of **ENTOMOLOGY**
 Vols. 10-16, 18 (1965-1971, 1973)
    20-30   (1975-1985)................$27.00/$30.00
 Vols. 31-32 (1986-1987)................$31.00/$34.00
 Vol. 33   (1988) ....$34.00/$38.00
 Vol. 34   (avail. Jan. 1989)..........$34.00/$38.00   Vol(s). _____   Vol. _____

Annual Review of **FLUID MECHANICS**
 Vols. 1-4, 7-17 (1969-1972, 1975-1985).......$28.00/$31.00
 Vols. 18-19 (1986-1987)................$32.00/$35.00
 Vol. 20   (1988) ....$34.00/$38.00
 Vol. 21   (avail. Jan. 1989)..........$34.00/$38.00   Vol(s). _____   Vol. _____

Annual Review of **GENETICS**
 Vols. 1-19   (1967-1985)................$27.00/$30.00
 Vols. 20-21 (1986-1987)................$31.00/$34.00
 Vol. 22   (1988) ....$34.00/$38.00
 Vol. 23   (avail. Dec. 1989)..........$34.00/$38.00   Vol(s). _____   Vol. _____

Annual Review of **IMMUNOLOGY**
 Vols. 1-3   (1983-1985)................$27.00/$30.00
 Vols. 4-5   (1986-1987)................$31.00/$34.00
 Vol. 6   (1988) ....$34.00/$38.00
 Vol. 7   (avail. April 1989)........$34.00/$38.00   Vol(s). _____   Vol. _____

Annual Review of **MATERIALS SCIENCE**
 Vols. 1, 3-12 (1971, 1973-1982).........$27.00/$30.00
 Vols. 13-17 (1983-1987)................$64.00/$67.00
 Vol. 18   (1988) ....$66.00/$70.00
 Vol. 19   (avail. Aug. 1989)..........$66.00/$70.00   Vol(s). _____   Vol. _____

Annual Review of **MEDICINE**
 Vols. 9, 11-15 (1958, 1960-1964)
    17-36   (1966-1985)................$27.00/$30.00
 Vols. 37-38 (1986-1987)................$31.00/$34.00
 Vol. 39   (1988) ....$34.00/$38.00
 Vol. 40   (avail. April 1989)........$34.00/$38.00   Vol(s). _____   Vol. _____